Multifunctional Nanostructured Coatings

Woodhead Publishing Series in
Electronic and Optical Materials

Multifunctional Nanostructured Coatings

Biomedical, Environmental, and Energy Applications

Edited by

Manviri Rani

Department of Chemistry, Malaviya National Institute of Technology, Jaipur, Rajasthan, India

Uma Shanker

Department of Chemistry, Dr B R Ambedkar National Institute of Technology, Jalandhar, Punjab, India

ELSEVIER

WP
WOODHEAD
PUBLISHING
An imprint of Elsevier

Woodhead Publishing is an imprint of Elsevier
50 Hampshire Street, 5th Floor, Cambridge, MA 02139, United States
125 London Wall, London EC2Y 5AS, United Kingdom

ISBN: 978-0-443-23683-9

For information on all Woodhead Publishing publications visit our website at https://www.elsevier.com/books-and-journals

Publisher: Matthew Deans
Acquisitions Editor: Stephen Jones
Editorial Project Manager: Emily Thomson
Production Project Manager: Surya Narayanan Jayachandran
Cover Designer: Greg Harris

Typeset by TNQ Technologies

Contents

Contributors xi

Section I Multifunctional nanostructured coatings: Fundamentals and basics 1

1 Fundamentals of multifunctional nanostructured coatings with recent updates 3

Maneesh Kumar Poddar and Vishnu Soman

1.1 Introduction 3
1.2 Types of nanocoatings 5
1.3 Salient features of nanocoatings and recent developments 8
1.4 Methods of synthesis of nanostructured coatings 14
1.5 Characterization of nanocoatings 20
1.6 Environmental impact and sustainability of nanocoatings 20
1.7 Conclusions 21
AI Disclosure 22
References 22

2 Recent updates in nanometal oxide—based corrosion-resistant coatings 37

Gauri Shukla, Manviri Rani and Uma Shanker

2.1 Introduction 37
2.2 Design of metal oxide—based nanocoatings 39
2.3 Metal/metal oxide incorporated surface coatings 41
2.4 Factors affecting the efficiency of metal/metal oxide nanoparticles as corrosion inhibitors 45
2.5 Mechanism of corrosion protection 48
2.6 Conclusion 49
References 50

Section II Synthetic approach of nanomaterials for coatings 55

3 Fundamentals and basics of synthetic approaches of nanostructured coatings 57

Subodh Barthwal and Sumit Barthwal

3.1 Introduction 57

3.2 Synthesis of nanocoatings **60**
3.3 Future scopes **85**
References **87**

4 Designing of multifunctional coatings: Recent updates **95**
Shikha Sharma, Manviri Rani and Uma Shanker
4.1 Introduction **95**
4.2 Preparation methods of nanocoatings **97**
4.3 Sputtering **101**
4.4 Spray coating **103**
4.5 Sol-gel method **103**
4.6 Electrospinning **105**
4.7 Self-assembly **105**
4.8 Dip coating **107**
4.9 Applications of nanomaterial coatings **108**
4.10 Antimicrobial applications **108**
4.11 Multifunctional smart nanocoatings for fire protection **110**
4.12 Wear-resistant coatings **111**
4.13 Smart anticorrosion coatings **112**
4.14 Multifunctional superhydrophobic coatings **113**
4.15 Nanocoatings for energy applications **114**
4.16 Conclusions **115**
References **115**

5 Fabrication strategies and surface tuning and structural modifications of MNCs **123**
Vinith Johnson, Sunil Vasu, Archana M and Uday Kumar S
5.1 Introduction **123**
5.2 Fabrication strategies for multifunctional nanostructured coatings **124**
5.3 Surface tuning **136**
5.4 Structure modifications of nanomaterials **143**
5.5 Conclusion **154**
References **154**

Section III Biomedical applications of multifunctional nanostructured coatings 169

6 Multifunctional cold-sprayed coatings as micro/nano-biointerfaces for biomedical applications **171**
Simone Russo and Giuseppe Vitiello
6.1 Functional micro/nano-structured biointerfaces for medicine **171**
6.2 High-performance coatings for biomedical field by cold spray route **172**

6.3 Fundamental and operational principles of CS technology **173**
6.4 Micro/nano-structured coatings for biomedical field: From materials chemistry to functional medical devices **177**
6.5 Conclusion and future perspectives **187**
References **188**

7 Surface-coated magnetic nanostructured materials for robust bio-catalysis and biomedical applications **195**
Amira Moustafa Kamel, Mona Moaness and Mostafa Mabrouk
7.1 Introduction **195**
7.2 Multifunctional coatings: Revolutionizing materials for diverse applications **196**
7.3 Magnetic nanostructured materials **200**
7.4 Surface coating techniques **208**
7.5 Bio-catalytic reactions and their significance in biomedical applications **221**
7.6 Surface-coated magnetic nanostructured materials for biomedical applications **223**
7.7 Biocompatibility and safety considerations **229**
7.8 Challenges and future perspectives **230**
7.9 Conclusion **231**
References **231**

8 Multifunctional mesoporous silica-based nanocomposites for biomedical applications **241**
Dzmitry Shcharbin, Burcu Önal Acet, Inessa Halets-Bui, Mehmet Odabaşı and Ömür Acet
8.1 Introduction **241**
8.2 A Brief development history and different categories of MSNs **242**
8.3 Importance of silica nanoparticles for biomedical applications **245**
8.4 Applications **248**
References **258**

9 Biomedical applications of multifunctional polymer based nanostructured coatings **267**
Burcu Önal Acet, Dzmitry Shcharbin, Inessa Halets-Bui, Mehmet Odabaşı and Ömür Acet
9.1 Introduction **267**
9.2 Important points for multifunctional nanocarriers **269**
9.3 Some multifunctional nanocarriers **270**
9.4 Usability of multifunctional polymer-based nanostructured coatings some diseases **272**
9.5 Applications **274**

9.6 Conclusions and some challenges **282**
References **283**

10 Nanostructured coatings for biosensing applications in surface plasmon−coupled emission (SPCE) interface **291**
Kalathur Mohan Ganesh, Seemesh Bhaskar and Sai Sathish Ramamurthy
10.1 Introduction **291**
10.2 Metal-enhanced fluorescence (MEF) **293**
10.3 SPR and SPCE **294**
10.4 Relevance of nanostructured coatings in SPCE **296**
10.5 Applications of nanostructured coatings for biosensing in SPCE technology **297**
10.6 Futuristic scope and perspectives **314**
10.7 Conclusions **315**
References **316**

Section IV Multifunctional nanostructured coatings for environmental applications **323**

11 Recent advances in bio-inspired multifunctional coatings for corrosion protection **325**
Subodh Barthwal and Sumit Barthwal
11.1 Introduction **325**
11.2 Functional surfaces in nature **327**
11.3 Bio-inspired multifunctional coatings **333**
11.4 Challenges and future scopes **355**
References **357**

12 Recent trends of micro and nanostructured conducting polymers for environmental applications **365**
Isabela Jasper, Vanessa Klobukoski, Andrei Elias Deller, Raquel Anastácio Amaral, Patricio Peralta Zamora and Marcio Vidotti
12.1 Introduction **365**
12.2 Electrochemical synthesis of conducting polymers **367**
12.3 Conducting polymers characterization **371**
12.4 Conducting polymers applied to degradation of environmental contaminants **377**
12.5 Conclusion **384**
References **384**

13 Recent developments in multifunctional nanofibrous membranes for oily wastewater treatment **397**
Fayez Uddin Ahmed and Debarun Dhar Purkayastha
13.1 Introduction **397**
13.2 Background of oil/water separation **399**
13.3 Background of electrospun nanofibrous membranes **403**
13.4 Oil/water separation by membrane technology **406**
13.5 Multifunctionality in nanofibrous membrane design **412**
13.6 Recent advances in nanofibrous membranes for oily wastewater treatment **421**
13.7 Application of nanofibrous membranes in oily wastewater treatment: Real-world impact **428**
13.8 Future perspectives and challenges **429**
13.9 Conclusion: The path forward in oily wastewater treatment **429**
References **430**

Section V Multifunctional nanostructured coatings for energy management 439

14 Multifunctional self-cleaning nanostructured coatings for PV panels, CSP mirrors, and related solar devices **441**
Bandana Bera, Sudipta Senapati and Yitzhak Mastai
14.1 Introduction **441**
14.2 Wetting ability: The science underlying self-cleaning mechanism **444**
14.3 Self-cleaning coating types **446**
14.4 Materials used for self-cleaning coating **447**
14.5 Durability **451**
14.6 Current challenges and outlook **452**
14.7 Conclusions **452**
References **453**

15 Applications of thermal spray–based coatings for renewable energy **457**
Durvesh Gautam and Yogendra K. Gautam
15.1 Introduction **457**
15.2 Thermal spray coating techniques **460**
15.3 Thermal spray coating materials **462**
15.4 Renewable energy technology's big threat **463**
15.5 Application of thermal spray coatings for renewable energy sources **467**
15.6 Conclusion **472**
Abbreviations **473**
Acknowledgments **473**
References **473**

16 Ternary metal-based composites for coating in electrochemical reactions **479**
Nikhil Thakur, Pawan Kumar, Sorabh Aggarwal, Bandna Bharti, Praveen Vummadisetty Naidu and Abhishek Kandwal
16.1 Introduction **479**
16.2 Design and synthesis of ternary metal-based composites for coating applications **480**
16.3 Characterization techniques for ternary metal-based composites **490**
16.4 Electrochemical performance and mechanisms of ternary metal-based coatings **497**
16.5 Applications of ternary metal-based composites in electrochemical reactions **498**
16.6 Future perspectives and challenges in the advancement of ternary metal-based coatings **500**
16.7 Conclusion **501**
AI Disclosure **503**
References **503**

17 Potential of conjugated polymers in nanostructured coatings for energy applications **511**
Maria Federica De Riccardis
17.1 Introduction **511**
17.2 Supercapacitor electrodes **515**
17.3 Electrocatalysts **520**
17.4 Thermoelectric devices **523**
17.5 Conclusions **527**
References **528**

Index **535**

Contributors

Burcu Önal Acet Faculty of Arts and Science, Chemistry Department, Aksaray University, Aksaray, Turkey

Ömür Acet Vocational School of Health Science, Pharmacy Services Program, Tarsus University, Tarsus, Turkey

Sorabh Aggarwal School of Core Engineering, Shoolini University, Solan, Himachal Pradesh, India

Fayez Uddin Ahmed Department of Physics, National Institute of Technology Nagaland, Chumukedima, Nagaland, India

Raquel Anastácio Amaral Universidade Federal do Paraná (UFPR), Departamento de Química, Curitiba, Brazil

Subodh Barthwal Department of Mechanical Engineering, Amity University, Greater Noida, Uttar Pradesh, India

Sumit Barthwal Nanomechatronics Lab, Kookmin University, Seoul, South Korea

Bandana Bera Institute for Nanotechnology and Advanced Materials, Bar-Ilan University, Ramat-Gan, Israel; Department of Chemistry, Bar-Ilan University, Ramat-Gan, Israel

Bandna Bharti Department of Chemistry, DAV University, Jalandhar, Punjab, India

Seemesh Bhaskar University of Illinois at Urbana-Champaign, Carl R. Woese Institute for Genomic Biology, Department of Electrical and Computer Engineering, Nick Holonyak Jr. Micro and Nanotechnology Laboratory (HMNTL), Urbana, IL, United States

Andrei Elias Deller Universidade Federal do Paraná (UFPR), Departamento de Química, Curitiba, Brazil

Maria Federica De Riccardis Italian National Agency for New Technologies, Energy and Sustainable Economic Development, Research Center of Brindisi, Brindisi, Italy

Debarun Dhar Purkayastha Department of Physics, National Institute of Technology Nagaland, Chumukedima, Nagaland, India

Kalathur Mohan Ganesh STAR Laboratory, Department of Chemistry, Sri Sathya Sai Institute of Higher Learning, Prasanthi Nilayam, Anantapur, Andhra Pradesh, India

Durvesh Gautam Smart Materials and Sensor Laboratory, Department of Physics, Chaudhary Charan Singh University, Meerut, Uttar Pradesh, India

Yogendra K. Gautam Smart Materials and Sensor Laboratory, Department of Physics, Chaudhary Charan Singh University, Meerut, Uttar Pradesh, India

Inessa Halets-Bui Institute of Biophysics and Cell Engineering of the National Academy of Sciences of Belarus, Minsk, Belarus

Isabela Jasper Universidade Federal do Paraná (UFPR), Departamento de Química, Curitiba, Brazil

Vinith Johnson Department of Chemical Engineering, Indian Institute of Technology Tirupati, Tirupati, Andhra Pradesh, India

Amira Moustafa Kamel Polymers and Pigments Department, National Research Centre, Cairo, Egypt

Abhishek Kandwal School of Physics and Materials Science, Shoolini University, Solan, Himachal Pradesh, India; School of Chips, Xi'an Jiaotong-Liverpool University, Taicang, Suzhou, China

Vanessa Klobukoski Universidade Federal do Paraná (UFPR), Departamento de Química, Curitiba, Brazil

Pawan Kumar School of Physics and Materials Science, Shoolini University, Solan, Himachal Pradesh, India

Uday Kumar S Department of Chemical Engineering, Indian Institute of Technology Tirupati, Tirupati, Andhra Pradesh, India

Archana M Department of Chemical Engineering, Indian Institute of Technology Tirupati, Tirupati, Andhra Pradesh, India

Mostafa Mabrouk Refractories, Ceramics and Building Materials Department, Advanced Materials Technology and Mineral Resources Research Institute, National Research Centre, Cairo, Egypt

Yitzhak Mastai Institute for Nanotechnology and Advanced Materials, Bar-Ilan University, Ramat-Gan, Israel; Department of Chemistry, Bar-Ilan University, Ramat-Gan, Israel

Mona Moaness Refractories, Ceramics and Building Materials Department, Advanced Materials Technology and Mineral Resources Research Institute, National Research Centre, Cairo, Egypt

Praveen Vummadisetty Naidu Interdisciplinary Centre for Security, Reliability and Trust, University of Luxembourg, Luxembourg

Mehmet Odabaşı Faculty of Arts and Science, Chemistry Department, Aksaray University, Aksaray, Turkey

Maneesh Kumar Poddar National Institute of Technology Karnataka, Surathkal, Chemical Engineering, Mangaluru, Karnataka, India

Sai Sathish Ramamurthy Sri Sathya Sai Institute of Higher Learning, Faculty, Department of Chemistry, Principle Investigator, STAR Lab, CRIF, Puttaparthi, India; University of Maryland Baltimore County, Center for Advanced Sensor Technology, Department of Chemical, Biochemical and Environmental Engineering, Baltimore, MD, United States

Manviri Rani Department of Chemistry, Malaviya National Institute of Technology, Jaipur, Rajasthan, India

Simone Russo University of Naples Federico II, Department of Chemical, Materials and Production Engineering (DICMaPI), Naples, Italy; CSGI, Center for Colloid and Surface Science, Florence, Italy

Sudipta Senapati Department of Chemistry, Bar-Ilan University, Ramat-Gan, Israel

Uma Shanker Department of Chemistry, Dr B R Ambedkar National Institute of Technology, Jalandhar, Punjab, India

Shikha Sharma Department of Chemistry, Dr B R Ambedkar National Institute of Technology, Jalandhar, Punjab, India

Dzmitry Shcharbin Institute of Biophysics and Cell Engineering of the National Academy of Sciences of Belarus, Minsk, Belarus

Gauri Shukla Department of Chemistry, Malaviya National Institute of Technology, Jaipur, Rajasthan, India

Vishnu Soman National Institute of Technology Karnataka, Surathkal, Chemical Engineering, Mangaluru, Karnataka, India

Nikhil Thakur School of Physics and Materials Science, Shoolini University, Solan, Himachal Pradesh, India

Sunil Vasu Department of Chemical Engineering, Indian Institute of Technology Tirupati, Tirupati, Andhra Pradcsh, India

Marcio Vidotti Universidade Federal do Paraná (UFPR), Departamento de Química, Curitiba, Brazil

Giuseppe Vitiello University of Naples Federico II, Department of Chemical, Materials and Production Engineering (DICMaPI), Naples, Italy; CSGI, Center for Colloid and Surface Science, Florence, Italy

Patricio Peralta Zamora Universidade Federal do Paraná (UFPR), Departamento de Química, Curitiba, Brazil

Section I

Multifunctional nanostructured coatings: Fundamentals and basics

Fundamentals of multifunctional nanostructured coatings with recent updates

1

Maneesh Kumar Poddar and Vishnu Soman
National Institute of Technology Karnataka, Surathkal, Chemical Engineering, Mangaluru, Karnataka, India

1.1 Introduction

It is well known that nanomaterials have revolutionized the field of material science with their exceptional properties and helped us tackle many prevalent challenges in the past years. Among the various revolutionary innovations in nanotechnology, the role of nanostructured coatings, often referred to as "nanocoatings", is unparalleled for surface enhancement and protection of materials. They are thin layers of materials with at least one dimension in the nanoscale (1−100 nm, where conventional coatings are a few thousand nanometers in thickness). A schematic diagram depicting how nanocoating is done on the surface of a substrate is as shown in Fig. 1.1. Just like other nanomaterials, these nanocoatings also have inherent properties like a high surface-to-volume ratio, enabling them to act as protectants against corrosion, wear and tear, microbial action, oil, UV rays, and so on. Nanocoatings are synthesized using various techniques like Chemical Vapor Deposition (CVD), Particle Vapor Deposition (PVD), electrodeposition, magnetron spluttering, laser cladding, spray coating, etc. The applicability of nanocoating is very diverse, including areas like mechanical, thermal, energy conservation, optical applications, etc. For instance, addressing the wear and tear in moving mechanical parts is essential in maintaining the operating efficiency and economic viability. The challenge is not only limited to wear and tear, but also the said components may be subjected to very high temperatures and mechanical load, depending upon the operating environment, resulting in the formation of cracks or peeling of the coating from the surface of the substrate. Hence, while synthesizing the coatings, it is crucial to optimize the properties of the nanocoatings, such as particle size, adhesion with the substrate, thickness, etc. The

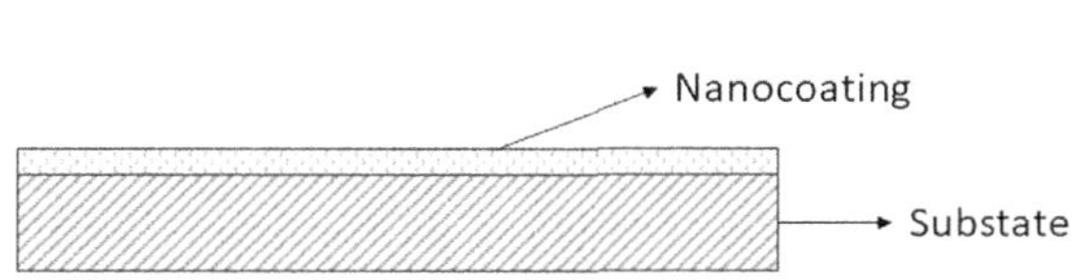

Figure 1.1 Schematic representation of a nanocoating on a substrate. The nanocoating (depicted in green [light gray in print version] dots) is shown as a thin layer uniformly covering the substrate surface (depicted in blue [gray in print version]). This diagram serves as a generalized illustration for all kinds of nanocoatings.

Multifunctional Nanostructured Coatings. https://doi.org/10.1016/B978-0-443-23683-9.00001-5

significant advantage of nanocoatings over conventional polymer-based coatings is the reduced use of Volatile Organic Compounds (VOCs), the use of which poses a severe threat to the environment. Nanocoatings have become the most flexible solution in the present day in different domains. These coatings have immense potential in the medical and dental science where inhibition of bacteria and biofilm formation is very crucial. Several mechanisms are responsible for their antimicrobial activity, ranging from biochemical reactions to physical damage of the microbial cells. Antimicrobial coatings formulated from materials like silver, copper, graphene, and chitosan are a hot topic of research. For example, a recent study on silver-substituted tricalcium phosphate aimed to prevent orthopedic infections and thereby increase the success rate of implants by virtue of its antibacterial activity (interfering with the adhesion of pathogens), antifungal activity and excellent biocompatibility (Graziani et al., 2021). An interesting research on Al-$SrTiO_3/TiO_2$ nanotube (Al-STNT) nanocoating as a promising material which offers patients with dental implants an option for self-treatment because this particular nanocoating along with ultrasound stimulation has the capability to produce reactive oxygen species which helps in the disruption bacterial biofilm, along with that these coatings increases the longevity of the implant by promoting osteointegration between the implant surface and alveolar bone (Pan et al., 2024). Nanocoatings have made it possible in the aeronautical engineering domain to greatly increase the resistance to corrosion of aluminum alloys that usually undergoes under extreme weather conditions. For example, the nanocoating of polyurethane nanocomposite coating containing graphene oxide and silanized zirconium nitride (PU-GO/AMDES-ZrN) is studied for protecting aluminum alloy structures from corrosion in aviation due to its high corrosion resistance, remarkable hydrophobic nature, and strong adhesive strength between the coating and the substrate (Jeeva, Thirunavukkarasu and Xavier, 2024). In transportation engineering, the application of photocatalytic coatings on road surfaces is a breakthrough in managing emission of exhaust gases. A recent study on nano g-C_3N_4 to be used as a photocatalytic coating on roads showed remarkable degradation efficiency for both water and air pollutants (Xie et al., 2024a). Similarly, a study on the integration of tungsten oxide and tungsten sulfide quantum dots into graphene oxide/TiO_2 proved to be an excellent photocatalytic material which could be used as pavement materials (Xie et al., 2024b). Marine engineering also focuses on polymer-based nanocoating in order to prevent biofouling; especially, the role of eco-friendly ceramic-based polymer composites for their antifouling properties is unparalleled (Selim et al., 2023). We shall look into more examples of each category in the coming sections.

These coatings are designed taking into account to manage the adhesion behavior and fouling release performance in view of the environmental conditions. There are several mechanisms by which the nanocoatings contribute to the wear resistance. One of such mechanisms is known as "*grain refinement*" where the incorporation of nanoparticles to the coating materials results in the formation of small "grains" which prevents the formation and propagation of cracks and hence resulting in improved strength. Also, this causes and increased surface area between the substrate and the coating which protects the material from the penetration of water and thereby protecting it from corroding. There are some other mechanisms, for example, zinc nanoparticles undergo electrochemical mechanisms on steel surface to prevent corrosion, by acting as a sacrificial anode. In the next sections, we will take a closer look at the various types of nanocoatings along with their salient features. We will also explore their unique qualities that make a significant impact on improving materials and addressing different issues.

1.2 Types of nanocoatings

There are several types of nanocoatings reported in the scientific literature, each having multitude of applications. Following are some of the most important types of nanocoatings used in the present day applications:

1. **Ceramic-based nanocoatings**
   a. **Oxide-based ceramics:** They are made up of metal oxides such as Alumina (Al_2O_3), the oxide form of aluminum commonly found in nature as corundum, and is a crystalline form of aluminum oxide, zirconia (ZrO_2), titania (TiO_2) etc. These oxides possess remarkable level of hardness making them suitable for applications where there is high amount of wear and tear, chances of corrosion and high temperatures are likely to be present. Alumina-based nanocoatings have also gained ample attention in the field of electronics, since they possess excellent dielectric properties by providing high dielectric strength, reducing dielectric loss, inhibiting irregular grain growth, preventing leakage current, and causing lattice distortion in materials (Ban et al., 2020). Alumina-based coating has also been tested in lithium-ion battery separators which help in increasing the electrolyte affinity, increasing mass uptake, and enhancing the thermal resistance (J. D. Xie et al., 2019). Zirconia-based nanocoating has been reported to be a suitable coating on titanium dental implants owing to their excellent biocompatibility, chemical inertness, impressive mechanical strength, esthetic value, antimicrobial activity against oral pathogens, etc (Das et al., 2016). A study demonstrated the combined potential of ZrO_2 and cerium oxide (CeO_2) nanoparticles in nanocoatings to enhance microhardness, wear resistance, and corrosion resistance leading to superior abrasion resistance and reduced adhesive wear compared to coatings containing only nickel or nickel with ZrO_2 nanoparticles(Y. Zhang ct al., 2024).
      i. **Nitride-based ceramics:** Nitrides are a class of inorganic compounds which are composed of nitrogen atom covalently or ionically bonded to one or more metal ions and as a result these compounds exhibit hardness, chemical stability, and high melting points, making them desirable for coating applications (Liu et al., 2012; Pogrebnjak et al., 2014; Robertson, 1991). Titanium nitride (TiN) stands out as the most widely used coating among the nitrides in industrial and medical contexts, attributable to their biocompatibility and wear resistance. TiN is also widely used a coating on medical devices such as orthopedic implants not only due to their biocompatibility and cytocompatibility but also due to the excellent tribological features such as reduced friction and improved hardness and a consequence of that life span of such medical devices may be increased (Liu et al., 2012; Santecchia et al., 2015; Subramanian et al., 2011; Van Hove et al., 2015). Another nitride-based coating which has gained substantial attention in the scientific community is aluminum nitride (AlN) known for its excellent thermal conductivity and electrical insulation properties in addition to the hardness, abrasion resistance, thermal and chemical stability, optoelectric properties, and low coefficient of thermal expansion making it a good choice in thermal and electrical applications (Balasubramanian et al., 2004). Studies have explored incorporating AlN into composite coatings to enhance surface properties without increasing material weight (Riquelme et al., 2020). Also, the studies on the

feasibility of forming aluminum nitride in aluminum alloys have been conducted which shows the potential to broaden the application of these coatings across different material systems (Hou, Mutharasan and Koczak, 1995). A study conducted on chromium aluminum nitride (CrAlN) which is a ternary nitride made by adding aluminum in the binary chromium nitride (CrN) coating resulted in excellent corrosion and wear and tear resistance (Yu et al., 2013).

ii. **Carbide-based ceramics:** The ceramic-based coatings have shown to be integral materials used in industries where high-wear environments and elevated temperatures prevail. These ceramics provide remarkable durability and long-life to tools and machinery by virtue of their abrasion resistance, exceptional hardness, and high thermal conductivity. Carbides can be binary and ternary, the former containing two elements, i.e., one metal and carbon forming a carbide and the latter is more complex and contains three elements including carbon. Titanium carbide (TiC), tungsten carbide (WC), silicon carbide (SiC), boron carbide (B4C), and hafnium carbide (HfC) are few examples of binary carbides and are extensively used as coating materials on tools (Kustas et al., 1997). The examples for tertiary carbides are tantalum aluminum carbide (Ta-Al-C) and titanium silicon carbide (Ti-Si-C). Tantalum aluminum carbide (Ta-Al-C) and titanium silicon carbide (Ti-Si-C) are used in high-temperature environments where materials need to withstand extreme heat and thermal cycling. They are used in aerospace components, turbine engines, and heat-resistant coatings for industrial furnaces (Sun et al., 2022). In one study, nanostructured zirconium carbide (ZiC) coatings were deposited on multiwalled carbon nanotubes which could be used as reinforcing fillers and adhesives (Vilkov et al., 2023). Another study on C/Si and C/W multilayers, where irradiation was done with Xe and Ar ions, resulting in the formation of carbides (SiC and WC) at the interfaces, finding excellent scope in the semiconductor manufacturing industry(Fogarassy et al., 2023). Ultrastretchable conductors composed of 2D $Ti_3C_2T_x$ nanosheets and single-walled carbon nanotubes have been developed for wearable antennas and electromagnetic interference shielding (Li et al., 2020).

2. **Metal-based nanocoatings:** These are thin layers of nanoparticles or nanoclusters one or more metals coated over a suitable substrate at nanoscale, synthesized using controlled deposition techniques like physical vapor deposition (PVD), CVD, electroplating, electrodeposition, etc (Farooq et al., 2022). These coatings have good conductivity and catalytic activity. In a study, the nickel coating in the trilayered composite material served to enhance conductivity, facilitate the surface grafting of keratin, and enable the metalization process. This helped achieve electromagnetic transparency of the material, thermal sensitivity, and deicing functionality, which are crucial for many military applications (Zhao et al., 2024). It is a well-known fact that silver nanoparticles have the ability to inhibit microbes, in light of that, a silver nanocoating was directly applied on dentine surface of teeth to prevent ailments such as dental caries, gingivitis, periodontitis, stomatitis, and peri-implantitis. Interestingly, this coating is compatible with biological fluids (Besinis, De Peralta and Handy, 2014). A study was conducted on the silver nanocoatings on polyester (PES) stents for minimizing the occurrence of coronary stent infections. The coating was composed of a combination of metallic silver as well as oxides of silver. The antimicrobial activity of these nanocoatings prevents bacterial colony formation, thereby preventing secondary infections (Rebelo et al., 2020). Nanocoating of copper oxide (CuO) on stainless steel has shown excellent corrosion resistance along with antibacterial activity against pathogens like *Staphylococcus aureus* and *Escherichia coli,* proves to be a good candidate for biomedical applications (Karabulut et al., 2023).

1. **Nanocomposite coatings with metal and polymer matrices**
   a. **Metal matrix nanocomposite-based coatings:** Here, a metal matrix is reinforced with nanoparticles for enhanced properties such as corrosion resistance, mechanical strength, wear resistance, etc. A research on aluminum oxide (Al_2O_3) and silicon carbide (SiC) nanoparticles incorporated into a matrix of nickel metal enhanced the corrosion resistance of the metal by providing a wide passivation range, meaning protecting the metal surface from further corrosion. Along with that, a high resistance to mechanical wear was also noted (Dehgahi, Amini and Alizadeh, 2016). Research studies conducted on Cu−Sn−Zn coatings with incorporated TiO_2 nanoparticles have shown enhanced durability, mechanical properties, and corrosion resistance, making them potentially suitable for decorative purposes (Gao et al., 2018). Incorporation of -α-Al_2O_3 particles in a nickel matrix resulted in enhanced surface hardness and wear resistance of approximately 2−3.5 times compared to unreinforced Ni coatings (Gül et al., 2009). Research on Ti_6Al_4V metal matrix composites studies the effect of incorporation TiC nanoparticles as a reinforcement via laser-engineered net shaping, resulted in the formation of Ti-Ni intermetallic compounds which contributed to the improvement in grain size and increased mechanical strength (Zheng et al., 2008).
   b. **Polymer matrix−based coatings:** Polymer matrix nanocoatings contain nanoparticles embedded homogeneously in a polymer matrix (depicted in Fig. 1.2) and find various application in fields such as packaging, automotive, and solar energy owing to their advantages such as improved stability, dissolution, and mechanical properties, to name a few. One study reported the use of epoxy resin along with nanochitosan as a reinforcing agent resulted in the improvement of the corrosion protection properties of the coating on mild steel substrates (Wonnie Ma et al., 2017). Another study introduces α-zirconium phosphate/polyurethane (ZrP/PU) nanocomposites for corrosion protection, with ZrP nanoplatelets synthesized via reflux method and modified with polyetheramine surfactant. The resulting coating has improves moisture barrier and reduced water vapor transmission compared to pure PU coatings (Huang et al., 2017). A study on epoxy-polysiloxane hybrid coatings modified with polyetheramines for anticorrosive properties found that coatings with shorter-chain polyetheramines (D230, T403) showed improved hardness and wet adhesion compared to their longer-chain counterparts (D400, THF100). Also, the corrosion resistance was found to increase with D230 and T403 but decrease with D400 and THF100, due to variations in cross-link density (Yao et al., 2014).

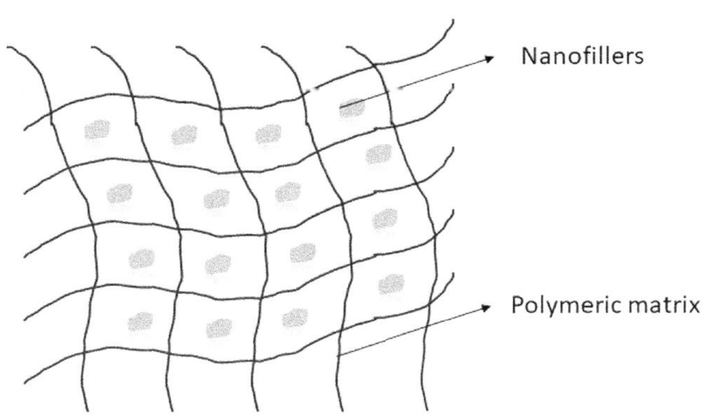

Figure 1.2 Schematic representation of polymer matrix−based nanocoatings. The diagram depicts a polymer matrix with embedded nanoparticles (shown as *green* [*light gray* in print version] dots) forming a nanocoating layer on a substrate surface.

1.3 Salient features of nanocoatings and recent developments

Nanostructured coatings exhibit a range of unique properties that set them apart from conventional coatings. These properties contribute to their enhanced performance in various applications. Here are some key properties of nanocoatings:

1.3.1 Enhanced mechanical strength

Nanostructured coatings often display improved hardness, toughness, and scratch resistance compared to conventional coatings. This is attributed to the nanoscale arrangement of materials, leading to better mechanical properties. Nanostructured coatings, such as those synthesized by Zhong et al., 2014, have been found to exhibit improved mechanical properties, wear resistance, and high-temperature corrosion resistance. These enhancements are attributed to the nanoscale arrangement of materials, which reduces defects and increases the density of the coatings. The use of nanostructured powders in plasma-sprayed coatings, as demonstrated by Li et al. (2004), has been shown to further improve wear resistance due to the reduction of micrometer-sized defects. Mishra (2020) also highlights the potential for toughness enhancement in nanostructured hard coatings through the use of nanocomposite and multilayer coatings, which can significantly increase toughness while maintaining hardness.

1.3.2 Superior wear resistance

Nanocoatings provide enhanced wear resistance, making them suitable for applications where surfaces are subjected to friction, abrasion, or contact with other materials. The wear resistance of nanocoatings is contributed by grain refinement which means increasing the number of grain boundaries and thereby preventing the formation and propagation of cracks. Also, the nanoeffects of the materials used promote atomic migration thereby preventing crack formation (Gu et al., 2020). This property is particularly valuable in industries like manufacturing and transportation. These coatings are particularly beneficial in transportation, aerospace, and manufacturing industries due to their superior mechanical properties and stability. Nanocoatings, particularly nanostructured and composite coatings, have been shown to significantly enhance wear resistance in various applications, including those involving severe operating conditions (Erdemir and Voevodin, 2010). The use of nanostructured coatings has also been found to improve wear resistance in machining and wear-resistant applications (Malshe, Jiang and Dhamdhere, 2002). Furthermore, the orientation of nanoparticles in these coatings can further enhance their wear resistance. They found out that using a nozzle spray coating method, halloysite nanotubes could be aligned in the in-plane and out-of-plane directions and in various randomly oriented states to enhanced the wear resistance (Song et al., 2016).

Adding to that, some of the recent research studies show the use of yttrium (III) oxide (Y_2O_3) as an additive in ceramic composite coatings to reduce the coefficient of friction as well as enhancing the tribological behavior including wear resistance, improved hardness, and toughness by virtue of their compact microstructure (Jia et al., 2023; Ma et al., 2024; C. Zhang et al., 2024).

1.3.3 Tailored surface functionality

The surface properties of nanocoatings can be precisely engineered at the nanoscale. This allows for the incorporation of specific functionalities, such as hydrophobicity, oleophobicity, antifouling, and self-cleaning, depending on the application requirements. The ability to engineer the surface properties of nanocoatings at the nanoscale has led to the development of coatings with specific functionalities such as hydrophobicity, oleophobicity, antifouling, and self-cleaning. Wei et al. (2016) demonstrated the fabrication of environmentally safe, transparent, and self-cleaning coatings with antifouling properties. Kamegawa, Irikawa and Yamashita (2017) further enhanced these properties by designing a multifunctional surface with superhydrophobic and self-cleaning properties. These are a few of the studies which collectively highlight the potential of nanocoatings in various applications, from self-cleaning glazing to antiicing surfaces. Table 1.1 summarizes a few nanocoatings for their surface properties and applicability.

Table 1.1 Nanocoatings with enhanced surface properties.

Nanocoating type	Surface property	Application	References
Laser-Induced periodic surface structures (LIPSS) combined with atomic layer deposition molybdenum disulfide (ALD MoS_2) nanocoatings	Enhanced friction and hydrophobic performance	On contact interfaces	Xing et al. (2020)
Poly(tannic acid)/Ceria nanozymes (PTA/CeO_2NZs)	Antioxidative activity	Bone repair, protection against oxidative stress	Yang et al. (2021)
Acid-catalyzed silica sol/ Hydrophobic silica nanoparticles (ACSS/ HSNs)	Superhydrophobicity	Optical devices, architectural glass, automotive windshields, and marine coatings	Wei et al. (2016)
Acid-catalyzed silica sol/ Solid silica nanoparticles (ACSS/ SSNs)	Superhydrophobicity, antifogging effects	Optical devices, architectural glass, automotive windshields, and marine coatings	Wang et al. (2021)

1.3.4 Improved corrosion resistance

Nanostructured coatings can offer superior corrosion protection by forming a more effective barrier against environmental factors. This property is essential in applications where materials are exposed to harsh or corrosive conditions. Nanostructured coatings, particularly organic nanocomposite coatings, have been shown to significantly enhance corrosion protection by forming a more effective barrier against environmental factors (Muresan, 2023). These coatings, which can be made of layers under 100 nm, have been found to improve the chemical, mechanical, and optical properties of traditional coatings, leading to better long-term corrosion resistance (Farag, 2020). The application of nanotechnology in corrosion protection has also been found to be effective in decreasing corrosion rates and improving the lifetime of materials and devices (Salam Hamdy, 2010). The development of "smart" and self-healing nanostructured coatings is an area of active research, with potential for further enhancing corrosion protection (Wang et al., 2019).

1.3.5 High thermal stability

Nanocoatings, particularly those with nanostructured superhard materials, exhibit excellent thermal stability, making them suitable for high-temperature applications in aerospace, automotive, and electronic industries. These coatings, with their fine grain size, also show superior mechanical properties at high temperatures, including high-temperature corrosion resistance (Gao, Liu and Li, 2001). The thermal stability of nanocoatings is a complex phenomenon and may be attributed to their nanoscale dimensions or nanocrystalline structure as previously mentioned. Some of those factors are interdiffusion control, oxidation resistance, energy input during growth, cooling rate, etc (Musil, Kos and Baroch, 2024). A research study on multilayer TiSiN/TiAlN nano-coating demonstrated thermal stability which may be attributed to its slower oxidation onset, dense microstructure with reduced grain size, and resistance to thermal fatigue, thereby enabling them withstand high-temperature oxidation while maintaining the structural integrity and resist thermal cracking and vaporization and are suitable for high speed machining applications (Zha et al., 2023). Similarly, a coating of silica-based nanocomposite on molybdenum ditellurides (MoT_2) showed the former due to its remarkable thermal stability acts as a heat barrier to MoT_2, without tampering the original two-dimensional structure of the crystals. This makes them suitable for electronic and optoelectronic applications (Heo et al., 2023). A study on nanocomposite Mo-B-C coatings demonstrate improved hardness and resistance to fracture after annealing, further highlighting their thermal stability (Gleich et al., 2018).

1.3.6 Enhanced adhesion

Nanocoatings often show improved adhesion to substrates due to their ability to conform at the molecular level. This property (Malshe, Jiang and Dhamdhere, 2002) is crucial for long-lasting and durable coatings. A nanocoating made from CsgA proteins found in biofilms is shown to have enhanced cell adhesion since it promotes rapid attachment, proliferation, and stabilization of the cytoskeleton on the substrate surface and has the potential for improving the performance of medical implants especially by preventing immune rejection (Yang et al., 2023). A nanocoating comprising tannic acid (TA)-functionalized hexagonal boron nitride nanosheets (TA-BNNS), TA, and polyvinyl alcohol (PVA) inspired by the interfacial hydrogen bonding found snails where the hydrogen bond between PVA and TA is dynamically regulated resulting in superior adhesion (Huang et al., 2024). A novel nanocoating consisting of porous coordination polymer based on zeolitic imidazole framework-8 (ZIF8) combined with cerium conversion showed superior adhesion due to decrease in delamination area and radius (Falahaty marvast et al., 2023). A polysaccharide-based nanocoating designed to encapsulate bacteria is attached to macrophages and helps maintain long-lasting stimulation and control over cells in vivo, enabling infused macrophages to reprogram tumor-associated macrophages and slow tumor in murine cancer cell model (An et al., 2024). A self-adhesive nanocoating made up of interconnected networks of MXene/cellulose nanofibers (MXene/CNF) may be used for fabricating multifunctional polydimethylsiloxane (PDMS) foam nanocomposites for use in intelligent healthcare monitoring and fire-safe thermal insulation (Chen et al., 2023). Nanoporous coatings have been shown to have excellent adhesion to polymer substrates (Poxson et al., 2012). The mechanical properties and adhesion of nanolayer coatings have also been found to be superior to monolayer coatings, further highlighting the importance of this property (Khlifi and Larbi, 2014). Additionally, the concept of nanocoupling has been proposed as a method to improve the interfacial adhesion of biodegradable polymer coatings on metal surfaces. The study explores enhancing adhesion and controlled drug release in drug-eluting metal stents by employing surface-initiated ring-opening polymerization of L-lactide, resulting in nanocoupling that significantly improves interfacial adhesion stability between the metal surface and the polymer layer (Choi et al., 2011).

1.3.7 Customizable optical properties

Nanocoatings can be designed to manipulate light interactions, leading to unique optical properties. This is exploited in applications such as optical filters, antireflective coatings, and coatings with specific optical functionalities. Kats et al. (2013) introduce a new type of optical coating based on interference effects in highly absorbing media, offering potential applications in photodetectors, solar cells, and optical filters. Askar et al. (2018) present a scalable technology for fabricating nanoparticle antireflection coatings on complex optical surfaces, which can significantly improve light transmission. Another study demonstrates the correlation between surface profiles of nanostructures and the behavior of scattered light, providing insight into the design of nanostructured antireflection coatings. These studies collectively underscore the potential of nanocoatings in various optical applications (Chao et al., 2011). A recent study on SiO_2 coated with TiO_2 showed antireflection and self-cleaning properties and finds application in solar cells and optical devices (Z. Li et al., 2023). A similar study done on hydrophobic antireflective nanocoatings on polycarbonate synthesized using dip-coating technique showed improved transmittance and mechanical stability and may be explored for optical and display applications (Yan et al., 2023). For improving the photovoltaic efficiency of crystalline silicon solar cells, TiO_2 honeycomb arrays coated on flat crystalline silicon (c-Si) wafers were studied and an enhancement in optical light-trapping effects was observed leading to a significant increase in efficiency up to 30% in planar interdigitated back-contact solar cells without compromising their electrical performance (Santos et al., 2023). An experimental study on superhydrophobic nanocoatings applied over the surface of photovoltaic panels to reduce the accumulation of dust and soiling showed lowering of the operating temperatures of the panels and as a result improvement in power generation efficiency and extended lifespan of the solar cells by reducing degradation caused by elevated temperatures could be achieved (Ehsan et al., 2022). A study on encapsulation of organic solar cells using poly(hexafluorobutyl acrylate) nanocoating deposited via initiated chemical vapor deposition technique proved to be efficient in mitigating the detrimental effects of oxygen and moisture which dramatically increases the stability and longevity of organic photovoltaic devices (Yurtdaş, Tozlu and Karaman, 2023). Pseudoboehmite nanoflakelets for superhydrophobic coatings on solar glass for PV panels showed high optical transmittance (>93%) and solar performance with sustained efficiency even after exposure to dust particles owing to the ultrahigh water contact angles (>175 degrees) and almost negligible roll-off angles (<1 degree) and hence providing excellent self-cleaning action (Pullithadathil et al., 2020).

1.3.8 Antimicrobial properties

Certain nanocoatings incorporate antimicrobial agents or nanomaterials with inherent antimicrobial properties. Nanocoatings with antimicrobial properties play a crucial role in maintaining hygiene in places like hospitals and food packaging (Mohammad and Ahmad, 2024; Yousaf et al., 2014). These coatings are designed to prevent the growth of harmful microorganisms. Recent research has shown that antimicrobial nanocoatings are effective against drug-resistant pathogens. Scientists have discovered that the unique structures of nanomaterials contribute to their antimicrobial abilities. For instance, a biogenic CuO/guava extract nanocoating acts as a barrier which helps to maintain the freshness of packaged fruits and extending their shelf life by preventing moisture loss and optimizing the ripening process (Sharma et al., 2024). Immunomodulatory IL-23 receptor antagonist peptide nanocoatings synthesized on titanium for dental implants is helpful in the reduction of inflammation by regulation cytokine production and preventing peri-implantitis and implant failure to a great extent (Pizarek, Fischer and Aparicio, 2023). A silver nanoparticle-based coatings has been developed with strong antibacterial effects and no harm to cells (Gallo et al., 2016). A study of silver nanocoating when applied onto electrospun polysulfone (PSU) membranes showed multifold increase in the antimicrobial activity and thus could be used in water treatment (Fiaschini et al., 2022). A nanocoating comprising poly(vinyl alcohol) (PVA) and chitosan-containing *Agrimonia eupatoria* L. extract showed great antibacterial activity of wound dressings by effectively inhibiting the growth of *S. aureus* and *Pseudomonas aeruginosa* along with maintaining biocompatibility with normal human dermal fibroblast cells (Mouro, Dunne and Gouveia, 2021).

1.3.9 Electrical conductivity or insulation

Depending on the composition, nanocoatings can exhibit electrical conductivity or insulation. This property is utilized in electronics, sensors, and other applications where electrical properties are critical. Nanocoatings are significant in changing how materials conduct or insulate electricity. Studies have revealed that these coatings can reduce electrical conductivity and enhance insulation in materials, for example, LDPE/Al_2O_3 nanocomposites and EVA/PP insulation (Montanari et al., 2004). Researchers found that nanocoatings are effective in improving insulation properties. Also, it has been found out that, in materials like cotton and silk fibers, nanocoatings containing carbon nanofillers greatly enhance electrical conductivity (Narayanan, Karpagam and Bhattacharyya, 2015). The widespread use of conductive nanocomposites highlights the important role of tiny conductive fillers and the potential of carbon-based nanostructures, such as carbon nanotubes and graphene, in achieving optimal electrical performance. Table 1.2 highlights a few recent nanocoatings with remarkable electrical properties.

1.3.10 Nano-scale thickness

Nanocoatings can achieve thin film thicknesses at the nanoscale, allowing for precise control over coating thickness. This can be advantageous in applications where minimal coating thickness is desired. One such approach involves nonline of sight (plasma) techniques, which enable the deposition of nanometer-scale coatings. Microwave irradiation is another method highlighted, showcasing the ability to achieve

Table 1.2 Electrical properties of nanocoatings.

Nanocoating	Electrical properties	Application	References
Ag nanocoating on silk fibroin (SF)	Improved electrical conductivity upon addition of Ag nanoparticle	Wearable electronic devices, biomedical applications	Shivananda et al. (2023)
Conductive polymer brushes (PMTM-co-P3MT)	High ohmic conductivity	Thin-film memristive coatings	Wolski et al. (2024)
ITO@MXene nanocoating on PET film	Low infrared-emissivity, high solar absorptivity and electrical conductivity	Infrared stealth, building daylighting, thermal management	Du et al. (2023)
CoO-nafion nanocoating on SWNTs	Improved specific capacitance, optimizing charge transfer	Next-generation supercapacitor technologies	Gong (2024)

ultrafast nanoscale polymer coatings on porous 3D structures. Additionally, methods like X-ray lithography, nanografting, plasma spray, electroplating, and electrodeposition, as explored by researchers, contribute to the formation of nanostructured coatings. These techniques utilize advanced processes as mentioned before, allowing for the precise application of coatings with thickness in the nanometer range. Table 1.3 gives an idea about the thickness of nanocoatings used in various applications.

Table 1.4 gives some of the recent advances in wear resistant nanocoatings from the literature.

1.4 Methods of synthesis of nanostructured coatings

Various synthesis techniques are employed to fabricate nanocoatings with distinct properties suitable for different applications. CVD is a widely used method for coating

Table 1.3 Nanocoating thickness and functions.

Nanocoating	Thickness (nm)	Function/Property/Use	References
TiO_2	50	Accelerate bone formation on zirconia implants	Li et al. (2022)
Covalent organic framework (COF)	100–300	Active and angle-independent structural coloration	Jiang et al. (2024)
SiO_2-modified APTMS	12.0–29.3	Amplify microwave absorption	Guo, Lu and Jian (2024)
Li1.3Al0.3Ti1.7$(PO_4)_3$ (LATP)	8	Improve interfacial stability between NCM811 cathode and Li5.5PS4.5Cl1.5 solid electrolyte in all-solid-state batteries	Zou et al. (2023)
Amorphous titanium (IV) oxide (A-TiO_2)	1–11	Acts as a barrier layer, inhibiting the diffusion of electrolyte ions into graphite. Protects the graphite from further reactions with electrolyte and improves cycling performance of the graphite negative electrodes.	Wu et al. (2023)
MoS_2/ZnO/PDMS superhydrophobic nano-antiicing coatings	5	Provides superhydrophobic properties, reduces adhesion strength to ice decreases icing thickness and area on wind turbine blades, and improves aerodynamic characteristics and safety performance of wind turbines.	Liu et al. (2023)

Table 1.4 Concise overview of different types of nanocoatings, their functions, and properties.

Chemical nature of nanocoating	Function/Properties	References
Ni, Ni−ZrO_2, Ni−ZrO_2−CeO_2 nanocoatings	Improved surface morphology, microhardness, abrasion resistance, corrosion resistance	Ma et al. (2024)
Epoxy resin coatings filled with MBH	Wear resistance, transparency	N. Li et al. (2023)
Ni/W-SiC nanocoatings	Improved surface characteristics, corrosion resistance, wear resistance	Xia et al. (2024)
Ni-SiC composite nanocoatings (CNCs)	Abrasion resistance, microhardness	Liu, Wang and Guo (2024)
TiCrN multilayer nanocoating	Enhanced wear resistance, hardness, corrosion resistance	Meysami et al. (2024)
MXene/black phosphorene nanocoatings (black phosphorus, Ti3C2Tx nanodispersion ink, poly(dimethylsiloxane))	Multifunctionality, fire safety, water resistance, heat response, flame retardancy, electromagnetic interference shielding effectiveness	W. Li et al. (2023)
Tetraethoxysilane (TEOS) base-catalyzed silica nanoparticles (SNs), acid-catalyzed silica sol (ACSS)	Antireflective properties, high transmittance, mechanical stability	Yan et al. (2023)
Pyrolytic carbon	Corrosion resistance, hydrophobicity, surface morphology, chemical composition, electrochemical behavior	Rashid et al. (2024)
Nb_2O_5 nanocoating	Improved electrochemical properties, enhanced thermal stability, maintained specific capacity at high temperatures	Zhang et al. (2021)
Multilayer coating of titanium silicon nitride (TiSiN) and titanium aluminum nitride (TiAlN) based nanocoating	Excellent thermal stability, oxidation resistance, resistance to ultrahigh frequency thermal fatigue shock	Zha et al. (2023)
Layered silica-based nanocomposite (SNC) coating	Thermal-protective, oxygen-resistant, preserves crystal structure of 2D materials during laser thinning process	Heo et al. (2023)
Copper oxide (CuO)	Selective solar absorption, thermal stability	Welegergs et al. (2021)
ZIF-8/Graphene oxide	Ultrafast response optical sensing, gas detection	Huang et al. (2022)
MnO_2 (birnessite-type manganese dioxide) and poly(3,4-ethylenedioxythiophene) (PEDOT) composite	Enhanced osteoblastic electrical stimulation, promote adhesion, proliferation, and differentiation of MC3T3-E1 osteoblastic cells under electrical stimulation	Liu et al. (2021)

metallic or ceramic compounds, which relies on vapor phase constituents introduced into a reaction chamber. Physical Vapor Deposition (PVD) is another technique performed under vacuum conditions, the principle of which is the evaporation of a solid or molten source. Spray coating involves the application of coating material through a nozzle onto a substrate, while the sol-gel process utilizes chemical reactions at low temperatures to create 3D networks. Self-assembly, involving the spontaneous arrangement of disordered components, includes static and dynamic processes, with examples like Langmuir—Blodgett (LB) films and Layer-by-Layer (LbL) deposition. Another prominent method is extrusion which is basically a polymer-processing technique employed for incorporating nanoparticles into polymeric compounds, especially in the production of nanocomposites. Each method has its unique advantages and applications and hence provides a diverse range of nanocoating solutions.

1.4.1 Chemical vapor deposition

CVD is a versatile method for producing high-quality thin films and coatings with precise control over structure and properties. It has found applications in a wide range of industries, including tribology, where it is used to create functional surfaces resistant to wear and corrosion. The method's unique features, such as its ability to coat complex-shaped components and fabricate nano-devices make it a valuable tool in materials processing (Carlsson and Martin, 2010; Makhlouf, 2011). This bottom-up manufacturing process can fabricate a variety of coatings, including monolayers, multilayers, nanocomposite layers, and functionally graded layers, at relatively low processing temperatures (McCarthy et al., 2014). The unique feature of CVD, its nonline-of-sight-deposition capability, allows for the coating of complex-shaped components and the fabrication of nanodevices, carbon—carbon composites, and ceramic matrix composites (Choy, 2003). The method can also be combined with techniques, such as sol—gel and plasma-enhanced CVD to produce functional nanostructured coatings with improved mechanical properties (Costacurta et al., 2011). CVD can be used to create controlled nanostructured silver coatings with a multitude of applications in catalysis, photo-activity, and biocidal surfaces (Sun, Yuan and Gao, 2021).

1.4.2 Physical vapor deposition

Physical Vapor Deposition (PVD) is a coating technique that operates under vacuum conditions. It involves the transfer of material on an atomic level. In contrast to CVD, PVD utilizes solid precursors for deposition. The process consists of evaporation of the material through a high-energy source, followed by the transport of vapor to the substrate. During transportation, a reaction occurs between the metal atoms and a reactive gas, leading to the deposition of the coating onto the substrate surface. While having positives such as improved coating properties and environmental friendliness, PVD has some short comings such difficulty in coating complex shapes, high costs, and process complexity (Makhlouf, 2011). Recent advances in physical vapor deposition (PVD) have been driven by the development of plasma-assisted processes, such as plasma-assisted physical vapor deposition. These processes have been enhanced by improved ionization systems, which have led to higher ionization efficiencies in commercial systems. Vacuum arc deposition (VAD) has also seen significant progress, with the development of filtered VAD and hot electrode vacuum arcs, leading to the production of high-quality, macroparticle-free films (Beilis, 2020; Geyao et al., 2020; Vladoiu et al., 2020). Directed vapor deposition is a novel PVD technology that combines low vacuum electron beam evaporation with vapor transport in a flowing gas stream, offering new possibilities for film structures and compositions (Schulz et al., 2020).

1.4.3 Spray coating

Spray coating is a widely used industrial technique which deposits materials on substrates through nozzles, enabling coating of diverse surfaces with organic lacquers. Thermal spraying, a common method, involves melting or heating coating materials for high deposition rates on large surfaces. Cold spraying, a low-temperature variation, requires high velocities for effective bonding, resulting in improved strength and adhesion. Methods like plasma spraying, high-velocity oxy-fuel spraying (HVOF), and high-velocity air fuel offer various coatings with properties influenced by parameters like powder size distribution, feed rate, distance to substrates, and spray velocity (Makhlouf, 2011; Nguyen-Tri et al., 2018; Yilmaz and Yorgancioglu, 2018). Cold spray is a solid-state process, resulting in unique coating characteristics, but it can lead to a loss of ductility in the coating (Karthikeyan, 2007). On the other hand, thermal spray processes, including flame spraying, wire spraying, plasma spray, HVOF, and detonation gun, can enhance wear and corrosion resistance, and have a wide range of applications (Cramer and Covino, 2003). The use of suspensions in thermal spraying can further enhance coating properties, but the economic efficiency of this process depends on various factors (Toma et al., 2015). Overall, thermal spray coating technology offers a range of benefits, including increased corrosion and wear resistance, protection against electromagnetic interference, and metal buildup.

1.4.4 Sol-gel process

The sol-gel process is yet another versatile method for the synthesis of nanocoatings where precise control over film thickness and composition can be achieved. It involves the transformation of a colloidal solution (sol) into a gel, which is then used to form a thin film on a substrate. The sol-gel method utilizes a chemically active precursor in a solvent, undergoing hydrolysis and polycondensation to form a sol, which is subsequently polymerized into a gel, dried, and heated to produce coatings with uniform properties. Despite advantages such as wear resistance and the ability to create nanocomposite coatings, challenges include thickness constraints, susceptibility to cracking, and drawbacks for large-scale production, such as time consumption, expensive raw materials, and the use of potentially hazardous organics (Gu et al., 2020; Jiang et al., 2019; Lavollee et al., 2016). Table 1.5 provides and overview of a few the recent advances in nanocoatings synthesized by sol-gel method.

Table 1.5 Nanocoating synthesized via Sol-gel method.

Nanocoating	Properties	References
Zirconium oxide (ZrO_2)	Hydrophobicity, improved surface morphology, water contact angle and surface free energy	Catauro et al. (2020)
Zirconium in bioactive glass	Improved sintering, enhanced mechanical properties, stimulated t-m transformation, biocompatibility	Soubelet et al. (2024)
Hydroxyapatite (HA)	Homogeneous surface, biocompatibility	Catauro et al. (2020)
Aluminum and zirconium oxide (sol-gel synthesis using N,N-dimethyloctylamine and acetylacetone)	Thermal stability	Trusova et al. (2018)
Bio-based epoxy coatings (vegetable oil-based epoxy resins from cardanol and soybean oil)	Mechanical and thermal stability, anticorrosive performance	Eral et al. (2023)

1.4.5 Self-assembly

Self-assembly refers to the formation of multilayered structures through the spontaneous arrangement of oppositely charged colloidal particles, particularly involving the LbL assembly of polyelectrolytes. This technique employs synthetic and natural polyelectrolytes, metal oxides, and clay nanoparticles for versatile applications like sensing and drug delivery. Driven by electrostatic attraction, it offers simplicity, adaptability to various surfaces, and controlled multilayer thickness, providing advantages over traditional coating methods. Self-assembly, a process of spontaneous arrangement, is exemplified by LB films and LbL deposition. These depositions are methods for constructing nanostructured films. In the Langmuir trough, amphiphilic compounds self-organize at the air—water interface, forming a monolayer that can be transferred onto a substrate to create LB films. LbL deposition involves the sequential adsorption of oppositely charged species onto a substrate, resulting in multilayered structures. LB films have been explored using various materials like polymers, phospholipids, enzymes, and peptides, with their organization and homogeneity evaluated through microscopy and spectroscopy (De Oliveira, De Barros and Ferreira, 2017). Table 1.6 gives an overview of recent advances in nanocoatings synthesizes using self-assembly method.

In summary, various synthesis methods as discussed above, such as CVD, Physical Vapor Deposition (PVD), spray coating, sol-gel process, and self-assembly, offer diverse approaches to fabricate nanocoatings with distinct properties. Each method has its advantages and applications, providing a wide range of solutions for specific coating needs, from wear-resistant films to controlled nanostructured layers. These

Table 1.6 Overview of recent advances in nanocoatings synthesizes using self-assembly method.

Nanocoating type	Synthesis method	Properties	References
Hybrid nanocoatings of self-assembled organic–inorganic amphiphiles for prevention of implant infections	Self-assembly of antimicrobial peptide (AMP) amphiphiles decorated with silver nanoparticles (AgNPs)	High in vitro antimicrobial potency against multiple bacteria strains. Relevant antimicrobial activity in vivo subcutaneous infection model in rats.	Ye et al. (2022)
Carboxymethyl chitosan (CCS) and modified montmorillonite (MMT)	One-step self-assembly	Well-arranged nacre-like hierarchical microstructure, high transparency and iridescence, super-efficient fire-safety for coated substrates, self-extinguishing in burning tests, remarkable decrease in heat release and smoke production	H. Xie et al. (2019)
Boron nitride nanosheets (BNNSs) as an inorganic–organic nanocoating on common polymer dielectrics	Self-assembly to form sandwich-like structure polymer composites	Wide bandgap, improved breakdown strength of polymer dielectrics, shallow trap levels facilitate dissipation of surface charges and increase flashover voltage, improved temperature stability and dielectric constant of polymer dielectric	Wang et al. (2022)
Graphene nanocoatings	Layer-by-layer stacking of graphene nanocoatings by marangoni self-assembly (for corrosion protection of stainless steel)	Good anticorrosion property, reduced self-corrosion current density, and increased self-corrosion potential of stainless steel, layered nanosheet structure enhances corrosion protection	Ye et al. (2021)
Tannic acid-modified MXene (TA-mxene) and synthesized P, N-co-doped cellulose nanocrystals (PA@ PANI@CNC)	Layer-by-layer self-assembly (mxene and P, N-co-doped cellulose nanocrystal nanocoating onto cotton fabrics)	High EMI shielding efficiency (SE), high flame-retardancy, reduced heat release rate and smoke production, high limited oxygen index	Mao, Wang and Fu (2022)

versatile techniques contribute to the advancement of nanotechnology across various industries, offering precise control over film composition, thickness, and properties for numerous applications.

1.5 Characterization of nanocoatings

Characterization techniques are essential for understanding and evaluating the properties of nanostructured coatings. Mechanical testing techniques, such as nanoscratch testing and microbeam bending with digital image correlation, are used assess the mechanical strength, durability, and failure behavior of nanocoatings (Barthelat et al., 2003; Beake, Harris and Liskiewicz, 2013). Microstructural and chemical characterization techniques, including TEM, SEM, XPS, and XRD elucidate the internal structure, composition, and crystallography of the nanocoatings. These analyses are crucial for ensuring the coatings meet the desired specifications and perform effectively in the practical scenario. These techniques contribute to a comprehensive understanding of the nanocoatings, which aid in quality control and optimization of coating processes; some of the techniques for the characterization of nanocoatings are highlighted in Table 1.7 as reported in recent literature.

1.6 Environmental impact and sustainability of nanocoatings

Nanocoatings are vital in promoting environmental sustainability by significantly reducing the use of VOCs. Integrating 'Green Chemistry' principles into nanoscience involves using naturally occurring, biodegradable materials as reducing and capping agents, thereby decreasing the toxicity of nanomaterials and their environmental impact. Nanoparticles in water-based varnish have been shown to improve VOC emissions, offering the potential for even more significant reductions when combined with UV irradiation. Additionally, using supercritical or liquid carbon dioxide as a carrier medium for coatings can significantly decrease VOC emissions (Agarwal et al., 2023; Aich et al., 2014; Hay and Khan, 2002; Zhu, Liu and Shen, 2016). Moreover, catalytic nanomaterials exhibit the potential to remove VOC, making them suitable for VOC-reducing coatings and paints to contribute to healthier buildings (Shah and Li, 2019). However, it is to be noted that the synthesis and application of nanostructured coatings have significant consequences for the environment, especially concerning their possible impact and risks to ecological systems. The environmental effects of nanocoatings involve various complexities that need careful consideration. Multiple factors like size, capping agent, and shape can complicate these implications, influencing nanomaterials' toxicity, bioaccumulation, and physical—chemical changes. As nanohybrids gain popularity in different uses, concerns arise about how they interact with the environment, including their fate, movement, and potential harm. The specific uses of nanostructured coatings, like in nanotextiles and façade coatings,

Table 1.7 Characterization of nanocoatings.

Nanocoating name	Characterization techniques	Purpose of the characterization	References
Highly pure metal (nb, Ta, V)	X-ray diffraction (XRD), field emission scanning electron microscopy (FESEM), energy-dispersive X-ray spectroscopy (EDS), atomic force microscopy (AFM)	To find out the crystalline structure, surface morphology, elemental composition and surface roughness of nanocoatings applied on stainless steel substrates.	Aldabagh, Alzubaydi and Alhuwaizi (2023)
Ni-AlN	DHM, SEM, TEM, XRD	To assess the microstructures, composition, and microhardness of nanocoatings on mild steel surfaces.	Xia et al. (2020)
Nickel–molybdenum (Ni-Mo) alloy	DC/Radio frequency sputtering, potentiodynamic polarization Curves, EIS	To evaluate the fabrication process, corrosion protection, and the electrochemical properties of bimetallic Ni-Mo nanocoatings on mild steel immersed in saline solution.	Zamani Meymian et al. (2020)

might unintentionally lead to environmental and health effects, posing risks during waste treatment and incineration due to possible hazardous effects, agglomeration, and stability issues (Aich et al., 2014; Musee et al., 2020; Som et al., 2011; dos Santos et al., 2022).

1.7 Conclusions

To conclude, nanocoatings (or nanostructured coatings) have revolutionized material science with their exceptional properties and diverse applications. Synthesized through various physio-chemical routes, nanocoatings provide solutions to challenges in mechanical, thermal, and optical domains. Notably, they offer remarkably high mechanical strength, superior wear resistance, tailorable surface functionality, improved corrosion resistance, high thermal stability, enhanced adhesion, and tailorable optical properties unlike conventional coatings. Their role spans from inhibiting bacteria in medical field to enhancing corrosion resistance in aeronautical engineering. Reduced use of VOC contributes to environmental sustainability. Characterization techniques ensure nanocoatings meet desired specifications, while environmental implications, involving factors like size and shape, require careful consideration for ecological systems. That having said, there are specific key issues which are to be addressed in order to explore the full potential of nanocoatings in future. Adaptation of cost-effective strategies for developing nanocoatings needs to be done for their widespread applicability. Also, additional efforts are needed to improve the adhesion between the coating and the substrate for better durability. Developing mathematical models to simulate coating processes and predict their impact on corrosion and tribological properties can aid in optimizing coating performance. These models can help understand the behavior of coatings under different conditions and refine their properties accordingly (Farooq et al., 2022). The biomimetic approach can be applied in the future to the development of nanocoatings such as the lotus leaf effect (Samaha, Tafreshi and Gad-el-Hak, 2012) to create water-repellent surfaces that are easy to clean and resistant to dirt and contaminants, shark skin effect (Chen et al., 2018) attributing to the presence of riblet structures can be used for reducing drag in fluid flow applications which can be used in improving aerodynamic performance in aviation and marine field, the mangrove leaf effect (Cui et al., 2019) to control salt and hence to create anticorrosive coatings and so on. In conclusion, nanocoatings are a noteworthy example in nanotechnology where state-of-the-art advancements coexist with a focus on environmental sustainability.

AI Disclosure

During the preparation of this work, the author(s) used OpenAI's ChatGPT to improve the readability and flow of the manuscript. After using this tool/service, the author(s) reviewed and edited the content as needed and take full responsibility for the content of the publication. After using this tool/service, the author(s) reviewed and edited the content as needed and take(s) full responsibility for the content of the publication.

References

Agarwal, N., Solanki, V.S., Pare, B., Singh, N., Jonnalagadda, S.B., 2023. Current trends in nanocatalysis for green chemistry and its applications- a mini-review. Curr. Opin. Green Sustainable Chem. 41. https://doi.org/10.1016/j.cogsc.2023.100788. Available from: http://www.journals.elsevier.com/current-opinion-in-green-and-sustainable-chemistry.

Aich, N., Plazas-Tuttle, J., Lead, J.R., Saleh, N.B., 2014. A critical review of nanohybrids: synthesis, applications and environmental implications. Environ. Chem. 11 (6), 609–623. https://doi.org/10.1071/EN14127. Available from: http://www.publish.csiro.au/nid/188.htm.

Aldabagh, D.J., Alzubaydi, T.L., Alhuwaizi, A.F., 2023. Surface characterization of stainless steel 316L coated with various nanoparticle types. International Journal of Biomaterials 2023. https://doi.org/10.1155/2023/3997281. Available from: http://www.hindawi.com/journals/ijbm/.

An, J.X., Han, Z.Y., Qin, Y.T., Li, C.X., He, J.L., Zhang, X.Z., 2024. Bacteria-based backpacks to enhance adoptive macrophage transfer against solid tumors. Adv. Mater. 36 (6). https://doi.org/10.1002/adma.202305384. Available from:. http://onlinelibrary.wiley.com/journal/10.1002/(ISSN)1521-4095.

Askar, K., Gu, Z., Leverant, C.J., Wang, J., Kim, C., Jiang, B., Jiang, P., 2018. Self-assembled nanoparticle antireflection coatings on geometrically complex optical surfaces. Opt Lett. 43 (21), 5238–5241. https://doi.org/10.1364/OL.43.005238. Available from: https://www.osapublishing.org/ol/abstract.cfm?uri=ol-43-21-5238.

Balasubramanian, C., Godbole, V.P., Rohatgi, V.K., Das, A.K., Bhoraskar, S.V., 2004. Synthesis of nanowires and nanoparticles of cubic aluminium nitride. Nanotechnology 15 (3), 370–373. https://doi.org/10.1088/0957-4484/15/3/024.

Ban, H., Fan, Y., Zhao, X., Du, A., Ma, R., Chen, K., Cao, X., Li, X., 2020. Effects of Al2O3 on dielectric properties of core–shell Ba0.7Sr0.3Ti0.9925Tm0.0103@Al2O3 ceramics. Mater. Chem. Phys. 241. https://doi.org/10.1016/j.matchemphys.2019.122376. Available from: http://www.journals.elsevier.com/materials-chemistry-and-physics.

Barthelat, F., Wu, Z., Prorok, B.C., Espinosa, H.D., 2003. Dynamic torsion testing of nanocrystalline coatings using high-speed photography and digital image correlation. Exp. Mech. 43 (3), 331–340. https://doi.org/10.1007/BF02410532. Available from: http://www.springer.com/dal/home/generic/search/results?SGWID=1-40109-70-36417762-0.

Beake, B.D., Harris, A.J., Liskiewicz, T.W., 2013. Review of recent progress in nanoscratch testing . Tribology - materials. Surface. Interfac. 7 (2), 87–96. https://doi.org/10.1179/1751584X13Y.0000000037. Available from: http://docserver.ingentaconnect.com/deliver/connect/maney/17515831/v7n2/s6.pdf?expires=1371289645&id=74573573&titleid=75001610&accname=Elsevier+BV&checksum=CE0BE311D38D0546E8AA0294851AA1E8.

Beilis, I., 2020. Vacuum Arc Plasma Sources. Thin Film Deposition. Springer Science and Business Media LLC, pp. 933–1001. https://doi.org/10.1007/978-3-030-44747-2_22.

Besinis, A., De Peralta, T., Handy, R.D., 2014. Inhibition of biofilm formation and antibacterial properties of a silver nano-coating on human dentine. Nanotoxicology 8 (7), 745–754. https://doi.org/10.3109/17435390.2013.825343. Available from: http://informahealthcare.com/nan.

Carlsson, J.O., Martin, P.M., 2010. Chemical vapor deposition handbook of deposition technologies for films and coatings. Elsevier Inc., United States, pp. 314–363. https://doi.org/10.1016/B978-0-8155-2031-3.00007-7. Available from: http://www.sciencedirect.com/science/book/9780815520313.

Catauro, M., Barrino, F., Blanco, I., Piccolella, S., Pacifico, S., 2020. Use of the sol-gel method for the preparation of coatings of titanium substrates with hydroxyapatite for biomedical application. Coatings 10 (3). https://doi.org/10.3390/coatings10030203. Available from: https://res.mdpi.com/d_attachment/coatings/coatings-10-00203/article_deploy/coatings-10-00203-v3.pdf.

Chao, Yen-C., Chen, Cheng-Y., Lin, Chin-A., He, Jr-H., 2011. Light scattering by nanostructured anti-reflection coatings. Energy Environ. Sci. 9, 3436–3441. Available from: https://pubs.rsc.org/en/content/articlelanding/2011/ee/c0ee00636j. http://xlink.rsc.org/?DOI=c0ee00636j.

Chen, D., Liu, Y., Chen, H., Zhang, D., 2018. Bio-inspired drag reduction surface from sharkskin. Biosurface and Biotribology 4 (2), 39–45. https://doi.org/10.1049/bsbt.2018.0006.

Chen, H.Y., Chen, Z.Y., Mao, M., Wu, Y.Y., Yang, F., Gong, L.X., Zhao, L., Cao, C.F., Song, P., Gao, J.F., Zhang, G.D., Shi, Y.Q., Cao, K., Tang, L.C., 2023. Self-adhesive polydimethylsiloxane foam materials decorated with MXene/cellulose nanofiber interconnected network for versatile functionalities. Adv. Funct. Mater. 33 (48). https://doi.org/10.1002/adfm.202304927. Available from: http://onlinelibrary.wiley.com/journal/10.1002/(ISSN)1616-3028.

Choi, J., Cho, S.B., Lee, B.S., Joung, Y.K., Park, K., Han, D.K., 2011. Improvement of interfacial adhesion of biodegradable polymers coated on metal surface by nanocoupling. Langmuir 27 (23), 14232–14239. https://doi.org/10.1021/la2030318.

Choy, K.L., 2003. Chemical vapour deposition of coatings. Prog. Mater. Sci. 48 (2), 57–170. https://doi.org/10.1016/S0079-6425(01)00009-3. Available from: https://www.sciencedirect.com/science/article/pii/S0079642501000093.

Costacurta, S., Falcaro, P., Vezzù, S., Colasuonno, M., Scopece, P., Zanchetta, E., Guglielmi, M., Patelli, A., 2011. Fabrication of functional nanostructured coatings by a combined sol–gel and plasma-enhanced chemical vapour deposition method. J. Sol. Gel Sci. Technol. 60 (340). https://doi.org/10.1007/s10971-011-2573-x. Available from: https://link.springer.com/article/10.1007/s10971-011-2573-x#citeas.

Cramer, S.D., Covino, B.S., 2003. Thermal Spray Coatings. ASM International, pp. 803–813. https://doi.org/10.31399/asm.hb.v13a.a0003690.

Cui, M., Wang, P.Y., Wang, Z., Wang, B., 2019. Mangrove inspired anti-corrosion coatings. Coatings 9 (11). https://doi.org/10.3390/coatings9110725. Available from: https://res.mdpi.com/d_attachment/coatings/coatings-09-00725/article_deploy/coatings-09-00725.pdf.

Das, I., Chattopadhyay, S., Mahato, A., Kundu, B., De, G., 2016. Fabrication of a cubic zirconia nanocoating on a titanium dental implant with excellent adhesion, hardness and biocompatibility. RSC Adv. 6 (64), 59030–59038. https://doi.org/10.1039/c6ra10661g. Available from: http://pubs.rsc.org/en/journals/journalissues.

De Oliveira, R.F., De Barros, A., Ferreira, M., 2017. Nanostructured Films: Langmuir-Blodgett (LB) and Layer-By-Layer (LbL) Techniques Nanostructures. Elsevier Inc., Brazil, pp. 105–123. https://doi.org/10.1016/B978-0-323-49782-4.00004-8. Available from: http://www.sciencedirect.com/science/book/9780323497824.

Dehgahi, S., Amini, R., Alizadeh, M., 2016. Corrosion, passivation and wear behaviors of electrodeposited Ni–Al2O3–SiC nano-composite coatings. Surf. Coating. Technol. 304, 502–511. https://doi.org/10.1016/j.surfcoat.2016.07.007. Available from: http://www.journals.elsevier.com/surface-and-coatings-technology/.

Du, X., Li, X., Zhang, Y., Guo, X., Li, Z., Cao, Y., Yang, Y., Wang, W., Wang, J., 2023. Visible transparent, infrared stealthy polymeric films with nanocoating of ITO@MXene enable efficient passive radiative heating and solar/electric thermal conversion. Nano Res. 16 (2), 3326–3332. https://doi.org/10.1007/s12274-022-4962-6. Available from: https://www.springer.com/journal/12274.

Ehsan, R.M., Simon, S.P., Kinattingal, S., Kumar, K.A., Sriharsha, T., 2022. Effects of nanocoatings on the temperature-dependent cell parameters and power generation of photovoltaic panels. Appl. Nanosci. 12 (12), 3945–3962. https://doi.org/10.1007/s13204-022-02633-0. Available from: https://www.springer.com/journal/13204.

Eral, S., Oktay, B., Dizman, C., Kayaman Apohan, N., 2023. Preparation of organic–inorganic bio-based epoxy coatings with high anti-corrosive performance. Polym. Bull. https://doi.org/10.1007/s00289-023-05037-4. Available from: https://www.springer.com/journal/289.

Erdemir, A., Voevodin, A.A., 2010. Nanocomposite coatings for severe applications handbook of deposition technologies for films and coatings. Elsevier Inc., United States, pp. 679–715. https://doi.org/10.1016/B978-0-8155-2031-3.00014-4. Available from: http://www.sciencedirect.com/science/book/9780815520313.

Falahaty marvast, M., Gholambargani, M., Ramezanzadeh, M., Ramezanzadeh, B., 2023. Enhancing self-healing and adhesion bonding properties: investigating the promising potential of Ce-ZIF8 MOF hybrid thin films as pH-responsive nanocoatings. Langmuir 39 (27), 9579–9594. https://doi.org/10.1021/acs.langmuir.3c01363. Available from: http://pubs.acs.org/journal/langd5.

Farag, A.A., 2020. Applications of nanomaterials in corrosion protection coatings and inhibitors. Corrosion Rev. 38 (1), 67–86. https://doi.org/10.1515/corrrev-2019-0011. Available from: https://www.degruyter.com/view/j/corrrev.

Farooq, S.A., Raina, A., Mohan, S., Singh, R.A., Jayalakshmi, S., Haq, M.I.U., 2022. Nanostructured coatings: review on processing techniques, corrosion behaviour and tribological performance. Nanomaterials 12 (8). https://doi.org/10.3390/nano12081323. Available from: https://www.mdpi.com/2079-4991/12/8/1324/pdf.

Fiaschini, N., Giuliani, C., Vitali, R., Tammaro, L., Valerini, D., Rinaldi, A., 2022. Design and manufacturing of antibacterial electrospun polysulfone membranes functionalized by Ag nanocoating via magnetron sputtering. Nanomaterials 12 (22). https://doi.org/10.3390/nano12223962. Available from: http://www.mdpi.com/journal/nanomaterials.

Fogarassy, Z., Kentsch, U., Panjan, P., Racz, A.S., 2023. Experimental and theoretical study on the production of carbide-rich composite nano-coatings. Hungary Surfaces and Interfaces 38. https://doi.org/10.1016/j.surfin.2023.102773. Available from: http://www.journals.elsevier.com/surfaces-and-interfaces.

Gallo, J., Panacek, A., Prucek, R., Kriegova, E., Hradilova, S., Hobza, M., Holinka, M., 2016. Silver nanocoating technology in the prevention of prosthetic joint infection. Czech Republic Materials 9 (5). https://doi.org/10.3390/ma9050337. Available from: http://www.mdpi.com/1996-1944/9/5/337/pdf.

Gao, W., Cao, D., Jin, Y., Zhou, X., Cheng, G., Wang, Y., 2018. Microstructure and properties of Cu-Sn-Zn-TiO2 nano-composite coatings on mild steel. Surf. Coating. Technol. 350, 801–806. https://doi.org/10.1016/j.surfcoat.2018.04.046. Available from: http://www.journals.elsevier.com/surface-and-coatings-technology/.

Gao, W., Liu, Z., Li, Z., 2001. Nano- and microcrystal coatings and their high-temperature applications. Adv. Mater. 13 (12–13), 1001–1004. https://doi.org/10.1002/1521 4095(200107)13:12/13%3C1001::AID-ADMA1001%3E3.0.CO;2-V. Available from: https://onlinelibrary.wiley.com/doi/10.1002/1521-4095(200107)13:12/13%3C1001::AID-ADMA1001%3E3.0.CO;2-V.

Geyao L. Yang D. Wanglin C. Chengyong W. 2020 1 2020/01 Procedia CIRP 10.1016/j.procir.2020.05.149 22128271 250 262 Elsevier B.V. China Development and application of physical vapor deposited coatings for medical devices: A review http://www.sciencedirect.com/science/journal/22128271 89.

Gleich, S., Breitbach, B., Peter, N.J., Soler, R., Bolvardi, H., Schneider, J.M., Dehm, G., Scheu, C., 2018. Thermal stability of nanocomposite Mo2BC hard coatings deposited by magnetron sputtering. Germany Surface and Coatings Technology 349, 378–383. https://

doi.org/10.1016/j.surfcoat.2018.06.006. Available from: http://www.journals.elsevier.com/surface-and-coatings-technology/.

Gong, K., 2024. Ionic-storage capacitance enhanced by a self-assembled CoO-Nafion nanocoating on single-walled carbon nanotubes. Electrochim. Acta 490. https://doi.org/10.1016/j.electacta.2024.144276. Available from: https://www.sciencedirect.com/science/journal/00134686.

Graziani, G., Barbaro, K., Fadeeva, I.V., Ghezzi, D., Fosca, M., Sassoni, E., Vadalà, G., Cappelletti, M., Valle, F., Baldini, N., Rau, J.V., 2021. Ionized jet deposition of antimicrobial and stem cell friendly silver-substituted tricalcium phosphate nanocoatings on titanium alloy. Bioact. Mater. 6 (8), 2629–2642. https://doi.org/10.1016/j.bioactmat.2020.12.019. Available from: http://www.keaipublishing.com/en/journals/bioactive-materials/.

Gu, Y., Xia, K., Wu, D., Mou, J., Zheng, S., 2020. Technical characteristics and wear-resistant mechanism of nano coatings: a review. Coatings 10 (3). https://doi.org/10.3390/coatings10030233. Available from: https://res.mdpi.com/d_attachment/coatings/coatings-10-00233/article_deploy/coatings-10-00233-v2.pdf.

Gül, H., Kiliç, F., Aslan, S., Alp, A., Akbulut, H., 2009. Characteristics of electro-co-deposited Ni-Al2O3 nano-particle reinforced metal matrix composite (MMC) coatings. Wear 267 (5–8), 976–990. https://doi.org/10.1016/j.wear.2008.12.022.

Guo, Y., Lu, H., Jian, X., 2024. SiO2–modified APTMS nanocoatings encapsulating FeNi: amplifying microwave absorption and corrosion resistance. Appl. Surf. Sci. 652. https://doi.org/10.1016/j.apsusc.2023.159286. Available from: https://www.sciencedirect.com/science/journal/01694332.

Hay, J.N., Khan, A., 2002. Environmentally friendly coatings using carbon dioxide as the carrier medium. J. Mater. Sci. 37 (22), 4743–4752. https://doi.org/10.1023/A:1020874532674.

Heo, S., Kim, S., Lee, S.Y., Kwak, I.H., Baik, J., Yang, H., Park, J.H., Cho, S., 2023. Thermal-protective and oxygen-resistant nanocoating using silica-nanocomposites for laser thinning of polymorphic molybdenum ditellurides. Appl. Surf. Sci. 638. https://doi.org/10.1016/j.apsusc.2023.157958. Available from: http://www.journals.elsevier.com/applied-surface-science/.

Hou, Q., Mutharasan, R., Koczak, M., 1995. Feasibility of aluminium nitride formation in aluminum alloys. Mater. Sci. Eng., A 195, 121–129. https://doi.org/10.1016/0921-5093(94)06511-x.

Huang, J., Yang, L., Wang, W., Yu, B., Cao, C., Bi, Z., Wang, J., Shen, J., Song, L., 2024. Biomimetic, highly adhesive, supramolecular nano-coatings for fire-safe polymeric foams. Prog. Org. Coating 186. https://doi.org/10.1016/j.porgcoat.2023.107995. Available from: http://www.elsevier.com/locate/porgcoat.

Huang, T.C., Lai, G.H., Li, C.E., Tsai, M.H., Wan, P.Y., Chung, Y.H., Lin, M.H., 2017. Advanced anti-corrosion coatings prepared from α-zirconium phosphate/polyurethane nanocomposites. RSC Adv. 7 (16), 9908–9913. https://doi.org/10.1039/c6ra27588e. Available from: http://pubs.rsc.org/en/journals/journalissues.

Huang, Y., Lin, W., Huang, T., Li, Z., Zhang, Z., Xiao, R., Yang, X., Lian, S., Pan, J., Ma, J., Wang, W., Sun, L.P., Li, J., Guan, B.O., 2022. Ultrafast response optical microfiber interferometric VOC sensor based on evanescent field interaction with ZIF-8/graphene oxide nanocoating. Adv. Opt. Mater. 10 (3). https://doi.org/10.1002/adom.202101561. Available from: http://onlinelibrary.wiley.com/journal/10.1002/(ISSN)2195-1071.

Jeeva, N., Thirunavukkarasu, K., Xavier, J.R., 2024. Multilayer functional polyurethane nanocomposite coating containing graphene oxide and silanized zirconium nitride for the protection of aluminum alloy structures in aerospace industries. J. Mater. Eng. Perform.

https://doi.org/10.1007/s11665-024-09392-9. Available from: https://www.springer.com/journal/11665.

Jia, D., Shi, W., Zhang, H., Wu, T., Diao, Y., Li, K., Lu, C., 2023. Effects of Y2O3 content on wear resistance and corrosion resistance of 316L/TiC coating fabricated by laser cladding. Coatings 13 (8). https://doi.org/10.3390/coatings13081348. Available from: http://www.mdpi.com/journal/coatings.

Jiang, Y., Shi, K., Tang, H., Wang, Y., 2019. Enhanced wettability and wear resistance on TiO2/PDA thin films prepared by sol-gel dip coating. Surf. Coating. Technol. 375, 334–340. https://doi.org/10.1016/j.surfcoat.2019.07.051. Available from: http://www.journals.elsevier.com/surface-and-coatings-technology/.

Jiang, Y., Chen, D., Zhang, Z., Wu, X., Tu, Y., Zheng, Z., Mao, L., Li, W., Ma, Y., Yang, X., Wang, W.J., Liu, P., 2024. Meta-structured covalent organic framework nanocoatings with active and angle-independent structural coloration. Adv. Mater. https://doi.org/10.1002/adma.202311784. Available from: http://onlinelibrary.wiley.com/journal/10.1002/(ISSN)1521-4095.

Kamegawa, T., Irikawa, K., Yamashita, H., 2017. Multifunctional surface designed by nano-composite coating of polytetrafluoroethylene and TiO2 photocatalyst: self-cleaning and superhydrophobicity. Sci. Rep. 7 (1). https://doi.org/10.1038/s41598-017-14058-9. Available from: http://www.nature.com/srep/index.html.

Karabulut, G., Beköz Üllen, N., Karakuş, S., Toruntay, C., 2023. Enhancing antibacterial and anticorrosive properties of 316L stainless steel with nanocoating of copper oxide nano-particles. Mater. Chem. Phys. 308. https://doi.org/10.1016/j.matchemphys.2023.128265. Available from: http://www.journals.elsevier.com/materials-chemistry-and-physics.

Karthikeyan, J., 2007. The Advantages and Disadvantages of the Cold Spray Coating Process the Cold Spray Materials Deposition Process: Fundamentals and Applications. Elsevier Ltd., United States, pp. 62–71. https://doi.org/10.1533/9781845693787.1.62. Available from: http://www.sciencedirect.com/science/book/9781845691813.

Kats, M.A., Blanchard, R., Genevet, P., Capasso, F., 2013. Nanometre optical coatings based on strong interference effects in highly absorbing media. Nat. Mater. 12 (1), 20–24. https://doi.org/10.1038/nmat3443.

Khlifi, K., Larbi, A.B.C., 2014. Mechanical properties and adhesion of TiN monolayer and TiN/TiAlN nanolayer coatings. J. Adhes. Sci. Technol. 28 (1), 85–96. https://doi.org/10.1080/01694243.2013.827094.

Kustas, F.M., Fehrehnbacher, L.L., Komanduri, R., 1997. Nanocoatings on cutting tools for dry machining. CIRP Annals 46 (1), 39–42. https://doi.org/10.1016/S0007-8506(07)60771-9. Available from: https://www.sciencedirect.com/science/article/abs/pii/S0007850607607719.

Lavollee, C., Gressier, M., Garcia, J., Sobrino, J.M., Reby, J., Menu, M.J., Rossi, S., Fedel, M., 2016. New architectured hybrid sol gel coatings for wear and corrosion protection of low-carbon steel. France Progress in Organic Coatings 99, 337–345. https://doi.org/10.1016/j.porgcoat.2016.06.015. Available from: http://www.elsevier.com/locate/porgcoat.

Li, J.F., Liao, H., Wang, X.Y., Normand, B., Ji, V., Ding, C.X., Coddet, C., 2004. Tribology International, vol 37. Elsevier. https://doi.org/10.1016/S0301-679X(03)00138-5. Available from: https://www.sciencedirect.com/science/article/pii/S0301679X03001385.

Li, N., Liu, Z.q., Yang, Y.x., Ying, Y.c., Zhou, Z.y., 2023. Enhanced wear resistance while maintaining transparency of epoxy resin coatings filled with novel one-dimensional magnesium borate hydroxide nanorods. Prog. Org. Coating 175. https://doi.org/10.1016/j.porgcoat.2022.107370. Available from: http://www.elsevier.com/locate/porgcoat.

Li, N., Liu, Z., Liu, G., Wang, Z., Guo, X., Guo, C., Han, J., 2022. TiO2Nanocoatings with controllable crystal type and nanoscale topography on zirconia implants to accelerate bone formation. Bioinorgan. Chem. Appl. 2022. https://doi.org/10.1155/2022/8650659. Available from: http://www.hindawi.com/journals/bca/.

Li, W., Yin, Z., Qi, L., Yu, B., Xing, W., 2023. Scalable production of bioinspired MXene/black phosphorene nanocoatings for hydrophobic and fire-safe textiles with tunable electromagnetic interference and exceeding thermal management. Chem. Eng. J. 460. https://doi.org/10.1016/j.cej.2023.141870. Available from: http://www.elsevier.com/inca/publications/store/6/0/1/2/7/3/index.htt.

Li, Y., Tian, X., Gao, S.P., Jing, L., Li, K., Yang, H., Fu, F., Lee, J.Y., Guo, Y.X., Ho, J.S., Chen, P.Y., 2020. Reversible crumpling of 2D titanium carbide (MXene) nanocoatings for stretchable electromagnetic shielding and wearable wireless communication. Adv. Funct. Mater. 30 (5). https://doi.org/10.1002/adfm.201907451. Available from: http://onlinelibrary.wiley.com/journal/10.1002/(ISSN)1616-3028.

Li, Z., He, H., Wang, X., Shou, C., Huang, M., Jin, S., Du, X., 2023. Robust SiO2@TiO2 nanocoatings with antireflection and photocatalytic self-cleaning properties by introducing commercial P25 TiO2. China Colloids and Surfaces A: Physicochemical and Engineering Aspects 664. https://doi.org/10.1016/j.colsurfa.2023.131176. Available from: http://www.elsevier.com/locate/colsurfa.

Liu, A., Deng, J., Cui, H., Chen, Y., Zhao, J., 2012. Friction and wear properties of TiN, TiAlN, AlTiN and CrAlN PVD nitride coatings. Int. J. Refract. Metals Hard Mater. 31, 82–88. https://doi.org/10.1016/j.ijrmhm.2011.09.010.

Liu, B., Liu, Z., Li, Y., Feng, F., 2023. A wind tunnel test of the anti-icing properties of MoS2/ZnO hydrophobic nano-coatings for wind turbine blades. Coatings 13 (4). https://doi.org/10.3390/coatings13040686. Available from: http://www.mdpi.com/journal/coatings.

Liu, H., Wang, H., Guo, X., 2024. Effect of electrodeposition method on the abrasion resistance of Ni-SiC composite nanocoatings. J. Mater. Eng. Perform. 33 (6), 2853–2859. https://doi.org/10.1007/s11665-023-08204-w. Available from: https://www.springer.com/journal/11665.

Liu, S., Li, K., Shen, Q., Shao, D., Huang, S., Xie, Y., Zheng, X., 2021. Electroactive MnO2-poly(3,4-ethylenedioxythiophene) composite nanocoatings enhance osteoblastic electrical stimulation. Appl. Surf. Sci. 545. https://doi.org/10.1016/j.apsusc.2020.148827. Available from: http://www.journals.elsevier.com/applied-surface-science/.

Ma, C., Wang, C., Xia, F., Wang, Q., Yan, P., Zhang, Y., 2024. Microstructure, wear and corrosion resistances of Ni–ZrO2–CeO2 nanocoatings. Ceram. Int. https://doi.org/10.1016/j.ceramint.2024.03.157. Available from: https://www.sciencedirect.com/science/journal/02728842.

Makhlouf, A.S.H., 2011. Current and advanced coating technologies for industrial applications Nanocoatings and Ultra-Thin Films. https://doi.org/10.1533/9780857094902.1.3. Available from: https://www.sciencedirect.com/science/article/abs/pii/B9781845698126500014.

Malshe, A.P., Jiang, W., Dhamdhere, A.R., 2002. Nanostructured coatings for machining and wear-resistant applications. Minerals, Metals and Materials Society, United States JOM 54 (9), 28–30. https://doi.org/10.1007/BF02709089. Available from: http://www.springer.com/materials/journal/11837.

Mao, Y., Wang, D., Fu, S., 2022. Layer-by-layer self-assembled nanocoatings of Mxene and P, N-co-doped cellulose nanocrystals onto cotton fabrics for significantly reducing fire hazards and shielding electromagnetic interference. Compos. Appl. Sci. Manuf. 153. https://doi.org/10.1016/j.compositesa.2021.106751. Available from: https://www.journals.elsevier.com/composites-part-a-applied-science-and-manufacturing.

McCarthy, M., Gerasopoulos, K., Maroo, S.C., Hart, A.J., 2014. Materials, fabrication, and manufacturing of Micro/Nanostructured surfaces for phase-change heat transfer enhancement. Nanoscale Microscale Thermophys. Eng. 18 (3), 288–310. https://doi.org/10.1080/15567265.2014.926436. Available from: http://www.tandfonline.com/toc/umte20/current.

Meysami, A., Najafabadi, R.A., Yosefnejad, T., Isfahani, T., 2024. Enhancing wear behavior and hardness of D5 cold work tool steel through TiCrN multilayer nanocoating via physical vapor deposition. Surf. Eng. Appl. Electrochem. 60 (1), 58–68. https://doi.org/10.3103/S1068375524010113. Available from: https://www.springer.com/journal/11987.

Mishra, S.K., 2020. Toughening of nanocomposite hard coatings. Rev. Adv. Mater. Sci. 59 (1), 553–585. https://doi.org/10.1515/rams-2020-0049. Available from: https://www.degruyter.com/view/j/rams.

Mohammad, Z.H., Ahmad, F., 2024. Nanocoating and its application as antimicrobials in the food industry: a review. Int. J. Biol. Macromol. 254. https://doi.org/10.1016/j.ijbiomac.2023.127906. Available from: https://www.sciencedirect.com/journal/international-journal-of-biological-macromolecules.

Montanari G.C. Fabiani D. Palmieri F. Kaempfer D. Thomann R. Mülhaupt R. 2004 10 2004/10 IEEE Transactions on Dielectrics and Electrical Insulation 10.1109/TDEI.2004.1349780 5 754 762 Italy Modification of Electrical Properties and Performance of EVA and PP Insulation through Nanostructure by Organophilic Silicates 11.

Mouro, C., Dunne, C.P., Gouveia, I.C., 2021. Designing new antibacterial wound dressings: development of a dual layer cotton material coated with poly(vinyl Alcohol)_Chitosan nanofibers incorporating Agrimonia eupatoria L. Extract. Molecules 26 (1), 83. https://doi.org/10.3390/molecules26010083.

Muresan, L.M., 2023. Nanocomposite coatings for anti-corrosion properties of metallic substrates. Materials 16 (14). https://doi.org/10.3390/ma16145092. Available from: http://www.mdpi.com/journal/materials.

Musee, N., Leareng, S., Kebaabetswe, L., Tubatsi, G., Mahaye, N., Thwala, M., 2020. Implications of Surface Coatings on Engineered Nanomaterials for Environmental Systems: Status Quo, Challenges, and Perspectives. Elsevier BV, pp. 399–416. https://doi.org/10.1016/b978-0-12-816787-8.00014-4.

Musil, J., Kos, Š., Baroch, P., 2024. Flexible hard nanocoatings with high thermal stability. Czech Republic Surface and Coatings Technology 476. https://doi.org/10.1016/j.surfcoat.2023.130195. Available from: http://www.journals.elsevier.com/surface-and-coatings-technology/.

Narayanan, S.C., Karpagam, K.R., Bhattacharyya, A., 2015. Nanocomposite coatings on cotton and silk fibers for enhanced electrical conductivity. Fibers Polym. 16 (6), 1269–1275. https://doi.org/10.1007/s12221-015-1269-1. Available from: http://link.springer.com/journal/volumesAndIssues/12221.

Nguyen-Tri, P., Nguyen, T.A., Carriere, P., Ngo Xuan, C., 2018. Nanocomposite coatings: preparation, characterization, properties, and applications. International Journal of Corrosion 2018. https://doi.org/10.1155/2018/4749501. Available from: http://www.hindawi.com/journals/ijc/.

Pan, Q., Zheng, Y., Zhou, Y., Zhang, X., Yuan, M., Guo, J., Xu, C., Cheng, Z., Kheraif, A.A.A., Liu, M., Lin, J., 2024. Doping engineering of piezo-sonocatalytic nanocoating confer dental implants with enhanced antibacterial performances and osteogenic activity. Adv. Funct. Mater. https://doi.org/10.1002/adfm.202313553. Available from: http://onlinelibrary.wiley.com/journal/10.1002/(ISSN)1616-3028.

Pizarek, J.A., Fischer, N.G., Aparicio, C., 2023. Immunomodulatory IL-23 receptor antagonist peptide nanocoatings for implant soft tissue healing. Dent. Mater. 39 (2), 204−216. https://doi.org/10.1016/j.dental.2023.01.001. Available from: https://www.journals.elsevier.com/dental-materials.

Pogrebnjak, A.D., Bagdasaryan, A.A., Yakushchenko, I.V., Beresnev, V.M., 2014. The structure and properties of high-entropy alloys and nitride coatings based on them. Chem. Rev. 83 (11), 1027−1061. https://doi.org/10.1070/RCR4407. Available from: http://iopscience.iop.org/0036-021X/83/11/1027/pdf/0036-021X_83_11_1027.pdf.

Poxson, D.J., Mont, F.W., Cho, J., Schubert, E.F., Siegel, R.W., 2012. Tailored nanoporous coatings fabricated on conformable polymer substrates. ACS Appl. Mater. Interfaces 4 (11), 6295−6301. https://doi.org/10.1021/am301882m.

Pullithadathil, B., Joghee, S.H., Uthandi, K.M., Singh, N., Katti, S., Kumar, P., Renganayagalu, R.K., 2020. Superhydrophobic coatings based on pseudoboehmite nanoflakelets for sustainable photovoltaic energy production. ACS Appl. Nano Mater. 3 (10), 9899−9911. https://doi.org/10.1021/acsanm.0c01956. Available from: https://pubs.acs.org/journal/aanmf6.

Rashid, M., Ul Haq, E., Abdul Karim, M.R., Shehzad, W., Ali, A., 2024. Polymer-infiltration-pyrolysis (PIP) inspired hydrophobic nano-coatings for improved corrosion resistance. J. Mater. Sci. 59 (3), 828−846. https://doi.org/10.1007/s10853-023-09226-4. Available from: https://www.springer.com/journal/10853.

Rebelo, R., Padrão, J., Fernandes, M.M., Carvalho, S., Henriques, M., Zille, A., Fangueiro, R., 2020. Aging effect on functionalized silver-based nanocoating braided coronary stents. Coatings 10 (12), 1−17. https://doi.org/10.3390/coatings10121234. Available from: https://www.mdpi.com/2079-6412/10/12/1234.

Riquelme, A., Rodrigo, P., Escalera-Rodriguez, M.D., García-Fogeda, P., Rams, J., 2020. Influence of the feed powder composition in mechanical properties of ALN-nano-reinforced aluminium composites coatings deposited by reactive direct laser deposition. Metals 10 (7), 1−18. https://doi.org/10.3390/met10070926. Available from: https://www.mdpi.com/2075-4701/10/7/926/pdf.

Robertson, J., 1991. Electronic structure of silicon nitride. Philos. Mag. B 63 (1), 47−77. https://doi.org/10.1080/01418639108224430.

Salam Hamdy, A., 2010. Corrosion protection performance via nano-coatings technologies. Recent Patents on Materials Sciencee 3 (3), 258−267. https://doi.org/10.2174/1874464811003030258.

Samaha, M.A., Tafreshi, H.V., Gad-el-Hak, M., 2012. Superhydrophobic surfaces: from the lotus leaf to the submarine. Compt. Rendus Mec. 340 (1−2), 18−34. https://doi.org/10.1016/j.crme.2011.11.002.

Santecchia, E., Hamouda, A.M.S., Musharavati, F., Zalnezhad, E., Cabibbo, M., Spigarelli, S., 2015. Wear resistance investigation of titanium nitride-based coatings. Ceram. Int. 41 (9), 10349−10379. https://doi.org/10.1016/j.ceramint.2015.04.152.

Santos, I.M., Alexandre, M., Mihailetchi, V.D., Silva, J.A., Mateus, T., Mouquinho, A., Boane, J., Vicente, A.T., Nunes, D., Menda, U.D., Águas, H., Fortunato, E., Martins, R., Mendes, M.J., 2023. Optically-boosted planar IBC solar cells with electrically-harmless photonic nanocoatings. Adv. Opt. Mater. 11 (15). https://doi.org/10.1002/adom.202300276. Available from: http://onlinelibrary.wiley.com/journal/10.1002/(ISSN)2195-1071.

dos Santos, O.A.L., Pizzorno Backx, B., Abumousa, R.A., Bououdina, M., 2022. Environmental implications associated with the development of nanotechnology: from synthesis to disposal. Nanomaterials 12 (23). https://doi.org/10.3390/nano12234319. Available from: http://www.mdpi.com/journal/nanomaterials.

Schulz, U., Nowotnik, A., Kunkel, S., Reiter, G., 2020. Effect of processing and interface on the durability of single and bilayer 7YSZ/gadolinium zirconate EB-PVD thermal barrier coatings. Germany Surface and Coatings Technology 381. https://doi.org/10.1016/j.surfcoat.2019.125107. Available from: http://www.journals.elsevier.com/surface-and-coatings-technology/.

Selim, M.S., Higazy, S.A., Hao, Z., Chen, X., 2023. Ceramic Polymer Nanocomposites as Eco-Friendly Marine Antifouling Coatings Advances in Nanotechnology for Marine Antifouling. Elsevier, China, pp. 117−140. https://doi.org/10.1016/B978-0-323-91762-9.00010-1. Available from: https://www.sciencedirect.com/book/9780323917629.

Sharma, B., Nigam, S., Verma, A., Garg, M., Mittal, A., Sadhu, S.D., 2024. A biogenic approach to develop guava derived edible copper and zinc oxide nanocoating to extend shelf life and efficiency for food preservation. J. Polym. Environ. 32 (1), 331−344. https://doi.org/10.1007/s10924-023-02972-1. Available from: https://www.springer.com/journal/10924.

Shivananda C.S. Nagi Reddy K.K. Deepika A.G. Madhu S. Hemalatha N. Lavanya S. Sahana R. 2023 1 2023/01 Materials Today: Proceedings 10.1016/j.matpr.2023.04.562 22147853 45 48 Elsevier Ltd India Preparation and characterization of conducting silk fibroin by Ag nanocoating for electrical applications https://www.sciencedirect.com/journal/materials-today-proceedings 89.

Som, C., Wick, P., Krug, H., Nowack, B., 2011. Environmental and health effects of nanomaterials in nanotextiles and façade coatings. Environ. Int. 37 (6), 1131−1142. https://doi.org/10.1016/j.envint.2011.02.013. Available from: http://www.elsevier.com/locate/envint.

Song, K., Chen, D., Polak, R., Rubner, M.F., Cohen, R.E., Askar, K.A., 2016. Enhanced wear resistance of transparent epoxy composite coatings with vertically aligned halloysite nanotubes. ACS Appl. Mater. Interfaces 8 (51), 35552−35564. https://doi.org/10.1021/acsami.6b11872. Available from: http://pubs.acs.org/journal/aamick.

Soubelet, C.G., Grillo, C.A., Suárez, G., Stabile, F.M., 2024. Zirconia/bioactive glass composites development through a particle nanocoating. Argentina Ceramics International 50 (9), 14347−14359. https://doi.org/10.1016/j.ceramint.2024.01.346. Available from: https://www.sciencedirect.com/science/journal/02728842.

Subramanian, B., Muraleedharan, C.V., Ananthakumar, R., Jayachandran, M., 2011. A comparative study of titanium nitride (TiN), titanium oxy nitride (TiON) and titanium aluminum nitride (TiAlN), as surface coatings for bio implants. Surf. Coating. Technol. 205 (21−22), 5014−5020. https://doi.org/10.1016/j.surfcoat.2011.05.004. Available from: http://www.journals.elsevier.com/surface-and-coatings-technology/.

Sun, L., Yuan, G., Gao, L., 2021. Chemical vapour deposition. Nature Reviews Methods Primers 1. https://doi.org/10.1038/s43586-020-00012-z. Available from: https://www.semanticscholar.org/paper/Chemical-vapour-deposition-Sun-Yuan/49e58b1ea62dea499098186bd04a3f18dcaebc6f.

Sun, X., Zhang, J., Pan, W., Wang, W., Tang, C., 2022. Research progress in surface strengthening technology of carbide-based coating. J. Alloys Compd. 905. https://doi.org/10.1016/j.jallcom.2022.164062. Available from: https://www.journals.elsevier.com/journal-of-alloys-and-compounds.

Toma, F.L., Potthoff, A., Berger, L.M., Leyens, C., 2015. Demands, potentials, and economic aspects of thermal spraying with suspensions: a critical review. J. Therm. Spray Technol. 24 (7), 1143−1152. https://doi.org/10.1007/s11666-015-0274-7.

Trusova, E.A., Koshcheev, A.P., Konovalov, A.A., Chipkin, P.I., Safronenko, M.G., Fortalnova, E.A., Khrustalev, V.N., 2018. Composition and thermal stability of Al- and Zr-containing gels prepared by a sol−gel synthesis using N,N-dimethyloctylamine and acetylacetone. Federation Inorganic Materials 54 (12), 1231−1237. https://doi.org/10.1134/S0020168518120166. Available from: http://www.springer.com/chemistry/inorganic+chemistry/journal/10789.

Van Hove, R.P., Sierevelt, I.N., Van Royen, B.J., Nolte, P.A., 2015. Titanium-nitride coating of orthopaedic implants: a review of the literature. BioMed Res. Int. 2015. https://doi.org/10.1155/2015/485975. Available from: http://www.hindawi.com/journals/biomed/.

Vilkov, I.V., Ob''edkov, A.M., Kaverin, B.S., Semenov, N.M., Kovylin, R.S., Dodonov, V.A., 2023. Synthesis and characterization of a hybrid nanomaterial based on multi-walled carbon nanotubes with zirconium carbide nanocoatings. Federation Russian Journal of General Chemistry 93, S844. https://doi.org/10.1134/S1070363223160247. Available from: https://www.springer.com/journal/11176.

Vladoiu, R., Tichý, M., Mandes, A., Dinca, V., Kudrna, P., 2020. Thermionic vacuum Arc-A versatile technology for thin film deposition and its applications. Coatings 10 (3). https://doi.org/10.3390/coatings10030211. Available from: https://res.mdpi.com/d_attachment/coatings/coatings-10-00211/article_deploy/coatings-10-00211-v2.pdf.

Wang, D., Li, Y., Wen, Y., Li, X., Du, X., 2021. Simple and low cost fabrication of large area nanocoatings with mechanical robustness, enhanced broadband transmittance and anti-fogging. Colloids Surf. A Physicochem. Eng. Asp. 629. https://doi.org/10.1016/j.colsurfa.2021.127522. Available from: http://www.elsevier.com/locate/colsurfa.

Wang, T.Y., Li, X.F., Liu, S.M., Liu, B.X., Liang, X.D., Li, S., Zhang, G.X., Liu, J.B., Dang, Z.M., 2022. Self-assembled wide bandgap nanocoatings enabled outstanding dielectric characteristics in the sandwich-like structure polymer composites. Nano Convergence 9 (1). https://doi.org/10.1186/s40580-022-00346-2. Available from: https://nanoconvergencejournal.springeropen.com.

Wang, Y., Guo, Q., Su, G., Cao, J., Liu, J., Zhang, X., 2019. Hierarchically structured self-healing actuators with superfast light- and magnetic-response. Adv. Funct. Mater. 29 (50). https://doi.org/10.1002/adfm.201906198. Available from: http://onlinelibrary.wiley.com/journal/10.1002/(ISSN)1616-3028.

Wei, C., Tang, Y., Zhang, G., Zhang, Q., Zhan, X., Chen, F., 2016. Facile fabrication of highly omniphobic and self-cleaning surfaces based on water mediated fluorinated nanosilica aggregation. RSC Adv. 6 (78), 74340–74348. https://doi.org/10.1039/C6RA13367C.

Welegergs, G.G., Gebretnisae, H.G., Tsegay, M.G., Nuru, Z.Y., Dube, S., Maaza, M., 2021. Thickness dependent morphological, structural and optical properties of SS/CuO nanocoatings as selective solar absorber. Infrared Phys. Technol. 113. https://doi.org/10.1016/j.infrared.2020.103619. Available from: https://www.journals.elsevier.com/infrared-physics-and-technology.

Wolski, K., Smenda, J., Świerz, W., Dąbczyński, P., Marzec, M., Zapotoczny, S., 2024. Self-templating copolymerization to produce robust conductive nanocoatings based on conjugated polymer brushes with implementable memristive characteristics. Small. https://doi.org/10.1002/smll.202309216. Available from: http://onlinelibrary.wiley.com/journal/10.1002/(ISSN)1613-6829.

Wonnie Ma, I.A., Sh, A., Rammesh, K., Vengadaesvaran, B., Ramesh, S., Arof, A.K., 2017. Anticorrosion properties of epoxy-nanochitosan nanocomposite coating. Prog. Org. Coating 113, 74–81. https://doi.org/10.1016/j.porgcoat.2017.08.014. Available from: http://www.elsevier.com/locate/porgcoat.

Wu, G.L., Cheong, K.K., Lee, W.J., Chang, C.C., 2023. Understanding the lithium-ion battery's aging mechanisms of mesophase graphite negative electrodes with/without amorphous titanium(IV) oxide nanocoatings by atomic layer deposition. J. Energy Storage 73. https://doi.org/10.1016/j.est.2023.108951. Available from: http://www.journals.elsevier.com/journal-of-energy-storage/.

Xia, F., Yan, P., Ma, C., Liu, Y., 2024. Effect of SiC nanoparticles on the structure and properties of Ni+W-SiC nanocoatings. J. Mater. Eng. Perform. https://doi.org/10.1007/s11665-024-09312-x. Available from: https://www.springer.com/journal/11665.

Xia, F., Li, Q., Ma, C., Guo, X., 2020. Preparation and characterization of Ni−AlN nanocoatings deposited by magnetic field assisted electrodeposition technique. Ceram. Int. 46 (2), 2500−2509. https://doi.org/10.1016/j.ceramint.2019.09.244. Available from: https://www.journals.elsevier.com/ceramics-international.

Xie, H., Lai, X., Wang, Y., Li, H., Zeng, X., 2019. A green approach to fabricating nacre-inspired nanocoating for super-efficiently fire-safe polymers via one-step self-assembly. J. Hazard Mater. 365, 125−136. https://doi.org/10.1016/j.jhazmat.2018.10.099. Available from: http://www.elsevier.com/locate/jhazmat.

Xie, J.D., Fu, C.C., Liao, C.C., Juang, R.S., Gandomi, Y.A., 2019. Alumina nanocoating of polymer separators for enhanced thermal and electrochemical performance of Li−ion batteries. Asia Pac. J. Chem. Eng. 14 (4). https://doi.org/10.1002/apj.2335. Available from: http://onlinelibrary.wiley.com/journal/10.1002/(ISSN)1932-2143.

Xie, T., He, Z., Yu, H., Huang, C., Ou, J., Dai, W., Ge, J., 2024. The preparation and performance analysis of pavement photocatalytic material based on quantum dot heterostructures technology. Construct. Build. Mater. 416. https://doi.org/10.1016/j.conbuildmat.2024.135066. Available from: https://www.sciencedirect.com/science/journal/09500618.

Xing, Y., Wu, Z., Yang, J., Wang, X., Liu, L., 2020. LIPSS combined with ALD MoS2 nanocoatings for enhancing surface friction and hydrophobic performances. Surf. Coating. Technol. 385. https://doi.org/10.1016/j.surfcoat.2020.125396. Available from: http://www.journals.elsevier.com/surface-and-coatings-technology/.

Yan, Y., Liu, J., Zhang, B., Xia, R., Zhang, Y., Guan, Z., 2023. Enhanced mechanical and hydrophobic antireflective nanocoatings fabricated on polycarbonate substrates by combined treatment of water and HMDS vapor. Materials 16 (10). https://doi.org/10.3390/ma16103850. Available from: http://www.mdpi.com/journal/materials.

Yang, C., Li, D., Zang, S., Zhou, Y., Zhang, L., Zhong, Z., 2023. Nanocoating of CsgA protein for enhanced cell adhesion and proliferation. Chin. Chem. Lett. 34 (4). https://doi.org/10.1016/j.cclet.2022.08.013. Available from: http://www.elsevier.com/wps/find/journaldescription.cws_home/997/description#description.

Yang, S., Ji, J., Luo, M., Li, H., Gao, Z., 2021. Poly(tannic acid) nanocoating based surface modification for construction of multifunctional composite CeO2NZs to enhance cell proliferation and antioxidative viability of preosteoblasts. Nanoscale 13, 16349−16361. https://doi.org/10.1039/D1NR02799A. Available from: https://pubs.rsc.org/en/content/articlelanding/2021/nr/d1nr02799a.

Yao, H., Hang, J., Sun, X., Jin, L., Shi, L., Zhang, J., 2014. Preparation and anticorrosive behavior of epoxy-polysiloxane hybrid coatings modified by polyetheramines. J. Adhes. Sci. Technol. 28 (12), 1103−1116. https://doi.org/10.1080/01694243.2013.875609. Available from: http://www.brill.nl/default.aspx?partid=10&pid=26935.

Ye, C., Zhu, Y., Sun, H., Chen, F., Sun, H., Dai, W., Wei, Q., Fu, L., Yu, A., Du, S., Yang, M., Huang, L.F., Yu, J., Jiang, N., Lin, C.T., 2021. Layer-by-layer stacked graphene nanocoatings by Marangoni self-assembly for corrosion protection of stainless steel. Chin. Chem. Lett. 32 (1), 501−505. https://doi.org/10.1016/j.cclet.2020.03.013. Available from: http://www.elsevier.com/wps/find/journaldescription.cws_home/997/description#description.

Ye, Z., Sang, T., Li, K., Fischer, N.G., Mutreja, I., Echeverría, C., Kumar, D., Tang, Z., Aparicio, C., 2022. Hybrid nanocoatings of self-assembled organic-inorganic amphiphiles for prevention of implant infections. Acta Biomater. 140, 338−349. https://doi.org/

10.1016/j.actbio.2021.12.008. Available from: http://www.journals.elsevier.com/acta-biomaterialia.

Yilmaz, O., Yorgancioglu, A., 2018. Nanocoatings: Preparation, Properties, and Biomedical Applications Polymeric Nanomaterials in Nanotherapeutics. Elsevier, Turkey, pp. 299–331. https://doi.org/10.1016/B978-0-12-813932-5.00008-X. Available from: http://www.sciencedirect.com/science/book/9780128139325.

Yousaf, S., Alhnan, M.A., Abdallah, A., Abdallah, B., Khan, I., Ahmed, W., 2014. Nanocoatings in Medicine: Antiquity and Modern Times Emerging Nanotechnologies for Manufacturing. United Kingdom Elsevier Science, United Kingdom, pp. 418–443. https://doi.org/10.1016/B978-0-323-28990-0.00016-6. Available from: http://www.sciencedirect.com/science/book/9780323289900.

Yu, C.Y., Wang, S.B., Li, T.B., Zhang, Z.X., 2013. Tribological behaviour of CrAlN coatings at 600°C. Surf. Eng. 29 (4), 318–321. https://doi.org/10.1179/1743294412y.0000000100.

Yurtdaş, S., Tozlu, C., Karaman, M., 2023. Vapor deposition of poly(hexafluorobutyl acrylate) nanocoating for encapsulation of organic solar cells. Prog. Org. Coating 177. https://doi.org/10.1016/j.porgcoat.2023.107443. Available from: http://www.elsevier.com/locate/porgcoat.

Zamani Meymian, M.R., Ghaffarinejad, A., Fazli, R., Kosari Mehr, A., 2020. Fabrication and characterization of bimetallic nickel-molybdenum nano-coatings for mild steel corrosion protection in 3.5% NaCl solution. Colloids Surf. A Physicochem. Eng. Asp. 593. https://doi.org/10.1016/j.colsurfa.2020.124617. Available from: http://www.elsevier.com/locate/colsurfa.

Zha, X., Wang, T., Guo, B., Chen, F., Lin, L., Zhang, T., Jiang, F., 2023. Research on the oxidation resistance and ultra-high frequency thermal fatigue shock failure mechanisms of the bilayer and multilayer nano-coatings on cemented carbide tools. Int. J. Refract. Metals Hard Mater. 110. https://doi.org/10.1016/j.ijrmhm.2022.106043. Available from: http://www.journals.elsevier.com/international-journal-of-refractory-metals-and-hard-materials/.

Zhang, C., Hu, H., Mao, L., Chen, H., Liao, H., Yu, Z., 2024. Effect of Y2O3 content on the microstructure and wear performance of plasma-sprayed Cr2O3–Y2O3 composite coatings. Ceram. Int. 50 (3), 4553–4562. https://doi.org/10.1016/j.ceramint.2023.11.194. Available from: https://www.sciencedirect.com/science/journal/02728842.

Zhang, W., Xiao, L., Zheng, J., Zhong, Y., Shi, B., Chen, H., Fu, H., 2021. Effect of Nb2O5 nanocoating on the thermal stability and electrochemical performance of LiNi0.6-Co0.2Mn0.2O2 cathode materials for lithium ion batteries. J. Alloys Compd. 880. https://doi.org/10.1016/j.jallcom.2021.160415. Available from: https://www.journals.elsevier.com/journal-of-alloys-and-compounds.

Zhang, Y., Xia, F., Yan, P., Ma, C., Wang, C., Wang, Q., 2024. Microstructure, wear and corrosion resistances of Ni–ZrO2–CeO2 nanocoatings. Ceram. Int. https://doi.org/10.1016/j.ceramint.2024.03.157.

Zhao, X., Xu, X., Lu, Y., Sun, Q., 2024. A tri-layer structural film containing nickel nanocoating for electromagnetic transmittance and joule heating. Applied Nano Materials. https://doi.org/10.1021/acsanm.4c00074. Available from: https://pubs.acs.org/journal/aanmf6.

Zheng, B., Smugeresky, J.E., Zhou, Y., Baker, D., Lavernia, E.J., 2008. Microstructure and properties of laser-deposited Ti6Al4V metal matrix composites using Ni-Coated powder. Metall. Mater. Trans. A 39 (5), 1196–1205. https://doi.org/10.1007/s11661-008-9498-1.

Zhong X.C. Liu Z.W. Wu Y.S. Liu M.T. Zeng D.C. 2014 1 2014/01 IOP Conference Series: Materials Science and Engineering 10.1088/1757-899X/60/1/012032 1757899X 1 Institute of Physics Publishing China Nano-composite coatings with improved mechanical properties and corrosion resistance by thermal spraying http://www.iop.org/EJ/journal/mse 60.

Zhu, X.D., Liu, Y., Shen, J., 2016. Volatile organic compounds (VOCs) emissions of wood-based panels coated with nanoparticles modified water based varnish. European Journal of Wood and Wood Products 74 (4), 601–607. https://doi.org/10.1007/s00107-016-1012-7. Available from:http://www.springer.com/life+sci/forestry/journal/107.

Zou, C., Zang, Z., Tao, X., Yi, L., Chen, X., Zhang, X., Yang, L., Liu, X., Wang, X., 2023. Stabilized cathode/sulfide electrolyte interface through conformally interfacial nanocoating for all-solid-state batteries. ACS Appl. Energy Mater. 6 (6), 3599–3607. https://doi.org/10.1021/acsaem.3c00255. Available from: http://pubs.acs.org/journal/aaemcq.

Recent updates in nanometal oxide–based corrosion-resistant coatings

2

Gauri Shukla [1], *Manviri Rani* [1] *and Uma Shanker* [2]
[1]Department of Chemistry, Malaviya National Institute of Technology, Jaipur, Rajasthan, India; [2]Department of Chemistry, Dr B R Ambedkar National Institute of Technology, Jalandhar, Punjab, India

2.1 Introduction

Corrosion is the spontaneous and irreversible process of transforming refined metals into more stable substances such as metal hydroxides, metal sulfides, or oxides. A sophisticated metal will naturally undergo corrosion to become a more chemically stable oxide. It is the process by which materials (often metals) gradually deteriorate due to chemical or electrochemical reactions with their surroundings. In a similar vein, when iron rusts, oxygen and moisture in the air combine to generate iron oxides. Corrosion affects practically every aspect of the global infrastructure and other assets that employ metals, with significant economic and environmental consequences. Corrosion not only seriously compromises public safety and causes considerable damage, but it also halts operations and necessitates substantial asset replacement and repair. Over $2.5 trillion, or 3.4% of the yearly global gross domestic product (GDP), is spent on corrosion damage worldwide (Koch et al., 2016). Corrosion costs the United States $276 billion annually, or 3.1% of GDP. By taking preventive action, corrosion expenses can be lowered to 15% from 35% (Rajendran et al., 2020). A Chinese study found that, out of all corrosion prevention techniques, companies spent 66.5% of their budgets on anticorrosion coatings. NACE International stated in 2016 that corrosion is thought to cost the world 2.5 trillion US dollars, or 3.4% of all goods produced (Baorong et al., 2017). These figures highlight the critical role that corrosion prevention techniques play in extending the useful life of metallic structures and lowering maintenance and repair costs in many sectors. A few methods for preventing corrosion include using appropriate design, anticorrosion coatings, corrosive inhibitors, cathodic protection, anodic passivation, salt removers, corrosion-resistant metals, and routine cleaning (Ghali et al., 2007; Honarvar Nazari et al., 2015). The use of coatings for corrosion protection is appealing as it is one of the most effective, adaptable, affordable, and simple methods available (Nazari et al., 2022). For metallic components, coating is one of the most useful and economical ways to increase mechanical strength and resistance to corrosion. The primary objective of corrosion-resistant coatings is to protect metals and alloys from corrosive organisms found in various environments,

Multifunctional Nanostructured Coatings. https://doi.org/10.1016/B978-0-443-23683-9.00002-7

including water, hot, corrosive fluids, soils, ultraviolet light, and air pollution. In particular, coatings are constantly exposed to the many chemicals and acid rain that are prevalent in the atmosphere, especially in industrial locations (Anjum et al., 2020). Depending on the kind of metallic substrate and its surroundings, a coating's level of protection varies. Anticorrosion coatings can offer two types of material protection. Polymeric paints and aluminum coatings are two instances of passive protection, where the coatings serve as a physical barrier against corrosive species (Yu et al., 2019; Luo et al., 2018; Pugazhenthi et al., 2018). When a coating is physically intact, passive protection may still be effective; however, this does not always mean the substrate is adequately shielded. Indeed, corrosive species can still reach the surface of the underlying substrate by passing via the pores, free volumes, and channels of an intact coating. Active protection is the second kind of protection that an anticorrosion coating offers. It works by limiting or stopping the electrochemical processes that promote corrosion by including a corrosion inhibitor or sacrificial element into the coating composition. Zinc-rich coatings and intelligent coatings are examples of active protective coatings that can effectively guard the underlying metal even after damage has occurred (Nazari and Shi, 2016).

Nanotechnology is becoming increasingly common in the coatings sectors, where researchers are looking at using nanofillers in polymer binders and employing various metal oxide-based coatings instead of conventional additives and pigments for efficient performance. Nanoparticles (NPs) or corrosion inhibitors are introduced into the coatings to better substrate protection and improve performance (Raps et al., 2009). Polymers are extensively used as barriers in corrosion-resistant coatings incorporated with nanofillers, which improves the mechanical, thermal, and physicochemical properties of polymers and also enhances the specific surface area of coatings (Tjong, 2006; Wetzel et al., 2003). Fillers also help in enhancing the barrier performance of organic polymer coatings by minimizing the pores that were created due to the evaporation of organic solvents and external environmental stresses (Liu et al., 2013a,b; Ramanathan et al., 2008). Various kinds of nanofillers, including graphene-based nanomaterials (Ramezanzadeh et al., 2018; Taheri et al., 2019), stacked doubled hydroxides (Alibakhshi et al., 2017; Hayatdavoudi and Rahsepar, 2017), inorganic nanomaterials (Xiao et al., 2016; Zhang et al., 2018), and clays (Majd et al., 2020), have been employed to create polymer nanocomposites with improved aspects. The characteristics of coatings can be influenced by the inherent qualities of nanofillers as well as their chemical functional groups, shape, size, and quantity (Ahmadi, 2019). Metal oxide-based coatings improve the coating's abrasion, scratch resistance, and conductivity and also enhance their corrosion resistance. Al_2O_3, MgO, TiO_2, ZnO, $CaCO_3$, SiO_2 and CeO_2 are the majorly used NPs in corrosion resistant coatings due to their intrinsic characteristics. For example, Al_2O_3 and SiO_2 are used to increase abrasion and scratch resistance; CeO_2 and MgO NPs are used as anti-corrosion and high-temperature oxidation-resistant materials; and TiO_2 and ZnO NPs are used as UV-blocking agents (Radhakrishnan et al., 2009). Si—O networks from SiO_2 polymer composites and silane coupling agents are also used for anticorrosive coatings, mostly because of the coating's increased crosslinking density and lower pinhole defect (Dhoke et al., 2009). The fact that commercially available

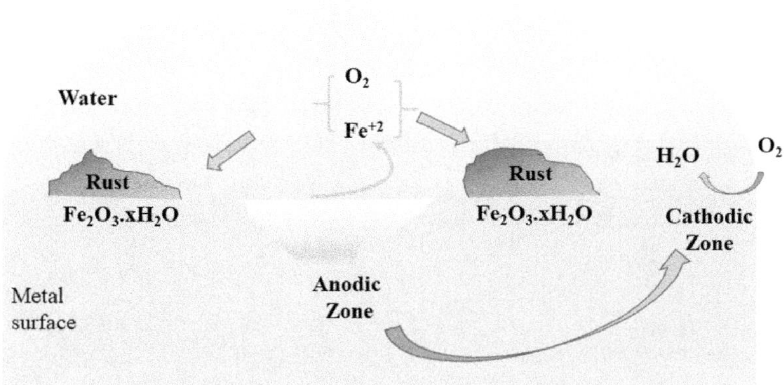

Figure 2.1 Schematic representation of the corrosion mechanism.

pigments and additives remain aggregated after grinding and mixing presents another difficulty in creating thin films since this changes the viscosity of coatings (Li et al., 2007). Less than 100 nm-sized metal oxide NPs can strengthen the polymer matrix without compromising the coatings' transparency or other crucial characteristics (Su and Chen, 2008). This chapter, therefore, highlights the incorporation of metal oxide NPs in corrosion-resistant coatings, their performance, efficiency, synthesis, factors, and mechanism for improving corrosion resistance. The brief mechanism of corrosion is described in Fig. 2.1.

2.2 Design of metal oxide—based nanocoatings

Nanoparticles were synthesized using different types of techniques that were classified as bottom-up and top-down methods. Building blocks are added to the substrate to generate nanostructures using the bottom-up method, which is a constructive nanosynthesis technique and includes various methods such as sol—gel, chemical vapor deposition (CVD), biosynthesis, etc., whereas top-down is a destructive approach in which bulk material reduces into nanometric-scale particles. Thermal breakdown and laser ablation are the most popular top-down techniques for creating nanoparticles. The various approaches of nanoparticle synthesis have been described in, and various types of coatings for corrosion resistance incorporating NPs have been summarized in Fig. 2.2.

2.2.1 Sol-gel method

Sol-gel method is frequently employed to create nanoparticles due to its reproducibility, accessibility, uniformity, and ease of use, whereby chemical solutions containing inorganic salts or metal alkoxides function as precursors. A colloidal sol solution was produced by partially condensing and hydrolyzing the precursors. The gel is

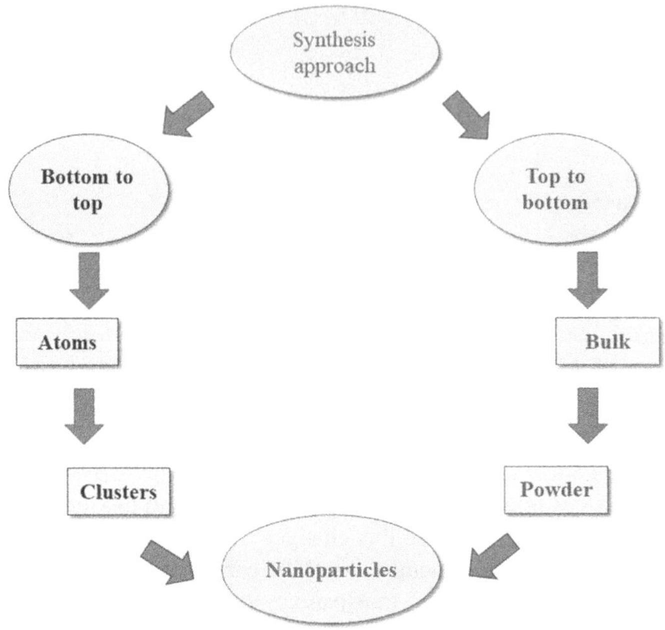

Figure 2.2 Various approaches of nanoparticle synthesis for coating applications.

created as sol particles continue to condense into a three-dimensional network (Akpan and Hameed, 2010). Using a variety of techniques, including centrifugation, filtering, and sedimentation, phase separation is used to recover the nanoparticles, and drying is used to eliminate any remaining moisture (Bokov et al., 2021).

2.2.2 Biosynthesis method

The synthesis of nanoparticles using biological materials and extracts from green plants is an economical, environmentally beneficial, and alternative approach, in which various biological resources such as bacteria, fungi, algae, plant extracts, yeast, and human cells are used as reducing/capping agents for producing nanoparticles. These organisms can undergo reduction because they include metabolites and proteins. Because of its simplicity, quick rate of synthesis, and environmental friendliness, the green synthesis process has special and improved qualities (Saravanan et al., 2021).

Fig. 2.3 represents various techniques of coating preparation.

2.2.3 Chemical vapor deposition method

This process involves the precursor occurring in the vapor phase and depositing on the surface of the substrate. Precursors are always supplied to the reactor as vapor; however, under ambient conditions, they might be solid, liquid, or gas. Thermal breakdown technique refers to CVD that uses a chemical reaction between the substrate surface

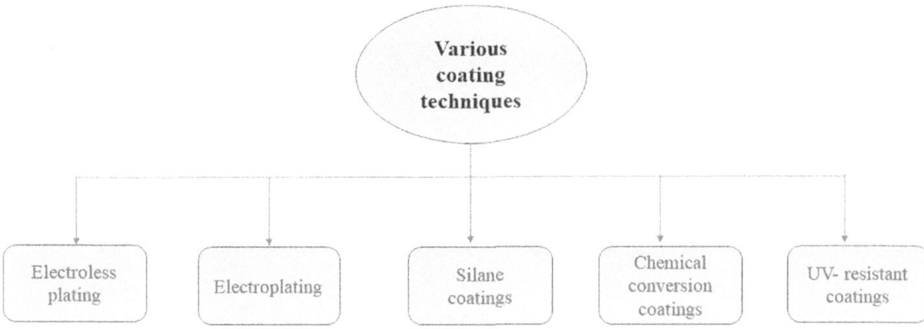

Figure 2.3 Various techniques of coating preparation.

and gaseous precursors; this reaction is triggered by temperature; if the process is accelerated by plasma, it is referred to as plasma-enhanced CVD (Cheng et al., 2022).

2.2.4 Laser ablation method

Laser ablation techniques are an effective and unique method of producing nanoparticles in a colloidal solution in a different solvent. Physically, laser ablation is usually thought of as a top-down method. It produces an enormous range of nanomaterials in a clean, scalable way (Kim et al., 2017). Using a bulk metal plate soaked in an appropriate solvent and plasma plume condensation, the laser ablation process creates nanomaterials. This technique uses water and an organic solvent to create nanoparticles (Chrzanowska et al., 2015).

2.2.5 Thermal decomposition method

An improved process to create suitable and monodispersed nanoparticles is endothermic thermal breakdown. Thermal breakdown is the process when a chemical substance separates into little fragments by heating and then does not reassemble when it cools. The temperature at which substances break down into their most basic chemical forms after being heated is known as the decomposition temperature (Singapati and Ravikumar, 2023).

2.3 Metal/metal oxide incorporated surface coatings

There are several advantages to incorporating the NPs into the coatings that encircle mechanical, optical, biological, and corrosion resistance properties. Many coating systems are available to which NPs are incorporated, such as electroless plating, electroplating, anodic films, organosilane coatings, and chemical conversion coatings. Fig. 2.4 offers schematic representation of nanoparticle-based coatings.

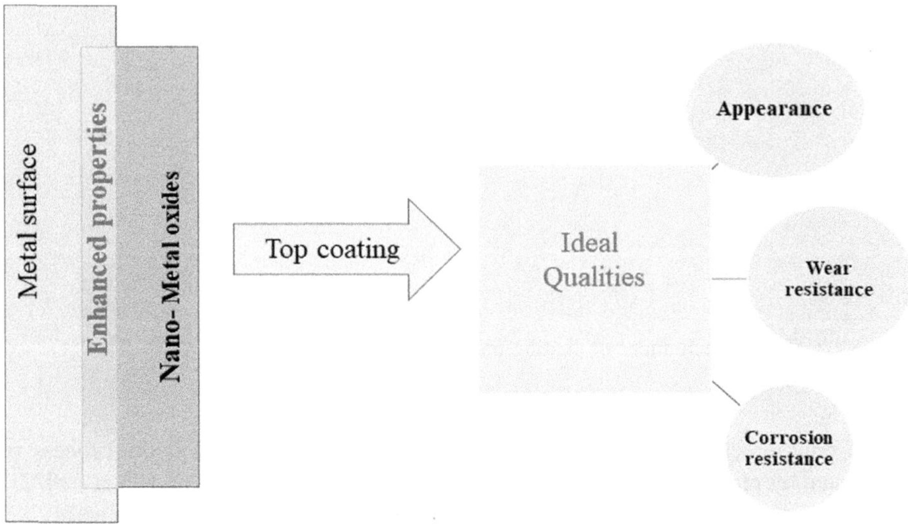

Figure 2.4 Schematic representation of nanoparticle-based coatings.

As different metal/metal oxide NPs have different chemical properties, which qualifies specific characteristics depending upon various employed NPs, metal/metal oxide NPs can be classified into three categories:

1. Metal/metal oxide NPs improve the barrier properties of the coating.
2. Metal/metal oxide NPs provide self-healing ability to coating.
3. Metal/metal oxide NPs improve resistance to photodegradation.

2.3.1 Metal oxide nanoparticles with barrier properties

Al_2O_3, ZrO_2, TiO_2, Y_2O_3, NiO, ZnO, Zn, and SnO_2 are the most frequently used NPs. These NPs could be utilized in various coating systems to augment the barrier properties of coating against different aggressive electrolyte. Electrodeposition and epoxy coatings are the most common techniques to deposit corrosion-resistant coatings that incorporate the NPs. However, NPs that will perform best need to have a deep understanding of the chemistry and the interaction between the NPs and coating components. Al_2O_3 and TiO_2, for instance, are frequently used in electrodeposited coatings and provide better mechanical as well as corrosion resistance properties. In epoxy coatings, metal oxide NPs not only improve the corrosion resistance but also increase the sustainability of coatings in more harsh environments. Many metal oxide NPs are utilized with epoxy coatings that help improve the corrosion resistance and cracking resistance of coatings. ZrO_2 NPs are abundantly employed to enhance the mechanical as well as corrosion resistance of epoxy coatings. The NPs such as Al_2O_3, Fe_2O_3, ZnO, and SiO_2 have also been utilized in epoxy coatings. The different concentrations of these NPs show different behavior. As the small amount of these NPs is added, only ZnO has shown the lowest corrosion current density. However, the increase in concentrations of these NPs has increased to 3%, and totally different behavior is observed, Al_2O_3 has shown the best performance, and, on the other hand, Fe_2O_3 and SiO_2 have shown the increase in corrosion current density. This increase in corrosion current density is caused by the agglomeration of NPs due to poor dispersibility in the epoxy solution.

2.3.2 Metal oxide nanoparticles with self-healing properties

Self-healing is the ability of coating, which depends upon the coating composition and type of inhibitors incorporated into the coatings. The phenomenon of self-healing keeps the coating intact and, in turn, protects the substrate. The self-healing mechanism for corrosion protection is an entirely automatic healing process without requiring external energy input. SiO_2, Cr_2O_3, and CeO_2 NPs show the ability to heal the defects and enhance barrier properties. Numerous coating systems employed these particles and obtained appreciably good results, and the self-healing ability is clearly observed. However, the use of Cr_2O_3 is restricted because of hexavalent chromium ions, which are toxic to the environment and humans. SiO_2 NPs can be incorporated into the coatings, which serve two purposes; first, it fills the pores and improves the barrier properties of coatings, and second, it acts as a cathodic inhibitor and suppresses the cathodic reactions. Furthermore, SiO_2 NP incorporation coatings' mechanical properties are better than the pristine coating. CeO_2 NPs are extensively used in many corrosion protection systems because of their outstanding barrier as well as self-healing properties. However, self-healing is highly related to the substrate material and coating system. In some cases, in steel, the self-healing property was not observed. Moreover, the CeO_2 NPs need to be activated with cerium nitrate to improve silane coatings' corrosion resistance properties. However, the self-healing and barrier properties were reported for Mg alloys, Al alloys, and, in some cases, galvanized steel and stainless steel. Primarily, metal oxide NPs are indirectly used in self-healing phenomena as nanocontainers. SiO_2 NPs, mixed oxide NPs (e.g., ZrO_2/CeO_2), and TiO_2 NPs are used as nanocontainers in sol-gel hybrid films. These nanocontainers can store organic and inorganic active chemicals and release them when environmental stimulus is sensed, consequently heals the defects.

2.3.3 Metal/metal oxide nanoparticles for photodegradation resistance

The long-term endurance against environmental actions and corrosion resistance is required for components in many outdoor applications. The huge source of degradation of protective coatings leads to the failure of components is ultraviolet (UV) radiation coming from the sun. The polymer matrix coatings undergo photodegradation when used in outdoor applications. Metal/metal oxide NPs with semiconducting behavior are found good for protecting coating against UV radiation. These NPs include ZnO, SiO_2, TiO_2, and Fe_2O_3; among these, TiO_2 and ZnO NPs are utilized frequently because they improve the photodegradation of polymer composite coatings. Even a small amount of TiO_2 and ZnO NPs improve the mechanical and photodegradation resistance without altering the optical properties of coatings. ZnO NPs are added in polyurethane nanocoatings to improve the photodegradation properties, proving effective. However, if the amount of ZnO NPs is increased from a critical value, then a negative effect on protective properties is observed. TiO_2 NPs are also performed well under accelerated weathering test. The acrylic polyurethane coatings are modified with a combination of rutile-TiO_2 NPs, which curtailed the photodegradation of acrylic polyurethane coatings. However, enhancement in photodegradation resistance is sensitive to the amount of NPs added. Briefly, metal oxide NPs are helpful to enhance the photodegradation resistance of organic coatings. However, it is consistently observed that the amount of NPs added into the coating plays a crucial role. If the amount of NPs exceeds a certain value, undesirable properties will result.

Metal oxide and related modified coatings offer a number of benefits depending on the specific characteristics of particular metal/metal oxide NPs. Based on these characteristics, metal/metal oxide NPs can be classified into various categories, such as barrier properties, enhanced resistance to photodegradation, self-healing ability, etc. Barrier-like characteristics of metal oxide nanoparticles are shown by different metal oxide NPs, including Al_2O_3 (Dhoke and Khanna, 2012), ZrO_2 (Lee et al., 2011), TiO_2 (Atta et al., 2017), Y_2O_3 (Popoola et al., 2016), NiO (Deepa and Venkatesha, 2018), ZnO (Torknezhad et al., 2018), and SnO_2 (Fayomi and Popoola, 2015). The self-healing properties of TiO_2 NP-based coatings have been shown in Fig. 2.5i. These

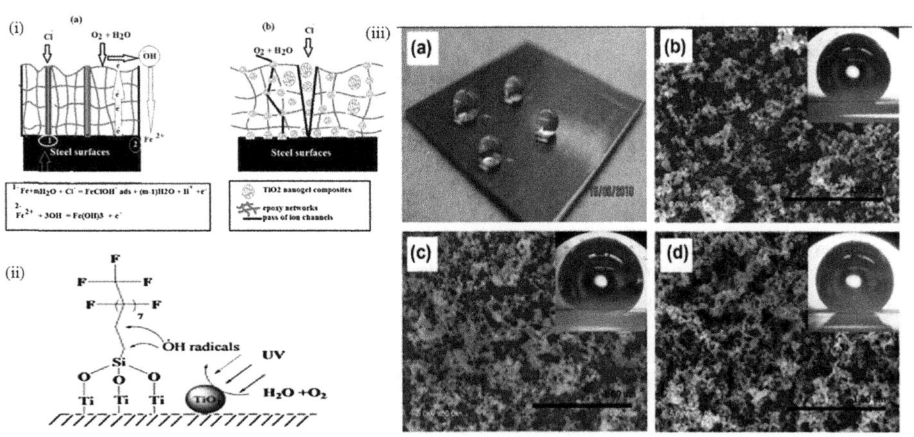

Figure 2.5 (i) Self-healing and corrosion mechanisms (a) in the absence and (b) in the presence of TiO_2 nanogel composites; (ii) photodegradation process of PTES-based superhydrophobic surface; (iii) SEM images of different layers of TiO_2 nanoparticles in steel surface: (a) Water droplets on the treated steel surface, (b) TiO_2 one layer $(TiO_2)*1$, (c) TiO_2 two layers $(TiO_2)*2$, and (d) TiO_2 three layers $(TiO_2)*3$.

(i) (With permission from Atta, A.M., Al-Lohedan, H.A., El-Saeed, A.M., Al-Shafey, H.I., Wahby, M.H. (2017). Epoxy embedded with TiO2 nanogel composites as promising self-healing organic coatings of steel. Prog. Org. Coating 105, 291–302; (iii) With permission from Isimjan, T.T., Wang, T., Rohani, S., (2012). A novel method to prepare superhydrophobic, UV resistance and anti-corrosion steel surface. Chem. Engin. J. 210, 182–187.

NPs might be added to different coating systems to improve the coating's barrier qualities against various hostile electrolytes. The incorporation of NPs has been carried out using electrodeposition (Al_2O_3 and TiO_2) and epoxy (ZrO_2, SiO_2) coatings as the most common techniques. While epoxy coatings boost coatings' durability and corrosion resistance in more aggressive settings, electrodeposited coatings offer superior mechanical and corrosion-resistant qualities. The performance of these NPs depends upon their concentration and quantity in the coatings (Haddadi et al., 2015; Low et al., 2006).

The corrosion prevention self-healing system functions entirely independently, requiring no outside energy input. In addition to improving barrier qualities, SiO_2, Cr_2O_3, and CeO_2 NPs can repair flaws. However, its usage is restricted due to hexavalent chromium ions' toxicity to both humans and the environment. Upon integration, SiO_2 NPs cover the pores, enhance the coatings' barrier qualities, and function as a cathodic inhibitor to stifle cathodic processes (Palanivel et al., 2003). CeO_2 NPs have excellent barrier and self-healing attributes, which make them widely employed in various corrosion prevention systems. However, magnesium alloys, aluminum alloys, and, occasionally, galvanized and stainless steels were claimed to have a barrier and self-healing qualities (Zhao et al., 2019). **The photodegradation efficiency of superhydrophobic PTES-based coating and the SEM images have been described in** Fig. 2.5.

Various semiconducting metal oxide nanoparticles (ZnO, SiO_2, TiO_2, and Fe_2O_3) have been discovered to be effective at shielding coatings from ultraviolet light. Among these, TiO_2 and ZnO NPs are used frequently as they enhance polymer composite coatings' mechanical and photodegradation efficiency without changing other properties, even when used in minute amounts (Nguyen et al., 2016). In polyurethane nanocoatings, ZnO NPs are added to enhance photodegradation capabilities, which are effective (Isimjan et al., 2012). The photodegradation efficiency of superhydrophobic PTES-based coating and the SEM images have been described in Fig. 2.5ii and iii. Various other metal oxide NPs used in corrosion-resistant coatings have been summarized in (Table 2.1).

2.4 Factors affecting the efficiency of metal/metal oxide nanoparticles as corrosion inhibitors

Corrosion protection ability of coating incorporating the metal/metal oxide NPs depends upon NPs' dispersibility in the coating, concentration, charge, and incorporation techniques. Photodegradation resistance improvement of polyurethane nanocoatings after 500 hours accelerated aging with ZnO nanoparticles. Photodegradation resistance improvement of acrylic polyurethane coatings modified with rutile TiO_2 nanoparticles.

Table 2.1 Various nanomaterials used, their efficiency, and applicability as corrosion inhibitors.

S. No.	Nanoparticles	Preparation methods	Characterization and corrosion measurement technique	Highest inhibition efficiency (%)	References
1.	CuO nanoparticle	Anodization technique potentiodynamic method	FESEM, EDXRD, XRD	86.2% in 3.5% NaCl and 74.5% 2 mg/L NH3	Mahmood et al. (2018)
2.	ZnO nanocomposite with poly (aniline-co-N-ethyl aniline)	Green synthesis and chemical precipitation	Potentiodynamic polarization measurements, electrochemical impedance spectroscopy FT-IR, XRD, SEM,	94.00% (0.1 M HCl) 82.46% (5% NaCl) 96% (distilled water	Mobin et al. (2017)
3.	Ag nanoparticles and poly (ethylene glycol) thiol	Sol−gel polarization method	Electrochemical impedance spectroscopy (EIS), TEM and DLS. UV−Vis spectroscopy	90.95% (1 M HCl)	Atta et al. (2013)
4.	Polydopamine (PDA) nanoparticles	Polymerization	Weight loss, potentiodynamic polarization, electrochemical impedance (EIS), scanning electron microscopy (SEM) and atomic force microscopy (AFM)	86.42% (at concentration of 40 mol/L) (1 M HCl)	Li et al. (2017)
5.	Iron oxide (FeO_2^3) with olive oil	Co-precipitation weight loss method,	Scanning electron microscopy (SEM)	80.88%	Palanisamy et al. (2014)
6.	MgO nanoparticle	Epoxy sol−gel weight loss technique	XRD, SEM	97% (acetic solution) 93.7% (saline solution)	Khodair et al. (2019)
7.	Mn_2O_3 nanoparticles	Co-precipitation	Potentiodynamic polarization, electrochemical impedance spectroscopy (EIS), XRD, TEM, FT-IR.	23.06%−55.6% (20 ppm) (0.5 M HCl)	Najjar et al. (2018)

8.	TiO2, MnO_2^3, and ZnO nanoparticles with polypyrrole	Electro polymerization	Potentiodynamic polarization, electrochemical impedance spectroscopy, SEM, TEM	98.09%-PPy/TiO2 (acidic solution)	Hosseini et al. (2017)
9.	SiO_2/acrylic nanoparticle	Eco-friendly methods	FT-IR, XRD, SEM–EDX, EIS and polarization measurement	45.38% in 1% HCl and 77.75% in 3.5% NaCl	Rajkumar and Vedhi (2019)
10.	ZnO nanoparticle	PMMA and resin chemical precipitation method	XRD, SEM, FT-IR, potentiodynamic polarization, EIS, weight loss method	99% in 3.5% NaCl at 25°C (at 10% PRZn conc.)	Rajendran et al. (2018)
11.	TiO_2, Ag, and Zn nanoparticles with polyaniline	Electrochemical method deposition	SEM–EDX, UV, FT-IR, EIS	97.54% (3.5% NaCl) (for PANI/Ag nanocomposite)	Ates and Topkaya (2015)

2.4.1 Dispersibility of metal oxide nanoparticles

Depending on the synthesis processes, NPs may end up perfectly dispersed in the coating or agglomerated particles. The ideal dispersion and distribution of NPs in the coating reflect the best possible enhancement in the properties. The dispersion of NPs may depend upon the following factors

- Charge on NPs
- Concentration
- Surface modification
- Temperature

Whenever NPs are added, as solute and solvent could be polar or nonpolar in nature, the charge on NPs determines their dispersion capability in a polar and nonpolar solution. TiO_2 NPs, for example, are dispersed in polar solvents such as water. However, nonpolar solvents such as n-hexane, acetone, plasticizer, and alcohol are unsuitable for TiO_2 NPs because of the absence of poles. Therefore, the charge on an NP should be a key to analyzing the dispersion capability in a particular solvent. The NPs can be treated with chemical compounds, which modify the surface of NPs and enhance their dispersion. The selection of surface-modifying compounds is based on the application and final goal of the component to which NPs are being added. For instance, TiO_2 NPs are treated with silane coupling agent KH-550 to improve the dispersion in organic solvents such as ethanol. The KH-550 was attached to the surface of TiO_2 and imparted the nonpolar characteristics to the surface of TiO_2 NPs. The concentration of NPs is also essential to get the uniform properties of the coating. An optimum concentration always exists beyond the agglomeration of NPs, which is more favored than dispersion; hence, undesirable properties will result. For instance, when incorporated in Ni–P coatings, the concentration of Y_2O_3 less than 1 g/L did not result in uniform coatings, and the corrosion resistance was the function of Y_2O_3 NP's concentration. Similarly, a modified PANI (polyaniline) coating with NPs of TiO_2 and ZnO revealed superior corrosion resistance. This exceptional improvement has been associated with the presence of NPs because they increase the surface area available for the liberation of the dopant. The dopant anion here acts similarly to a corrosion inhibitor and helps passivate the steel surfaces, even if the coating is damaged. The bath temperature is also essential for the uniform dispersion of NPs. The surface modification of NPs may provide the best dispersion when heated to a specific temperature. For example, KH-550-modified TiO_2 NPs show better dispersion even after 24 hours if the bath temperature is maintained at 80°C.

2.5 Mechanism of corrosion protection

NPs incorporated into the various coatings have been explained in previous sections. The mechanism behind the improvement in corrosion protection is that making coating self-cleaning, antifouling, or filling the cracks and pores present in the coatings with

NPs. The detailed explanation of all of these mechanisms is beyond the scope of this chapter; however, a brief description of each mechanism is given in the following sections.

2.5.1 Self-healing

Self-healing coatings have been considered promising in the industries in recent years. The objective is that once the defects form on the surface and the protective coating is damaged, they can mend themselves and continue to protect the substrate without external aid. Metal/metal oxide NPs are utilized as inhibitors containing reservoirs, that is, hollow SiO_2 particles, mesoporous TiO_2, mesoporous silica, CeO_2 NPs, and TiO_2 porous particles. These particles may comprise the organic inhibitors to be released at active sites to heal the produced defects. It is also observed that metal/metal oxide NPs act as inhibitors, that is, cerium, cerium (III), and lanthanum(III). However, in this case, the containers that are carrying these inhibitors may be of nonmetallic in nature.

2.5.2 Antifouling

Fouling is a severe issue that arises when the materials are in contact with water. The unwanted biological substances grow over the material constantly and could produce microfouling or macrofouling by attacking bacterial species, barnacles, tubeworms, bryozoans, and algae. The metal/metal oxide NPs such as Ag, TiO2, AgO, and ZnO had been utilized as antimicrobial agents for cleansing of freshwater and in biomedical applications. ZnO nanorods incorporated coatings were performed well to protect the Mg alloy substrate from the maritime microfouling- and macrofouling-causing species. Similarly, various metal oxide NPs are studied to develop cheap and more fouling-resistant coatings. For instance, TiO_2, Nb_2O_5, and SiO_2 NPs were studied; among them, TiO_2 NPs exhibited enhanced antifouling capability. Fig. 2.6 describes the brief mechanism of corrosion inhibition: (a) Self-healing and (b) active corrosion resistance.

2.6 Conclusion

Nanotechnology can protect metals and alloys from corrosion and harsh environmental conditions. Here, organic and inorganic coatings made by a range of methods such as electroplating, sol-gel, anodization, and electroless plating, among others, have been applied to metal/metal oxide NPs in a broad range of coating systems have been explored. They restrict corrosion rate through the mechanism of self-healing, self-cleaning, antifouling, barrier protection, photodegradation efficiency, etc. The efficacy of barrier coatings to prevent corrosion depends on several variables, including pH, concentration, temperature, nature, charge, and others, which affect the dispersibility of the NPs in the coatings. In corrosion-resistant coatings, various metal oxide NPs

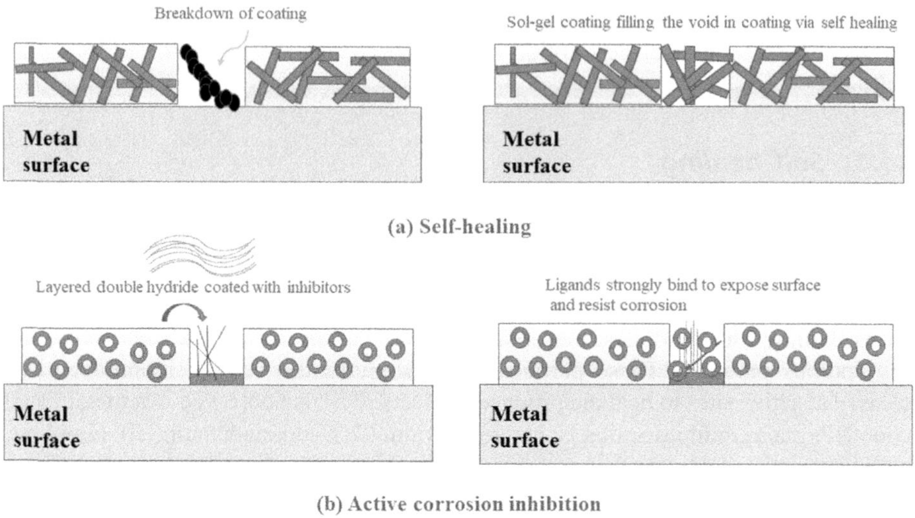

Figure 2.6 The brief mechanism of corrosion inhibition: (a) self-healing and (b) active corrosion resistance.

such as CuO, ZnO, Fe_2O_3, MgO, Mn_2O_3, SiO_2, TiO_2, and others. A comparison and discussion of several coating methods to improve corrosion resistance were conducted. We also suggest using hybrid composites as coating materials instead of just one filler since NP combinations can improve performance.

References

Ahmadi, Z., 2019. Epoxy in nanotechnology: a short review. Progr. Organ. Coat. 132, 445–448.

Akpan, U.G., Hameed, B.H., 2010. The advancements in sol–gel method of doped-TiO2 photocatalysts. Appl. Catal. Gen. 375 (1), 1–11.

Alibakhshi, E., Ghasemi, E., Mahdavian, M., Ramezanzadeh, B., 2017. A comparative study on corrosion inhibitive effect of nitrate and phosphate intercalated Zn-Al-layered double hydroxides (LDHs) nanocontainers incorporated into a hybrid silane layer and their effect on cathodic delamination of epoxy topcoat. Corrosion Sci. 115, 159–174.

Anjum, M.J., Ali, H., Khan, W.Q., Zhao, J., Yasin, G., 2020. Metal/metal oxide nanoparticles as corrosion inhibitors. In: Corrosion Protection at the Nanoscale. Elsevier, pp. 181–201.

Ates, M., Topkaya, E., 2015. Nanocomposite film formations of polyaniline via TiO2, Ag, and Zn, and their corrosion protection properties. Prog. Org. Coating 82, 33–40.

Atta, A.M., Allohedan, H.A., El-Mahdy, G.A., Ezzat, A.R.O., 2013. Application of stabilized silver nanoparticles as thin films as corrosion inhibitors for carbon steel alloy in 1 M hydrochloric acid. J. Nanomater. 2013.

Atta, A.M., Al-Lohedan, H.A., El-saeed, A.M., Al-Shafey, H.I., Wahby, M.H., 2017. Epoxy embedded with TiO2 nanogel composites as promising self-healing organic coatings of steel. Prog. Org. Coat. 105, 291–302.

Baorong, H., Li, X., Xiumin, M., Cuiwei, D., Zhang, D., Zheng, M., Fubin, M., 2017. The cost of corrosion in China. Npj Mater. Degradat. 1 (1).

Bokov, D., Turki Jalil, A., Chupradit, S., Suksatan, W., Javed Ansari, M., Shewael, I.H., Kianfar, E., 2021. Nanomaterial by sol-gel method: synthesis and application. Adv. Mater. Sci. Eng. 2021, 1–21.

Cheng, Y., Wang, K., Qi, Y., Liu, Z., 2022. Chemical vapor deposition method for graphene fiber materials. Acta Phys.-Chim. Sin 38 (2), 2006046.

Chrzanowska, J., Hoffman, J., Małolepszy, A., Mazurkiewicz, M., Kowalewski, T.A., Szymanski, Z., Stobinski, L., 2015. Synthesis of carbon nanotubes by the laser ablation method: effect of laser wavelength. Phys. Status Solidi (B) 252 (8), 1860–1867.

Deepa, K., Venkatesha, T., 2018. Synthesis of NiO-ZrO2 mixed metal oxide nanoparticles and their application in Zncomposite coating on mild steel. Anal. Bioanal. Electrochem 10, 890–900.

Dhoke, S.K., Khanna, A.S., 2012. Electrochemical impedance spectroscopy (EIS) study of nano-alumina modified alkyd based waterborne coatings. Prog. Org. Coating 74 (1), 92–99.

Dhoke, S.K., Khanna, A.S., Sinha, T.J.M., 2009. Effect of nano-ZnO particles on the corrosion behavior of alkyd-based waterborne coatings. Prog. Org. Coat. 64 (4), 371–382.

Fayomi, O.S.I., Popoola, A.P.I., 2015. Anti-corrosion and tribo-mechanical properties of co-deposited Zn–SnO 2 composite coating. Acta Metall. Sin. 28, 521–530.

Ghali, E., Sastri, V.S., Elboujdaini, M., 2007. Corrosion Prevention and Protection: Practical Solutions. John Wiley & Sons.

Haddadi, S.A., Mahdavian, M., Karimi, E., 2015. Evaluation of the corrosion protection properties of an epoxy coating containing sol–gel surface modified nano-zirconia on mild steel. RSC Adv. 5 (36), 28769–28777.

Hayatdavoudi, H., Rahsepar, M., 2017. Smart inhibition action of layered double hydroxide nanocontainers in zinc-rich epoxy coating for active corrosion protection of carbon steel substrate. J. Alloys Compd. 711, 560–567.

Honarvar Nazari, M., Bergner, D., Shi, X., 2015. Managing metallic corrosion on winter maintenance equipment assets. In: Environmental Sustainability in Transportation Infrastructure, pp. 61–76.

Hosseini, M., Fotouhi, L., Ehsani, A., Naseri, M., 2017. Enhancement of corrosion resistance of polypyrrole using metal oxide nanoparticles: potentiodynamic and electrochemical impedance spectroscopy study. J. Colloid Interface Sci. 505, 213–219.

Isimjan, T.T., Wang, T., Rohani, S., 2012. A novel method to prepare superhydrophobic, UV resistance and anti-corrosion steel surface. Chem. Eng. J. 210, 182–187.

Khodair, Z.T., Khadom, A.A., Jasim, H.A., 2019. Corrosion protection of mild steel in different aqueous media via epoxy/nanomaterial coating: preparation, characterization and mathematical views. J. Mater. Res. Technol. 8 (1), 424–435.

Kim, M., Osone, S., Kim, T., Higashi, H., Seto, T., 2017. Synthesis of nanoparticles by laser ablation: a review. KONA Powder Particl. J. 34, 80–90.

Koch, G., Varney, J., Thompson, N., Moghissi, O., Gould, M., Payer, J., 2016. NACE International Impact Report: International Measures of Prevention, Application, and Economics of Corrosion Technologies Study. NACE Int., Houston, TX, USA, Tech. Rep.

Lee, K.M., Shin, K.R., Namgung, S., Yoo, B., Shin, D.H., 2011. Electrochemical response of ZrO_2-incorporated oxide layer on AZ91 Mg alloy processed by plasma electrolytic oxidation. Surf. Coating. Technol. 205 (13–14), 3779–3784.

Li, H., Liu, G., Liu, B., Chen, W., Chen, S., 2007. Dielectric properties of polyimide/Al2O3 hybrids synthesized by in-situ polymerization. Mater. Lett. 61 (7), 1507–1511.

Li, Z., Yang, Y., Wang, Z., Zhang, X., Chen, Q., Qian, X., Ji, Y., 2017. Polydopamine nanoparticles doped in liquid crystal elastomers for producing dynamic 3D structures. J. Mater. Chem. A 5 (14), 6740–6746.

Li, G.L., Zheng, Z., Möhwald, H., Shchukin, D.G., 2013a. Silica/polymer double-walled hybrid nanotubes: synthesis and application as stimuli-responsive nanocontainers in self-healing coatings. ACS Nano 7 (3), 2470–2478.

Li, X.W., Song, R.G., Jiang, Y., Wang, C., Jiang, D., 2013b. Surface modification of TiO2 nanoparticles and its effect on the properties of fluoropolymer/TiO2 nanocomposite coatings. Appl. Surf. Sci. 276, 761–768.

Low, C.T.J., Wills, R.G.A., Walsh, F.C., 2006. Electrodeposition of composite coatings containing nanoparticles in a metal deposit. Surf. Coating. Technol. 201 (1–2), 371–383.

Luo, X., Zhong, J., Zhou, Q., Du, S., Yuan, S., Liu, Y., 2018. Cationic reduced graphene oxide as self-aligned nanofiller in the epoxy nanocomposite coating with excellent anticorrosive performance and its high antibacterial activity. ACS Appl. Mater. Interfaces 10 (21), 18400–18415.

Mahmood, M.H., Suryanto, S., Al Hazza, M.H., Haidera, F.I., 2018. Developing of corrosion resistance nano copper oxide coating on copper using anodization in oxalate solution. Int. J. Eng. 31 (3), 450–455.

Majd, M.T., Davoudi, M., Ramezanzadeh, M., Ghasemi, E., Ramezanzadeh, B., Mahdavian, M., 2020. Construction of a smart active/barrier anti-corrosion system based on epoxy-ester/zinc intercalated kaolin nanocontainer for steel substrate. Construct. Build. Mater. 247, 118555.

Mobin, M., Aslam, J., Alam, R., 2017. Corrosion protection of poly (aniline-co-N-ethylaniline)/ZnO nanocomposite coating on mild steel. Arabian J. Sci. Eng. 42, 209–224.

Najjar, R., Abdel-Gaber, A.M., Awad, R., 2018. Electrochemical corrosion behaviour of carbon steel in acidic media in presence of Mn2O3 nanoparticles synthesized at different pH. Int. J. Electrochem. Sci. 13 (9), 8723–8735.

Nazari, M.H., Shi, X., 2016. Polymer-based nanocomposite coatings for anticorrosion applications. In: Industrial Applications for Intelligent Polymers and Coatings, pp. 373–398.

Nazari, M.H., Zhang, Y., Mahmoodi, A., Xu, G., Yu, J., Wu, J., Shi, X., 2022. Nanocomposite organic coatings for corrosion protection of metals: a review of recent advances. Prog. Org. Coating 162, 106573.

Nguyen, T.V., Nguyen, T.A., Dao, P.H., Nguyen, A.H., Do, M.T., 2016. Effect of rutile titania dioxide nanoparticles on the mechanical property, thermal stability, weathering resistance and antibacterial property of styrene acrylic polyurethane coating. Adv. Nat. Sci. Nanosci. Nanotechnol. 7 (4), 045015.

Palanisamy, K., Devabharathi, V., Sundaram, N.M., 2014. Corrosion inhibition studies of mild steel with carrier oil stabilized of iron oxide nanoparticles incorporated into a paint. Environ. Times 1, 2.

Palanivel, V., Zhu, D., van Ooij, W.J., 2003. Nanoparticle-filled silane films as chromate replacements for aluminum alloys. Prog. Org. Coating 47 (3–4), 384–392.

Popoola, A.P.I., Aigbodion, V.S., Fayomi, O.S.I., 2016. Anti-corrosion coating of mild steel using ternary Zn-ZnO-Y2O3 electro-depositon. Surf. Coating. Technol. 306, 448–454.

Pugazhenthi, I., Mohammed Safiullah, S., Anver Basha, K., 2018. UV and corrosion protective behavior of polymer hybrid coating on mild steel. J. Appl. Polym. Sci. 135 (16), 46175.

Radhakrishnan, S., Siju, C.R., Mahanta, D., Patil, S., Madras, G., 2009. Conducting polyaniline–nano-TiO2 composites for smart corrosion resistant coatings. Electrochim. Acta 54 (4), 1249–1254.

Rajendran, A., Sivalingam, T., Narayanan, P., Murugavel, S.C., 2018. Nano composite polymer resin coating to control mild steel corrosion in marine environment. Am. J. Appl. Chem. 6 (3), 102–125.

Rajendran, S., Nguyen, T.A., Kakooei, S., Yeganeh, M., Li, Y. (Eds.), 2020. Corrosion Protection at the Nanoscale. Elsevier.

Rajkumar, R., Vedhi, C., 2019. A study of corrosion protection efficiency of silica nanoparticles acrylic coated on mild steel electrode. Vacuum 161, 1–4.

Ramanathan, T., Abdala, A.A., Stankovich, S., Dikin, D.A., Herrera-Alonso, M., Piner, R.D., Brinson, L.C., 2008. Functionalized graphene sheets for polymer nanocomposites. Nat. Nanotechnol. 3 (6), 327–331.

Ramezanzadeh, B., Shamshiri, M., Sari, M.G., 2018. Designing a multi-functionalized clay lamellar-co-graphene oxide nanosheet system: an inventive approach to enhance mechanical characteristics of the corresponding epoxy-based nanocomposite coating. Prog. Org. Coating 116, 7–20.

Raps, D., Hack, T., Wehr, J., Zheludkevich, M.L., Bastos, A.C., Ferreira, M.G.S., Nuyken, O.J.C.S., 2009. Electrochemical study of inhibitor-containing organic–inorganic hybrid coatings on AA2024. Corrosion Sci. 51 (5), 1012–1021.

Saravanan, A., Kumar, P.S., Karishma, S., Vo, D.V.N., Jeevanantham, S., Yaashikaa, P.R., George, C.S., 2021. A review on biosynthesis of metal nanoparticles and its environmental applications. Chemosphere 264, 128580.

Singapati, A.Y., Ravikumar, C., 2023. Mechanism of nanoparticle formation in the liquid-phase thermal decomposition method. Langmuir 39 (27), 9325–9342.

Su, H.W., Chen, W.C., 2008. High refractive index polyimide–nanocrystalline-titania hybrid optical materials. J. Mater. Chem. 18 (10), 1139–1145.

Taheri, N.N., Ramezanzadeh, B., Mahdavian, M., 2019. Application of layer-by-layer assembled graphene oxide nanosheets/polyaniline/zinc cations for construction of an effective epoxy coating anti-corrosion system. J. Alloys Compd. 800, 532–549.

Tjong, S.C., 2006. Structural and mechanical properties of polymer nanocomposites. Mater. Sci. Eng. R Rep. 53 (3–4), 73–197.

Torknezhad, Y., Khosravi, M., Assefi, M., 2018. Corrosion protection performance of nanoparticle incorporated epoxy paint assessed by linear polarization and electrochemical impedance spectroscopy. Mater. Corros. 69 (4), 472–480.

Wetzel, B., Haupert, F., Zhang, M.Q., 2003. Epoxy nanocomposites with high mechanical and tribological performance. Compos. Sci. Technol. 63 (14), 2055–2067.

Xiao, X., Wang, D., Li, Y., Jackson, E., Fang, Y., Zhang, Y., Shi, X., 2016. Investigation into the synergistic effect of nano-sized materials on the anti-corrosion properties of a waterborne epoxy coating. Int. J. Electrochem. Sci. 11, 6023–6042.

Yu, Y., Zuo, Y., Zhang, Z., Wu, L., Ning, C., Zuo, C., 2019. Al2O3 coatings on zinc for anticorrosion in alkaline solution by electrospinning. Coatings 9 (11), 692.

Zhang, Y., Zhao, M., Zhang, J., Shao, Q., Li, J., Li, H., Guo, Z., 2018. Excellent corrosion protection performance of epoxy composite coatings filled with silane functionalized silicon nitride. J. Polym. Res. 25, 1–13.

Zhao, Y., Zhang, Z., Shi, L., Zhang, F., Li, S., Zeng, R., 2019. Corrosion resistance of a self-healing multilayer film based on SiO2 and CeO2 nanoparticles layer-by-layer assembly on Mg alloys. Mater. Lett. 237, 14–18.

Section II

Synthetic approach of nanomaterials for coatings

Fundamentals and basics of synthetic approaches of nanostructured coatings

3

Subodh Barthwal[1] *and Sumit Barthwal*[2]
[1]Department of Mechanical Engineering, Amity University, Greater Noida, Uttar Pradesh, India; [2]Nanomechatronics Lab, Kookmin University, Seoul, South Korea

3.1 Introduction

A coating is used on a substrate to achieve protection, coloration, or various functionalities. Its composition comprises resins, solvents, pigments, fillers, and additives. Pigments contribute color or corrosion resistance to the coating, while fillers play a role in shaping mechanical properties and imparting specific additional characteristics to the substrate. In recent decades, the incorporation of nanoparticles has garnered significant attention within the coatings industry due to their ability to enhance and introduce novel properties to paints (Gaur and Khanna, 2015; Pilotek and Tabellion, 2005). Nanotechnology has captured the attention of scientists worldwide across various disciplines such as chemistry, physics, and biology. It holds immense promise as a technology due to its broad range of applications, benefits, and advantages. The coatings industry stands out as a sector where nanoparticles are anticipated to make substantial contributions. Demonstrated through various studies, the integration of nanoparticles into coatings not only enhances their performance but also introduces novel functionalities, enabling the creation of multipurpose coatings (Barthwal and Barthwal, 2023). Examples of benefits derived from nanomodified paints include enhanced chemical resistance, increased resilience to erosion and abrasion, protection against UV light, and antifouling properties, among others (Khanna, 2008).

Nanotechnology has experienced significant growth in coatings applications recently, driven by two key factors: the increased availability of nanoscale materials, such as various nanoparticles, and advancements in processes enabling precise control of coating structures at the nanoscale. This surge is also fueled by nanotechnology's potential to address diverse performance challenges across a wide array of products and structures that rely on coatings.

The term "nanoparticles" refers to solid particles with dimensions falling within the range of 1−100 nm, typically existing as a solid powder or through dispersion in a liquid solvent. Additionally, it is essential to maintain a specific separation between these nanoparticles to prevent agglomeration, ensuring the stability of dispersions. This discussion focuses on the utilization of nanofillers and nanopigments in the

Multifunctional Nanostructured Coatings. https://doi.org/10.1016/B978-0-443-23683-9.00003-9

coatings industry. A coating earns the designation of "nanocoating" when it incorporates a nanocomponent (Pilotek and Tabellion, 2005).

Coatings find application in various sectors, including residential and commercial buildings, transportation vehicles, electronics, biomedical devices, and industrial maintenance. Ranging from interior and exterior paints to specialized coatings for high-rise buildings, automobiles, airplanes, bridges, and marine vessels, coatings serve essential functions such as enhancing aesthetics, protecting substrates from damage (e.g., scratches, corrosion, weathering), and providing specialized functionalities like conductivity, insulation, and water repellency.

Nanotechnology has particularly revolutionized coatings by offering possibilities to enhance their protective capabilities and specialized functionalities. Coatings can now be engineered with precision, with thicknesses ranging from hundreds of micrometers to less than 100 nm, catering to diverse needs from antiskid coatings on aircraft carriers to insulating coatings in microchips. This innovation has opened up avenues for coatings to exhibit improved resistance to scratches, impact, corrosion, weathering, and bio-fouling, while also enabling the incorporation of specialized features such as conductivity, insulation, water repellency, and heat reflection.

Nanoparticles inherently possess the properties of the materials from which they are composed. In the coatings sector, one of the pivotal characteristics of nanoparticles is their larger surface area to volume ratio, resulting in abundant surface activity and increased interactions within the system. Consequently, surface properties take precedence over bulk properties. This unique attribute allows achieving comparable results for a specific property at lower concentrations than those required for the bulk material.

The coating thickness depends on the fillers and pigment particle size. Consequently, the use of smaller fillers and pigments results in the formation of thinner films, requiring less paint for a given surface area, and facilitating the coating of miniature objects. According to market research, the nanocoatings market is projected to achieve a valuation of $20.1 billion by 2030, with a substantial increase from $10.7 billion in 2020 (Nanocoatings market size, share, competitive landscape and trend analysis report by type (Anti-fingerprint, antimicrobial, self-cleaning (bionic & photocatalytic), anti-corrosion, conductive, and others) and application (electronics, energy, food & packaging, construction, marine, automotive, healthcare, others): Global Opport. Anal. Ind. Forec., 2021-2030).

Among the various types of nanocoatings, antimicrobial variants are the most widely utilized, closely followed by self-cleaned and antifouling coatings. Antimicrobial nanocoatings find predominant application in the healthcare and medical domains, as well as in manufacturing, food packaging, and water treatment. Self-cleaned and antifouling coatings are employed across diverse sectors, including automotive industries, marine, electronics, and food manufacturing. Projections indicate an anticipated growth in these sectors between 2013 and 2019 (Nanocoatings Market_ product (Anti-fingerprint, Antimicrobial, Anti-fouling & Easy-to-clean, Self-cleaning (Bionic and Photocatalytic) and Specialty coatings) Applications (Medical & Healthcare, Food Manufacturing, Packaging, Marine, Water Treatment Electronics, Building & Construction, Automotive, and Energy and Textiles, Leathers, Industrial Engineering)_ Global Industry Analysis, Size, Share, Growth, Trends and Forecast 2013_2019). The marine and automotive industries emerge as lucrative segments for the application of nanocoatings. With the automotive sector experiencing growth, the potential utilization of this technology within the industry is on the rise. The marine industry is very interested in antifouling nanocoatings because fouling formation has a significant negative influence on marine vehicle performance. Estimates suggest that fouling may lead to an excess fuel consumption of about 40% and a reduction in the speed of marine vehicles.

Nanocoatings have found diverse applications across various industries due to their unique properties and functionalities. Some relevant examples include:

1. Self-Cleaning Coatings: Nanocoatings can be engineered to create self-cleaning surfaces by incorporating nanoparticles with photocatalytic properties, such as titanium dioxide (TiO_2). These coatings can break down organic contaminants when exposed to sunlight or artificial light, making them ideal for applications in building exteriors, glass surfaces, and automotive coatings. An example is self-cleaning glass used in windows and facades, which reduces the need for manual cleaning and maintenance.
2. Anticorrosion Coatings: Nanoparticle-based coatings can provide excellent corrosion protection by forming a barrier against moisture and corrosive agents. Zinc oxide (ZnO) and zinc nanoparticles, for instance, have been incorporated into coatings for steel structures, pipelines, and marine vessels to prevent corrosion and extend their lifespan.
3. Antifouling Coatings: Nanocoatings can deter the adhesion of organisms like algae, barnacles, and bacteria, thereby reducing bio-fouling on surfaces. Copper nanoparticles, for example, have been used in marine coatings to prevent the growth of marine organisms on ship hulls, reducing drag and fuel consumption.
4. Antiscratch Coatings: Nanostructured coatings can enhance the scratch resistance of surfaces by creating a tough, protective layer. Silica nanoparticles, for instance, have been incorporated into automotive clear coats to improve scratch resistance and maintain the appearance of vehicles over time.
5. Thermal Barrier Coatings: Nanocoatings can provide insulation against heat transfer, making them useful for applications requiring thermal management. Ceramic nanoparticles, such as alumina (Al_2O_3) or zirconia (ZrO_2), have been used in thermal barrier coatings for aerospace components, gas turbines, and automotive exhaust systems to reduce heat transfer and improve efficiency.
6. Conductive Coatings: Nanoparticle-based coatings can impart electrical conductivity to surfaces, enabling applications in electronics, sensors, and electromagnetic shielding. Silver nanoparticles, for example, have been incorporated into conductive coatings for touchscreens, printed circuit boards, and flexible electronics.
7. Anti-icing Coatings: Nanocoatings engineered to prevent the formation of ice on surfaces have applications in aviation, transportation, and infrastructure. By incorporating nanoparticles with ice-repellent properties, these coatings can reduce the risk of ice accumulation on aircraft wings, wind turbines, and power lines, enhancing safety and efficiency.
8. Antimicrobial Coatings: Nanocoatings containing antimicrobial nanoparticles, such as silver or copper, have been developed to inhibit the growth of bacteria and fungi on surfaces. These coatings find applications in healthcare facilities, food packaging, and consumer products, contributing to hygiene and infection control.

In essence, nanotechnology has propelled coatings beyond traditional boundaries, empowering them to meet evolving performance demands across a myriad of applications, from everyday consumer products to critical infrastructure and advanced technological devices.

3.2 Synthesis of nanocoatings

Through the application of specific synthesis methods, it becomes feasible to create nanomaterials customized for particular purposes. Essentially, two distinct approaches for the synthesis of nanocoatings exist: A "top-down" strategy and a "bottom-up" strategy (Fig. 3.1) (Barthwal et al., 2022). Both strategies are flexible and can be used together to create complex hierarchical structures. Although top-down approaches are generally quicker and more scalable, bottom-up approaches offer better control over the composition of the material and can make it easier to incorporate a variety of features. The particulars of the application and the required properties of the superhydrophobic surface will determine which of these methods is suitable for selection.

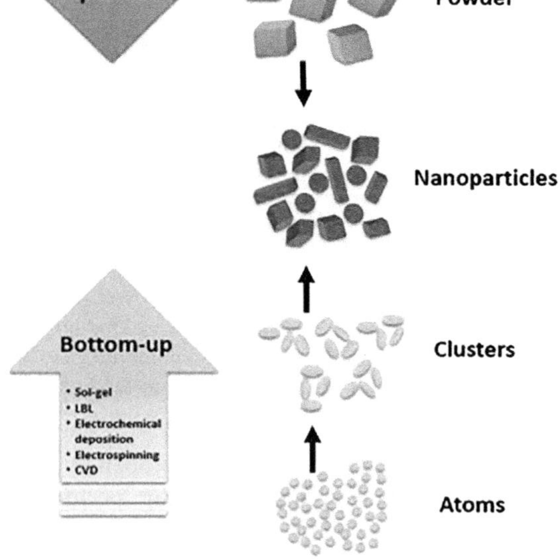

Figure 3.1 Schematic description of synthesis of nanomaterials.

3.2.1 Top-down synthesis

For the purpose of creating nanocoatings, top-down synthesis entails making the necessary macroscopic modifications of existing structures or materials. This method usually starts with a bulk material, and then the surface is modified to create micro- and nanostructures using different techniques like lithography, etching, and abrasion. The precise control over nanostructure shape, size, and arrangement is made possible by the top-down synthesis approach. Top-down synthesis has advantages in terms of precision and repeatability, but it may not be as cost-effective or scalable for large-scale manufacturing.

3.2.1.1 Lithography technique

The lithography technique stands out as a sophisticated and precise method for nanocoating, leveraging the principles of light and pattern transfer to form micro- and nanostructures on a substrate. In lithography, a photosensitive material is exposed to light through a mask or photomask that carries the desired pattern. This exposure induces chemical or physical changes in the material, producing a positive or negative replica of the pattern on the substrate.

Various forms of the lithography-based technique exist, depending on the light source employed. These encompass electron beam lithography (EBL) (Feng et al., 2011), nanoimprinting (Baji et al., 2022), photolithography using X-ray and ultraviolet (UV) sources (Shirtcliffe et al., 2004; Marquez-Velasco et al., 2010), laser lithography (Zheng et al., 2023), and nanosphere lithography (NSL) (Dev et al., 2014). The precision and versatility of lithography render it a vital tool in the advancement of sophisticated materials for various applications (Zhang et al., 2023; Guo et al., 2023; Liu et al., 2023). Ji et al. (2021) designed various diverse bio-inspired micro-/nano-scaled cylindrical patterns by manipulating their width and interspacing on a silicon substrate through 3D photo/colloidal lithography techniques. Subsequently, the bio-inspired surface textures were coated with polydimethylsiloxane, serving as a soft negative template, and then overlaid with polyaniline (PANI) to explore the morphology's impact in supercapacitor applications. Findings indicated that the bio-inspired texture featuring micro-/nano-architectured surfaces with two distinct widths and interspacing of 5 μm and 500 nm exhibited the most favorable PANI capacitance performance, reaching 429 F/g. In comparison, other single and nonpatterned (plain) templates demonstrated capacitance performances of 364 F/g and 316 F/g, respectively, at 5 mV/second. A collaborative performance yielding 614 F/g was witnessed at 1 A/g, accompanied by outstanding capacitive retention of 85% even after 10,000 cycles. Remarkably, the flexibility of the system was maintained over extended cycling periods without any deterioration in performance. Zhu et al. (2024) investigated a novel edge lithography technique based on aluminum reactive ion etching for the creation of silicon nanostructures. Unlike other methods, this approach eliminates the need for an additional film deposition step. The oxidation of the aluminum film itself during the oxygen plasma ashing process results in the formation of an aluminum oxide layer on the Al sidewalls. After two rounds of aluminum etching,

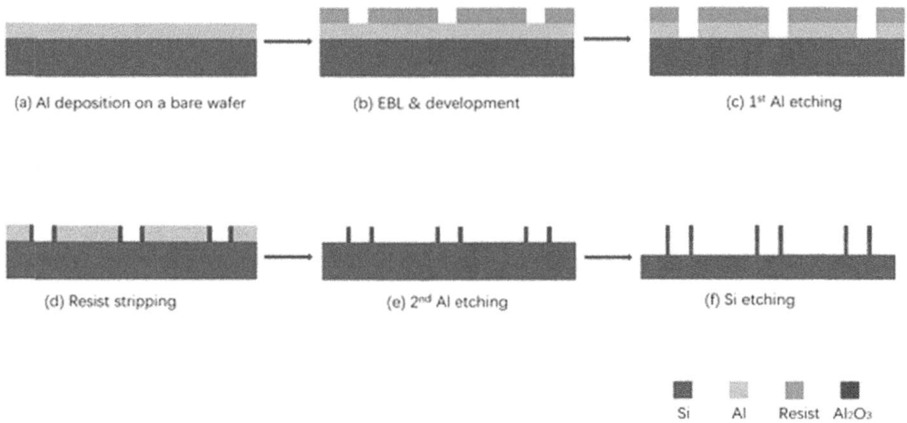

Figure 3.2 Process flow diagram of Al edge lithography (Zhu et al., 2024).
From Zhu, C., Pan, A., Zhu, X., Zheng, S., Cui, B., 2024. Edge lithography based on aluminum dry etching. Micro Nano Eng. 22, 100233. https://doi.org/10.1016/j.mne.2023.100233.

the thin aluminum oxide structure is selectively kept, and a nonswitching mixed gas Si etching procedure is used to transfer the pattern onto a silicon substrate (Fig. 3.2). Findings show that oxygen plasma ashing greatly accelerates the development of an aluminum oxide surface layer. Moreover, the aluminum film's deposition technique had a significant impact on the result. Table 3.1 shows various findings and applications of using lithographical techniques for different substrates and coating materials.

3.2.1.2 Chemical etching

Chemical etching emerges as a highly efficient technique to fabricate nanocoated surfaces, which involves carefully regulating a chemical reaction to precisely remove material and create complex micro- and nanostructures on a substrate. This procedure starts with the application of a particular etchant solution, which interacts with the material surface and causes reactions that produce customized surface textures. The final surface morphology is significantly influenced by the choice of etchant (acid, like H_2SO_4 and HCl, and a base, like NaOH), as well as by variables like concentration and duration. Chemical etching is a flexible method for producing nanocoated surfaces because it has many benefits, such as affordability, scalability, and suitability for a wide variety of materials (Barthwal et al., 2019; Zhu et al., 2023). He et al. (2023) used submicron silicon particles and a two-step copper-assisted chemical etching procedure to develop porous structures. Asphalt was subsequently used as a carbon source to create carbon-coated copper-modified porous silicon (pSi/Cu@C) composite anode material. Excellent electrochemical performance is demonstrated by the pSi/Cu@C. The discharge specific capacity is still 820 mAh g^{-1} after 200 cycles at a rate of 0.5C (1C = 4.2 A g^{-1}). In contrast, raw silicon anode material (Si-raw) has a low discharge specific capacity of 229 $mAhg^{-1}$. They observed in their study that the electrode material's entire conductivity can be efficiently enhanced by the external carbon

Table 3.1 Recent work by using lithographical technique.

Fabrication technique	Coating material	Substrate	Findings/objective	Applications	References
Lithography	Silica nanospheres (SNs)	Silicon substrate	The fabrication of ordered arrays of silicon nanoparticles (SiNPs) has been accomplished through a combination of nanosphere lithography and metal-assisted chemical etching. The presence of silicon nanocrystals adorned on the sidewalls of SiNPs gives rise to the photoluminescence phenomenon, attributed to the quantum confinement effect.	Solar cell, lithium ion battery, surface enhanced raman scattering (SERS), and functional micro/nanodevices	(Van Minh et al., 2022)
	Diamond silver nanoparticle (DSN)	SERS substrates	SPL technology offers a straightforward and cost-effective approach for achieving large-scale, sensitive, and uniform surface enhanced raman scattering (SERS) substrates.	Environmental, biological, food safety	(Wang et al., 2019)
	Polystyrene (PS)	Silicon wafer	The PS honeycomb mesh holds potential utility as a mask or etch-resist layer in various applications, enabling selective surface treatments for the creation of a periodic and manageable surface topography in diverse contexts.	Solar, antireflection application	(Sharma et al., 2022)
	ZnO	Si and SiO_2 substrates	This study provides a thorough exploration of the integration of purely chemical, bottom-up solution-based synthesis with advanced electron-beam lithography for the systematic generation of periodic arrays of ZnO nanostructures on Si and SiO_2 substrates.	Nano-LEDs, single-wire transistors, nano-harvesters	(Papageorgiou et al., 2020)
	SnO_2 nanowires	Si or Al_2O_3 substrate	In contrast to comparable nanowire integration methods, the nanosensors detailed in this description exhibit enhanced responsiveness and a reduced response time.	Gas sensors, electronics	(Samà et al., 2019)

layer, while the copper particles are dispersed throughout the porous structure. Torralba et al. (2024) conducted a study to test the feasibility of electroless deposition of bimetallics and to characterize the deposits from a compositional, structural, and morphological perspective. They also reported on the changes in surface morphology brought about by Si dissolution. The extent to which these various factors may be accurately controlled to create the desired bimetallic catalysts has been the subject of their investigation. Furthermore, XPS has been employed to examine the potential dependence of the electronic effects on the bimetal composition and the deposition parameters. Lastly, they have attempted to investigate the causes and particularities of the deposition of bimetallic complexes of Cu and Pd and deposition on Si in a variety of manners when left alone, based on the experimental data and band bending simulations at the micro- and nanoscale.

Barthwal and Barthwal (2024) successfully developed a durable, cost-efficient, and easy-to-process superhydrophobic surface on an aluminum base using a two-stage chemical etching method. Initially, the process generated a hierarchical microporous (HMP) structure through chemical etching. This was followed by depositing a self-assembled layer of PFOTS onto the aluminum, resulting in exceptional superhydrophobic and oleophobic properties. The surface boasts an impressive water contact angle of 162 degrees and effectively repels a variety of oils with surface tensions ranging from 21.5 to 63.3 mN/m.

Thorough examinations of the surface morphology and chemical composition were conducted, alongside investigations into how surface morphology affects wettability, chemical stability, and antibacterial properties. Mechanical assessments demonstrated that the surface maintained its superhydrophobicity even after enduring linear abrasion for 120 cm under an applied pressure of 12.25 kPa. Additionally, the surface exhibited self-cleaning properties and resistance to fouling, along with notable antibacterial activity against *Staphylococcus aureus*, resulting in a 23 mm zone of inhibition.

Yao et al. (2021) introduced an innovative and simple approach for producing wafer-scale arrays of uniformly vertically aligned silicon nanowires (Si-NWs) within the diameter range of 50–200 nm. The fundamental concept involves the fabrication of thin gold (Au) nanostructures through magnetron sputtering, followed by the dewetting of Au and subsequent realization of wafer-scale, highly uniform Si-NW arrays through metal-assisted chemical etching at a low temperature (2°C), all achieved without the need for complex photolithography (Fig. 3.3). The outstanding properties of these Si-NW arrays were also demonstrated numerically, showcasing a maximum exciton generation rate of 1.76×1024 and a maximum energy generation rate of 28 for Si-NW arrays with a diameter of 200 nm. Some other findings and applications of using the chemical etching method for different substrates and coating materials are listed in Table 3.2.

3.2.1.3 Plasma treatment

Manufacturers frequently seek to apply plasma coating to their components and products to enhance overall performance and reliability. Plasma coating represents a fundamental technology capable of driving improvements in product functionality and

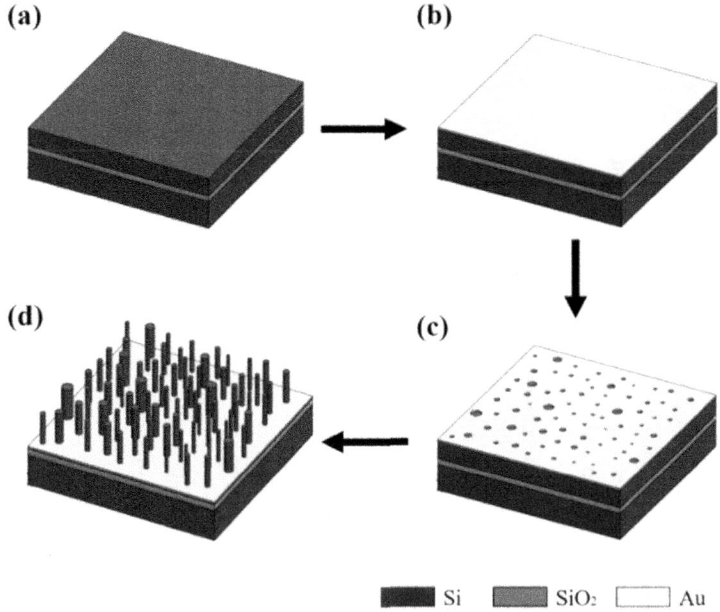

Figure 3.3 Schematic illustration of fabrication of Si-NWs by metal-assisted chemical etching (MACE) (Yao et al., 2021).
From Yao, C., Zhao, Y., Zhang, X., Li, H., Xie, C., 2021. Design and fabrication of wafer-scale highly uniform silicon nanowire arrays by metal-assisted chemical etching for antireflection films. Results Phys. 31, 105018. https://doi.org/10.1016/j.rinp.2021.105018.

dependability. The incorporation of plasma coating across a product line has the potential to elevate performance and open avenues for entirely new products that were previously unattainable. This method involves exposing a substrate material to a plasma discharge, stimulating diverse chemical and physical transformations on its surface. The precise exposure to plasma enables the generation of micro- and nanostructures, playing a key role in establishing nanocoated surfaces. Plasma-based surface modification primarily encompasses plasma etching, plasma sputtering, and plasma polymerization (Jafari et al., 2013). Dimitrakellis et al. (2021) introduced the concept of atmospheric pressure plasma etching/nanotexturing of ZnO/PMMA nanocomposite films with a constant nanoparticle load to selectively eliminate organic material from the surface. This improved surface coverage of ZnO particles and aggregates results in increased antibacterial activity against Gram-negative bacteria *E. coli*. He/O_2 combination was used in an RF-driven dielectric barrier discharge (DBD) source to apply plasma treatment, allowing quick polymer etching in an open-air setting. ZnO nanoparticles on the surface of the nanocomposite were gradually revealed and aggregated as a result of oxidative etching. Using this technique resulted in a nearly 30% coverage of the composite surface area by ZnO particles and a quick rise in ZnO aggregate surface coverage with plasma etching lasting up to 2 minutes. The ZnO coverage saturates between 55% and 60% of the entire surface

Table 3.2 Recent work by using chemical etching method.

Fabrication technique	Coating material	Substrate	Findings/objective	Applications	References
Metal-assisted chemical etching (MACE)	Cu100-xPdx nanoparticles	Silicon	This study introduces a novel investigation into the direct synthesis of Cu100-xPdx nanoparticles on silicon through metal-assisted chemical etching (MACE). The provided method enables meticulous regulation of the bimetallic composition and structure of the nanoparticles.	Glucose detection, water purification, electrocatalysis, CO2 reduction	(Torralba et al., 2024)
	Ag nanoparticles (AgNPs)	Silicon substrate	The MACE process proves to be a proficient technique for producing diverse optically active silicon nanostructures. These nanostructures hold promise as potential candidates for solar cell applications, particularly serving as effective antireflection coatings.	Photonics, optoelectronics, energy storage, photovoltaics, chemical sensor, biological sensor	(Salem et al., 2020)
	Au	N and P-type silicon	The findings indicate that high aspect ratio nanostructures can be effectively produced using both N and P-type silicon as feasible alternatives. While these alternatives demonstrate distinct etch rates, they exhibit similarities in terms of etch direction and porosity.	Solar cell	(Ohlin et al., 2023)
	Pt/Au	GaN substrates	Utilizing a patterned metal-assisted chemical etching process, nanowire arrays are generated on GaN thin films. The explicit elucidation of the formation mechanism behind these nanowire arrays is provided.	Field effect transistors, chemical and biochemical sensing, nano LED, solar cell	(Wang et al., 2016)
	AgNO3	Silicon	Examining the impact of both dopant concentration and species on the formation rate of silicon nanowires (SiNWs) in bulk silicon and silicon-on-insulator (SOI) samples, the objective is to gain a deeper understanding and enhance the optimization of the silver-assisted chemical etching process.	Photovoltaics, battery technology, thermo-electrics, chemical and biological sensors, electronics	(Canevali et al., 2015)

area for extended etching durations. Dynamic incubation and in-solution tests for the antibacterial study showed a decrease in the microbial load with plasma etching time. An almost exponential decrease in the number of bacterial cells was observed when ZnO surface percentage was correlated with ZnO surface coverage and the ensuing antibacterial activity. Even with brief plasma etching phases of 2–4 minutes, effective bacterial death is achieved in 4 hours of incubation, with the ZnO aggregates remaining on the surface.

Spyrides et al. (2019) studied that oxygen and argon plasma on UHMWPE plates cause distinct etching mechanisms, and these differences may be understood by combining the results of AFM and FTIR (Fig. 3.4). Surface etching is connected to the roughness created by argon plasma, which tends to increase with treatment duration. By using FTIR, this etching is detected as a decrease in the strength of the peaks associated with hydrogen suppression. However, following oxygen plasma treatment, roughness rises as a result of unidirectional surface etching and then falls as treatment duration increases as a result of surface amorphization. After 5 minutes of treatment, there is a further rise in roughness that is correlated with surface degradation. In this instance, the ordering-related peak's FTIR intensity falls while the coupling-related peak's increases in the a-direction. Some other findings and applications of using the plasma spraying method for different substrates and coating materials are listed in Table 3.3.

3.2.1.4 Template method

The template technique has grown into an easy and flexible way to create nanocoated surfaces using premade templates or molds to give well-defined micro- and nanostructures onto a substrate. Using this technique, surface characteristics from the naturally or artificially developed templates are replicated onto a selected material to produce complex designs. Three separate stages make up the template method: creating the template, molding it, and demolding it.

This technique is frequently used for fabricating polymeric surfaces because it is scalable, repeatable, and affordable. Furthermore, the template technique allows for the flexible creation of surfaces with multifunctional features by including a variety of materials and functionalities (Kaur et al., 2022). Consequently, the template approach presents itself as a viable way to create relevant nanocoated surfaces for certain industrial and technical uses. Tao et al. (2016) introduced a soft-templating method to synthesize shape-controlled hollow silica nanoparticles (HSNs), such as wormlike hollow silica nanotubes (HSNTs) and hollow silica nanospheres (HSNSs), by adjusting the composition of the template, poly(acrylic acid) (PAA). The research delved into the potential formation mechanism of HSNs. The outcomes revealed that the hydrophobic modification resulted in a water contact angle reaching 153 degrees on the coating. Upon calcination at temperatures ranging from 300 to 450°C, the refractive indices of the corresponding films varied from 1.10 to 1.12. The coated K9 glass exhibited significantly higher transmittance than the blank K9 glass. These multifunctional coatings demonstrate promising applications in opto- and

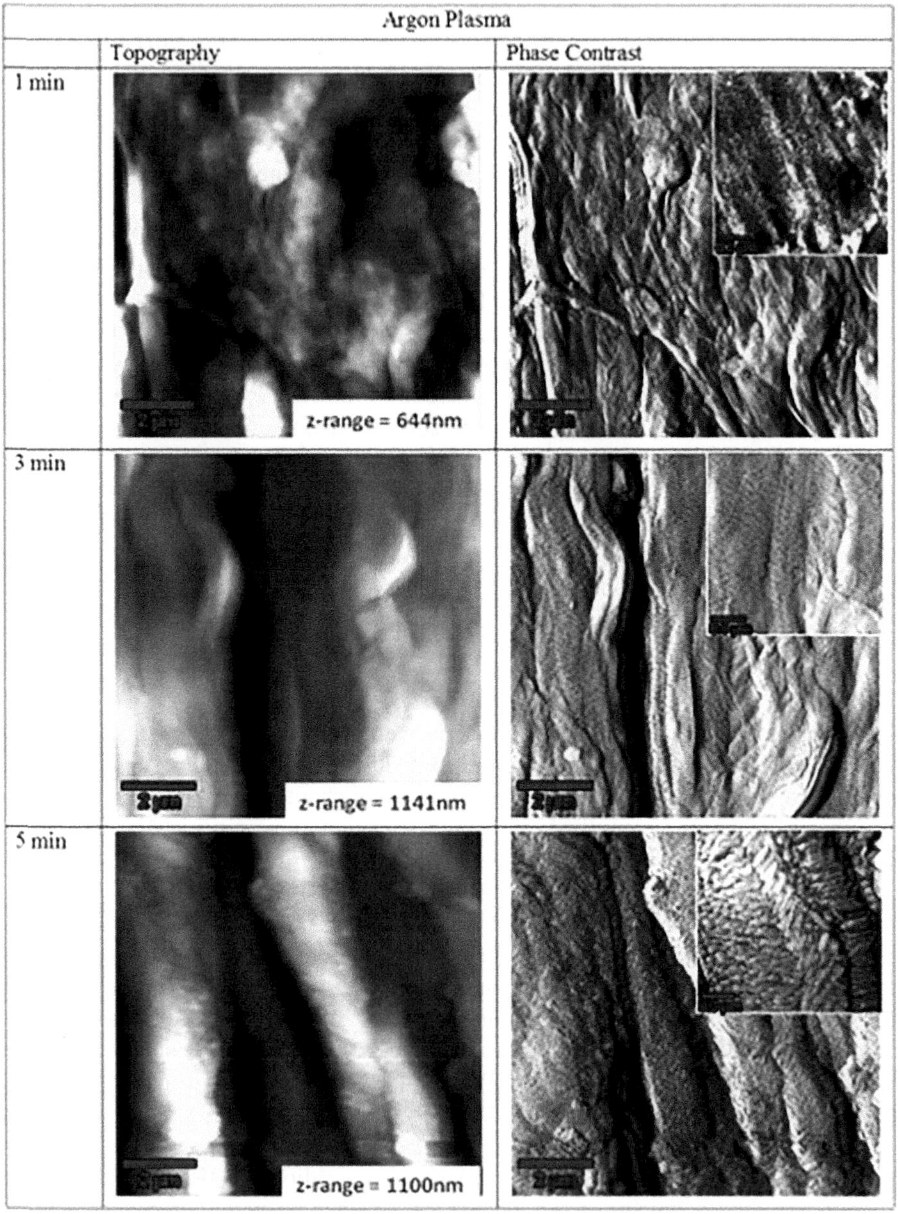

Figure 3.4 AFM topography and phase contrast pictures of UHMW-PE following various argon plasma treatments (Spyrides et al., 2019).
From Spyrides, S.M.M., Alencastro, F.S., Guimaraes, E.F., Bastian, F.L., Simao, R.A., 2019. Mechanism of oxygen and argon low pressure plasma etching on polyethylene (UHMWPE). Surf. Coat. Technol. 378. https://doi.org/10.1016/j.surfcoat.2019.124990.

Table 3.3 Recent work by using plasma spraying method.

Fabrication technique	Coating material	Substrate	Findings/objective	Applications	References
Plasma spraying	Ag−TiO2	SS304 stainless steel	For the atmospheric plasma spray (ASPS) process, a suspension comprising AgNO3 solution and TiO2 particles (pure anatase) was created and applied through plasma spraying onto an SS304 stainless steel substrate to produce the Ag−TiO2 coating.	Biomedical	(Wang et al., 2024)
	HA, BAG, BAG-1CNT and BAG-2CNT	Ti	The emphasis of this investigation is on improving the characteristics of plasma-sprayed bioceramic coatings through the incorporation of carbon nanotubes. The deposited coatings encompass hydroxyapatite, baghdadite, and baghdadite reinforced with carbon nanotubes.	Biomedical	(Singh et al., 2024)
	Cr7C3−CrSi2−Al2O3	TC4 and NiCrAlY	This research delves into the microstructure, formation mechanism, and properties of two Cr7C3−CrSi2−Al2O3 composite coatings achieved through plasma spraying. Additionally, the investigation explores the reaction mechanism within the Al−Cr2O3−SiC system.	Wear resistance, corrosion resistance, oxidation resistance	(Ru et al., 2022)
	Hydroxyapatite (HA)	Ti6Al4V	Hydroxyapatite (HA) coatings, reinforced with and without graphene nanosheets (GNSs), were manufactured on Ti6Al4V substrates through plasma spray. Subsequently, these coatings underwent heat and hydrothermal treatments.	Biomedical	(Ren et al., 2020)
	ZrO2	Ti−6Al−4V (TC4) titanium alloy	Plasma spraying was employed to create porous nanostructured ZrO2 coatings on the TC4 titanium alloy surface. The incorporation of MDPE and PEEK as pore-forming agents in the ZrO2 powder contributed to enhancing the porosity of the coatings.	Temperature resistance, wear resistance, corrosion resistance	(Cui et al., 2019)

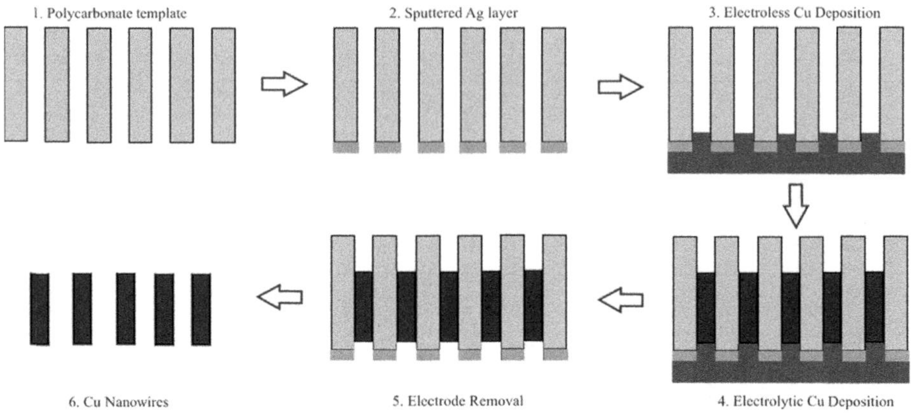

Figure 3.5 Schematic illustration of preparation of electrode layer for template-assisted electrodeposition of nanowires (Graves et al., 2018).
From Graves, J.E., Bowker, M.E.A., Summer, A., Greenwood, A., Ponce de León, C., Walsh, F.C., 2018. A new procedure for the template synthesis of metal nanowires. Electrochem. Commun. 87, 58–62. https://doi.org/10.1016/j.elecom.2017.11.022.

microelectronic fields due to their combination of high transmittance, ultralow refractive indices, and superhydrophobicity.

Graves et al. (2018) described a new, simple procedure for preparing the electrode layer for template-assisted electrodeposition of nanowires (Fig. 3.5). Electroless copper deposition can be started with a tiny ($\leq$15 nm) inexpensive silver seed layer, and a thicker (300–500 nm) homogeneous, highly conductive copper layer can be applied for further nanowire development. Reduced vacuum processing times, minimal consumption of precious metals, and the ease with which the electrode layer may be removed using a chemical etchant are among the benefits of this innovative technique. The electrodeposited nanowires exhibit consistent development, strong electroless layer adhesion to the electroplated layer, and a structure like that of conventional growth methods. Table 3.4 shows various findings and applications of using the template method for different substrates and coating materials.

3.2.2 Bottom-up synthesis

Bottom-up synthesis is a process used to create nanocoated surfaces by creating materials or nanostructures from smaller molecules or components. With bottom-up synthesis, the required structures are assembled from components starting at the molecular or nanoscale level, as opposed to top-down techniques, which begin with bigger structures and decrease them. This method frequently depends on molecular interactions, such as hydrogen bonds, van der Waals forces, or chemical reactions, to drive self-assembly processes. Through meticulous planning and manipulation of these interactions, scientists may fabricate nanostructures possessing certain characteristics that enhance nanocoating. Because bottom-up techniques are beneficial for their capability to form complex and

Table 3.4 Recent work by using template method.

Fabrication technique	Coating material	Substrate	Findings/objective	Applications	References
Templated synthesis	Chalcogenide glass (ChG) nanospheres	Silicon	A technique referred to as liquid-phase template (LPT) is employed, leveraging the advantages of both liquid-phase and template methods. This method facilitates the production of uniform ChG (Chalcogenide glass) nanospheres.	Environmental remediation, nanocontainer, biomedicine	(He et al., 2023)
	Mesoporous polystyrene (PS)	Quartz glass	This study presents the templated synthesis of SiO2 films on quartz glass through the utilization of mesoporous polystyrene (PS) templates.	Solar cells, photovoltaic devices	(Krishnan et al., 2023)
	PI-*b*-P2VP	Cellulose fiber	The combination of cellulose fiber materials with the PI-b-P2VP system was undertaken to explore its impact on cross-linking, pore formation, and the stability of block copolymers (BCP) within a cellulose fiber substrate.	Membranes or sensors, sustainable microfluidics	(Gemmer et al., 2022)
	Hierarchical porous ZIF-8	Aluminum	A hierarchical porous ZIF-8-coated aluminum foil was synthesized through an in-situ growth approach, and its performance for the thermal flow-mediated extraction (TFME) of selected chlorophenols was subsequently assessed.	Separation technologies	(Ghani et al., 2021)
	Cobalt nanoporous carbon (Co-NPC)	Borosilicate glass tube	In this study, Co-NPC was produced through a one-step direct carbonization process of ZIF-67, without the need for additional carbon precursors. These coatings were then applied to the surface of a stir bar for use in stir bar sorptive extraction. The resulting coatings demonstrated notable extraction efficiency for the chosen analytes.	Biological samples testing	(Ghani et al., 2019)

precisely controlled nanostructures, offering a high degree of customization for nanocoated surface qualities. Standard bottom-up methodologies employed for the fabrication of nanocoated surfaces include the hydrothermal technique, chemical vapor deposition, sol-gel processes, electrochemical deposition, and layer-by-layer deposition.

3.2.2.1 Sol–gel method

The sol-gel process provides control over surface texture and characteristics, making it a flexible and popular technology for creating superhydrophobic surfaces. In this procedure, precursor chemicals such as metal alkoxides or organometallic inorganic salts are hydrolyzed and condensed to create a sol, which is a colloidal suspension of nanoparticles. After applying the sol to a substrate, a three-dimensional network is created via gelation. To create a solid film on the substrate, the resultant gel is dried and cured. The sol-gel approach is notable for its scalability, affordability, and ease of use. Since it has been used on a variety of substrates, such as metal, glass, and polymers, it is a potential method for creating nanocoated surfaces with customized functions for a range of uses. Unfortunately, there are a number of issues with this technology. First, it may be time-consuming, especially for thick coatings. Secondly, it can be difficult to get exact control over purity and microstructure, which can impair repeatability. Other restrictions include the possibility of cracking and shrinkage while drying and compatibility problems with specific substrates (Al-luhaibi and Sendi, 2022; Pratap et al., 2023).

Xiong et al. (2019) fabricated a GS nanocomposite functionalized with amino silane and corrosion inhibitor BTA to enhance the protective qualities of sol-gel coatings (Fig. 3.6). In the synthetic fGS-BTA nanocomposite, amino silane, which functions as a nanogate on the pore outlets and inhibitor BTA is loaded into the mesoporous channels, regulates the release of BTA.

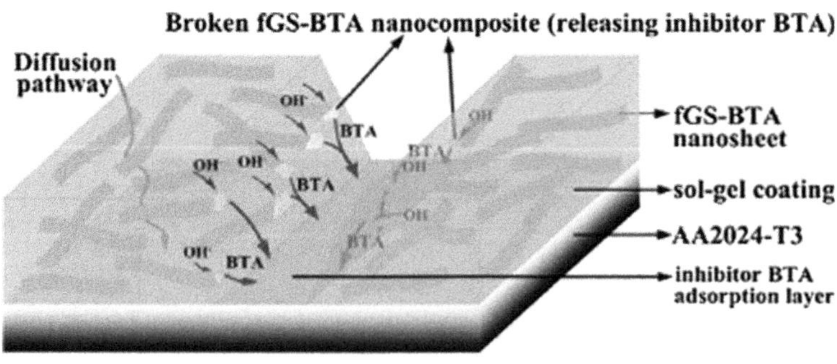

Figure 3.6 Schematic drawing describing the mechanism of fGS-BTA/sol-gel coating (Xiong et al., 2019).
From Xiong, L., Liu, J., Li, Y., Li, S., Yu, M., 2019. Enhancing corrosion protection properties of sol-gel coating by pH-responsive amino-silane functionalized graphene oxide-mesoporous silica nanosheets. Prog. Org. Coat. 135, 228–239. https://doi.org/10.1016/j.porgcoat.2019.06.007.

The two-dimensional structure and good compatibility of fGS with sol solution are confirmed by AFM and sedimentation tests, which helps to improve the barrier effect of sol-gel coating. Furthermore, during the immersion procedure, the fGS-BTA samples maintain the greatest impedance (above $1 \times 10^5\ \Omega\ cm^2$) in comparison to the other samples rapid fall in impedance. Furthermore, following 170 hours of neutral salt spray or 168 hours of immersion testing, no localized pits or products were seen on the samples. These show that there was a noticeable improvement in the protective property of the fGS-BTA coating. It also results from the controlled release inhibitor's enhanced barrier capacity and inhibitive quality.

Akbarzadeh et al. (2022) reported that to improve the corrosion protection of AA2024, the duplex system's performance was evaluated by introducing various quantities of sodium montmorillonite (Na-MMT) distributed throughout the silane coating. In this work, AA2024 was coated with an optimized PEO layer using a solution comprising sodium silicate and potassium hydroxide solution as the electrolyte. This was followed by the application of a hybrid sol-gel layer (30% V/V) made from tetraethoxysilane (TEOS) and 3-glycidoxypropyltrimethoxysilane (GPTMS) sol-gel solution. The corrosion protection effectiveness of the PEO layer on the AA2024 substrate in a 0.1 M NaCl solution was shown to be significantly impacted by the sol-gel/clay nanocomposite layer, according to electrochemical impedance spectroscopy (EIS). When it came to the sealing ability of the sol-gel coatings, the low-frequency impedance of the various coating systems after 5 weeks of exposure to the aggressive solution revealed that the silane coating modified with 1000 ppm of clay nanoparticles had the highest corrosion resistance among them. Some other findings and applications of using the sol-gel method for different substrates and coating materials are listed in Table 3.5.

3.2.2.2 Electrospinning

The electrospinning technique is a flexible method used to create nanocoated surfaces. It operates by applying an electric field to a polymeric solution, thereby generating fine fibers. This procedure involves applying a high voltage to a syringe holding the polymer solution. This causes a charged jet to develop, which then elongates and hardens into ultrafine fibers as a result of solvent evaporation. Surface tension, conductivity, and viscosity of the polymeric solution, as well as the electrospinning process parameters (applied voltage, tip-to-collector distance, and solution flow rate), all have an impact on the morphology of the electrospun fibers (Su et al., 2024; Sas et al., 2012; Barthwal et al., 2024). The electrospinning technology is a potential way to create nanocoated surfaces because it gives exquisite control over the surface properties and fiber shape. Tian et al. (2016) prepared PU nanofiber mats by the electrospinning method and utilized them as a substrate to create conducting nanocomposites. Fig. 3.7 depicted the method for fabricating polyurethane/polyaniline fibrous nanocomposites. Aniline monomers were chemically polymerized on the surface of an electrospun PU nanofiber substrate to create an electrospun polyurethane (core)/ polyaniline (skin) nanofibrous mat. PANi/PU mats measured about 8 mm in thickness, which is even smaller than a single cotton fiber of 20 mm. They also showed good elasticity (114%) and an average tensile strength (93 MPa).

Table 3.5 Recent work by using sol-gel method.

Fabrication technique	Coating material	Substrate	Findings/objective	Applications	References
Sol-gel method	Monoclinic vanadium dioxide (VO2(M))	Polyethylene terephthalate (PET)	This study introduces a straightforward and scalable sol-gel technique coupled with postannealing to produce spherical VO2(M) nanoparticles. The synthesis involves the careful control of parameters such as the concentration of H2O2 solution, oxygen intake, annealing time, and annealing temperature.	Optical sensing devices, optical switching devices, smart windows infrared modulators	(Li et al., 2024)
	Zirconium	Glass	In this study, a nanozirconia sol was synthesized through the sol-gel method and subsequently subjected to hydrophobic modification using hexamethyldisilazane (HMDS). The resulting sol was applied as a coating on a glass substrate, and its wetting behavior was subsequently analyzed.	Self-cleaning, antireflectivity	(Dogra et al., 2023)
	1.2Nd3+ 80SiO2−20LaF3	Glass-slides	Transparent oxyfluoride glass-ceramic (OxGCs) coatings, with a composition of 1.2Nd3+ 80SiO2−20LaF3, were synthesized using a sol-gel method referred to as the "precrystallized nanoparticles route." OxGCs are characterized as transparent materials consisting of one or more fluoride nanocrystals embedded in a glass matrix.	Tunable lasers, solid-state lasers, fiber amplifiers	(Cruz et al., 2023)
	PDMS modified MWCNTs/ZnO	Soft (paper, sponge, filter paper, fabric) and hard materials (glass and aluminum surface)	A cost-effective and fluorine free method is outlined to create a durable superhydrophobic composite coating by using a mixture of multiwalled carbon nanotubes (MWCNTs) and zinc oxide (ZnO) that's been modified with a polydimethylsiloxane (PDMS) coating. The MWCNTs/ZnO composite is synthesized through a sol-gel technique. This composite coating can be easily applied onto different surfaces to achieve superhydrophobic properties.	Corrosion inhibitor	(Barthwal et al., 2020)
	Silver	AA2024-T3	The investigation focused on exploring the effectiveness of silanol-type hybrid polymers that had been enhanced with silver nanoparticles encapsulated within SiO2 for the purpose of protecting 2024-T3 aluminum alloy against biocorrosion. This was achieved through employing electrochemical characterization and surface analysis techniques.	Corrosion resistance	(González et al., 2019)

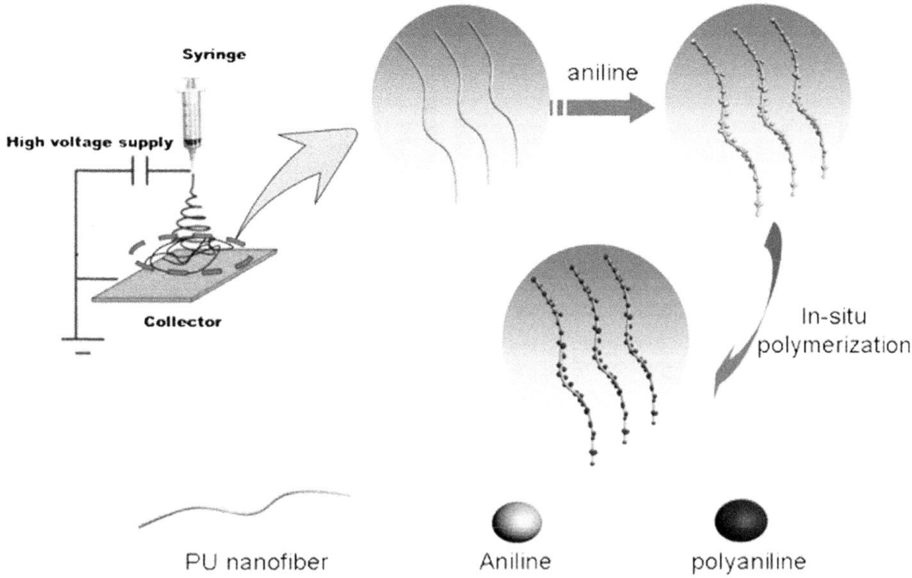

Figure 3.7 Schematic illustration of fabrication process of polyurethane/polyaniline fibrous nanocomposites by electrospinning method (Tian et al., 2016).
From Tian, M., Wang, Y., Qu, L., Zhu, S., Han, G., Zhang, X., Zhou, Q., Du, M., Chi, S., 2016. Electromechanical deformation sensors based on polyurethane/polyaniline electrospinning nanofibrous mats. Synthetic Metals 219, 11–19. https://doi.org/10.1016/j.synthmet.2016.05.005.

The emeraldine oxide state lasts for 120 minutes during the polymerization process, and the PU/PANi nanofibrous mats that are produced have exceptional electrical conductivity (0.43 S/cm). The average strain gauge factor for the stretching measurement instance was 17.15, which was more than eight times more than the value of a conventional polymer strain gauge. PU/PANi nanomats demonstrated a highly strain sensitivity of $(R-R_0)/R_0 = 0.75$ (curvature $= 0.4\ cm^{-1}$) for the bending measurement situation, together with cyclically reproducible sensor responses. Since the PU/PANi conductive nanofiber had more specific surface area, porosity, and flexibility than the traditional conductive fiber, its application in wearable electronics, robotic systems, and electron skin was anticipated.

Aydogdu et al. (2019) used electrospinning to encapsulate gallic acid in a combination of hydroxypropyl methylcellulose (HPMC) and polyethylene oxide (PEO). This study explores the potential use of nanofibers as active packaging materials. The nanofibers exhibited robust antioxidant activity and effectively absorbed gallic acid. With the increase in quantity of gallic acid, its degradation temperature (T_d) dropped from 275 to 250°C as a result of the crystalline structure alteration, and the TGA curves moved from one stage to two stage degradation. Gallic acid effectively incorporated itself into the HPMC/PEO nanofiber, as evidenced by the interaction between them. Gallic acid-loaded nanofibers reduced walnut oxidation while it was being

Table 3.6 Recent work by using electrospinning method.

Fabrication technique	Coating material	Substrate	Findings/objective	Applications	References
Electrospinning	Lanthanide-doped strontium aluminate (NLSA)	GNF@PU	An easy method was employed to create transparent, UV-resistant, photochromic, hydrophobic, and energy-efficient smart coatings for dark lighting applications. To formulate photoluminescent coatings (GNF@PU), lanthanide-doped strontium aluminate nanoparticles (NLSA) were incorporated within the GNF@PU composite.	Anticounterfeiting labeling, safety signs, glow in the dark coatings	(Abdu et al., 2024)
	TiO2–SiO2	Glass	Transparent and hydrophilic thin films of silica (SiO2) modified titania (TiO2) nanofibers were successfully synthesized on glass substrates using the electrospinning technique. These films are designed for self-cleaning applications in solar cells.	Solar cell	(Batool et al., 2024)
	BaTiO3 and SrTiO3	Ti–25Zr	This study involved individually immersing active nano-ceramic compounds, including piezoelectric BaTiO3 and perovskite SrTiO3, into a blended polymer solution of PCL/Chitosan. These solutions were then utilized in an electrospinning process to coat the surface of the Ti–25Zr alloy. The resulting composite nano-fiber aims to enhance cell viability and provide resistance against bacterial infections.	Orthodontic and orthopedic bone regeneration applications	(Al-Khateeb et al., 2023)
	Graphene nanoplate-carbon nanotube-polydimethylsiloxane (GCP)	Polyvinyl alcohol (PVA)	In this research, a GCP composite was utilized as a printable, flexible, and electrically conductive collector, allowing for both 2D and 3D conformal electrospinning on a nonconductive substrate.	Artificial ear-cartilage implants	(Song et al., 2022)
	PAN-NFs/BTA	Fe	In this study, a core-shell fiber was effectively produced using PAN as the shell and BTA as the core through the coaxial electrospinning technique. The resulting core-shell fiber was combined with EP to create a composite coating system named PAN-NFs/BTA/EP.	Corrosion resistance	(Zhao et al., 2020)

stored. Overall, it was found that the electrospinning approach was an effective way to encapsulate bioactive substances, and the findings of testing gallic acid-loaded nanofibers were encouraging, showing that they had a great potential as active packaging materials. Some other findings and applications of using the electrospinning method for different substrates and coating materials are listed in Table 3.6.

3.2.2.3 Layer-by-layer deposition

Layer-by-layer (LBL) deposition is an easy, inexpensive, and substrate-independent method used to adsorb layers of materials with opposing charges one after the other onto a substrate in order to fabricate nanocoated surfaces. This method creates a multilayered thin film by dipping the substrate alternately into solutions containing positively and negatively charged organisms. After each dipping, the coated film is cleaned with water and allowed to dry (Richardson et al., 2016). A variety of materials, including polymers and nanoparticles, can be used to fabricate these films, depending on the required nanocoated surface qualities. The deposited film thickness, constituents, and surface morphology may all be precisely controlled in the layering process. Layer-by-layer deposition is a versatile and scalable method of customizing nanocoated surfaces that may be used for a wide range of applications (Yang et al., 2023; Ding et al., 2023; Sethi et al., 2023; Guo et al., 2017; Zheng et al., 2019; Shao et al., 2023; Syed et al., 2017). Nevertheless, this method's main drawback is that it takes a lot of time and is appropriate for small quantities of materials.

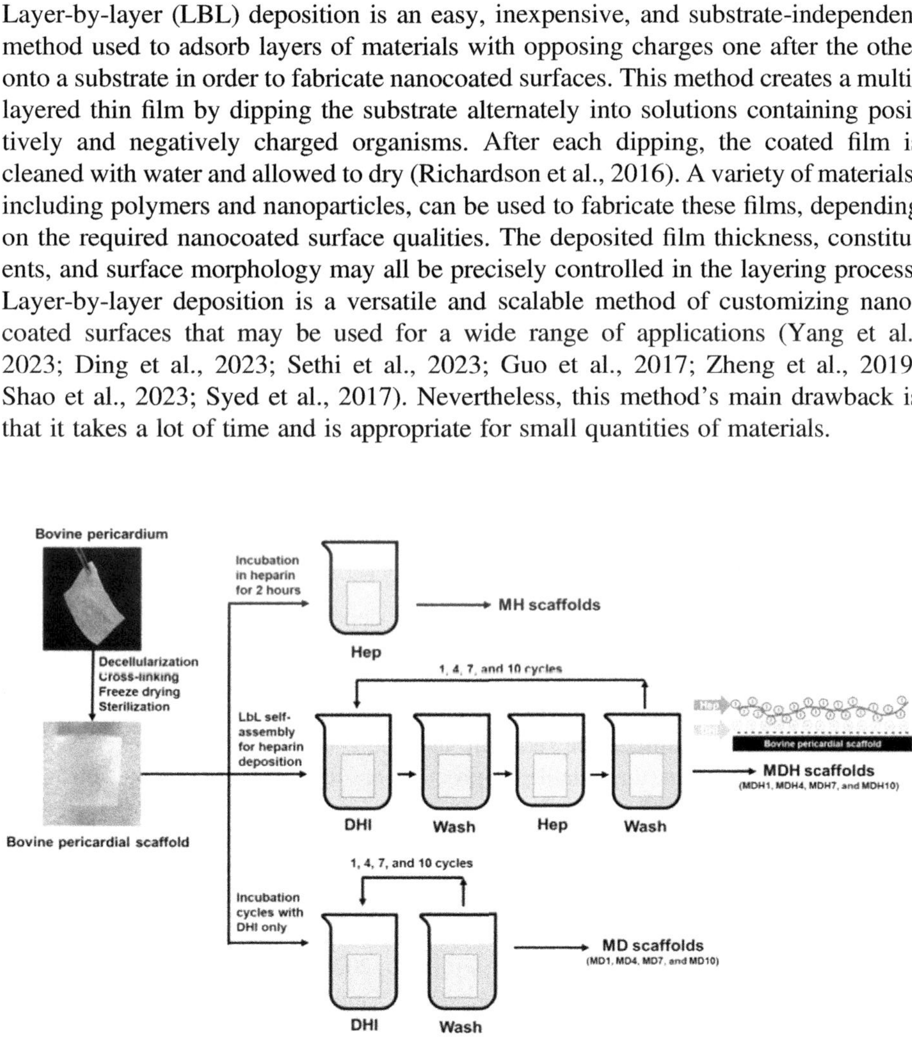

Figure 3.8 Schematic description of preparation of DHI/heparin immobilized scaffolds by LBL self-assembly deposition (Nguyen and Tran, 2022).
From Nguyen, M.T.N., Tran, H.L.B., 2022. Heparinization of the bovine pericardial scaffold by layer-by-layer (LBL) assembly technique. J. Sci. Adv. Mater. Dev. 7 (2). https://doi.org/10.1016/j.jsamd.2021.11.007.

Nguyen and Tran (2022) used the layer-by-layer self-assembly approach to establish a bovine pericardial scaffold immobilized with heparin. The bovine pericardial scaffold was created in this work using the processes of decellularization, cross-linking, and freeze-drying. Using the DHI/heparin complex formation process, which involves the interaction between DHI ions and ECM as well as DHI ions and heparin molecules, the LBL assembly approach effectively incorporated heparin onto the bovine pericardial scaffold (Fig. 3.8).

The heparinized scaffold demonstrated excellent antithrombosis in vitro, according to the results. The surface of the bovine pericardial scaffold exhibited heparin residue with anticoagulant action even after 30 days of in vitro release. The DHI/heparinized bovine pericardial scaffold that was used in this study has potential as a cardiovascular patch material.

Shi et al. (2024) suggested the integration of the spin-coating technique with the LBL assembly of nanocellulose and nanochitin, a practical and environmentally friendly approach for the manufacture and application of amphoteric nanochitin to create fog water collectors. In order to successfully spin-coat the hydrophilic nanocellulose (TEMPO-oxidized cellulose nanofiber, TOCN) and nanochitin (partially deacetylated chitin nanofibers, DEChN) on the hydrophobic polylactic acid (PLA) film substrate, the oxalic acid-hydrolyzed nanochitin (OAChN) was applied as a transition layer. Wherein a LBL assembly of DEChN (amino-rich positive surface charge) and TOCN (carboxyl-rich negative surface charge) was intended to create a polysaccharide coating over PLA with adjustable surface properties and thickness. Table 3.7 shows various findings and applications of using the layer-by-layer (LBL) deposition method for different substrates and coating materials.

3.2.2.4 Chemical Vapor Deposition (CVD)

Chemical Vapor Deposition (CVD) is a vacuum deposition technique that relies on chemical reactions involving gaseous precursors occurring on the surface of a substrate. This process forms a thin, homogeneous layer of the target material by reacting precursor gases containing the necessary components on the substrate surface. The substrate is heated to initiate the reaction, and the chemical reactions take place in the gas phase. After that, the volatile by-products are removed, leaving the substance that was deposited behind.

Although CVD offers fine control over film properties, including composition, thickness, and crystal structure, it is an expensive and time-consuming procedure. One significant issue with CVD is that it is not as easy to get larger samples. Despite these disadvantages, CVD is a favored technique for depositing a variety of materials due to its versatility (Zheng et al., 2022; Fu et al., 2022; Şakalak et al., 2020; Zhang et al., 2023; Siddiqui et al., 2020; Cai et al., 2018; Yang et al., 2020; Peng et al., 2024).

Saringer et al. (2023) used chemical vapor deposition to fabricate $Ti_{1-x}Al_xN$ coating with Al contents till $x > 0.8$ in an FCC crystal structure. It was discovered that a somewhat wide parameter window allows for the deposition of mostly face-centered cubic coatings. The coating texture can be changed to 001,101, and 111 inside this timeframe, but the grain size is mostly unaffected. The synthesis of w-AlN may be entirely

Table 3.7 Recent work by using layer-by-layer (LBL) deposition method.

Fabrication technique	Coating material	Substrate	Findings/objective	Applications	References
Layer-by-Layer coating	Ag	Acrylonitrile Butadiene Styrene (ABS)	A spray-assisted quasi-layer-by-layer (quasi-LBL) technique has been developed to form metallic coatings that are both pure and uniform on plastic surfaces.	Solar cells, electromagnetic interference shielding, nanophotonic, flexible electronic, plastic metalization, electrocatalytic electrodes, luminescent semiconducting	(Chen et al., 2019)
	SPH/(PAH/BG)7	Mg	A unique layer-by-layer coating, incorporating SPH and (PAH/BG)n through the dip technique, is employed to create a biomimetic $CaHPO_4.H_2O$ coating. This aims to enhance the corrosion resistance of magnesium alloy.	Corrosion resistance, biomedical applications	(Khaled & Santhiya, 2024)
	Diamond nanoparticles (DNPs)	Si	An approach involving electrostatic layer-by-layer deposition of DNPs onto substrate surfaces is implemented using colloidal suspensions of DNPs with varying zeta potentials and ionic strength.	Antireflective, antibacterial, coating applications	(Yoshikawa et al., 2022)
	Chitin nanofibers (ChNFs)	Glass	This work suggests development of omnidirectional antireflective coatings using charged chitin nanofibers (ChNFs) through the layer-by-layer (LBL) self-assembly technology.	Antireflective coatings	(Hu et al., 2023)
	Carbon nanotubes (CNTs)	Stainless-steel tube	This study aimed to experimentally explore the impact of a layer-by-layer (LBL)-assembled carbon nanotube (CNT) coating on the dropwise condensation heat transfer mechanism.	Heat exchanger component	(Kim et al., 2021)

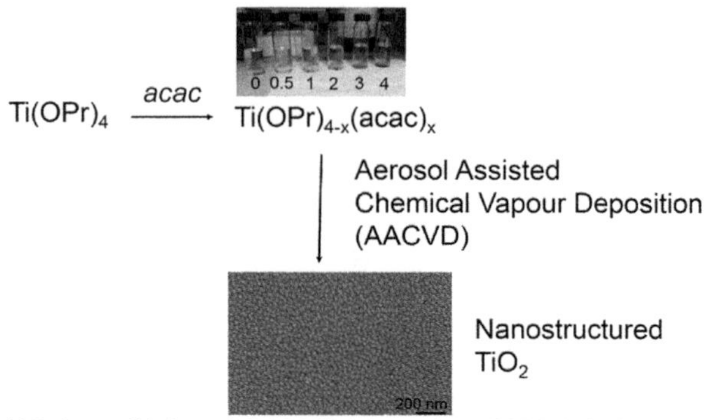

Figure 3.9 Process flow diagram of synthesis of titanium dioxide coatings using aerosol assisted chemical vapor deposition (AACVD) process (Taylor et al., 2020).
From Taylor, M., Pullar, R.C., Parkin, I.P., Piccirillo, C., 2020. Nanostructured titanium dioxide coatings prepared by Aerosol Assisted Chemical Vapor Deposition (AACVD). J. Photochem. Photobiol. A Chem. 400. https://doi.org/10.1016/j.jphotochem.2020.112727.

prevented at a deposition temperature of 730°C and a pressure of 54−70 mbar. These parameters, together with a rotating speed of 4 minute^{-1}, are recommended for the deposition of TiAlN; however, the precise values may vary for various deposition systems, as obtaining a high hardness usually involves decreasing the quantity of w-AlN.

Taylor et al. (2020) used the Aerosol Assisted Chemical Vapor Deposition (AACVD) process to fabricate titanium dioxide coatings employing complexes that are formed between acetyl acetone (acac) and titanium isopropoxide (TIPP) as precursors (Fig. 3.9). Different ratios of TIPP to acac were used in deposition studies to evaluate the impact on the coatings composition, morphology, and photocatalytic activity. Results revealed that the use of acac produced nanostructured titanium dioxide, or nanoparticles with a diameter of around 10−25 nm. The material with the highest performance in photocatalytic tests was found to have an acac:TIPP ratio of 1, as demonstrated by the FQE value being approximately three times higher than that of the coating made with TIPP alone. Table 3.8 shows various findings and applications of using the chemical vapor deposition (CVD) method for different substrates and coating materials.

3.2.2.5 Electrochemical deposition

Electrochemical deposition offers a versatile and precise method for fabricating nanocoated surfaces by electroplating materials onto conductive substrates (Shrama et al., 2014). With this technique, the material to be deposited dissolves in an electrolyte solution, and the substrate serves as the cathode. A solid layer forms on the substrate when an electric current is applied because the metal ions in the solution are reduced at the cathode. By adjusting the deposition conditions, the method is not only quick, scalable, and repeatable, but it also makes it possible to fabricate a variety of surface morphologies, including fibers, cones, tubes, rods, sheets, and needles (Darband et al., 2020). The electrochemical deposition method presents a favorable pathway for the creation of tailored superhydrophobic surfaces endowed with enhanced functionalities.

Liu et al. (2024) suggest a gradient ultrasonic-assisted electrodeposition method to examine its influence on the mechanical, tribological, and microstructural

Table 3.8 Recent work by using chemical vapor deposition (CVD) method.

Fabrication technique	Coating material	Substrate	Findings/objective	Applications	References
Chemical vapor deposition	Al2O3	Silica powders	This study focuses on depositing nanometric alumina layers onto easily fluidizable silica powders through the FBCVD process, with a comprehensive exploration of the deposition parameters.	Microelectronic, energy storage applications	(Aslam et al., 2023)
	EGDA, PFDA, and initiator TBP	Glass slides and silicon wafers	Transparent superhydrophobic polymer nanocone array coatings with outstanding anti-icing performance were successfully produced through a simple one-step iCVD process.	Anti-icing, antifrosting applications	(Huang et al., 2023)
	Silicon oxynitride thin films	Silicon	SiOxNy thin films were deposited successfully onto silicon wafers through the PECVD technique at a low temperature, catering to applications in nano optical/ photonics on-chip interconnects.	Optoelectronics, microelectronics, nanoelectronics, biomedical applications	(Sharma et al., 2013)
	Carbon nanotubes (CNTs)	Aluminum	The objective of this study is to employ thermal chemical vapor deposition (TCVD) to grow carbon nanotubes (CNTs) on aluminum substrates (sheets) to facilitate the cladding of aluminum sheets together. Experimental findings indicate that the mechanical properties of the composite are suitable across varying CNT weight percentages. Simulation analyses were carried out and compared with existing numerical and experimental data in the literature, demonstrating favorable correlations.	Anticorrosion, wear resistance applications	(Morovvati et al., 2023)
	SiC ($dp = .2\ \mu m$) with pyrolytic carbon (pyC)	Alumina	With the objective of coating fine SiC powders ($dp = 0.2\ \mu m$) with pyrolytic carbon (pyC) through CVD, a fluidized-bed reactor was designed and constructed. This reactor operates efficiently at low pressures and high temperatures for long durations, ensuring the deposition of sufficiently thick coatings.	Aircraft and automobile braking, atmospheric reentry, spatial propulsion, ground and aircraft turbines, industrial hot components	(El Mansouri et al., 2023)

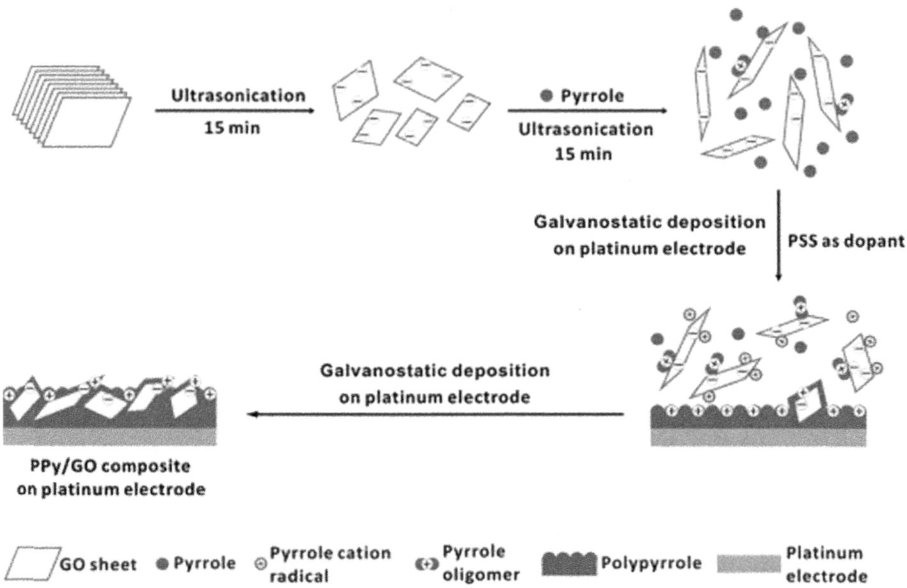

Figure 3.10 Schematic illustration of electrochemical deposition of PPy/GO coating on the Pt electrode site (Deng et al., 2011).
From Deng, M., Yang, X., Silke, M., Qiu, W., Xu, M., Borghs, G., Chen, H., 2011. Electrochemical deposition of polypyrrole/graphene oxide composite on microelectrodes toward tuning the electrochemical properties of neural probes. Sens. Actuat. B Chem. 158 (1), 176–184. https://doi.org/10.1016/j.snb.2011.05.062.

characteristics of Ni–Co coatings. Meanwhile, wear simulations are carried out to clarify the deterministic link between microstructure, process, and characteristics through the integration of surface topography and the dynamic modification of the wear model's coefficient of friction (COF). Results showed that maximum hardness (about 489 HV) and minimal COF (approximately 0.17) are achieved with a high Co content (Ni:Co = 5:5) and with gradient-increased ultrasonic power (30–100 W). This indicates a 72% drop in COF and a 1.73-fold increase in hardness compared to Ni coating.

Deng et al. (2011) used a one-step electrochemical technique to create polypyrrole–graphene oxide (PPy–GO) composite sheets. They achieved this by taking advantage of the electrostatic interaction that occurs between positively charged pyrrole cation radicals and negatively charged graphene oxide (GO) sheets (Fig. 3.10). Additionally, the effects of the electrolyte GO concentration and polymerization current density were investigated with regard to the formation of PPy–GO coatings on platinum neural microelectrode sites. The PPy–GO coatings showed a rougher surface characteristic with micrometer-scale bulges in comparison to the pure PPy film. Results showed that at the clinically relevant 1 kHz, the impedance of the PPy–GO-coated Pt electrode is only around 10% of the naked Pt electrode. However, compared to the bare Pt electrode, the charge capacity density is more

Table 3.9 Recent work by using electrochemical deposition method.

Fabrication technique	Coating material	Substrate	Findings/objective	Applications	References
Electrochemical deposition	Carboxylic multiwalled carbon nanotubes (CMWNT)	Ti6Al4V	In this study, electrochemical deposition was employed to graft carboxylic multiwalled carbon nanotubes (CMWNT) coatings onto the surface of Ti6Al4V alloys, aiming to improve their biotribological properties.	Bone implants	(Cao et al., 2023)
	CaP/nano-sized SiO2	Titanium	In this study, creation of SiO2-containing CaP coatings on a titanium substrate was successfully achieved through ultrasound-assisted electrochemical deposition (ECD).	Biomedical applications	(Fathyunes, 2021)
	Ni–P–Al2O3	Q235 low carbon steel sheet	The successful preparation of Ni–P–Al2O3 nanocomposite coatings was achieved in this study by integrating a jet electrochemical deposition process with pulse-assisted technology.	Corrosion resistance	(Li et al., 2022)
	Platinum nanoparticles (PtNPs), selenium nanoparticles (SeNPs) and nanocomposites of PtNPs-SeNPs	Fluorine doped tin oxide (FTO)	The development of electrochemical sensors for hydrogen peroxide detection involved development of PtNPs-SeNPs-FTO nanocomposites using an electrochemical deposition method.	Optoelectronic device, electrochemical sensor, catalysis, electronic	(Dumore and Mukhopadhyay, 2022)
	Cu–Fe–P	Nickel foam (NF)	This study proposes a straightforward and rapid approach for cost-effective electrochemical hydrogen production.	Urea oxidation, hydrogen production	(Song et al., 2022)

than two orders of magnitude. Furthermore, while applying a neural probe, the PPy–GO-coated Pt electrodes performed better than the PPy-coated electrodes. Table 3.9 shows various findings and applications of using the electrochemical deposition method for different substrates and coating materials.

3.2.2.6 Hydrothermal technique

The hydrothermal method is recognized as an economical, consistent, and eco-friendly approach for synthesizing and altering materials within an aqueous environment under meticulously regulated high-temperature and high-pressure conditions. In this technique, precursor materials, which are usually metal salts or metal organic compounds, are dissolved in a solution. They are then placed in a sealed reaction vessel and subjected together with a substrate to high temperatures and pressures. The hydrothermal environment's harsh circumstances encourage chemical reactions as well as the nucleation and development of materials on the substrate. Hydrothermal methods are widely used in many scientific and industrial applications, such as the development of crystals, the synthesis of nanomaterials, and surface modification to accomplish certain functions (Barthwal and Lim, 2021; Bai et al., 2023).

Yigit et al. (2020) effectively used the hydrothermal approach to create the nanohydroxyapatite (nHA) matrix coatings on the Ti6Al4V alloy doped with various weight percentages of graphene nanosheets (GNSs). Results showed that it was possible to create a largely porous structure with nanoscale HA nucleation in the coating, and as the GNS weight% in hybrid structures increased, so did the number of nHA nucleations in the layers. As the fraction of GNS weight in the coating layer increased, so did the hybrid coatings' micro Vickers hardness. Electrochemical results

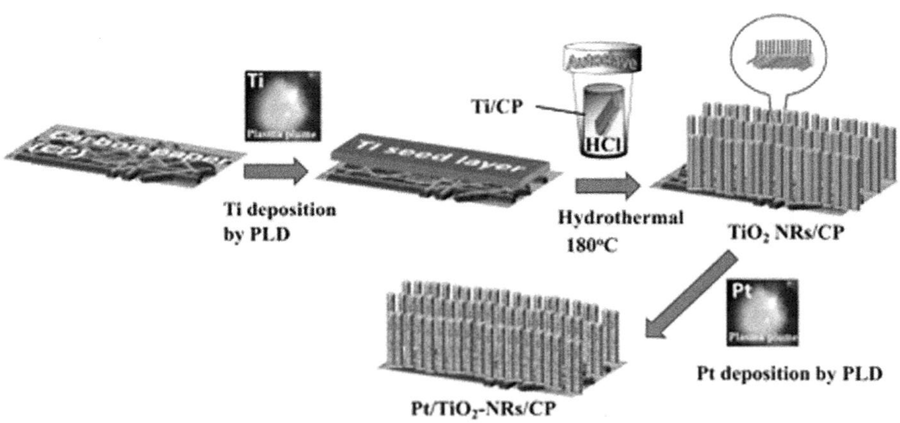

Figure 3.11 Schematic description of the production of Pt/TiO_2-NRs on CP substrate by hydrothermal technique (Mohammadi et al., 2023).
From Mohammadi, N., Zuria, A.M., Mohamedi, M., 2023. Hierarchical TiO2 nanorods grown on carbon paper via the combination of pulsed laser deposited Ti seed layer and hydrothermal technique as advanced hybrid electrocatalyst support. Hybrid Adv. 2, 100021. https://doi.org/10.1016/j.hybadv.2023.100021.

revealed that the nHA/GNS hybrid structure improves the hydrothermally treated Ti6Al4V surfaces' in vitro corrosion resistance. Due to the superior surface quality than the others, nHA/5GNS coating has the largest surface area and the strongest corrosion resistance.

Mohammadi et al. (2023) reported the hydrothermal synthesis of hierarchical TiO_2 nanorods (NRs) is achieved by growing them from a pulsed laser-deposited Ti seed layer on a carbon paper (CP) substrate, eliminating the need for templates or binders (Fig. 3.11). XRD analysis reveals the crystallization of TiO_2-NRs in the rutile phase. The resulting TiO_2-NRs significantly enhance the electroactivity and stability of Pt in ethanol oxidation reactions. Chronoamperometry durability studies indicate a specific mass activity of 47.92 $mAmg^{-1}$ for the Pt/TiO_2-NRs/CP catalyst, which is seven times higher than that of the commercial Pt (6.72 $mAmg^{-1}$). When used as an anode in a membraneless microfluidic ethanol fuel cell, the Pt/TiO_2-NRs/CP catalyst achieves a maximum current density of 56.4 $mAcm^{-2}$, 5.1 $mWcm^{-2}$, and durability comparable to a commercial Pt anode with 16.7 times higher Pt loading. The freestanding nature, low platinum loading, and high stability position TiO_2-NRs/CP as a promising catalyst support for fuel cell technologies. Some other findings and applications of using the hydrothermal method for different substrates and coating materials are listed in Table 3.10.

3.3 Future scopes

Future scopes in the synthesis of nanomaterials for coatings encompass several promising areas of research and development:

1. **Precise Control of Nanoparticle Morphology**: Advancements in synthesis techniques can enable precise control over the size, shape, and surface properties of nanoparticles. This control allows for tailoring the nanomaterials to specific coating applications, optimizing performance characteristics such as adhesion, dispersion, and coverage.
2. **Scalable and Cost-effective Production Methods**: Developing scalable synthesis methods that are economically viable for large-scale production is essential for commercial adoption of nanocoatings. Research focused on improving the efficiency of synthesis processes, reducing raw material costs, and minimizing energy consumption can make nanomaterials more accessible to industries.
3. **Functionalization for Enhanced Properties**: Functionalizing nanoparticles with organic or inorganic molecules can impart additional properties to nanocoatings, such as improved adhesion, chemical resistance, or self-healing capabilities. Future research may explore novel functionalization strategies to enhance the performance and versatility of nanomaterial-based coatings.
4. **Integration of Sustainable Materials**: Investigating sustainable sources of raw materials and eco-friendly synthesis routes is a growing area of interest in nanomaterial synthesis. Researchers are exploring bio-based precursors, green chemistry approaches, and recycling methods to develop environmentally friendly nanomaterials for coatings, aligning with global efforts toward sustainability.

Table 3.10 Recent work by using hydrothermal method.

Fabrication technique	Coating material	Substrate	Findings/objective	Applications	References
Hydrothermal technique	Bi2S3	Nickel	This study centers on evaluating the energy storage efficiency of Bi2S3 nanoparticles and employs various methods for their characterization. The nanoparticles were synthesized through a hydrothermal technique, and an analysis of their structural properties was conducted using powder X-ray diffraction and Rietveld refinement.	Energy storage applications	(Rizan et al., 2024)
	Ag-doped ZnO nanorods	SS-400	A straightforward method is outlined for producing Ag-doped ZnO nanorods through a fast hydrothermal process, aimed at achieving a superamphiphilic surface on SS-400 substrates. Even after extended periods of storage, as well as rigorous thermal and mechanical assessments, the resulting surface consistently demonstrates outstanding superamphiphilic properties.	Oil separator, applications in a variety of fields	(Barthwal et al., 2016)
	SnS2	SiNWs	A hydrothermal method was employed to create a SnS2 core−shell heterostructure on vertically aligned silicon nanowires (SiNWs), aiming to enhance the optoelectronic performance across a broad spectral range spanning from ultraviolet (UV) to near-infrared (NIR) regions.	Photodetector	(Das et al., 2023)
	ZnO and CuO	Glass substrates	In this research, nanostructured oxide ceramics of ZnO and CuO were synthesized on microglass substrates seeded with ZnO (referred to as ZOZO) and CuO (referred to as ZOCO), respectively. The study investigated their surface properties and gas-sensing capabilities. The deposition of thin film seed layers was carried out via dip-coating, followed by the hydrothermal synthesis of both ZnO and CuO nanostructures under comparable conditions.	Gas sensing	(Ajjaq et al., 2023)
	TiO2 nanorod films	Titanium	This research focuses on producing dense TiO2 nanorod films with customized morphology on Ti substrates. This is achieved through a hydrothermal process, complemented by annealing treatment of the Ti substrate and the incorporation of alkali metal chloride in the hydrothermal solution.	Orthopedic, dental implants, possible photocatalysis	(Dong et al., 2011)

5. **Nanostructured Composite Materials**: Combining different types of nanomaterials or incorporating nanoparticles into composite structures can lead to synergistic effects and enhanced properties in coatings. Future research may focus on developing innovative composite materials with tailored nanostructures for improved mechanical strength, thermal stability, or barrier properties.
6. **Responsive and Adaptive Coatings**: Designing nanocoatings that respond to external stimuli, such as changes in temperature, humidity, or pH, can enable adaptive functionalities for dynamic surface protection. Future efforts may explore stimuli-responsive nanomaterials and smart coating systems capable of self-repair, antifouling, or controlled release of active agents in response to environmental cues.
7. **Multifunctional Coatings for Emerging Applications**: As new challenges and opportunities arise in industries such as electronics, healthcare, and renewable energy, there is a demand for multifunctional coatings tailored to specific applications. Future research may focus on developing nanocoatings with advanced functionalities, such as conductive, antireflective, or bioactive properties, to address emerging needs in diverse sectors.
8. **Cross-disciplinary Collaboration and Innovation**: Collaborative efforts between materials scientists, chemists, engineers, and industry partners can drive innovation in nanomaterial synthesis for coatings. Future research initiatives may involve interdisciplinary approaches, drawing insights from fields such as nanotechnology, surface science, polymer chemistry, and materials engineering to push the boundaries of coating technology.

By addressing these future scopes, researchers can unlock new possibilities in the synthesis of nanomaterials for coatings, paving the way for next-generation surface treatments with enhanced performance, sustainability, and functionality.

References

Abdu, M.T., Khattab, T.A., Abdelrahman, M.S., Abdelrahman, M.S., 2024. Electrospinning of glass nanofibers to roughen a waterborne polyurethane coating toward smart applications. Ceram. Int. 50, 11518–11527.

Ajjaq, A., Barin, Ö., Çağırtekin, A.O., Soltabayev, B., Acar, S., 2023. ZnO seed-mediated hydrothermal growth of advanced 1-D ZnO and 2-D CuO nanostructured oxide ceramics for gas sensing applications. Ceram. Int. 49 (24), 40853–40865.

Akbarzadeh, S., Santos, L.S., Vitry, V., Paint, Y., Olivier, M.G., 2022. Improvement of the corrosion performance of AA2024 alloy by a duplex PEO/clay modified sol-gel nanocomposite coating. Surf. Coating. Technol. 434, 128168.

Al-Khateeb, A., Al-Hassani, E.S., Jabur, A.R., 2023. Active nanoceramic compound dipped in biopolymers to create composite coating for metallic implant surface. Heliyon 9 (9).

Al-luhaibi, A.A., Sendi, R.K., 2022. Synthesis, potential of hydrogen activity, biological and chemical stability of zinc oxide nanoparticle preparation by sol–gel: a review. J. Radiat. Res. Appl. Sci. 15 (3), 238–254.

Aslam, S., Sekkat, A., Vergnes, H., Esvan, J., Pugliara, A., Samélor, D., Eshraghi, N., Vahlas, C., Auvergniot, J., Caussat, B., 2023. A new route to apply nanometric alumina coating on powders by fluidized bed chemical vapor deposition. Chem. Eng. J. Adv. 16, 100554.

Aydogdu, A., Sumnu, G., Sahin, S., 2019. Fabrication of gallic acid loaded Hydroxypropyl methylcellulose nanofibers by electrospinning technique as active packaging material. Carbohydr. Polym. 208, 241–250.

Bai, C., Hu, C., Ni, P., Zhang, X., Zhang, W., Zhang, S., Tang, J., Li, T., Li, Y., 2023. Fabrication of robust, superhydrophobic-superoleophilic PVA sponge by one-pot hydrothermal method for oil-water separation. Surface. Interfac. 37, 102679.

Baji, A., Gangadoo, S., Truong, V.K., Abtahi, M., Budiman, A., Oopath, S.V., 2022. Fabrication of superhydrophobic polyvinylidene fluoride-co-hexafluoropropylene films enabled by nanoimprint lithography. Mater. Lett. 311, 131555.

Barthwal, S., Barthwal, S., 2023. A highly robust, nonfluorinated, and economical PDMS based superhydrophobic flexible surface with repairable and flame retardant properties. J. Appl. Polym. Sci. 140 (17), e53766.

Barthwal, S., Barthwal, S., 2024. Engineering a robust, multifunctional superhydrophobic/oleophobic microporous aluminum surface via a two-step chemical etching process. Surface. Interfac. 46, 103933.

Barthwal, S., Lim, S.H., 2021. A durable, fluorine-free, and repairable superhydrophobic aluminum surface with hierarchical micro/nanostructures and its application for continuous oil-water separation. J. Membr. Sci. 618, 118716.

Barthwal, S., Kim, Y., Ahn, J., Lim, S.H., 2016. Fabrication of a superamphiphilic SS-400 oil separator surface using a Ag-doped ZnO nanorod coating. Sci. Adv. Mater. 8 (8), 1595–1602.

Barthwal, S., Lee, B., Lim, S.H., 2019. Fabrication of robust and durable slippery anti-icing coating on textured superhydrophobic aluminum surfaces with infused silicone oil. Appl. Surf. Sci. 496, 143677.

Barthwal, S., Barthwal, S., Singh, B., Singh, N.B., 2020. Multifunctional and fluorine-free superhydrophobic composite coating based on PDMS modified MWCNTs/ZnO with self-cleaning, oil-water separation, and flame retardant properties. Colloids Surf. A Physicochem. Eng. Asp. 597, 124776.

Barthwal, S., Barthwal, S., Singh, N.B., 2022. Bioinspired superhydrophobic nanocomposite materials: introduction, design, and applications. In: Nanocomposites. Jenny Stanford Publishing, pp. 429–481.

Barthwal, S., Uniyal, S., Barthwal, S., 2024. Nature-inspired superhydrophobic coating materials: drawing inspiration from nature for enhanced functionality. Micromachines 15 (3), 391.

Batool, I., Shahzad, N., Shahzad, R., Satti, A.N., Liaquat, R., Waqas, A., Shahzad, M.I., 2024. Self-cleaning study of SiO_2 modified TiO_2 nanofibrous thin films prepared via electrospinning for application in solar cells. Sol. Energy 268, 112271.

Cai, Z., Lin, J., Hong, X., 2018. Transparent superhydrophobic hollow films (TSHFs) with superior thermal stability and moisture resistance. RSC Adv. 8 (1), 491–498.

Canevali, C., Alia, M., Fanciulli, M., Longo, M., Ruffo, R., Mari, C.M., 2015. Influence of doping elements on the formation rate of silicon nanowires by silver-assisted chemical etching. Surf. Coating. Technol. 280, 37–42.

Cao, H., Tian, P., Deng, J., Li, Y., Wang, C., Han, S., Zhao, X., 2023. Electrochemical deposition multi-walled carbon nanotube coatings on the surface of Ti6Al4V alloy for enhancing its biotribological properties. J. Mech. Behav. Biomed. Mater. 142, 105825.

Chen, D., Kang, Z., Hirahara, H., Li, W., 2019. Quasi layer by layer spray deposition of high-quality Ag coatings with excellent micro mechanical properties. Appl. Surf. Sci. 481, 1213–1219.

Cruz, M.E., Fernández, J., Durán, A., Balda, R., Castro, Y., 2023. Optically active nano-glass-ceramic coatings of Nd^{3+}doped-$80SiO_2$-$20LaF_3$ prepared by the pre-crystallized nanoparticles sol-gel route. J. Non-Cryst. Solids 601, 122050.

Cui, Y.H., Hu, Z.C., Ma, Y.D., Yang, Y., Zhao, C.C., Ran, Y.T., Gao, P.Y., Wang, L., Dong, Y.C., Yan, D.R., 2019. Porous nanostructured ZrO_2 coatings prepared by plasma spraying. Surf. Coating. Technol. 363, 112–119.

Darband, G.B., Aliofkhazraei, M., Khorsand, S., Sokhanvar, S., Kaboli, A., 2020. Science and engineering of superhydrophobic surfaces: review of corrosion resistance, chemical and mechanical stability. Arab. J. Chem. 13 (1), 1763–1802.

Das, S., Pal, S., Larsson, K., Mandal, D., Giri, S., Banerji, P., Chandra, A., Basori, R., 2023. Hydrothermally grown SnS_2/Si nanowire core-shell heterostructure photodetector with excellent optoelectronic performances. Appl. Surf. Sci. 624, 157094.

Deng, M., Yang, X., Silke, M., Qiu, W., Xu, M., Borghs, G., Chen, H., 2011. Electrochemical deposition of polypyrrole/graphene oxide composite on microelectrodes towards tuning the electrochemical properties of neural probes. Sensor. Actuator. B Chem. 158 (1), 176–184.

Dev, A., Dev Choudhury, B., Abedin, A., Anand, S., 2014. Fabrication of periodic nanostructure assemblies by interfacial energy driven colloidal lithography. Adv. Funct. Mater. 24 (29), 4577–4583.

Dimitrakellis, P., Kaprou, G.D., Papavieros, G., Mastellos, D.C., Constantoudis, V., Tserepi, A., Gogolides, E., 2021. Enhanced antibacterial activity of ZnO-PMMA nanocomposites by selective plasma etching in atmospheric pressure. Micro Nano Eng. 13, 100098.

Ding, D., Wu, Q., Wang, J., Chen, Y., Li, Q., Hou, L., Zhao, L., Xu, Y.Y., 2023. Superhydrophobic encapsulation of flexible Bi_2Te_3/CNT coated thermoelectric fabric via layer-by-layer assembly. Compos. Commun. 38, 101509.

Dogra, V., Kishore, C., Mishra, A., Gaur, A., Verma, A., 2023. Sol-Gel preparation and wetting behaviour analysis of hydrophobic Zirconium based nano-coating: implications for solar panel coating. Chem. Eng. J. Adv. 15, 100507.

Dong, L., Cheng, K., Weng, W., Song, C., Du, P., Shen, G., Han, G., 2011. Hydrothermal growth of rutile TiO_2 nanorod films on titanium substrates. Thin Solid Films 519 (15), 4634–4640.

Dumore, N.S., Mukhopadhyay, M., 2022. Development of novel electrochemical sensor based on PtNPs-SeNPs-FTO nanocomposites via electrochemical deposition for detection of hydrogen peroxide. J. Environ. Chem. Eng. 10 (2), 107058.

El Mansouri, A., Guette, A., Bertrand, N., Vignoles, G.L., 2023. Fluidized bed chemical vapor deposition of pyrocarbon on various types of powders: heat and mass transfer analyses and nanotexture characterization. Chem. Eng. J. 468, 143544.

Fathyunes, L., 2021. Effect of ultrasonic waves on the electrochemical deposition of calcium phosphate/nano-sized silica composite coating. J. Mater. Res. Technol. 14, 2345–2356.

Feng, J., Tuominen, M.T., Rothstein, J.P., 2011. Hierarchical superhydrophobic surfaces fabricated by dual scale electron beam lithography with well ordered secondary nanostructures. Adv. Funct. Mater. 21 (19), 3715–3722.

Fu, J., Sun, Y., Ji, Y., Zhang, J., 2022. Fabrication of robust ceramic based superhydrophobic coating on aluminum substrate via plasma electrolytic oxidation and chemical vapor deposition methods. J. Mater. Process. Technol. 306, 117641.

Gaur, S., Khanna, A., 2015. Functional coatings by incorporating nanoparticles. Nano Res. 1 (7).

Gemmer, L., Hu, Q., Niebuur, B.J., Kraus, T., Balzer, B.N., Gallei, M., 2022. A block copolymer templated approach for the preparation of nanoporous polymer structures and cellulose fiber hybrids by ozone treatment. Polym. Chem. 13 (27), 4028–4046.

Ghani, M., Ghoreishi, S.M., Shahin, M., Azamati, M., 2019. Zeolitic imidazole framework templated synthesis of nanoporous carbon as a coating for stir bar sorptive extraction of fluorouracil and phenobarbital in human body fluids. Microchem. J. 146, 798–806.

Ghani, M., Raoof, J.B., Masoum, S., 2021. Template-directed synthesis of zeolitic imidazolate framework-8 derived Zn–Al layered double oxides decorated on the electrochemically anodized nanoporous aluminum substrate for thin film microextraction of chlorophenols followed by determination with high-performance liquid chromatography. J. Chromatogr. A 1656, 462550.

González, E.A., Leiva, N., Vejar, N., Sancy, M., Gulppi, M., Azócar, M.I., Gomez, G., Tamayo, L., Zhou, X., Thompson, G.E., Páez, M.A., 2019. Sol–gel coatings doped with encapsulated silver nanoparticles: inhibition of biocorrosion on 2024-T3 aluminum alloy promoted by *Pseudomonas aeruginosa*. J. Mater. Res. Technol. 8 (2), 1809–1818.

Graves, J.E., Bowker, M.E.A., Summer, A., Greenwood, A., de León, C.P., Walsh, F.C., 2018. A new procedure for the template synthesis of metal nanowires. Electrochem. Commun. 87, 58–62.

Guo, X.J., Xue, C.H., Li, M., Li, X., Ma, J.Z., 2017. Fabrication of robust, superhydrophobic, electrically conductive and UV-blocking fabrics via layer-by-layer assembly of carbon nanotubes. RSC Adv. 7 (41), 25560–25565.

Guo, C., Liu, K., Zhang, T., Sun, P., Liang, L., 2023. Development of flexible photothermal superhydrophobic microarray by photolithography technology for anti-icing and deicing. Prog. Org. Coating 182, 107675.

He, J., Yang, W., Jin, C., Wan, Y., Xu, G., Wang, C., Zhou, L., Yue, Z., 2023. Study on copper-assisted chemical etching of porous Si/Cu@C anode materials. Solid State Ionics 401, 116352.

He, Y., Zhao, R., He, Y., Chen, X., Tao, G., Hou, C., 2023. Chalcogenide glass nanospheres with tunable morphology by liquid-phase template approach. iScience 26 (3).

Hu, Z., Shang, J., Wang, P., Zhang, L., Zhou, J., 2023. Omnidirectional antireflective coatings prepared with chitin nanofibers via layer-by-layer self-assembly. J. Colloid Interface Sci. 650, 676–685.

Huang, X., Sun, M., Shi, X., Shao, J., Jin, M., Liu, W., Zhang, R., Huang, S., Ye, Y., 2023. Chemical vapor deposition of transparent superhydrophobic anti-Icing coatings with tailored polymer nanoarray architecture. Chem. Eng. J. 454, 139981.

Jafari, R., Asadollahi, S., Farzaneh, M., 2013. Applications of plasma technology in development of superhydrophobic surfaces. Plasma Chem. Plasma Process. 33, 177–200.

Ji, W.F., Ahmed, M.M., Bibi, A., Lee, Y.C., Yeh, J.M., 2021. Graphene-based PANI composite coatings with fine-controllable 3D hierarchical structures prepared from bio-inspired photo-/colloidal-lithography technique for flexible supercapacitor application. Electrochim. Acta 390, 138890.

Kaur, A., Bajaj, B., Kaushik, A., Saini, A., Sud, D., 2022. A review on template assisted synthesis of multi-functional metal oxide nanostructures: status and prospects. Mater. Sci. Eng., B 286, 116005.

Khaled, N.I., Santhiya, D., 2024. Multifunctional poly (allylamine hydrochloride)/bioactive glass layer by layer surface coating on magnesium alloy for biomedical applications. Prog. Org. Coating 186, 108059.

Khanna, A.S., 2008. Nanotechnology in high performance paint coatings. Asian J. Exp. Sci. 21 (2), 25–32.

Kim, T., Shin, D., Lee, J., Kim, S.J., 2021. Effect of layer-by-layer assembled carbon nanotube coatings on dropwise condensation heat transfer associated with non-condensable gas effect. Int. J. Heat Mass Tran. 175, 121345.

Krishnan, M.R., Rajendran, V., Alsharaeh, E., 2023. Anti-reflective and high-transmittance optical films based on nanoporous silicon dioxide fabricated from templated synthesis. J. Non-Cryst. Solids 606, 122198.

Li, Y., Liu, M., Xu, X., Qu, Z., Zhao, H., Zhang, M., Zheng, L., 2022. Effects of duty cycle on properties of Ni–P–Al_2O_3 nanocomposite deposited layer prepared by pulse-assisted jet electrochemical deposition. J. Mater. Res. Technol. 21, 3556–3569.

Li, M., Fang, C., Cheng, Y., Zhang, X., Han, H., Zhao, J., Zhang, Y., 2024. A facile pathway to fabricate VO_2(M) nanoparticles via sol-gel method for flexible thermochromic films with efficient infrared stealth. Vacuum 221, 112885.

Liu, D., Liu, R., Cao, L., Saeed, S., Bryanston-Cross, P.J., Wang, Z., 2023. Fabrication of periodic hierarchical structures with anti-icing performance by direct laser interference lithography and hydrothermal treatment. Surf. Coating. Technol. 471, 129819.

Liu, H., He, M., Li, J., Chen, K., Zhang, H., 2024. Additive manufacturing of high-performance Ni-Co coatings for micro/nanomold applications using an advanced gradient ultrasonic electrochemical deposition process. Addit. Manuf. 79, 103949.

Marquez-Velasco, J., Vlachopoulou, M.E., Tserepi, A., Gogolides, E., 2010. Stable super-hydrophobic surfaces induced by dual-scale topography on SU-8. Microelectron. Eng. 87 (5–8), 782–785.

Mohammadi, N., Zuria, A.M., Mohamedi, M., 2023. Hierarchical TiO_2 nanorods grown on carbon paper via the combination of pulsed laser deposited Ti seed layer and hydrothermal technique as advanced hybrid electrocatalyst support. Hybrid Adv. 2, 100021.

Morovvati, M.R., Mollaei-Dariani, B., Lalehpour, A., Toghraie, D., 2023. Fabrication and finite element simulation of aluminum/carbon nanotubes sheet reinforced with Thermal Chemical Vapor Deposition (TCVD). J. Mater. Res. Technol. 23, 1887–1902.

Nanocoatings market size, share, competitive landscape and trend analysis report by type (Anti-fingerprint, antimicrobial, self-cleaning (bionic & photocatalytic), anti-corrosion, conductive, and others) and application (electronics, energy, food & packaging, construction, marine, automotive, healthcare, others). Global Opport. Anal. Ind. Forec., 2021-2030 Retrieved from https://www.alliedmarketresearch.com.

Nanocoatings Market_ product (Anti-fingerprint, Anti-microbial, Anti-fouling & Easy-to-clean, Self-cleaning (Bionic and Photocatalytic) and Specialty coatings) Applications (Medical & Healthcare, Food Manufacturing, Packaging, Marine, Water Treatment Electronics, Building & Construction, Automotive, and Energy and Textiles, Leathers, Industrial Engineering)_ Global Industry Analysis, Size, Share, Growth, Trends and Forecast 2013_2019. Retrieved from: ,https://www.transparencymarketresearch.com.

Nguyen, M.T.N., Tran, H.L.B., 2022. Heparinization of the bovine pericardial scaffold by layer-by-layer (LBL) assembly technique. J. Sci. Adv. Mater. Dev. 7 (2), 100405.

Ohlin, H., Frisk, T., Sychugov, I., Vogt, U., 2023. Comparing metal assisted chemical etching of N and P-type silicon nanostructures. Micro and Nano Eng. 19, 100178.

Papageorgiou, G.P., Karydas, A.G., Papageorgiou, G., Kantarelou, V., Makarona, E., 2020. Controlled synthesis of periodic arrays of ZnO nanostructures combining e beam lithography and solution-based processes leveraged by micro X-ray fluorescence spectroscopy. Micro Nano Eng. 8, 100063.

Peng, Y., Hu, J., Huan, Y., Zhang, Y., 2024. Chemical vapor deposition growth of graphene and other nanomaterials with 3D architectures towards electrocatalysis and secondary batteries related applications. Nanoscale (16), 7734–7751.

Pilotek, S., Tabellion, F., 2005. Nanoparticles in coatings. Eur. Coat. J. 4, 170–176.

Pratap, R., Chowdhury, S.R., Singh, A.K., Giri, R., 2023. Study on structural, microstructural and dielectric properties of zinc oxide nanostructure synthesized by Sol-gel technique. Mater. Today: Proc. 82, 369–374.

Ren, J., Zhao, D., Qi, F., Liu, W., Chen, Y., 2020. Heat and hydrothermal treatments on the microstructure evolution and mechanical properties of plasma sprayed hydroxyapatite coatings reinforced with graphene nanoplatelets. J. Mech. Behav. Biomed. Mater. 101, 103418.

Richardson, J.J., Cui, J., Bjornmalm, M., Braunger, J.A., Ejima, H., Caruso, F., 2016. Innovation in layer-by-layer assembly. Chem. Rev. 116 (23), 14828−14867.

Rizan, M., Samaje, V., Kumara, K., Ravindra, K.N., Giresha, A.S., Thiyagaraj, S., 2024. Hydrothermal synthesis of Bi_2S_3 nanoparticle as a prospective electrode for supercapacitor. Materials 2, 100081.

Ru, P.W., Lei, W.A.N.G., Yong, Y.A.N.G., Gao, P.Y., Wang, Y.W., Cui, Y.H., Sun, W.W., Wang, X.Y., Liang, W.A.N.G., 2022. Microstructure, formation mechanism and properties of plasma-sprayed Cr_7C_3—$CrSi_2$—Al_2O_3 coatings. Trans. Nonferrous Metals Soc. China 32 (11), 3635−3649.

Şakalak, H., Yılmaz, K., Gürsoy, M., Karaman, M., 2020. Roll-to roll initiated chemical vapor deposition of super hydrophobic thin films on large-scale flexible substrates. Chem. Eng. Sci. 215, 115466.

Salem, A.M.S., Harraz, F.A., El-Sheikh, S.M., Shah, S.I., 2020. Novel Si nanostructures via Ag-assisted chemical etching route on single and polycrystalline substrates. Mater. Sci. Eng., B 262, 114793.

Samà, J., Domènech-Gil, G., Gràcia, I., Borrisé, X., Cané, C., Barth, S., Steib, F., Waag, A., Prades, J.D., Romano-Rodríguez, A., 2019. Electron beam lithography for contacting single nanowires on non-flat suspended substrates. Sensor. Actuator. B Chem. 286, 616−623.

Saringer, C., Tkadletz, M., Thurner, J., Czettl, C., Schalk, N., 2023. Influence of deposition parameters on microstructure and mechanical properties of chemical vapor deposited Ti1-xAlxN coatings. Int. J. Refract. Metals Hard Mater. 113, 106203.

Sas, I., Gorga, R.E., Joines, J.A., Thoney, K.A., 2012. Literature review on superhydrophobic self-cleaning surfaces produced by electrospinning. J. Polym. Sci. B Polym. Phys. 50 (12), 824−845.

Sethi, S.K., Kadian, S., Gogoi, R., Manik, G., 2023. Layer-by-layer fabrication of self-cleaning superhydrophobic surface made from Carboxymethylcellulose and ZnO quantum dots: a combined experimental and computational study. Surface. Interfac. 37, 102752.

Shao, C., Jiang, M., Zhang, J., Zhang, Q., Han, L., Wu, Y., 2023. Construction of a superhydrophobic wood surface coating by layer-by-layer assembly: self-adhesive properties of polydopamine. Appl. Surf. Sci. 609, 155259.

Sharma, S.K., Barthwal, S., Singh, V., Kumar, A., Dwivedi, P.K., Prasad, B., Kumar, D., 2013. PECVD based silicon oxynitride thin films for nano photonic on chip interconnects applications. Micron 44, 339−346.

Sharma, A., Jaiswal, A., Jaiswal, R.P., 2022. Fabrication of a tailorable polystyrene nanoscale mesh of honeycomb morphology using nano sphere lithography. Thin Solid Films 763, 139567.

Shi, Z., Liu, L., Chen, H., Tang, C., Yu, J., Fan, Y., 2024. Preparation of Janus film for fog water collection via layer-by-layer assembling of nanocellulose and nanochitin on PLA. Carbohydr. Polym. 323, 121369.

Shirtcliffe, N.J., Aqil, S., Evans, C., McHale, G., Newton, M.I., Perry, C.C., Roach, P., 2004. The use of high aspect ratio photoresist (SU-8) for super-hydrophobic pattern prototyping. J. Micromech. Microeng. 14 (10), 1384.

Shrama, S.K., Saurakhiya, N., Barthwal, S., Kumar, R., Sharma, A., 2014. Tuning of structural, optical, and magnetic properties of ultrathin and thin ZnO nanowire arrays for nano device applications. Nanoscale Res. Lett. 9, 1−17.

Siddiqui, A.R., Maurya, R., Katiyar, P.K., Balani, K., 2020. Superhydrophobic, self-cleaning carbon nanofiber CVD coating for corrosion protection of AISI 1020 steel and AZ31 magnesium alloys. Surf. Coating. Technol. 404, 126421.

Singh, S., Kumar, A., Kamboj, M., Das, B., Rakha, K., Singh, H., 2024. Corrosion behaviour of plasma-sprayed baghdadite bioceramic coatings reinforced with carbon nanotubes. J. Alloys Compd. 976, 173094.

Song, W., Zhu, C., Safikhani, A., 2022. Binder-free electrochemical deposition of 3-D superhydrophilic Cu-Fe-P nanostructure for improving urea oxidation and hydrogen evolution reaction. J. Electroanal. Chem. 927, 116993.

Song, J.Y., Yun, H.J., Kim, J.H., Yi, C.C., Park, S.M., 2022. Electroconductive, flexible, and printable graphene nanoplate-carbon nanotube-polydimethylsiloxane composite collectors for three-dimensional conformal electrospinning. Compos. Sci. Technol. 227, 109629.

Spyrides, S.M.M., Alencastro, F.S., Guimaraes, E.F., Bastian, F.L., Simao, R.A., 2019. Mechanism of oxygen and argon low pressure plasma etching on polyethylene (UHMWPE). Surf. Coating. Technol. 378, 124990.

Su, W., Chang, Z., Yuyu, E., Feng, Y., Yao, X., Wang, M., Ju, Y., Wang, K., Jiang, J., Li, P., Lei, F., 2024. Electrospinning and electrospun polysaccharide-based nanofiber membranes: a review. Int. J. Biol. Macromol., 130335

Syed, J.A., Tang, S., Meng, X., 2017. Super-hydrophobic multilayer coatings with layer number tuned swapping in surface wettability and redox catalytic anti-corrosion application. Sci. Rep. 7 (1), 4403.

Tao, C., Yan, H., Yuan, X., Yao, C., Yin, Q., Zhu, J., Ni, W., Yan, L., Zhang, L., 2016. Synthesis of shape-controlled hollow silica nanostructures with a simple soft-templating method and their application as superhydrophobic antireflective coatings with ultralow refractive indices. Colloids Surf. A Physicochem. Eng. Asp. 501, 17−23.

Taylor, M., Pullar, R.C., Parkin, I.P., Piccirillo, C., 2020. Nanostructured titanium dioxide coatings prepared by aerosol assisted chemical vapour deposition (AACVD). J. Photochem. Photobiol. Chem. 400, 112727.

Tian, M., Wang, Y., Qu, L., Zhu, S., Han, G., Zhang, X., Zhou, Q., Du, M., Chi, S., 2016. Electromechanical deformation sensors based on polyurethane/polyaniline electrospinning nanofibrous mats. Synth. Met. 219, 11−19.

Torralba, E., Le Gall, S., Chaliyawala, H., Chopard, É., Rahhou, Z.A., Muller-Bouvet, D., Cachet-Vivier, C., Bastide, S., 2024. Metal-Assisted Chemical etching for the direct synthesis of bimetallic Cu-Pd nanoparticles on silicon. Appl. Surf. Sci. 649, 159117.

Van Minh, N., Van Hieu, D., Van Tuan, P., Dung, N.D., Hoang, C.M., 2022. Characteristics of silicon nano-pillars fabricated by nano-sphere lithography and metal assisted chemical etching. Mater. Sci. Semicond. Process. 142, 106483.

Wang, K.C., Yuan, G.D., Wu, R.W., Lu, H.X., Liu, Z.Q., Wei, T.B., Wang, J.X., Li, J.M., Zhang, W.J., 2016. GaN nanowire arrays by a patterned metal-assisted chemical etching. J. Cryst. Growth 440, 96−101.

Wang, Q., Zhang, C., Gong, T., Kong, W., Yue, W., Chen, W., Xie, Z., Su, Y., Li, L., 2019. Large-scale diamond silver nanoparticle arrays as uniform and sensitive SERS substrates fabricated by surface plasmon lithography technology. Opt Commun. 444, 56−62.

Wang, Q., Tang, Z., Herout, R., Liu, C., Yu, K., Lange, D., Godin, R., Kizhakkedathu, J.N., Troczynski, T., Wang, R., 2024. Axial suspension plasma sprayed Ag-TiO_2 coating for enhanced photocatalytic and antimicrobial properties. Surface. Interfac., 103856

Xiong, L., Liu, J., Li, Y., Li, S., Yu, M., 2019. Enhancing corrosion protection properties of sol-gel coating by pH-responsive amino-silane functionalized graphene oxide-mesoporous silica nanosheets. Prog. Org. Coating 135, 228−239.

Yang, R., Liang, Y., Hong, S., Zuo, S., Wu, Y., Shi, J., Cai, L., Li, J., Mao, H., Ge, S., Xia, C., 2020. Novel low-temperature chemical vapor deposition of hydrothermal delignified wood for hydrophobic property. Polymers 12 (8), 1757.

Yang, N., Yuan, R., Li, W., Tan, X., Liu, Z., Zhang, Q., Li, C., Ge, L., 2023. Durable LBL microcapsule-based sponges with superhydrophobic/super-oleophilic and photothermal behaviors for oil–water separation. Separ. Purif. Technol. 327, 124915.

Yao, C., Zhao, Y., Zhang, X., Li, H., Xie, C., 2021. Design and fabrication of wafer-scale highly uniform silicon nanowire arrays by metal-assisted chemical etching for antireflection films. Results Phys. 31, 105018.

Yigit, O., Dikici, B., Senocak, T.C., Ozdemir, N., 2020. One-step synthesis of nano-hydroxyapatite/graphene nanosheet hybrid coatings on Ti6Al4V alloys by hydrothermal method and their in-vitro corrosion responses. Surf. Coating. Technol. 394, 125858.

Yoshikawa, T., Tokuda, N., Makino, Y., Tsurui, A., Ieki, R., Kojima, R., Liu, M., Mahiko, T., Nishikawa, M., 2022. Electrostatic layer-by-layer deposition of diamond nanoparticles onto substrate surfaces. Carbon Trends 9, 100202.

Zhang, H., Gan, J., Wu, Y., Wu, Z., 2023. Biomimetic high water adhesion superhydrophobic surface via UV nanoimprint lithography. Appl. Surf. Sci. 633, 157610.

Zhang, M., Zhou, T., Li, H., Liu, Q., 2023. UV-durable superhydrophobic ZnO/SiO2 nanorod arrays on an aluminum substrate using catalyst-free chemical vapor deposition and their corrosion performance. Appl. Surf. Sci. 623, 157085.

Zhao, X., Yuan, S., Jin, Z., Zhu, Q., Zheng, M., Jiang, Q., Song, H., Duan, J., 2020. Fabrication of composite coatings with core-shell nanofibers and their mechanical properties, anticorrosive performance, and mechanism in seawater. Prog. Org. Coating 149, 105893.

Zheng, L., Su, X., Lai, X., Chen, W., Li, H., Zeng, X., 2019. Conductive superhydrophobic cotton fabrics via layer-by-layer assembly of carbon nanotubes for oil-water separation and human motion detection. Mater. Lett. 253, 230–233.

Zheng, J., Yang, J., Cao, W., Huang, Y., Zhou, Z., Huang, Y.X., 2022. Fabrication of transparent wear-resistant superhydrophobic SiO_2 film via phase separation and chemical vapor deposition methods. Ceram. Int. 48 (21), 32143–32151.

Zheng, J., Liu, R., Liu, D., Weng, Z., Song, G., Li, W., Wang, Z., 2023. Slippery liquid infused porous surfaces with anti-icing performance fabricated by direct laser interference lithography. Prog. Org. Coating 175, 107308.

Zhu, S., Deng, W., Su, Y., 2023. Recent advances in preparation of metallic superhydrophobic surface by chemical etching and its applications. Chin. J. Chem. Eng. 61, 221–236.

Zhu, C., Pan, A., Zhu, X., Zheng, S., Cui, B., 2024. Edge lithography based on aluminum dry etching. Micro Nano Eng. 22, 100233.

Designing of multifunctional coatings: Recent updates

4

Shikha Sharma[1], *Manviri Rani*[2] *and Uma Shanker*[1]
[1]Department of Chemistry, Dr B R Ambedkar National Institute of Technology, Jalandhar, Punjab, India; [2]Department of Chemistry, Malaviya National Institute of Technology, Jaipur, Rajasthan, India

4.1 Introduction

A thin film or coating refers to applying a deposited layer onto a surface substrate in order to modify the substrate's chemical, thermal, optical, or environmental characteristics (Abu-Thabit and Makhlouf, 2020). By modifying the chemical composition, structure, and physical qualities through surface coatings, the desired surface finish can be achieved. Coatings are created by adding useful materials to surfaces (Mozetič, 2019). Coating technology offers a straightforward technique to change surfaces or matrices, yielding properties such as UV protection, water resistance, corrosion resistance, antibacterial activity, and more in industries such as textiles, food, and paper packaging.

Cotton and leather are widely used in clothing, home furnishings, and other industrial products because of their skin-friendliness, breathability, and flexibility. However, due to their natural texture, these goods are sensitive to bacteria and wear, necessitating antibacterial and wear-resistant coating technologies these surfaces (Fan et al., 2018). Coatings can preserve, color, or serve other purposes to the substrate. They are made up of solvents, pigments, resins, and fillers. Pigments offer the coating color or protection from corrosion, while fillers affect mechanical qualities and supply substrate properties. The addition of nanoparticles enhances paints and introduces new qualities, making them particularly interesting to the coatings industry (Vidales-Herrera and López, 2020).

Nanocoatings involve adding a nanometer-scale layer to a surface or incorporating a tiny item into a mixture. Coatings with two or more distinct components, at least one of which must be nanometer-sized, are referred to as nanocomposite coatings. As a result, the final materials' qualities are frequently improved by the interaction of several components (Yilmaz and Yorgancioglu, 2019). Nanostructured materials for coatings are attractive due to their remarkable mechanical, electrical, magnetic, and optical properties.

Introducing nanostructured elements in coatings enhances electrical properties, permeability, antireflection, abrasion resistance, corrosion resistance, flame retardancy, wear resistance, mechanical properties, and surface qualities (Gu et al., 2020).

Multifunctional Nanostructured Coatings. https://doi.org/10.1016/B978-0-443-23683-9.00004-0

For instance, utilizing plasma-assisted procedures, it was usual practice to insert carbides, nitrides, borides, and oxides of transition metals from groups III and IV elements into a thermoset resin to create hard and superhard covering. The coating's gas permeability can be increased by adding layered structure materials such as nanoclay and double-layered oxides, while the coating's thermal resistance can be increased by utilizing carbonaceous materials such as black carbon, carbon nanotubes, carbon fiber, or graphene-based materials (Ielo et al., 2021). To enhance the coating's photocatalytic behavior, several photocatalytic additives such as ZnO and TiO_2 were used into the coating's formulation (Nardi et al., 2015).

Nanoparticles retain their intrinsic qualities even at small scale. There is still silica at small scales, titanium dioxide still has a high refractive index, and zinc oxide, even though its particles are nanoscale, still absorbs UV radiations (Butt et al., 2017). Coatings preserve these properties. The change in light scattering as particle size decreases is likely the most remarkable characteristic, as particles small enough to avoid light scattering can create transparent coatings. Nanoparticles' high surface area/volume ratio increases surface activity and system interaction, and their small size reduces light-scattering, making coatings more transparent (Vidales-Herrera and López, 2020).

Nanofillers and nanopigments are used in the coatings industry, making them "nanocoating" (Pilotek and Tabellion, 2005). The dimension of the pigments and fillers in a coating determines how thick it can be. Smaller fillers and pigments result in thinner films, which reduces the amount of paint needed to cover a given surface area and even allows for the coating of smaller objects (Nikolic et al., 2015). The design of nanocoatings requires control of many factors, such as interface volume, size of particles, single-layer thickness, interfacial energy, texture of the surface and substrate, selection of materials, application method, and processing parameters (Fernando, 2009).

The presence of multifunctional nanofillers in today's nanocomposite coatings allows them to perform additional functions beyond protecting components. Notable examples include antibacterial and sustainable energy-based coatings. Nanofillers-based silver nanoparticles and related products hold great promise for biological coatings with strong antibacterial activity (Barabadi et al., 2023). As challenges with traditional energy sources grow, research into sustainable energy conversion is expected to increase, both theoretically and experimentally, and nanocomposite coatings will play a role in this trend. Self-cleaning or "easy-to-clean" coatings on structures, protective substrates, and glass can reduce energy and water consumption in cleaning facilities (Tayel et al., 2022). Nanocomposite coating insulation reduces energy loss, saving billions of dollars in winter home maintenance (Tri et al., 2019).

There are numerous preparative methods available to carry out atomistic/molecular deposition on a certain substrate, including electroplating, chemical vapor deposition, physical deposition technique, anodization, molecular beam epitaxy, sol-gel, spinning, Langmuir Blodgett, spinning, electrophoresis, etching, plasma polymerization, dipping, atomic layer deposition (Miikkulainen et al., 2013), cathodic arc evaporation, ball impact deposition (Yazdani and Zakeri, 2017), ion-assisted deposition, layer-by-layer assembly (Tettey et al., 2010), ion plating sputtering, thermal evaporation, self-assembly (Wang et al., 2014), and pulsed laser deposition. Each technique demonstrates certain benefits and disadvantages; the suitable approach depends on the specific use case.

This chapter delves into the design of multifunctional coatings, advantages and disadvantages of coating preparation methods, exploring the diverse applications, materials, and techniques that drive innovations in this field.

4.2 Preparation methods of nanocoatings

4.2.1 Chemical vapor deposition

A common procedure for coating metallic or ceramic materials, including element metals and their alloys and intermetallic compounds, is CVD. In CVD, the components of a vapor phase are delivered into a reaction chamber and adsorbed on a heated substrate surface. This adsorption causes a chemical reaction that forms a solid coating. During deposition, homogeneous gas-phase reactions occur in the gas phase, while heterogeneous chemical reactions take place near the heated surface, producing thin films or coatings (Gu et al., 2020). The critical substrate temperature can significantly influence the occurrence of these reactions (Makhlouf, 2011). Numerous CVD processes exist, including low-pressure CVD, plasma-assisted CVD, atmospheric pressure CVD, laser CVD, metal-organic CVD, photochemical vapor deposition, chemical beam epitaxy, chemical vapor infiltration, and plasma-enhanced CVD (Yates et al., 2008). Fig. 4.1a depicts the schematic diagram illustrating the CVD technique. In this methodology, a gaseous carbon source (hydrocarbons, such as methane, carbon monoxide, or acetylene) is utilized, and energy is transferred to a carbon molecule in the gas phase through the application of plasma or a resistively heated coil. During CVD, a substrate is subjected to elevated temperatures of around 700°C while being coated with metal catalysts, including nickel, cobalt, iron, or their various combinations. The initiation of growth occurs subsequent to the introduction of two gases into the chamber, namely hydrocarbon gas and a carrier gas such as nitrogen, hydrogen, or argon.

The CVD process offers several advantages, including low power consumption, operation at low temperatures, relatively high purity, and most notably, the potential for process scalability. Table 4.1 lists the advantages and disadvantages of the CVD method. The CVD process can produce both multiwalled carbon nanotubes (MWCNTs) and single-walled carbon nanotubes (SWCNTs), with the specific type of nanotube generated depending on the temperature used. It has been observed that producing SWCNTs requires a higher temperature compared with producing MWCNTs (Khan et al., 2016).

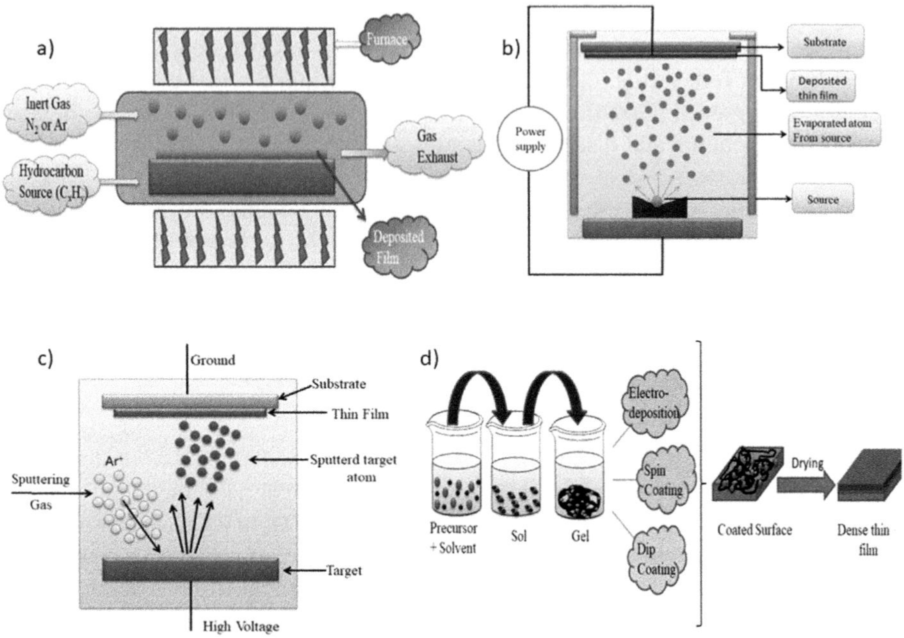

Figure 4.1 Pictorial representation of (a) Chemical vapour deposition. (b) Physical vapour deposition. (c) Sputtering. (d) Sol-gel method.

Traditionally, ceramics have been regarded as heavy, bulky, and brittle materials that are prone to breaking when subjected to external forces, resulting in changes to their shape and structure. However, recent advancements have shown that reducing the dimensions and scale of ceramics can transform them into highly flexible compounds. These flexible ceramics can undergo various deformations, including bending and twisting, without exhibiting visible cracks or signs of failure. To achieve this transformation, a reliable synthesis method is required, allowing precise control over the formation of desired morphological features and accurate realization of defined geometries. CVD has emerged as an ideal approach for this purpose. CVD offers excellent control over structural modifications at the subnanometer scale and allows for precise tuning of the chemical and electronic composition. The wide range of parameters available in the CVD experimental setup facilitates the efficient production of flexible ceramics with specific characteristics (Gurylev, 2023).

Huang et al. (2023) showcase a straightforward and efficient approach to produce transparent superhydrophobic coatings with customized nanocone array structures. They achieved this through a one-step method utilizing initiated chemical vapor deposition (iCVD). Transparent superhydrophobic coatings are extremely sought-after in outdoor optoelectronic devices due to their advantageous features such as resistance to wetting, self-cleaning properties, and antiicing capabilities.

Table 4.1 Lists the advantages and disadvantages of the CVD method.

S. No.	Method	Advantages	Disadvantages	References
1.	CVD	(1) Significant manufacturing efficiency; (2) Superior performance; (3) Straightforward management; (4) Pure materials; (5) Processing expense.	(1) Production of toxic gases. (2) The production of multicomponent materials poses significant challenges. (3) Chances of contamination.	Gu et al. (2020), Abbasi et al. (2019), varshney et al. (2016)
2.	PVD	(1) The coating exhibits uniformity in its distribution. (2) The deposition occurs locally, targeting specific areas and at high rates. (3) Adjustable thin film deposition. (4) The process is environmentally friendly and does not contribute to pollution. (5) The process demonstrates a high level of repeatability. (6) No harm to environment.	(1) Complicated technology (2) Expensive technique (3) Decomposition on high temperature. (4) Amorphous coatings are produced by rapid cooling.	Fotovvati et al. (2019), Helmersson et al. (2006), Gu et al. (2020)
3.	Sol-gel	(1) Simple, homogeneous, affordable, trustworthy, repeatable, and controllable. (2) Uniform coatings of high purity. (3) Suitable for deposition on large surface area. (4) Simple to carry out in a lab (5) Uniform and high purity coatings.	(1) Expensive production (2) Long fabrication period (3) The gel contains a large number of micropores. (4) Only thin layer production.	Li et al., (2020), Priyadarshini et al. (2019), Sukhodub et al. (2014), Gu et al. (2020)

Continued

Table 4.1 Lists the advantages and disadvantages of the CVD method.—cont'd

S. No.	Method	Advantages	Disadvantages	References
4.	Thermal spraying	**(1)** The procedure is straightforward **(2)** Low cost production **(3)** Highly efficient technique **(4)** High deposition rates. **(5)** There is a wide range of coating thickness variation. **(6)** Simple composite coating construction.	**(1)** Extreme heat induces decomposition, oxide inclusion, which affect coating properties. **(2)** Amorphous coatings are produced by rapid cooling.	Fotovvati et al. (2019); Li et al., (2017), Gu et al. (2020)
5.	Sputtering	**(1)** Beneficial for substrates susceptible to heat. **(2)** High rate of growth **(3)** A thick, poreless, uniform covering **(4)** Strong adherence.	**(1)** Ionized particle harm coating. **(2)** low rate of deposition **(3)** Amorphous coatings are produced by low deposition rates **(4)** The selective supply rate of elements.	Uzakbaiuly et al. (2021); Varshney et al. (2016); Sukhodub et al. (2014); Priyadarshini et al. (2019)

4.2.2 Physical vapor deposition

Physical vapor deposition is a process that sputters atoms from a source target under vacuum utilizing either the evaporation of a solid or molten source or the intense gaseous ions in plasma. There are four steps in the PVD procedure: (1) the coating material is evaporated by a high energy source, such as an electron beam or ions; (2) the evaporated atoms or molecules are transported to the coating substrate; (3) during the transport stage, the metal atoms react with the appropriate reactive gas (such as methane, nitrogen, or oxygen); and (4) the coating is deposited on the substrate surface (Fig. 4.1b). Almost all types of inorganic materials and several types of organic materials can be employed in the PVD process. In comparison to many other procedures, such as electroplating, the technique is environmentally friendly. PVD does, however, have several drawbacks, such as difficult shape fabrication, high cost, and low production (Yilmaz and Yorgancioglu, 2019). Table 4.1 lists the advantages and disadvantages of the method. In this method, techniques such as thermal evaporation, sputter deposition, pulsed laser deposition, E−beam evaporation could be used (Uzakbaiuly et al., 2021).

The initial application of physical vapor deposition was in the field of tools and molds. The application of TiC coating on molds significantly increases their lifespan. Coating high-speed steel tools enhances their wear resistance, chip resistance, and cutting speed. Moreover, the coated tools exhibit exceptional properties such as high hardness, chemical stability, toughness, and low friction coefficient (Shafyei and Ashiri, 2019). Wang et al. (2016) conducted research on the wear and corrosion characteristics of various metal and nanostructured coatings in diverse environments using vapor deposition. The wear rate of $Al_2O_3-3TiO_2$ nanocoatings was discovered to be between 30% and 50% lower than that of conventional coatings. Prikryl et al. (2018) prepared SiO_x nanocoatings on specific polymer matrices to enhance their resistance to harmful chemicals seeping through. These nanocoatings are made by vacuum-based thermal deposition of SiO using the Plasma-Assisted Physical Vapor Deposition (PA-PVD) method or vacuum-based chemical vapor deposition of hexamethyldisiloxane (HMDSO) using the PECVD (plasma-enhanced chemical vapor deposition) method. Taran et al. (2019) used vacuum-arc deposition (PVD) to create monoclinic m-ZrO_2 thin films on Nd-Fe-B magnet surfaces. In the context of orthodontics, vacuum-arc deposition (PVD) was employed to apply thin ZrO_2 coatings on the surfaces of Nd-Fe-B permanent magnets. These magnets serve as repelling devices and are utilized as an effective and biologically safe force-generating system for orthodontic tooth movement.

4.3 Sputtering

Sputtering is a significant technique in the field of physical vapor deposition (PVD), wherein the process involves the expulsion of atoms from a surface through the transfer of momentum from atomic-sized energetic particles that bombard the surface. The technique of sputter deposition offers enhanced control over the composition of films containing multiple elements, thereby providing increased flexibility in the selection of materials for deposition (Chi, 2010). It is one of the earliest and most widely used methods of thin-film deposition for industrial uses, such as the semiconductor industry's fabrication of integrated circuit processing and computer hard discs. Antireflection coatings are applied on glass through sputtering for solar thermal (Selvakumar and Barshilia, 2012) and optical applications (Mazur et al., 2013). Both photovoltaic solar cells and optical waveguides are made using sputtering. Thin-contact metals for thin-film transistors can be deposited using the low-temperature deposition feature. On glass used in double-window assemblies, magnetron sputtering is used to apply low-emissivity coatings (such as silver, zinc oxide, tin oxide, and titanium oxide) (Sun et al., 2017). Thin films of metals, metal alloys, refractory materials (such oxides and nitrides), and nonmetals such as polytetrafluorethylene (PTFE) can all be deposited by sputtering (Lim et al., 2019). Sputtering is a method for depositing thin films that has benefits such as consistent thickness across a vast area, simple thickness control, and a long target lifetime. In the case of alloy sputtering, the alloy composition (i.e., the alloy composition's stoichiometry) is preserved in comparison to the initial target composition.

Sputtering is a glow-discharge process, which means that glow discharge produces the energetic particles, which are utilized to attack the target. As a result of its reliance on the ejection of surface atoms from the electrode surface by momentum transfer from bombardment ions to surface atoms, the sputtering process is sometimes regarded as an etching process (Mattox, 2010). An exchange of momentum occurs when energetic ions—typically inert gas species such as Ar—collide with target material atoms (Fig. 4.1c.). These "incident ions" cause collision cascades in the target. An atom will be ejected if a collision cascade encounters the target's surface and its leftover energy exceeds the target's surface binding energy (Abu-Thabit and Makhlouf, 2020).

Magnetrons are near targets during sputtering. Accelerated ionic gas is driven into the vacuum chamber, exploding the target and releasing atomized particles that will be violently pushed onto the substrate. Electron beam evaporation uses a cathode target with an evaporation source containing the material to be evaporated. The discharged particles will collide with gas molecules injected into the reactor to accelerate them and generate plasma through the deposition chamber, which will be more intense in the middle of the reactor. Particles will form a constantly compacted layer on the substrate, allowing a good film to adhere and produce a coating (Gu et al., 2020).

Li et al. (2021) created strontium nanolayer on zirconia implant by magneton sputtering to enhance osteogenesis. There are several therapeutic uses for zirconia implants, but because of its biological inertness and lack of osteoinductive characteristics, these uses are restricted. Superior biocompatibility and excellent osteogenic characteristics are found in strontium. Utilizing a neon-argon-based gas injection magnetron sputtering process, Wicher et al. (2022) created thin diamond-like carbon (DLC)/TiO_2 layer interference coatings on glass screen protectors using magnetron sputtering approach. The interference shift effect is brought about by the deposition of a thin TiO_2 layer on the surface, whereas DLC coatings are taken into consideration for applications that boost the durability of commercially available safety glasses.

4.4 Spray coating

In this process, a coating substance is sprayed onto a substrate to create a layer; it is done so through a nozzle. Spray coating is frequently employed in industry, particularly for organic lacquers, for coatings of metals, glass, and wood surfaces, or in finishing fabrics and leather (Kuroda et al., 2008). In this method of nanocoating, nanocomposite coatings can be applied to substrates in molten, solution, or dispersion state. Following the application, the coating is typically dried and/or cooled to solidify.

In the thermal spraying process, particles are partially melted and propelled at high speeds using a flame or an electric arc. Heating of the coating materials is commonly achieved through electrical or chemical methods, and the materials used for spraying can include metals, ceramics, or polymers (Leroux et al., 2008). The plasma spraying method is also utilized for coatings made of powder materials, in which the powders are injected into the plasma jet at a temperature of about 10,000 K and sprayed over the substrate while still molten. The energy inputs, gas flow rate, torch geometry, type of coating powder, and distance from the substrate are some of the variables that determine the coating's ultimate qualities (Makhlouf, 2011).

Wang et al. (2017) prepared TiO_2-oleic acid (OA) superhydrophobic nanocomposite coating using novel vacuum cold spray (VCS) technique. These newly created superhydrophobic nanocoatings may be able to withstand intense abrasion. Baghdan et al. (2019), prepared nanospray-dried antimicrobial dental implants. Xu et al. (2023) created polyvinylidene difluoride composite membrane by repeatedly spraying Fe^{3+} and tannic acid (TA) to build a TA—Fe complex on the hydrophobic PVDF membrane, giving it superhydrophilicity.

4.5 Sol-gel method

The sol-gel method uses a chemical reaction at a low temperature to create things, films, fibers, particles, or composites. A 3D network is often created through a series of reactions between inorganic precursors, such as metal salts or organometallic compounds. A stiff colloid component or polymer suspension in a liquid is referred to as a sol. These particles typically range in size from 1 to 100 nm and can be either amorphous or crystalline. A gel is made up of a continuous, three-dimensional network that surrounds a liquid phase in a colloidal gel; the network is constructed from an aggregation of colloidal particles.

The sol-gel method can be aqueous or alcohol-based. Alcohol-based methods avoid water buildup until hydrolysis, while aqueous-based systems use water. However, solvent-free nonhydrolytic sol-gel methods also exist. Sol-gel precursors are generally alkoxides or nonalkoxides. This method prefers alkoxides over nonalkoxides due to their higher volatility. Additionally, metal salts can be used. Group I and Group II elements commonly have nonvolatile solid alkoxides with low solubility (Percy et al., 2000).

Typically, the sol-gel coating procedure entails four phases:

1. The liquid medium is used to disperse colloidal particles, creating a sol.
2. The sol solution is applied to the substrates through methods such as spraying, dipping, or spinning, resulting in the formation of coatings.
3. The particles within the sol undergo polymerization as the stabilizing components are removed, leading to the formation of a gel with a continuous network structure.
4. The remaining organic or inorganic components are subjected to heat treatments, which result in the pyrolysis of these components and the formation of either an amorphous or crystalline coating (Troczynski and Yang, 2001; Olding et al., 2001). Fig. 4.1d depicts the schematic diagram illustrating the sol-gel method.

Metal alkoxides are suitable sol-gel initiators. All metals form metal alkoxides with general formula $M(OR)_x$. Here x represents the oxidation state or valence of the metal, R represents an alkyl group, and M represents a metal (Choi et al., 2020). The hydrolysis and condensation reactions of metal alkoxides to produce bigger metal oxide molecules are the most prevalent methods of creating sol-gel networks (Brinker and Scherer, 2013).

Hydrolysis:

$$M(OR)_4 + H_2O \longrightarrow OH - M(OR)_3 + ROH$$

$$OH - M(OR)_3 + ROH \longrightarrow M(OH)_4 + 4ROH$$

Condensation:

$$M(OR)_4 + H_2O \longrightarrow OH - M(OR)_3 + ROH$$

$$OH - M(OR)_3 + ROH \longrightarrow M(OH)_4 + 4ROH$$

After removing the template, sol-gel procedures are frequently utilized in casting techniques to create ceramics with controlled morphologies and hybrid materials. This technique has been used to cover inorganic particles with sizes ranging from nanometers to micrometers successfully. Monodisperse particles with inorganic surface functionality but lower density than completely inorganic particles are produced by coating organic spherical templates with varied inorganic layers. Hollow inorganic materials are made possible by the capacity to remove the inner core and have a variety of uses, including biomedicine. Mohammed and Hussein (2019), prepared pure hydroxyapatite (HA) and HA/TiO_2 composite bioceramic coatings by sol-gel method and efficiently coated on the surface of Ti–Nb–Zr alloy system. Composite HA/TiO_2 film has outstanding wear and corrosion properties for hard tissue applications. Nanotechnology's innovative manufacturing technologies make nanoceramics manufacture straightforward and affordable. Subramaniyan (2023) utilized a straightforward sol-gel technique to synthesize nanopowders and introduced a simple approach for creating nonuniform nanocoatings. The focus of the study was on the preparation of a $SiO_2-Al_2O_3$ composite. Nano silicate (SiO_2) was employed as a filler material to enhance the mechanical properties of the composite, while aluminum oxide (Al_2O_3) served as a secondary material to provide additional properties such as wear resistance and corrosion resistance.

4.6 Electrospinning

A practical construction method for controlled, fibrous 2D or 3D nanostructures is electrospinning. In order to create desired structures, the electrospinning process involves charging polymeric solutions that contain high-voltage metallic precursors or ceramic particles. The charged solution is then pulled by an electric field from a nozzle onto a collector. Numerous ceramic materials can be electrospun to resemble the nanoscale structures of biological systems, such as the collagen fibrils found in bone and the fibrous proteins formed in the native extracellular matrix. Electrospinning provides a versatile method for coating nanofibers onto textile fabrics and implant stents, offering flexibility in the process (Kuraishi et al., 2009; Lee and Obendorf, 2007).

Due to the submicrometric fibrous structure with interconnected pores that could mimic the extracellular matrix, polymeric fibers produced by electrospinning techniques are suitable for applications in tissue engineering (scaffolds and/or wound healing materials). This encourages the adhesion and growth of particular tissue cells. However, that specific morphology is also amenable to adherence of harmful microbes, which can produce a protective assemblage known as a biofilm (made of extracellular DNA, proteins, and polysaccharides). The development of new electrospun composite materials with antibacterial characteristics is becoming more popular as a result. While numerous organic compounds exhibit antibacterial properties, nanoparticles have emerged as a promising alternative due to their ability to introduce novel functionalities and enhance existing ones. Consequently, the subsequent sections will initially focus on the design of composite fibers through the utilization of electrospinning and comparable methodologies. Subsequently, the discussion will shift toward the investigation of the most extensively studied antimicrobial nanoparticles and their impact on the fabrication process and ultimate properties of (bio) polymeric fibers (Quirós et al., 2015; De Faria et al., 2015; Rodríguez-Tobías et al., 2019).

In order to improve the biocompatibility and corrosion resistance of AZ31 Mg alloys substrate, which is utilized as a promising biodegradable implant material in orthopedic clinical settings, the electrospinning technique is employed to fabricate a fiber coating composed of polycaprolactone/collagen/cerium-hydroxyapatite (PCL/Col/Ce-HA) (Chen et al., 2023). Ghiasi et al. (2014) employed the electrospinning method to produce a silk fibroin nanocoated textured silk yarn, with the aim of utilizing it as a scaffold for tendon and ligament applications.

4.7 Self-assembly

Self-assembly is a reversible process whereby disorganized parts are organized in predetermined pattern. It falls into the dynamic or static categories. The ordered state that develops when a system is in equilibrium and doesn't lose energy is known as static self-assembly. When the ordered state demands energy dissipation, dynamic self-assembly occurs. Weather patterns, solar systems, and self-assembling monolayers are a few examples of self-assembling systems (Surekha and Sundararajan, 2014). Thin organic coatings can be created on top of a substrate in an organized fashion using this method (Padma et al., 2014).

Langmuir was the first to provide a scientific explanation for a technique that allows for the creation of a single-layer coating on solid surfaces. Blodgett further developed this technique, now known as the Langmuir–Blodgett (LB) Technique, to enable the production of multilayer coatings (Langmuir, 1917; Blodgett, 1935; Blodgett and Langmuir, 1937). These coatings are formed by arranging monolayers of long alkyl chain surfactant molecules, which possess a specific attraction to particular substrates and are firmly attached through chemisorption. Self-assembled monolayers (SAMs) offer a practical, stable, and uncomplicated approach to modify the interfacial properties of metals, semiconductors, and metal oxides (Lee et al., 2015).

SAMs are organic structures that form when molecules from a gas or solution phase are adsorbed onto solid surfaces or orderly arranged on liquid surfaces (Fig. 4.2a.). These adsorbates naturally come together to create crystalline or semicrystalline structures. The molecules in SAMs have a specific chemical functionality, known as the head group, at one end, which exhibits a particular chemical attraction to a specific substrate (Mahapatro, 2015). However, SAMs have limitations such as low loading

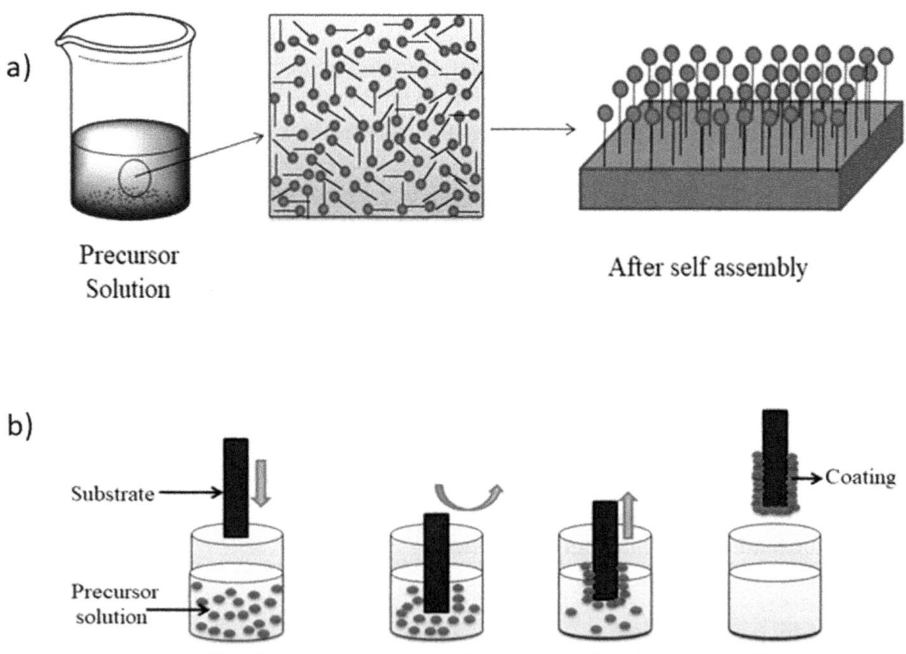

Figure 4.2 Pictorial representation of (a) self-assembly method. (b) Dip coating.

efficiency and applicability to only a restricted range of surfaces. On the other hand, LB coatings require costly equipment and are primarily suitable only for encapsulating amphiphilic components. Consequently, in recent years, the utilization of layer-by-layer (LbL) self-assembling techniques has grown significantly in the deposition of molecules on various material surfaces. LbL deposition is a straightforward and adaptable process with extensive applications in materials science, including biosensors, implant coatings, biomotors, fiberoptics, superhydrophobic surfaces, semiconductors, and drug delivery devices (Keeney et al., 2015).

Combining different techniques allows users to customize films with precise control over parameters such as film thickness, concentration of biomolecules, film stability, release mechanisms, and duration. The LbL technique uses dip coating, spin coating, and spray coating to create a multilayered coating. The LbL self-assembly method involves immersing a charged substrate in an oppositely charged polyelectrolyte, leading to electrostatic attraction and the formation of the first monolayer, washing it, depositing the second layer followed by washing, and repeating the four steps to achieve a multilayered coating (Sakr and Borchard, 2013).

This method offers the potential to create ultrathin multilayer films with a level of precision finer than 1 nanometer, allowing for the precise control of molecular composition. This coating method controls medication release, stabilizes pharmaceuticals, and improves excipient characteristics. The medicine or excipient stays intact since shell formation does not entail covalent binding. Thus, the medicine or excipient is not a new compound. LbL assembly initially used electrostatic interactions. Due to the absence of covalent bonding during shell formation, the drug or excipient remains unchanged, thereby preventing it from being identified as a distinct chemical entity (Cloutier et al., 2015). Mao et al. (2022), prepared nanocoatings based on 2D MXene and 1D cellulose nanocrystals (CNCs) on cotton fabrics using the layer-by-layer (LbL) self-assembly method. These nanocoatings greatly lower fire risks and block electromagnetic interference. Shao et al. (2023), treated wood surface with a superhydrophobic coating prepared from PDA (polydopamine) and PFDTS (1, 1, 2H, 2H-perfluorodecyl trichlorosilane) using layer-by-layer assembly techniques. A superhydrophobic coating helps to stop the biodeterioration of wood caused by moisture deformation, severe mildew, water swelling, and decay when used outdoors.

4.8 Dip coating

The simplest and fastest way to prepare thin films is by dip coating. When a substrate is submerged in a solution, gravity, capillary rise, and viscous drag all work together to evenly and naturally spread the solution throughout the surface of the substrate. At a steady pace, the substrate is withdrawn vertically (Fig. 4.2b.). The final layer is then solidified through evaporation (Mishra et al., 2019). A homogeneous liquid layer forms on the substrate's surface once the substrate is removed from the solution. The volatile solvents will be removed after drying at room temperature, and potential chemical reactions will take place, producing a thin film of coating. The film typically requires heat treatment for hardening or chemical change after drying in a water atmosphere. Mesoporosity can be produced by adding tiny nanoparticles (a few nanometers to tens of nanometers) to the precursor sol, combining the powder with a volatile substance, such as hydrogen peroxide or naphthalene, or using polymethyl methacrylate (PMMA), which, when deposited in a compact, crack-free layer, produces films with pores between 2 and 50 nm in diameter (the IUPAC description for mesoporosity). The preparation of complex architectures is also made possible by the dip coating technique. The viscosity of the solution, the velocity of solvent evaporation, and the angle at which the substrate is removed from the coating can all affect the coating thickness. The solvent's properties affect how much evaporation occurs (Neacşu et al., 2016).

The high-performance disc ceramic membrane was made using a printing and dip-coating approach for effective separation of the oil-water emulsion. A thin, error-free membrane separation layer should be created directly on the large pore size support surface for effective oil-water separation, with the membrane operating in a dynamic filtration mode. Without alumina particle leakage, the printing layer effectively isolates the support flaws; however, the printing layer's surface developed many holes. In order to achieve a smooth membrane surface and a narrow pore size distribution, the dip-coating layer eliminates the hole defects in the printing layer with a negligibly higher membrane thickness (Liu et al., 2023).

4.9 Applications of nanomaterial coatings

Fig. 4.3 illustrates the diverse applications of nanocoatings. These coatings provide benefits such as infection prevention, scratch resistance, moisture protection, self-cleaning surfaces, and stain repellence, significantly improving product performance and durability across various industries. Some of the applications are explained as follows:

4.10 Antimicrobial applications

Researchers in pharmacology are exploring new antimicrobial agents using nanotechnology due to its high surface area-to-volume ratio and unique properties (Wani and Lone, 2019). Controlling harmful microorganism growth is crucial for improving human livelihoods and maintaining natural balance. Uncontrolled growth can lead to health issues, contaminating surfaces such as wound dressings, medical devices, food packages, and industrial pipes, posing a global threat to life and society (Mauter et al., 2011). In favorable settings, these bacteria stick to these devices and produce biofilms that are hard to remove. Biofilms provide safe havens for bacteria to infect, grow, and disseminate by allowing them to metabolize with enhanced antibiotic and human immune defense (Lichter et al., 2009). To contain the infection, these devices are discarded and replaced, which is inconvenient and expensive.

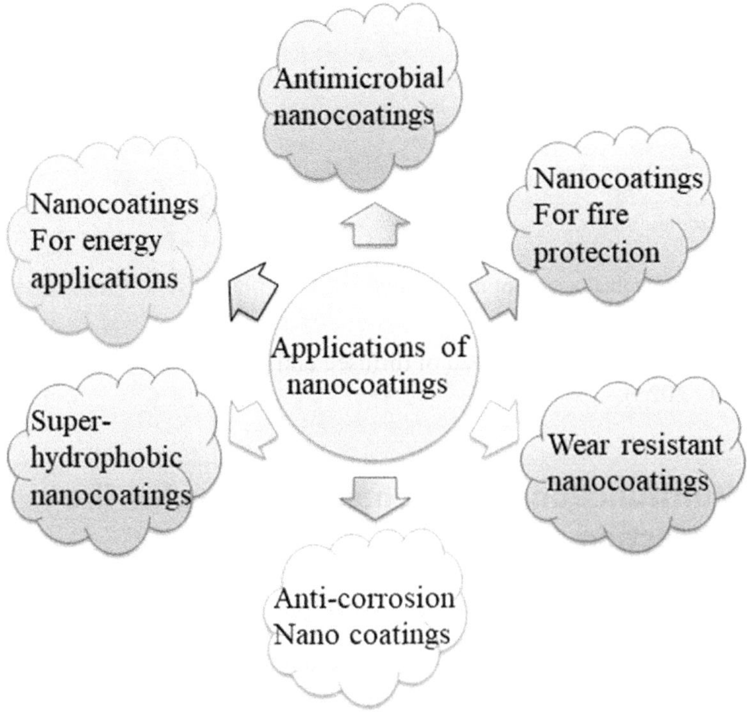

Figure 4.3 Applications of nanocoatings.

Nearly 80% of microbial infections that raise patient morbidity and medical costs are biofilm-related (Hall-Stoodley et al., 2004). After secreting adhesion proteins, bacteria become firmly adhered to the surface and form surface films. These bacteria settle on surfaces, grow, and form colonies inside peptidoglycan envelopes, forming mature biofilms. Biofilm creation makes these germs inaccessible to antimicrobial agents and the immune system, causing a persistent infection that spreads throughout the body. These biofilms are dangerous and resistant to standard antibiotics. Thus, high-performance antimicrobial surfaces that kill bacteria or prevent biofilm are needed. Due to their size and shape-dependent physicochemical properties and huge surface-to-area ratio, nanoparticles are evaluated for antibacterial activity (Guo et al., 2013). Nanoparticle-coated surfaces (NPs) have antifouling qualities that prevent these germs from adhering to the surfaces by either killing the microbial cells directly or damaging the proteins that the microbes make while forming a biofilm (Fig. 4.4) (Wani and Lone, 2019). Ag nanoparticles (Machado et al., 2019; Boccalon et al., 2021), TiO_2 nanoparticles (Ganguli and Chaudhuri, 2021; Chen et al., 2022) and ZnO nanoparticles (Solano et al., 2020) are widely used for their antimicrobial applications. Antimicrobial nanocoatings have a substantial impact on safeguarding surfaces from bacterial colonization in the future, regardless of whether those surfaces are environmental or implanted. Antimicrobial nanocoatings offer advantages over traditional antibiotics

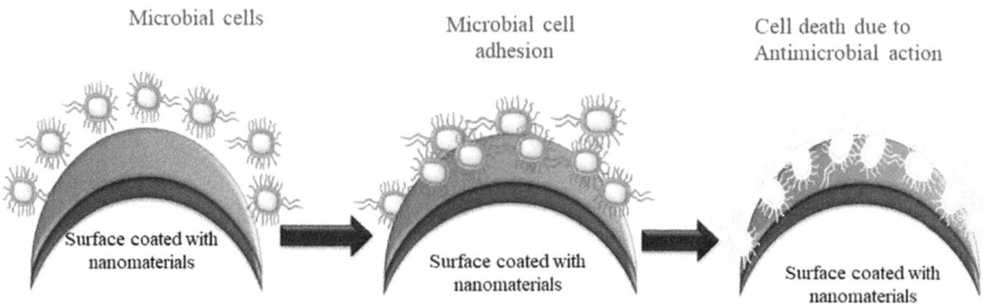

Figure 4.4 Antimicrobial action of nanocoatings.

by providing targeted effects instead of diffused and perhaps undesired consequences (Butler et al., 2023).

4.11 Multifunctional smart nanocoatings for fire protection

Polymers are extensively utilized as lightweight materials owing to their exceptional functional properties, cost-effectiveness, and favorable processability. The majority of these substances consist primarily of carbon and hydrogen constituents, rendering them susceptible to combustion and posing a significant hazard to both human life and property. Consequently, it is essential to improve the fire-resistant properties of lightweight materials. For the time being, there are three primary techniques used to enhance the flame retardancy of lightweight materials: (1) The incorporation of flame-retardant elements (e.g., P, N) or groups into the molecular chains of materials through chemical reactions is a common approach. (2) The physical blending technique involves the integration of flame retardants into the matrix. (3) The application of flame-retardant coatings onto the surface of materials is a method employed to impede the transfer of heat and oxygen, thereby reducing the risk of combustion. The third approach is regarded as a viable option as it has demonstrated the ability to enhance flame retardancy without compromising the substrate's processability and mechanical characteristics. Cheng et al. employed a one-step electrolyte deposition method to fabricate a phytic acid/polyethylene imine coating on the surface of cotton fabric. The fabric's Limiting Oxygen Index (LOI) exhibited an increase from 23.6% to 36.8%. Furthermore, the aforementioned textile material retained its flame-retardant properties even after undergoing 10 cycles of washing (Cheng et al., 2019). The research team led by Sun developed a nanocoating with properties similar to nacre by utilizing montmorillonite (MMT) and polyvinyl alcohol through a simple one-step assembly process. The resulting coating exhibited remarkable mechanical strength, superior barrier properties, and resistance to flames (Sun et al., 2018). The application of nanocoating to diverse substrates exhibited auspicious prospects for its potential utilization. Nonetheless, the aforementioned coatings are susceptible to external forces over extended periods of use, leading to a decline in their flame-retardant properties. To increase the lifespan of flame-retardant coatings, it is crucial to provide them with a self-healing feature.

In light of the swift advancement of the Internet of Things, there has been a surge in demands for materials that exhibit superior fire protection and multifunctional capabilities. Equipping combustible substances with prompt fire-detection ability is deemed as a viable approach to enhance their fire-prevention measures. Wu and his research team assembled a multilayered graphene oxide (GO)/silicone structure onto several flammable substrate materials to produce hierarchical coatings. The coatings that are produced demonstrate a noticeable alteration in electrical resistance in response to changes in temperature, making them effective in detecting elevated environmental temperatures and serving as effective early warning sensors. This capability enables the prevention of fires before the ignition temperature of combustible materials is reached. The multilayered GO/silicone structure exhibits a remarkable flame detection response time of 2–3 s and exceptional self-extinguishing retardancy following a flame attack. This performance is attributed to the synergistic transformation of the structure into a multiscale graphene/nanosilica protection layer. The coatings that have been developed in a hierarchical manner show great potential for use in fire prevention and protection in a range of high-risk fire scenarios and other hazardous situations (Wu et al., 2018).

Zeng et al. (2022) developed a multifunctional fire protection nanocoating (MFNC) through the process of dip-coating Ti_3C_2TX (MXene), 2-ureido-4[1H]-pyrimidinone-containing cellulose (UPC), and montmorillonite (MMT) nanosheets onto the surface of flammable materials using a layer-by-layer assembly technique. The nanocoating was equipped with precise temperature sensing ability through the incorporation of the MXene/UPC layer, which was attributed to the remarkable conductivity and distinctive thermoelectric properties of MXene. Upon ignition, MFNC demonstrated the ability to activate the fire-alert mechanism in a time frame of under 4 seconds. The exceptional flame retardancy of MFNC was attributed to the barrier and catalytic carbonization of MMT, facilitated by the MMT/UPC layer.

4.12 Wear-resistant coatings

The utilization of polymers in various facets of daily existence is progressively expanding. Regrettably, polymer surfaces exhibit inadequate scratch and abrasion resistance and necessitate the application of scratch-resistant coatings to preserve their aesthetically pleasing appearance for extended durations. Consequently, there is an increasing interest in the development of coatings that can provide protection to polymer surfaces. Barna et al. (2005) have emphasized the utilization of nanoparticles such as silicon dioxide (SiO_2), titanium oxide (TiO_2), aluminum oxide (Al_2O_3), and zirconium oxide (ZrO_2) in order to enhance the hardness and mechanical characteristics of coatings, ultimately leading to an improvement in their resistance to wear and scratches. Furthermore, owing to their diminutive dimensions, nanoparticles do not exert any influence on the transparency or glossiness of the coatings. Nanocoatings have the potential to preserve the surface aesthetics and longevity of parquet floorings and windowpanes. Li et al. (2023) reported 1-D inorganic magnesium borate hydroxide (MBH) nanorods as a novel nanocoating filler to increase the transparency and wear resistance of epoxy resin (EP) coatings.

Polysiloxane coatings must have scratch resistance in order to be used in the numerous industries that demand both long-term security and aesthetics. Therefore, the production of a polysiloxane coating that is both highly transparent and scratch-resistant is needed. This is achieved through the incorporation of modified silica nanoparticles (m-SiO_2) that possess epoxy groups into polysiloxane coatings through a homogeneous dispersion achieved through an in situ polymerization process. The uniform distribution of m-SiO_2 particles confers advantageous properties to the scratch resistance of polysiloxane coatings, while exerting minimal impact on their transparency (Li et al., 2020).

4.13 Smart anticorrosion coatings

Corrosion occurs when metals and their environments react chemically, changing metal properties and causing breakdowns in metal function (wei et al., 2015). Coating is a potential way to protect carbon steel and other metals. Polymer covering prevents corrosive species from penetrating metal substrates (Shchukin et al., 2008). The implementation of nanocoatings with anticorrosive properties has had a significant impact on safeguarding metals and alloys in diverse settings. The development of these feasible coatings has advanced from the production of materials on a laboratory scale to their practical implementation.

Magnesium (Mg) alloys have garnered significant attention in various industries including aerospace, communication, automotive, and portable microelectronic fields due to their numerous benefits (Xu et al., 2019). These benefits include low density, high specific strength, and exceptional cast performance. Nonetheless, the susceptibility of magnesium alloys to corrosion continues to pose a significant obstacle, thereby impeding their widespread utilization. It is a well-established fact that when an alloy is subjected to corrosive environments, it can lead to the occurrence of electrochemically induced corrosion. Therefore, it is imperative to identify and expeditiously remediate the initial stages of corrosion. Protective coatings are widely acknowledged as an efficacious approach to mitigate the corrosion of highly chemically reactive magnesium alloys (Zang et al., 2017). A novel smart composite coating has been formulated by Huangfu et al. (2023), through the integration of ionic liquid grafted carboxylated carbon nanotube and phenolphthalein into a micro-arc oxidation coating (MAO/PP/CCNT-IL) with the aim of providing protection to magnesium alloy. The results demonstrate that the composite coating of MAO/PP/CCNT-IL confers significant enhancements in both corrosion resistance and self-repairing capabilities to the Mg alloy.

4.14 Multifunctional superhydrophobic coatings

Superhydrophobic surfaces are those that have a sliding angle (SA) of less than 5 and a water contact angle (CA) larger than 150. Natural biological surfaces with superhydrophobic and self-cleaning qualities include the lotus leaves, the rice leaves, the butterfly wings, the gecko, the namib desert beetle, and the water strider. The observed superhydrophobicity of these biological surfaces has been attributed to their rough surface with hierarchical micro/nanostructure and hydrophobic chemical composition (low surface energy) (Li and Amirfazli, 2007). Researchers have tried to change and manipulate the wettability of various materials in an effort to better understand the inherent mechanism of water repulsion based on the surface features of lotus leaves. Superhydrophobic nanomaterials have emerged as one of the most popular topics across a variety of areas due to the rapid advancement of science and technology. These are widely used in industry, agriculture, defense, medicine, and other industries due to their unique features, including as superhydrophobicity, antiicing, and corrosion resistance (Fig. 4.5). Therefore, the creation of superhydrophobic materials with excellent functionality, affordable costs, useful features, and eco-friendly attributes is crucial for both industrial development and environmental conservation (Wang et al., 2023).

Researchers are working to discover efficient ways to remove oil from water because increased discharge of industrial oily wastewater threatens both the environment and human health. Superhydrophobic surfaces with oil-water separation

Figure 4.5 Multifunctional superhydrophobic nanocoatings.

capabilities have drawn interest recently. Barthwal et al. (2022) described a technique for creating a superhydrophobic/superoleophilic sponge by manufacturing hydrophobic metal organic framework (MOF-5), then modifying it with polydimethylsiloxane (PDMS) using a straightforward dip-coating technique. This sponge's superior water repellency, oil/organic solvent-water separation, antisticking, and antibacterial activity were all outstanding. The hydrophobic PDMS coating serves as a binding layer between the nanoparticles and substrates, which helps to create a superhydrophobic surface, while the hydrophobic MOF-5 nanoparticles embellished on the sponge surface reduce surface roughness. Superhydrophobic PDMS@H-MOF-5-coated sponge offers high separation efficiency of up to 98% for diverse oil/water mixture, good recycling, and antibacterial activity against *S. aureus*. To create superhydrophobic surfaces, the PDMS@H-MOF-5 suspension can also be sprayed onto a variety of surfaces.

Chen et al. (2023) fabricated superhydrophobic sand through the self-assembly of polytetrafluorethylene (PTFE) nanocoatings. Superhydrophobic and superoleophilic characteristics were simultaneously present in the as-prepared sands. Compared with unprocessed desert sand, the as-prepared sand demonstrates good water storage capacity. With a separation efficiency of >96%, the as-prepared sand layer can be employed as an adsorbent or filter to effectively separate both light and heavy oil/water combinations, demonstrating strong application potential in the industrial sector.

4.15 Nanocoatings for energy applications

Solar energy is a promising technique addressing global energy and environmental issues. Solar photovoltaic (PV) applications are developing in urban and rural locations, including residential, industrial, and commercial buildings, to use solar radiation. The goal of PV technology development is to improve PV system performance and reduce costs by increasing module cover glass transmittance and lowering maintenance costs. Solar irradiance harvest depends on PV system performance and climate. Sunlight passes through the environment and the PV module cover before being transformed into electricity. The PV cover acts as a light "gate" for internal energy conversion components such as solar cells. Therefore, cover transmittance is crucial (Al Shehri et al., 2016; Sun et al., 2018). However, dust/particle accumulation on the PV module cover surface is a major transmittance hindrance factor. Dust reduces photovoltaic (PV) panel efficiency (Aljdaeh et al., 2021).

To address this problem, Aljdaeh et al. (2021) suggested a new hydrophobic silicon dioxide-based nanoparticle coating for the PV panel. By lowering dust buildup and the price of cleaning cycles, the suggested PV coating improved the characteristics of PV panels. The coated panels have 13% more output power generation, even without routine surface cleaning. Additionally, the coated panels used half as much water as the uncoated panels. Furthermore, the coated panels do not need any cleaning instruments that are necessary when using traditional cleaning techniques. Tayel et al.

(2022) fabricated PDMS/SiO2-based hydrophobic nanocoatings to reduce dust deposition on PV solar panels.

Dogra et al. (2023) prepared hydrophobic Zirconium based nanocoating, which can be used as optical coatings for self-cleaning and antireflectivity applications. Nanozirconia sol was prepared using sol-gel method, which is further modified hydrophobically with the help of hexamethyldisilazane (HMDS).

4.16 Conclusions

This chapter provides an overview of the processes involved in the fabrication of thin films and nanostructured coatings, as well as an exploration of their various applications. The potential applications of nanocoatings encompass significant sectors including the high-tech industry, civil industry, and national defense. In comparison to conventional coatings, the nanostructured coating exhibits significant enhancements in terms of strength, toughness, wear resistance, and thermal barrier properties. Hence, the implementation of nanostructured coatings is expected to facilitate the advancement of conventional industries due to their commendable performance and long-lasting characteristics. Consequently, these coatings have the potential to supplant traditional technologies, resulting in favorable societal and economic outcomes. The ongoing progress and advancement in surface nanotechnology will inevitably lead to the broader utilization of nanostructured coatings in industrial production.

References

Abbasi, H., Antunes, M., Velasco, J.I., 2019. Effects of graphene nanoplatelets and cellular structure on the thermal conductivity of polysulfone nanocomposite foams. Polymers 12 (1), 25.

Abu-Thabit, N.Y., Makhlouf, A.S.H., 2020. Fundamental of smart coatings and thin films: synthesis, deposition methods, and industrial applications. In: Advances in Smart Coatings and Thin Films for Future Industrial and Biomedical Engineering Applications. Elsevier, pp. 3−35.

Al Shehri, A., Parrott, B., Carrasco, P., Al Saiari, H., Taie, I., 2016. Impact of dust deposition and brush based dry cleaning on glass transmittance for PV modules applications. Sol. Energy 135, 317−324.

Aljdaeh, E., Kamwa, I., Hammad, W., Abuashour, M.I., Sweidan, T.E., Khalid, H.M., Muyeen, S.M., 2021. Performance enhancement of self-cleaning hydrophobic nanocoated photovoltaic panels in a dusty environment. Energies 14 (20), 6800.

Baghdan, E., Raschpichler, M., Lutfi, W., Pinnapireddy, S.R., Pourasghar, M., Schäfer, J., Bakowsky, U., 2019. Nano spray dried antibacterial coatings for dental implants. Eur. J. Pharm. Biopharm. 139, 59−67.

Barabadi, H., Jounaki, K., Pishgahzadeh, E., Morad, H., Bozorgchami, N., Vahidi, H., 2023. Bioengineered metal-based antimicrobial nanomaterials for surface coatings. In: Antiviral and Antimicrobial Smart Coatings. Elsevier, pp. 489−539. https://doi.org/10.1016/B978-0-323-99291-6.00012-8.

Barna, E., Bommer, B., Kürsteiner, J., Vital, A., Trzebiatowski, O.V., Koch, W., Graule, T., 2005. Innovative, scratch proof nanocomposites for clear coatings. Compos. Appl. Sci. Manuf. 36 (4), 473–480.

Barthwal, S., Jeon, Y., Lim, S.H., 2022. Superhydrophobic sponge decorated with hydrophobic MOF-5 nanocoating for efficient oil-water separation and antibacterial applications. Sustain. Mater. Technol. 33, e00492.

Blodgett, K.B., 1935. Films built by depositing successive monomolecular layers on a solid surface. J. Am. Chem. Soc. 57 (6), 1007–1022.

Blodgett, K.B., Langmuir, I., 1937. Built-up films of barium stearate and their optical properties. Phys. Rev. 51 (11), 964.

Boccalon, E., Nocchetti, M., Pica, M., Romani, A., Sterflinger, K., 2021. Hydrogels: a 'stepping stone'towards new cleaning strategies for biodeteriorated surfaces. J. Cult. Herit. 47, 1–11.

Brinker, C.J., Scherer, G.W., 2013. Sol-gel Science: The Physics and Chemistry of Sol-Gel Processing. Academic press.

Butler, J., Handy, R.D., Upton, M., Besinis, A., 2023. Review of antimicrobial nanocoatings in medicine and dentistry: mechanisms of action, biocompatibility performance, safety, and benefits compared to antibiotics. ACS Nano 17 (8), 7064–7092.

Butt, M.A., Fomchenkov, S.A., Kazanskiy, N.L., Ullah, A., Ali, R.Z., Habib, M., April 2017. Infrared reflective coatings for building and automobile glass windows for heat protection. In: Optical Technologies for Telecommunications 2016, vol 10342. SPIE, pp. 199–204.

Chen, M.C., Koh, P.W., Ponnusamy, V.K., Lee, S.L., 2022. Titanium dioxide and other nanomaterials based antimicrobial additives in functional paints and coatings. Prog. Org. Coating 163, 106660.

Chen, Z., Zhang, Z., Ouyang, Y., Chen, Y., Yin, X., Liu, Y., Yang, W., 2023. Electrospinning polycaprolactone/collagen fiber coatings for enhancing the corrosion resistance and biocompatibility of AZ31 Mg alloys. Colloids Surf. A Physicochem. Eng. Asp. 662, 131041.

Nanotechnology. In: Chi, L. (Ed.), 2010. Nanostructured Surfaces, vol. 8. John Wiley & Sons.

Cheng, X.W., Tang, R.C., Yao, F., Yang, X.H., 2019. Flame retardant coating of wool fabric with phytic acid/polyethyleneimine polyelectrolyte complex. Prog. Org. Coating 132, 336–342.

Choi, G., Choi, A.H., Evans, L.A., Akyol, S., Ben-Nissan, B., 2020. A review: recent advances in sol-gel-derived hydroxyapatite nanocoatings for clinical applications. J. Am. Ceram. Soc. 103 (10), 5442–5453.

Cloutier, M., Mantovani, D., Rosei, F., 2015. Antibacterial coatings: challenges, perspectives, and opportunities. Trends Biotechnol. 33 (11), 637–652.

De Faria, A.F., Perreault, F., Shaulsky, E., Arias Chavez, L.H., Elimelech, M., 2015. Antimicrobial electrospun biopolymer nanofiber mats functionalized with graphene oxide–silver nanocomposites. ACS Appl. Mater. Interfaces 7 (23), 12751–12759.

Dogra, V., Kishore, C., Mishra, A., Gaur, A., Verma, A., 2023. Sol-Gel preparation and wetting behaviour analysis of hydrophobic Zirconium based nano-coating: implications for solar panel coating. Chem. Eng. J. Adv. 15, 100507.

Fan, Q., Ma, J., Xu, Q., An, W., Qiu, R., 2018. Multifunctional coatings crafted via layer-by-layer spraying method. Prog. Org. Coating 125, 215–221.

Fernando, R.H., 2009. Nanocomposite and Nanostructured Coatings: Recent Advancements. https://doi.org/10.1021/bk-2009-1008.ch001.

Fotovvati, B., Namdari, N., Dehghanghadikolaei, A., 2019. On coating techniques for surface protection: a review. J. Manufact. Mater. Proc. 3 (1), 28.

Ganguli, P., Chaudhuri, S., 2021. Nanomaterials in antimicrobial paints and coatings to prevent biodegradation of man-made surfaces: a review. Mater. Today: Proc. 45, 3769–3777.

Ghiasi, M., Naghashzargar, E., Semnani, D., 2014. Silk fibroin nano-coated textured silk yarn by electrospinning method for tendon and ligament scaffold application. Nano Hybrids 7, 35–51.

Gu, Y., Xia, K., Wu, D., Mou, J., Zheng, S., 2020. Technical characteristics and wear-resistant mechanism of nano coatings: a review. Coatings 10 (3), 233.

Guo, L., Yuan, W., Lu, Z., Li, C.M., 2013. Polymer/nanosilver composite coatings for antibacterial applications. Colloids Surf. A Physicochem. Eng. Asp. 439, 69–83.

Gurylev, V., 2023. Chemical vapor deposition processing and its relevance to build flexible ceramics materials. In: Advanced Flexible Ceramics. Elsevier, pp. 171–191.

Hall-Stoodley, L., Costerton, J.W., Stoodley, P., 2004. Bacterial biofilms: from the natural environment to infectious diseases. Nat. Rev. Microbiol. 2 (2), 95–108.

Helmersson, U., Lattemann, M., Bohlmark, J., Ehiasarian, A.P., Gudmundsson, J.T., 2006. Ionized physical vapor deposition (IPVD): a review of technology and applications. Thin Solid Films 513 (1–2), 1–24.

Huang, X., Sun, M., Shi, X., Shao, J., Jin, M., Liu, W., Ye, Y., 2023. Chemical vapor deposition of transparent superhydrophobic anti-Icing coatings with tailored polymer nanoarray architecture. Chem. Eng. J. 454, 139981.

Huangfu, H., Guo, X., Li, N., Xiong, Y., Huang, Y., Zhang, J., Wang, L., 2023. A smart composite coating with self-reporting and self-healing functions to enhance corrosion protection for magnesium alloys. Prog. Org. Coating 181, 107598.

Ielo, I., Giacobello, F., Sfameni, S., Rando, G., Galletta, M., Trovato, V., Plutino, M.R., 2021. Nanostructured surface finishing and coatings: functional properties and applications. Materials 14 (11), 2733.

Keeney, M., Jiang, X.Y., Yamane, M., Lee, M., Goodman, S., Yang, F., 2015. Nanocoating for biomolecule delivery using layer-by-layer self-assembly. J. Mater. Chem. B 3 (45), 8757–8770.

Khan, W., Sharma, R., Saini, P., 2016. Carbon nanotube-based polymer composites: synthesis, properties and applications. In: Carbon Nanotubes-Current Progress of Their Polymer Composites.

Kuraishi, K., Iwata, H., Nakano, S., Kubota, S., Tonami, H., Toda, M., Taki, W., 2009. Development of nanofiber-covered stents using electrospinning: in vitro and acute phase in vivo experiments. J. Biomed. Mater. Res. Part B: Appl. Biomater. 88 (1), 230–239.

Kuroda, S., Kawakita, J., Watanabe, M., Katanoda, H., 2008. Warm spraying—a novel coating process based on high-velocity impact of solid particles. Sci. Technol. Adv. Mater. 9.

Langmuir, I., 1917. The constitution and fundamental properties of solids and liquids. II. Liquids. J. Am. Chem. Soc. 39 (9), 1848–1906.

Lee, S., Obendorf, S.K., 2007. Use of electrospun nanofiber web for protective textile materials as barriers to liquid penetration. Textil. Res. J. 77 (9), 696–702.

Lee, J., Bong, J., Ha, Y.G., Park, S., Ju, S., 2015. Durability of self-assembled monolayers on aluminum oxide surface for determining surface wettability. Appl. Surf. Sci. 330, 445–448.

Leroux, F., Campagne, C., Perwuelz, A., Gengembre, L., 2008. Fluorocarbon nano-coating of polyester fabrics by atmospheric air plasma with aerosol. Appl. Surf. Sci. 254 (13), 3902–3908.

Li, W., Amirfazli, A., 2007. Microtextured superhydrophobic surfaces: a thermodynamic analysis. Adv. Colloid Interface Sci. 132 (2), 51–68.

Li, L., Yao, L., Wang, H., Shen, X., Lou, W., Huang, C., Wu, G., 2021. Magnetron sputtering of strontium nanolayer on zirconia implant to enhance osteogenesis. Mater. Sci. Eng. C 127, 112191.

Li, T., Liu, Y., Liu, B., Guo, W., Xu, L., 2017. Microstructure and wear behavior of FeCoCr-NiMo0. 2 high entropy coatings prepared by air plasma spray and the high velocity oxy-fuel spray processes. Coatings 7 (9), 151.

Li, N., Liu, Z.Q., Yang, Y.X., Ying, Y.C., Zhou, Z.Y., 2023. Enhanced wear resistance while maintaining transparency of epoxy resin coatings filled with novel one-dimensional magnesium borate hydroxide nanorods. Prog. Org. Coating 175, 107370.

Li, Y., Zhang, L., Li, C., 2020. Highly transparent and scratch resistant polysiloxane coatings containing silica nanoparticles. J. Colloid Interface Sci. 559, 273–281.

Lichter, J.A., Van Vliet, K.J., Rubner, M.F., 2009. Design of antibacterial surfaces and interfaces: polyelectrolyte multilayers as a multifunctional platform. Macromolecules 42 (22), 8573–8586.

Lim, J.E., Yoon, S., Hwang, B.U., Lee, N.E., Kim, H.K., 2019. Self-connected Ag nanoporous sponge embedded in sputtered polytetrafluoroethylene for highly stretchable and semi-transparent electrodes. Adv. Mater. Interfac. 6 (8), 1801936.

Liu, G., Yang, Y., Liu, H., Wang, Q., Wang, Y., Zhou, J.E., Chang, Q., 2023. Preparation of disc ceramic membrane by a printing and dip-coating method for oil-water separation. Sep. Purif. Technol. 315, 123552.

Machado, G.E., Pereyra, A.M., Rosato, V.G., Moreno, M.S., Basaldella, E.I., 2019. Improving the biocidal activity of outdoor coating formulations by using zeolite-supported silver nanoparticles. Mater. Sci. Eng. C 98, 789–799.

Mahapatro, A., 2015. Bio-functional nano-coatings on metallic biomaterials. Mater. Sci. Eng. C 55, 227–251.

Makhlouf, A.S.H., 2011. Current and advanced coating technologies for industrial applications. In: Nanocoatings and Ultra-thin Films. Woodhead Publishing, pp. 3–23.

Mao, Y., Wang, D., Fu, S., 2022. Layer-by-layer self-assembled nanocoatings of Mxene and P, N-co-doped cellulose nanocrystals onto cotton fabrics for significantly reducing fire hazards and shielding electromagnetic interference. Compos. Appl. Sci. Manuf. 153, 106751.

Mattox, D.M., 2010. Physical sputtering and sputter deposition (sputtering). In: Handbook of Physical Vapor Deposition (PVD) Processing, pp. 237–286.

Mauter, M.S., Wang, Y., Okemgbo, K.C., Osuji, C.O., Giannelis, E.P., Elimelech, M., 2011. Antifouling ultrafiltration membranes via post-fabrication grafting of biocidal nanomaterials. ACS Appl. Mater. Interfaces 3 (8), 2861–2868.

Mazur, M., Wojcieszak, D., Domaradzki, J., Kaczmarek, D., Song, S., Placido, F., 2013. TiO2/SiO2 multilayer as an antireflective and protective coating deposited by microwave assisted magnetron sputtering. Opto-Electron. Rev. 21 (2), 233–238.

Miikkulainen, V., Leskelä, M., Ritala, M., Puurunen, R.L., 2013. Crystallinity of inorganic films grown by atomic layer deposition: overview and general trends. J. Appl. Phys. 113 (2), 2.

Mishra, A., Bhatt, N., Bajpai, A.K., 2019. Nanostructured superhydrophobic coatings for solar panel applications. In: Nanomaterials-Based Coatings. Elsevier, pp. 397–424.

Mohammed, M.T., Hussein, S.M., 2019. Synthesis and characterization of nanocoatings derived by sol–gel onto a new surgical titanium surface. Mater. Res. Exp. 6 (7), 076426.

Mozetič, M., 2019. Surface modification to improve properties of materials. Materials 12 (3), 441.

Nardi, T., Rtimi, S., Pulgarin, C., Leterrier, Y., 2015. Antibacterial surfaces based on functionally graded photocatalytic Fe_3O_4@ TiO_2 core–shell nanoparticle/epoxy composites. RSC Adv. 5 (127), 105416–105421.

Neacşu, I.A., Nicoară, A.I., Vasile, O.R., Vasile, B.Ş., 2016. Inorganic micro-and nano-structured implants for tissue engineering. In: Nanobiomaterials in Hard Tissue Engineering. William Andrew Publishing, pp. 271–295.

Nikolic, M., Lawther, J.M., Sanadi, A.R., 2015. Use of nanofillers in wood coatings: a scientific review. J. Coating Technol. Res. 12, 445–461. https://doi.org/10.1007/s11998-015-9659-2.

Olding, T., Sayer, M., Barrow, D., 2001. Ceramic sol–gel composite coatings for electrical insulation. Thin Solid Films 398, 581–586.

Padma, N., Saxena, V., Sudarsan, V., Raval, H., Sen, S., 2014. Disordered self assembled monolayer dielectric induced hysteresis in organic field effect transistors. J. Nanosci. Nanotechnol. 14 (6), 4418–4423.

Percy, M.J., Bartlett, J.R., Spiccia, L., West, B.O., Woolfrey, J.L., 2000. The influence of β-diketones on the hydrolysis and growth of particles from zirconium (IV) n-propoxide. J. Sol. Gel Sci. Technol. 19, 315–319.

Pilotek, S., Tabellion, F., 2005. Nanoparticles in coatings. Eur. Coat. J. 4, 170–176.

Prikryl, R., Otrisal, P., Obsel, V., Svorc, L., Karkalic, R., Buk, J., 2018. Protective properties of a microstructure composed of barrier nanostructured organics and SiO x layers deposited on a polymer matrix. Nanomaterials 8 (9), 679.

Priyadarshini, B., Rama, M., Chetan, Vijayalakshmi, U., 2019. Bioactive coating as a surface modification technique for biocompatible metallic implants: a review. J. Asian Ceram. Soc. 7 (4), 397–406.

Quirós, J., Borges, J.P., Boltes, K., Rodea-Palomares, I., Rosal, R., 2015. Antimicrobial electrospun silver-, copper-and zinc-doped polyvinylpyrrolidone nanofibers. J. Hazard Mater. 299, 298–305.

Rodríguez-Tobías, H., Morales, G., Grande, D., 2019. Comprehensive review on electrospinning techniques as versatile approaches toward antimicrobial biopolymeric composite fibers. Mater. Sci. Eng. C 101, 306–322.

Sakr, O.S., Borchard, G., 2013. Encapsulation of enzymes in Layer-by-Layer (LbL) structures: latest advances and applications. Biomacromolecules 14 (7), 2117–2135.

Selvakumar, N., Barshilia, H.C., 2012. Review of physical vapor deposited (PVD) spectrally selective coatings for mid-and high-temperature solar thermal applications. Sol. Energy Mater. Sol. Cell. 98, 1–23.

Shao, C., Jiang, M., Zhang, J., Zhang, Q., Han, L., Wu, Y., 2023. Construction of a super-hydrophobic wood surface coating by layer-by-layer assembly: self-adhesive properties of polydopamine. Appl. Surf. Sci. 609, 155259.

Shafyei, H., Ashiri, R., 2019. Electron beam assisted physical vapor deposition of very hard TiCN coating with nanoscale characters. Ceram. Int. 45 (12), 14821–14828.

Shchukin, D.G., Lamaka, S.V., Yasakau, K.A., Zheludkevich, M.L., Ferreira, M.G.S., Möhwald, H., 2008. Active anticorrosion coatings with halloysite nanocontainers. J. Phys. Chem. C 112 (4), 958–964.

Solano, R., Patino-Ruiz, D., Herrera, A., 2020. Preparation of modified paints with nano-structured additives and its potential applications. Nanomater. Nanotechnol. 10, 1847980420909188.

Subramaniyan, A.L., 2023. A Simple Method of Synthesizing Nano Ceramic Coatings on Engine Pistons.

Sukhodub, L.F., Yanovska, G.O., Sukhodub, L.B., Kuznetsov, V.M., Stanislavov, O.S., 2014. Nanocomposite apatite-biopolymer materials and coatings for biomedical applications. J. Nano Electr. Phys.

Sun, G., Cao, X., Zhou, H., Bao, S., Jin, P., 2017. A novel multifunctional thermochromic structure with skin comfort design for smart window application. Sol. Energy Mater. Sol. Cell. 159, 553–559.

Sun, K., Lu, L., Jiang, Y., Wang, Y., Zhou, K., He, Z., 2018. Integrated effects of PM2. 5 deposition, module surface conditions and nanocoatings on solar PV surface glass transmittance. Renew. Sustain. Energy Rev. 82, 4107–4120.

Surekha, K., Sundararajan, S., 2014. ÃDepartment of mechanical engineering science, University of Johannesburg, South Africa;† ndizani civil works, durban, South Africa. Anti-Abras. Nanocoat.: Curr. Future Applic. 81.

Taran, A., Garkusha, I., Taran, V., Timoshenko, A., Misiruk, I., Starikov, V., Mamalis, A.G., 2019. Nanostructured ZrO2 ceramic PVD coatings on Nd-Fe-B permanent magnets. Nanotechnol. Percept 15 (1), 13–20.

Tayel, S.A., Abu El-Maaty, A.E., Mostafa, E.M., Elsaadawi, Y.F., 2022. Enhance the performance of photovoltaic solar panels by a self-cleaning and hydrophobic nanocoating. Sci. Rep. 12 (1), 21236. https://doi.org/10.1038/s41598-022-25667-4.

Tettey, K.E., Yee, M.Q., Lee, D., 2010. Photocatalytic and conductive MWCNT/TiO2 nanocomposite thin films. ACS Appl. Mater. Interfaces 2 (9), 2646–2652.

Tri, P.N., Nguyen, T.A., Rtimi, S., Plamondon, C.M.O., 2019. Nanomaterials-based coatings: an introduction. In: Nanomaterials-Based Coatings. Elsevier, pp. 1–7.

Troczynski, T., Yang, Q., 2001. U.S. Patent No. 6. U.S. Patent and Trademark Office, Washington, DC, pp. 284–682.

Uzakbaiuly, B., Mukanova, A., Zhang, Y., Bakenov, Z., 2021. Physical vapor deposition of cathode materials for all solid-state Li ion batteries: a review. Front. Energy Res. 9, 625123.

Varshney, G., Kanel, S.R., Kempisty, D.M., Varshney, V., Agrawal, A., Sahle-Demessie, E., Nadagouda, M.N., 2016. Nanoscale TiO2 films and their application in remediation of organic pollutants. Coord. Chem. Rev. 306, 43–64.

Vidales-Herrera, J., López, I., 2020. Nanomaterials in coatings: an industrial point of view. In: Handbook of Nanomaterials for Manufacturing Applications. Elsevier, pp. 51–77.

Wani, I.A., Lone, M.N., 2019. Antibacterial and antifungal. In: Nanomaterials-Based Coatings. Elsevier, pp. 41–71.

Wang, C., Wang, Y., Wang, L., Hao, G., Sun, X., Shan, F., Zou, Z., 2014. Nanocomposite lanthanum zirconate thermal barrier coating deposited by suspension plasma spray process. J. Therm. Spray Technol. 23, 1030–1036.

Wang, P., Deng, S., He, Y., Liu, C., Zhang, J., 2016. Oxidation and hot corrosion behavior of Al2O3/YSZ coatings prepared by cathode plasma electrolytic deposition. Corrosion Sci. 109, 13–21.

Wang, Z., Chen, X., Gong, Y., Zhang, B., Li, H., 2017. Superhydrophobic nanocoatings prepared by a novel vacuum cold spray process. Surf. Coating. Technol. 325, 52–57.

Wang, Z., Ren, Y., Wu, F., Qu, G., Chen, X., Yang, Y., Lu, P., 2023. Advances in the research of carbon-, silicon-, and polymer-based superhydrophobic nanomaterials: synthesis and potential application. Adv. Colloid Interface Sci. 318, 102932.

Wei, H., Wang, Y., Guo, J., Shen, N.Z., Jiang, D., Zhang, X., Guo, Z., 2015. Advanced micro/nanocapsules for self-healing smart anticorrosion coatings. J. Mater. Chem. A 3 (2), 469–480.

Wicher, B., Chodun, R., Skowroński, Ł., Trzcinski, M., Kulikowski, K., Zdunek, K., 2022. Design of thin DLC/TiO2 film interference coatings on glass screen protector using a neon–argon-based gas injection magnetron sputtering technique. Diam. Relat. Mater. 123, 108859.

Wu, Q., Gong, L.X., Li, Y., Cao, C.F., Tang, L.C., Wu, L., Mai, Y.W., 2018. Efficient flame detection and early warning sensors on combustible materials using hierarchical graphene oxide/silicone coatings. ACS Nano 12 (1), 416–424.

Xu, J., Xie, A., Sun, H., Wu, Y., Li, C., Xue, C., Pan, J., 2023. Construction of tannic acid-Fe complex coated PVDF membrane via simple spraying method for oil/water emulsion separation. Colloids Surf. A Physicochem. Eng. Asp. 671, 131621.

Xu, T., Yang, Y., Peng, X., Song, J., Pan, F., 2019. Overview of advancement and development trend on magnesium alloy. J. Magnesium Alloys 7 (3), 536–544.

Yates, H.M., Brook, L.A., Sheel, D.W., Ditta, I.B., Steele, A., Foster, H.A., 2008. The growth of copper oxides on glass by flame assisted chemical vapour deposition. Thin Solid Films 517 (2), 517–521.

Yazdani, A., Zakeri, A., 2017. Fabrication and characterization of Ni-SiC nanocomposite coatings on Al substrates by ball impact deposition method. Metall. Mater. Trans. 48, 4180–4192.

Yilmaz, O., Yorgancioglu, A., 2019. Nanocoatings: preparation, properties, and biomedical applications. In: Polymeric Nanomaterials in Nanotherapeutics. Elsevier, pp. 299–331.

Zang, D., Zhu, R., Zhang, W., Yu, X., Lin, L., Guo, X., Jiang, L., 2017. Corrosion-resistant superhydrophobic coatings on Mg alloy surfaces inspired by lotus seedpod. Adv. Funct. Mater. 27 (8), 1605446.

Zeng, Q., Zhao, Y., Lai, X., Jiang, C., Wang, B., Li, H., Chen, Z., 2022. Skin-inspired multifunctional MXene/cellulose nanocoating for smart and efficient fire protection. Chem. Eng. J. 446, 136899.

Fabrication strategies and surface tuning and structural modifications of MNCs

5

Vinith Johnson, Sunil Vasu, Archana M and Uday Kumar S
Department of Chemical Engineering, Indian Institute of Technology Tirupati, Tirupati, Andhra Pradesh, India

5.1 Introduction

In the domain of nanotechnology, precise control and manipulation of the morphology and size of nanoparticles are of utmost importance due to their significant impact on various performance properties such as electrical conductivity, bioactivity, catalytic reactivity, light interaction, and magnetic behavior. The choice of understanding nanomaterials synthesis is important as it gives unique physicochemical, structural, and morphological characteristics to nanoparticles, rendering them promising properties suitable for a myriad of applications across optoelectronics, optics, electrochemistry, electronics, environmental science, and biomedicine. Advancements in nanoparticle synthesis methods empower researchers to precisely tailor and modulate the shape and size of nanoparticles to meet specific application requirements. Nanofabrication techniques are commonly categorized into two principal methodologies: the top-down approach and bottom-up approaches. Within this domain, the concept of "nano-space" serves as a pivotal nexus linking atomic and molecular scales with macroscopic entities. Nanomaterials, characterized by their small sizes and expansive interfaces, manifest distinctive properties that distinguish them from bulk materials. Analogous to the hierarchical scales of meters and microns, nanometers represent a fundamental unit of length. Nanomaterials are defined as materials wherein at least one dimension falls within the nanometer size range (1−100 nm). Globally, there exists a consensus that the nanoscale encompasses dimensions ranging from 1 to 100 nm, while dimensions spanning from 100 to 1000 nm constitute a submicron regime. Entities smaller than 1 nm are typically classified as atomic clusters (Xia and Jiang, 2008; Tan and Saltzman, 2004; Pulido-Reyes et al., 2017). Some of the important properties of nanomaterials that are often explored for diverse applications include the quantum confinement effect, high surface area to volume ratio, superior mechanical, optical, electrical, and magnetic properties. As the size of nanoparticles is very small, that is, in the range of 1−100 nm, most of the atoms that constitute the nanomaterial are present on the surface, resulting in heightened surface-free energy. Consequently, nanoparticles demonstrate a propensity for spontaneous agglomeration, oxidation, or adsorption phenomena,

Multifunctional Nanostructured Coatings. https://doi.org/10.1016/B978-0-443-23683-9.00005-2

effectively mitigating the presence of unstable surface atoms. These inherent surface properties of nanomaterials offer significant opportunities for precise modulation, making them highly versatile for a wide range of applications across various fields (Liu et al., 2010). Despite notable progress made by numerous research groups over the past 2 decades in developing nanoparticle and nanostructure synthesis strategies, there remains a gap in the literature concerning a comprehensive and detailed review that thoroughly assesses and compares various fabrication strategies, surface tuning techniques, and methods for structure modification. This book chapter aims to bridge this gap by offering a thorough and comprehensive exploration of different methods employed in fabrication techniques, surface tuning properties, and structure modification methods of multifunctional nanostructure coatings (MNCs).

This book chapter is structured into three main sections, each focusing on a crucial aspect of multifunctional nanostructure coatings (MNCs) development. The first section outlines various fabrication techniques from conventional methods like chemical vapor deposition (CVD) and physical vapor deposition (PVD) to more advanced approaches such as electrospinning and laser cladding. Following this, the second section delves into surface tuning techniques, essential for customizing the properties of nanostructured coatings. This section covers a spectrum of methods including chemical functionalization, nanoparticle coating, hydrophobic/hydrophilic modification, and layered functionalization, all aimed at enhancing surface functionalities and overall performance. The final section of the chapter focuses on structure modification methods, offering in-depth discussions on innovative approaches such as template-assisted methods, strain engineering, gradient coatings, plasma treatment, ion beam irradiation, nanotexturing, and nanomaterials doping. Through comprehensive exploration of these techniques, the chapter highlights the potential for synergy and innovation in MNCs research, providing readers with a comprehensive understanding of how to enhance the structural characteristics of nanostructured coatings for various applications.

5.2 Fabrication strategies for multifunctional nanostructured coatings

The fabricating approaches used to produce MNCs involve creating functional coatings at the nanoscale with a range of desired properties. These coatings are tailored to possess a blend of attributes including enhanced mechanical strength, increased durability, resistance to corrosion, antifouling properties, self-cleaning abilities, optical effects, and electrical conductivity. The fabrication methods entail meticulously selecting and incorporating nanostructures, nanoparticles, or nanocomposites into the coating matrix through techniques such as physical or chemical deposition, sol-gel processes, electrodeposition, self-assembly, etc. The objective is to develop coatings capable of meeting diverse requirements across various biomedical and industrial applications. Some of the conventional strategies for the fabrication of MNCs are as follows.

5.2.1 Chemical vapor deposition

The dissociation of gaseous or vaporous material above the heated surface followed by deposition on the surface is referred to as CVD. An illustrative representation of CVD is shown in Fig. 5.1.

The CVD method of deposition takes place in a series of steps during the deposition process. The fundamental processes for the CVD of thin films are as follows:

- First, the gaseous precursors are generated via sublimation or evaporation.
- The generated gaseous precursors are introduced to the chamber or reactor via forced convection.
- The gaseous precursors react with each other to produce reactive intermediates.
- These reactants are transported to the surface of the substrate through diffusion.
- The surface of the heated substrate favors the chemical reaction at the solid–liquid interface, leading to the adsorption of the reactants.
- The chemical reaction promotes surface processes such as chemical decomposition, site migration, and site incorporation of the reactants, etc., supporting the growth of thin film on the surface of the substrate.
- The gaseous by-products formed and the unreacted precursors are removed from the chamber or reactor by diffusion or convection.

CVD is widely used to coat ceramic or metallic compositions, that is, the substrate used to be coated is ceramic or metallic materials. The crucial point is the temperature of the substrate surface, which influences the adhesion of the coatings to the substrate

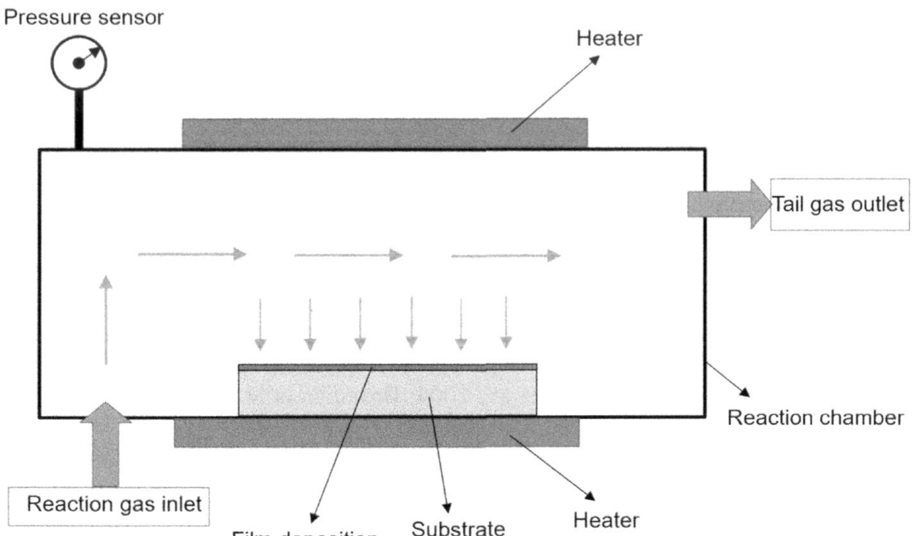

Figure 5.1 Basic components involved in chemical vapor deposition process.
From Ramezani, M., Mohd Ripin, Z., Pasang, T., Jiang, C.P., 2023. Surface engineering of metals: techniques, characterizations and applications. Metals 13 (7). https://doi.org/10.3390/met13071299.

(Makhlouf, 2011). Multidirectional deposition in the CVD method is possible due to its inherent nature (Behera et al., 2020). This implies that substrates with varying sizes and geometries can be instantly coated by this method. The CVD method is utilized in a variety of industrial applications, which includes the deposition of refractory materials on turbine blades, etc. The CVD process demands substrates with higher temperature capacity and lower precipitation rates, which makes it expensive for large surface coatings.

An example of CVD is the synthesis of graphene thin film over a polycrystalline copper substrate. The growth of the graphene film at 850°C using ethanol as the chemical precursor for 5–6 min exhibits anticorrosive properties (Abdeen et al., 2019). Characterization by potentiodynamic polarization and electrochemical impedance spectroscopy (EIS) validates that, in contrast to uncoated copper substrates, the graphene coating yields a reduced corrosion rate and increased impedance.

5.2.2 Physical vapor deposition

The deposition of target material by converting it into the gaseous phase and depositing it on the surface of the substrate is known as PVD. The conversion of the target material into the gaseous phase occurs through various means such as laser energy, thermal energy, electron beams, high-energy particles, resistive heating, etc. (van Lente and van Til, 2008). The basic principle behind the PVD process involves:

- Evaporation of the target material to be coated on the surface of the substrate via various energy sources.
- Transport of the evaporated material to the surface of the substrate.
- Condensation and deposition of the target material on the surface once it reaches the cold substrate (Baptista et al., 2018) Fig. 5.2

The PVD methods are distinguished based on their principle of deposition (Gu et al., 2020b). Some examples of PVD methods include cathodic arc deposition, laser pulsed deposition, magnetron sputtering, and thermal evaporation (Nguyen-Tri et al., 2018). For example, novel nanostructured carbon-based coatings, including fullerene-like carbon nitride and carbon fluoride (FL-CNx and FL-CFx) nanostructured coatings, were formed through the process of magnetron sputtering, which exhibits less absorption of water compared to conventionally used amorphous carbon films due to the lack of dangling bonds. Therefore, FL-CNx coatings prove to be an appropriate protective overlay for magnetic storage devices like computer disk drive systems (Gueorguiev et al., 2011; Hellgren et al., 2004; Broitman et al., 2008).

The structure of nanostructured coatings, along with their mechanical, wear, and corrosion properties, is significantly influenced by experimental factors, namely processing parameters. For example, in the case of titanium-aluminum-nitride (Ti–Al–N) films produced through reactive magnetron sputtering, the substrate bias plays a crucial role in affecting aspects such as hardness, microstructure, roughness, mechanical stress, and wear performance. Ti–Al–N films exhibit increased residual stress and hardness with an increase in substrate bias, whereas they exhibit decreased wear rate and surface roughness (Chu et al., 2006).

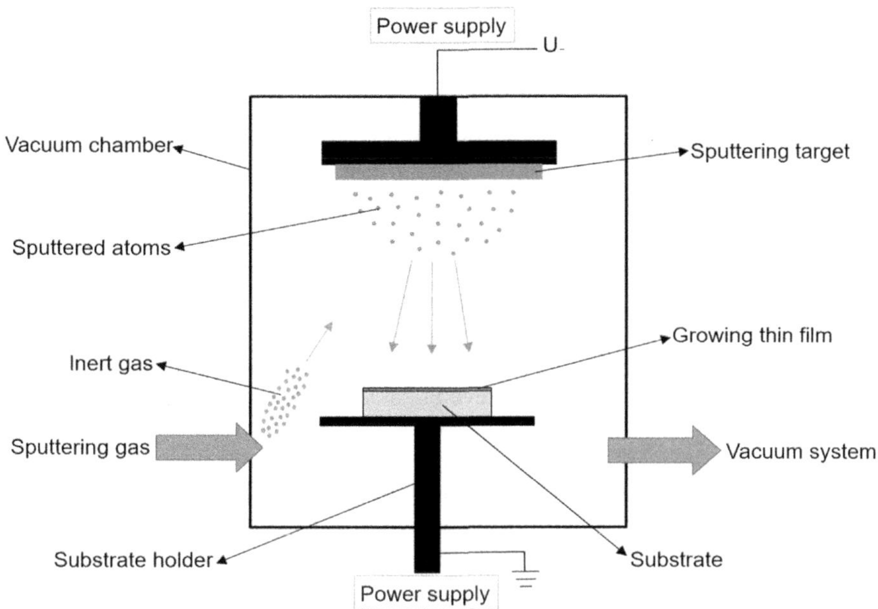

Figure 5.2 Basic components involved in physical vapor deposition process.
From Ramezani, M., Mohd Ripin, Z., Pasang, T., Jiang, C.P., 2023. Surface engineering of metals: techniques, characterizations and applications. Metals 13 (7). https://doi.org/10.3390/met13071299.

Likewise, the wear performance, deposition rate, and hardness of Cr—Al—N nanocoating are influenced by the dynamic glancing angle in DC magnetron sputtering. Dynamic glancing angle deposition (DGLAD) is a method that involves oscillating substrates in front of targets with continuous variations in angle. For instance, employing an oscillatory range of $\pm 10^{\circ}$ leads to approximately three times greater wear resistance in coatings compared to those deposited through conventional methods (Avila et al., 2021).

In the PVD method, based on the principle, it restricts the substrate coating to line-of-sight deposition. This technique allows the deposition of various inorganic materials and certain organic materials onto substrates. In terms of environmental impact, PVD is considered more eco-friendly compared to other deposition methods like electroplating. However, the requirement of a vacuum chamber in PVD imposes limitations on the size of surfaces that can be coated. Compared to the CVD method, the PVD process is unable to uniformly coat substrates with intricate or complex surface geometries.

5.2.3 Layer by layer assembly

Layer-by-layer (LbL) assembly serves as a prevalent method in engineering materials and coatings. Layer-by-layer assembly offers a practical approach to coating, which is ideal for safeguarding consumer products like paints for corrosion-resistant paints or antigraffiti coatings (Andreeva et al., 2008; Zheludkevich et al., 2007; Andreeva et al., 2010). For example, the deposition of alternatingly charged (PEI/PSS) polyelectrolyte (PE) multilayers onto aluminum surfaces by Andreeva et al. resulted in the prevention of corrosion processes on the aluminum surfaces, attributed to the pH-buffering capacity of the polyelectrolyte-based LbL coatings, Fig. 5.3 (Zhao et al., 2019).

The composite multilayers created through LbL offers efficient storage and extended release of the inhibitor. Likewise, SiO_2 particles coated with LbL multilayers containing inhibitors were utilized as nanocontainers, enabling simultaneous achievement of self-healing and anticorrosion properties in composite coatings (Shchukin et al., 2006; Zheludkevich et al., 2007). The key element of the self-healing anticorrosion coatings consists of capsules arranged in flat layers. Employing the LbL assembly technique proves to be a powerful method for producing capsule shells that manage the release of corrosion inhibitors as needed. These capsules facilitate the controlled release of corrosion inhibitors precisely where needed, confined only to the corroded region Fig. 5.3. Such controlled release serves as a localized trigger for the repair mechanism that heals the defects.

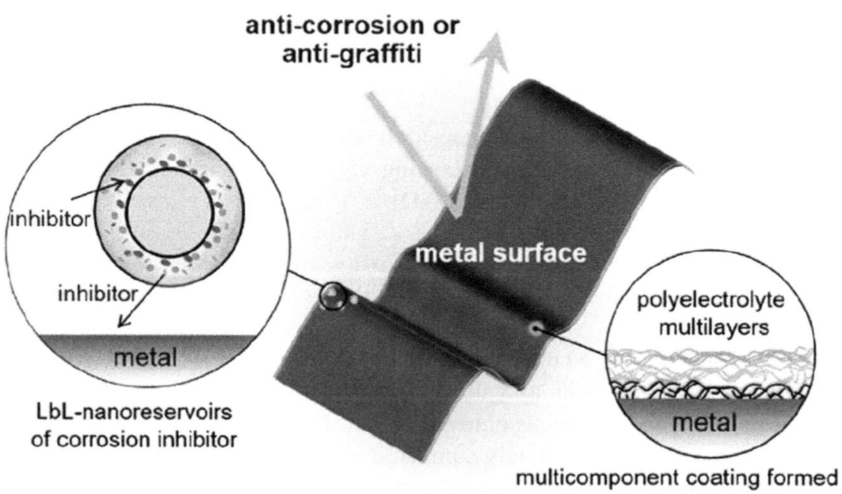

Figure 5.3 Protective coating using the layer-by-layer assembly method. The polyelectrolyte-based coatings provide corrosion prevention on aluminum surfaces through their pH-buffering capacity.
From Zhao, S., Caruso, F., Dahne, L., Decher, G., De Geest, B.G., Fan, J., Feliu, N., Gogotsi, Y., Hammond, P.T., Hersam, M.C., Khademhosseini, A., Kotov, N., Leporatti, S., Li, Y., Lisdat, F., Liz-Marzan, L.M., Moya, S., Mulvaney, P., Rogach, A.L., … Parak, W.J. (2019). The future of layer-by-layer assembly: a tribute to ACS nano associate editor helmuth mohwald. ACS Nano 13 (6), 6151–6169. https://doi.org/10.1021/acsnano.9b03326.

Layer-by-layer assembly facilitates the utilization of diverse materials as components for shell formation, employing predominantly weak electrostatic forces for their assembly. The nature of the "smart" materials introduced into the container shell determines the various stimuli capable of inducing reversible and irreversible modifications in the shell, such as pH, temperature, ionic strength, electromagnetic fields and ultrasonic treatment. The range of observed responses can vary from subtle effects like adjustable permeability to more significant outcomes such as complete rupture of the container shell. These diverse behaviors are contingent upon the composition of the polyelectrolyte multilayers, such as interactions based on weak polyanion-weak polycation or strong polyanion-weak polycation.

5.2.4 Sol–gel method

The chemically active precursors dissolved in the solvent undergo the reaction mechanism of hydrolysis with polycondensation to form a system of sol, which slowly forms a gel finally by polymerization. This is known as the sol–gel method. The gel formed is then dried and formed as a coating by heating (Gu et al., 2020b; Wang, 2020). Through this process, the coating with uniform chemical and physical properties is achieved. The major factors involved in the coatings by sol–gel methods are based on the nature of the catalyst, precursor materials involved in the formation of the sol, thickness, temperature, pH, etc. These factors are crucial and affect the hydrolysis speed and density of the coating. Wear-resistant coatings are produced by this method (Gu et al., 2020b; Lavollee et al., 2016; Aliofkhazraei, 2011).

An example of a sol–gel method–based coating is the arrangement of ultrathin hydrophilic photo-active TiO_2 nanoporous membrane layered onto a nanoporous hydrophobic hybrid silica coating developed to create antireflective, self-cleaning properties by the sol–gel method. This system promotes prolonged stability in water, with optical characteristics that are customizable through the modification of silica layer thickness and porosity (Faustini et al., 2010).

This method has also been employed for the development of nanocomposite coatings with a thickness of micron scale. However, coatings produced through this method face limitations in thickness and are prone to cracking (Nguyen-Tri et al., 2018). The presence of micropores in gels leads to the escape of gases and organics during the drying process, resulting in shrinkage.

There are some drawbacks associated with sol–gel processes, especially when applied to large-scale production (i.e., industrial production). They are:

- The process is time-consuming and typically demands an extended duration (days or even weeks) for preparation.
- The precursors required for the sol–gel process are costly.
- Some of the organics utilized in the sol–gel process pose hazards to both the environment and human health.

5.2.5 Electrodeposition

Electrodeposition is a process that employs depositing a metal or alloy coating onto a conductive substrate surface through the application of an electric current (electrolysis). This occurs by immersing it in an electrolyte solution containing a salt precursor of the metal or alloy intended for coating. The electrolyte solution, which contains the metal salt, is known as the bath. By varying the experimental parameters like applied potential, current density, time of deposition, and plating solution composition, modifications in the composition, morphology, and texture of the deposited film coating can be achieved (Milan and Schlesinger, 2010; Basile et al., 2010). Electrodeposition is one of the most commonly utilized techniques for producing nanostructured metallic or alloy coatings. It is classified into two categories based on the type of current employed: direct current (DC) electrodeposition and pulse electrodeposition.

An example of an electrodeposition-based coating is fabricating a superhydrophobic film of nickel hydroxide/nickel stearate on the surface of an aluminum alloy through the electrodeposition technique utilizing an ethanol solution containing stearic acid and nitrate hexahydrate as an electrolyte (Gurrappa and Binder, 2008).

5.2.6 Spray coating

Spray coating is a technique wherein a target material is forced through a nozzle onto a substrate surface, forming a coating upon impact (Yilmaz and Yorgancioglu, 2018). The substance being sprayed can exist in various states, including liquid, molten, or softened particles (Makhlouf, 2011). Industries commonly use spray coating techniques to apply organic lacquers and binders to coat irregularly shaped surfaces like glass, metals, wood, and textiles. Spray coating can be classified into two types: thermal spraying and cold spraying.

Thermal spraying is a widely utilized method in spray coating, where the coating material is melted or heated before being sprayed onto the surface of the substrate (see Fig. 5.4). The heating of the coated material is done by an electrical or chemical process. Thermal spraying offers higher deposition rates on substrates with expansive surface areas compared to conventional methods like CVD and PVD.

Cold spraying is a variant method of spray coating that facilitates the application of coatings at temperatures below those specified in thermal spray methods. However, achieving high velocities is crucial in cold spraying, as the impingement velocity plays a pivotal role in determining the bonding between the coating material and substrates. Coatings produced through cold spraying exhibit enhanced strength, low porosity, and superior adhesion capabilities (Nguyen-Tri et al., 2018).

Additional methods of spray coating include plasma spraying, high-velocity oxy-fuel spraying (HVOF), high-velocity air fuel spraying (HVAF), etc. In these processes, the ultimate properties of the sprayed coating depend on factors such as the size distribution of coating powders, distance between the spray gun and substrates, powder feed rate, spray velocity, etc.

An example of a spray coating method is a superhydrophobic coating utilizing a multifunctional composite containing polydimethylsiloxane mixed with graphene,

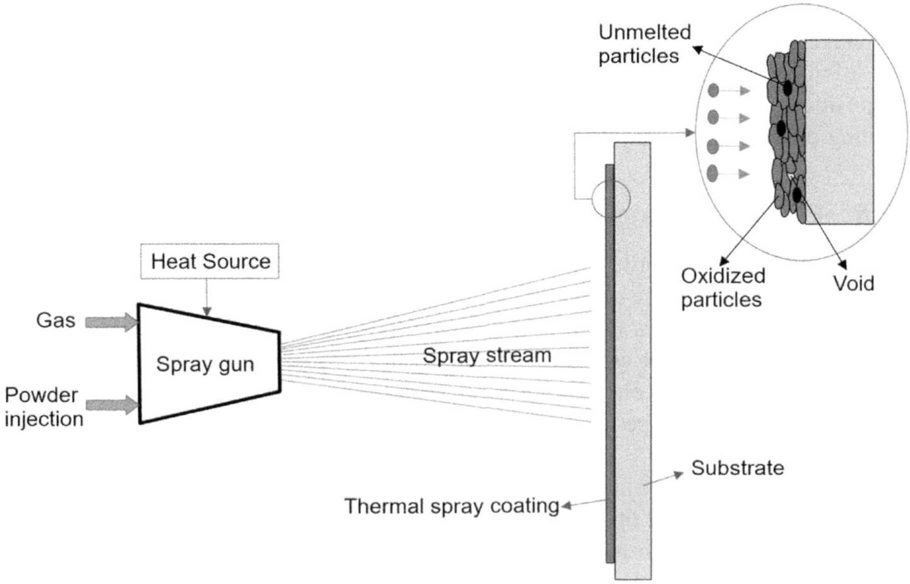

Figure 5.4 Basic components involved in thermal spray coating technique.
From Ramezani, M., Mohd Ripin, Z., Pasang, T., Jiang, C.P., 2023. Surface engineering of metals: techniques, characterizations and applications. Metals, 13 (7). https://doi.org/10.3390/met13071299.

titanium dioxide nanoparticles, and micro/nanoparticles of diatomaceous earth, which was coated using spray coating. This coating displayed self-cleaning capabilities alongside strong mechanical resilience and corrosion resistance, attributed to its molecular diffusion barrier properties. Properties of the sprayed coating are depending on factors such as the size distribution of coating powders, distance between the spray gun and substrates, powder feed rate, spray velocity, etc (Nine et al., 2015).

5.2.7 Laser cladding

Laser cladding involves the deposition of nanostructured coatings by melting either powdered or wire feedstock target material through a high energy laser onto the substrate surface (Gu et al., 2020b). As the molten target material solidifies on the surface, the substrate and the target material exhibit a strong metallurgical bond. For example, Nazari et al. (2018) innovated a novel Ti–Fe composite coating using laser cladding. This coating marked the first instance of applying a Ti–Fe coating onto a titanium substrate utilizing titanium powder obtained from mechanical processing. This durable coating promises to greatly enhance the wear resistance of titanium alloy, rendering it suitable for applications in the aerospace and national defense sectors.

The quality of the coating is substantially influenced by laser process parameters such as laser power, power feed, and scan speed (Gu et al., 2020a; Rahman Rashid et al., 2019; Zou et al., 2020). Laser cladding provides several advantages, including low dilution rates, high cooling rates, and ease of automation when compared to other alternative methods. This technique allows the use of various powders for coating deposition. However, drawbacks such as poor metallurgical quality, nonuniform composition, and the occurrence of cracks in coatings are associated with the laser cladding process (Gu et al., 2020b).

5.2.8 Electrospinning

Electrospinning is a versatile and simple technique for depositing extremely fine polymeric fibers across a range of sizes from submicron to nanoscale, through the utilization of electrostatic forces (Rivero et al., 2018). Moreover, this method serves as a highly effective means of creating materials with elevated surface areas (Frenot and Chronakis, 2003). These electrospun nanofibers (referred to as ENFs) are fabricated using various polymer types including synthetic, natural, biodegradable, copolymers, or polymer blends.

The fabrication of nanofibers involves applying high voltage to a liquid polymeric precursor contained at the tip of a conductive capillary tube (needle/spinneret) (López-Rubio et al., 2019). Upon the application of an electric field, the electrostatic forces balance the surface tension of the polymeric liquid, resulting in the elongation of the droplets in a characteristic cone shape at the tip of the needle known as the "Taylor cone." This results in a continuous jet strand propelled from the dispensing needle toward a typically grounded collector (Bhardwaj and Kundu, 2010) as illustrated in Fig. 5.5. During the travel from the tip to the grounded collector, the solvent evaporates from the polymeric jet strand and is deposited as electrospun nanofibers (ENFs) onto the grounded collector.

The electrospinning process is sensitive to various parameters for the generation of polymeric fibers with the desired morphology and diameter, instead of beads or droplets. The initial set encompasses characteristics of the polymeric precursor, including molecular weight, surface tension, chain entanglement, electrical conductivity, viscosity, solvent nature, molar concentration, and others (Jiang et al., 2015). The second set of parameters pertains to operational factors like applied voltage, flow rate of the polymeric solution, and distance between the capillary tip (needle/spinneret) and collector (Pillay et al., 2013). The third set of parameters relates to environmental conditions such as relative humidity, temperature, and airflow within the electrospinning chamber, influencing solvent evaporation rate.

The electrospinning technique facilitates the fabrication of porous coatings that mimic the extracellular matrix of the human body, supporting cell growth for tissue regeneration or serving in prolonged drug release for the control of infection.

Synthetic polymers such as conducting polymers, water-soluble polymers, hydrophobic polymers, and superhydrophobic polymers are used for the fabrication of nanofibers with self-healing protective properties and anticorrosion properties. The design of Al_2O_3 coatings on zinc alloys improves the corrosion resistance in alkaline solutions. This approach offers an efficient alternative for metal-air batteries within renewable energy devices, given zinc's prevalent use as electrodes. A schematic illustration of the fabrication process is shown in Fig. 5.6. Additionally,

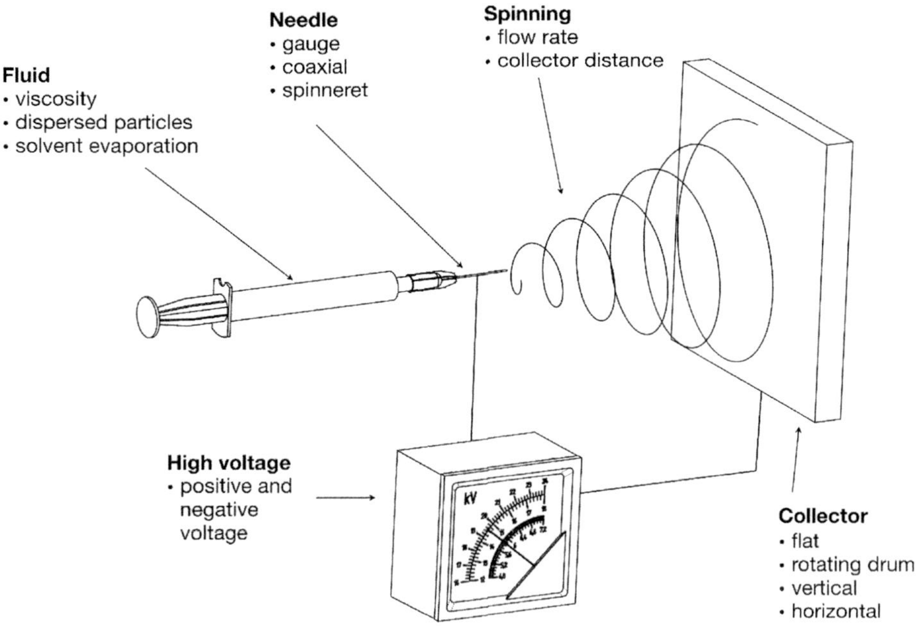

Figure 5.5 Schematic representation of electrospinning process. Basic components and parameters involved in electrospinning technique are illustrated.
From Rivero, P.J., Garcia, J.A., Quintana, I., Rodriguez, R., 2018. Design of nanostructured functional coatings by using wet-chemistry methods. Coatings 8 (2). https://doi.org/10.3390/coatings8020076.

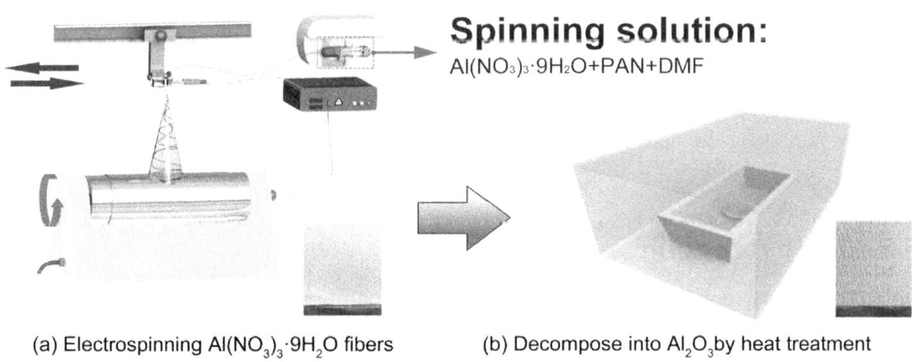

Figure 5.6 Fabrication of Al_2O_3 coating. (a) electrospun $Al(NO_3)_3 \cdot 9H_2O$ fibers on the zinc substrate. (b) Heat treatment of the electrospun films in muffle furnace in an atmospheric environment to obtain Al_2O_3 coatings.
From Yu, Y., Zuo, Y., Zhang, Z., Wu, L., Ning, C., Zuo, C., 2019. Al2O3 coatings on zinc for anti-corrosion in alkaline solution by electrospinning. Coatings 9 (11). https://doi.org/10.3390/coatings9110692.

electrospun coatings possess self-healing capabilities through the release of corrosion inhibitors encapsulated within micro/nanocontainers integrated within the electrospun fibers.

5.2.9 Dip coating

Dip coating is a simple method for coating the surface with various precursors. Although straight forward, this coating method becomes more versatile with the grasp of extensive microscopic processes occurring during dip coating. This understanding allows for modification of the final films by utilizing the coating process as a crucial intermediary for connecting the solution or sol's structure to the microstructure of the deposited film.

The process dip coating is a combination of three important stages:

1. Immersion and dwell time: the substrate is immersed into the precursor solution and allowed to remain in the solution for a certain duration or dwell time to enhance the interaction between the substrate and the coating precursor solution, ensuring complete wetting.
2. Drainage and deposition: In continuation of the dwell time in the precursor solution, the substrate is withdrawn at a constant rate, typically ranging from 1 to 10 mm $second^{-1}$ within the Landau–Levich draining regime. As the substrate is withdrawn, a fine layer of solution is drawn from the reservoir due to surface wetting and draining, that is, film deposition
3. Evaporation: after the deposition the, the coated surface was heat-dried to remove the solvent from the solution on the substrate forming an initial thin film. Following this, the coating will be undergone additional heat treatment for the removal any remaining organic residues present and encourage the crystallization of functional oxides (Brinker, 2013).

The thickness of the initial liquid layer and the resulting layers on the thin film can be controlled by adjusting the withdrawal speed (Landau and Levich, 1988). The ability to adjust thickness to predetermined gradients or more intricate profiles is simply by modulating the withdrawal speed Fig. 5.7.

The dip coating process is applicable to various types of precursor solutions such as sol–gel, metal organic decomposition (MOD), hybrid, etc. By utilizing sol-gel solutions, the greatest potential possibilities to alter film properties are achieved. This is done by adjusting the size and structure of the inorganic species within the sol along with the solvent(s).

5.2.10 Spin coating

Spin coating is a technique used for depositing or coating uniform thin films of organic materials in the size range of nanometers to micrometers on the surface of flat substrates. The typical procedure for coating involves dropping the liquid materials either at a slow spinning speed or under static conditions onto the surface of the substrate. Then, the substrate is rotated at a critical speed, allowing the coating material to be evenly distributed by centrifugal force. The substrate continues to spin continuously until the desired thickness is reached, and any excess fluid spins off the boundaries of the substrate.

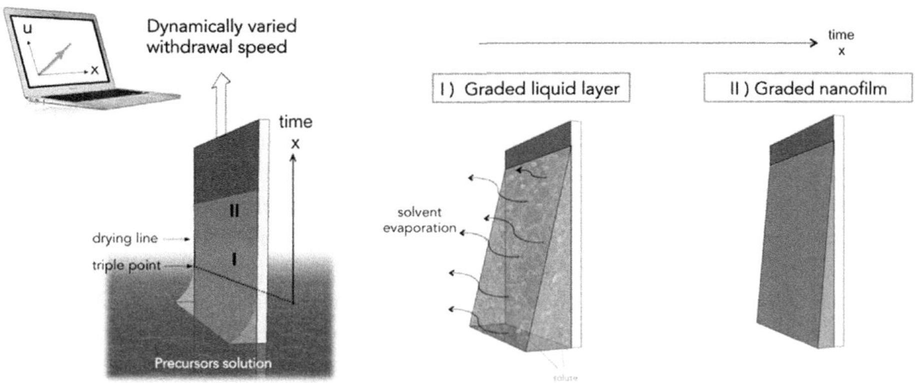

Figure 5.7 This figure represents the process of dip coating in the mode of acceleration for the formation of graded nanofilms in the controlled manner. Identified by two zones: Zone I, below the drying with the graded liquid layer; zone II, above the graded line with the graded nanofilm. From Faustini, M., Ceratti, D.R., Louis, B., Boudot, M., Albouy, P.A., Boissière, C., Grosso, D., 2014. Engineering functionality gradients by dip coating process in acceleration mode. ACS Appl. Mater. Interfaces 6 (19), 17102–17110. https://doi.org/10.1021/am504770x.

Spin coating dynamics can be outlined in four stages: deposition, spin-up, spin-off, and solvent evaporation.

- Deposition involves applying the coating solution onto the substrate, either statically or dynamically, where centrifugal force aids in dispersion.
- Spin-up initiates substrate rotation, ejecting excess fluid.
- Spin-off follows, spreading the remaining fluid and often leaving droplets at the edges.
- During evaporation, solvent evaporation occurring throughout the process dictates the final film thickness, with thicker films resulting from longer evaporation times due to increased concentration and viscosity. Thus, immediate evaporation at room temperature is crucial.

Deposition, spin-up, and spin-off typically occur sequentially, while evaporation persists throughout the entire preparation process. Both spin-off (flow controlled) and the evaporation step hold significant importance as they exert the most substantial influence on the final coating thickness.

The factors influencing the coating thickness include spin speed, acceleration, and the time of rotation. This technique is considered a versatile method to coat the surface of the substrate with various materials in the forms of polymers, nanoparticles, and biomaterials (Salehi et al., 2018). For example, spin coating of polymeric solutions containing HPMC and PVA on the PMMA substrate surface increases the hydrophobicity (contact angle) and surface free energy of the substrate (Kitsara et al., 2014).

5.2.11 Hybrid approaches

It is an interesting fabricating approach to implement multifunctionalized surface coatings utilizing the combination of different conventional methods to enhance surface properties like corrosion resistance, bioactivity, hydrophobicity, etc. Some examples of hybrid approaches using the combination of electrospinning and other conventional methods were described in this section.

Cellulose acetate (CA) nanofibers were deposited onto a metallic surface (304 stainless steel) by the electrospinning technique. Following this, the deposition of a composite consisting of chitosan (CHI) and hydroxyapatite (HAP) nanoparticles is achieved through dip-coating, resulting in CA nanofibers being fully incorporated within the HAP/CHI coating. This combines to form a densely packed uniform layer on the substrate. Subsequent electrochemical polarization and impedance tests were conducted in simulated body fluid (SBF), confirming that the presence of CA nanofibers significantly enhanced the corrosion resistance of the metallic substrate. Moreover, this type of coating shows promise for implementation in medical implants, offering potential stability in biomedical applications (Zhong et al., 2015).

Another novel approach is employing the combination of sol−gel coatings and electrospinning as a hybrid approach. In this approach the electrospinned poly vinyly alcohol (PVA) nanofibers infused with ceria particles and ceria nitrate combined with the sol−gel matrix composed on organic-inorganic precursors such as 3-glycidoxypropyltrimethoxysilane (GPTMS) and tetra n-propoxyzirconium (TPOZ). This combination enhances the barrier properties of the surfaces of the coated metal substrate aluminum alloy (AA2024-T3), resulting in enhanced corrosion resistance of the substrate to the saline environment owing to the prolonged exposure (Covelo et al., 2015).

5.3 Surface tuning

Surface degradation of a coating as a result of physiochemical reactions between its surface and the surrounding environment alters functional properties and hampers its industrial application. Hence, improved surface properties are substantial for several industrial applications and the long-term stability of nanostructured coatings. To accomplish this, surface tuning has been widely adopted and proven to be efficient in enhancing the integrity, stability, and other functional properties of the coating. Furthermore, contemporary advancements in nanotechnology have enabled surface tuning at the nanoscale, giving rise to advanced functional coatings.

Surface tuning can be simply understood as a method of introducing a functional moiety to a coating to tailor its properties for specific applications. It is vital to ensure that surface tuning manipulates surface properties without modifying the other structural properties of a coating. Surface tuning imparts superior properties that offer a significant advantage in numerous applications across various fields, including antiadhesion coatings, protective coatings, thin-film technology, technical fabrics, biomaterials, sensing (chemical and biosensing), and microelectronics. Herein, various surface tuning strategies such as chemical functionalization, nanoparticle functionalization, polymer modification, and layered material (LM) functionalization are briefly summarized in this section.

5.3.1 Chemical functionalization

Chemical functionalization refers to attaching a functional group to the surface of MNCs to tailor its properties for various applications. It is carried out using either covalent or noncovalent approaches (Xia and Jiang, 2008). Covalent functionalization relies on a covalent linkage between the chemical modifier and the coating, while noncovalent functionalization relies on physical interactions (Tan and Saltzman, 2004). Covalent functionalization is a widely adopted strategy as it results in stable surface modification. However, it may also affect other properties of the MNCs, so the choice of chemical functionalization should be based on the type of material and its application (Tan and Saltzman, 2004). Thiol and amine functionalization are widely studied chemical functionalization methods, which are briefly summarized as follows:

5.3.1.1 Thiol functionalization

Thiol functionalization is a postfunctionalization process in which the sulfur containing compounds are deposited on the nanostructured coating. Self-assembled monolayer is the well-established and attractive strategy for thiol functionalization. It is a simple bottom-up approach that involves formation of ordered assemblies of an active surfactant on the solid surface (Ulman, 1996). The common compounds utilized in the SAM based thiolation are n-Alkanethiol, α, ω-alkanedithiol, ω-mercaptoalkanol, ω-mercaptoalkane carboxylic acid, 4-methyl-4-mercaptobiphenyl, and 1,1-dialkyl-disulfide (Nsude et al., 2023). Thiol-ene coupling is another strategy for thiolation of nanostructured coatings. It is a click chemistry reaction that involves anti-Markovnikov addition of thiols to an alkene. Thiol-Michael addition and free-radical thiol-ene addition are two main thiol-ene coupling reactions. Of these, free-radical thiol-ene, particularly photo-initiated (UV light) thiol-ene coupling has been gaining prominence due to simple and rapid reaction kinetics (Lowe, 2014).

Thiolation utilizing these methods has been used for multiple applications. For instance, F. Rezaei et al. utilized thiol-ene coupling strategy to enhance corrosion and abrasion resistance of a low-emissivity film substrate (Rezaei et al., 2019). Thiolation of coatings also enables tuning the hydrophobicity/hydrophilicity of MNCs for specific applications. On the other hand, thiolation has also found applications in drug delivery and diagnostic applications (biosensors). In addition to this, thiolation has also increased the adhesive properties, stability, cellular uptake, and so on, making it an ideal functionalization strategy for variety of biomedical applications (Hock et al., 2022).

Thiol functionalization enhances various properties of nanostructured coatings; however, it presents some constraints by impacting the thermal and mechanical properties in some applications (Trey et al., 2011). Furthermore, the selection of ideal thiols and functionalization strategy is an intricate process due to the availability of several techniques and its crucial to select an appropriate method for efficient surface tuning of MNCs (Yaseen et al., 2021).

5.3.1.2 Amine functionalization

Amine functionalization is also a postfunctionalization strategy implemented for surface tuning of MNCs, primarily inorganic nanostructured coatings. Chemical grafting via covalent coupling is a widely adopted method for grafting amines on the surface of MNCs. It relies on several parameters such as the type of solvent, temperature, and type of amine precursor. While both natural and synthetic amines are used, 3-aminopropyltriethoxysilane (APTES), diethanolamine (DEA), and branched polyethyleneimine (PEI) have proved to be highly efficient for amine functionalization (Baptista et al., 2018). Like thiol functionalization, amine functionalization also contributes to stability, catalytic, optical, and electrical properties (Gu et al., 2020b).

The improvement in properties by amines is primarily attributed to an increase in the activated sites on the surface of nanostructured coatings. Amine-functionalized coatings have found potential applications in water treatment, gas capture, and biomedical coatings due to increased surface reactivity. In this regard, Bouazizi et al. (2019) employed APTES for the functionalization of graphene oxide sheets and demonstrated the active role of APTES in enhancing the catalytic activity for the reduction of 4-nitrophenol. Amines exhibit excellent gas sorption ability, making them applicable for the capture of CO_2 and other harmful gases. Apart from this, amine functionalization has also been found to be effective in improving the antibacterial activity of the coatings (Bouazizi et al., 2022).

Despite improving several properties, amine functionalization also poses some limitations. It impacts the morphological properties of the coating altering its functionality. In some instances, it also reduces surface area and pore size leading to poor gas sorption (Sunder et al., 2022). Hence, these challenges needed to be addressed for implantation of amine functionalized nanostructured coatings for wider and large-scale applications.

5.3.2 Polymer modification

Polymer-based functionalization offers numerous advantages by enhancing the optical, electrical, and catalytical properties of MNCs. There are three common strategies employed in polymer functionalization. This includes grafting to, grafting through, and grafting from (Hellgren et al., 2004). The grafting to method involves the covalent attachment of previously synthesized polymers to the surface. In contrast, the grafting from approach initiates polymerization of the monomer from the nanostructured surfaces. While grafting through involves radical copolymerization of a low-molecular-weight monomer with a functionalized macromonomer. Of these methods, grafting from has been widely utilized as it allows precise control of the graft polymer for tailored applications. Graft polymerization can be classified into three types based on the route of polymerization. They are chemical graft polymerization, plasma-induced graft polymerization, and photoinduced and radiation polymerization.

5.3.2.1 Chemical graft polymerization

In chemical graft polymerization, either free radical initiator or ionic initiator is used for polymer grafting. Initiator plays a critical role in these types of grafting as it determines the route of polymerization. In free radical graft polymerization, initiator such as ammonium persulfate (APS), benzoyl peroxide (BPO), potassium permanganate ($KMnO_4$) are used to generate free radicals, which then react with monomers and initiates the polymerization (Chu et al., 2006). While in ionic initiated graft polymerization, ionic initiators such as Lewis's acids, strong based, and metal carbonyls are used to initiate either cationic or anionic graft polymerization (Choudhary et al., 2020). Ionic grafting demands strict reaction conditions (inert, anhydrous and low reaction temperatures) whereas free radical grafting does not require such stringent reaction conditions. Therefore, free radical grafting has been widely used for polymer grafting (Šebenik, 1998).

5.3.2.2 Plasma induced graft polymerization

Plasma is a state of matter that majorly consists of charged particles. It can be divided into thermal and nonthermal. Since thermal plasma operates at high temperature, it might possibly degrade the material surface. Hence, nonthermal plasma which generated by corona discharge, dielectric barrier discharges, and so on is widely adopted for polymer grafting of MNCs (Desmet et al., 2009). In plasma-induced grafting, surfaces of MNCs are plasma activated by treating with the plasma at desired conditions. Subsequently, monomer will be polymerized on to the plasma activated surface to produce polymer grafted MNCs. This method is confined to a depth of few nanometers and does not interfere with other functional properties of MNCs (Gupta et al., 2002a,b).

5.3.2.3 Photografting

As name suggests, photografting utilizes UV radiation for polymer grafting. Some MNCs are inert making it difficult to modify their surfaces. In these cases, photografting is employed to break the chemical bonds on the surface of MNCs, thus making it reactive for further polymer grafting. Photografting can proceed either with photoinitiator or without photoinitiator depending upon the type of polymer and surface. For instance, in case of light reactive or sensitive surfaces like, grafting occurs without the need of photoinitiator. However, non-light-sensitive surfaces mandate the use of photoinitiators for photografting (Nsude et al., 2023).

In addition to photografting, radiation grafting, and electron beam grafting are also used for polymer grafting as shown in Fig. 5.8. Radiation grafting typically employs gamma radiation to irradiate the surface resulting in the generation of surface free radicals and peroxide groups due to molecule ionization (Biazar et al., 2010). Subsequently, surface can be grafted with desired polymer by using appropriate monomer. Similarly, electron beam grafting utilizes electron beams to generate free radicals to initiate the polymerization.

The main advantage with photo-grafting is the ability to control the thickness of the graft polymer. By changing the reaction parameters, one can precisely control the

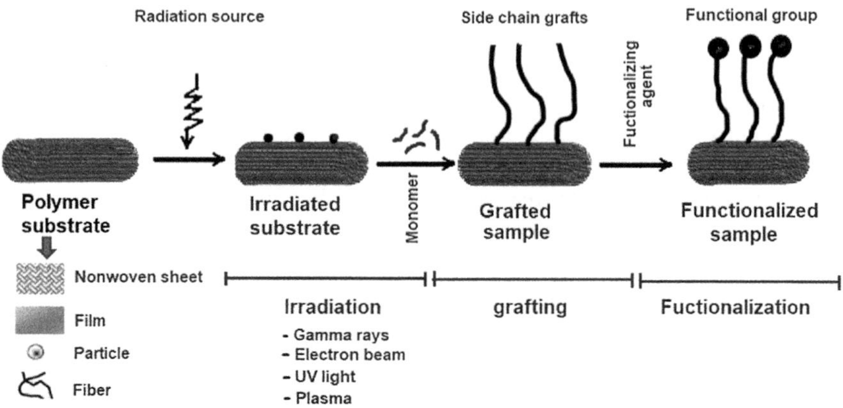

Figure 5.8 Strategies in photografting of polymers. Schematic diagram illustrating the various radiation sources for initiating photografting, monomer grafting, and functionalization.
From Yusefi, M., Nasef, M.M., Tareq, M.A., Gupta, B., Shameli, K., Ali, R.R., Ting, T.M., El Enshasy, H.A., 2022. Bibliometrics of functional polymeric biomaterials with bioactive properties prepared by radiation-induced graft copolymerization: a review. Polymers 14 (22). https://doi.org/10.3390/polym14224831.

depth of the polymer coating on the surface for tailored applications (Nasef and Hegazy, 2004). For further reading on the detailed mechanisms involved in each method, we direct readers to refer (Saldívar-Guerra et al., 2013; Purohit et al., 2023).

Polymers used in functionalization can be categorized as natural and synthetic polymers. Natural polymers include chitosan, dextran, and cyclodextrin, while synthetic includes polyethylene glycol (PEG), polyvinyl alcohol (PVA), polyethyleneimine (PEI), poly-vinylpyrrolidone (PVP), polyaniline (PANI), polypyrrole (PPY), and so on. Given that these polymers exist in diverse forms, selection of polymers is crucial. For instance, molecular weight of polymer is a critical factor that influences various aspects of the polymer such as glass transition temperature (T_g), solubility, viscosity, and film-forming characteristics. Higher molecule weight polymer often results in improved adhesion, toughness, and durability of the coatings (Avila et al., 2021). However, other factors such as UV stability, hardness, chemical resistance, resistance to wear, and corrosion also need to be considered for optimal selection of the polymer. Hence, it is essential to find a balance of all desired properties of polymer to achieve efficient functionalization of nanostructured coatings.

The polymer-based surface functionalization has been employed for wide range application spanning from corrosion resistance to biomedical applications. Insulating polymers have been employed for antiwetting, self-cleaning, and antifouling applications (Steele et al., 2012). Likewise, conducting polymers have demonstrated high oxidation stability, photostability, thermal stability, and biocompatibility, making them promising for MNCs. These polymers have found applications in corrosion-resistant coatings, drug delivery systems, energy storage systems, and so on (Deshpande et al., 2014). On the other hand, biopolymers have employed with specific focus on biomedical applications such as drug delivery, tissue engineering, and wound dressing applications (Kumar and Sharma, 2022). Stimuli-responsive polymers have been also incorporated into the nanostructured coatings for self-healing properties. These polymers respond to mechanical stress or any other deformation and initiates the healing process of the coating (Cho et al., 2009).

The application of polymer functionalization for practical applications has been hindered by various limitations associated with it. Firstly, mechanical and tensile strength of polymer are less compared to that of metal and other functionalization moieties (Fu et al., 2019). Scaling up is also a formidable challenge as it is difficult to control molecular weight and properties of polymers with existing technologies (Makvandi et al., 2021). Furthermore, selection of polymer is an intricate process and crucial to tailor the nanostructured surfaces for specific applications. For instance, conducting polymers can be used as because of their conjugated backbone, while porous polymers cannot be used for corrosion resistant coatings as they promote corrosion (Kumar and Sharma, 2022). Hence, these factors need to be considered for employing polymer functionalization for nanostructured coatings.

5.3.3 Nanoparticles functionalization

Nanoparticles can substantially modify the properties of nanostructured coatings, leading to a wide range of applications. The common strategies employed for nanoparticle-based functionalization of nanostructured coatings include sol-gel, dip coating, electrodeposition, sputtering, vapor deposition, etc (Yilmaz and Yorgancioglu, 2018). Functionalization can be achieved either by depositing the presynthesized nanoparticles onto the coating or via in-situ synthesis of nanoparticles on the surface of the coatings. The latter approach provides homogeneous distribution of nanoparticles on the coatings. However, it is limited by the control of the size and morphologies of the nanoparticles. The former approach provides better size control but suffers from other bottlenecks such as nonuniform deposition and the use of molecule linkers. Thus, the choice of method should be based on the intended application. Various nanoparticles have been investigated for their surface tuning capabilities and can be grouped into metal and metal oxide nanoparticles.

Metallic nanoparticles are the most promising nanoparticles for surface tuning because of the unique optical, catalytical, and electronic properties. In addition, all these properties of metal nanoparticles can be modified by controlling the size and shape of the nanoparticles offering countless applications for nanostructured coatings. Noble metals like gold, silver, palladium, and platinum have been widely studied followed by other metal nanoparticles. The high catalytical activities of metal nanoparticles synergy well with that of the nanostructured coatings and improve the several electrical, mechanical, and optical properties enabling multiple applications. For instance, noble metal nanoparticles functionalization has enhanced the selectivity and sensitivity of chemical and gas sensors. Additionally, metal nanoparticles also improved photovoltaic performances of solar cells by preventing electron hole recombination processes (Chen et al., 2009). Beyond these applications, metal nanoparticles have also been employed for anticorrosion (Cai et al., 2020), biocompatibility, and antimicrobial properties of nanostructured coatings (Li et al., 2021). Furthermore, functionalization with plasmonic nanoparticles such as gold and silver enable biomolecule conjugation without the need of complex chemical reactions such as EDC-NHS cross linking, thiol-maleimide conjugation, and click chemistry.

Metal oxides (MOs) are another popular choice of nanomaterials for surface tuning of nanostructured coatings broadening the application of the coatings. MOs possess high chemical stability, photostability, low toxicity, biodegradability, broad range of radiation absorption, and high biocompatibility. Zinc oxide, titanium dioxide, iron oxides, aluminum oxide, and copper oxide are widely applied for surface tuning of the coatings. The particle size and crystal structure specifically oxygen vacancies are primary parameters of MOs that influences the properties of the surface coatings (Yang et al., 2005). MOs exhibits high catalytical activity identical to metal nanoparticles which have led to improved antimicrobial, anticorrosive, anti-UV degradation, and antistatic properties of the coating. Furthermore, adhesion and self-healing capabilities of coatings also have been improved by MOs as a result of chemical interactions and metal oxide reinforcement effect (Rahman, 2020). The improved properties have led to applications of MO-functionalized coatings in sensing, photocatalysis, energy storage, and electronics. The semiconducting nature of metal oxides enables electron transfer under photoirradiation resulting in a photocatalytic chemical reaction and also enhances the photostability of nanostructured coatings (Sandua et al., 2022). In few instances, MO functionalization can also result in formation of heterojunctions that improves sensitivity in chemical sensing and photocatalytic efficiency by controlling electron-hole recombination process as highlighted in this review (Laurenti et al., 2017).

Both metal nanoparticles and MO-based functionalization have been possible in academic pursuits, and practical applications are severely limited due to the following factors: biocompatibility and toxicity, difficulty in controlling the morphology of nanoparticles, scaling up complexity, and high cost incurred in the synthesis of some nanoparticles (López-Rubio et al., 2019).

5.3.4 Layered materials functionalization

Layered materials are solid materials that exhibit highly anisotropic bonding. These materials consist of two-dimensional sheets strongly bonded internally and weakly bonded with adjacent sheets. Although there are 1800 LMs, only few LMs which includes graphene, hexagonal boron nitride (hBN), black phosphorus, transition-metal dichalcogenide, layered double hydroxides (LDH), and MXenes have been studied thus far (Montblanch et al., 2023). Of these, graphene has been widely used for tuning the nanostructured coatings.

Graphene is a 2D-layered material consisting of single layer of carbon atoms arranged in a hexagonal lattice nanostructure. Graphene is hydrophobic, while its derivatives namely graphene oxide (GO) and reduced graphene oxide (rGO) are hydrophilic, hence applied for broader applications. The addition of graphene has resulted in superior mechanical properties which includes fracture toughness, tensile strength, and young's modulus fatigue resistance. The improved performance was attributed to high specific surface area, enhanced interfacial adhesion, and the planar geometry of the graphene (Lin et al., 2019). In addition, graphene also acts as an impermeable barrier and prevents air oxidation enhancing the anticorrosion properties of the coatings (Yu et al., 2014). In contrast to graphene, its derivatives GO and rGO consists of functional groups (carboxyl and epoxy) that enable them to react chemically with the nanostructured coatings improving electromagnetic radiation absorption and electrocatalytical properties (Karami et al., 2023).

Beyond graphene, other layered materials also have been employed functionalization of nanostructured coatings. For instance, hBN, which is also known as "white graphene" have been utilized to prevent galvanic corrosion. Likewise, all other layered materials such as TMDs, MXenes, phosphorene, and LDH have also been increasingly applied for improving corrosion resistance, adhesion, lubrication, wear resistance, and other properties of coatings for wider range of applications. Layered material–based functionalization has been used to modify the properties of filtration membranes such as hydrophilicity, pore size, and charge effects for water treatment applications. It also been demonstrated that rejection of various toxic substances has improved due to the functionalization (Chu et al., 2017). It also been employed for enhancing photocatalytic performance by increasing absorption of reactants due to large surface area (Jing et al., 2013). Electrocatalytic properties of coatings have also been improved considerably enabling applications in chemical sensing.

However, as indicated in various research articles, research on layered materials other than graphene is still in its early stages, and further research on optimizing the properties of these layered materials is essential for their application in surface tuning of nanostructured coatings. Also, large-scale nonuniformity affecting consistency of nanostructured coating, manufacturing challenges, scaling up issues, and high surface energy are some limitations associated with layered material–based functionalization (Choi et al., 2022).

In summary, each functional moiety has their own advantages and disadvantages. These can be mitigated by integrating aforementioned functional moieties that could solve many of these challenges due to synergistic properties. However, it may also undermine certain properties of nanostructured coatings. Hence, selection of suitable method is essential for accomplishing effective surface tuning of nanostructured coatings. The further exploration can be directed to find how these surface functionalization strategies can be manipulated using existing technologies or developing novel technologies for precise control over surface tuning for intended applications and also to extend its application in industrial scale.

5.4 Structure modifications of nanomaterials

The surface properties of synthesized nanomaterials often lack desired characteristics such as wettability, biocompatibility, catalytic activity, adhesion properties, electrical conductivity, and magnetic properties. Consequently, they require modification before application or further processing, such as coating with functional materials. Modification involves altering the chemical structure, composition, and morphological properties to achieve the desired surface finish. This can be achieved through specific processing techniques and assembly methods tailored to the type of nanomaterial, with the aim of improving, enhancing, and modifying their surface properties (Lord et al., 2010; Guerrini et al., 2018; Nedela et al., 2017; Variola et al., 2011). Due to the unique surface characteristics of nanomaterials, modification becomes essential during the preparation or dispersion of nanoparticles to enhance stability. Surface modification significantly contributes to the preparation, adjustment, and preservation of nanoparticles. Some important structure modification methods include template-assisted methods, strain engineering, gradient coatings, plasma treatment, ion beam irradiation, nanotexturing, and nanomaterials doping, among others.

5.4.1 Template-assisted modification

The template-assisted method enables precise control of morphology and tunable properties in metal oxide nanostructures. This is achieved by incorporating soft and hard templates, leveraging their structural and characteristic features. In the template-assisted method, the porous surface of the template serves as a directing agent during the fabrication process, restricting the growth of crystals and nucleation period to produce mesoporous materials ranging from microsized to nanosized dimensions (Kaur et al., 2022).

The emergence of template-assisted nanostructure synthesis introduces an intriguing and innovative synthetic technique known for its easy handling, implementation, and resilience to reaction conditions. Literature highlights that the presence of a template affects the particle size and crystal structure of metal-oxide nanoparticles. Additionally, this synthetic strategy provides enhanced control over shape, size, and morphology by restricting the growth of the crystal structure and nucleation period during synthesis. Table 5.1, compiling literature studies, underscores the potential of the template-assisted method not only to impact particle size but also to influence the phase/crystal structure compared to template-free synthesis.

The template-assisted methodology has the potential to generate numerous morphologies, including nanowires, nanofibers, nanotubes, and nanoleaves, depending on the chosen template material. This technique basically involves three steps. The first step entails preparing a template, such as anodic aluminum oxide. In the second step, synthesis techniques like hydrothermal, sonochemical, electrochemical deposition, and CVD are employed for template-mediated synthesis of metal-oxide nanostructures. Finally, in the third step, pure nanostructured materials are obtained by removing the template through physical and chemical methods such as etching, sintering, and dissolution (Xie et al., 2016). The distinct steps involved in the template-assisted methodology for mesoporous material synthesis have been illustrated in Fig. 5.9.

The template-assisted synthesis of nanostructured metal oxides can be achieved through both top-down and bottom-up methods, as depicted in Fig. 5.10. Various techniques, including sol–gel, microwave, electroless deposition, CVD, and hydrothermal/solvothermal methods, can be employed to prepare metal-oxide nanostructures with specific target responses. However, literature findings suggest that, in most cases, the template-assisted synthesis technique is predominantly carried out using

Table 5.1 Summary of the template-assisted synthesis method impact on the particle size and crystal structure in chosen MO nanomaterials.

	Particle size		Phase/crystal structure		
Synthesized MO nanomaterial	**Without template**	**With template**	**Without template**	**With template**	**References**
NiO	31 nm	8 and 26 nm	Face centered cubic	Hexagonal	(Cao et al., 2007; Abbasi et al., 2019)
TiO_2	Approx. 100 nm	30–40 nm	Rutile	Anatase	(El-Naggar et al., 2016; Koh et al., 2008)
In_2O_3	20–30 nm	15 nm	Rhombohedral	Cubic	(Han et al., 2014; Xiao et al., 2016)
CeO_2	15–36	4 nm	Face centered cubic	Cubic fluorite	(Khan et al., 2011; Sifontes et al., 2011)
ZnO	28 nm	20 nm	Hexagonal	Wurtzite	(Sharma et al., 2010; Rawal, 2015)
CO_3O_4	20–37 nm	10 nm	Face centered cubic	Cubic	(Khalil et al., 2020; Khun et al., 2015)
SnO_2	18 nm	6–15 nm	Tetragonal rutile	Tetragonal rutile	(Honarmand et al., 2019; Wang et al., 2005b)
Mn_3O_4	25 nm	5 nm	Tetragonal	Tetragonal hausmannite	(Xue and Zhang, 2019; Salavati-Niasari et al., 2008)

From Kaur, A., Bajaj, B., Kaushik, A., Saini, A., Sud, D., 2022. A review on template assisted synthesis of multi-functional metal oxide nanostructures: status and prospects. Mater. Sci. Eng. B 286. https://doi.org/10.1016/j.mseb.2022.116005.

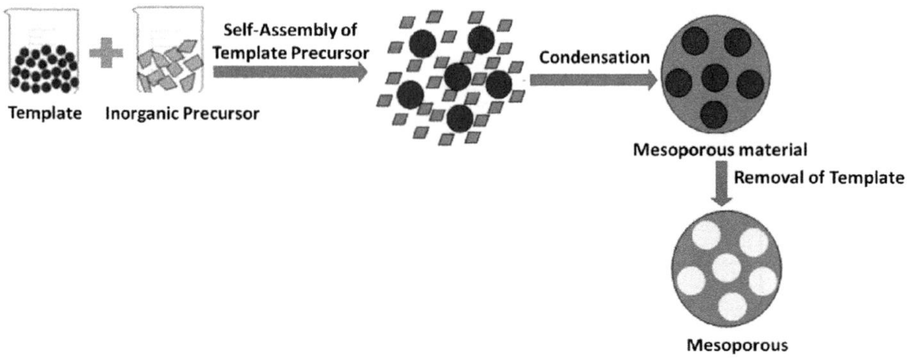

Figure 5.9 Illustration of steps involved in template-assisted synthesis method for the fabrication of metal oxide nanostructures (Kaur et al., 2022).
From Kaur, A., Bajaj, B., Kaushik, A., Saini, A., Sud, D., 2022. A review on template assisted synthesis of multi-functional metal oxide nanostructures: status and prospects. Mater. Sci. Eng. B 286. https://doi.org/10.1016/j.mseb.2022.116005. Reproduced with permission.

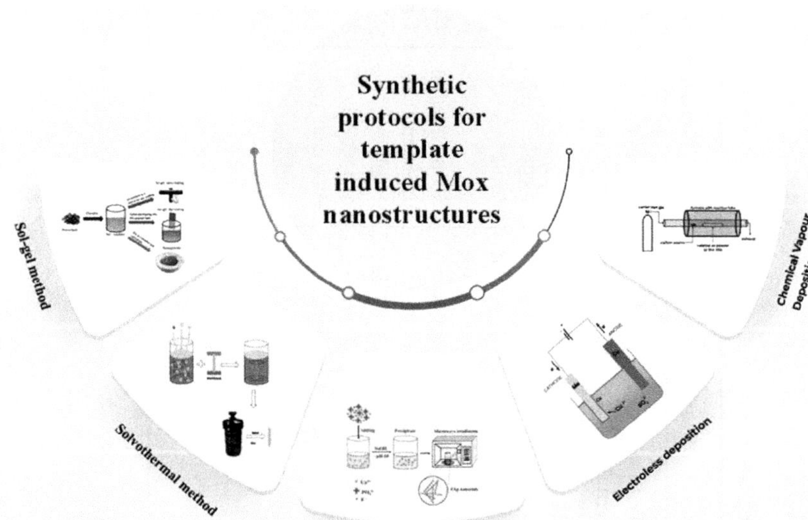

Figure 5.10 Different synthetic methods for the fabrication of template-assisted metal-oxide nanostructures (Kaur et al., 2022).
From Kaur, A., Bajaj, B., Kaushik, A., Saini, A., Sud, D., 2022. A review on template assisted synthesis of multi-functional metal oxide nanostructures: status and prospects. Mater. Sci. Eng. B 286. https://doi.org/10.1016/j.mseb.2022.116005. Reproduced with permission.

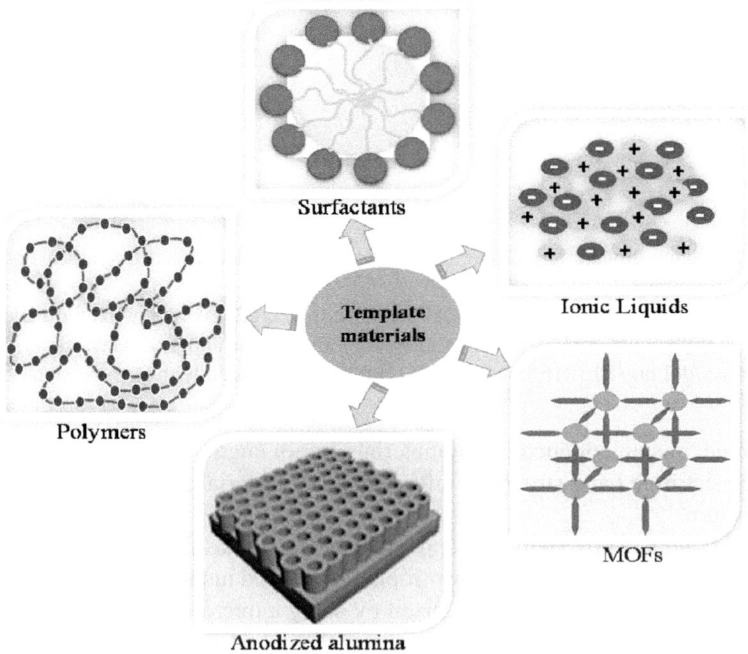

Figure 5.11 Illustration of the different template materials employed in the synthesis of MOx nanostructures.
From Kaur, A., Bajaj, B., Kaushik, A., Saini, A., Sud, D., 2022. A review on template assisted synthesis of multi-functional metal oxide nanostructures: status and prospects. Mater. Sci. Eng. B 286. https://doi.org/10.1016/j.mseb.2022.116005. Reproduced with permission.

hydrothermal and sol–gel techniques. Fig. 5.11 illustrates the various template materials employed, such as surfactants, ionic liquids, MOFs, anodized alumina, polymers, etc., for the synthesis of metal-oxide nanostructures. Furthermore, Fig. 5.12 represents the polymeric template used for the synthesis of metal-oxide nanostructures.

5.4.2 Strain-engineered modification

Strain engineering involves intentionally introducing mechanical strain or stress to nanomaterials to modify their properties and enhance their performance for specific applications. This modification technique is based on altering the structural arrangement or lattice parameters of nanomaterials that can bring about changes in their electronic, mechanical, and optical properties. The strain engineering can be achieved through various methods such as mechanical deformation, application of external pressure, or interfacing with substrates with different lattice constants. By inducing strain can effectively tune the bandgap, catalytic activity, conductivity, and other critical characteristics of nanomaterials. This approach is particularly relevant in the development of advanced materials for electronics, photonics, catalysis, and energy storage.

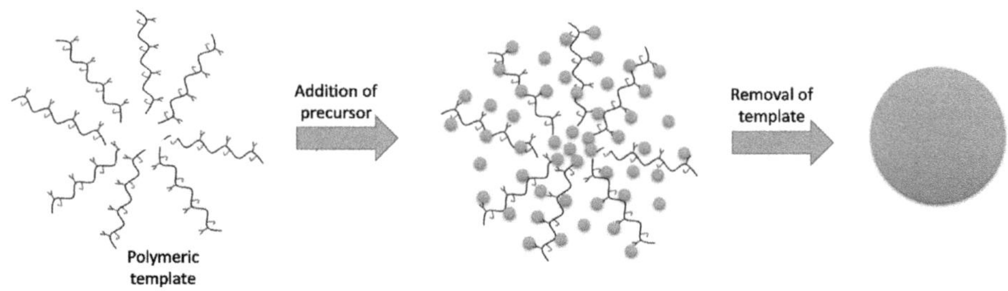

Figure 5.12 Diagrammatic representation of polymeric template assisted synthesis of metal-oxide nanostructures.
From Kaur, A., Bajaj, B., Kaushik, A., Saini, A., Sud, D., 2022. A review on template assisted synthesis of multi-functional metal oxide nanostructures: status and prospects. Mater. Sci. Eng. B, 286. https://doi.org/10.1016/j.mseb.2022.116005. Reproduced with permission.

For example, in semiconductor nanomaterials, strain engineering can be employed to tailor the electronic band structure, influencing the material's conductive or semiconductive nature.

A study conducted by Haibo Hu and coworkers suggests that the hierarchical nanostructures of Ag–ZnO, featuring anisotropic growth and tunable properties induced by strain, have been successfully synthesized by using a microwave synthesis technique. The surface morphology of ZnO can be engineered, and strain can be induced in the material by adjusting the microwave irradiation period. The strained ZnO nanospikes facilitate the integration of numerous Ag nanoparticles due to the expanded surface area and greater number of active sites on the hierarchical ZnO surfaces. These hierarchical Ag–ZnO nanostructures exhibit outstanding sensing capabilities for detrimental organic molecules, glucose, and pharmaceutical waste, achieving detection limits of 10^{-18} M, 10^{-10} M, and 10^{-13} M, respectively (Hu et al., 2011).

5.4.2.1 Top of form

The anisotropic growth and strain-induced hierarchical nanostructure of ZnO can serve as an effective matrix for embedding a substantial quantity of Ag noble nanoparticles, facilitating the generation of hot spots and strong interactions with electromagnetic radiations (Anker, 2008). As a result, numerous toxic molecules are adsorbed on the surface, significantly enhancing Raman signals. Fig. 5.13 illustrates the strain-induced changes in the nanostructure, leading to a gradual shift in Raman peaks. Consequently, the SERS-based hierarchical nanostructure demonstrates the capability to detect ultra-low concentrations of pollutants and chemical species in water bodies by modifying the metal oxide matrix of nanomaterials (Hu et al., 2011).

5.4.3 Gradient coating modification

Extensive research has been conducted on nanostructured (NS) materials in both scientific research and industrial domains due to their unique mechanical and chemical

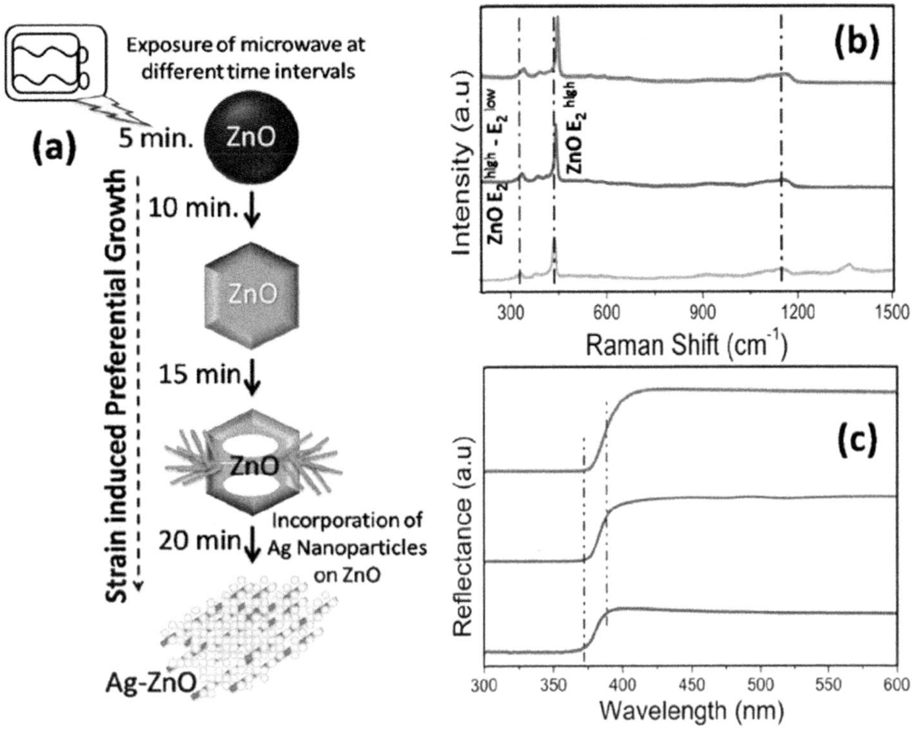

Figure 5.13 Schematic representation of strain induced structure modification. The structural transformation of ZnO nanostructures was investigated at various microwave exposure durations (a), and the induced strain in the nanostructures was analyzed through Raman spectroscopy, revealing a gradual shift in the Raman spectrum (b). Additionally, UV–Vis diffuse reflectance spectroscopy (UV–vis DRS) was employed to examine the strain-induced changes, demonstrating a progressive shift in the wavelength (c) (Ajay Rakkesh et al., 2020). From Ajay Rakkesh, R., Durgalakshmi, D., Karthe, P., Balakumar, S., 2020. Anisotropic growth and strain-induced tunable optical properties of Ag–ZnO hierarchical nanostructures by a microwave synthesis method. Mater. Chem. Phys. 244. https://doi.org/10.1016/j.matchemphys.2020.122720. Reproduced with permission.

properties compared to conventional coarse grain counterparts (Gleiter, 2000; Jeong et al., 2001; Kumar et al., 2003; Lei, 2004). These materials find extensive applications in bulk applications and as coatings on engineered substrates. However, coatings used for thermal, wear, or corrosion resistance, as well as in microelectronics, may encounter stress concentrations at the interface due to property disparities between the coating and the substrate, potentially leading to interface failure.

One potential solution to this challenge is the incorporation of gradient materials, known to enhance material lifespan (Kokini and Choules, 1995; Wang et al., 2005a,b). Gradient materials offer a method for improving material properties by modifying characteristics as the composition and structure change from the bottom to the surface. To meet advanced demands, it is essential to fabricate an ideal material through simple processes. The electroplating technique, renowned for its simplicity and versatility, proves to be a useful method for producing nanostructured materials (Erb, 1995; Koch et al., 2005). Various nanostructured metals, including nickel (Kumar et al., 2003; Gu et al., 2006), cobalt (Jeong et al., 2001), and copper (Lei, 2004; Zhang et al., 2008), have been manufactured and investigated for their mechanical behavior and deformation mechanisms.

The introduction of varied concentrations of grain-refining agents, such as saccharin or sodium citrate, into the electrolyte during electrodeposition facilitates the creation of nanostructures with different grain sizes (Abraham et al., 2002). Earlier literature studies emphasized the significant impact of grain size and distribution on the thermal stability, mechanical, corrosion behavior, and tribological properties of nanostructured materials (Lei, 2004; Wang et al., 2005a,b; Shen et al., 2008; Mishra and Balasubramaniam, 2004). However, there exists a gap in the literature studies regarding nanostructured coatings produced with different saccharin concentrations during electrodeposition, leading to a gradual shift in grain size from conventional crystalline structure to nanocrystalline structure.

For instance, a layered nanostructured-Ni was generated by intermittently introducing 2-butyne-1,4-diol (acting as the grain refiner) into the electrolyte during direct-current electrodeposition process (Gu et al., 2007). This layered nanostructured-Ni sample consists of alternating layers of ultrafine and nano-sized grains, resulting in a material with strength and ductility positioned between those of ultrafine-grained and nano-sized Ni (Qin et al., 2008).

5.4.4 Plasma treatment modification

Plasma treatment modification of nanomaterials involves subjecting them to a plasma environment to induce changes in their surface properties, structure, or chemical composition. Plasma, the fourth state of matter, is characterized by ionized gases containing a mixture of ions, electrons, and neutral species. This process, commonly known as plasma modification or plasma etching, holds significant promise for various applications.

Coating medical implant surfaces with silver can effectively prevent biofilm formation and many bacterial infections (Taglietti et al., 2014). It is crucial to ensure precise control over biomaterial surfaces as they play a vital role in the interaction between bacteria, body tissue, and medical devices.

Plasma treatment methods offer an interesting platform for modifying material surfaces through the deposition of thin films. These films enable chemical surface modifications on various materials, even on intricately shaped and flexible scaffolds, with no risk of cracking. Importantly, these structural modifications can be easily controlled concerning the density and type of functional groups without altering the inherent bulk properties of the substrate (Hamerli et al., 2003; Yang et al., 2002; Duque et al., 2010).

Moreover, plasma treatment methods, which are solvent-free, facilitate surface functionalization without the use of hazardous solvents (Morent et al., 2011). Specifically, plasma-enhanced CVD (PECVD) is acknowledged as an effective technique for producing thin films that are adherent, sufficiently flexible, and biocompatible, minimizing the static forces on adjacent tissues (Lewis et al., 2000). PECVD enables plasma polymerization (PFM), resulting in covalently bonded reactive polymeric thin films on surfaces. These PFM-modified surfaces, adaptable to materials like polydimethylsiloxane (PDMS), can undergo micro- and nanostructuring. They exhibit an affinity for interacting with amine-bearing molecules, leading to the development of bioactive and biocompatible surfaces with antibacterial properties (Duque et al., 2010; Horna et al., 2012; Langer and Tirrell, 2004).

5.4.5 Ion beam irradiation

As the demand for energy storage devices increases, there is continual exploration of various material approaches to improve performance, including increasing surface area, introducing surface defects, and enhancing the conductivity of active electrode materials. Several studies have demonstrated enhancements in the charge storage capabilities of MnO_2 nanoparticles through the application of low-energy (5 keV) ion beam irradiation. This process induces alterations in the morphology of the nanoparticles and introduces a notable quantity of surface defects.

Conductivity measurements reveal an overall improvement in the electrical conductivity of the altered nanoparticles, coupled with an expansion in the specific surface area resulting from ion beam irradiation (Wang et al., 2022b; Gupta et al., 2022a,b; Fu et al., 2023; Das et al., 2021). Recent research has extensively examined energy storage systems, with modern supercapacitor technology presenting advantages over traditional batteries. These advantages include high power density, extended life cycles, and exceptionally rapid charge-discharge rates (Winter and Brodd, 2004).

Supercapacitors are categorized into different types, such as electrochemical double-layer capacitors (EDLCs), pseudocapacitors, and hybrid-type capacitors, which vary based on their charge storage mechanisms and electrode materials. Carbon-based materials are commonly utilized in EDLCs, providing a high surface area and excellent cyclic stability but with relatively low specific capacitance values (Yang et al., 2014). Transition metal oxides, like MnO_2, find frequent application in pseudocapacitors. In this context, faradic reactions store charges, yielding high specific capacitance values, albeit with lower stability when compared to EDLCs (Cottineau et al., 2006; Lang et al., 2011; Wu et al., 2014; Wang et al., 2015; An et al., 2019). The drawbacks of low surface area and poor electrical conductivity in metal oxides necessitate innovative tools for enhancing nanomaterial properties (Wang et al., 2019, 2022a).

Defect and morphology engineering, as a novel tool, prove essential for improving the conductivity interaction sites and surface area of metal oxides to reduce charge transfer resistance. Ion beam irradiation has emerged as a novel technique in recent years for altering the physical properties and surface morphology of various nanomaterials (Majhi et al., 2022). Adjusting parameters such as ion quantity, ion types, range of energy, and precise control over the process is achievable. Energetic ions, penetrating the surface of the material, dissipate their energy through electronic and nuclear-stopping power mechanisms, leading to events like defect creation, bond breakage, carbonization, sputtering, ion beam mixing, and cross-linking (Abe et al., 2015; Smith et al., 2017; Sarmah and Kumar, 2019).

In the field of energy storage applications, ion beam irradiation has proven effective in enhancing charge transfer properties and surface area by creating porous structures and incorporating defects into nanomaterials (Wang et al., 2022b; Gupta et al., 2022a,b; Fu et al., 2023; Das et al., 2021). The ion beam irradiation approach is particularly promising, offering a means to alter the surface of nanostructured transition metal oxides favorably without the need for further functionalization or chemicals. This method introduces defects and alters surface morphology, thereby improving properties related to electrical charge storage (Maharana et al., 2023).

5.4.6 Nanotexturing

Nanotexturing refers to the process of creating specific patterns, structures, or surface features on materials at the nanoscale. In the area of the modern food industry, plastic packaging plays a vital role in ensuring the preservation of food quality and safety. These packaging products are categorized as high-touch surfaces (HTSs), given their extensive use in daily life. Consequently, they have the potential to serve as reservoirs for various pathogens, posing a risk of infection spread (Liu et al., 2022; Casini et al., 2018; Birkett et al., 2022). The ongoing global pandemic emphasizes the pressing need for antibacterial HTS to mitigate and prevent the emergence of future diseases. Conventional approaches to achieve HTS with inherent antibacterial effects often involve using antibiotics or coatings of metals and metal oxides (Vincent et al., 2018; Pedroni et al., 2021; Jann et al., 2021; Fan et al., 2021). However, challenges persist due to inadequate active agent concentration, antimicrobial resistance (AMR) in bacteria, and the potential contamination of food products with toxins. Recognized as a significant global cause of death, surpassing even HIV and malaria, the longstanding threat of AMR requires innovative approaches to curb bacterial pathogen transmission. Unintended food contamination can pose substantial public health risks. Nature-inspired surfaces, exemplified by nanostructures with diverse shapes and dimensions found on cicada and dragonfly wings, offer a potential solution to these issues.

The surfaces of these organisms utilize a contact-killing mechanism, where mechanical stress is applied to bacterial cells on the external surface of the packaging, not in direct contact with the food (Zhou et al., 2021a; Modaresifar et al., 2019; Luo et al., 2020). Recent research emphasizes the efficacy of such nanotextured surfaces against both bacteria and viruses (Zhou et al., 2021b; Ma et al., 2021). Various methods, including lithography, hydrothermal synthesis, and plasma etching, have been documented in the literature for reproducing antibacterial bio-inspired surfaces on different substrate materials (Dundar Arisoy et al., 2018; Michalska et al., 2021; Jaggessar and Yarlagadda, 2020; Bright et al., 2022; Ma et al., 2022; Akdoğan and Şirin, 2021; Hirano et al., 2021). Plasma surface engineering techniques are solvent-free and suitable for thermally sensitive materials such as polymers (Dimitrakellis et al., 2021; Palumbo et al., 2019). The absence of chemical solvents reduces the risk of contamination, and the adaptability of plasma technology permits diverse configurations, enhancing potential for industrial-scale applications (Weltmann et al.,

2019). A particular study demonstrates the use of a plasma-based approach to achieve a nanotextured antibacterial surface for poly(butylene succinate) (PBS), a promising biodegradable polymer for food packaging. PBS is appealing due to its thermal and mechanical properties comparable to conventional nonbiodegradable polymers like polyethylene (PE) and polypropylene (PP) (Petchwattana et al., 2016; Wei et al., 2020). Furthermore, PBS is applied in biomedical contexts where nanotextured bactericidal surfaces can address challenges associated with antimicrobial resistance (AMR) (Ma et al., 2022; Rafiqah et al., 2021).

5.4.7 Nanomaterials doping

The process of incorporating dopant or impurity elements into interstitial sites or the lattice of host photocatalytic materials, typically semiconductors, is known as the doping process. Dopants include a variety of metal ions, such as alkali ($M = K+$, $Na+$), alkaline-earth metals ($M = Ca2+$, $Mg2+$, $Sr2+$), transition metals ($M = Ag+$, $Zn2+$, $Cu2+$, $Al3+$, $Zr4+$), and inner transition ($M = Ce4+$, $Nb5+$) metal ions. Additionally, nonmetals like phosphorus (P), sulfur (S), nitrogen (N), oxygen (O), and carbon (C) can serve as dopants in photocatalysts. Various methods, such as co-precipitation, solvothermal, hydrolysis sol−gel, spin coating and thermal decomposition, can be utilized for the doping process (Ismael, 2019). The incorporation of dopants enhances the physico-chemical properties of host photocatalysts, making them effective catalysts for hydrogen production from water (Suresh et al., 2023).

Various methods, including functionalization with noble metal nanoparticles, metal doping, incorporation of carbonaceous nanostructures, utilization of different metal oxide nanostructures, UV activation, and electron-beam (e-beam) irradiation, can be employed to enhance the sensing response in ZnO. Doping stands out as one of the crucial factors in increasing the sensing response of ZnO nanostructures (Chang et al., 2014). Notably, transition metals like Ni, Co, Fe, and Cu have been extensively utilized as dopants in ZnO nanostructures, demonstrating a significant impact (Zhang et al., 2017a,b; Maswanganye et al., 2017; Bai et al., 2014; Gong et al., 2006). These dopants play a dual role by increasing the activity and modifying the resistance of ZnO. Additionally, they contribute to the reduction of the sensor's working temperature while enhancing selectivity, stability, and rapid response to the target gas (Bhati et al., 2018a,b).

A notable observation relates to the alteration in the surface morphology of ZnO nanostructures upon the incorporation of metal dopants (Iqbal et al., 2013). The introduction of dopants leads to smaller particle sizes compared to pristine ZnO nanostructures, attributed to the restricted movement of crystallites at the boundaries between the dopant and the host material (Maciel, 2003). This limitation inhibits crystal growth, resulting in improved gas sensing capabilities. The enhanced gas sensing performance is ascribed to the increased surface area of small-sized particles, facilitating a higher number of chemisorbed oxygen ions and barrier height. A study by Bhati et al. (2020) involved the deposition of Ni-doped ZnO nanostructures (0, 2, 4, 6% Ni) on a silicon substrate, illustrating the practical application of metal-doped ZnO in gas sensing.

5.5 Conclusion

This chapter offers a comprehensive overview of the various strategies employed in fabricating nanostructured coatings and their multifunctional capabilities. Through meticulous surface tuning and structural modifications, these coatings can be customized to showcase desired properties such as improved durability, biocompatibility, corrosion resistance, and catalytic activity. By carefully selecting and optimizing fabrication techniques, precise control over coating thickness, morphology, porosity, and composition can be achieved, thereby facilitating the creation of coatings tailored to specific application requirements. Future research efforts are directed toward integrating advanced materials such as 2D materials, nanoparticles, and biomolecules into nanostructured coatings to unlock new functionalities and applications. Additionally, the development of smart and responsive coatings capable of adapting to changing environmental conditions or stimuli presents an exciting frontier for innovation. In conclusion, this chapter highlights the diverse array of techniques and approaches available for creating and enhancing nanostructured coatings with multifunctional capabilities. By fostering interdisciplinary collaboration, innovation, and ongoing advancements in materials science and nanotechnology, researchers are poised to unlock the full potential of nanostructured coatings, driving technological progress and societal impact in the near future.

References

Abbasi, B.A., Iqbal, J., Mahmood, T., Ahmad, R., Kanwal, S., Afridi, S., 2019. Plant-mediated synthesis of nickel oxide nanoparticles (NiO) via Geranium wallichianum: characterization and different biological applications. Mater. Res. Express 6 (8). https://doi.org/10.1088/2053-1591/ab23e1.

Abdeen, D.H., El Hachach, M., Koc, M., Atieh, M.A., 2019. A review on the corrosion behaviour of nanocoatings on metallic substrates. Materials 12 (2). https://doi.org/10.3390/ma12020210.

Abe, H., Tokuhira, S., Uchida, H., Ohshima, T., 2015. Surface modifications of hydrogen storage alloy by heavy ion beams with keV to MeV irradiation energies. Nucl. Instrum. Methods Phys. Res. Sect. B Beam Interact. Mater. Atoms 365, 214–217. https://doi.org/10.1016/j.nimb.2015.07.085.

Abraham, M., Holdway, P., Thuvander, M., Cerezo, A., Smith, G.D.W., 2002. Thermal stability of electrodeposited nanocrystalline nickel. Surf. Eng. 18 (2), 151–156. https://doi.org/10.1179/026708401225002749.

Ajay Rakkesh, R., Durgalakshmi, D., Karthe, P., Balakumar, S., 2020. Anisotropic growth and strain-induced tunable optical properties of Ag–ZnO hierarchical nanostructures by a microwave synthesis method. Mater. Chem. Phys. 244. https://doi.org/10.1016/j.matchemphys.2020.122720.

Akdoğan, E., Şirin, H.T., 2021. Plasma surface modification strategies for the preparation of antibacterial biomaterials: a review of the recent literature. Mater. Sci. Eng. C 131. https://doi.org/10.1016/j.msec.2021.112474.

Aliofkhazraei, M., 2011. Synthesis, Processing and Application of Nanostructured Coatings Engineering Materials. Springer Science and Business Media B.V., Iran, pp. 1–28. https://doi.org/10.1007/978-3-642-17966-2_1.

An, C., Zhang, Y., Guo, H., Wang, Y., 2019. Metal oxide-based supercapacitors: progress and prospectives. R. Soc. Chem. China Nanoscale Adv. 1 (12), 4644–4658. https://doi.org/10.1039/c9na00543a.

Andreeva, D.V., Fix, D., Möhwald, H., Shchukin, D.G., 2008. Buffering polyelectrolyte multilayers for active corrosion protection. J. Mater. Chem. 18 (15), 1738–1740. https://doi.org/10.1039/b801314d.

Andreeva, D.V., Skorb, E.V., Shchukin, D.G., 2010. Layer-by-layer polyelectrolyte/inhibitor nanostructures for metal corrosion protection. ACS Appl. Mater. Interfaces 2 (7), 1954–1962. https://doi.org/10.1021/am1002712.

Anker, J.N., 2008.

Avila, P.R.T., Apolinário, R.C., Rodrigues, A.M., Fernandes, J.V., Menezes, R.R., Neves, G.d.A., Pinto, H.C., 2021. On improving wear resistance of cr-al-n coatings using dynamic glancing angle dc magnetron sputtering. Nanomaterials 11 (9). https://doi.org/10.3390/nano11092187.

Bai, S., Guo, T., Zhao, Y., Sun, J., Li, D., Chen, A., Liu, C.C., 2014. Sensing performance and mechanism of Fe-doped ZnO microflowers. Sens. Actuators B 195, 657–666. https://doi.org/10.1016/j.snb.2014.01.083.

Baptista, A., Silva, F.J.G., Porteiro, J., Míguez, J.L., Pinto, G., Fernandes, L., 2018. On the physical vapour deposition (PVD): evolution of magnetron sputtering processes for industrial applications. Procedia Manuf. 17, 746–757. https://doi.org/10.1016/j.promfg.2018.10.125.

Basile, F., Benito, P., Fornasari, G., Monti, M., Scavetta, E., Tonelli, D., Vaccari, A., 2010. A novel electrochemical route for the catalytic coating of metallic supports. Studies in Surface Science and Catalysis. Elsevier Inc, pp. 51–58.

Behera, A., Mallick, P., Mohapatra, S.S., 2020. Nanocoatings for Anticorrosion Corrosion Protection at the Nanoscale. Elsevier, India, pp. 227–243. https://doi.org/10.1016/B978-0-12-819359-4.00013-1.

Bhardwaj, N., Kundu, S.C., 2010. Electrospinning: a fascinating fiber fabrication technique. Biotechnol. Adv. 28 (3), 325–347. https://doi.org/10.1016/j.biotechadv.2010.01.004.

Bhati, V.S., Ranwa, S., Fanetti, M., Valant, M., Kumar, M., 2018a. Efficient hydrogen sensor based on Ni-doped ZnO nanostructures by RF sputtering. Sens. Actuators B 255, 588–597. https://doi.org/10.1016/j.snb.2017.08.106.

Bhati, V.S., Ranwa, S., Rajamani, S., Kumari, K., Raliya, R., Biswas, P., Kumar, M., 2018b. Improved sensitivity with low limit of detection of a hydrogen gas sensor based on rGO-loaded Ni doped ZnO nanostructures. ACS Appl. Mater. Interfaces 10 (13), 11116–11124. https://doi.org/10.1021/acsami.7b17877.

Bhati, V.S., Hojamberdiev, M., Kumar, M., 2020. Enhanced sensing performance of ZnO nanostructures-based gas sensors: a review. Energy Rep. 6, 46–62. https://doi.org/10.1016/j.egyr.2019.08.070.

Biazar, E., Zeinali, R., Montazeri, N., Pourshamsian, K., Behrouz, M.J., Asefnejad, A., Khoshzaban, A., Shahhosseini, G., Najafabadi, M.S., Abyani, R., Jamalzadeh, H., Fouladi, M., Hagh, S.R.F., Khamaneh, A.S., Kabiri, S., Keshel, S.H., Mansourkiaei, A., 2010. Cell engineering: nanometric grafting of poly-N-isopropylacrylamide onto polystyrene film by different doses of gamma radiation. Int. J. Nanomed. 5 (1), 549–556. https://doi.org/10.2147/ijn.s8269.

Birkett, M., Dover, L., Cherian Lukose, C., Wasy Zia, A., Tambuwala, M.M., Serrano-Aroca, Á., 2022. Recent advances in metal-based antimicrobial coatings for high-touch surfaces. Int. J. Mol. Sci. 23 (3). https://doi.org/10.3390/ijms23031162.

Bouazizi, N., Vieillard, J., Bargougui, R., Couvrat, N., Thoumire, O., Morin, S., Ladam, G., Mofaddel, N., Brun, N., Azzouz, A., Le Derf, F., 2019. Entrapment and stabilization of iron nanoparticles within APTES modified graphene oxide sheets for catalytic activity improvement. J. Alloys Compd. 771, 1090–1102. https://doi.org/10.1016/j.jallcom.2018.08.240.

Bouazizi, N., Vieillard, J., Samir, B., Le Derf, F., 2022. Advances in amine-surface functionalization of inorganic adsorbents for water treatment and antimicrobial activities: a review. Polymers 14 (3), 378. https://doi.org/10.3390/polym14030378.

Bright, R., Hayles, A., Wood, J., Ninan, N., Palms, D., Visalakshan, R.M., Burzava, A., Brown, T., Barker, D., Vasilev, K., 2022. Bio-inspired nanostructured Ti-6Al-4V alloy: the role of two alkaline etchants and the hydrothermal processing duration on antibacterial activity. Nanomaterials 12 (7). https://doi.org/10.3390/nano12071140.

Brinker, C.J., 2013. Dip coating. Chemical Solution Deposition of Functional Oxide Thin Films. Springer-Verlag, Wien, United States, pp. 233–261. https://doi.org/10.1007/978-3-211-99311-8_10.

Broitman, E., Gueorguiev, G.K., Furlan, A., Son, N.T., Gellman, A.J., Stafström, S., Hultman, L., 2008. Water adsorption on fullerene-like carbon nitride overcoats. Thin Solid Films 517 (3), 1106–1110. https://doi.org/10.1016/j.tsf.2008.07.022.

Cai, Z., Ou, Y., Wang, J., Xiao, R., Fu, L., Yuan, Z., Zhan, R., Sun, Y., 2020. Chemically resistant Cu–Zn/Zn composite anode for long cycling aqueous batteries. Energy Storage Mater. 27, 205–211. https://doi.org/10.1016/j.ensm.2020.01.032.

Cao, Y., Cao, J., Zheng, M., Liu, J., Ji, G., 2007. Synthesis, characterization, and electrochemical properties of ordered mesoporous carbons containing nickel oxide nanoparticles using sucrose and nickel acetate in a silica template. J. Solid State Chem. 180 (2), 792–798. https://doi.org/10.1016/j.jssc.2006.10.024.

Casini, B., Righi, A., De Feo, N., Totaro, M., Giorgi, S., Zezza, L., Valentini, P., Tagliaferri, E., Costa, A.L., Barnini, S., Baggiani, A., Lopalco, P.L., Malacarne, P., Privitera, G.P., 2018. Improving cleaning and disinfection of high-touch surfaces in intensive care during carbapenem-resistant acinetobacter baumannii endemo-epidemic situations. Int. J. Environ. Res. Publ. Health 15 (10). https://doi.org/10.3390/ijerph15102305.

Chang, C.J., Lin, C.Y., Chen, J.K., Hsu, M.H., 2014. Ce-doped ZnO nanorods based low operation temperature NO2 gas sensors. Ceram. Int. 40 (7), 10867–10875. https://doi.org/10.1016/j.ceramint.2014.03.080.

Chen, Z.H., Tang, Y.B., Liu, C.P., Leung, Y.H., Yuan, G.D., Chen, L.M., Wang, Y.Q., Bello, I., Zapien, J.A., Zhang, W.J., Lee, C.S., Lee, S.T., 2009. Vertically aligned ZnO nanorod arrays sentisized with gold nanoparticles for Schottky barrier photovoltaic cells. J. Phys. Chem. C 113 (30), 13433–13437. https://doi.org/10.1021/jp903153w.

Cho, S.H., White, S.R., Braun, P.V., Feb. 2009. Self-healing polymer coatings. Adv. Mater. 21 (6), 645–649. https://doi.org/10.1002/adma.200802008.

Choi, S.H., Yun, S.J., Won, Y.S., Oh, C.S., Kim, S.M., Kim, K.K., Lee, Y.H., 2022. Large-scale synthesis of graphene and other 2D materials towards industrialization. Nat. Commun. 13 (1). https://doi.org/10.1038/s41467-022-29182-y.

Choudhary, S., Sharma, K., Sharma, V., Kumar, V., Polymers, G., 2020. Reactive and Functional Polymers Volume Two: Modification Reactions, Compatibility and Blends, 2, pp. 199–243. https://doi.org/10.1007/978-3-030-45135-6_8.

Chu, K., Shum, P.W., Shen, Y.G., 2006. Substrate bias effects on mechanical and tribological properties of substitutional solid solution (Ti, Al)N films prepared by reactive magnetron sputtering. Mater. Sci. Eng. B 131 (1–3), 62–71. https://doi.org/10.1016/j.mseb.2006.03.036.

Chu, K.H., Fathizadeh, M., Yu, M., Flora, J.R.V., Jang, A., Jang, M., Park, C.M., Yoo, S.S., Her, N., Yoon, Y., 2017. Evaluation of removal mechanisms in a graphene oxide-coated ceramic ultrafiltration membrane for retention of natural organic matter, pharmaceuticals, and inorganic salts. ACS Appl. Mater. Interfaces 9 (46), 40369–40377. https://doi.org/10.1021/acsami.7b14217.

Cottineau, T., Toupin, M., Delahaye, T., Brousse, T., Bélanger, D., 2006. Nanostructured transition metal oxides for aqueous hybrid electrochemical supercapacitors. Appl. Phys. Mater. Sci. Process 82 (4), 599–606. https://doi.org/10.1007/s00339-005-3401-3.

Covelo, A., Genescá, J., Barba, A., Menchaca, C., Uruchurtu, J., Hernández, M., 2015. Corrosion behavior of hybrid sol-gel films reinforced with electrospun nanofibers. Solid State Phenomena.

Das, P., Das, S., Ratha, S., Chakraborty, B., Chatterjee, S., 2021. Ion beam engineered hydrogen titanate nanotubes for superior energy storage application. Electrochim. Acta 371. https://doi.org/10.1016/j.electacta.2021.137774.

Deshpande, P.P., Jadhav, N.G., Gelling, V.J., Sazou, D., 2014. Conducting polymers for corrosion protection: a review. J. Coating Technol. Res. 11 (4), 473–494. https://doi.org/10.1007/s11998-014-9586-7.

Desmet, T., Morent, R., De Geyter, N., Leys, C., Schacht, E., Dubruel, P., 2009. Nonthermal plasma technology as a versatile strategy for polymeric biomaterials surface modification: a review. Biomacromolecules 10 (9), 2351–2378. https://doi.org/10.1021/bm900186sBelgium.

Dimitrakellis, P., Ellinas, K., Kaprou, G.D., Mastellos, D.C., Tserepi, A., Gogolides, E., 2021. Bactericidal action of smooth and plasma micro-nanotextured polymeric surfaces with varying wettability, enhanced by incorporation of a biocidal agent. Macromol. Mater. Eng. 306 (4). https://doi.org/10.1002/mame.202000694.

Dundar Arisoy, F., Kolewe, K.W., Homyak, B., Kurtz, I.S., Schiffman, J.D., Watkins, J.J., 2018. Bioinspired photocatalytic shark-skin surfaces with antibacterial and antifouling activity via nanoimprint lithography. ACS Appl. Mater. Interfaces 10 (23), 20055–20063. https://doi.org/10.1021/acsami.8b05066.

Duque, L., Queralto, N., Francesch, L., Bumbu, G.G., Borros, S., Berger, R., Förch, R., 2010. Reactions of plasma-polymerised pentafluorophenyl methacrylate with simple amines. Plasma Process. Polym. 7 (11), 915–925. https://doi.org/10.1002/ppap.201000058.

El-Naggar, M.E., Shaheen, T.I., Zaghloul, S., El-Rafie, M.H., Hebeish, A., 2016. Antibacterial activities and UV protection of the in situ synthesized titanium oxide nanoparticles on cotton fabrics. Ind. Eng. Chem. Res. 55 (10), 2661–2668. https://doi.org/10.1021/acs.iecr.5b04315.

Erb, U., 1995. Electrodelqslted nanocrystals: syntresis, properties and indusiiual applications. Nanostruct. Mater. 6, 533–538.

Fan, X., Yahia, L., Sacher, E., 2021. Antimicrobial properties of the Ag, Cu nanoparticle system. Biology 10 (2), 1–37. https://doi.org/10.3390/biology10020137.

Faustini, M., Nicole, L., Boissière, C., Innocenzi, P., Sanchez, C., Grosso, D., 2010. Hydrophobic, antireflective, self-cleaning, and antifogging sol-gel coatings: an example of multifunctional nanostructured materials for photovoltaic cells. Chem. Mater. 22 (15), 4406–4413. https://doi.org/10.1021/cm100937e.

Frenot, A., Chronakis, I.S., 2003. Polymer nanofibers assembled by electrospinning. Curr. Opin. Colloid Interface Sci. 8 (1), 64–75. https://doi.org/10.1016/S1359-0294Ž03.00004-9.

Fu, S., Sun, Z., Huang, P., Li, Y., Hu, N., 2019. Some basic aspects of polymer nanocomposites: a critical review. Nano Mater. Sci. 1 (1), 2–30. https://doi.org/10.1016/j.nanoms.2019.02.006.

Fu, S., Qiao, D., Luo, Y., Jiao, Z., Cheng, L., 2023. Heteroatom modified TiO2/graphene aerogel composites by electron beam irradiation in supercapacitors. ACS Appl. Nano Mater. 6 (7), 6312–6322. https://doi.org/10.1021/acsanm.3c00709.

Gleiter, H., 2000. Nanostructured materials: basic concepts and microstructure. Acta Mater. 48 (1), 1–29.

Gong, H., Hu, J.Q., Wang, J.H., Ong, C.H., Zhu, F.R., 2006. Nano-crystalline Cu-doped ZnO thin film gas sensor for CO. Sens. Actuators B 115 (1), 247–251. https://doi.org/10.1016/j.snb.2005.09.008.

Gu, C., Lian, J., Jiang, Z., Jiang, Q., 2006. Enhanced tensile ductility in an electrodeposited nanocrystalline Ni. Scripta Mater. 54 (4), 579–584. https://doi.org/10.1016/j.scriptamat.2005.10.041.

Gu, C., Lian, J., Jiang, Q., 2007. Layered nanostructured Ni with modulated hardness fabricated by surfactant-assistant electrodeposition. Scripta Mater. 57 (3), 233–236. https://doi.org/10.1016/j.scriptamat.2007.04.005.

Gu, Y., Yu, S., Mou, J., Wu, D., Zheng, S., 2020a. Research progress on the collaborative drag reduction effect of polymers and surfactants. Materials 13 (2). https://doi.org/10.3390/ma13020444.

Gu, Y., Xia, K., Wu, D., Mou, J., Zheng, S., 2020b. Technical characteristics and wear-resistant mechanism of nano coatings: a review. Coatings 10 (3). https://doi.org/10.3390/coatings10030233.

Gueorguiev, G.K., Goyenola, C., Schmidt, S., Hultman, L., 2011. CFx: a first-principles study of structural patterns arising during synthetic growth. Chem. Phys. Lett. 516 (1–3), 62–67. https://doi.org/10.1016/j.cplett.2011.09.045.

Guerrini, L., Alvarez-Puebla, R.A., Pazos-Perez, N., 2018. Surface modifications of nanoparticles for stability in biological fluids. Materials 11 (7). https://doi.org/10.3390/ma11071154.

Gupta, B., Plummer, C., Bisson, I., Frey, P., Hilborn, O., 2002a. Plasma-induced Graft Polymerization of Acrylic Acid onto Poly(ethylene Terephthalate) Films: Characterization and Human Smooth Muscle Cell Growth on Grafted Films, p. 2002.

Gupta, D., Chauhan, V., Upadhyay, S., Koratkar, N., Singh, F., Kumar, S., Mahajan, A., Chandra, R., Kumar, R., 2022b. Defects engineering and enhancement in optical and structural properties of 2D-MoS2 thin films by high energy ion beam irradiation. Mater. Chem. Phys. 276. https://doi.org/10.1016/j.matchemphys.2021.125422.

Gurrappa, I., Binder, L., 2008. Electrodeposition of nanostructured coatings and their characterization—a review. Sci. Technol. Adv. Mater. 9 (4). https://doi.org/10.1088/1468-6996/9/4/043001.

Hamerli, P., Weigel, T., Groth, T., Paul, D., 2003. Surface properties of and cell adhesion onto allylamine-plasma-coated polyethylenterephtalat membranes. Biomaterials 24 (22), 3989–3999. https://doi.org/10.1016/S0142-9612(03)00312-0.

Han, D., Song, P., Zhang, H., Yang, Z., Wang, Q., 2014. Cu2O template-assisted synthesis of porous In2O 3 hollow spheres with fast response towards acetone. Mater. Lett. 124, 93–96. https://doi.org/10.1016/j.matlet.2014.03.045.

Hellgren, N., Berlind, T., Gueorguiev, G.K., Johansson, M.P., Stafström, S., Hultman, L., 2004. Fullerene-like B-C-N thin films: a computational and experimental study. Mater. Sci. Eng. B 113 (3), 242–247. https://doi.org/10.1016/S0921-5107(04)00430-1.

Hirano, M., Hashimoto, M., Miura, K., Ohtsu, N., 2021. Fabrication of antibacterial nanopillar surface on AISI 316 stainless steel through argon plasma etching with direct current discharge. Surf. Coating. Technol. 406. https://doi.org/10.1016/j.surfcoat.2020.126680.

Hock, N., Racaniello, G.F., Aspinall, S., Denora, N., Khutoryanskiy, V.V., Bernkop-Schnürch, A., 2022. Thiolated nanoparticles for biomedical applications: mimicking the workhorses of our body. Adv. Sci. 9 (1). https://doi.org/10.1002/advs.202102451.

Honarmand, M., Golmohammadi, M., Naeimi, A., 2019. Biosynthesis of tin oxide (SnO2) nanoparticles using jujube fruit for photocatalytic degradation of organic dyes. Adv. Powder Technol. 30 (8), 1551–1557. https://doi.org/10.1016/j.apt.2019.04.033.

Horna, D., Ramírez, J.C., Cifuentes, A., Bernad, A., Borrós, S., González, M.A., 2012. Efficient cell reprogramming using bioengineered surfaces. Adv. Healthcare Mater. 1 (2), 177–182. https://doi.org/10.1002/adhm.201200017.

Hu, H., Wang, Z., Wang, S., Zhang, F., Zhao, S., Zhu, S., 2011. ZnO/Ag heterogeneous structure nanoarrays: photocatalytic synthesis and used as substrate for surface-enhanced Raman scattering detection. J. Alloys Compd. 509 (5), 2016–2020. https://doi.org/10.1016/j.jallcom.2010.10.125.

Iqbal, J., Jan, T., Ronghai, Y., 2013. Effect of Co doping on morphology, optical and magnetic properties of ZnO 1-D nanostructures. J. Mater. Sci. Mater. Electron. 24 (11), 4393–4398. https://doi.org/10.1007/s10854-013-1415-8.

Ismael, M., 2019. Highly effective ruthenium-doped TiO2 nanoparticles photocatalyst for visible-light-driven photocatalytic hydrogen production. New J. Chem. 43 (24), 9596–9605. https://doi.org/10.1039/c9nj02226k.

Jaggessar, A., Yarlagadda, P.K.D.V., 2020. Modelling the growth of hydrothermally synthesised bactericidal nanostructures, as a function of processing conditions. Mater. Sci. Eng. C 108. https://doi.org/10.1016/j.msec.2019.110434.

Jann, J., Drevelle, O., Chen, X.G., Auclair-Gilbert, M., Soucy, G., Faucheux, N., Fortier, L.C., 2021. Rapid antibacterial activity of anodized aluminum-based materials impregnated with quaternary ammonium compounds for high-touch surfaces to limit transmission of pathogenic bacteria. RSC Adv. 11 (60), 38172–38188. https://doi.org/10.1039/d1ra07159a.

Jeong, D.H., Gonzalez, F., Palumbo, G., Aust, K.T., Erb, U., 2001. The effect of grain size on the wear properties of electrodeposited nanocrystalline nickel coatings. Scripta Mater. 44 (3), 493–499. https://doi.org/10.1016/s1359-6462(00)00625-4.

Jiang, T., Carbone, E.J., Lo, K.W.H., Laurencin, C.T., 2015. Electrospinning of polymer nanofibers for tissue regeneration. Prog. Polym. Sci. 46, 1–24. https://doi.org/10.1016/j.progpolymsci.2014.12.001.

Jing, L., Zhou, W., Tian, G., Fu, H., 2013. Surface tuning for oxide-based nanomaterials as efficient photocatalysts. Chem. Soc. Rev. 42 (24), 9509–9549. https://doi.org/10.1039/c3cs60176e.

Karami, M.R., Jaleh, B., Eslamipanah, M., Nasri, A., Rhee, K.Y., 2023. Design and optimization of a TiO2/RGO-supported epoxy multilayer microwave absorber by the modified local best particle swarm optimization algorithm. Nanotechnol. Rev. 12 (1). https://doi.org/10.1515/ntrev-2023-0121.

Kaur, A., Bajaj, B., Kaushik, A., Saini, A., Sud, D., 2022. A review on template assisted synthesis of multi-functional metal oxide nanostructures: Status and prospects. Mater. Sci. Eng. B 286. https://doi.org/10.1016/j.mseb.2022.116005.

Khalil, A.T., Ovais, M., Ullah, I., Ali, M., Shinwari, Z.K., Maaza, M., 2020. Physical properties, biological applications and biocompatibility studies on biosynthesized single phase cobalt oxide (Co3O4) nanoparticles via Sageretia thea (Osbeck.). Arab. J. Chem. 13 (1), 606–619. https://doi.org/10.1016/j.arabjc.2017.07.004.

Khan, S.B., Faisal, M., Rahman, M.M., Jamal, A., 2011. Exploration of CeO2 nanoparticles as a chemi-sensor and photo-catalyst for environmental applications. Sci. Total Environ. 409 (15), 2987–2992. https://doi.org/10.1016/j.scitotenv.2011.04.019.

Khun, K., Ibupoto, Z.H., Liu, X., Beni, V., Willander, M., 2015. The ethylene glycol template assisted hydrothermal synthesis of Co3O4 nanowires; structural characterization and their application as glucose non-enzymatic sensor. Mater. Sci. Eng. B 194, 94–100. https://doi.org/10.1016/j.mseb.2015.01.001.

Kitsara, M., Nwankire, C.E., Walsh, L., Hughes, G., Somers, M., Kurzbuch, D., Zhang, X., Donohoe, G.G., O'Kennedy, R., Ducrée, J., 2014. Spin coating of hydrophilic polymeric films for enhanced centrifugal flow control by serial siphoning. Microfluid. Nanofluidics 16 (4), 691–699. https://doi.org/10.1007/s10404-013-1266-x.

Koch, C.C., Youssef, K.M., Scattergood, R.O., Murty, K.L., 2005. Breakthroughs in optimization of mechanical properties of nanostructured metals and alloys. Adv. Eng. Mater. 7 (9), 787–794. https://doi.org/10.1002/adem.200500094.

Koh, J.H., Park, J.T., Roh, D.K., Seo, J.A., Kim, J.H., 2008. Synthesis of TiO2 nanoparticles using amphiphilic POEM-b-PS-b-POEM triblock copolymer template film. Colloids Surf. A Physicochem. Eng. Asp. 329 (1–2), 51–57. https://doi.org/10.1016/j.colsurfa.2008.06.055.

Kokini, K., Choules, B.D., 1995. Surface thermal fracture of functionally graded ceramic coatings: effect of architecture and materials. Compos. Eng. 5 (7), 865–877. https://doi.org/10.1016/0961-9526(95)00045-o.

Kumar, N., Sharma, A., 2022. Surface coatings and functionalization strategies for corrosion mitigation. ACS (Am. Chem. Soc.) Symp. Ser. 1418, 291–316. https://doi.org/10.1021/bk-2022-1418.ch014.

Kumar, K.S., Van Swygenhoven, H., Suresh, S., 2003. Mechanical behavior of nanocrystalline metals and alloys. Acta Mater. 51 (19), 5743–5774. https://doi.org/10.1016/j.actamat.2003.08.032.

Landau, L., Levich, B., 1988. Dragging of a Liquid by a Moving Plate. Elsevier BV, pp. 141–153. https://doi.org/10.1016/b978-0-08-092523-3.50016-2.

Lang, X., Hirata, A., Fujita, T., Chen, M., 2011. Nanoporous metal/oxide hybrid electrodes for electrochemical supercapacitors. Nat. Nanotechnol. 6 (4), 232–236. https://doi.org/10.1038/nnano.2011.13.

Langer, R., Tirrell, D.A., 2004. Designing materials for biology and medicine. Nature 428 (6982), 487–492. https://doi.org/10.1038/nature02388.

Laurenti, M., Stassi, S., Canavese, G., Cauda, V., 2017. Surface engineering of nanostructured ZnO surfaces. Adv. Mater. Interfac. 4 (2). https://doi.org/10.1002/admi.201600758.

Lavollee, C., Gressier, M., Garcia, J., Sobrino, J.M., Reby, J., Menu, M.J., Rossi, S., Fedel, M., 2016. New architectured hybrid sol-gel coatings for wear and corrosion protection of low-carbon steel. Prog. Org. Coating 99, 337–345. https://doi.org/10.1016/j.porgcoat.2016.06.015.

Lei, L., Yongfeng, S., Xianhua, C., Lihua, Q., Lu. K., 2004. Ultrahigh strength and high electrical conductivity in copper. Science 304, 422–426. https://doi.org/10.1126/science.1092905.

Lewis, H.G.P., Edell, D.J., Gleason, K.K., 2000. Pulsed-PECVD films from hexamethylcyclotrisiloxane for use as insulating biomaterials. Chem. Mater. 12 (11), 3488–3494. https://doi.org/10.1021/cm0003370.

Li, J., Zhou, P., Attarilar, S., Shi, H., 2021. Innovative surface modification procedures to achieve micro/nano-graded ti-based biomedical alloys and implants. Coatings 11 (6). https://doi.org/10.3390/coatings11060647.

Lin, Y.T., Don, T.M., Wong, C.J., Meng, F.C., Lin, Y.J., Lee, S.Y., Lee, C.F., Chiu, W.Y., 2019. Improvement of mechanical properties and anticorrosion performance of epoxy coatings by the introduction of polyaniline/graphene composite. Surf. Coating. Technol. 374, 1128–1138. https://doi.org/10.1016/j.surfcoat.2018.01.050.

Liu, X., Chu, P.K., Ding, C., 2010. Surface nano-functionalization of biomaterials. Mater. Sci. Eng. R. Rep. 70, 275–302.

Liu, M., Bauman, L., Nogueira, C.L., Aucoin, M.G., Anderson, W.A., Zhao, B., 2022. Antimicrobial polymeric composites for high-touch surfaces in healthcare applications. Curr. Opin. Biomed. Eng. 22. https://doi.org/10.1016/j.cobme.2022.100395.

López-Rubio, A., Fabra, M.J., Martínez-Sanz, M., 2019. Food packaging based on nanomaterials. Nanomaterials 9 (9). https://doi.org/10.3390/nano9091224.

Lord, M.S., Foss, M., Besenbacher, F., 2010. Influence of nanoscale surface topography on protein adsorption and cellular response. Nano Today 5 (1), 66–78. https://doi.org/10.1016/j.nantod.2010.01.001.

Lowe, A.B., 2014. Thiol-ene "click" reactions and recent applications in polymer and materials synthesis: a first update. Polym. Chem. 5 (17), 4820–4870. https://doi.org/10.1039/c4py00339j.

Luo, L., Zhou, Y., Xu, X., Shi, W., Hu, J., Li, G., Qu, X., Guo, Y., Tian, X., Zaman, A., Hui, D., Zhou, Z., 2020. Progress in construction of bio-inspired physico-antimicrobial surfaces. Nanotechnol. Rev. 9 (1), 1562–1575. https://doi.org/10.1515/ntrev-2020-0089.

Ma, C., Nikiforov, A., De Geyter, N., Dai, X., Morent, R., Ostrikov, K., 2021. Future antiviral polymers by plasma processing. Prog. Polym. Sci. 118. https://doi.org/10.1016/j.progpolymsci.2021.101410.

Ma, C., Nikiforov, A., De Geyter, N., Morent, R., Ostrikov, K., 2022. Plasma for biomedical decontamination: from plasma-engineered to plasma-active antimicrobial surfaces. Curr. Opin. Chem. Eng. 36. https://doi.org/10.1016/j.coche.2021.100764.

Maciel, A.P., 2003. Microstructural and morphological analysis of pure and Ce-doped tin dioxide nanoparticles. J. Eur. Ceram. Soc. 23 (5), 707–713.

Maharana, B., Rajbhar, M.K., Sanyal, G., Chakraborty, B., Jha, R., Chatterjee, S., 2023. Defect-engineered MnO2 nanoparticles by low-energy ion beam irradiation for enhanced electrochemical energy storage applications. Electrochim. Acta 464. https://doi.org/10.1016/j.electacta.2023.142868.

Majhi, R., Rajbhar, M.K., Das, P., Elliman, R.G., Chatterjee, S., 2022. Low energy ion beam-induced joining of TiO2 nanoparticles. J. Alloys Compd. 924. https://doi.org/10.1016/j.jallcom.2022.166440.

Makhlouf, A.S.H., 2011. Current and Advanced Coating Technologies for Industrial Applications. Elsevier BV, pp. 3–23. https://doi.org/10.1533/9780857094902.1.3.

Makvandi, P., Iftekhar, S., Pizzetti, F., Zarepour, A., Zare, E.N., Ashrafizadeh, M., Agarwal, T., Padil, V.V.T., Mohammadinejad, R., Sillanpaa, M., Maiti, T.K., Perale, G., Zarrabi, A., Rossi, F., 2021. Functionalization of polymers and nanomaterials for water treatment, food packaging, textile and biomedical applications: a review. Environ. Chem. Lett. 19 (1), 583–611. https://doi.org/10.1007/s10311-020-01089-4.

Maswanganye, M.W., Rammutla, K.E., Mosuang, T.E., Mwakikunga, B.W., 2017. The effect of Co and in combinational or individual doping on the structural, optical and selective sensing properties of ZnO nanoparticles. Sens. Actuators B 247, 228–237. https://doi.org/10.1016/j.snb.2017.02.039.

Michalska, M., Laney, S.K., Li, T., Portnoi, M., Mordan, N., Allan, E., Tiwari, M.K., Parkin, I.P., Papakonstantinou, I., 2021. Bioinspired multifunctional glass surfaces through regenerative secondary mask lithography. Adv. Mater. 33 (43). https://doi.org/10.1002/adma.202102175.

Milan, P., Schlesinger, M., 2010. Modern Electroplating. Wiley, p. 2010.

Mishra, R., Balasubramaniam, R., 2004. Effect of nanocrystalline grain size on the electrochemical and corrosion behavior of nickel. Corrosion Sci. 46 (12), 3019–3029. https://doi.org/10.1016/j.corsci.2004.04.007.

Modaresifar, K., Azizian, S., Ganjian, M., Fratila-Apachitei, L.E., Zadpoor, A.A., 2019. Bactericidal effects of nanopatterns: a systematic review. Acta Biomater. 83, 29–36. https://doi.org/10.1016/j.actbio.2018.09.059.

Montblanch, A.R.P., Barbone, M., Aharonovich, I., Atatüre, M., Ferrari, A.C., 2023. Layered materials as a platform for quantum technologies. Nat. Nanotechnol. 18 (6), 555–571. https://doi.org/10.1038/s41565-023-01354-x.

Morent, R., De Geyter, N., Desmet, T., Dubruel, P., Leys, C., 2011. Plasma surface modification of biodegradable polymers: a review. Plasma Process. Polym. 8 (3), 171–190. https://doi.org/10.1002/ppap.201000153.

Nasef, M.M., Hegazy, E.S.A., 2004. Preparation and applications of ion exchange membranes by radiation-induced graft copolymerization of polar monomers onto non-polar films. Prog. Polym. Sci. 29 (6), 499–561. https://doi.org/10.1016/j.progpolymsci.2004.01.003.

Nazari, K.A., Rahman Rashid, R.A., Palanisamy, S., Xia, K., Dargusch, M.S., 2018. A novel Ti-Fe composite coating deposited using laser cladding of low cost recycled nano-crystalline titanium powder. Mater. Lett. 229, 301–304. https://doi.org/10.1016/j.matlet.2018.07.038.

Nedela, O., Slepicka, P., Švorcík, V., 2017. Surface modification of polymer substrates for biomedical applications. Materials 10 (10). https://doi.org/10.3390/ma10101115.

Nguyen-Tri, P., Nguyen, T.A., Carriere, P., Ngo Xuan, C., 2018. Nanocomposite coatings: preparation, characterization, properties, and applications. Int. J. Corrosion 2018. https://doi.org/10.1155/2018/4749501.

Nine, M.J., Cole, M.A., Johnson, L., Tran, D.N.H., Losic, D., 2015. Robust superhydrophobic graphene-based composite coatings with self-cleaning and corrosion barrier properties. ACS Appl. Mater. Interfaces 7 (51), 28482–28493. https://doi.org/10.1021/acsami.5b09611.

Nsude, H.E., Nsude, K.U., Obodo, R.M., Nwanya, A.C., Ahmad, I., Maaza, M., Ezema, F., 2023. Surface Modification and Functionalization of Ceramic Composite Using Self-Assembled Monolayer and Graft Polymerization Surface Modification and Functionalization of Ceramic Composites. Elsevier, Nigeria, pp. 21–44. https://doi.org/10.1016/B978-0-323-85883-0.00006-5.

Palumbo, F., Lo Porto, C., Favia, P., 2019. Plasma nano-texturing of polymers for wettability control: why, what and how. Coatings 9 (10). https://doi.org/10.3390/coatings9100640.

Pedroni, M., Vassallo, E., Aloisio, M., Brasca, M., Chen, H., Firpo, G., Ghezzi, F., Morandi, S., Pietralunga, S.M., Silvetti, T., Virgili, T., 2021. Plasma sputtered tungsten oxide thin film on poly(Lactic acid) for food packaging applications. Coatings 11 (11). https://doi.org/10.3390/coatings11111281.

Petchwattana, N., Covavisaruch, S., Wibooranawong, S., Naknaen, P., 2016. Antimicrobial food packaging prepared from poly(butylene succinate) and zinc oxide. Measurement 93, 442–448. https://doi.org/10.1016/j.measurement.2016.07.048.

Pillay, V., Dott, C., Choonara, Y.E., Tyagi, C., Tomar, L., Kumar, P., Du Toit, L.C., Ndesendo, V.M.K., 2013. A review of the effect of processing variables on the fabrication of electrospun nanofibers for drug delivery applications. J. Nanomater. 2013. https://doi.org/10.1155/2013/789289.

Pulido-Reyes, G., Leganes, F., Fernández-Piñas, F., Rosal, R., 2017. Bio-nano interface and environment: a critical review. Environ. Toxicol. Chem. 36 (12), 3181–3193. https://doi.org/10.1002/etc.3924.

Purohit, P., Bhatt, A., Mittal, R.K., Abdellattif, M.H., Farghaly, T.A., 2023. Polymer Grafting and its chemical reactions. Front. Bioeng. Biotechnol. 10. https://doi.org/10.3389/fbioe.2022.1044927.

Qin, L., Xu, J., Lian, J., Jiang, Z., Jiang, Q., 2008. A novel electrodeposited nanostructured Ni coating with grain size gradient distribution. Surf. Coating. Technol. 203 (1–2), 142–147. https://doi.org/10.1016/j.surfcoat.2008.08.044.

Rafiqah, S.A., Khalina, A., Harmaen, A.S., Tawakkal, I.A., Zaman, K., Asim, M., Nurrazi, M.N., Lee, C.H., 2021. A review on properties and application of bio-based poly(Butylene succinate). Polymers 13 (9). https://doi.org/10.3390/polym13091436.

Rahman Rashid, R.A., Nazari, K.A., Barr, C., Palanisamy, S., Orchowski, N., Matthews, N., Dargusch, M.S., 2019. Effect of laser reheat post-treatment on the microstructural characteristics of laser-cladded ultra-high strength steel. Surf. Coating. Technol. 372, 93–102. https://doi.org/10.1016/j.surfcoat.2019.05.021.

Rahman, M.M., 2020. Polyurethane/zinc oxide (PU/ZnO) composite-synthesis, protective property and application. Polymers 12 (7). https://doi.org/10.3390/polym12071535.

Rawal, I., 2015. Facial synthesis of hexagonal metal oxide nanoparticles for low temperature ammonia gas sensing applications. RSC Adv. 5 (6), 4135–4142. https://doi.org/10.1039/c4ra12747a.

Rezaei, F., Dickey, M.D., Hauser, P.J., Aug. 2019. Corrosion resistant coating based on thiol-ene polymeric system. Prog. Org. Coating 133, 350–356. https://doi.org/10.1016/j.porgcoat.2019.04.073.

Rivero, P.J., Garcia, J.A., Quintana, I., Rodriguez, R., 2018. Design of nanostructured functional coatings by using wet-chemistry methods. Coatings 8 (2). https://doi.org/10.3390/coatings8020076.

Salavati-Niasari, M., Davar, F., Mazaheri, M., 2008. Synthesis of Mn3O4 nanoparticles by thermal decomposition of a [bis(salicylidiminato)manganese(II)] complex. Polyhedron 27 (17), 3467–3471. https://doi.org/10.1016/j.poly.2008.04.015.

Saldívar Guerra, E., Bonilla Cruz, J., Dehonor, M., González-Montiel, A., 2013. Polymer Modification: Functionalization and Grafting Handbook of Polymer Synthesis, Characterization, and Processing. John Wiley and Sons, Mexico, pp. 205–223. https://doi.org/10.1002/9781118480793.ch10.

Salehi, A., Naderi, P., Boroumand, F.A., Dunbar, A., 2018. Fabrication and characterization of hybrid photovoltaic devices based on N-type GaAs and polymer composites. In: Proceedings of the 2nd International Conference of Energy Harvesting, Storage, and Transfer (EHST'18).

Sandua, X., Rivero, P.J., Esparza, J., Fernández-Palacio, J., Conde, A., Rodríguez, R.J., 2022. Design of photocatalytic functional coatings based on the immobilization of metal oxide particles by the combination of electrospinning and layer-by-layer deposition techniques. Coatings 12 (6). https://doi.org/10.3390/coatings12060862.

Sarmah, D., Kumar, A., 2019. Ion beam modified molybdenum disulfide-reduced graphene oxide/polypyrrole nanotubes ternary nanocomposite for hybrid supercapacitor electrode. Electrochim. Acta 312, 392–410. https://doi.org/10.1016/j.electacta.2019.04.174.

Šebenik, A., 1998. Living free-radical block copolymerization using thio-iniferters. Prog. Polym. Sci. 23 (5), 875–917. https://doi.org/10.1016/s0079-6700(98)00001-x.

Sharma, D., Rajput, J., Kaith, B.S.K.M., Sharma, S., 2010. Synthesis of ZnO nanoparticles and study of their antibacterial and antifungal properties. Thin Solid Films 519, 1224–1229.

Shchukin, D.G., Zheludkevich, M., Yasakau, K., Lamaka, S., Ferreira, M.G.S., Möhwald, H., 2006. Layer-by-layer assembled nanocontainers for self-healing corrosion protection. Adv. Mater. 18 (13), 1672–1678. https://doi.org/10.1002/adma.200502053.

Shen, X., Lian, J., Jiang, Z., Jiang, Q., 2008. High strength and high ductility of electrodeposited nanocrystalline Ni with a broad grain size distribution. Mater. Sci. Eng. A 487 (1–2), 410–416. https://doi.org/10.1016/j.msea.2007.10.018.

Sifontes, A.B., Gonzalez, G., Ochoa, J.L., Tovar, L.M., Zoltan, T., Cañizales, E., 2011. Chitosan as template for the synthesis of ceria nanoparticles. Mater. Res. Bull. 46 (11), 1794–1799. https://doi.org/10.1016/j.materresbull.2011.07.049.

Smith, K.A., Savva, A.I., Deng, C., Wharry, J.P., Hwang, S., Su, D., Wang, Y., Gong, J., Xu, T., Butt, D.P., Xiong, H., 2017. Effects of proton irradiation on structural and electrochemical charge storage properties of TiO2 nanotube electrodes for lithium-ion batteries. J. Mater. Chem. A 5 (23), 11815–11824. https://doi.org/10.1039/c7ta01026e.

Steele, A., Bayer, I., Loth, E., Jul. 2012. Adhesion strength and superhydrophobicity of polyurethane/organoclay nanocomposite coatings. J. Appl. Polym. Sci. 125 (Suppl. 1). https://doi.org/10.1002/app.36312.

Sunder, N., Fong, Y.Y., Bustam, M.A., Suhaimi, N.H., 2022. Development of amine-functionalized metal-organic Frameworks hollow fiber mixed matrix membranes for CO2 and CH4 separation: a review. Polymers 14 (7). https://doi.org/10.3390/polym14071408.

Suresh, R., Gnanasekaran, L., Rajendran, S., Soto-Moscoso, M., 2023. Doped nanomaterials: application in hydrogen production via photocatalytic water splitting. Fuel 348. https://doi.org/10.1016/j.fuel.2023.128528.

Taglietti, A., Arciola, C.R., D'Agostino, A., Dacarro, G., Montanaro, L., Campoccia, D., Cucca, L., Vercellino, M., Poggi, A., Pallavicini, P., Visai, L., 2014. Antibiofilm activity of a monolayer of silver nanoparticles anchored to an amino-silanized glass surface. Biomaterials 35 (6), 1779–1788. https://doi.org/10.1016/j.biomaterials.2013.11.047.

Tan, J., Saltzman, W.M., 2004. Biomaterials with hierarchically defined micro- and nanoscale structure. Biomaterials 25 (17), 3593–3601. https://doi.org/10.1016/j.biomaterials.2003.10.034.

Trey, S.M., Kristofer Gamstedt, E., Mäder, E., Jönsson, S., Johansson, M., 2011. Glass fiber reinforced high glass transition temperature thiol-ene networks. Compos. Appl. Sci. Manuf. 42 (11), 1800–1808. https://doi.org/10.1016/j.compositesa.2011.08.003.

Ulman, A., 1996. Formation and structure of self-assembled monolayers. Chem. Rev. 96 (4), 1533–1554. https://doi.org/10.1021/cr9502357.

van Lente, H., van Til, J.I., 2008. Articulation of sustainability in the emerging field of nanocoatings. J. Clean. Prod. 16 (8–9), 967–976. https://doi.org/10.1016/j.jclepro.2007.04.020.

Variola, F., Brunski, J.B., Orsini, G., Tambasco De Oliveira, P., Wazen, R., Nanci, A., 2011. Nanoscale surface modifications of medically relevant metals: state-of-the art and perspectives. Nanoscale 3 (2), 335–353. https://doi.org/10.1039/c0nr00485e.

Vincent, M., Duval, R.E., Hartemann, P., Engels-Deutsch, M., 2018. Contact killing and antimicrobial properties of copper. J. Appl. Microbiol. 124 (5), 1032–1046. https://doi.org/10.1111/jam.13681.

Wang, L., Gao, Y., Xue, Q., Liu, H., Xu, T., 2005a. Graded composition and structure in nanocrystalline Ni-Co alloys for decreasing internal stress and improving tribological properties. J. Phys. Appl. Phys. 38 (8), 1318–1324. https://doi.org/10.1088/0022-3727/38/8/033.

Wang, Y., Lee, J.Y., Zeng, H.C., 2005b. Polycrystalline SnO2 nanotubes prepared via infiltration casting of nanocrystallites and their electrochemical application. Chem. Mater. 17 (15), 3899–3903. https://doi.org/10.1021/cm050724f.

Wang, Y., Guo, J., Wang, T., Shao, J., Wang, D., Yang, Y.W., 2015. Mesoporous transition metal oxides for supercapacitors. Nanomaterials 5 (4), 1667–1689. https://doi.org/10.3390/nano5041667.

Wang, M., Wu, G., Cao, 2019. Multi-segmented nanowires: a high tech bright future. Materials 12 (23), 3908. https://doi.org/10.3390/ma12233908.

Wang, D., Mukhtar, A., Humayun, M., Wu, K., Du, Z., Wang, S., Zhang, Y., 2022a. A critical review on nanowire-motors: design, mechanism and applications. Chem. Rec. 22 (8). https://doi.org/10.1002/tcr.202200016.

Wang, Y., Cheng, X., Zhang, K., Chen, G., Wang, R., Zhang, J., 2022b. Ion-irradiation of catalyst and electrode materials for water electrolysis/photoelectrolysis cells, rechargeable batteries, and supercapacitors. Mater. Adv. 3 (20), 7384–7405. https://doi.org/10.1039/d2ma00547f.

Wang, X., 2020. Preparation, synthesis and application of Sol-gel method. Vidyasirimedhi Inst. Sci. Technol. 25, 1–30.

Wei, Z., Gu, J., Ye, Y., Fang, M., Lang, J., Yang, D., Pan, Z., 2020. Biodegradable poly(butylene succinate) nanofibrous membrane treated with oxygen plasma for superhydrophilicity. Surf. Coating. Technol. 381. https://doi.org/10.1016/j.surfcoat.2019.125147.

Weltmann, K.D., Kolb, J.F., Holub, M., Uhrlandt, D., Šimek, M., Ostrikov, K., Hamaguchi, S., Cvelbar, U., Černák, M., Locke, B., Fridman, A., Favia, P., Becker, K., 2019. The future for plasma science and technology. Plasma Process. Polym. 16 (1). https://doi.org/10.1002/ppap.201800118.

Winter, M., Brodd, R.J., 2004. What are batteries, fuel cells, and supercapacitors? Chem. Rev. 104 (10), 4245–4269. https://doi.org/10.1021/cr020730k.

Wu, H., Li, D., Zhu, X., Yang, C., Liu, D., Chen, X., Song, Y., Lu, L., 2014. High-performance and renewable supercapacitors based on TiO2 nanotube array electrodes treated by an electrochemical doping approach. Electrochim. Acta 116, 129–136. https://doi.org/10.1016/j.electacta.2013.10.092.

Xia, F., Jiang, L., 2008. Bio-inspired, smart, multiscale interfacial materials. Adv. Mater. 20 (15), 2842–2858. https://doi.org/10.1002/adma.200800836China.

Xiao, B., Wang, F., Zhai, C., Wang, P., Xiao, C., Zhang, M., 2016. Facile synthesis of In2O3 nanoparticles for sensing properties at low detection temperature. Sens. Actuators B 235, 251–257. https://doi.org/10.1016/j.snb.2016.05.082.

Xie, Y., Kocaefe, D., Chen, C., Kocaefe, Y., 2016. Review of research on template methods in preparation of nanomaterials. J. Nanomater. 2016. https://doi.org/10.1155/2016/2302595.

Xue, Q., Zhang, Q., 2019. Agar hydrogel template synthesis of Mn3o4 nanoparticles through an ion diffusion method controlled by ion exchange membrane and electrochemical performance. Nanomaterials 9 (4). https://doi.org/10.3390/nano9040503.

Yang, J., Bei, J., Wang, S., 2002. Enhanced cell affinity of poly (D,L-lactide) by combining plasma treatment with collagen anchorage. Biomaterials 23 (12), 2607–2614. https://doi.org/10.1016/S0142-9612(01)00400-8.

Yang, L.H., Liu, F.C., Han, E.H., 2005. Effects of P/B on the properties of anticorrosive coatings with different particle size. Prog. Org. Coating 53 (2), 91–98. https://doi.org/10.1016/j.porgcoat.2005.01.003.

Yang, P., Ding, Y., Lin, Z., Chen, Z., Li, Y., Qiang, P., Ebrahimi, M., Mai, W., Wong, C.P., Wang, Z.L., 2014. Low-cost high-performance solid-state asymmetric supercapacitors based on MnO2 nanowires and Fe2O3 Nanotubes. Nano Lett. 14 (2), 731–736. https://doi.org/10.1021/nl404008e.

Yaseen, M., et al., 2021. Preparation, functionalization, modification, and applications of nanostructured gold: a critical review. Energies 14 (5). https://doi.org/10.3390/en14051278.

Yilmaz, O., Yorgancioglu, A., 2018. Nanocoatings: preparation, properties, and biomedical applications polymeric nanomaterials in nanotherapeutics. Elsevier, Turkey, pp. 299–331. https://doi.org/10.1016/B978-0-12-813932-5.00008-X.

Yu, Y.H., Lin, Y.Y., Lin, C.H., Chan, C.C., Huang, Y.C., 2014. High-performance polystyrene/graphene-based nanocomposites with excellent anti-corrosion properties. Polym. Chem. 5 (2), 535–550. https://doi.org/10.1039/c3py00825h.

Zhang, H., Jiang, Z., Lian, J., Jiang, Q., 2008. Bulk nanostructured Cu with high strength and good ductility. Adv. Eng. Mater. 10 (1–2), 41–45. https://doi.org/10.1002/adem.200700237.

Zhang, L., Dong, B., Lin, X., Zhang, X., Chen, J., Sun, X., Xu, H., Zhang, T., Bai, X., Zhang, S., Song, H., 2017a. Three-dimensional ordered ZnO–Fe3O4 inverse opal gas sensor toward trace concentration acetone detection. Sensor. Actuator. B Chem. 252, 367–374. https://doi.org/10.1016/j.snb.2017.05.167.

Zhang, X., Dong, Z., Liu, S., Shi, Y., Dong, Y., Feng, W., 2017b. Maize straw-templated hierarchical porous ZnO:Ni with enhanced acetone gas sensing properties. Sens. Actuators B 243, 1224–1230. https://doi.org/10.1016/j.snb.2016.12.076.

Zhao, S., Caruso, F., Dahne, L., Decher, G., De Geest, B.G., Fan, J., Feliu, N., Gogotsi, Y., Hammond, P.T., Hersam, M.C., Khademhosseini, A., Kotov, N., Leporatti, S., Li, Y., Lisdat, F., Liz-Marzan, L.M., Moya, S., Mulvaney, P., Rogach, A.L., Roy, S., Shchukin, D.G., Skirtach, A.G., Stevens, M.M., Sukhorukov, G.B., Weiss, P.S., Yue, Z., Zhu, D., Parak, W.J., 2019. The future of layer-by-layer assembly: a Tribute to ACS nano associate editor Helmuth Mohwald. ACS Nano 13 (6), 6151–6169. https://doi.org/10.1021/acsnano.9b03326.

Zheludkevich, M.L., Shchukin, D.G., Yasakau, K.A., Möhwald, H., Ferreira, M.G.S., 2007. Anticorrosion coatings with self-healing effect based on nanocontainers impregnated with corrosion inhibitor. Chem. Mater. 19 (3), 402–411. https://doi.org/10.1021/cm062066k.

Zhong, Z., Qin, J., Ma, J., 2015. Cellulose acetate/hydroxyapatite/chitosan coatings for improved corrosion resistance and bioactivity. Mater. Sci. Eng. C 49, 251–255. https://doi.org/10.1016/j.msec.2015.01.020.

Zhou, C., Koshani, R., O'Brien, B., Ronholm, J., Cao, X., Wang, Y., 2021a. Bio-inspired mechano-bactericidal nanostructures: a promising strategy for eliminating surface foodborne bacteria. Curr. Opin. Food Sci. 39, 110–119. https://doi.org/10.1016/j.cofs.2020.12.021.

Zhou, Y., Fletcher, N.F., Zhang, N., Hassan, J., Gilchrist, M.D., 2021b. Enhancement of antiviral effect of plastic film against SARS-CoV-2: combining nanomaterials and nanopatterns with scalability for mass manufacturing. Nano Lett. 21 (24), 10149–10156. https://doi.org/10.1021/acs.nanolett.1c02266.

Zou, Y., Ma, B., Cui, H., Lu, F., Xu, P., 2020. Microstructure, wear, and oxidation resistance of nanostructured carbide-strengthened cobalt-based composite coatings on Invar alloys by laser cladding. Surf. Coating. Technol. 381. https://doi.org/10.1016/j.surfcoat.2019.125188.

Section III

Biomedical applications of multifunctional nanostructured coatings

Multifunctional cold-sprayed coatings as micro/nano-biointerfaces for biomedical applications

6

Simone Russo [1,2] *and Giuseppe Vitiello* [1,2]
[1]University of Naples Federico II, Department of Chemical, Materials and Production Engineering (DICMaPI), Naples, Italy; [2]CSGI, Center for Colloid and Surface Science, Florence, Italy

6.1 Functional micro/nano-structured biointerfaces for medicine

The design and creation of functional interfaces is a decisive point in developing advanced tools and devices for application in the biomedical field. Indeed, the spread and transmission of diseases and infections are facilitated by the ability of bacteria, fungi, and viruses to adhere to surfaces and proliferate (Donlan, 2002). Hence, treatments and cleaning protocols are essential to avoid spread of diseases and to reduce the risk of postsurgery infection (Jaggessar et al., 2017). In addressing these problems, it is crucial to consider also all those issues related to mechanical/structural stability, bioactivity, and biocompatibility of biomedical implants and devices (Kumar et al., 2020).

Micro-/nanomaterials can be used to solve these problems, since they possess excellent antipathogenic properties and because of they can mimic parts of real bone, tissues, and/or other human biological environments (Jaggessar et al., 2017; Neacşu et al., 2016; Wang et al., 2013, 2016). Since nanoparticles have a high surface-to-volume ratio, they are more likely to come in contact with bacteria, break through their cell walls and membranes, interfering with bacteria vital chemical processes (Wang et al., 2013). Moreover, micro-/nano metal oxides possess photocatalytic properties leading to the formation of reactive oxygen species (ROS), which cause oxidative damages to bacteria and viruses structures inducing their killing (Paiva and Bozza, 2014; Russo et al., 2024). At the same time, given that bone exists in three structural levels, such as nano-, micro- and macro-structure (Rho et al., 1998), the use of nanomaterials can enhance the biocompatibility, bioactivity, and osseointegration. Furthermore, the complex osseointegration process begins with the recruitment of cells on the biointerfaces, making cell adhesion a central aspect (Bagno et al., 2007). Studies demonstrated that nanoscale films, by imitating the structural elements of naturally occurring body proteins that resemble bone, can be used to enhance cell adhesion and encourage osteoblast activity (Amantea et al., 2023). Indeed, it was thought that

Multifunctional Nanostructured Coatings. https://doi.org/10.1016/B978-0-443-23683-9.00006-4

the nano-scaled films improved cell adhesion and interface spreading and also physiological activity. Also, nanostructured interfaces encourage osteoblast activities due to their capacity to mimic the structural components of naturally occurring body proteins that resemble bone.

Definitely, the rational design of micro/nano biointerfaces with exceptionals antipathogenic (i.e., antibacterial, antiviral, and antifungal) properties and remarkable bioactivity, adhesivity, and biocompatibility is a key issue to improve the performances of biomedical devices directly interacting with biological environments. Their significant influence on implant function, infection prevention, and general biocompatibility highlight their essential role in pushing the boundaries of biomedical technology.

6.2 High-performance coatings for biomedical field by cold spray route

The demand for high-performance multifunctional coatings onto biomedical devices is continuously increasing, and there are various fabrication techniques used today to produce them. In particular, a significant scientific interest has been focused on creating high-performance biointerfaces in which the functionality of materials for medical use is enhanced by immobilizing different biomolecules, creating multilayer coatings, and/or incorporating surface micro- and nanostructures with different chemical features. However, achieving a perfect medical device coating as bioactive interfaces able to control all the biological responses is not currently feasible, considering the inherent complexity and hierarchical structure of biological systems. Among the wide range of techniques for the production of micro/nano coatings for medical devices, the casting and spraying ones represent a valid opportunity, not requiring chemical reactions during the deposition process and favoring the realization of multifunctional surfaces with improved biological performances.

Specifically, the cold spray (CS) route emerged in the 1980s as a result of advancements in thermal spray processes. During their study on a model of a gas flow carrying solid particles at supersonic speeds, scientists observed that solid particles were deposited onto a surface when the velocity of impact reached a critical value (Ramakrishna et al., 2001). In the late 1990s, CS equipment became commercially available, leading to a surge in activities and patents related to this technique (Irissou et al., 2008; Ramakrishna et al., 2001). Reports highlighting the potentialities of cold sprayed tantalum coatings for biomedical applications or titanium and bio-ceramic coatings used in bioengineering (Ghelichi and Guagliano, 2009; Singh et al., 2013) are dated starting from 2006. The significance of CS technology in the biomedical field is evident from the patent literature, including a filing in 2009 related to the fabrication of coatings for dental implants and prostheses (Liao et al., 2022). However, it is crucial to consider the biological performance of these micro- and nano-coatings, as there has been insufficient reporting of related studies over the past decade. The aim of this chapter is to focus on the most significant examples of biomedical nanostructured coatings with different functionalities, already fabricated through the CS way. As schematized

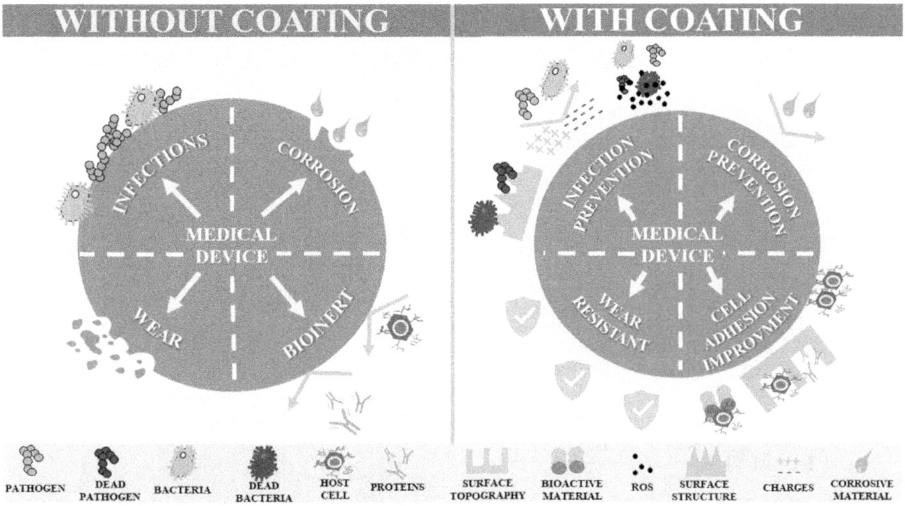

Figure 6.1 Illustrative description of the main applications of micro-/nano coatings obtained through CS technology as bioactive interfaces for medical devices.

in Fig. 6.1, the biomedical applications address various issues, such as pathogen infection, enhanced cytocompatibility, corrosion and abrasive wear, and biological inertness.

6.3 Fundamental and operational principles of CS technology

CS, also referred to as "cold gas dynamic spray," "kinetic spray," and other equivalent terms, offers benefits over conventional thermal spray processes for biomedical and biological applications, making it competitive within industrial sectors (Ghelichi and Guagliano, 2009; Vilardell et al., 2015). In this technique, the solid feedstock is suspended in a pressurized gas stream (N_2, He, or air) that flows through a converging-diverging nozzle (i.e., De Laval nozzle). The design of this nozzle forced the stream to accelerate the speed of sound in the throat area by squeezing it in the convergent portion. The divergence expands to atmospheric pressure and accelerates to a supersonic velocity upon traveling through it. This can be seen in Fig. 6.2 and Eq. (6.1), that is the result of the combination of mass conservation, momentum conservation, isentropic flow relation, and the ideal gas law equation (Huang et al., 2021; Yadav et al., 2022).

$$\left(1 - \mathrm{M}^2\right) * \frac{d_v}{V} = -\frac{d_A}{A} \tag{6.1}$$

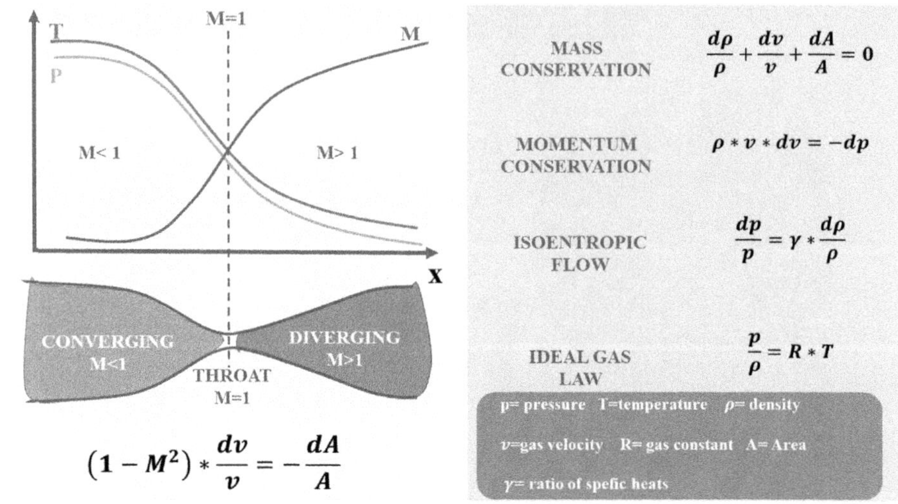

Figure 6.2 Summary of the key equations behind CS functioning mechanism together with the corresponding graphical representations.

where V is the velocity of the gas, A is the cross-sectional flow area, and M is the Mach number (the ratio between the gas velocity and the sound speed). In the converging section, the flow is subsonic, as indicated by $M < 1$, suggesting an increase in the velocity as the cross-section decreases. At the throat, $M = 1$, marking the transition to supersonic flow in the gas ($M > 1$), where the velocity increases while the cross section increases.

Inert gases such as helium and nitrogen are preferable to air since they provide a higher powder exit velocity, enabling momentum transfer to powder particles (Kumar et al., 2020). The possibility of reaching such high velocities allows us to obtain high kinetic energies, even at low temperatures. Indeed, CS operates at temperatures lower than the melting points of several coating materials, ranging from ambient temperature to 1000°C. This feature makes it suitable for depositing oxygen-sensitive metals, heat-sensitive organic molecules, and biomolecules, as it generates deposits with identical feedstock compositions (Prashar and Vasudev, 2021). The formation of the CS coatings is a two-step process. Initially, there is a bond between the micro-/nanoparticles and the substrate, followed by a particle—particle bond that determines the thickness of the coating (Singh et al., 2021). However, an important parameter that has to be considered in CS coatings is the critical velocity (v_c) that is the lowest velocity required to minimize surface abrasion while yet achieving effective deposits (Liao et al., 2022). This means that successful adhesion is obtained only at velocities higher than which depends on the properties and morphologies of the coating materials (Ganesan et al., 2012). Critical velocity is determined by various factors such as feedstock powder, substrate thermomechanical properties, physical attributes, carrier gas, operating conditions (temperature and pressure), and nozzle design (Singh et al., 2016). The critical velocity can be predicted through mathematical models or laser Doppler anemometry, but these two methods are time consuming and require extensive knowledge of the properties of materials (Assadi et al., 2003). Raletz et al. suggested two simple methods to determine the critical velocities. In the first method, the dispersion of the sprayed particles before impact is determined using an imaging technique. The second approach measured deposit efficiency and used a one-dimensional isentropic model to calculate the particle impact velocity (Raletz, Vardelle and Ezo'o, 2006). This study compared the predictions of a one-dimensional isentropic model with findings from the literature and concluded that their methods were dependable and credible. They suggested that using smaller particles and more advanced detection technologies could improve the measurement accuracy. The mechanism of adhesion between coatings and substrates is still unclear owing to the high complexity of particle-particle and gas carrier interactions. For example, high velocities are believed to cause adiabatic shear instability when working with metallic microparticles (Singh et al., 2021). This instability arises from high strain rate deformation processes and leads to the formation of metal jets (Hussain et al., 2009). Nevertheless, Hassani et al. argued that large interfacial strain for bonding may not need shear instability and proposed that the rapid impact generates pressure waves, causing hydrodynamic plasticity at particle edges, facilitating bonding without shear instability (Hassani-Gangaraj et al., 2018). The authors proved that adhesion does not require adiabatic shear instability and proposed a proportionality between the critical velocity and bulk speed of sound for optimal coating. However, they did not consider the effects of substrate conditions such as temperature, roughness, and hardness, factors that may influence the CS process (Adaan-Nyiak and Tiamiyu, 2023). CS technique can be divided into two subcategories: **low-pressure cold spray (LPCS)** and **high-pressure cold spray (HPCS)**, as schematized in Fig. 6.3. It is important to note that LPCS stands out from HPCS in two significant ways: first, the LPCS employs 5−10 bars of gas pressure, whereas the HPCS uses 25−30 pressures; second, the LPCS injects powder radially, that is absent in HPCS.

In LPCS, the aerodynamics of the gas heater are maximized by using preheated air or nitrogen at a pressure of 5−10 bar. The gas then flows through a "DeLaval" or comparable nozzle, which accelerates it to 300−600 m/second. After being inserted radially downstream of the supersonic nozzle's throat portion, solid powder particles are propelled in the direction of the substrate. Because of the low pressure, the system is more portable, has more automation flexibility, enhanced operational safety, and has a longer nozzle service life. Nevertheless, it presents different drawbacks such as limited deposition efficiency and particle path constraints (Singh et al., 2013).

In HPCS, helium or nitrogen is preheated up to 1000°C and pressurized at 25−30 bar to achieve aerodynamics, instead of particle melting. It is imperative to preheat the pressurized gas as the speed of sound is directly proportional to temperature.

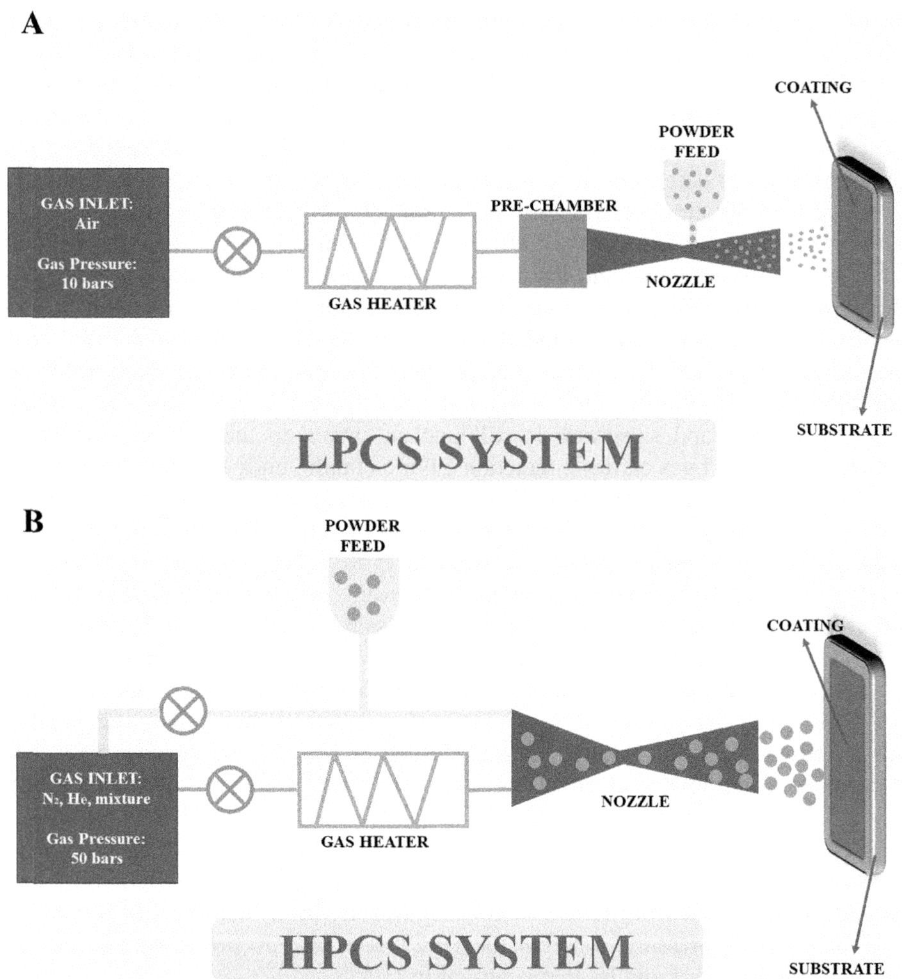

Figure 6.3 Schematic diagrams showing the operating parameters of (a) low-pressure cold spray (LPCS) and (b) high-pressure cold spray (HPCS). Both HPCS and LPCS systems include a gas inlet, gas heater, de Laval nozzle, and powder feeder. The two systems differ how the powder feed is arranged, where LPCS uses a prechamber and radial feeding, while HPCS uses axial feeding.

Therefore, elevating the gas temperature leads to increased gas velocity. Furthermore, raising the temperature of the gas results in a significant rise in particle impact temperatures and a consequential drop in particle critical velocity (Liao et al., 2022; Schmidt et al., 2006). Solid powder particles are mixed in the prechamber and then axially injected into the gas stream. With HPCS up to 90% spray efficiency can be achieved, which is significantly higher than the 50% efficiency achieved with low-pressure systems (Singh et al., 2013).

6.4 Micro/nano-structured coatings for biomedical field: From materials chemistry to functional medical devices

A wide range of materials, such as metals, ceramics, composites, and polymers, has proven to be effectively deposited through CS technology, thus offering a valuable opportunity to develop multifunctional coatings for biomedical devices (Fig. 6.4).

The desirable properties of the feedstock powder, such as its chemical characteristics, amorphous structure, biocompatibility and bioactivity, hardness, corrosion, and wear resistance, can be preserved by optimizing the CS process parameters. As a result, the substrate materials are effectively functionalized by the as-sprayed coatings. Indeed, to favorably design orthopedic devices, the properties of the bulk material and micro-/nanostructured coatings are decisive to confer wear resistance and to favor the interactions with the biological environment (Dosta et al., 2018). In this way, surface chemical modification and functionalization of prosthetic devices aims to enhance their osseointegration and mitigate allergic reactions (Liao et al., 2022; Vilardell et al., 2015). Similarly, antibiotic resistance, hospital infections, and contamination of clinical materials are a serious treat for human health. Although the hygiene and surface-cleaning protocols are already increasing, the risk of infections is actually considered a high importance level (Lucas et al., 2020). The use of engineered micro-/nanomaterials with intrinsic antiviral/antibacterial activity or loaded with active drugs is an intriguing way to coat surfaces conferring an extraordinary potential to deactivate germs (Fu and Gray, 2021).

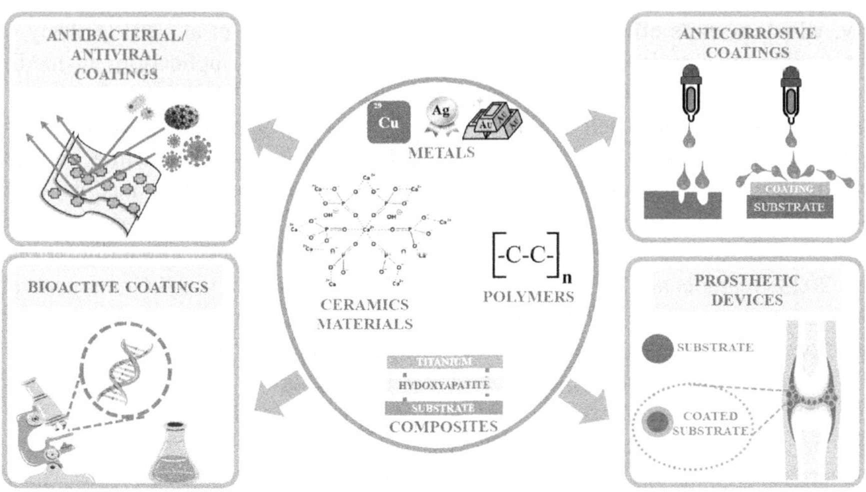

Figure 6.4 Schematic representation of the main classes of materials in terms of chemical composition mostly used to create multifunctional CS coatings as biointerfaces for biomedical applications.

For these aims, CS represents one of the greatest methods for creating resistant, antipathogenic, and self-cleaning surfaces, since it allows for the fabrication of thick micro-/nano coatings while maintaining the chemical and physical characteristics of the raw material.

The selection of the materials as function of their chemical nature and structural features plays a crucial role in achieving successful coatings with high deposition efficiency. Since CS is a solid-state process, particles with sufficient kinetic energy and plasticity upon impact are the ones that contribute to the production of a successful CS deposit. From a kinetic energy perspective, lighter particles with lower density can be more easily accelerated, while heavier particles present more challenges in acceleration. Additionally, particle velocity is inversely proportional to particle size for a given material. Consequently, it is common to use CS particles smaller than 100 μm, depending on the material's density. However, for particle sizes below 10 μm or for nanoparticles ($\leq$100 nm), it is important to consider the bow shock effect and particle-particle self-aggregation. These phenomena can lead to nozzle clogging and poor deposition, thus determining the formation of not stable or inhomogeneous biointerfaces. In this context, several examples of micro-/nanostructured coatings with different chemical composition and physicochemical properties are already realized by CS technology in the last years, as schematized in Table 6.1.

6.4.1 Metals

The extraordinary qualities of metallic micro-/nano coatings can include many beneficial attributes (Fig. 6.5). First, their mechanical strength guarantees their toughness and resistance to a range of environmental factors (Oliver et al., 2019). Furthermore, metal-based coatings show an excellent wear resistance, which makes them extremely durable and resistant to abrasion and exhibit excellent thermal and electrical conductivity, which permits effective heat/electrical transfer (Dosta et al., 2018). They can exert a significant antipathogenic activity, promoting many applications in medical field (Guillem-Marti et al., 2019; Lucas et al., 2020).

For example, metals, such as Ti, Zr, Co, Cr, Ni, and their alloys, perform a crucial function as fundamental components of dental and orthopedic everlasting implantation. Similarly, stainless steel coatings deposited on aluminum substrates, which are used to create surfaces with excellent corrosion resistance, wear resistance, and mechanical properties after the contact with simulated bodily fluids (Rashidy Ahmady et al., 2023; Wang et al., 2020). Additionally, alloys based on Co and Cr are generally employed as appropriate metallic bio-inert materials developed with the slightest negative impact on host tissue following the transplantation. All these kinds of coatings have the ability to provide a smooth surface, which enhances the functioning of articulating surfaces and biointerfaces, thus enhancing for example the implants tribological characteristics (Trelka et al., 2023).

Alternatively, various metals, such as copper and silver, show good antimicrobial/antiviral abilities (Ali et al., 2023; Ciacotich et al., 2018; Mitra et al., 2020); for this reason, they are used to create bioactive interfaces against different kind of pathogens. Indeed, these metals prevent the development of infections, thanks their

Table 6.1 Summary of the most recent micro/nano coatings by CS route in terms of chemical composition, particle morphology and size, and biomedical functionality.

Chemical composition of micro/nano coatings	Particle morphology	Particle size	Biomedical activity	References
Au	Spherical	5–30 nm	Plasmonic biosensors	Moonga and Jen (2019)
Cu/Ag	Undefined	–	Antimicrobial	Lucas et al. (2020)
Cu	Spherical	20–50 μm	Antimicrobial	da Silva et al. (2019)
Cu	Irregular	–	Antimicrobial, antiviral	Saha et al. (2023), Hutasoit et al. (2020)
Al/ZnO	Undefined	45 μm	Antimicrobial	Sanpo et al. (2008)
Zn–Al–Mg–TiO_2	Undefined	–	Antimicrobial, antifouling	Wang et al. (2020)
Stainless steel	Spherical	5–25 μm	Anticorrosive	Dikici and Topuz (2018)
Hydroxyapatite (HA)	Spherical	15–65 μm	Biocompatible	Vilardell et al. (2019)
		75–150 μm	Bioactive, biocompatible	Paterlini et al. (2022)
Ti/HA (Ti 80%-HA 20%)	Sponge	–	Biocompatible	Judd et al. (2019)
Ta/HA	Sponge	35 μm (Ta) 10 μm (HA)	Biocompatible, bioactive	Tang et al. (2019)
Ti/HA	Undefined	70–110 nm (Ti) 40–60 nm (HA)	Bioactive, bone-integrated	Paterlini et al. (2022)
HA/Gentamicin sulfate (GS)	Rodlike	20–100 nm	Antimicrobial, bioactive	Li et al. (2017)
Glass/Polyetheretherketone (PEEK)	Spherical and porous	40.5 μm (glass) 22.5 μm (PPEK)	Bioactive, wear resistant	Garrido et al. (2022)
Mg/HA	Spherical	4 μm	Biocompatible, biodegradable, bioactive	Noorakma et al. (2013)

Continued

Table 6.1 Continued

Chemical composition of micro/nano coatings	Particle morphology	Particle size	Biomedical activity	References
Al/CNT	Near-spherical (Al) Rodlike (CNT)	4 μm(Al) 1–5 μm (length); 15–30 nm (diameter)	Biocompatible	Xie et al. (2021)
HA/graphene-nanosheet (GN)	Rodlike (HA) Wrinkled-paper-like flakes (GN)	20–45 nm (length), 10 nm (diameter), 0.8–1.1 nm (height) <4 μm (lateral face)	Cells proliferation	Liu et al. (2014)
HA-Ag/PEEK	Irregular	45 μm	Antibacterial	Sanpo et al. (2009)
Silk fibroin	Cylindrical fibrils	40–80 nm (diameter)	Biocompatible, cell adhesion	Du et al. (2006)
CNT/Biopolymers	Tubular (CNT)	100 nm-1 cm (length) 2–50 nm (diameter)	Antiangiogenic, biocompatible, biosensing, cell adhesion	Rogala-Wielgus and Zieliński (2023)

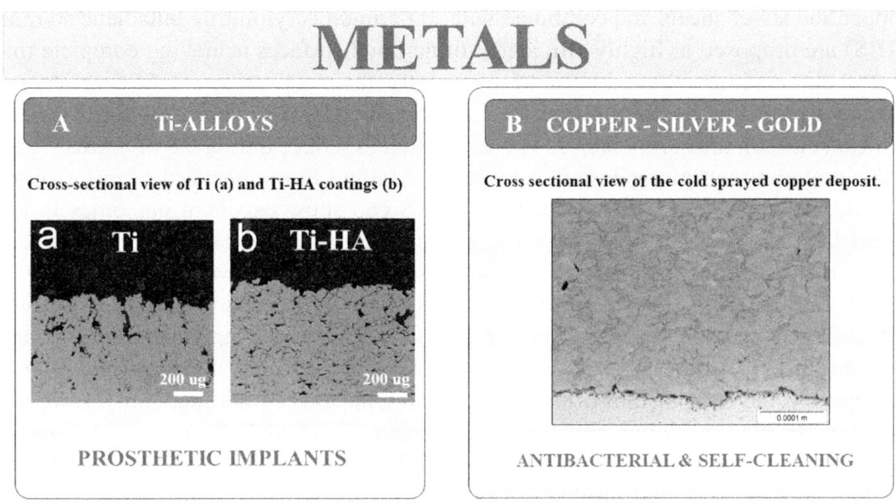

Figure 6.5 Schematic summary of the main metals forming micro/nano coatings for biomedical devices (A) cross-sectional view of Ti-Alloys coating obtained via cold spray technique for prostethic implants, (B) Cross sectional view of copper deposit obtained via cold spray technique for antibacterial and self-cleaning applications.
Modified From Guillem-Marti, J., Cinca, N., Punset, M., Cano, I.G., Gil, F.J., Guilemany, J.M., Dosta, S., 2019. Porous titanium-hydroxyapatite composite coating obtained on titanium by cold gas spray with high bond strength for biomedical applications. Coll. Surf. B Biointerf. 180, 245–253. https://doi.org/10.1016/j.colsurfb.2019.04.048 (panel on the left). Modified from From Champagne, V.K., Helfritch, D.J., 2013. A demonstration of the antimicrobial effectiveness of various copper surfaces. J. Biol. Eng. 7 (1), 8–8. https://doi.org/10.1186/1754-1611-7-8. Open Access article distributed under the terms of the Creative Commons Attribution License (panel on the right)

antipathogenic functions, and the use reduces the impact of cleaning procedures on the clinical environment and can contribute to stop the transmission of dangerous germs and viruses (Fu and Gray, 2021; Saha et al., 2023). Furthermore, the use of these metals to maintain a sterile environment reduces the risk of hospital-acquired infections and raises the sanitation standards (Dosta et al., 2018; Ghosh et al., 2020; Lucas et al., 2020). Copper and its alloys, as Cu–Ag for example, are combined with stainless or carbon steel to produce surface having enhanced antimicrobial characteristics against for distinct microorganisms after very short time (da Silva et al., 2019). Similarly, Zn–Al–Mg–TiO_2 coatings (Wang et al., 2020) proposed via CS are proposed as antimicrobial and antifouling agent for medical field under light irradiation, thanks to their improved photocatalytic properties (Chung et al., 2007; Chung et al., 2008; Tsou and Hsieh, 2017). Efficient antimicrobial metal-based coatings are lately proposed by combining the CS and 3D printing technologies (Lucas et al., 2020) to address the issues of antibiotic resistance and hospital-acquired illnesses. For these coatings, the authors, used copper and silver metals which have antimicrobial properties and as substrates they used copper sheets and. In particular, CS coatings formed by copper and silver sheets and combined with 3D printed acrylonitrile butadiene styrene (ABS) are proposed as highly efficient antimicrobial surfaces achieving complete microbial elimination within minutes. More recently, the inactivation of SARS-CoV-2 on contact is proposed by using copper coatings obtained by CS preparation route (Sousa and Cote, 2020), thus proposing as viable material to reduce the spread of SARS-CoV-2 also providing intriguing design concepts for coatings that suppress the virus. Consequently, the metal micro-/nano coatings by CS can show excellent outcomes in the biomedical and medical fields for projects, thus aiming to prevent the spread of bacterial and viral infections as well as for prosthetic implant development. Finally, a recent study proposes the possibility of depositing gold nanoparticles by CS process with tailored operating parameters for plasmonic biosensor applications (Moonga and Jen, 2019). Although CS is used to deposit nanoparticles onto surfaces, limited examples in the field of thin films for biosensing appears in the literature. However, it represents an intriguing perspective to realize functional plasmonic biosensing thin films with finely tuned coatings thickness (less than 100 m), surface and interface roughness, grain size, and residual stresses.

6.4.2 Ceramics

Ceramic coatings generate a great interest because of their antibacterial and anticorrosive properties (Liao et al., 2022) (Fig. 6.6). Using a bioactive ceramic coating has advantages in the biomedical industry since bioactive fixation creates a stronger binding than basic mechanical fixation (Vilardell et al., 2015). For instance, the main biocompatible and bioactive material that might enhance the osseointegration of titanium implants is hydroxyapatite (HA), but to that is important that the chemical properties of HA are preserved during the synthesis (Vilardell et al., 2019).

To create HA coatings with greater density and regulated crystallinity, CS represent a promising strategy in comparison to other low-temperature methods for its straightforward and affordable method of producing coatings at low temperatures with refined coatings' microstructure.

Also, metal oxides are investigated as antibacterial coatings. For example, ZnO nanopowders and their Al-mixed composite deposited on glass substrates via CS methodology display excellent bacteriostatic activity against *Escherichia coli* (Sanpo et al., 2008).

Magnesium (Mg) has garnered considerable attention as a potential substitute biomaterial due to its facile biodegradability and mechanical properties that closely resemble those of bone. Indeed, Mg-based alloys possess remarkable qualities that render them suitable in constructing porous scaffolds for bone healing, bone fixation materials, and cardiovascular stents (Rahman et al., 2020). Nevertheless, the main limitation to the medical application is their rapidly and localized corrosion behavior. Studies have not been developed for its high temperatures that could melt magnesium substrate and decompose HA in other calcium phosphate phases, and the crystallinity of HA may also be lowered due to rapid solidifications. CS has offered solution to both problems (Noorakma et al., 2013). Indeed, to overcome this issue, an effective

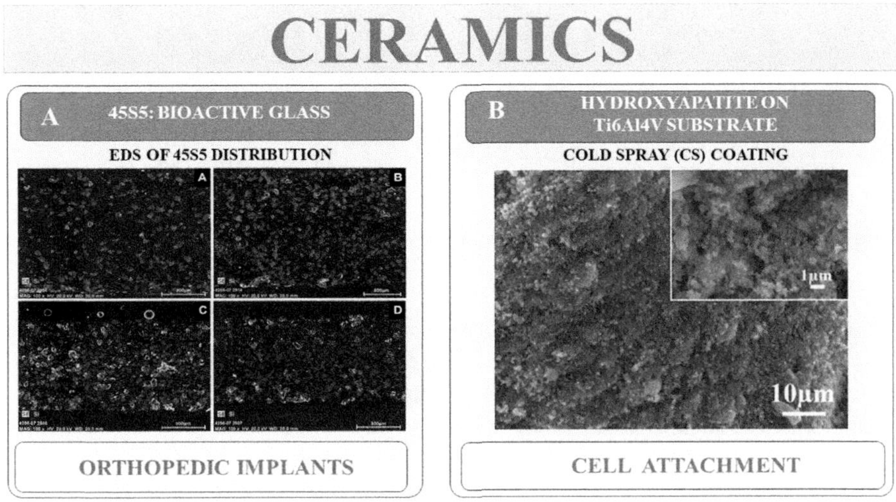

Figure 6.6 Schematic summary of the main ceramics forming micro/nano coatings for biomedical devices. (A) EDS of a bioglass/PEEK composite coating obtained via cold spray technique for orthopedic implants, (B) FESEM micrographs of the top surface areas of hydroxyapatite coating obtained via cold spray techinque
Modified from Garrido, B., Albaladejo-Fuentes, V., Cano, I.G., et al., 2022. Development of bioglass/PEEK composite coating by cold gas spray for orthopedic Implants. J Therm Spray Tech 31, 186−196. https://doi.org/10.1007/s11666-021-01312-w. Open Access article distributed under the terms of the creative commons attribution Licence (panel on the left). Modified from Vilardell, A., Cinca, N., Garcia-Giralt, N., Dosta, S., Cano, I., Nogués, X., Guilemany, J., 2020. In-vitro comparison of hydroxyapatite coatings obtained by cold spray and conventional thermal spray technologies. 107, https://doi.org/10.1016/j.msec.2019.110306 (panel on the right).

approach involves coating the Mg alloy with HA to decelerate the degradation rate, which would otherwise occur rapidly.

6.4.3 Composites

Another important class of material to realize functional biointerfaces for biomedical applications is represented by ceramic/metal and organic/inorganic micro-/nanostructured composites. They are able to exert a high biocompatibility and a great ability to defend against bacterial pathogens, which are very useful for tissue and prothesis engineering, among other medical equipment, Fig. 6.7.

Specifically, metal/ceramic composites are commonly employed to create bioactive coating by CS technique with the aim to prevent the ceramic sample from breaking down related to the poor ductileness metals counteracting ceramics and to improve the functional properties. Thick and dens nanocoatings of titanium/hydroxyapatite (Ti/HA) (Judd et al., 2019) and microcoating of tantalum/hydroxyapatite (Ta/HA) (Tang et al., 2019) are proposed to avoid HA decomposition and to confer excellent

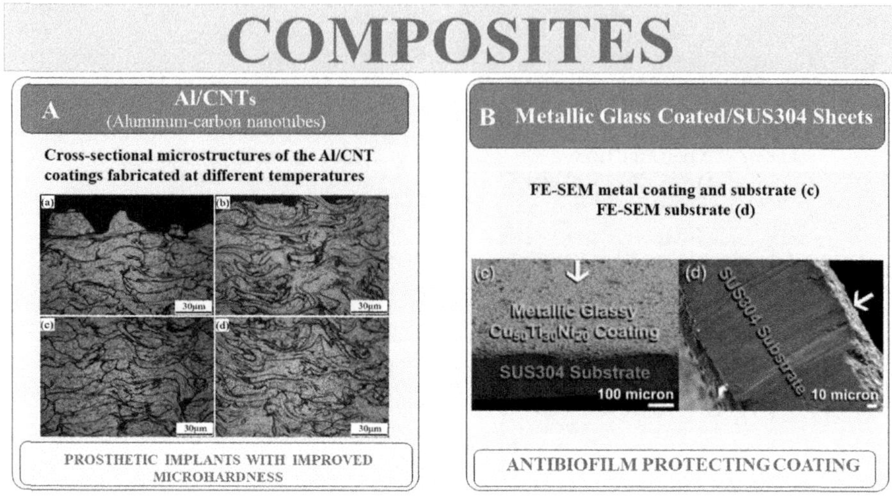

Figure 6.7 Schematic summary of the main composites materials forming micro-/nanocoatings for biomedical devices. (A) Optical micrographs illustrating etched cross-sectional microstructures of the carbon nanotubes/Al composite coatings obtained via cold spray technique, (B) Images of a two-faced coated sample of SUS304 utilizing cold spray metallic glassy $Cu_{52}Ti_{17}Ni_{31}$ powders.
Modified From Xie, X., Tan, Z., Chen, C., Xie, Y., Wu, H., Yan, X., Gao, S., Li, Z., Ji, G., Liao, H., 2021. Synthesis of carbon nanotube reinforced Al matrix composite coatings via cold spray deposition. Surf. Coat. Technol. 405, 126676−126676. https://doi.org/10.1016/j.surfcoat.2020.126676 (panel on the left). Modified From Aldhameer, A., El-Eskandarany, M., Kishk, M., Alajmi, F., Banyan, M., 2022. Mechanical alloying integrated with cold spray coating for fabrication Cu50(Ti50−xNix), x; 10, 20, 30, and 40 at.% antibiofilm metallic glass coated/SUS304 sheets. Nanomaterials 12 (10), 1681 https://doi.org/10.3390/nano12101681. Open Access article distributed under the terms of the Creative Commons Attribution License (panel on the right)

resistance to corrosion, biocompatibility and osteoconductivity, as useful for dental and orthopedic implants. Through a simulated body fluid (SBF) test, the formation of a bone-like apatite layer on the coating is obtained, thus confirming a great biocompatibility to be integrated with the host tissue. Certainly, these coatings determine increased levels of cell adhesion, proliferation, and differentiation, thus enhancing their tissue regeneration properties (Lotfi et al., 2013; Naqvi and McNamara, 2020; Watanabe and Osada, 2017). Paterlini et al. propose nanocoatings able to maximized bone osseointegration of metallic implants by using stoichiometric hydroxyapatite and biomimetic nanocrystalline apatite powders. Utilizing a low-pressure CS system, different powders are deposited onto Ti6Al4V alloy substrates, showing high bioactivity and mechanical stability after distinct surface pretreatments (Paterlini et al., 2022).

Organic/inorganic micro-/nanostructured composites are also proposed to furnish antimicrobial and antiviral ability to the coatings. A mixed hydroxyapatite/gentamicin sulfate (HA/GS) nanocoating on titanium substrate is obtained via CS method (Li et al., 2017) which, thanks to the low operating temperatures, allows to maintain undamaged the chemical and antibacterial characteristics against *E. coli* of GS. The designed HA/GS biointerfaces demonstrate exceptional biocompatibility with human osteoblast cells and long-term GS release. Similarly, biodegradable stents formed by coatings of Fe and SS 316L composite material onto mild carbon steel cylindrical rod are also produced by CS way, thus resulting in coatings with improved tensile strength, ductility, bioactivity, and a slower corrosion rate, thanks the tuning of the weight percentage of Fe powder in the feedstock, the grain sizes (in the submicron range) of sprayed particles, as well as the thermal treatments with excellent strength and fatigue resistance (Frattolin et al., 2019). Another important class of micro-/nano-structured interfaces for biomedical applications is represented by coatings including on carbon nanotubes (CNTs) and graphene-like materials, which are highly considered for implants in implantology, cardiology, and neurology. For example, Al/CNT composite powders deposited onto stainless steel plates by CS are used to fabricate hybrid composite coatings exerting an enhanced improvement in the microhardness if compared to pure Al coatings. The resulted strengthening derives from grain boundary hardening due to nanocrystalline structure and by CNTs inclusion (Xie et al., 2021), thus proposing as stable and resistant interfaces to imitate biomaterials placed in touch with biological environments. Similarly, hydroxyapatite/graphene—nanosheet composite coatings deposited by vacuum CS show a markedly enhanced attachment and proliferation of the osteoblast cells. It is ascribable to the nanotexture or nanoroughness determined by graphene nanosheets, which favor a fast adsorption of key serum proteins-like fibronectin with an elongated stretching conformation and length of ~100—135 nm, thus leading to the formation of a biomimetic interface (Liu et al., 2014).

6.4.4 Polymers

Polymeric materials are utilized due to their convenient manufacturability, flexibility, and compatibility, as well as their extensive array of mechanical, thermal, electrical, and chemical properties when coupled with various materials to form composites (Prashar and Vasudev, 2023) for application in biomedical field Fig. 6.8. An emerging field in biomedical devices is shape memory, where materials can undergo deformation based on specific features. Currently, there are numerous obstacles in utilizing shape memory polymers as medical implants and devices, such as the production of polymer fibers and the activation of polymers by different stimuli. In this context, CS represents a good strategy to fabricate bioactive interfaces.

The thermoplastic polymer PEEK shows great potential as a biomaterial in the field of orthopedics due to its Young's modulus of 4—24 GPa, which closely matches that of cortical bone. Thanks to its good biocompatibility as well as both chemical and physically stability, PEEK is a suitable alternative to metal implant component. For this reason, 45S5/PEEK bioactive coatings with enhanced wear resistance are easily designed through CS technique (Garrido et al., 2022). PEEK is a biocompatible polymer with elasticity comparable to that of bone, but it does not show a bioactivity

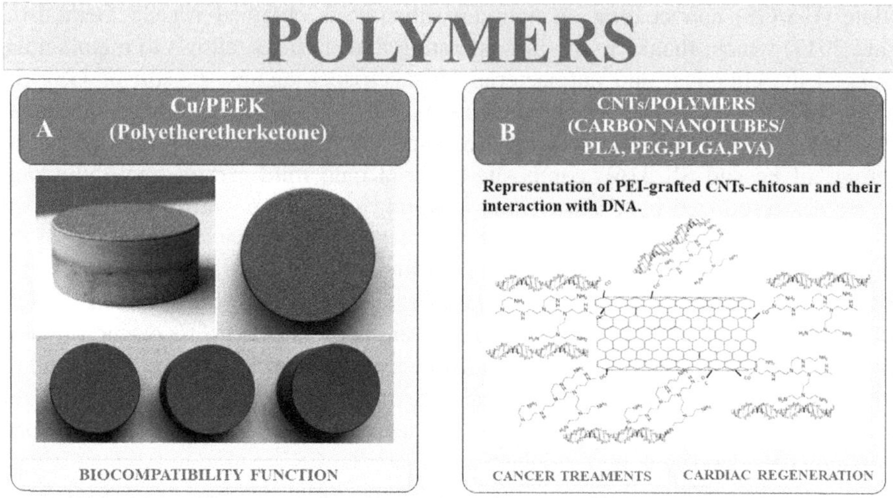

Figure 6.8 Schematic summary of the main polymers forming micro-/nanocoatings for biomedical devices. (A) Cu coatings on cylindrical PEEK cylinders obtained via cold spray techinque. (B) Description of the interaction of polymers/carbon nanotubes composites with DNA.
Modified from Raoelison, R., Lalu Koithara, L., Costil, S., 2021. Cold spray coating of PEEK surface by copper deposition: Interfacial adhesion at high deposition efficiency and bonding strength. CIRP J. Manuf. Sci. Technol. 35, 63–68 https://doi.org/10.1016/j.cirpj.2021.05.008 (panel on the left). Modified From Arumugam, S., Ju, Y., 2021. Carbon nanotubes reinforced with natural/synthetic polymers to mimic the extracellular matrices of bone – a review. Mater. Today Chem. 20, 100420–100420. https://doi.org/10.1016/j.mtchem.2020.100420 (panel on the right).

(Almasi et al., 2016). In order to produce orthopedic implants, it may be advantageous to modify PEEK using a bioactive glass. CS technique is proposed as helpful for the production of medical implants by using blends of PEEK and 45S5 powders. The presence of PEEK in the blend facilitates the deposition of glass particles, which are homogeneously distributed through the coatings and on the surface at the end of the process. Moreover, the obtained coatings with the glass showed higher wear resistance and hardness then pure PEEK coatings, thus forming a bond with bone tissue and promote bone regrowth, although with a poor mechanical quality (Fan et al., 2004; Ur Rehman et al., 2018).

Silk fibroin polymers possess a biological origin and represent another noteworthy material for application in the biomedical industry, particularly in the development of medical equipment (Du et al., 2006). They are used to enhance cell adhesion, also serving as components for bone regeneration, exerting a significant ability to accelerate the rate of healing (Algarrahi et al., 2015). Similarly, the integration of CNTs within polymeric scaffolds is promising for cardiac regeneration, cancer treatment, and bone or nerve restoration. Indeed, composited of CNT reinforced with chitosan, poly(lactic acid) (PLA), poly(lactide-*co*-glycolide) (PLGA), poly(ethylene glycol) (PEG), polyvinyl alcohol (PVA), and polycaprolactone (PCL) can mimic the extracellular matrices of bone, showing increased scaffold strength and flexibility, improved biocompatibility, retardation of cancer cells division, and enhancement of angiogenesis. They show antimicrobial and antiadhesive behavior and can be opportunely functionalized to fabricate various biosensors for biomolecular detection (Rogala-Wielgus and Zieliński, 2023).

6.5 Conclusion and future perspectives

The demand for functional coatings for biomedical applications is rising steadily. Here, CS processes are starting to play a very important role in regard to future developments, thus proposing as an option to engineer customized coatings with well-controlled microstructures and compositions, which are key requirements in biomedical devices where several properties often have to be incorporated into the design. Although the CS process is a relatively new coating technology, many studies have shown the potential it has in tailoring surface properties such as roughened surfaces for better biocompatibility, extremely dense surfaces for corrosion resistance, porous surfaces for enhanced mechanical compatibility, and antimicrobial surfaces retaining long-term effectiveness. In addition, coatings consisting of composite materials can be produced by either a single CS step or by hybrid surface treatments. The favorable and tunable process parameters can guarantee the opportunity to exploit this technology to cover a range of functionalities providing a superior implant coating, combining antibacterial activity, anticorrosion activity, bioactivity, and biocompatibility.

In this chapter, a general overview of the main examples of CS coatings designed as micro-/nanobio-interfaces have been proposed in terms of the chemical, dimensional, and structural features of the materials used for their production and as function of their applications into biomedical devices for personalized medicine. In the paragraph 8.1, the concept of a rationale design of the functional micro-/nanostructured biointerfaces for medicine was introduced. In the paragraphs 8.2 and 8.3, the high-performance biomedical coatings obtained by CS technology as well as its fundamental and operational principles were proposed. In the paragraph 8.4, the main significant examples of micro-/nanostructured coatings for biomedical field were finally described, also focusing on the links between the chemical features of adopted materials (e.g., metals, ceramics, composites, and polymers) and the functional properties (e.g., antimicrobial/antiviral, biocompatibility, biosensing, and cells adhesion) for fabricating advanced medical devices.

Therefore, in the next future, the CS technique is expected to play a significant role in the 3D-printed patient-specific implant industry, based on its ability to efficiently deposit thick, low cost, and high-quality coatings. The use of CS technique to embed particles into soft substrate materials such as polymers is predicted to be of extraordinary interest going forward, as this provides access to unique multifunctional materials and properties. At this point in time, a great interest was focused on the investigation of production design as well as mechanical properties and functionalities of biocompatible coatings, while only few attempts have been made to understand and control the bio-interfacial interactions occurring between these coatings and the biological environment. This represents a point on which a great of interest will be focused in the next future.

A long road still has to be covered. As the technology will mature, it will also become apparent that the technology will require tight control of the multiple parameters playing a key role for the resulting coatings, due to the need to adhere to regulatory requirements related to biomedical applications and devices. However, excellent opportunities for new intellectual property and scientific advances will be able to open up.

References

Adaan-Nyiak, M.A., Tiamiyu, A.A., 2023. Recent advances on bonding mechanism in cold spray process: a review of single-particle impact methods. J. Mater. Res. 38 (1), 69–95. https://doi.org/10.1557/s43578-022-00764-2.

Algarrahi, K., Franck, D., Ghezzi, C.E., Cristofaro, V., Yang, X., Sullivan, M.P., Chung, Y.G., Affas, S., Jennings, R., Kaplan, D.L., Estrada, C.R., Mauney, J.R., 2015. Acellular bi-layer silk fibroin scaffolds support functional tissue regeneration in a rat model of onlay esophagoplasty. Biomaterials 53, 149–159. https://doi.org/10.1016/j.biomaterials.2015.02.092.

Ali, A., Petrů, M., Azeem, M., Noman, T., Masin, I., Amor, N., Militky, J., Tomková, B., 2023. A comparative performance of antibacterial effectiveness of copper and silver coated textiles. J. Ind. Textil. 53. https://doi.org/10.1177/15280837221134990.

Almasi, D., Iqbal, N., Sadeghi, M., Sudin, I., Kadir, M.R.A., Kamarul, T., 2016. Preparation methods for improving PEEK's bioactivity for orthopedic and dental application: a review. Int. J. Biomater. 2016, 1–12. https://doi.org/10.1155/2016/8202653.

Amantea, F., Antignani, G., Pota, G., Cascone, E., Parisi, S., Alfè, M., Gargiulo, V., Luciani, G., Pezzella, A., D'Errico, G., Di Capua, R., Vitiello, G., 2023. Hybrid eumelanin/graphene-like/TiO2 bio-interface for mouse embryonic stem cell (mESC) growth: compatibility toward neurons and beating cardiomyocytes differentiation. Appl. Surf. Sci. 633. https://doi.org/10.1016/j.apsusc.2023.157608.

Assadi, H., Gärtner, F., Stoltenhoff, T., Kreye, H., 2003. Bonding mechanism in cold gas spraying. Acta Mater. 51 (15), 4379–4394. https://doi.org/10.1016/S1359-6454(03)00274-X.

Bagno, A., Piovan, A., Dettin, M., Chiarion, A., Brun, P., Gambaretto, R., Fontana, G., Di Bello, C., Palù, G., Castagliuolo, I., 2007. Human osteoblast-like cell adhesion on titanium substrates covalently functionalized with synthetic peptides. Bone 40 (3), 693–699. https://doi.org/10.1016/j.bone.2006.10.007.

Chung, C.J., Lin, H.I., Tsou, H.K., Shi, Z.Y., He, J.L., 2008. An antimicrobial TiO_2 coating for reducing hospital-acquired infection. J. Biomed. Mater. Res. B Appl. Biomater. 85B (1), 220–224. https://doi.org/10.1002/jbm.b.30939.

Chung, C.J., Lin, H.I., He, J.L., 2007. Antimicrobial efficacy of photocatalytic TiO2 coatings prepared by arc ion plating. Surf. Coating. Technol. 202 (4–7), 1302–1307. https://doi.org/10.1016/j.surfcoat.2007.07.077.

Ciacotich, N., Ud Din, R., Sloth, J.J., Møller, P., Gram, L., 2018. An electroplated copper–silver alloy as antibacterial coating on stainless steel. Surf. Coating. Technol. 345, 96–104. https://doi.org/10.1016/j.surfcoat.2018.04.007.

Dikici, B., Topuz, M., 2018. Production of annealed cold-sprayed 316L stainless steel coatings for biomedical applications and their in-vitro corrosion response. Protect. Met. Phys. Chem. Surface 54 (2), 333–339. https://doi.org/10.1134/S2070205118020168.

Donlan, R.M., 2002. Biofilms: microbial life on surfaces. Emerg. Infect. Dis. 8 (9), 881–890. https://doi.org/10.3201/eid0809.020063.

Dosta, S., Cinca, N., Vilardell, A.M., Cano, I.G., Maria Guilemany, J., 2018. Cold Spray Coatings for Biomedical Applications Cold-Spray Coatings. Springer International Publishing, Cham, pp. 533–557. https://doi.org/10.1007/978-3-319-67183-3_19.

Du, N., Liu, X.Y., Narayanan, J., Li, L., Lim, M.L.M., Li, D., 2006. Design of superior spider silk: from nanostructure to mechanical properties. Biophys. J. 91 (12), 4528–4535. https://doi.org/10.1529/biophysj.106.089144.

Fan, J.P., Tsui, C.P., Tang, C.Y., Chow, C.L., 2004. Influence of interphase layer on the overall elasto-plastic behaviors of HA/PEEK biocomposite. Biomaterials 25 (23), 5363–5373. https://doi.org/10.1016/j.biomaterials.2003.12.050.

Frattolin, J., Roy, R., Rajagopalan, S., Walsh, M., Yue, S., Bertrand, O.F., Mongrain, R., 2019. A manufacturing and annealing protocol to develop a cold-sprayed Fe-316L stainless steel biodegradable stenting material. Acta Biomater. 99, 479–494. https://doi.org/10.1016/j.actbio.2019.08.034.

Fu, H., Gray, K.A., 2021. The key to maximizing the benefits of antimicrobial and self-cleaning coatings is to fully determine their risks. Curr. Opin. Chem. Eng. 34. https://doi.org/10.1016/j.coche.2021.100761, 100761–100761.

Ganesan, A., Affi, J., Yamada, M., Fukumoto, M., 2012. Bonding behavior studies of cold sprayed copper coating on the PVC polymer substrate. Surf. Coating. Technol. 207, 262–269. https://doi.org/10.1016/j.surfcoat.2012.06.086.

Garrido, B., Albaladejo-Fuentes, V., Cano, I.G., Dosta, S., 2022. Development of Bioglass/PEEK composite coating by cold gas spray for orthopedic implants. J. Therm. Spray Technol. 31 (1–2), 186–196. https://doi.org/10.1007/s11666-021-01312-w.

Ghelichi, R., Guagliano, M., 2009. Coating by the cold spray process: a state of the art. Frat. Ed. Integrità Strutt. 3 (8), 30–44. https://doi.org/10.3221/IGF-ESIS.08.03.

Ghosh, M., Roy, A., Ghosh, A., Kumar, H., Saha, G., 2020. Antibacterial and antimicrobial coatings on metal substrates by cold spray technique: present and future perspectives. In: Green Approaches in Medicinal Chemistry for Sustainable Drug Design. Elsevier, pp. 15–45. https://doi.org/10.1016/B978-0-12-817592-7.00002-2.

Guillem-Marti, J., Cinca, N., Punset, M., Cano, I.G., Gil, F.J., Guilemany, J.M., Dosta, S., 2019. Porous titanium hydroxyapatite composite coating obtained on titanium by cold gas spray with high bond strength for biomedical applications. Colloids Surf. B Biointerfaces 180, 245–253. https://doi.org/10.1016/j.colsurfb.2019.04.048.

Hassani-Gangaraj, M., Veysset, D., Champagne, V.K., Nelson, K.A., Schuh, C.A., 2018. Adiabatic shear instability is not necessary for adhesion in cold spray. Acta Mater. 158, 430–439. https://doi.org/10.1016/j.actamat.2018.07.065.

Huang, C., Wang, T., Zhang, Z., 2021. Research progress in numerical simulation of Laval nozzle. J. Phys. Conf. 2012 (1). https://doi.org/10.1088/1742-6596/2012/1/012090.

Hussain, T., McCartney, D.G., Shipway, P.H., Zhang, D., 2009. Bonding mechanisms in cold spraying: the contributions of metallurgical and mechanical components. J. Therm. Spray Technol. 18 (3), 364–379. https://doi.org/10.1007/s11666-009-9298-1.

Hutasoit, N., Kennedy, B., Hamilton, S., Luttick, A., Rahman Rashid, R.A., Palanisamy, S., 2020. Sars-CoV-2 (COVID-19) inactivation capability of copper-coated touch surface fabricated by cold-spray technology. Manuf. Lett. https://doi.org/10.1016/j.mfglet.2020.08.007.

Irissou, E., Legoux, J.-G., Ryabinin, A.N., Jodoin, B., 2008. Christian moreau, review on cold spray process and technology: Part I—intellectual property. J. Therm. Spray Technol. 17 (4), 495–516. https://doi.org/10.1007/s11666-008-9203-3.

Jaggessar, A., Shahali, H., Mathew, A., Prasad, K., Yarlagadda, D.V., 2017. Bio-mimicking nano and micro-structured surface fabrication for antibacterial properties in medical implants. J. Nanobiotechnol. 15 (1). https://doi.org/10.1186/s12951-017-0306-1, 64–64.

Judd, K.G., Sharma, M.M., Eden, T.J., 2019. Multifunctional bioceramic composite coatings deposited by cold spray. Key Eng. Mater. 813, 228–233. https://doi.org/10.4028/www.scientific.net/KEM.813.228.

Kumar, S., Nehra, M., Kedia, D., Dilbaghi, N., Tankeshwar, K., Kim, K.-H., 2020. Nanotechnology-based biomaterials for orthopaedic applications: recent advances and future prospects. Mater. Sci. Eng. C 106. https://doi.org/10.1016/j.msec.2019.110154.

Kumar, S., Kumar, M., Jindal, N., 2020. Overview of cold spray coatings applications and comparisons: a critical review. World J. Eng. 17 (1), 27–51. https://doi.org/10.1108/WJE-01-2019-0021.

Li, D., Gong, Y., Chen, X., Zhang, B., Zhang, H., Jin, P., Li, H., 2017. Room-temperature deposition of hydroxyapatite/antibiotic composite coatings by vacuum cold spraying for antibacterial applications. Surf. Coating. Technol. 330, 87–91. https://doi.org/10.1016/j.surfcoat.2017.09.085.

Liao, T.-Y., Biesiekierski, A., Berndt, C.C., King, P.C., Ivanova, E.P., Thissen, H., Kingshott, P., 2022. Multifunctional cold spray coatings for biological and biomedical applications: a review. Prog. Surf. Sci. 97 (2). https://doi.org/10.1016/j.progsurf.2022.100654.

Liu, Y., Dang, Z., Wang, Y., Huang, J., Hua, L., 2014. Hydroxyapatite/graphene-nanosheet composite coatings deposited by vacuum cold spraying for biomedical applications: inherited nanostructures and enhanced properties. Carbon 67, 250–259. https://doi.org/10.1016/j.carbon.2013.09.088.

Lotfi, M., Nejib, M., Naceur, M., 2013. Cell adhesion to biomaterials: concept of biocompatibility. In: Advances in Biomaterials Science and Biomedical Applications. InTech. https://doi.org/10.5772/53542.

Lucas, M.D.I., Botef, I., Reid, R.G., van Vuuren, S.F., 2020. Laboratory-based study of novel antimicrobial cold spray coatings to combat surface microbial contamination. Infect. Control Hosp. Epidemiol. 41 (12), 1378–1383. https://doi.org/10.1017/ice.2020.335.

Mitra, D., Kang, E.-T., Gee Neoh, K., 2020. Antimicrobial copper-based materials and coatings: potential multifaceted biomedical applications. ACS Appl. Mater. Interfaces 12 (19), 21159–21182. https://doi.org/10.1021/acsami.9b17815.

Moonga, K.H., Jen, T.-C., 2019. Techniques to investigate the application of cold spray in fabrication of nanoplasmonic films: parameter identification. Procedia Manuf. 35, 327–336. https://doi.org/10.1016/j.promfg.2019.05.048.

Naqvi, S.M., McNamara, L.M., 2020. Stem cell mechanobiology and the role of biomaterials in governing mechanotransduction and matrix production for tissue regeneration. Front. Bioeng. Biotechnol. 8. https://doi.org/10.3389/fbioe.2020.597661.

Neacşu, I.A., Nicoară, A.I., Vasile, O.R., Vasile, B.Ş., 2016. Inorganic micro- and nano-structured implants for tissue engineering. In: Nanobiomaterials in Hard Tissue Engineering. Elsevier, pp. 271–295. https://doi.org/10.1016/B978-0-323-42862-0.00009-2.

Noorakma, A.C.W., Zuhailawati, H., Aishvarya, V., Dhindaw, B.K., 2013. Hydroxyapatite-coated magnesium-based biodegradable alloy: cold spray deposition and simulated body fluid studies. J. Mater. Eng. Perform. 22 (10), 2997–3004. https://doi.org/10.1007/s11665-013-0589-9.

Oliver, J.-anne N., Su, Y., Lu, X., Kuo, P.-H., Du, J., Zhu, D., 2019. Bioactive glass coatings on metallic implants for biomedical applications. Bioact. Mater. 4, 261–270. https://doi.org/10.1016/j.bioactmat.2019.09.002.

Paiva, C.N., Bozza, M.T., 2014. Are reactive oxygen Species always detrimental to pathogens? Antioxidants Redox Signal. 20 (6), 1000–1037. https://doi.org/10.1089/ars.2013.5447.

Paterlini, A., Alexis, J., Balcaen, Y., Bertrand, G., 2022. Cold spraying of thick biomimetic and stoichiometric apatite coatings for orthopaedic implants. Coatings 12 (6). https://doi.org/10.3390/coatings12060722.

Prashar, G., Vasudev, H., 2021. A comprehensive review on sustainable cold spray additive manufacturing: state of the art, challenges and future challenges. J. Clean. Prod. 310. https://doi.org/10.1016/j.jclepro.2021.127606, 127606–127606.

Prashar, G., Vasudev, H., 2023. Deposition and fabrication of biomaterials using cold spray technique: a review on the application of biomedical-implants. Appl. Surf. Sci. Adv. 18. https://doi.org/10.1016/j.apsadv.2023.100533.

Rahman, M., Li, Y., Wen, C., 2020. HA coating on Mg alloys for biomedical applications: a review. J. Magnesium Alloys 8 (3), 929–943. https://doi.org/10.1016/j.jma.2020.05.003.

Raletz, F., Vardelle, M., Ezo'o, G., 2006. Critical particle velocity under cold spray conditions. Surf. Coating. Technol. 201 (5), 1942–1947. https://doi.org/10.1016/j.surfcoat.2006.04.061.

Ramakrishna, S., Mayer, J., Wintermantel, E., Leong, K.W., 2001. Biomedical applications of polymer-composite materials: a review. Compos. Sci. Technol. 61 (9), 1189–1224. https://doi.org/10.1016/S0266-3538(00)00241-4.

Rashidy Ahmady, A., Ekhlasi, A., Nouri, A., Haghbin Nazarpak, M., Gong, P., Solouk, A., 2023. High entropy alloy coatings for biomedical applications: a review. Smart Mater. Manuf. 1. https://doi.org/10.1016/j.smmf.2022.100009.

Rho, J.-Y., Kuhn-Spearing, L., Zioupos, P., 1998. Mechanical properties and the hierarchical structure of bone. Med. Eng. Phys. 20 (2), 92–102. https://doi.org/10.1016/S1350-4533(98)00007-1.

Rogala-Wielgus, D., Zieliński, A., 2023. Preparation and properties of composite coatings, based on carbon nanotubes, for medical applications. Carbon Lett. https://doi.org/10.1007/s42823-023-00626-9.

Russo, S., Muscetta, M., Amato, P., Venezia, V., Verrillo, M., Rega, R., Lettieri, S., Cocca, M., Marotta, R., Vitiello, G., 2024. Humic substance/metal-oxide multifunctional nanoparticles as advanced antibacterial antimycotic agents and photocatalysts for the degradation of PLA microplastics under UVA/solar radiation. Chemosphere 346. https://doi.org/10.1016/j.chemosphere.2023.140605.

Saha, D.C., Boegel, S.J., Tanvir, S., Nogueira, C.L., Aucoin, M.G., Anderson, W.A., Jahed, H., 2023. Antiviral and antibacterial cold spray coating application on rubber substrate, disruption in disease transmission chain. J. Therm. Spray Technol. 32 (4), 818–830. https://doi.org/10.1007/s11666-023-01553-x.

Sanpo, N., Saraswati, Lu, T.M., Cheang, P., 2008. Anti-bacterial property of cold sprayed ZnO-Al coating. In: International Conference on BioMedical Engineering and Informatics. IEEE, pp. 488–491. https://doi.org/10.1109/BMEI.2008.291.

Sanpo, N., Tan, M.L., Cheang, P., Khor, K.A., 2009. Antibacterial property of cold-sprayed HA-Ag/PEEK coating. J. Therm. Spray Technol. 18 (1), 10–15. https://doi.org/10.1007/s11666-008-9283-0.

Schmidt, T., Gärtner, F., Assadi, H., Kreye, H., 2006. Development of a generalized parameter window for cold spray deposition. Acta Mater. 54 (3), 729–742. https://doi.org/10.1016/j.actamat.2005.10.005.

da Silva, F.S., Cinca, N., Dosta, S., Cano, I.G., Guilemany, J.M., Caires, C.S.A., Lima, A.R., Silva, C.M., Oliveira, S.L., Caires, A.R.L., Benedetti, A.V., 2019. Corrosion resistance and antibacterial properties of copper coating deposited by cold gas spray. Surf. Coating. Technol. 361, 292–301. https://doi.org/10.1016/j.surfcoat.2019.01.029.

Singh, H., Sidhu, T.S., Kalsi, S.B.S., Karthikeyan, J., 2013. Development of cold spray from innovation to emerging future coating technology. J. Braz. Soc. Mech. Sci. Eng. 35 (3), 231–245. https://doi.org/10.1007/s40430-013-0030-1.

Singh, R., Singh, G., Singh Sidhu, H., 2016. The cold spray coating process: a future technique in material deposition. Asian J. Eng. Appl. Technol. 5 (1), 1–3. https://doi.org/10.51983/ajeat-2016.5.1.776.

Singh, S., Singh Raman, R.K., Berndt, C.C., Singh, H., 2021. Influence of cold spray parameters on bonding mechanisms: a review. Metals 11 (12). https://doi.org/10.3390/met11122016.

Sousa, B.C., Cote, D.L., 2020. Antimicrobial copper cold spray coatings and SARS-CoV-2 surface inactivation. MRS Adv. 5 (56), 2873–2880. https://doi.org/10.1557/adv.2020.366.

Tang, J., Zhao, Z., Liu, H., Cui, X., Wang, J., Xiong, T., 2019. A novel bioactive Ta/hydroxyapatite composite coating fabricated by cold spraying. Mater. Lett. 250, 197–201. https://doi.org/10.1016/j.matlet.2019.04.123.

Trelka, A., Góral, A., Żórawski, W., Gancarz, T., Petrzak, P., 2023. The mechanical and tribological properties of cold-sprayed cermet coatings—Al alloy substrate systems. J. Therm. Spray Technol. 32 (6), 1714–1727. https://doi.org/10.1007/s11666-023-01590-6.

Tsou, H.K., Hsieh, P.Y., 2017. Anticorrosive, antimicrobial, and bioactive titanium dioxide coating for surface-modified purpose on biomedical material. In: Application of Titanium Dioxide. InTech. https://doi.org/10.5772/intechopen.68854.

Ur Rehman, M.A., Bastan, F.E., Nawaz, Q., Goldmann, W.H., Maqbool, M., Virtanen, S., Boccaccini, A.R., 2018. Electrophoretic deposition of lawsone loaded bioactive glass (BG)/chitosan composite on polyetheretherketone (PEEK)/BG layers as antibacterial and bioactive coating. J. Biomed. Mater. Res. 106 (12), 3111–3122. https://doi.org/10.1002/jbm.a.36506.

Vilardell, A.M., Cinca, N., Concustell, A., Dosta, S., Cano, I.G., Guilemany, J.M., 2015. Cold spray as an emerging technology for biocompatible and antibacterial coatings: state of art. J. Mater. Sci. 50 (13), 4441–4462. https://doi.org/10.1007/s10853-015-9013-1.

Vilardell, A.M., Cinca, N., Dosta, S., Cano, I.G., Guilemany, J.M., 2019. Feasibility of using low pressure cold gas spray for the spraying of thick ceramic hydroxyapatite coatings. Int. J. Appl. Ceram. Technol. 16 (1), 221–229. https://doi.org/10.1111/ijac.13088.

Wang, F., Shi, L., He, W.-X., Han, D., Yan, Y., Niu, Z.-Y., Shi, S.-G., 2013. Bioinspired micro/nano fabrication on dental implant–bone interface. Appl. Surf. Sci. 265, 480–488. https://doi.org/10.1016/j.apsusc.2012.11.032.

Wang, K., Wang, S., Xiong, T., Wen, D., Wang, G., Liu, W., Du, H., 2020. Properties of Zn-Al-Mg-TiO2 coating prepared by cold spraying. Surf. Coating. Technol. 387. https://doi.org/10.1016/j.surfcoat.2020.125549.

Wang, Q., Yan, J., Yang, J., Li, B., 2016. Nanomaterials promise better bone repair. Mater. Today 19 (8), 451–463. https://doi.org/10.1016/j.mattod.2015.12.003.

Watanabe, N., Osada, H., 2017. Cell Proliferation and Differentiation Bioprobes. Springer Japan, Tokyo, pp. 11–35. https://doi.org/10.1007/978-4-431-56529-1_2.

Xie, X., Tan, Z., Chen, C., Xie, Y., Wu, H., Yan, X., Gao, S., Li, Z., Ji, G., Liao, H., 2021. Synthesis of carbon nanotube reinforced Al matrix composite coatings via cold spray deposition. Surf. Coating. Technol. 405. https://doi.org/10.1016/j.surfcoat.2020.126676.

Yadav, K., Daware, D., Akolkar, V., Padalwar, P., 2022. A study on concepts de-laval nozzle using CFD tool. In: Graduate Research in Engineering and Technology, pp. 65–69. https://doi.org/10.47893/GRET.2022.1132.

Surface-coated magnetic nanostructured materials for robust bio-catalysis and biomedical applications

7

Amira Moustafa Kamel[1], *Mona Moaness*[2] *and Mostafa Mabrouk*[2]
[1]Polymers and Pigments Department, National Research Centre, Cairo, Egypt; [2]Refractories, Ceramics and Building Materials Department, Advanced Materials Technology and Mineral Resources Research Institute, National Research Centre, Cairo, Egypt

7.1 Introduction

Over the past few decades, nanotechnology has emerged as a revolutionary force, transforming diverse fields from electronics to energy. One particularly impactful branch of biomedicine is the development of nanostructured coatings. These minute but mighty materials, meticulously crafted at the nanoscale, offer a plethora of advantages over traditional materials, paving the way for ground-breaking advancements in medical diagnosis, treatment, and therapeutics (Sardo, 2023). At the heart of this revolution lies the unique nature of nanostructured coatings. Unlike their bulk counterparts, these materials possess features sculpted at the atomic level, resulting in remarkable properties specifically tailored for biomedical applications. Some of the key advantages of nanostructured coatings include enhanced biocompatibility and surface versatility. Traditional materials often face compatibility challenges within the body, triggering inflammatory responses and hindering integration with tissues. Nanostructured coatings can mimic the natural architecture of tissues at the nanoscale, promoting seamless integration and minimizing unwanted reactions (García-Merino et al., 2022).

The surface of these coatings acts as a blank canvas, readily amenable to functionalization with diverse molecules. This lets the coating's properties be changed to fit different needs, like adding bioactive molecules for targeted drug delivery or antimicrobial agents to stop infections (Witika et al., 2020). Nanostructured coatings also offer precise control over surface area and topography due to their nanoscale dimensions, providing an immense surface area-to-volume ratio that can significantly enhance performance in various applications. In addition, the precise control over surface topography allows for engineering-specific interactions with biological molecules and cells (Zheng et al., 2021).

Multifunctional Nanostructured Coatings. https://doi.org/10.1016/B978-0-443-23683-9.00007-6

The diverse benefits of nanostructured coatings translate into a vast array of potential applications across various aspects of biomedicine (Barhoum and Makhlouf, 2018). Here are some key examples:

For example, coating implants with nanostructured materials can significantly improve their biocompatibility, reducing the risk of rejection and infection. We can also functionalize these coatings to boost bone growth and tissue integration, resulting in quicker healing and enhanced functionality. These coatings can also be functionalized to enhance bone growth and tissue integration, leading to faster healing and improved functionality (Haleem et al., 2023; Khan et al., 2019; Thakur et al., 2022). It can also design nanostructured coatings for drug delivery, which encapsulate and release drugs in a controlled manner to target specific cells or tissues. This targeted delivery system minimizes side effects, improves therapeutic efficacy, and enables personalized medicine (Faraji et al., 2018; Akl et al., 2023). In addition, utilizing the unique interactions between nanostructured surfaces and biological molecules to develop highly sensitive biosensors that can detect biomarkers in diseases like cancer leads to early diagnosis and improved treatment outcomes (Stephanie et al., 2021).

In regenerative medicine and tissue engineering, nanostructured coatings can also serve as supports for cell growth and differentiation. This makes it easier for damaged organs and tissues to heal, like when bones break or cartilage wears down (Pina et al., 2019).

7.2 Multifunctional coatings: Revolutionizing materials for diverse applications

Multifunctional coatings are emerging as transformative tools, reshaping a wide range of fields, from medicine and energy to aerospace and environmental protection. These meticulously engineered materials transcend the limitations of traditional coatings, offering a symphony of functionalities within a single layer (Pal et al., 2023). This essay delves into the latest advancements in these remarkable materials, exploring their key characteristics, burgeoning applications, and potential benefits that beckon on the horizon (Bagale et al., 2019; Pal et al., 2023). With their numerous capabilities, multifunctional coatings are actively transforming a variety of fields, no longer just a vision of the future. To truly understand their potential, we must delve deeper into the specific types, and techniques used to create them, their remarkable uses, and the challenges that still lie ahead.

7.2.1 Types of multifunctional coatings

The versatility of these coatings is astonishing, stemming from a multitude of material combinations and functionalities (Amirtharaj Mosas et al., 2022). Scientists have developed various types of these coatings, each serving specific purposes. Self-cleaning coatings, for example, leverage photocatalytic properties or superhydrophobic surfaces to repel dirt, grime, and pathogens, making them ideal for use on solar panels, building facades, and medical devices. Materials like titanium dioxide (TiO2), silanes, and fluoro polymers are commonly used in these coatings (Hosseini et al., 2020; Adak et al., 2017).

Antimicrobial coatings, on the contrary, incorporate embedded nanoparticles or biocidal agents to actively combat bacteria and viruses, finding applications in hospitals, medical instruments, implants, food packaging, and public transportation. One of the most commonly used materials for antimicrobial coatings is silver nanoparticles, copper oxide (CuO), and quaternary ammonium compounds (QACs). For instance, Asri et al. created an antibacterial coating that adapts to shapes, using immobilized QACs attached to hyperbranched polyurea. This coating demonstrated effective antibacterial properties against both Gram-positive and Gram-negative bacteria (Perni and Prokopovich, 2020; Meier et al., 2023).

Energy-harvesting coatings utilize thin-film solar cells or piezoelectric materials to convert ambient light or mechanical energy into electricity, powering sensors, wearable electronics, and remote devices (Adak et al., 2017; Rashid et al., 2023).

Self-healing coatings are designed with polymers containing reversible bonds or microfluidic channels filled with healing agents, enabling them to repair minor scratches and cracks, thereby extending the lifespan of materials used in aerospace, infrastructure, and marine environments (Ekeocha et al., 2021; Zhang et al., 2018)

Lastly, biocompatible coatings are tailored for seamless integration with living tissue, making them suitable for use in implants, prosthetics, and drug delivery systems. These examples demonstrate the diverse range of multifunctional coatings and their applications across various industries (Saleh et al., 2016; Jayaraj and Pius, 2018).

7.2.2 Techniques for multifunctional coatings

The artistry behind these coatings lies in meticulously choosing materials and employing various techniques to achieve desired functionalities. One such technique is Layer-by-Layer (LbL) deposition. This technique precisely builds up the coating layer by layer, allowing for fine-tuning of functionalities and properties. The main advantages of this technique are precise control over coating thickness and functionality, and multimaterial versatility. Furthermore, various applications like drug delivery systems, sensors, and biocompatible coatings can utilize this technique (Fan et al., 2018).

Another technique is the sol–gel process, in which precursors form a gel network that solidifies into a coating, providing high flexibility, customizable surface properties, and strong adhesion. Applications include antireflective coatings for optical enhancement, scratch-resistant surfaces for material durability, and the production of ceramic membranes for diverse applications such as filtration. The sol–gel process is versatile, providing tailored properties to meet industry demands (Dehghanghadikolaei et al., 2018; Wu et al., 2022).

Chemical vapor deposition (CVD) is a precise process where gaseous precursors react to form thin, uniform coatings directly onto a substrate. This method is advantageous for creating conformal coatings that adhere closely to substrate contours. Widely used in semiconductor manufacturing, CVD produces thin films for integrated circuits and protective coatings for metals, enhancing durability and performance. Its versatility and precision make CVD valuable for tailoring coatings with specific properties across diverse materials and applications (Chemical vapour deposition (CVD): advances, technology and applications, 2019; Khlyustova et al., 2020).

Electrochemical deposition involves using an electric current to drive ions from a solution onto the substrate. This process enables precise control over the thickness and composition of the coating. By carefully manipulating the electric current, this method allows for the creation of coatings with specific properties tailored to meet desired requirements. The technique finds applications in various industries where controlled deposition is essential for achieving optimal performance and functionality in coated materials (Mozumder et al., 2019; Walsh et al., 2020).

Lastly, laser-assisted techniques involve the use of lasers for various applications, such as etching patterns, activating specific material properties, and directly writing functional structures onto surfaces. The precision and versatility of lasers enable intricate modifications and customization of material surfaces. This technique finds widespread use in diverse fields, ranging from manufacturing to electronics, where the unique capabilities of lasers contribute to the development of advanced materials and functional surfaces with specific and tailored properties (Badiceanu et al., 2022).

Multifunctional coatings exhibit diverse applications across industries. In aerospace, they serve in anti-icing coatings on wings, corrosion-resistant coatings on aircraft bodies, and self-healing coatings for critical components. The energy sector benefits from self-cleaning solar panels, energy-harvesting coatings for smart buildings, and anticorrosive coatings for offshore wind turbines (Mozumder et al., 2019). In healthcare, multifunctional coatings contribute to biocompatible coatings for implants, antimicrobial coatings for medical instruments, and drug-releasing coatings for targeted therapy (Bagale et al., 2019). Environmental applications include air-purifying coatings for buildings, self-cleaning surfaces for public spaces, and antifouling coatings for marine vessels, promoting sustainability (Idumah et al., 2020). The textile industry utilizes multifunctional coatings for wrinkle-resistant fabrics, self-cleaning clothes, and antimicrobial coatings for protective gear (Petrova et al., 2021). The broad range of applications underscores the versatility and adaptability of multifunctional coatings, addressing challenges and enhancing performance across various fields. Fig.7.1 demonstrates different coatings to enhance biocompatibility.

Biomedical applications can tailor multifunctional coatings to seamlessly integrate with tissues, reducing inflammation and promoting tissue growth. This integration is crucial for implants, prosthetics, and biosensors, leading to better patient outcomes and longer device lifespans. For example, coatings with specific biomolecules can mimic the natural extracellular matrix, encouraging cell adhesion and reducing rejection (Saleh et al., 2016; Jayaraj and Pius, 2018).

These coatings also contribute to enhanced infection control by combating bacteria and viruses on medical devices and surfaces, significantly reducing healthcare-associated infections (HAIs). Nanoparticles embedded in the coating, biocidal agents, or even light-activated reactions can achieve this (Teixeira-Santos et al., 2021).

In addition, it can design multifunctional coatings for targeted drug delivery and controlled release, which can release specific drugs directly to the target site, minimize side effects, and optimize treatment efficacy. This can involve stimuli-responsive

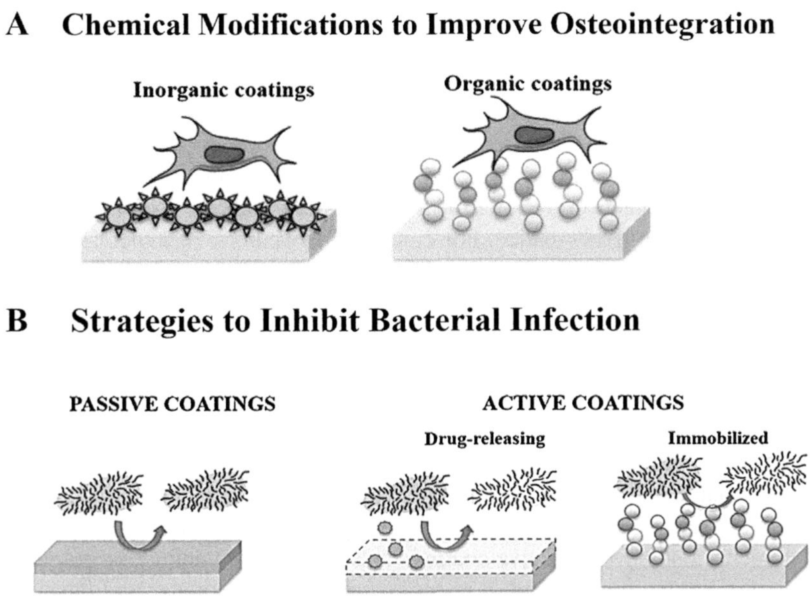

Figure 7.1 Improvement of osteointegration as one of biomedical applications by chemical methods, which are based on inorganic (e.g., calcium phosphate) or organic (e.g., peptide and protein) coatings. (b) The strategies to inhibit bacterial infection are divided into passive (e.g., antiadhesive) or active (e.g., drug-eluting or immobilized) coatings.
Reproduced from Mas-Moruno, C., Su, B., Dalby, M.J., 2019. Multifunctional coatings and nanotopographies: toward cell instructive and antibacterial implants. Adv. Healthcare Mater. 8 (1). https://doi.org/10.1002/adhm.201801103/open access.

materials that release drugs in response to changes in pH, temperature, or other environmental factors (Liang et al., 2018; Chang et al., 2023).

Furthermore, these coatings play a role in personalized medicine and diagnostics, enabling the creation of biosensors that detect specific biomarkers in real time, enabling personalized medicine and early disease diagnosis. For instance, coatings with antibodies can capture specific molecules from the patient's blood or tissue, providing valuable diagnostic information (Tang et al., 2021).

Lastly, in regenerative medicine and tissue engineering, coatings promote tissue regeneration and repair by providing a supportive environment for cell growth and differentiation, promising advancements in treating injuries, burns, and neurodegenerative diseases (Muzzio et al., 2021) Fig. 7.1.

7.2.3 Real-world biomedical applications for multifunctional coatings

Multifunctional coatings have a wide range of real-world biomedical applications, contributing to advancements in medical devices, diagnostics, and therapies. One prominent application is in developing implantable medical devices, such as pacemakers, stents, and orthopedic implants. These coatings can improve the biocompatibility of the devices, reducing the risk of rejection and infection, and promoting integration with surrounding tissues. In the real world, there are plenty of biomedical applications for multifunctional coatings like AcuvueOasys Max with HydraLuxe tear-like lubricant coating and light-filtering technology. This product, featuring a tear-like lubricant coating and light-filtering technology, aims to enhance comfort, reduce dryness, and provide protection from harmful blue light, ultimately promoting eye health and optimizing visual performance. For contact lens wearers, AcuvueOasys Max has improved wearability, increased productivity, and reduced eye fatigue.

Biocompatible titanium hip replacements with an antibacterial coating, like silver nanoparticles, are another example. These are meant to lower the risk of implant-associated infections, which are a big problem in joint replacement surgeries (see Fig.7.2). The inclusion of antibacterial elements seeks to improve patient outcomes, promoting faster healing and reducing healthcare costs linked to infection-related complications. This approach not only addresses immediate infection challenges but also has the potential to enhance the overall efficiency and success of joint replacement procedures (Diez-Escudero and Hailer, 2021; Xi et al., 2021). Ti−6Al−4V titanium dental implants also have bioactive coatings and surfaces that help bone osseointegration. This means that they are more stable, work better for longer, and have a lower chance of failing. This innovation enhances overall patient outcomes, offering improved aesthetics and chewing function in dental implant procedures, as shown in Fig.7.2. In essence, these implants advance both the technical and patient-centric aspects of dental implantology (Nicholson, 2020; De la Garza-Ramos et al., 2020) Fig. 7.2.

7.3 Magnetic nanostructured materials

Magnetic materials, characterized by aligned magnetic moments, respond uniquely to external magnetic fields based on atomic and molecular arrangements. Understanding these properties is crucial for diverse applications. These materials fall into three main types—ferromagnetic, paramagnetic, and diamagnetic—based on their response to magnetic fields (De la Garza-Ramos et al., 2020). These materials are categorized into three main types: ferromagnetic, paramagnetic, and diamagnetic, depending on their interactions with magnetic fields (see Figs.7.3−7.5). Ferromagnetic materials, such as iron, cobalt, nickel, and steel alloys, display a strong attraction to magnets and retain their magnetization even in the absence of an external magnetic field. This behavior is responsible for magnetic domains, which are regions within the material where atomic magnetic moments align. When subjected to a magnetic field, these domains align, contributing to the material's overall magnetic behavior. Transformers, electromagnets, and magnetic storage devices utilize ferromagnetic materials due to their high magnetic permeability, which enhances responsiveness and efficiency in various technological applications (Rasaili et al., 2022; Krishnan, 2016).

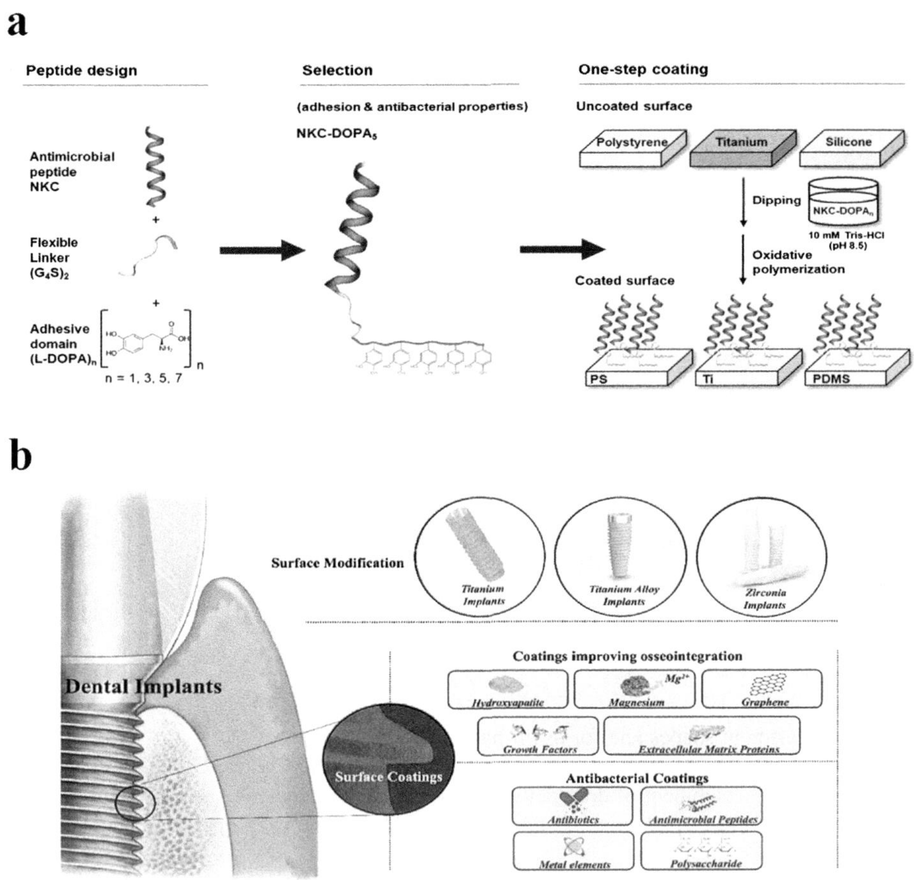

Figure 7.2 Applicable coated materials for (a) antimicrobial coating for medical applications and (b) surface modifies and functional coating of dental implants.
Reproduced from Dong, H., Liu, H., Zhou, N., Li, Q., Yang, G., Chen, L., Mou, Y., 2020. Surface modified techniques and emerging functional coating of dental implants. Coatings 10 (11), 1012. https://doi.org/10.3390/coatings10111012/open access.

Paramagnetic materials exhibit a weak attraction to magnets, becoming magnetized in the direction of an applied magnetic field. Unlike ferromagnetic substances, paramagnetic materials align their magnetic moments when exposed to the field, but they revert to random orientations upon removal of the field. Examples include aluminum, platinum, and oxygen, with a small positive magnetic susceptibility indicating a tendency to align with an external magnetic field, though less pronounced than in ferromagnetic materials. Paramagnetic materials find applications in scientific and technological contexts where controlled magnetic responses are crucial (Abrikosov et al., 2016). Diamagnetic materials, on the contrary, demonstrate a weak repulsion when exposed to magnets, displaying a negative susceptibility. When an external

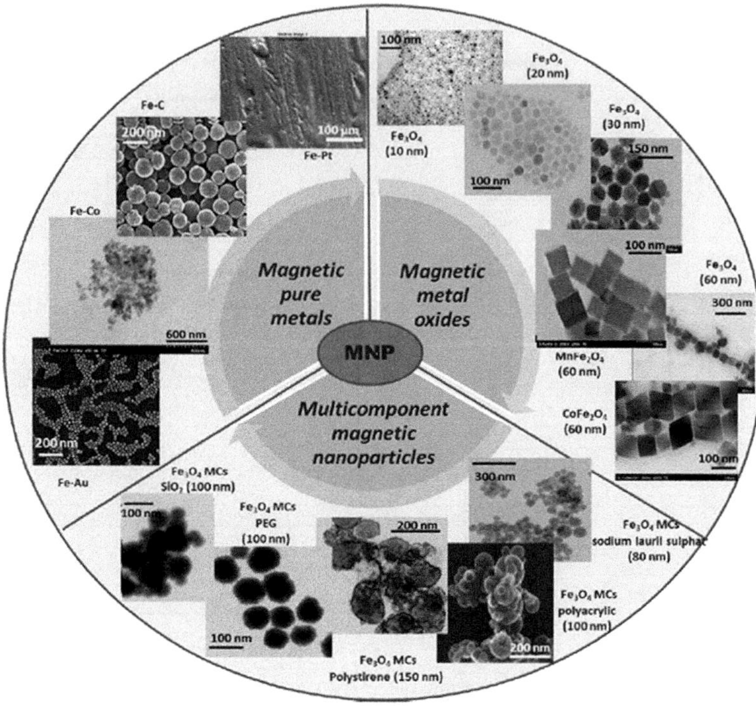

Figure 7.3 Different shapes and forms of magnetic nanoparticles.
From Pusta, A., Tertis, M., Crăciunescu, I., Turcu, R., Mirel, S., Cristea, C., 2023. Recent advances in the development of drug delivery applications of magnetic nanomaterials. Pharmaceutics 15 (7). https://doi.org/10.3390/pharmaceutics15071872-open access.

magnetic field is present, the atomic moments in diamagnetic materials align in the opposite direction, creating a weak opposing force. Common examples include water, copper, gold, and organic compounds. Diamagnetic materials contribute to scientific applications and enhance our understanding of magnetic properties across different material classes, distinguished by their negative susceptibility (Yamato and Kimura, 2020) Fig. 7.3.

7.3.1 Synthesis of magnetic nanoparticles

Innovative approaches to synthesize MNPs (magnetic nanoparticles) have garnered significant attention. Precision in their preparation and surface modification is crucial, as it can impact their physical and chemical traits, stability, mobility, and effectiveness in removing pollutants (Ali et al., 2021). Generally, the synthesis of MNPs can be categorized into three main methods: (a) physical methods, (b) chemical methods, and (c) biological or microbial methods (Kudr et al., 2017). Physical methods adopt a top-down approach, initiating with bulk material and reducing it to create NPs.

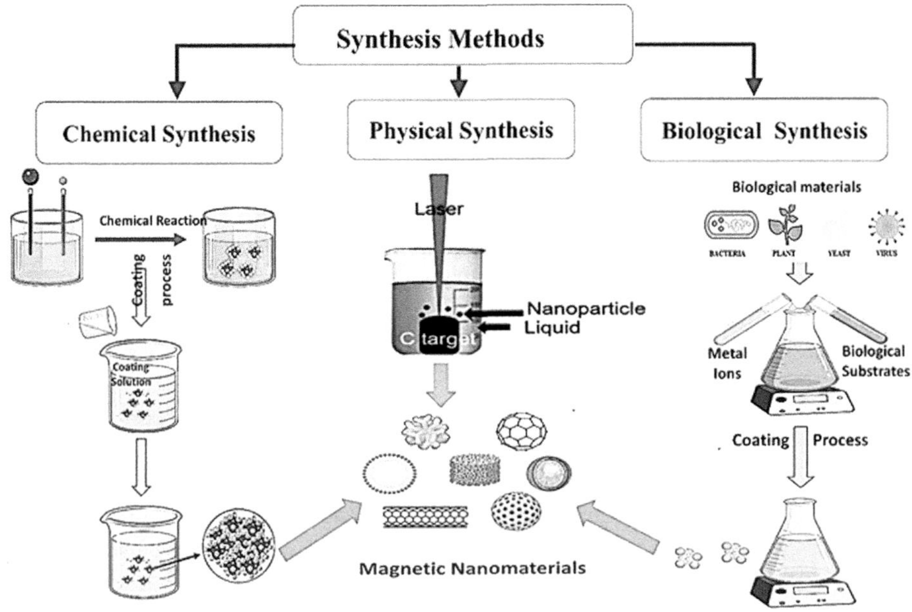

Figure 7.4 Illustrates the synthesis methods of magnetic nanoparticles (NPs); to select one of these methods there are needs to compare between them as emphasizing the broader applications of chemical methods compared with physical ones.

Conversely, chemical and biological methods use a bottom-up approach, assembling atoms or molecules to form NPs of varying sizes.

7.3.1.1 Physical methods of synthesis

Mechanical milling

The mechanical milling method is a simple and organized way to make MNP (magnetic nanoparticle) powder. It makes it easy to make different materials from different starting compounds and to make metallic granular alloys. This milling process employs various mills, such as shaker and planetary mills (Shukla et al., 2021). Several parameters play a role in MNP production via mechanical milling, including the morphology and enhancement achieved at temperatures between 700 and 900°C. Researchers found that the magnetic nanopowders had ferromagnetic properties, including low coercivity (68.68 Oe), remanence (0.840 emu/g), and saturation magnetization (6.517 emu/g). In a different study, activated carbon and Fe_3O_4 nanoparticles were mixed and then milled in a ball mill to make carbonaceous magnetic adsorbents. They found it environmentally friendly, highly efficient, cost-effective, and operationally convenient (Song et al., 2023).

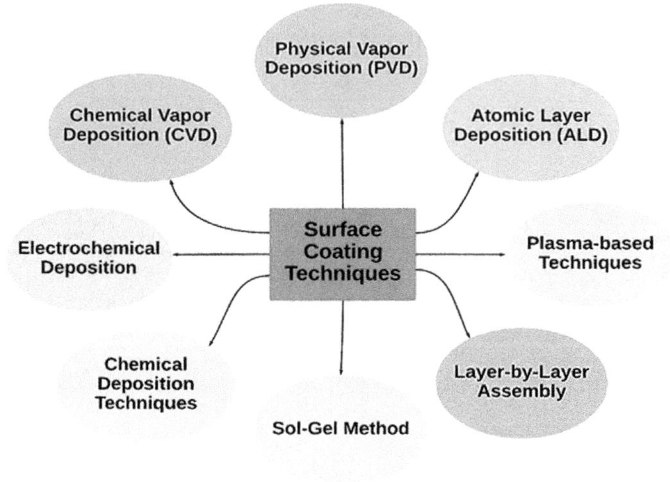

Figure 7.5 Different surface coating techniques for magnetic nanomaterials.

Electron beam lithography

Electron beam lithography (EBL) uses an electron beam (e-beam) to transform iron particles into iron oxides (Fe_3O_4) by generating nanoscale iron oxide nanoparticles (NPs). These nanosized iron oxide particles form as a result of the controlled emission of the e-beam onto an iron-coated surface. This technique has gained recognition for its cost-effectiveness, ability to produce stable nanoparticles, and adaptability (Wang et al., 2019).

Gas-phase deposition

The gas-phase deposition technique involves a catalyst-supported CVD process utilizing molecular precursors, such as $Sn(OBut)_4$ or $[Fe(OBut)_3]_2$, on alumina substrates coated with gold. This method facilitates the creation of one-dimensional iron oxide nanostructures (Ali et al., 2021). Gas-phase synthesis has been widely adopted by numerous researchers for efficiently and economically producing magnetic nanoparticles (MNPs) (Thiam et al., 2020; Khan et al., 2020; Lerner et al., 2016)

Vapor deposition and patterning

This method involves creating magnetic particles by forming a seamless film and filling voids within a given template. Vapor deposition methods, such as laser ablation, sputtering, evaporation, and electrodeposition, deposit the substance that forms these particles. This technique has demonstrated the ability to generate high-quality nanoparticles (NPs), lower alloying temperatures, and has the potential for large-scale application (Wang et al., 2018).

Electrical explosion of wires

The electrical explosion of wires technique, although not widely used for synthesizing MNPs, involves the vaporization of a metal wire using a high-intensity electric current. The initial purity of the metal wire constrains the resulting metal particles, which exhibit exceptional purity and possess a perfectly spherical shape. Notably, this method is considered environmentally friendly as it generates no waste (Lerner et al., 2016).

7.3.1.2 Chemical methods of synthesis

Spray pyrolysis

Spray pyrolysis is the process of applying a solution to reactors to form a solid. The process starts with solvent evaporation from aerosol droplets, followed by solute condensation within the droplets. Subsequently, molecules in the precipitated particle dry and decompose at high temperatures. This technique has notably enhanced the properties of magnetic nanoparticles (MNPs), including magnetization and photocatalytic degradation capabilities, as indicated by studies (Emil Kaya et al., 2020).

Laser pyrolysis

The laser pyrolysis method employs continuous CO_2 laser heat applied to a gas mixture to initiate and sustain chemical reactions. This technique induces localized heating and cooling effects, effectively influencing precursor heating for reactions and nucleation. Through this method, notable enhancements in chemical and morphological characteristics have been observed, including high surface area, crystalline structure, stability, electrical conductivity, coercivity, and magnetic saturation levels reaching up to 70 emu/g (Nizamuddin et al., 2018; Dutu et al., 2015).

Co-precipitation

The co-precipitation method is a widely employed synthesis technique for producing magnetic nanoparticles, such as magnetite (Fe_3O_4) or maghemite (γ-Fe_2O_3). This approach concurrently precipitates ferrous and ferric ions within a solution by adding an alkaline substance, typically sodium hydroxide (NaOH), to a mixture of ferrous and ferric salts. Subsequent transformations transform the resulting iron oxide nanoparticles, and the synthesized material undergoes washing and drying processes to obtain a nanoparticle powder. This method is known for its simplicity, cost effectiveness, and versatility. It allows for the customization of nanoparticle properties by adjusting parameters such as precursor concentrations and pH levels. This versatility makes it applicable in various fields, including biomedicine and environmental science, where magnetic nanoparticles find diverse applications (Darwish et al., 2019; Saragi et al., 2018).

Sol-gel method

The sol-gel method is an extensively researched and widely used technique for nanoparticle synthesis, particularly for metal oxides via a wet chemical route. This method involves adding hydroxyl groups to a molecular precursor and then condensing it in water. This makes a nanoparticle "sol" (Khan et al., 2021). After that, condensation and polymerization create a three-dimensional network of metal oxides (a wet gel), which needs more heat treatments to get to the crystalline structure that is wanted. Notably, this technique offers advantages such as lower working temperatures and improved control over reaction kinetics by varying chemical constituents (Subha et al., 2018).

Polyol method

While the Sol-Gel technique utilizing surfactants can lead to surface charge and morphology alterations in nanoparticles (NPs), the polyol method serves as an alternative approach. This method uses polyols, such as propylene glycol and polyethylene glycol, to control the growth of particles. This keeps newly formed nanoparticles crystal clear and stops particles from sticking together (Vega-Chacón et al., 2016). Widely adopted, especially in biomedical applications such as magnetic resonance imaging, this method involves the direct production of fine metallic particles. It achieves this by reducing dissolved metallic salts and subsequently precipitating metals. The polyol method demonstrates remarkable efficacy in synthesizing bimetallic clusters and nanocrystalline alloys (Lim et al., 2022).

Nonthermal plasma methods

Plasma methods offer eco-friendly alterations without solvents, presenting advantages over traditional reduction processes in terms of speed, environmental safety, adaptability, and energy efficiency. Conventional plasma methods display high gas temperatures exceeding 4000 K, leading to significant energy consumption (Wang et al., 2020). However, researchers are exploring nonthermal plasma techniques due to their enhanced chemical activity, selectivity, and superior energy conversion efficiency. These methods feature lower gas temperatures around 3000 K and higher electron energies surpassing 1 eV, making them more environmentally friendly for synthesizing magnetic nanoparticles (Smolkováet al., 2019).

7.3.1.3 Biological methods

Biological methods are gaining prominence in the synthesis of eco-friendly nanoparticles (NPs) as alternatives to traditional techniques. Utilizing microbes and plants, particularly for magnetic nanoparticles (MNPs) like zero-valent iron, magnetite, and maghemite, has become a focus due to its rapid, environmentally friendly nature. The biocompatibility of MNPs, studied extensively, highlights the pathways offered by microbes and plants, utilizing genetic diversity and enzymes for synthesis. Microbes demonstrate unique morphological characteristics in producing inorganic nanoproducts, while plants use various parts such as extracts, barks, tissues, and exudates for effective metal NP synthesis (Niculescu et al., 2022). These biological approaches offer advantages including higher yield, enhanced reproducibility, scalability, and precise control over NP size and properties. Biological methods present advantages such as higher yield, enhanced reproducibility, scalability, and greater control over the size and properties of resulting NPs (Ahmadi et al., 2021).

7.3.1.4 Comparison between physical, chemical and biological synthesis approaches

Fig. 7.1 illustrates the synthesis methods of magnetic nanoparticles (NPs). To select one of these methods, one needs to compare them, emphasizing the broader applications of chemical methods compared with physical ones. However, concerns about potential adverse impacts, such as the sorption and desorption of toxic substances during chemical synthesis, have led to environmental and toxicity worries. Both physical and chemical methods pose drawbacks, including high costs, hazardous chemical usage, and environmental threats (Kudr et al., 2017; Shukla et al., 2021). Technological advancements have reduced the prevalence of traditional methods, with an increasing focus on biological or 'green synthesis.' Biological methods, which align more with natural processes, use enzymes like nitrate reductase as reducing agents instead of synthetic chemical solutions. Raw materials for biological methods, such as plant concentrates or microorganisms, are more accessible than those for chemical synthesis. Significant secretory components facilitate extracellular NP synthesis in fungi, enabling them to produce larger quantities of NPs naturally. To develop and implement sustainable and environmentally friendly synthesis methods, comprehensive risk and toxicity assessments are crucial (Ali et al., 2021)

7.3.2 Properties and characterization of MNPs

Magnetic nanoparticles (MNPs) possess unique physicochemical properties that are essential for their various applications. These properties include size, morphology, surface charge, surface chemistry, and crystalline structure. The size of MNPs is a critical factor influencing their magnetic behavior and other characteristics. Small-sized nanoparticles often exhibit superparamagnetism, making them suitable for biomedical applications like magnetic resonance imaging (MRI) contrast agents. Morphology, referring to the shape of the nanoparticles, also plays a crucial role; different shapes, such as spheres, rods, or cubes, can influence the magnetic and surface properties of MNPs. Surface charge, determined by the presence of functional groups and coatings, affects the colloidal stability of MNPs in solution. This stability is crucial for applications such as drug delivery, where the nanoparticles need to remain dispersed. Surface chemistry involves the composition of the outer layer of MNPs, which can be modified for specific functionalities. For instance, coatings with polymers, silica, or metal oxides are often employed to enhance stability and biocompatibility (Avval et al., 2020).

Another physicochemical property of MNPs is their crystalline structure, which can consist of various materials such as iron oxides (magnetite or maghemite), cobalt, or nickel. The crystalline structure influences the magnetic behavior and thermal stability of MNPs. We can analyze the crystal structure and composition of MNPs using techniques like X-ray diffraction (XRD). Understanding and tailoring these physicochemical properties is crucial for designing MNPs with the desired characteristics for specific applications. For instance, in biomedical applications, controlling the size and surface properties of MNPs is essential to ensuring biocompatibility, cellular uptake, and targeted drug delivery. In addition, in catalysis or environmental remediation, optimizing the physicochemical properties of MNPs can enhance their performance and efficiency in various processes (Zamani Kouhpanji and Stadler, 2020).

7.4 Surface coating techniques

Surface coating techniques for nanostructured materials play a crucial role in tailoring the properties and functionality of nanomaterials for various applications. These techniques are essential to modify the surface characteristics, improve stability, enhance performance, and enable specific functionalities.

7.4.1 Surface coating techniques for nanostructured materials

7.4.1.1 Chemical vapor deposition (CVD)

CVD is the process of dissociating gaseous or vaporous material through chemical reactions near a heated substrate surface. Subsequently, the coating is deposited on the heated substrate surface, which is commonly used to coat metallic or ceramic compositions. Typically, metallic or ceramic materials serve as substrates, and the substrate temperature is critical because it influences coating adhesion. CVD allows multidirectional deposition, making it suitable for coating substrates with complex geometries and varying sizes. CVD is widely used in industrial applications, such as depositing refractory materials on turbine blades. It requires higher substrate temperatures and has a slow precipitation rate, making it expensive for large surface coatings. CVD methods include atmospheric pressure CVD, low-pressure CVD, laser CVD, plasma-assisted CVD, plasma-enhanced CVD, and atomic layer deposition (Xia, 2020).

7.4.1.2 Physical vapor deposition (PVD)

Physical vapor deposition (PVD) transforms the target material into a gaseous phase, followed by its deposition onto a substrate surface. Unlike CVD, PVD does not involve chemical reactions between the gaseous substance and the substrate. Steps in the PVD process include target material evaporation, transportation of the evaporated material to the substrate surface, and deposition of the target material as a coating. Various deposition methods fall under PVD, such as thermal evaporation, cathodic arc deposition, magnetron sputtering, and laser pulsed deposition. Compared with electroplating, PVD is environmentally friendly, but its requirement for a vacuum chamber limits the dimensions of coated surfaces. Unlike CVD, the line-of-sight deposition principle of PVD limits its ability to uniformly coat substrates with intricate or complex surface geometries (Growth and Transport in Nanostructured Materials: Reactive Transport in PVD, CVD, 2016)

7.4.1.3 Spray coating

Spray coating involves passing a target material through a nozzle, depositing it on a substrate surface to create a coating. Spraying can apply the substance in liquid, molten, or softened particle form. Spray coating is commonly used in industry to coat irregularly shaped surfaces like glass, metals, wood, and textiles with organic lacquers and binders. It includes methods like thermal spraying, plasma spraying, high-velocity oxy-fuel spraying (HVOF), and high-velocity air fuel (HVAF). Thermal spraying allows for higher deposition rates on substrates with large surface areas compared with CVD and PVD, but it is limited to line-of-sight deposition (Yin et al., 2018).

7.4.1.4 Sol-gel process

The sol-gel method entails using a solvent containing a chemically active component as a precursor, followed by hydrolysis and polycondensation to form a sol system. Slow polymerization transforms the sol into a gel, followed by drying and heating to form a coating. Sol-gel coatings exhibit uniform chemical and physical properties, but factors such as catalyst nature, initial material, temperature, thickness, and pH affect the coating's characteristics. Despite the advantages of producing wear-resistant coatings, sol—gel processes have drawbacks, including time-consuming preparation, expensive raw materials, and potential environmental and health hazards associated with some organics used in the process (Bokov et al., 2021).

7.4.1.5 Electrodeposition

Electrodeposition is the process of depositing a metal or alloy coating material using electrolysis over a conductive substrate surface, immersed in an electrolyte containing a salt of the target metal or alloy. Adjusting experimental parameters such as applied potential, current density, deposition time, and plating solution composition allows for altering the film coating's composition, morphology, and texture. Electrodeposition is widely used for producing nanostructured metallic or alloy coatings, with subclassifications like direct current (DC) electrodeposition and pulse electrodeposition based on the type of current used (Zamani Kouhpanji and Stadler, 2020; Nasirpouri et al., 2020).

7.4.1.6 Laser cladding

Laser cladding deposits nanostructured coatings by melting powdered or wire feedstock target material using a high-energy laser on the substrate surface. Process parameters such as laser power, power feed, and scan speed all have a significant impact on coating quality. Laser cladding offers advantages such as high cooling rates, low dilution rates, and ease of automation, but it suffers from disadvantages like poor metallurgical quality, nonuniform composition, and cracks in coatings (Zhu et al., 2021).

7.4.1.7 Plasma-based techniques

Plasma-based techniques are a versatile and powerful tool for coating the surface of nanostructured materials. These techniques utilize plasma, which is an ionized gas consisting of free electrons, ions, and neutral species, to modify the surface properties of materials.

There are a few common plasma-based surface coating techniques, such as:

Various fields such as the fabrication of microelectronic devices and the production of optical coatings, employ Plasma enhanced chemical vapor deposition (PECVD) to deposit thin films of silicon, silicon dioxide, or silicon nitride. Electron beam evaporation (EBE) is utilized for the deposition of metals and alloys. Reactive ion etching (RIE) is employed in the fabrication of nanostructures and the creation of high-aspect-ratio features. Plasma polymerization is utilized to create thin, pinhole-free films with good adhesion to the substrate. Atmospheric pressure plasma (APP) is employed for surface cleaning, activation, and modification. Magnetron sputtering is utilized for depositing thin films of metals, alloys, and compounds. Electron cyclotron Resonance (ECR) plasma proves useful for the deposition of high-quality films with low impurity levels (Phan et al., 2017). When using plasma-based techniques for the surface coating of nanostructured materials, careful control of process parameters such as plasma conditions, substrate temperature, and gas composition is crucial to achieving the desired surface properties and nanostructures (De Souza et al., 2022).

7.4.1.8 Other approaches

Additional methods for nanocoating synthesis include layer-by-layer assembly, electroless deposition, and self-assembly coatings. These offer alternative techniques for creating coatings with specific properties. Fig.7.2 illustrates key parameters influencing nanocoating formation, and Table 7.1 summarizes various preparation methods with their process parameters. These techniques' variations highlight the flexibility of surface coatings for nanostructured materials. Researchers can choose methods based on specific application needs in electronics, energy storage, biomedical devices, or emerging technologies. Ongoing research aims to advance these techniques for innovative coating processes. Table 7.1 presents a guide for selecting the appropriate technique, taking into account the advantages and disadvantages of each method. Moreover, in Table 7.2, there is a comparison between them based on the used applications. In addition, the suitable materials and the process parameters are described in Table 7.3.

7.4.2 Integration of nanostructured coatings with 3D printing

The integration of nanostructured coatings with 3D printing technology represents a convergence of advanced manufacturing techniques, offering a synergistic approach to enhance the functionality, durability, and versatility of printed objects. Nanostructured coatings, characterized by features at the nanoscale, are applied to 3D-printed surfaces to impart specific properties, augmenting the performance of the printed

Table 7.1 Comparison between surface coating techniques.

Technique	Advantages	Disadvantages	Applications
Chemical vapor deposition (CVD)	Uniform coating Precise control over thickness Compatibility with various substrates	High equipment cost Limited scalability High-temperature processing required	Semiconductors Thin-film coatings
Physical vapor deposition (PVD)	High purity coatings Control over thickness and composition Versatility	Limited to certain materials Coating may not be uniform Limited scalability	Optics Sensors Electronics
Chemical deposition techniques	Cost-effective Simple Suitable for large-area coatings	Coating uniformity may vary Limited to certain materials Processing conditions critical	Photovoltaics Catalysis Corrosion protection
Atomic layer deposition (ALD)	Excellent conformality Precise thickness control Uniformity	Relatively slow deposition process Limited scalability Complex equipment	Microelectronics Energy storage Catalysis
Sol-gel method	Versatile Cost-effective Good for thin films and coatings	Limited to certain materials Processing conditions critical Film uniformity may vary	Optics Sensors Biomedical coatings
Layer-by-layer assembly	Customizable Precise control over film composition Versatility	Limited to certain materials Assembly time can be lengthy Limited scalability	Drug delivery Sensors Biomaterials
Plasma-based techniques	Enhanced adhesion Improved hardness and wear resistance	Complex equipment Limited scalability	Aerospace Automotive Medical implants
Electrochemical deposition	Uniform coating Simple Cost-effective	Limited to conductive substrates Electrolyte composition critical Limited scalability	Corrosion protection Sensors Microelectronics

Table 7.2 Uses of surface coating techniques in biomedical and biocatalyst applications.

Technique	Biomedical applications	Biocatalyst applications
Chemical vapor deposition (CVD)	Limited direct use due to high-temperature processing Potential for coating implantable medical devices Can modify surface properties for biocompatibility	Limited use for enzyme immobilization Substrates may not be suitable for all enzymes May require additional surface functionalization May not be ideal for delicate biomolecules
Physical vapor deposition (PVD)	Used in some biomedical device coatings Limited scalability for large area coatings High purity coatings for medical sensors	Can provide thin, uniform coatings on substrates May not be suitable for all enzymes Functionalization methods for surface coating suitable for certain immobilization strategies
Chemical deposition techniques	Biocompatible coatings for medical Cost-effective and versatile for various materials Used in drug delivery systems	Good for biomolecule immobilization Suitable for enzyme coatings Can be applied to a variety of substrates
Atomic layer deposition (ALD)	Excellent conformal coatings with precise control Biocompatible coatings for implants Used in controlled drug delivery systems	Uniform coatings on complex structures Good for immobilizing enzymes on surfaces May require additional surface functionalization Suitable for delicate biomolecules
Sol-gel method	• Biocompatible thin films for implants • Versatile, suitable for various substrates • Can encapsulate biomolecules for controlled release	• Immobilization of enzymes in sol-gel matrices • Encapsulation of biomolecules • May require optimization for enzyme stability
Layer-by-layer assembly	Customizable for multilayered coatings Potential for drug delivery systems Biomolecule incorporation in multilayer structures	Good for enzyme layer-by-layer immobilization Controlled assembly of enzyme-containing layers May require optimization for enzyme activity

Table 7.2 Continued

Technique	Biomedical applications	Biocatalyst applications
Plasma-based techniques	Surface modification for biocompatibility Used in medical device manufacturing Plasma treatment for improved cell adhesion	Plasma polymerization for enzyme coatings Enhanced adhesion for enzyme immobilization Can modify surface properties for enzyme stability.
Electrochemical deposition	Biocompatible coatings on conductive substrates Simple and cost-effective process Used in the fabrication of bioelectrodes	Electrodeposition of enzymes on surfaces May require optimization for enzyme activity Controlled thickness and composition of coatings

objects in various applications. This amalgamation holds significant promise in diverse fields such as medicine, electronics, aerospace, and beyond. While integration brings numerous advantages, including improved mechanical strength, tailored surface properties, and enhanced functionalities, challenges such as compatibility issues, optimization of coating deposition techniques, and process scalability warrant comprehensive exploration (Xu et al., 2021). This synthesis of nanostructured coatings with 3D printing stands at the forefront of cutting-edge manufacturing, heralding a new era of intricately engineered, high-performance objects with applications spanning from customized medical implants to advanced electronic components.

The integration of nanostructured coatings with 3D printing opens up a broad spectrum of applications across various industries. The integration of nanostructured coatings with 3D printing has a wide range of applications across industries, offering transformative possibilities. Notable applications include the enhancement of biomedical implants, where nanostructured coatings on 3D-printed medical implants improve biocompatibility, reduce inflammation, and promote osseointegration. In drug delivery systems, these coatings control drug release and enhance targeting. In electronics and sensors, nanostructured coatings on 3D-printed components enhance conductivity, protect against environmental factors, and allow miniaturization (Aguirre-Cortés et al., 2023) as shown in Fig.7.3. Aerospace components benefit from nanostructured coatings on 3D-printed parts, improving wear resistance and thermal stability in extreme conditions. Coating energy storage devices, like 3D-printed batteries, can improve their energy density and performance. Nanostructured coatings benefit microfluidic devices by improving surface wettability and fluid manipulation precision. Environmental sensors, optical components, and electronic device encapsulation also find applications.

Table 7.3 The process parameters and coating materials for surface coating techniques.

Technique	Process parameters	Coating materials
Chemical vapor deposition (CVD)	High temperature Controlled precursor gas concentration Vacuum conditions for some processes Substrate heating for film growth	Metals (e.g., Ti, Pt) Semiconductors (e.g., Si, GaAs) Oxides and nitrides
Physical vapor deposition (PVD)	Vacuum conditions Substrate temperature control Sputtering gas composition Deposition rate control	Metals (e.g., Au, Ag) Alloys and compounds Transparent conductive oxides (e.g., ITO) Superconductors
Chemical deposition techniques	Temperature and solution chemistry Deposition time and solution concentration pH control for some methods Substrate choice and preparation	Metals (e.g., Au, Ag) Metal oxides (e.g., TiO_2, ZnO) Polymers (e.g., PLLA, PLGA) Hydroxyapatite for biomedical applications
Atomic layer deposition (ALD)	Precursor exposure cycles Temperature control Vacuum conditions for some processes Substrate surface functionalization	Metal oxides (e.g., Al_2O_3, TiO_2) Nitrides (e.g., TiN) Organic-inorganic hybrid materials Ferroelectric materials (e.g., PZT)
Sol-gel method	Solution chemistry and pH Hydrolysis and condensation rates Temperature for gel formation drying and calcination conditions	Silicates and aluminates Hybrid organic-inorganic materials Bioactive glasses Metal oxides (e.g., SiO_2, TiO_2)
Layer-by-layer assembly	pH and ionic strength of solutions Immersion time and rinsing steps Surface functionalization Temperature control during assembly DNA and other biomolecules	Polyelectrolytes (e.g., polystyrene sulfonate) Proteins and enzymes Nanoparticles (e.g., gold)
Plasma-based techniques	Plasma gas composition and pressure Substrate bias and temperature Treatment time and power density Gas flow rate and composition	Plasma polymerized films Diamond-like carbon coatings Nitride and carbide coatings Hydrophilic and hydrophobic coatings.

Table 7.3 Continued

Technique	Process parameters	Coating materials
Electrochemical deposition	Electrolyte composition and pH Applied potential and current density Deposition time and temperature Substrate choice and preparation	Metals (e.g., cu, Pt) Alloys and composites Conductive polymers (e.g., polypyrrole) Bioactive coatings (e.g., hydroxyapatite).

Nanostructured coatings improve the hardness, wear resistance, and longevity of 3D-printed industrial tools. Furthermore, these coatings help to lighten structural components without compromising strength. The diverse applications showcase the versatility and transformative potential of integrating nanostructured coatings with 3D printing across various technological and manufacturing fields (Du Toit et al., 2020). Moreover, in biomedical applications, the integration of nanostructured coatings with 3D printing offers benefits such as improved biocompatibility, controlled drug delivery, antimicrobial properties, and enhanced tissue engineering. For biocatalyst uses, the technology makes it easier for enzymes to stick to 3D-printed supports, bind to substrates, work with cofactors, and stay stable. It also allows for better reactor design, reusability, and recycling of biocatalytic systems, which makes them useful and long-lasting in many catalytic reactions (Shen et al., 2022). Using nanostructured coatings along with 3D printing in biomedical and biocatalyst applications shows a variety of ways to deal with problems and improve performance in these important areas. These benefits underscore the potential of this combined technology to revolutionize medical treatments, diagnostics, and sustainable catalytic processes.

7.4.3 Functionalization methods for surface coating

Integrating functionalization techniques into surface coating processes. This practice involves purposeful alterations to material surfaces, imparting specific characteristics tailored to precise criteria. The benefits range from enhanced adhesion and biocompatibility to controlled release mechanisms and catalytic capabilities. Functionalization is indispensable across diverse applications, such as biomedical devices and catalytic systems (Wu et al., 2015). Despite its merits, challenges exist, necessitating a comprehensive examination for informed decision-making in coating technologies. This introduction aims to elucidate the importance, advantages, and potential drawbacks of functionalization, contributing to a scholarly understanding of this intricate process. Surface functionalization is a critical aspect of coating technologies, enabling the modification of material surfaces for specific properties or functionalities, with various methods employed based on desired outcomes and substrate material nature.

Here are some common functionalization methods for surface coating:

(a) Chemical Modification:

Chemical functionalization is a process that involves modifying surface chemistry by forming covalent bonds with functional groups. For instance, we employ silane coupling agents like aminopropyltriethoxysilane (APTES) to modify the surfaces of materials like glass and silica. This modification enables the introduction of specific functionalities, such as hydrophobic or hydrophilic groups, onto the surfaces. An illustrative example involves the use of APTES to functionalize glass surfaces, where covalent bonds are formed, leading to enhanced adhesion and facilitating the attachment of biomolecules (Ahangaran and Navarchian, 2020).

(b) Polymer Coating:

Polymer coating, a versatile technique, applies a thin polymer layer to a substrate, providing specific functionalities. Industries employ it for functionalization, to alter surface energy, and to offer protective features. Tailoring polymer coatings achieves the desired adhesion, wettability, biocompatibility, and corrosion resistance. Examples include healthcare (biocompatible coatings), automotive (corrosion-resistant coatings), and electronics (dielectric coatings) (Mao et al., 2020). For instance, polyethylene glycol (PEG) coatings enhance biocompatibility on medical devices like catheters. Polymer coating is an effective method for diverse industrial and scientific needs. Polymer thin films enhance mechanical properties and introduce biocompatibility or corrosion resistance. Layer-by-layer assembly, using polyelectrolytes, provides precise control for applications such as controlled drug release in biomedicine (Abd Ellah and Abouelmagd, 2017).

(c) Plasma Treatment:

Plasma treatment is a method utilizing plasma to modify surface properties by introducing functional groups or altering surface energy. This can enhance wettability, adhesion, and biocompatibility. Activating gaseous precursors through plasma is what plasma polymerization does. This creates polymerlike coatings with specific properties. An example includes the plasma polymerization of tetrafluoroethylene (TFE) to create a thin, nonstick fluoropolymer coating with enhanced chemical resistance (Primc, 2020).

(d) Biological Functionalization:

Biological functionalization is the process of immobilizing biomolecules such as proteins, enzymes, or DNA onto surfaces to form bioactive interfaces. Peptide functionalization, specifically designing and attaching peptides to surfaces, promotes cell adhesion, proliferation, and tissue integration. An example includes immobilizing antibodies onto surfaces for biosensing applications, enabling the selective detection of target molecules with high sensitivity (Xiao et al., 2017).

(e) Nanoparticle Functionalization:

Nanoparticle functionalization integrates nanoparticles with tailored surface chemistry into coatings, providing functionalities such as antimicrobial properties, enhanced catalysis, or improved optical characteristics. Quantum dot functionalization is the process of attaching quantum dots with specific ligands to surfaces for applications in sensing, imaging, and electronic devices. Gold nanoparticles functionalized with thiol groups, for instance, can provide antimicrobial properties in medical instruments or touch surfaces (Ahmad et al., 2022).

(f) Self-Assembled Monolayers (SAMs):

Self-assembled monolayers (SAMs) involve the immobilization of molecules on surfaces, altering properties for various applications. Alkanethiol-based SAMs on metal surfaces enhance corrosion resistance and sensing, whereas phospholipid-based SAMs provide biocompatible and antifouling properties. An example includes forming a monolayer of alkylphosphonic acids on metal oxide surfaces to control wettability and enhance corrosion resistance (Sánchez-Bodón et al., 2022).

(g) Electrochemical Functionalization:

Electrochemical functionalization encompasses techniques like electrografting and electrochemical reduction to modify surfaces. Using an electric potential to create covalently bonded polymer coatings is what electrografting is all about. For example, polyaniline was electrografted onto conducting substrates to help supercapacitors store more charge (Altay et al., 2017).

(h) Hydrothermal and Solvothermal Processes:

Hydrothermal and solvothermal processes involve treating substrates in high-temperature, high-pressure environments to promote the growth of crystalline coatings. For example, hydrothermal coating uses water to grow zinc oxide (ZnO) nanorods on a substrate. This creates a photocatalytic coating that cleans water and surfaces that clean themselves (Ndlwana et al., 2021). These functionalization methods offer versatility in tailoring surface properties for specific applications, ranging from medical devices and sensors to corrosion protection and advanced materials. As illustrated in Table 7.4, the choice of method depends on the substrate material, desired functionality, and intended application of the coated surface.

According to Table 7.4, The choice of functionalization method for biomedical and biocatalyst applications depends on the application's specific requirements, the substrate's nature, and the desired surface properties. There is no one-size-fits-all solution, and researchers often select a method based on the characteristics needed for a particular use case. However, in these applications, some methods are more commonly used or have specific advantages, such as:

Polymer coating techniques, such as PEG, are widely used in biomedicine for versatile, biocompatible surfaces on medical devices that prevent protein adsorption. We can functionalize nanoparticles for targeted drug delivery or coat them with biomolecules for imaging. Techniques such as antibody immobilization create specific surfaces in biomolecular immobilization, which is critical for biosensors and tissue engineering (Buxadera-Palomero et al., 2017).

Several methods, such as physical adsorption, covalent bonding, and nanoparticle functionalization, can be used to immobilize enzymes in biocatalyst applications. These methods improve stability and reusability. Engineered nanoparticles make biocatalysts work better, and self-assembled monolayers (SAMs) give catalytic reactions controlled environments. This makes it possible to precisely tailor surface chemistry for different uses. It's essential to note that ongoing research in both biomedical and biocatalyst fields often explores combinations of these methods or hybrid approaches to capitalize on the unique advantages offered by each technique.

7.4.4 Advantages and disadvantages of functionalization for surface coating

Functionalization for surface coating has several advantages, including tailored properties, improved biocompatibility, enhanced adhesion, biocatalyst immobilization, versatility for different materials, controlled release in drug delivery, antifouling

Table 7.4 Comparison between functionalization methods for surface coating.

Technique	Advantage	Disadvantage	Applications
Chemical functionalization	**Tailored properties:** Allows precise customization of surface properties. **Versatility:** Applicable to various substrates and materials. **Diverse applications:** Widely used for biomolecule attachment, improving adhesion	**Complexity:** May require expertise and specialized equipment. **Cost:** Certain reagents can be expensive. **Limited durability:** Coatings may have limited durability under certain conditions.	Biomedical: Used for biomolecule immobilization on surfaces (e.g., DNA, proteins). Biocatalyst: Suitable for anchoring enzymes or catalytic molecules onto surfaces.
Polymer Coating	**Biocompatibility:** Suitable for medical devices due to biocompatible nature. **Versatility:** Adaptable to different substrates and applications. **Improved adhesion:** Enhances coating durability.	**Cost:** Specialized polymers can be expensive. **Processing Complexity:** May require controlled conditions for application. **Environmental impact:** Certain polymers may have environmental considerations.	Biomedical: Biocompatible polymer coatings for medical devices, and drug delivery systems. Biocatalyst: Can encapsulate enzymes for improved stability and catalytic performance.
Plasma treatment:	**Surface activation:** Improves adhesion and wettability. **Versatile:** Effective for modifying a range of materials. **Enhanced bonding:** Can improve coating-substrate bonding.	**Equipment cost:** Plasma treatment equipment can be expensive. **Energy consumption:** Requires energy-intensive processes. **Specificity:** May have limited specificity for certain applications	Biomedical: Surface activation for improved cell adhesion in implants. Biocatalyst: Can modify surface properties to enhance enzyme binding.

Table 7.4 Continued

Technique	Advantage	Disadvantage	Applications
Biological functionalization	**Biocompatibility:** Useful for biomedical applications. **Selective interactions:** Enables specific binding of biomolecules. **Biosensing:** Important for creating biosensors.	**Expertise needed:** Requires bioconjugation expertise. **Cost:** Biomolecules can be expensive. **Limited to biological applications:** Specific to biomolecular interactions.	Biomedical: Immobilization of antibodies for biosensors, tissue engineering. Biocatalyst: Attachment of enzymes for catalytic transformations.
Nanoparticle functionalization	**Enhanced properties:** Improves antimicrobial, catalytic, or optical properties. **High surface area:** Provides a large surface area for functionality. **Drug delivery:** Used for controlled drug delivery systems.	**Synthesis Complexity:** Nanoparticle synthesis can be complex. **Cost:** Production of certain nanoparticles can be expensive. **Potential toxicity:** Concerns about the environmental and health impact.	Biomedical: Drug delivery carriers, imaging agents, antimicrobial coatings. Biocatalyst: Nanoparticles as catalyst supports for enhanced biocatalysis.
Self-assembled monolayers (SAMs)	**Precision:** Allows precise control over surface properties. **Corrosion resistance:** Used for corrosion-resistant coatings. **Wettability control:** Can control surface wettability.	**Preparation time:** Time-consuming preparation process. **Thickness Constraints:** SAMs may have thickness limitations. **Susceptible to Cracking:** SAMs can be susceptible to cracking.	Biomedical: Controlling surface properties for implants or biosensors. Biocatalyst: SAMs can be designed to facilitate enzyme immobilization.

Continued

Table 7.4 Continued

Technique	Advantage	Disadvantage	Applications
Electrochemical functionalization	**Versatility:** Suitable for metallic substrates in various applications. **Controlled deposition:** Allows controlled deposition of coatings. **Enhanced properties:** Can improve electrical and catalytic properties	**Energy Intensive:** The process may be energy-intensive. **Equipment cost:** Electrochemical setups can be expensive. **Limited to conductive substrates:** Primarily applicable to conductive materials.	Biomedical: Electrografting for modifying surfaces of medical implants. Biocatalyst: Electrochemical methods for enzyme immobilization on electrodes.
Hydrothermal/ Solvothermal processes	**Crystalline coatings:** Effective for producing crystalline coatings. **Controlled synthesis:** Allows controlled synthesis of nanomaterials. **Photocatalysis:** Applicable for photocatalytic applications.	**Moderate Complexity:** Requires controlled conditions. **Solvent considerations:** Impact on the environment due to solvent use. **Limited to line-of-sight deposition:** Coating is limited to line-of-sight.	Biomedical: Hydrothermal coatings for controlled drug release. Biocatalyst: Synthesis of nanomaterials for enzyme immobilization.

properties, and precision in selective interactions. However, there are disadvantages for surface coating, such as; functionalization methods can be complex and costly, with limited durability; potential environmental impact; challenges in achieving surface uniformity; concerns about long-term stability; specificity limitations, compatibility issues with certain materials, and regulatory considerations in biomedical applications (Gao, 2019; Thorn et al., 2021). It's important to note that the advantages and disadvantages can vary significantly depending on the specific functionalization method, the materials involved, and the intended application. Researchers and engineers carefully evaluate these factors to select the most suitable approach for their particular needs.

7.5 Bio-catalytic reactions and their significance in biomedical applications

Biocatalytic reactions are chemical transformations mediated by enzymes. These remarkable molecules act as matchmakers, bringing specific substrates (reactant molecules) together in a precise dance to form desired products. Unlike traditional chemical catalysts, enzymes are highly selective, ensuring minimal side reactions and unwanted by-products. Biocatalytic reactions have transformative potential in healthcare, impacting drug development, gene therapy, biosensing, and tissue engineering. Enzymes can synthesize drugs with improved efficacy, treat genetic disorders through DNA repair, act as disease detectors for early diagnosis, and contribute to regenerative medicine by building biocompatible scaffolds for organ repair. Advantages of biocatalytic reactions include high specificity, mild reaction conditions, sustainability, and versatility in biomedicine applications (Thorn et al., 2021). However, challenges persist. Enzyme stability is a concern due to its susceptibility to denaturation by factors like temperature and pH. Large-scale production of pure enzymes can be cost-prohibitive, and delivering enzymes to specific tissues while ensuring prolonged activity remains challenging. Despite these hurdles, the potential of biocatalytic reactions to revolutionize healthcare underscores the need for addressing these challenges for future advancements (Casadey et al., 2020). Biocatalytic reactions have made a lot of progress recently in a few key areas. These include enzyme engineering, sustainable synthesis, multi-enzyme systems, AI in enzyme design, non-natural substrate transformations, biocatalysis in green chemistry, and industrial scale-up. Efforts bridge laboratory-scale and industrial bio-catalysis for applications in pharmaceuticals, chemicals, sustainable manufacturing, bioremediation, and renewable fuels.

7.5.1 Bio-catalysis applications

Surface-coated magnetic nanostructured materials have emerged as promising tools for robust bio-catalysis applications. These materials, which consist of magnetic nanoparticles enveloped in biocompatible and catalytically active surface layers, allow for efficient catalyst separation and recovery using external magnetic fields. The surface coatings keep things stable, stop them from sticking together, and give biocatalysts a place to stay. This improves the catalysts' ability to work, their selectivity, and their ability to be used again and again. The magnetic nature enables easy manipulation and recyclability, finding applications in pharmaceuticals, environmental remediation, and biofuel production (Uthaman et al., 2015). The core of these materials consists of magnetic nanoparticles like iron oxides, providing the magnetic properties necessary for separation and recovery. Surface coating, crucial for functionality, stabilizes the cores, ensures biocompatibility, and furnishes catalytic functionality. These magnetic nanostructures with a coating on the surface improve catalytic activity, create better microenvironments around catalysts, and make it possible for biocatalysts to stay in place.

The materials' magnetic nature facilitates easy separation and recovery, enabling reuse in multiple catalytic cycles without significant efficiency loss. Applications span pharmaceuticals, environmental remediation, and biofuel production, making them versatile in enzyme immobilization and catalytic transformations. In the biomedical realm, biocatalytic reactions offer exciting possibilities. Enzymes catalyze drug synthesis with enhanced therapeutic profiles, gene therapy corrects genetic defects at their root, biosensors detect disease markers for early diagnosis, and biocatalytic reactions guide tissue engineering for regenerative medicine. These reactions bring advantages of exquisite specificity, gentle reaction conditions, eco-friendliness, and a diverse enzyme arsenal, expanding the possibilities in biomedicine (Bilal et al., 2022).

7.5.2 Nanostructured coatings role in bio-catalysis

Nanostructured coatings play a crucial role in enhancing biocatalysis through various mechanisms. First, they enhance enzyme immobilization by increasing the surface area, which maximizes active sites for catalysis. Tailored structures also improve substrate accessibility, thereby enhancing catalytic efficiency (Bilal et al., 2022). Secondly, these coatings act as protective layers, shielding enzymes from harsh conditions and creating a stable microenvironment that minimizes enzyme leaching and enhances overall stability. Tuning enzyme properties via Surface functionalization with specific chemical groups or biomolecules tailors enzyme interactions. This enhances enzyme-substrate affinity, selectivity, and catalytic performance. Some coatings have the ability to modulate enzyme conformation, influencing catalytic properties. Third, adding certain chemical groups or biomolecules to the surface of an enzyme lets its properties be changed, which improves its ability to bind and select substrates and work as a catalyst. In addition, some coatings can modulate enzyme conformation, influencing catalytic properties and facilitating cofactor regeneration. Nanostructured coatings aid in cofactor immobilization, which is critical for continuous and efficient regeneration in enzyme-catalyzed reactions.

Moreover, these coatings ensure minimal interference with enzyme functionality, which is crucial for applications like enzyme immobilization, due to their biocompatibility. Certain coatings also provide controlled enzyme release or gradual degradation, which is beneficial in drug delivery systems. Finally, nanostructured coatings make tiny spaces that are perfect for certain reaction conditions, which improves biocatalytic reactions (Bilal et al., 2022). Biocatalysts work better and last longer when they are coated with nanostructured materials that can be used in many ways to immobilize and change enzymes. Their ability to fine-tune enzyme properties and create tailored environments expands the possibilities for efficient bio-catalysis in various industrial and biomedical applications. Magnetic nanostructured materials (MNSMs) with a surface coating have become useful tools for immobilizing enzymes, which allows for a variety of biocatalytic reactions each with their own benefits. One key advantage is their high surface area, which results from the multitude of tiny pores and crevices in MNSMs. This feature gives enzymes a lot of room to stick to the surface, which increases the number of catalytic sites and makes the reaction more efficient compared with using free enzymes (Bilal et al., 2022).

In addition, the magnetic core of MNSMs allows for easy biocatalyst separation and recovery using an external magnet. This simplifies downstream processing and enables enzyme reuse, making the process more cost-effective and environmentally friendly (Bilal et al., 2021). Moreover, we can precisely modify the surface of MNSMs with various functional groups to create specific interactions with enzymes and substrates. Customizing the surface chemistries improves biocatalytic performance and selectivity, which is like making docking stations just right for enzymes and substrates to make sure they fit perfectly.

7.6 Surface-coated magnetic nanostructured materials for biomedical applications

Due to their special qualities and potential for a wide range of uses, including biocatalysis and biomedical applications, surface-coated magnetic nanostructured materials have attracted a lot of attention recently. Typically, these materials consist of a magnetic core, such as iron oxide or cobalt, and a functional coating that provides stability, biocompatibility, and specific functions for their intended applications. Medical fields rarely use magnetic nanoparticles containing transition metals like Co, Ni, and Mn due to their severe toxicity. In their free form, they generate reactive oxygen species (ROS) that cause oxidative stress and cell damage (Kudr et al., 2017). Accordingly, biocompatible inorganic/organic materials must be used to functionalize these systems to reduce their cytotoxicity. The body's iron storage system can incorporate the iron ions produced during metabolism, making iron oxide nanoparticles biocompatible (Weissleder et al., 1989). Magnetic nanoparticles are hazardous at varying doses. Magnetic nanoparticles are generally nontoxic in small doses.

Magnetic nanostructured materials with a surface coating offer a plethora of biological applications. Their magnetic core allows for the controlled release of therapeutic drugs in the presence of an external magnetic field, making them suitable for targeted medication delivery. Besides providing specific targeting features such as cell-specific recognition ligands, we can engineer the functional coating to ensure the stability of the therapeutic payload. Bioimaging techniques like magnetic resonance imaging (MRI) can also employ magnetic nanostructures. By functionalizing the surface coating with contrast agents, these materials can improve the visibility of certain tissues or cells, allowing for better disease detection and monitoring. Researchers have also investigated the use of surface-coated magnetic nanostructured materials in tissue engineering and regenerative medicine. They can act as scaffolds for cell proliferation and differentiation, affording mechanical support and regulating the release of bioactive chemicals to promote tissue regeneration. Overall, magnetic properties and changed surface coatings work together to make long-lasting bio-catalysis systems and flexible biomedical tools. These materials show considerable promise for enhancing a variety of sectors, including industrial biotechnology, medication delivery, diagnostics, and regenerative medicine.

The unique application criteria, such as stability, biocompatibility, and intended functionality, undoubtedly determine the selection of functional coatings for magnetic nanostructured materials. Here are some popular forms of functional coatings (Bilal et al., 2022):

- **Polymers**: Polymer coatings are versatile and can be customized to give various functionality. For example, biocompatible polymers such as polyethylene glycol (PEG) can be utilized to enhance the stability and biocompatibility of magnetic nanoparticles. Other polymers, such as chitosan or alginate, can control medication release or improve cell adherence in tissue engineering applications.
- **Silica (SiO_2)** coatings are commonly used due to their high biocompatibility and simplicity of functionalization. Silica can form a stable and protective coating surrounding the magnetic core, reducing aggregation and protecting the nanoparticles' integrity. Furthermore, silica coatings can be changed with functional groups such as amino or carboxyl groups, permitting for additional functionalization and particular bioconjugation.
- **Metal-Organic Frameworks (MOFs):** Metal ions and organic ligands are coupled to form a family of porous materials known as MOFs. Because of their large surface areas and adjustable pore architectures, they can be used in gas storage and medication administration, among other applications. To add other functions, such targeted distribution or controlled release of encapsulated medications, MOFs can be coated onto magnetic nanoparticles.
- **Lipids and Liposomes:** Lipid-based coatings are especially beneficial in applications that include cell membranes or lipid bilayers. Lipid coatings can improve the durability and biocompatibility of magnetic nanoparticles, allowing them to interact more easily with biological systems. Liposomes, which are lipid vesicles, can contain medications or contrast agents and have controlled release properties.
- **Protein and Enzyme Coatings:** Magnetic nanoparticles can be coated with proteins or enzymes to add specific capabilities or catalytic activity. Antibodies or aptamers, for example, can be immobilized on the surface to allow for precise biomarker detection and binding. Enzymes can be immobilized on nanoparticles to improve catalytic activity or facilitate biocatalytic processes.
- **Biomimetic Coatings:** Biomimetic coatings try to mimic the composition or characteristics of biological materials found in nature. Coatings that are modeled after extracellular matrix or cell membranes, for example, can improve biocompatibility and replicate the natural milieu of cells or tissues.

Surface-coated magnetic nanostructured materials have shown significant potential for various biomedical applications via combination of magnetic properties with functional coatings enables precise manipulation, targeting, and imaging of biological systems. Here are some key biomedical applications of these materials as presented in Fig.7.7 and as following: Fig. 7.7

(a) Magnetic-targeted drug delivery

Therapeutic medicine-containing coatings can functionalize magnetic nanostructured materials. The magnetic core allows for controlled medicine release, resulting in targeted delivery to specific tissues or cells. An external magnetic field can direct drug-loaded nanoparticles to their intended location, improving their efficacy and reducing systemic side effects. Thomsen et al. attempted to breach the blood-brain barrier (BBB) by magnetic targeting. They chose iron oxide nanoparticles as the carrier, and an in vitro model demonstrated that using an external magnet significantly increased the amount of medication that crossed the BBB (Thomsen et al., 2013). In a recent study, Mushtaq et al. coated cobalt ferrite nanoparticles with polymers and injected doxorubicin into the polymer layer. It was discovered that the release was significantly higher in the presence of a magnetic field. Furthermore, in vitro investigations confirmed the nanoparticles' safety (Mushtaq et al., 2017).

Thus far, we have only discussed in vitro research using a two-dimensional cell culture. In a three-dimensional model, Child et al. investigated the magnetic targeting potential of nanoparticles in cells cultured. Their results supported the use of magnetically targeted administration in a three-dimensional model, which is more like the system that works in real life (Child et al., 2011). Chertok et al. employed polymer-coated iron oxide nanoparticles (IONPs) to target brain tumors. The polymer, starch or gum Arabic, underwent treatment with polyethyleneimine (PEI). This formulation was injected into rats with orthotopicglio sarcoma using the intracarotid route. Magnetic targeting allowed the particles to efficiently collect in the tumor (Chertok et al., 2010). Aryan et al. used superparamagnetic iron oxide nanoparticles and a magnetic field from the outside to successfully deliver therapeutic drugs to the right branch in a 3D carotid bifurcation. They determined that the particle size and magnet configuration were the primary characteristics that needed to be fine-tuned to obtain maximum efficiency in magnetic medication delivery (Aryan et al., 2022).

Mah et al. developed a novel technique for delivering recombinant AAV-2 (recombinant adeno-associated virus) for targeted gene therapy by employing magnetic microspheres that are reversibly linked to AAV vectors. The researchers intramuscularly injected both free and magnetic microsphere-bound rAAV into the right gastrocnemius of 129/SvJ mice. Delivery of the rAAV vector using magnetic microspheres demonstrated higher transduction efficiencies compared with using a free vector alone (kianfar, 2021). Another study by Shen et al. created a highly complicated nanoformulation containing folic acid (FA), iron oxide, CdTe quantum dots, a tetrapeptide, chitosan, and camptothecin (CPT). Shen et al. used quantum dots as a fluorescence probe, CPT as an anticancer medication, and folic acid as a targeted ligand due to the overexpression of the folate receptor in various malignancies. Iron oxide and quantum dots were combined to create a hybrid nanogel using chitosan as a gelation material. The tetrapeptide GFFG, also known as LGPV, or Leu-Gly-Pro-Val, linked folic acid to the core. Mice afflicted with sarcoma were given this hybrid formulation. With the use of a magnetic field, the particles were successfully contained inside the tumor and were specific for cancer cells.

(b) Magnetic Hyperthermia:

The process generates heat to destroy nearby cells. This focused type of therapy is currently receiving a lot of attention. Because tumor cells have a faster metabolic rate than normal cells and are therefore more susceptible to heat, heat therapy is a very effective treatment for cancer (Tran and Webster, 2010). When an alternating magnetic field is applied to magnetic nanoparticles, either hysteresis losses or relaxation mechanisms (Neel or Brownian) absorb energy and subsequently release it as heat (Laurent et al., 2011). The heat generated depends upon the nanoparticle (size and composition) and the magnetic field (frequency and amplitude) (Psimadas et al., 2014). For ferromagnetic materials, heat is generated through hysteresis losses, but these are almost negligible for superparamagnetic materials. For superparamagnetic material, the heat generation is due to Neel and Brownian relaxation mechanisms. Neel relaxation produces heat when the particle size of a superparamagnetic nanoparticle is less than 20 nm, while Brownian relaxation and hysteresis losses produce heat for larger particles. To ensure a safe treatment for the patients, the optimum size range for treatment should be around 10−100 nm, and the product of frequency and amplitude of the external magnetic field should not exceed this value. IONPs, owing to their biocompatibility, nontoxicity, and high saturation magnetization, have been well explored for hyperthermia applications employing core−shell formulations (Kefeni et al., 2020).

Hysteresis losses cause heat generation in ferromagnetic materials, however, they hardly ever occur in superparamagnetic materials. Neel and Brownian relaxation mechanisms are responsible for heat generation in superparamagnetic materials. Heat is created by Neel relaxation in superparamagnetic nanoparticles less than 20 nm, and by Brownian relaxation and hysteresis losses in particles larger than that. The best size range for treatment, ensuring patient safety, is approximately 10−100 nm. In addition, the product of the frequency and amplitude of the external magnetic field should not exceed the value of 5× 109 a.m/second. Researchers have thoroughly studied IONPs for hyperthermia applications using core-shell formulations due to their high saturation magnetization, nontoxicity, and biocompatibility. An earlier investigation loaded IONPs with either (a) curcumin or (b) curcumin plus folate after coating them with the polymer O-carboxyl methyl chitosan. The system's ability to generate heat was assessed with an external magnetic field (frequency = 236 kHz, amplitude = 60 oe). The mice with sarcoma-180 tumors were given injections of both formulations, with a dosage of 20 g/kg. When the folate was affixed to the particles, both systems produced heat efficiently and showed improved biodistribution (Thu Huong et al., 2016).

Hardiansyah et al. synthesized iron oxide nanoparticles coated in citric acid, which they then loaded into liposomes containing the anticancer medication doxorubicin. The magnetic nanocarriers did not produce any cytotoxicity in the cell line L-929, according to the MTT experiment. However, an examination of the particles in the colorectal cell line CT-26 revealed a synergistic impact between chemotherapy and magnetic hyperthermia (Hardiansyah et al., 2014). Researchers have also researched Cobalt ferrite nanoparticles (CFNPs) from the perspective of hyperthermia, discovering that they demonstrate even more effective heat generation than Fe_3O_4. They have significant saturation magnetization and magnetic anisotropy (Gandhi and Roy, 2019). Carbon nanoparticles (CFNPs) covered with silica, measuring 14 and 40 nm, were made. It was found that the size affected the amount of heat generated, with smaller sizes producing more heat (Cano et al., 2014). Balakrishnan et al. compared the efficacy of cube-shaped iron oxide (IONCs) and cobalt ferrite (Co-Fe NCs) nanoparticles in vivo for tumor regression and heat production. Co-Fe NCs demonstrated improved cytotoxic effects due to inherent cobalt toxicity, moderate hyperthermia, and their unique capacity to form expanding chains at tumor sites when exposed to an alternating magnetic field (AMF) (Balakrishnan et al., 2020).

(c) Magnetic Resonance Imaging (MRI)

Magnetic resonance imaging (MRI) is a useful tool in radiology for diagnosing diseases and anomalies. The magnetic field impacts hydrogen atoms (protons), which are ubiquitous, differently depending on their position and the atoms surrounding them in the body. A strong magnetic field aligns and processes these protons, followed by a radiofrequency pulse that perturbs them. The disturbed protons return to equilibrium via two distinct relaxation pathways: T1 (longitudinal or spin-lattice) and T2 (transverse or spin-spin) relaxation. The variance in these processes is tracked and used to generate images displaying a difference between tissues with distinct anatomy/pathology. We commonly use exogenous contrast agents, which generally reduce (sometimes increase) relaxation time, to improve visibility and obtain sharper images. Traditional gadolinium (Gd)-based contrast agents are paramagnetic and have relatively low relaxivity values. At low levels, Gd-chelates can shorten T1 relaxation time, but large amounts are necessary for T2 shortening, which can be hazardous (Terreno et al., 2010; Rümenapp et al., 2012; Rümenapp et al., 2012). Superparamagnetic nanoparticles are a relatively new family of MRI contrast agents that can reduce T1 and T2 relaxation times at low concentrations/doses when compared to paramagnetic agents. They can be sized, functionalized, and modified according to the target tissue (Terreno et al., 2010; Rümenapp et al., 2012).

Chee et al. created a library of 86 short peptides and ligands to modify the properties of iron oxide nanoparticles and create USPIONs, which are very small and magnetic. The ligand chosen for IONPs has long-term stability in physiological settings, binding to living cells, and low cytotoxicity at large dosages, making it ideal for in-vivo use (Fig.7.6). IONPs functionalized with this ligand outperformed commercial MRI contrast agents (Magnevist and Resovist) in distinguishing between liver tumors and healthy liver tissue (Chee et al., 2018).

(d) Magnetic Particle Imaging (MPI)

MPI is a sensitive tracer imaging technique that accurately measures the location and concentration of SPIONs in vivo. MPI imaging captures the SPION electronic moment, which is 22 million times stronger than nuclear MRI moments. It utilizes the nonlinear magnetization of SPIONs. When a time-varying excitation field is provided during signal creation, SPIONs quickly "flip" and generate a signal in the receiver coil. To localize a signal, a large magnetic gradient is used to create a field-free point (FFP), which has a magnetic field of zero. Magnetic hyperthermia can cause unintended accumulation of SPIONs in off-target tissues, including the liver and spleen, leading to collateral heat injury. Tay et al. created an image-guided theranostic platform that combines MPI and magnetic hyperthermia. MPI-guided magnetic hyperthermia was shown to effectively target tumors while minimizing collateral damage to the liver at a distance of 1–2 cm (Tay et al., 2018).

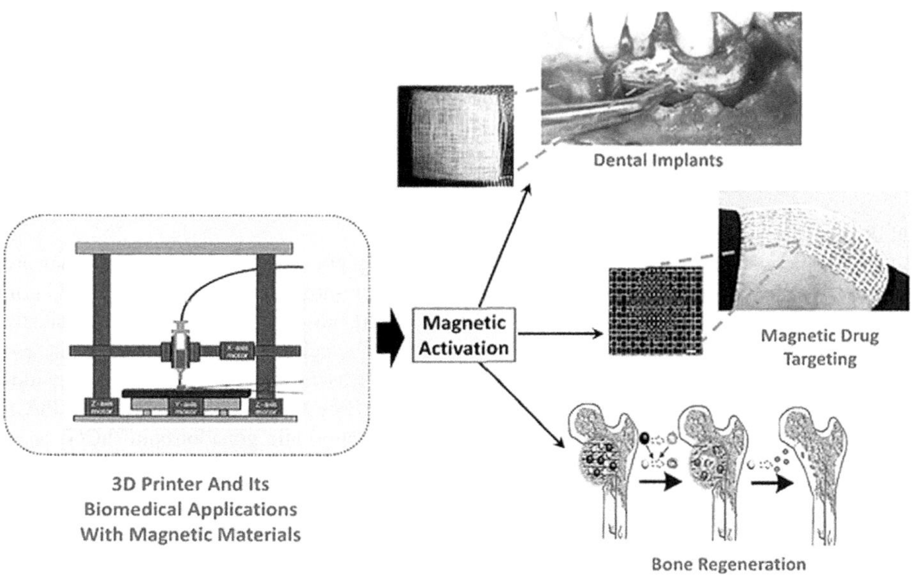

Figure 7.6 3D printer and its biomedical applications with magnetic nanoparticles.

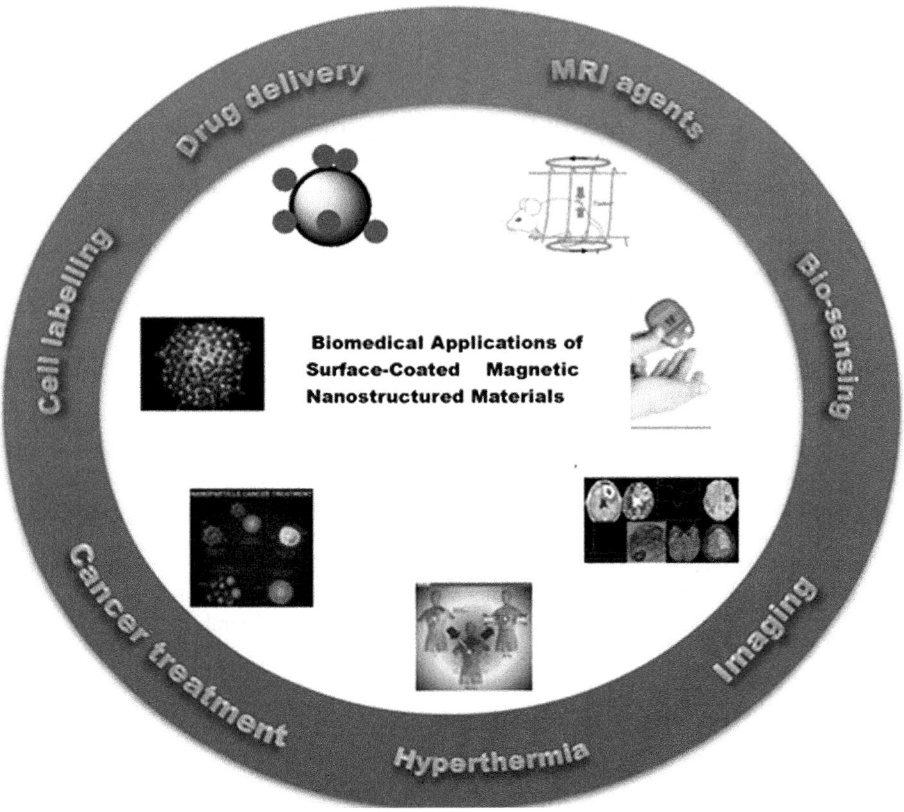

Figure 7.7 Biomedical applications of surface-coated magnetic nanostructured materials. From Natarajan, S., Harini, K., Gajula, G.P., Sarmento, B., Neves-Petersen, M.T., Thiagarajan, V., 2019. Multifunctional magnetic iron oxide nanoparticles: diverse synthetic approaches, surface modifications, cytotoxicity toward biomedical and industrial applications. BMC Mater. 1 (1). https://doi.org/10.1186/s42833-019-0002-6.

(e) Biosensors.

Magnetic nanoparticle-based biosensors are gaining popularity in biomedicine. They are known to be far more sensitive, have a greater signal-to-noise ratio, have a lower detection limit, and require less processing time than standard biosensors. Biomolecules typically functionalize magnetic nanoparticles to detect viruses, cells, proteins, nucleic acids, enzymes, and other substances (Rocha-Santos, 2014). It was discovered by Kim et al. that adding platinum to the surface of iron oxide nanoparticles (Fe_3O_4—Pt/core-shell NPs or MPt/CS NPs) made them more active like enzymes. Human chorionic gonadotropin (hCG) is an important pregnancy indicator. Using hCG/Ab-MPt CS NPs, which are anti-hCG antibody-modified nanoparticles, in LFIA strips made them twice as sensitive as regular gold strips (Kim et al., 2017). LFIA strips combined with magnetic nanoparticles can improve point-of-care testing results.

(f) Tissue Engineering

Tissue engineering (TE) is an interdisciplinary field that has successfully combined and translated basic technical, biological, and chemical ideas to make materials and tools that can replace, preserve, and restore biological tissues(Howard et al., 2008). Tissue engineering is one of the intriguing uses of magnetic nanoparticles. By binding to the cell membrane or internalizing within the cell, magnetic nanoparticles can regulate and control biological activities using an external magnetic field. They can also be used with three-dimensional scaffolds, which are biomaterials like films, hydrogels, and fibers that are known for being able to mimic the extracellular matrix of tissues and provide support for cells to attach and grow. Cartmell et al. discovered that human osteoblasts adhered to RGD-coated CrO_2 magnetic nanoparticles, which they then exposed to a daily magnetic field for 21 days. After 21 days, osteoblasts were viable and proliferated, and the bone matrix calcified compared with the control group (without a magnetic field) (Cartmell et al., 2002).

7.7 Biocompatibility and safety considerations

7.7.1 Biocompatibility of surface-coated magnetic nanostructured materials in biomedical applications

The biocompatibility of surface-coated magnetic nanostructured materials is critical in biomedical applications, playing a pivotal role in the safety and success of therapeutic and diagnostic interventions. Biocompatible coatings ensure favorable interactions with cells, tissues, and physiological fluids, minimizing adverse reactions. In medical imaging, these coatings enable the use of magnetic nanostructured materials as contrast agents in MRI without inducing harmful effects on tissues (Thanka Rajan et al., 2022). For targeted drug delivery, biocompatible coatings on magnetic nanoparticles serve as safe carriers, navigating through the biological environment for the secure transport of therapeutic agents. In hyperthermia treatment, crucial for localized heating, biocompatible coatings ensure that magnetic nanoparticles generate heat without harming surrounding healthy tissues. Biocompatible magnetic nanostructured materials are indispensable for biosensors, implantable devices, and reducing toxicity, contributing to stability, long-term durability, immune tolerance, and regulatory compliance in various biomedical applications (kianfar, 2021). Overall, biocompatibility is foundational for the seamless integration of these materials with biological systems, ensuring safety, efficacy, and minimal adverse effects in the dynamic field of nanomedicine and magnetic nanotechnology.

7.7.2 Safety considerations and regulatory aspects of surface-coated magnetic nanostructured materials

Safety considerations and regulatory aspects surrounding surface-coated magnetic nanostructured materials are pivotal for their responsible development and use across various applications. Biocompatibility is a key focus, requiring an assessment of potential health impacts, toxicity, and long-term effects on human health (Nel et al.,

2011). Thorough in vivo studies should examine toxicity, biodistribution, stability, and biodegradability, while ecotoxicity studies assess environmental impact. Immunogenicity and ethical/social implications, especially in medical applications, necessitate investigation. Regulatory compliance with medical device regulations, risk–benefit analyses, and adherence to established standards are crucial. Material classification, accurate labeling, and preclinical/clinical testing align with regulatory requirements. Compliance with environmental regulations and international collaboration for standard harmonization are essential. Postmarket surveillance and public engagement strategies ensure ongoing safety monitoring and transparent communication, contributing to the responsible development and deployment of these materials (Ray et al., 2009; Paradise, 2011). To ensure the responsible development and deployment of surface-coated magnetic nanostructured materials, it is crucial to address safety considerations and navigate the regulatory landscape, thereby realizing their benefits and minimizing potential risks to human health and the environment.

7.7.3 Nanotoxicity and safety assessment of nanostructured coatings

The nanotoxicity and safety assessment of nanostructured coatings are critical aspects that require a thorough investigation to ensure their responsible use and application. A comprehensive approach is required to consider the nanotoxicity of surface-coated magnetic nanostructured materials, taking into account factors such as particle size, surface area, chemical composition, and surface chemistry. Size-dependent toxicity, cellular interactions, and biodistribution in vivo are crucial aspects requiring thorough assessment. Additional considerations include inflammatory responses, genotoxicity, oxidative stress, and neurotoxicity (Gupta and Xie, 2018).

The safety assessment involves a risk–benefit analysis and regulatory compliance, especially in medical and consumer applications. Rigorous biocompatibility testing, addressing long-term effects and environmental impact, ensures responsible deployment. Adherence to standards, clear product labeling, and postmarket surveillance all contribute to continuous safety monitoring. Public awareness and interdisciplinary collaboration are key for the safe and responsible development of these materials (Wong and Tong, 2018). Understanding nanotoxicity and conducting thorough safety assessments are integral to the responsible development and application of nanostructured coatings. These considerations help mitigate potential risks while maximizing the benefits of innovative nanotechnologies.

7.8 Challenges and future perspectives

When making and using surface-coated magnetic nanostructured materials, it can be hard to make sure they are biocompatible, standardize the methods used for characterizing them, deal with concerns about stability, and make more of them. Achieving precise delivery in medical applications, optimizing magnetic properties, and navigating regulatory pathways are additional complexities. Long-term biocompatibility, environmental impact, and sustainability add dimensions to the challenges. Overcoming these requires interdisciplinary collaboration and continuous innovation. Surface-coated magnetic nanostructured materials are a dynamic field. Some of the latest trends in this field are combining smart materials to make systems that respond to stimuli, combining precision medicine with personalized uses, and using AI to improve treatments and diagnoses. Environmental applications, advancements in nanomanufacturing, and the exploration of novel coating materials further shape the future. Continuous research in enzyme engineering, drug delivery with nanocarriers, and personalized medicine holds transformative potential in bio-catalysis and biomedical applications. Future advancements anticipate personalized medicine, immuno-engineering strategies, AI-driven drug discovery, 3D bioprinting, CRISPR-Cas9 gene editing, and advanced biosensors for real-time monitoring. Nanoparticles in nanomedicine enable targeted drug delivery and biomaterials for implants, whereas synthetic biology and the human microbiome contribute to sustainable biocatalytic processes. The evolving landscape holds promise for innovative therapies, environmentally friendly chemistry, and enhanced disease management.

7.9 Conclusion

In conclusion, exciting trends such as the integration of smart materials, personalized medicine, AI-driven applications, environmental solutions, nanomanufacturing advancements, and the exploration of novel coating materials mark the future of surface-coated magnetic nanostructured materials. These trends collectively contribute to the evolving landscape of nanotechnology, promising ground-breaking developments with far-reaching implications across diverse scientific and industrial domains. Potential advancements and breakthroughs in bio-catalysis and biomedical applications shape a dynamic landscape characterized by interdisciplinary collaboration and technological innovation. These developments have the capacity to redefine disease treatment, drug development, and sustainable chemical processes, promising a future where precision, customization, and sustainability take center stage in healthcare and industry.

References

Abd Ellah, N.H., Abouelmagd, S.A., 2017. Surface functionalization of polymeric nanoparticles for tumor drug delivery: approaches and challenges. Expet Opin. Drug Deliv. 14 (2), 201–214. https://doi.org/10.1080/17425247.2016.1213238.

Abrikosov, I.A., Ponomareva, A.V., Steneteg, P., Barannikova, S.A., Alling, B., 2016. Recent progress in simulations of the paramagnetic state of magnetic materials. Curr. Opin. Solid State Mater. Sci. 20 (2), 85–106. https://doi.org/10.1016/j.cossms.2015.07.003.

Adak, D., Ghosh, S., Chakrabarty, P., Mondal, A., Saha, H., Mukherjee, R., Bhattacharyya, R., 2017. Self-cleaning V-TiO2:SiO2 thin-film coatings with enhanced transmission for solar glass cover and related applications. Sol. Energy 155, 410–418. https://doi.org/10.1016/j.solener.2017.06.014.

Aguirre-Cortés, J.M., Moral-Rodríguez, A.I., Bailón-García, E., Davó-Quiñonero, A., Francisco Pérez-Cadenas, A., Carrasco-Marín, F., 2023. 3D printing in photocatalysis: methods and capabilities for the improved performance. Appl. Mater. Today 32, 101831. https://doi.org/10.1016/j.apmt.2023.101831.

Ahangaran, F., Navarchian, A.H., 2020. Recent advances in chemical surface modification of metal oxide nanoparticles with silane coupling agents: a review. Adv. Colloid Interface Sci. 286, 102298. https://doi.org/10.1016/j.cis.2020.102298.

Ahmad, F., Salem-Bekhit, M.M., Khan, F., Alshehri, S., Khan, A., Ghoneim, M.M., Wu, H.F., Taha, E.I., Elbagory, I., 2022. Unique properties of surface-functionalized nanoparticles for bio-application: functionalization mechanisms and importance in application. Nanomaterials 12 (8). https://doi.org/10.3390/nano12081333.

Ahmadi, S., Fazilati, M., Nazem, H., Mojtaba Mousavi, S., Ramoji, K., 2021. Green synthesis of magnetic nanoparticles using Satureja hortensis essential oil toward superior antibacterial/fungal and anticancer performance. BioMed Res. Int. 2021, 1–14. https://doi.org/10.1155/2021/8822645.

Akl, M.A., Kamel, A.M., El-Ghaffar, M.A.A., 2023. Biodegradable functionalized magnetite nanoparticles as binary-targeting carrier for breast carcinoma. BMC Chem. 17 (1). https://doi.org/10.1186/s13065-023-00915-4.

Ali, A., Shah, T., Ullah, R., Zhou, P., Guo, M., Ovais, M., Tan, Z., Rui, Y.K., 2021. Review on recent progress in magnetic nanoparticles: synthesis, characterization, and diverse applications. Front. Chem. 9. https://doi.org/10.3389/fchem.2021.629054.

Altay, C., Senay, R.H., Eksin, E., Congur, G., Erdem, A., Akgol, S., 2017. Development of amino functionalized carbon coated magnetic nanoparticles and their application to electrochemical detection of hybridization of nucleic acids. Talanta 164, 175–182. https://doi.org/10.1016/j.talanta.2016.11.012.

Amirtharaj Mosas, K.K., Raja Chandrasekar, A., Dasan, A., Pakseresht, A., Galusek, D., 2022. Recent advancements in materials and coatings for biomedical implants. Gels 8 (5), 323. https://doi.org/10.3390/gels8050323.

Aryan, H., Beigzadeh, B., Siavashi, M., 2022. Euler-Lagrange numerical simulation of improved magnetic drug delivery in a three-dimensional CT-based carotid artery bifurcation. Comput. Methods Progr. Biomed. 219, 106778. https://doi.org/10.1016/j.cmpb.2022.106778.

Avval, Z.M., Malekpour, L., Raeisi, F., Babapoor, A., Mousavi, S.M., Hashemi, S.A., Salari, M., 2020. Introduction of magnetic and supermagnetic nanoparticles in new approach of targeting drug delivery and cancer therapy application. Drug Metabol. Rev. 52 (1), 157–184. https://doi.org/10.1080/03602532.2019.1697282.

Badiceanu, M., Anghel, S., Mihailescu, N., Ioana Visan, A., Mihailescu, C.N., Mihailescu, I.N., 2022. Coatings functionalization via laser versus other deposition techniques for medical applications: acomparative review. Coatings 12 (1), 71. https://doi.org/10.3390/coatings12010071.

Bagale, U.B., Sonawane, S.H., Bhanvase, B., Hakke, V.S., Kakunuri, M., Manickam, S., Sonawane, S.S., 2019. Multifunctional coatings based on smart nanocontainers. In: Advances in Smart Coatings and Thin Films for Future Industrial and Biomedical Engineering Applications. Elsevier, India, pp. 135–162. https://doi.org/10.1016/B978-0-12-849870-5.00014-8.

Balakrishnan, P.B., Silvestri, N., Fernandez-Cabada, T., Marinaro, F., Fernandes, S., Fiorito, S., Miscuglio, M., Serantes, D., Ruta, S., Livesey, K., Hovorka, O., Chantrell, R.,

Pellegrino, T., 2020. Exploiting unique alignment of cobalt ferrite nanoparticles, mild hyperthermia, and controlled intrinsic cobalt toxicity for cancer therapy. Adv. Mater. 32 (45). https://doi.org/10.1002/adma.202003712.

Barhoum, Makhlouf, A.S., 2018. Emerging Applications of Nanoparticles and Architecture Nanostructures. Elsevier.

Bilal, M., Hussain, N., Américo-Pinheiro, J.H.P., Almulaiky, Y.Q., Iqbal, H.M.N., 2021. Multi-enzyme co-immobilized nano-assemblies: bringing enzymes together for expanding bio-catalysis scope to meet biotechnological challenges. Int. J. Biol. Macromol. 186, 735–749. https://doi.org/10.1016/j.ijbiomac.2021.07.064.

Bilal, M., Iqbal, H.M.N., Adil, S.F., Shaik, M.R., Abdelgawad, A., Hatshan, M.R., Khan, M., 2022. Surface-coated magnetic nanostructured materials for robust bio-catalysis and biomedical applications-A review. J. Adv. Res. 38, 157–177. https://doi.org/10.1016/j.jare.2021.09.013.

Bokov, D., Jalil, A.T., Chupradit, S., Suksatan, W., Ansari, M.J., Shewael, I.H., Valiev, G.H., Kianfar, E., Wang, Z., 2021. Nanomaterial by sol-gel method: synthesis and application. Adv. Mater. Sci. Eng. 2021, 1–21. https://doi.org/10.1155/2021/5102014.

Buxadera-Palomero, J., Calvo, C., Torrent-Camarero, S., Gil, F.J., Mas-Moruno, C., Canal, C., Rodríguez, D., 2017. Biofunctional polyethylene glycol coatings on titanium: an in vitro-based comparison of functionalization methods. Colloids Surf. B Biointerf. 152, 367–375. https://doi.org/10.1016/j.colsurfb.2017.01.042.

Cano, M.E., Medina, R.H., Fernández, V.V.A., García-Casillas, P.E., 2014. Magnetic heating ability of silica-cobalt ferrite nanoparticles. Revista Mexicana de Ingeniera Quimica 13 (2), 555–561.

Cartmell, S.H., Dobson, J., Verschueren, S.B., El Haj, A.J., 2002. Development of magnetic particle techniques for long-term culture of bone cells with intermittent mechanical activation. IEEE Trans. NanoBiosci. 1 (2), 92–97. https://doi.org/10.1109/TNB.2002.806945.

Casadey, R., Broglia, M., Barbero, C., Criado, S., Rivarola, C., 2020. Controlled release systems of natural phenolic antioxidants encapsulated inside biocompatible hydrogels. React. Funct. Polym. 156, 104729. https://doi.org/10.1016/j.reactfunctpolym.2020.104729.

Chang, H.H.R., Chen, K., Lugtu-Pe, J.A., AL-Mousawi, N., Zhang, X., Bar-Shalom, D., Kane, A., Wu, X.Y., 2023. Design and optimization of a nanoparticulate pore former as a multifunctional coating excipient for pH transition-independent controlled release of weakly basic drugs for oral drug delivery. Pharmaceutics 15 (2). https://doi.org/10.3390/pharmaceutics15020547.

Chee, H.L., Gan, C.R.R., Ng, M., Low, L., Fernig, D.G., Bhakoo, K.K., Paramelle, D., 2018. Biocompatible peptide-coated ultrasmall superparamagnetic iron oxide nanoparticles for in vivo contrast-enhanced magnetic resonance imaging. ACS Nano 12 (7), 6480–6491. https://doi.org/10.1021/acsnano.7b07572.

Chemical vapour deposition (CVD), 2019. Advances, Technology and Applications. CRC Press.

Chertok, B., David, A.E., Yang, V.C., 2010. Polyethyleneimine-modified iron oxide nanoparticles for brain tumor drug delivery using magnetic targeting and intra-carotid administration. Biomaterials 31 (24), 6317–6324. https://doi.org/10.1016/j.biomaterials.2010.04.043.

Child, H.W., Del Pino, P.A., De La Fuente, J.M., Hursthouse, A.S., Stirling, D., Mullen, M., McPhee, G.M., Nixon, C., Jayawarna, V., Berry, C.C., 2011. Working together: the combined application of a magnetic field and penetratin for the delivery of magnetic nanoparticles to cells in 3D. ACS Nano 5 (10), 7910–7919. https://doi.org/10.1021/nn202163v.

Darwish, K., Lee, R., Lee, Y., 2019. Synthesis of magnetic ferrite nanoparticles with high hyperthermia performance via a controlled Co-precipitation method. Nanomaterials 9 (8), 1176. https://doi.org/10.3390/nano9081176.

De la Garza-Ramos, M.A., Estupiñan-Lopez, F.H., Gaona-Tiburcio, C., Beltrán-Novelo, L.G., Zambrano-Robledo, P., Cabral-Miramontes, J., Almeraya-Calderón, F., 2020. Electrochemical behavior of Ti6Al4V alloy used in dental implants immersed in Streptococcus gordonii and fusobacterium nucleatum solutions. Materials 13 (18). https://doi.org/10.3390/ma13184185.

De Souza, F.M., Gupta, R.K., Yasin, G., Nguyen, T.A., 2022. Plasma Technology for Nanostructures Plasma at the Nanoscale. Elsevier Applied Science, United States, pp. 265–281. https://doi.org/10.1016/B978-0-323-89930-7.00006-6.

Dehghanghadikolaei, A., Ansary, J., Ghoreishi, R., 2018. Sol-gel process applications: a mini-review. Proc. Nat. Res. Soc. 2. https://doi.org/10.11605/j.pnrs.201802008.

Diez-Escudero, A., Hailer, N.P., 2021. The role of silver coating for arthroplasty components. Bone Joint J. 103 (3), 423–429. https://doi.org/10.1302/0301-620X.103B3.BJJ-2020-1370.R1.

Du Toit, L.C., Kumar, P., Choonara, Y.E., Pillay, V., 2020. Advanced 3D-printed systems and nanosystems for drug delivery and tissue engineering: a volume in woodhead publishing series in biomaterials. In: Advanced 3D-Printed Systems and Nanosystems for Drug Delivery and Tissue Engineering. Elsevier, South Africa, pp. 1–302. https://doi.org/10.1016/B978-0-12-818471-4.01001-3.

Dutu, E., Dumitrache, F., Fleaca, C.T., Morjan, I., Gavrila-Florescu, L., Morjan, I.P., Sandu, I., Scarisoreanu, M., Luculescu, C., Niculescu, A.M., Vasile, E., 2015. Metallic tin-based nanoparticles synthesis by laser pyrolysis: parametric studies focused on the decreasing of the crystallite size. Appl. Surf. Sci. 336, 290–296. https://doi.org/10.1016/j.apsusc.2014.12.092.

Ekeocha, J., Ellingford, C., Pan, M., Wemyss, A.M., Bowen, C., Wan, C., 2021. Challenges and opportunities of self-healing polymers and devices for extreme and hostile environments. Adv. Mater. 33 (33). https://doi.org/10.1002/adma.202008052.

Emil Kaya, E., Kaya, O., Alkan, G., Gürmen, S., Stopic, S., Friedrich, B., 2020. New proposal for size and size-distribution evaluation of nanoparticles synthesized via ultrasonic spray pyrolysis using search algorithm based on image-processing technique. Materials 13 (1), 38. https://doi.org/10.3390/ma13010038.

Fan, Q., Ma, J., Xu, Q., An, W., Qiu, R., 2018. Multifunctional coatings crafted via layer-by-layer spraying method. Prog. Org. Coating 125, 215–221. https://doi.org/10.1016/j.porgcoat.2018.09.015.

Faraji, H., Nedaeinia, R., Nourmohammadi, E., Malaekeh-Nikouei, B., Sadeghnia, H.R., Ziapour, S.P., Sarkarizi, H.K., Oskuee, R.K., 2018. A review on application of novel solid nanostructures in drug delivery. J. Nano Res. 53, 22–36. https://doi.org/10.4028/www.scientific.net/JNanoR.53.22.

Gandhi, S., Roy, I., 2019. Methylene blue loaded, silica coated cobalt ferrite nanoparticles with potential for combination therapy. Mater. Res. Express 6 (7). https://doi.org/10.1088/2053-1591/ab187a.

Gao, F., 2019. An overview of surface-functionalized magnetic nanoparticles: preparation and application for wastewater treatment. ChemistrySelect 4 (22), 6805–6811. https://doi.org/10.1002/slct.201900701.

García-Merino, B., Bringas, E., Ortiz, I., 2022. Synthesis and applications of surface-modified magnetic nanoparticles: progress and future prospects. Rev. Chem. Eng. 38 (7), 821–842. https://doi.org/10.1515/revce-2020-0072.

Growth and Transport in Nanostructured Materials: Reactive Transport in PVD, CVD, 2016.

Gupta, R., Xie, H., 2018. Nanoparticles in daily life: applications, toxicity and regulations. J. Environ. Pathol. Toxicol. Oncol. 37 (3), 209–230. https://doi.org/10.1615/JEnvironPatholToxicolOncol.2018026009.

Haleem, A., Javaid, M., Singh, R.P., Rab, S., Suman, R., 2023. Applications of nanotechnology in medical field: a brief review. Glob. Health J. 7 (2), 70–77. https://doi.org/10.1016/j.glohj.2023.02.008.

Hardiansyah, A., Huang, L.Y., Yang, M.C., Liu, T.Y., Tsai, S.C., Yang, C.Y., Kuo, C.Y., Chan, T.Y., Zou, H.M., Lian, W.N., Lin, C.H., 2014. Magnetic liposomes for colorectal cancer cells therapy by high-frequency magnetic field treatment. Nanoscale Res. Lett. 9 (1), 1–13. https://doi.org/10.1186/1556-276X-9-497.

Hosseini, M.S., Ebratkhahan, M., Shayegan, Z., Niaei, A., Salari, D., Rostami, A., Raeisipour, J., 2020. Investigation of the effective operational parameters of self-cleaning glass surface coating to improve methylene blue removal efficiency; application in solar cells. Sol. Energy 207, 398–408. https://doi.org/10.1016/j.solener.2020.06.109.

Howard, D., Buttery, L.D., Shakesheff, K.M., Roberts, S.J., 2008. Tissue engineering: strategies, stem cells and scaffolds. J. Anat. 213 (1), 66–72. https://doi.org/10.1111/j.1469-7580.2008.00878.x.

Idumah, C.I., May Obele, C., Emmanuel, E.O., Hassan, A., 2020. Recently emerging nanotechnological advancements in polymer nanocomposite coatings for anti-corrosion, anti-fouling and self-healing. Surface. Interfac. 21, 100734. https://doi.org/10.1016/j.surfin.2020.100734.

Jayaraj, K., Pius, A., 2018. Biocompatible Coatings for Metallic Biomaterials Fundamental Biomaterials: Metals. Elsevier, India, pp. 323–354. https://doi.org/10.1016/B978-0-08-102205-4.00016-7.

Kefeni, K.K., Msagati, T.A.M., Nkambule, T.T., Mamba, B.B., 2020. Spinel ferrite nanoparticles and nanocomposites for biomedical applications and their toxicity. Mater. Sci. Eng. C 107. https://doi.org/10.1016/j.msec.2019.110314.

Khan, I., Saeed, K., Khan, I., 2019. Nanoparticles: properties, applications and toxicities. Arab. J. Chem. 12 (7), 908–931. https://doi.org/10.1016/j.arabjc.2017.05.011.

Khan, S., Katsumata, K.-I., Rodríguez-González, V., Terashima, C., Fujishima, A., 2020. Gas-phase Synthesis for Mass Production of TiO2 Nanoparticles for Environmental Applications. Springer Science and Business Media LLC, pp. 1–21. https://doi.org/10.1007/978-3-030-11155-7_10-1.

Khan, I., Morishita, S., Higashinaka, R., Matsuda, T.D., Aoki, Y., Kuzmann, E., Homonnay, Z., Katalin, S., Pavić, L., Kubuki, S., 2021. Synthesis, characterization and magnetic properties of ε-Fe2O3 nanoparticles prepared by sol-gel method. J. Magn. Magn Mater. 538, 168264. https://doi.org/10.1016/j.jmmm.2021.168264.

Khlyustova, A., Cheng, Y., Yang, R., 2020. Vapor-deposited functional polymer thin films in biological applications. J. Mater. Chem. B 8 (31), 6588–6609. https://doi.org/10.1039/d0tb00681e.

kianfar, E., 2021. Magnetic nanoparticles in targeted drug delivery: a review. J. Supercond. Nov. Magnetism 34 (7), 1709–1735. https://doi.org/10.1007/s10948-021-05932-9.

Kim, M.S., Kweon, S.H., Cho, S., An, S.S.A., Kim, M.I., Doh, J., Lee, J., 2017. Pt-decorated magnetic nanozymes for facile and sensitive point-of-care bioassay. ACS Appl. Mater. Interfaces 9 (40), 35133–35140. https://doi.org/10.1021/acsami.7b12326.

Krishnan, K.M., 2016. Fundamentals and Applications of Magnetic Materials. Oxford University Press. https://doi.org/10.1093/acprof:oso/9780199570447.001.0001.

Kudr, J., Haddad, Y., Richtera, L., Heger, Z., Cernak, M., Adam, V., Zitka, O., 2017. Magnetic nanoparticles: from design and synthesis to real world applications. Nanomaterials 7 (9), 243. https://doi.org/10.3390/nano7090243.

Laurent, S., Dutz, S., Häfeli, U.O., Mahmoudi, M., 2011. Magnetic fluid hyperthermia: focus on superparamagnetic iron oxide nanoparticles. Adv. Colloid Interface Sci. 166 (1−2), 8−23. https://doi.org/10.1016/j.cis.2011.04.003.

Lerner, M.I., Glazkova, E.A., Lozhkomoev, A.S., Svarovskaya, N.V., Bakina, O.V., Pervikov, A.V., Psakhie, S.G., 2016. Synthesis of Al nanoparticles and Al/AlN composite nanoparticles by electrical explosion of aluminum wires in argon and nitrogen. Federat. Powder Technol. 295, 307−314. https://doi.org/10.1016/j.powtec.2016.04.005.

Liang, H., Zhou, B., Li, J., Liu, X., Deng, Z., Li, B., 2018. Engineering multifunctional coatings on nanoparticles based on oxidative coupling assembly of polyphenols for stimuli-responsive drug delivery. J. Agric. Food Chem. 66 (26), 6897−6905. https://doi.org/10.1021/acs.jafc.8b01208.

Lim, J., Park, S., Cho, H., Lee, Y., Ha, I.H., Kim, Y., Hwang, E., Lee, H., Shin, J., Kwon, J., Ko, S.H., Hong, S., 2022. Monolithic digital patterning of polyimide by laser-induced pyrolytic jetting. Chem. Eng. J. 428, 131050. https://doi.org/10.1016/j.cej.2021.131050.

Mao, S., Zhang, D., Zhang, Y., Yang, J., Zheng, J., 2020. A universal coating strategy for controllable functionalized polymer surfaces. Adv. Funct. Mater. 30 (40). https://doi.org/10.1002/adfm.202004633.

Meier, P., Clement, P., Altenried, S., Reina, G., Ren, Q., Züst, R., Enger, O., Choi, F., Nestle, N., Deisenroth, T., Neubauer, P., Wick, P., 2023. Quaternary ammonium-based coating of textiles is effective against bacteria and viruses with a low risk to human health. Sci. Rep. 13 (1). https://doi.org/10.1038/s41598-023-47707-3.

Mozumder, M.S., Mourad, A.H.I., Pervez, H., Surkatti, R., 2019. Recent developments in multifunctional coatings for solar panel applications: a review. Sol. Energy Mater. Sol. Cell. 189, 75−102. https://doi.org/10.1016/j.solmat.2018.09.015.

Mushtaq, M.W., Kanwal, F., Batool, A., Jamil, T., Zia-ul-Haq, M., Ijaz, B., Huang, Q., Ullah, Z., 2017. Polymer-coated CoFe2O4 nanoassemblies as biocompatible magnetic nanocarriers for anticancer drug delivery. J. Mater. Sci. 52 (16), 9282−9293. https://doi.org/10.1007/s10853-017-1141-3.

Muzzio, N., Moya, S., Romero, G., 2021. Multifunctional scaffolds and synergistic strategies in tissue engineering and regenerative medicine. Pharmaceutics 13 (6), 792. https://doi.org/10.3390/pharmaceutics13060792.

Nasirpouri, F., Alipour, K., Daneshvar, F., Sanaeian, M.R., 2020. Electrodeposition of Anti-corrosion Nanocoatings Corrosion Protection at the Nanoscale. Elsevier, Iran, pp. 473−497. https://doi.org/10.1016/B978-0-12-819359-4.00024-6.

Ndlwana, L., Raleie, N., Dimpe, K.M., Ogutu, H.F., Oseghe, E.O., Motsa, M.M., Msagati, T.A.M., Mamba, B.B., 2021. Sustainable hydrothermal and solvothermal synthesis of advanced carbon materials in multidimensional applications: a review. Materials 14 (17). https://doi.org/10.3390/ma14175094.

Nel, A., Grainger, D., Alvarez, P.J., Badesha, S., Vincent, C., Ferrari, M., Godwin, H., Grodzinski, P., Morris, J., Savage, N., Scott, N., Wiesner, M., 2011. Nanotechnology Environmental, Health, and Safety Issues. Springer Science and Business Media LLC, pp. 159−220. https://doi.org/10.1007/978-94-007-1168-6_5.

Nicholson, J.W., 2020. Titanium alloys for dental implants: areview. Prosthesis 2 (2), 100−116. https://doi.org/10.3390/prosthesis2020011.

Niculescu, A.G., Chircov, C., Grumezescu, A.M., 2022. Magnetite nanoparticles: synthesis methods −a comparative review. Methods 199, 16−27. https://doi.org/10.1016/j.ymeth.2021.04.018.

Nizamuddin, S., Siddiqui, M.T.H., Mubarak, N.M., Baloch, H.A., Mazari, S.A., Tunio, M.M., Griffin, G.J., Srinivasan, M.P., Tanksale, A., Riaz, S., 2018. Advanced nanomaterials synthesis from pyrolysis and hydrothermal carbonization: a review. Curr. Org. Chem. 22 (5), 446−461. https://doi.org/10.2174/1385272821666171026153215.

Pal, N., Chakraborty, D., Cho, E.B., Seo, J.G., 2023. Recent developments on the catalytic and biosensing applications of porous nanomaterials. Nanomaterials 13 (15). https://doi.org/10.3390/nano13152184.

Paradise, J., 2011. Reassessing safety for nanotechnology combination products: what do biosimilars add to regulatory challenges for the FDA. Louis ULJ 56.

Perni, S., Prokopovich, P., 2020. Nanostructured Coatings for Antimicrobial Applications Advances in Nanostructured Materials and Nanopatterning Technologies: Applications for Healthcare, Environmental and Energy. Elsevier, United Kingdom, pp. 115−140. https://doi.org/10.1016/B978-0-12-816865-3.00005-6.

Petrova, L., Kozlova, O., Vladimirtseva, E., Smirnova, S., Lipina, A., Odintsova, O., 2021. Development of multifunctional coating of textile materials using silver microencapsulated compositions. Coatings 11 (2), 159. https://doi.org/10.3390/coatings11020159.

Phan, L., Yoon, S., Moon, M.-W., 2017. Plasma-based nanostructuring of polymers: areview. Polymers 9 (12), 417. https://doi.org/10.3390/polym9090417.

Pina, S., Ribeiro, V.P., Marques, C.F., Maia, F.R., Silva, T.H., Reis, R.L., Oliveira, J.M., 2019. Scaffolding strategies for tissue engineering and regenerative medicine applications. Materials 12 (11). https://doi.org/10.3390/ma12111824.

Primc, G., 2020. Recent advances in surface activation of Polytetrafluoroethylene (PTFE) by gaseous plasma treatments. Polymers 12 (10), 2295. https://doi.org/10.3390/polym12102295.

Psimadas, D., Baldi, G., Ravagli, C., Comes Franchini, M., Locatelli, E., Innocenti, C., Sangregorio, C., Loudos, G., 2014. Comparison of the magnetic, radiolabeling, hyperthermic and biodistribution properties of hybrid nanoparticles bearing CoFe2O4and Fe304 metal cores. Nanotechnology 25 (2). https://doi.org/10.1088/0957-4484/25/2/025101Greece.

Rasaili, P., Kumar Sharma, N., Bhattarai, A., 2022. Comparison of ferromagnetic materials: past work, recent trends, and applications. Conden. Matt. 7 (1), 12. https://doi.org/10.3390/condmat7010012.

Rashid, A., Zubair, U., Ashraf, M., Javid, A., Abid, H.A., Akram, S., 2023. Flexible piezoelectric coatings on textiles for energy harvesting and autonomous sensing applications: a review. J. Coating Technol. Res. 20 (1), 141−172. https://doi.org/10.1007/s11998-022-00690-2.

Ray, P.C., Yu, H., Fu, P.P., 2009. Toxicity and environmental risks of nanomaterials: challenges and future needs. J. Environ. Sci. Health, Part C. 27 (1), 1−35. https://doi.org/10.1080/10590500802708267.

Rocha-Santos, T., 2014. Sensors and biosensors based on magnetic nanoparticles. TrAC Trends Anal. Chem. 62. https://doi.org/10.1016/j.trac.2014.06.016.

Rümenapp, C., Gleich, B., Haase, A., 2012. Magnetic nanoparticles in magnetic resonance imaging and diagnostics. Pharmaceut. Res. 29 (5), 1165−1179. https://doi.org/10.1007/s11095-012-0711-y.

Saleh, M.M., Saleh, M.M., Touny, A.H., Al-Omair, M.A., 2016. Biodegradable/biocompatible coated metal implants for orthopedic applications. Bio Med. Mater. Eng. 27 (1), 87–99. https://doi.org/10.3233/BME-161568.

Sánchez-Bodón, J., Andrade del Olmo, J., Alonso, J.M., Moreno-Benítez, I., Luis Vilas-Vilela, J., Pérez-Álvarez, L., 2022. Bioactive coatings on titanium: areview on hydroxylation, self-assembled monolayers (SAMs) and surface modification strategies. Polymers 14 (1), 165. https://doi.org/10.3390/polym14010165.

Saragi, T., Depi, B.L., Butarbutar, S., Permana, B., Risdiana, 2018. The impact of synthesis temperature on magnetite nanoparticles size synthesized by co-precipitation method. J. Phys. Conf. 1013 (1), 012190. https://doi.org/10.1088/1742-6596/1013/1/012190.

Sardo, C., 2023. Nanobiomaterials revolutionizing biomedical science: special issue—" nanotechnology for biomedical applications". Appl. Sci. 13 (10), 6123. https://doi.org/10.3390/app13106123.

Shen, J., Zhang, S., Fang, X., Salmon, S., 2022. Advances in 3D gel printing for enzyme immobilization. Gels 8 (8), 460. https://doi.org/10.3390/gels8080460.

Shukla, S., Khan, R., Daverey, A., 2021. Synthesis and characterization of magnetic nanoparticles, and their applications in wastewater treatment: a review. Environ. Technol. Innovat. 24, 101924. https://doi.org/10.1016/j.eti.2021.101924.

Smolková, B., Uzhytchak, M., Lynnyk, A., Kubinová, Š., Dejneka, A., Lunov, O., 2019. A critical review on selected external physical cues and modulation of cell behavior: magnetic nanoparticles, non-thermal plasma and lasers. J. Funct. Biomater. 10 (1), 2. https://doi.org/10.3390/jfb10010002.

Song, N., Chen, Z., Yu, Y., Yang, B., Guo, Y., Chu, Y., Chen, T., Bi, Y., Zhang, Q., Feng, C., Jiang, J., 2023. Electrolytic manganese residue to metal phosphides via ball milling: enhancing flame retardancy, mechanical properties, and electrical conductivity of epoxy resin. Appl. Surf. Sci. 638, 158154. https://doi.org/10.1016/j.apsusc.2023.158154.

Stephanie, R., Kim, M.W., Kim, S.H., Kim, J.K., Park, C.Y., Park, T.J., 2021. Recent advances of bimetallic nanomaterials and its nanocomposites for biosensing applications. TrAC, Trends Anal. Chem. 135. https://doi.org/10.1016/j.trac.2020.116159.

Subha, A., Shalini, M.G., Sahu, B., Sahoo, S.C., 2018. Structural transformation and magnetic properties of copper ferrite nanoparticles prepared by sol–gel method. J. Mater. Sci. Mater. Electron. 29 (24), 20790–20799. https://doi.org/10.1007/s10854-018-0221-8.

Tang, K., Wang, W., Song, Z., Luo, X., 2021. Multifunctional nano-biosensor based on metal-organic framework for enhanced fluorescence imaging of intracellular miRNA-122 and synergistic chemo-photothermal therapy of tumor cells. Anal. Chim. Acta 1176, 338779. https://doi.org/10.1016/j.aca.2021.338779.

Tay, Z.W., Chandrasekharan, P., Chiu-Lam, A., Hensley, D.W., Dhavalikar, R., Zhou, X.Y., Yu, E.Y., Goodwill, P.W., Zheng, B., Rinaldi, C., Conolly, S.M., 2018. Magnetic particle imaging-guided heating in vivo using gradient fields for arbitrary localization of magnetic hyperthermia therapy. ACS Nano 12 (4), 3699–3713. https://doi.org/10.1021/acsnano.8b00893.

Teixeira-Santos, R., Lima, M., Gomes, L.C., Mergulhão, F.J., 2021. Antimicrobial coatings based on chitosan to prevent implant-associated infections: a systematic review. iScience 24 (12), 103480. https://doi.org/10.1016/j.isci.2021.103480.

Terreno, E., Castelli, D.D., Viale, A., Aime, S., 2010. Challenges for molecular magnetic resonance imaging. Chem. Rev. 110 (5), 3019–3042. https://doi.org/10.1021/cr100025t.

Thakur, A., Kumar, A., Kaya, S., Marzouki, R., Zhang, F., Guo, L., 2022. Recent advancements in surface modification, characterization and functionalization for enhancing the biocompatibility and corrosion resistance of biomedical implants. Coatings 12 (10), 1459. https://doi.org/10.3390/coatings12101459.

Thanka Rajan, S., Subramanian, B., Arockiarajan, A., 2022. A comprehensive review on biocompatible thin films for biomedical application. Ceram. Int. 48 (4), 4377−4400. https://doi.org/10.1016/j.ceramint.2021.10.243.

Thiam, Z., Abou-Hamad, E., Dereli, B., Liu, L., Emwas, A.H., Ahmad, R., Jiang, H., Isah, A.A., Ndiaye, P.B., Taoufik, M., Han, Y., Cavallo, L., Basset, J.M., Eddaoudi, M., 2020. Extension of surface organometallic chemistry to metal−organic frameworks: development of a well-defined single site [(≡Zr− O−)W(=O)(CH2TBu)3] olefin metathesis catalyst. J. Am. Chem. Soc. 142 (39), 16690−16703. https://doi.org/10.1021/jacs.0c06925.

Thomsen, L.B., Linemann, T., Pondman, K.M., Lichota, J., Kim, K.S., Pieters, R.J., Visser, G.M., Moos, T., 2013. Uptake and transport of superparamagnetic iron oxide nanoparticles through human brain capillary endothelial cells. ACS Chem. Neurosci. 4 (10), 1352−1360. https://doi.org/10.1021/cn400093z.

Thorn, C.R., Howell, P.L., Wozniak, D.J., Prestidge, C.A., Thomas, N., 2021. Enhancing the therapeutic use of biofilm-dispersing enzymes with smart drug delivery systems. Adv. Drug Deliv. Rev. 179. https://doi.org/10.1016/j.addr.2021.113916.

Thu Huong, L.T., Nam, N.H., Doan, D.H., My Nhung, H.T., Quang, B.T., Nam, P.H., Thong, P.Q., Phuc, N.X., Thu, H.P., 2016. Folate attached, curcumin loaded Fe3O4 nanoparticles: a novel multifunctional drug delivery system for cancer treatment. Mater. Chem. Phys. 172, 98−104. https://doi.org/10.1016/j.matchemphys.2015.12.065.

Tran, N., Webster, T.J., 2010. Magnetic nanoparticles: biomedical applications and challenges. J. Mater. Chem. 20 (40), 8760−8767. https://doi.org/10.1039/c0jm00994f.

Uthaman, S., Lee, S.J., Cherukula, K., Cho, C.S., Park, I.K., 2015. Polysaccharide-coated magnetic nanoparticles for imaging and gene therapy. BioMed Res. Int. 2015. https://doi.org/10.1155/2015/959175.

Vega-Chacón, J., Picasso, G., Avilés-Félix, L., Jafelicci, M., 2016. Influence of synthesis experimental parameters on the formation of magnetite nanoparticles prepared by polyol method. Adv. Nat. Sci. Nanosci. Nanotechnol. 7 (1), 015014. https://doi.org/10.1088/2043-6262/7/1/015014.

Walsh, F.C., Wang, S., Zhou, N., 2020. The electrodeposition of composite coatings: diversity, applications and challenges. Curr. Opin. Electrochem. 20, 8−19. https://doi.org/10.1016/j.coelec.2020.01.011.

Wang, F., Shifa, T.A., Yu, P., He, P., Liu, Y., Wang, F., Wang, Z., Zhan, X., Lou, X., Xia, F., He, J., 2018. New Frontiers on van der Waals Layered Metal Phosphorous Trichalcogenides. Adv. Funct. Mater. 28 (37). https://doi.org/10.1002/adfm.201802151.

Wang, Y., Pan, J.A., Wu, H., Talapin, D.V., 2019. Direct wavelength-selective optical and electron-beam lithography of functional inorganic nanomaterials. ACS Nano 13 (12), 13917−13931. https://doi.org/10.1021/acsnano.9b05491.

Wang, C., Li, D., Lu, Z.S., Song, M., Xia, W., 2020. Synthesis of carbon nanoparticles in a non-thermal plasma process. Chem. Eng. Sci. 227, 115921. https://doi.org/10.1016/j.ces.2020.115921.

Weissleder, R., Stark, D.D., Engelstad, B.L., Bacon, B.R., Compton, C.C., White, D.L., Jacobs, P., Lewis, J., 1989. Superparamagnetic iron oxide: pharmacokinetics and toxicity. Am. J. Roentgenol. 152 (1), 167−173. https://doi.org/10.2214/ajr.152.1.167.

Witika, B.A., Makoni, P.A., Matafwali, S.K., Chabalenge, B., Mwila, C., Kalungia, A.C., Nkanga, C.I., Bapolisi, A.M., Walker, R.B., 2020. Biocompatibility of biomaterials for nanoencapsulation: current approaches. Nanomaterials 10 (9), 1−40. https://doi.org/10.3390/nano10091649.

Wong, J., Tong, R., 2018. Handbook of Medical Device Regulatory Affairs in Asia. CRC Press.

Wu, G., Li, P., Feng, H., Zhang, X., Chu, P.K., 2015. Engineering and functionalization of biomaterials via surface modification. J. Mater. Chem. B 3 (10), 2024–2042. https://doi.org/10.1039/c4tb01934b.

Wu, J., Tu, J., Hu, K., Xiao, X., Li, L., Yu, S., Xie, Y., Wu, H., Yang, Y., 2022. Sol–gel-derived bayberry-like SiO2@TiO2 multifunctional antireflective coatings for enhancing photovoltaic power generation. Colloids Surf. A Physicochem. Eng. Asp. 654, 130173. https://doi.org/10.1016/j.colsurfa.2022.130173.

Xi, W., Hegde, V., Zoller, S.D., Park, H.Y., Hart, C.M., Kondo, T., Hamad, C.D., Hu, Y., Loftin, A.H., Johansen, D.O., Burke, Z., Clarkson, S., Ishmael, C., Hori, K., Mamouei, Z., Okawa, H., Nishimura, I., Bernthal, N.M., Segura, T., 2021. Point-of-care antimicrobial coating protects orthopaedic implants from bacterial challenge. Nat. Commun. 12 (1). https://doi.org/10.1038/s41467-021-25383-z.

Xia, L., 2020. Importance of nanostructured surfaces. In: Bioceramics: From Macro to Nanoscale. Elsevier, China, pp. 5–24. https://doi.org/10.1016/B978-0-08-102999-2.00002-8.

Xiao, M., Chen, Y.M., Biao, M.N., Zhang, X.D., Yang, B.C., 2017. Bio-functionalization of biomedical metals. Mater. Sci. Eng. C 70, 1057–1070. https://doi.org/10.1016/j.msec.2016.06.067.

Xu, W., Jambhulkar, S., Ravichandran, D., Zhu, Y., Kakarla, M., Nian, Q., Azeredo, B., Chen, X., Jin, K., Vernon, B., Lott, D.G., Cornella, J.L., Shefi, O., Miquelard-Garnier, G., Yang, Y., Song, K., 2021. 3D printing-enabled nanoparticle alignment: areview of mechanisms and applications. Small 17 (45). https://doi.org/10.1002/smll.202100817.

Yamato, M., Kimura, T., 2020. Magnetic processing of diamagnetic materials. Polymers 12 (7), 1491. https://doi.org/10.3390/polym12071491.

Yin, S., Chen, C., Suo, X., Lupoi, R., 2018. Cold-sprayed metal coatings with nanostructure. Adv. Mater. Sci. Eng. 2018, 1–19. https://doi.org/10.1155/2018/2804576.

Zamani Kouhpanji, M.R., Stadler, B.J.H., 2020. A guideline for effectively synthesizing and characterizing magnetic nanoparticles for advancing nanobiotechnology: areview. Sensors 20 (9), 2554. https://doi.org/10.3390/s20092554.

Zhang, F., Ju, P., Pan, M., Zhang, D., Huang, Y., Li, G., Li, X., 2018. Self-healing mechanisms in smart protective coatings: a review. Corrosion Sci. 144, 74–88. https://doi.org/10.1016/j.corsci.2018.08.005.

Zheng, J., Cheng, X., Zhang, H., Bai, X., Ai, R., Shao, L., Wang, J., 2021. Gold nanorods: the most versatile plasmonic nanoparticles. Chem. Rev. 121 (21), 13342–13453. https://doi.org/10.1021/acs.chemrev.1c00422.

Zhu, L., Xue, P., Lan, Q., Meng, G., Ren, Y., Yang, Z., Xu, P., Liu, Z., 2021. Recent research and development status of laser cladding: a review. Opt Laser. Technol. 138, 106915. https://doi.org/10.1016/j.optlastec.2021.106915.

Multifunctional mesoporous silica-based nanocomposites for biomedical applications

8

Dzmitry Shcharbin[1], *Burcu Önal Acet*[2], *Inessa Halets-Bui*[1], *Mehmet Odabaşı*[2] *and Ömür Acet*[3]
[1]Institute of Biophysics and Cell Engineering of the National Academy of Sciences of Belarus, Minsk, Belarus; [2]Faculty of Arts and Science, Chemistry Department, Aksaray University, Aksaray, Turkey; [3]Vocational School of Health Science, Pharmacy Services Program, Tarsus University, Tarsus, Turkey

8.1 Introduction

Nanotechnology is a developing area of study focused on utilizing scientific progress to manipulate material structure on a smaller scale. It presents remarkable prospects for medical field, mainly through the using of hybrid nanoparticles at the subcellular and cellular levels. The arrangement and characterization of products for nanotechnology can be effectively achieved using multifunctional mesoporous silica-based nanostructures (MSNs). MSNs possess a flexible mesoporous structure, making them highly versatile among other nanomaterials (Burns et al., 2006). Additionally, they have large pore volumes ($0.5-1.5\ cm^3/g$) (Boissière et al., 2003). MSNs also have substantial surface area, which acts a noteworthy role in facilitating adsorption and boosting the loading capacity of numerous therapeutic substances (Beck et al., 1992; Kresge et al., 1992). Despite having low hemolytic activity, MSNs exhibit exceptional thermal and chemical stability (Yang et al., 2005). MSNs in nanoformulations consistently exhibit several advantageous characteristics compared to their larger counterparts. These include enhanced in vivo bioavailability, decreased toxicity, and heightened selectivity (Mitchell et al., 2021).

Nanoformulations are typically categorized into two primary groups: organic and inorganic nanoformulations. In a broader sense, these nanoformulations are extensively studied and analyzed (Mitchell et al., 2021; Yang et al., 2005). Doxil[1], the pioneering nanoliposomal drug formulation, incorporates organic nanoformulations like liposomes and polymers. It is the initial nanoliposomal drug formulation approved by the American Food and Drug Administration (FDA) (Barenholz, 2012). Moreover, it has demonstrated its potential to enhance the efficacy of two lipid nanoparticle-based COVID-19 mRNA vaccines (Anselmo and Mitragotri, 2021). While formulations utilizing organic nanoparticles may surpass those based on inorganic nanoparticles in terms of biocompatibility and safety, the latter outshine the former regarding stability and drug delivery (DD) performance. It is essential to highlight that the remarkable

Multifunctional Nanostructured Coatings. https://doi.org/10.1016/B978-0-443-23683-9.00008-8

qualifications possessed by inorganic nanoparticles, including optical, ultrasonic, magnetic, and catalytic features, have paved the way for developing novel nanoparticle-based therapies (Xu et al., 2023). Optimistically, inorganic-based nanoparticles are gradually advancing toward the clinical stages, showcasing promising progress. Around 25 inorganic nanomedicines have received approval for clinical applications, further highlighting their potential in the field (Huang et al., 2020).

In the past few years, silica-based technologies in conjunction with various nanomaterials have emerged, enabling the practical synthesis of multiple materials within a single nanounit. As a result of the advanced features exhibited by these engineered nanostructures, they can be effectively incorporated with the drug encapsulation capabilities of mesoporous silica materials, thereby generating considerable attention from both clinical and academic communities. Hence, the notion of incorporating various attributes into a solitary framework may appear unattainable. Nevertheless, the challenge lies in their integration into composite substances, similar to the safety concerns and constraints associated with unprocessed nanomaterials (Castillo and Vallet-Regí, 2019).

MSNs that possess convenient surface functionalization are favored because of their specific aptness for transporting materials that are not soluble in water (Noreen et al., 2022). The utilization of MSNs enables the production of significant hybrid materials, including quantum dots (QDs) (Wang et al., 2012), noble metals (Davidson et al., 2018), and magnetic nanoparticles (Li et al., 2011), through the manipulation of other nanoparticles that are coated on the surfaces of the pores. In addition, MSNs have the capability to undergo surface functionalization in order to facilitate cell-specific targeting and delivery investigations. The potential for advanced drug delivery applications using these novel hybrid nanostructures has garnered significant interest among researchers (Noreen et al., 2022). The surface and entrance of MSNs possess the necessary characteristics to be modified with stimuli-responsive groups, inorganic nanostructures, supramolecules, and proteins that are able to serve as lids and gatekeepers. By utilizing the encapsulation technique, the release of drugs could be precisely controlled and triggered in response to internal or external stimuli such as pH, temperature, redox potential, light, and enzymes. Additionally, MSNs that have been modified with fluorescent dye and paramagnetic metal complex have been employed as probes. In this perspective, we explore the most novel research on multifunctional MSNs designed for biomedical implementations (Lee et al., 2011). Some applications of MSNs in biomedicine are presented in Fig. 8.1.

8.2 A Brief development history and different categories of MSNs

Silica structures exhibit a tetrahedral framework, comprising a silicon atom and four oxygen atoms that are connected through covalent bonds. MSNs belong to a category of porous materials described by a pore size distribution ranging from 2 to 50 nm. Soluble silica precursors are able to be incorporated into liquid crystalline mesophases

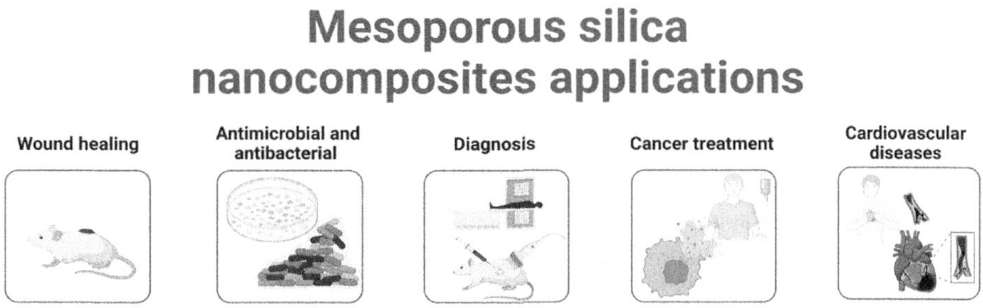

Figure 8.1 Some applications of MSNs in biomedicine.
Created with BioRender.com.

during the preparation of mesoporous silica by introducing block copolymers and/or amphiphilic surfactants as agents that direct the structure (Croissant et al., 2018; Yang et al., 2012).

Various mesoporous phases, including hexagonal, cubic, layered, and disordered structures, can be obtained through the condensation of silanes and subsequent removal of structure-driving substances. This can be achieved using techniques like solvent extraction or calcination. The end result is the formation of amorphous mesoporous silica structures (Croissant et al., 2020; Xu et al., 2023). The research on mesoporous silica has a historical background in the 1960s, during which certain United States patents made references to the synthesis of mesoporous silica. However, it was not until the 1990s that the scientific community began to show a growing interest in studying mesoporous silica (Xu et al., 2023). In the year 1992, researchers at the Mobil Research and Development Corporation successfully created a new type of mesoporous substance known as Mobil Composition of Matter No. 41 (MCM-41). This particular material belongs to a group of silicate-based mesoporous molecular sieves called M41S (Kresge et al., 1992). In addition to MCM-41, Santa Barbara Amorphous-15 (SBA-15) has also shown great potential as a nanomaterial in the field of biomedicine. By 1998, this exceptionally stable mesoporous silica sieve was discovered by researchers from the University of California at Santa Barbara (Zhao et al., 1998). MCM-41 has been proven to possess the capability of loading and dispensing the anti-inflammatory medication, ibuprofen, at a weight percentage ratio of 30% (Vallet-Regi et al., 2001). In the year of the same occurrence, two separate research groups published reports indicating that the size of MCM-41 particles is be able to be adjusted to the nanoscale. On the other hand, its structure may be meticulously tailored. These discoveries provided significant motivation for scientists to delve into the potential biomedical uses of MSNs.

Silica-based nanostructures have found extensive applications across various fields (Acet et al., 2019; Noma et al., 2020; Önal et al., 2019). Upon examining the literature, it becomes evident that MSNs research encompasses various facets of biomedical area. Undoubtedly, as the development of secure and effective MSN-based nanocomposites progresses, along with the systematic investigation of their in vivo biological processes, MSNs will progressively transition from fundamental investigation to clinical application. This transition will undoubtedly contribute to the advancement of nanomedicine (Xu et al., 2023). Typically, the production of MSNs involves the surfactant-templating sol–gel technique. Various factors, including surfactants, silica origins, catalyst substances, and some external circumstances such as pH and temperature, exert an influence on the structure and morphology of these materials (Costa and Paranhos, 2020; Narayan et al., 2018; Wan and Zhao, 2007).

As a result of the wide range of surfactants available, researchers have developed numerous MSNs with distinct structures, which have garnered significant interest. Currently, the sorting of MSNs is based on the mesoporous materials family, including the M41S groups, SBA groups, Fudan University (FDU) groups, and Korea Institute of Technology (KIT) groups, among others. The M41S group materials are commonly distinguished by a significant abundance of silanol groups (Si–OH) present on both the interior pores and surface. This characteristic facilitates the surface-functionalization process for specific bioapplications. Furthermore, mesophase organizations, pore sizes, particle morphologies, and dimensions of these materials are able to be conveniently modified by altering the synthesis circumstances. The MCM-41, MCM-48, and MCM-50 are the most representative materials within the M41S series, and their synthesis involves checking the ratios of surfactant substances to silica sources (Xu et al., 2023). MCM-41, a thoroughly examined member of nanostructured mesoporous materials, exhibits a symmetrical arrangement of unidirectional mesoporous pores in a two-dimensional (2D) hexagonal pattern, characterized by the P6mm space group symmetry. In contrast to MCM-41, MCM-48 exhibits a cubic arrangement with Ia3d space-group symmetry, along with a greater specific pore volume, specific surface area, and thermal stability (Schumacher et al., 1999). MCM-48 possesses a three-dimensional pore architecture and exhibits high degree of porosity, which renders it highly beneficial in the realm of DDSs (Pajchel and Kolodziejski, 2018; Raza et al., 2021; Zeleňák et al., 2018). The uncalcined form of the lamellar phase-MCM-50 exhibits a p2 space-group symmetry, wherein a surfactant layer separates to create a sandwich-like structure. At elevated temperatures, the removal of the surfactant from MCM-50 causes destabilization of its lamellar structure. Consequently, this leads to the formation of dense phases characterized by minor structural organization and porosity (Tanev and Pinnavaia, 1998).

The initial discovery of SBA-series mesoporous materials was made by scientists affiliated with the University of California at Santa Barbara (Zhao et al., 1998). These materials are characterized by a framework composed of silica and possess a well-organized mesoporous structure. Notably, the pore size of these materials can be adjusted, and they demonstrate a substantial specific surface area and exceptional thermal stability. There exists a wide range of silica-origin mesoporous substances within the SBA group, SBA-1 through SBA-16. Nevertheless, among the various materials available, SBA-15 and SBA-16 have gained significant popularity in the realm of biomedical implementations. Conversely, other species of MSNs like SBA-1, SBA-2, SBA-3, SBA-8, SBA-11, and SBA-12 find their primary utility in the domains of adsorption and catalysis. In contrast to MCM-41, SBA-15 exhibits enhanced thermal and mechanical stability due to its thicker pore walls. Notably, the remarkable combination of a high specific surface area and larger pore size in SBA-15 renders it an exceptional candidate for facilitating the effective hosting of large molecule drugs in the biomedical area (Xu et al., 2023).

The mesoporous silica known as FDU series primarily consists of FDU-1, FDU-2, FDU-5, and FDU-12. These silica materials were initially introduced by Zhao's research group at Fudan University (Yu et al., 2000). All these materials exhibit 3D porous structures, meticulously arranged pores, noncrystalline pore walls, and exceptional thermal and mechanical durability (Wan and Zhao, 2007). FDU-1, the initial MSN of the FDU-series, was successfully synthesized in the year 2000. This particular MSN possesses an Im3m space-group symmetry and showcases a mesoporous structure that closely resembles that of SBA-16 (Wan and Zhao, 2007). Unlike the MSNs of SBA and MCM groups, the FDU groups exhibit limited bio-implementations, and only a small number of reports on FDU-12 have been witnessed in the drug delivery field (Carmona et al., 2014; Liu and Che, 2015). FDU-12 exhibits a 3D cubic mesostructure architecture characterized by Fm3m space-group symmetry. It has a spacious cavity with dimensions ranging from 10 to 12.3 nm, and the entrance sizes of this cavity can be adjusted within the approximate range of $\approx$ 4–9 nm (Fan et al., 2003).

The Korea Advanced Institute of Science and Technology (KAIST) initially discovered the KIT series, introduced by Ryoo's group (Ryoo et al., 1996). The KIT series encompasses various materials such as KIT-1, KIT-5, and KIT-6. KIT-1, in particular, exhibits irregular mesopores and amorphous pore walls. However, despite these structural characteristics, KIT-1 possesses a uniform and adjustable pore size. Furthermore, scientific literature has highlighted its superior thermal stability compared to MCM-41 (Ryoo et al., 1996). In the meantime, KIT-5 and KIT-6 exhibit meticulously arranged mesostructures, characterized by the space-group symmetries of Fm3m and Ia3d, respectively. Both materials exhibit high surface areas, monotonous and uniform size distributions, and impressive stability, making them outstanding catalyst supports. And also, KIT-6 could be utilized to design DDSs (Xu et al., 2023).

In addition to these types, various structures with mesoporous architectures such as (IBN)-series, (AMS)-series, (HMS)-series, and (MSU)-series have also been reported. These materials exhibit comparable structural features to the SBA- and MCM-series MSNs, consequently making them highly versatile in various domains such as separation, adsorption, and catalysis. However, studies on these structures in the biomedical field are limited (Xu et al., 2023).

8.3 Importance of silica nanoparticles for biomedical applications

Silica has been acknowledged by the FDA as a biocompatible material with a safe grade, rendering it highly appealing for pharmaceutical applications. The FDA's approval of silica as a safe grade has been in place for more than 5 decades, enabling the utilization of silica-based medications that are deemed safe due to variations in their composition and contents (Downing and Jain, 2020). Silica nanostructures possess crucial attributes, including remarkable levels of biocompatibility, absence of toxicity, and biodegradability, which render them particularly advantageous for therapeutic applications. Extensive research has substantiated the biocompatibility of MSNs within the human body. While certain nanostructures may exhibit acute toxicity, literature indicates that MSNs exhibit toxicity that is dependent on the dosage administered (Croissant et al., 2018).

Irrespective of their morphologies, the surface modifications of MSNs act a significant function in determining their performance and safety. Furthermore, while relying solely on the "enhanced permeability and retention" (EPR) effect may boost the performance of free drugs, it may not always be adequate for tailoring personalized treatments or enhancing the effectiveness of current therapies (Castillo and Vallet-Regí, 2019). And also, it is crucial to consider the fact that exogenous nanoparticles experience the deposition of serum proteins due to their interaction with the immune system. This interaction, known as opsonization, results in a significant alteration of the particles' surface—a protein corona (Pisani et al., 2017a,b)—causing a disruption in their ability to be recognized and ultimately leading to their elimination. Thankfully, the effects of opsonization can be reduced by incorporating highly hydrophilic polymers or fragments like polyethylene glycol or zwitterion structures, which can create robust hydration layers on the particle's surface. Fig. 8.2 represents protein corona formation of MSNs.

The medical applications of MSNs are highly advantageous due to the significant difference between the therapeutic dose and the toxic dose required for their delivery. This characteristic renders MSNs as an ideal choice for medical purposes. Indeed, a significant portion of the biocompatibility of these structures stems from the potential formation of a protein corona on the surface areas of silica nanostructures (Casals et al., 2010; Cedervall et al., 2007; Lundqvist et al., 2008; Monopoli et al., 2011b; Pino et al., 2014; Tenzer et al., 2013). Both chemical and biological characteristics of MSNs offer significant prospects for pharmaceutical chemists to enhance the medical utility of MSNs. Groups situated on the surfaces are among the foremost reactive entities that can induce toxicity reliant on MSN. By making substantial changes to the surface

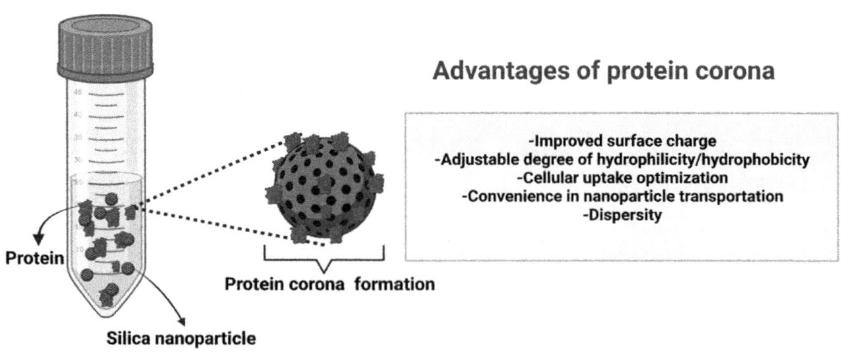

Figure 8.2 Protein corona formation of MSNs.
Created with BioRender.com.

chemistry, it is possible to modify the toxicity and enhance the medical usability of MSNs. The biodistribution of MSNs in the body can be altered through the process of surface modification. Achieving optimal biodistribution for effective medical treatment requires precise targeting of tissues and cells. By enhancing the precision of delivering nanostructures to the appropriate cell(s), the therapeutic dosage is substantially decreased due to reduced medication consumption. Surface modifications for MSNs have facilitated the accomplishment of this objective, enabling the focus of research on particular tissues or diseases. For instance, implementing PEI-PEG copolymer on the surface of MSN caused a notable enhancement in the proportion of MSNs that successfully reached the desired tumor cells, surpassing the range of less than 1%−10%. The enhanced level of specificity, coupled with an elevated degree of specificity, will enable the attainment of equivalent outcomes with reduced dosage rates (Downing and Jain, 2020).

The protein corona contributes to the exceptional chemical and biological surface modifications of MSNs, making them stand out as remarkable structures. Due to proteins adhering to the mesoporous surface, the nanoparticles can acquire a protective shell known as the protein corona structure (Dobrovolskaia et al., 2009; Pisani et al., 2017b). The interaction between the body and the MSN may be influenced by the characteristics of the proteins that bind to the exterior of the MSN. The formation of a protein corona could potentially mitigate the hemolytic impacts of silica nanostructure surfaces (Martinez et al., 2015; Paula et al., 2012). It can also exert a significant effect on the internalization process of cells (Lesniak et al., 2012).

It is crucial to acknowledge that this can also conceal surface alterations, particularly in fabricating nanostructures based on silica. Furthermore, it can potentially facilitate the passage of protein corona structures across the blood−brain barrier (Kreuter et al., 2002; Monopoli et al., 2011a) or be taken up by monocytes or macrophages (Botella et al., 2011). The manipulation of surface characteristics, pore structure, shape, and hydrophobicity/hydrophilicity properties can exert a certain level of control over protein corona formation. In vivo, protein coronas exhibit dynamic behavior, making it challenging to subject MSNs to any pretreatment involving protein structures before their application. Additional investigation into the interaction between proteins and MSNs could potentially present a valuable avenue for enhancing selectivity and diminishing toxicity (Downing and Jain, 2020).

The protein corona architecture refers to a layer of proteins that can adhere to the external surface of silica nanostructures, effectively concealing its presence while it traverses through the human body. MSNs, being composed entirely of SiO_2, exhibit enhanced biodegradability over an extended period within the human body. The phenomenon of biodegradability is associated with a hydrolysis reaction that takes place in the surface regions of MSNs within the human body, resulting in the decomposition of silica nanostructures into silicic acid. The event here hiders the toxic accumulation of MSNs in the body. The surface chemistry of MSNs can be adjusted to exploit the protein corona, thereby enhancing crucial factors like MSN toxicity and stability (Downing and Jain, 2020).

8.4 Applications

8.4.1 Antimicrobial and antibacterial mesoporous nanocomposites

One of the main directions in synthesis of biomedical mesoporous silica nanocomposites is the creation of antimicrobial and antibacterial scaffolds. Jiang et al. synthesized MSNs with poly(lactic acid) films composed with Chanterelles. During storage, it was discovered that Chanterelles packaged with films exhibited more stable bacterial flora changes and possessed superior properties. They controlled growth and life of Enterobacteriaceae spp. (Jiang et al., 2023). Shishmakova et al. developed the antibacterial scaffold based on mesoporous silica nanocontainers loaded with the antioxidant quercetin and accessed its activity in vitro (Fig. 8.3). The data indicated the high release of an antioxidant using mesoporous nanocomposites (Shishmakova et al., 2023).

Tabriz et al. proposed the nanocomposite based on MSNs, silver, and antibacterial drugs imipenem and meropenem. The proposed nanocomposites were effective against Gram-positive (*Staphylococcus aureus*) and Gram-negative (*Klebsiella pneumoniae* and *Escherichia coli)* bacteria and were safe for human cells. The main mechanism of action was the generation of reactive oxygen species (Tabriz et al., 2023). In similar research, Sánchez-Salcedo et al. printed using 3D printer mesoporous silica scaffolds with silver nanoparticles. The high efficiency of nanoscaffolds against MC3T3-E1 preosteoblastic cells and bacteria *Staphylococcus aureus* and *Escherichia coli* was found. The authors proposed that nanoscaffolds will have application for bone tissue regeneration (Sánchez-Salcedo et al., 2023). Rasool et al. synthesized nanocomposites based on MSNs with cerium oxide which as it known has an enhanced

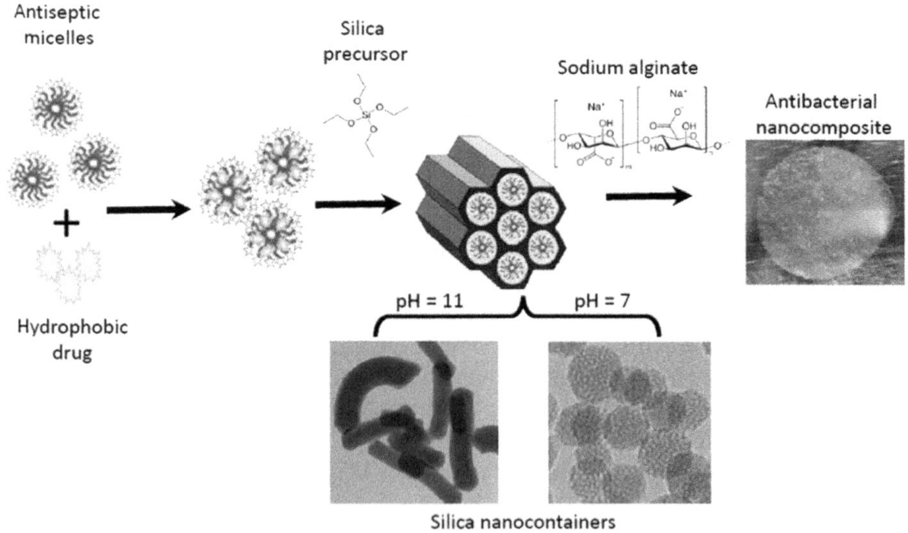

Figure 8.3 Antibacterial scaffold based on mesoporous silica nanocontainers (Shishmakova et al., 2023).

antibacterial activity. *Staphylococcus aureus* was effectively inhibited by 99.9%, while *Escherichia coli* was inhibited by 81%. In contrast, they did not induce hemolysis in erythrocytes (Rasool et al., 2022). Karaman et al. developed ceria nanoparticles as core coated with a mesoporous silica shell to create a complex nanostructure. In vitro studies showed a significant effect on *Escherichia coli*. Studies in vivo indicated that nanocomposites inhibited the bacterial proliferation in the intestine of bacteria-fed *Drosophila melanogaster* larvae (Şen Karaman et al., 2022). Su et al. created mesoporous silica nanocomposites modified by nano zinc oxides, silver nanoparticles, and polydopamine as effective antibacterial scaffolds. Due to synergism of components, the nanocomposites showed an enhanced antibacterial activity even at small concentration of 313 ppm (Su et al., 2022). For water disinfection, Furlan et al. (Fig. 8.4) proposed a complex systems based on mesoporous silica scaffolds and Fe_3O_4, SiO_2, and Ag(II). The nanocomposites inhibited 99.99% of bacteria *E. coli* with a silver release of less than 0.1 ppb. Moreover, this system allowed the multiple repeated uses over the significant time due to presence of Fe_3O_4 (Furlan et al., 2021).

Motealleh et al. using 3D-printer printed mesoporous organosilica nanocomposites with tetracycline (as antibacterial agent) and poly-D-lysine (cell-adhesive bipolymer). The nanocomposites had a remarkable antibacterial impact toward Gram-negative bacteria *P. aeruginosa* but has no effect on Gram-positive bacteria *S. aureus* and *E. faecalis* (Motealleh et al., 2021).

8.4.2 Mesoporous silica nanocomposites for diabetes diagnosis and treatment

Sarkar et al. created mesoporous silica nanocomposites as drug carriers for delivery of insulin, GLP-1, exenatide, DPP-4 inhibitor, and plasmid-containing GLP-1 genes to treat diabetes mellitus. As result, nanocomposites allowed the controlled delivery

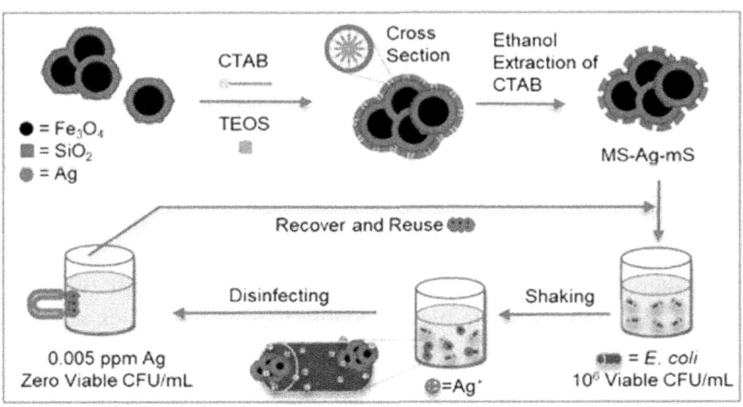

Figure 8.4 The nanocomposite based on mesoporous silica scaffolds and Fe_3O_4, SiO_2, and Ag(II)for water disinfection (Furlan et al., 2021).

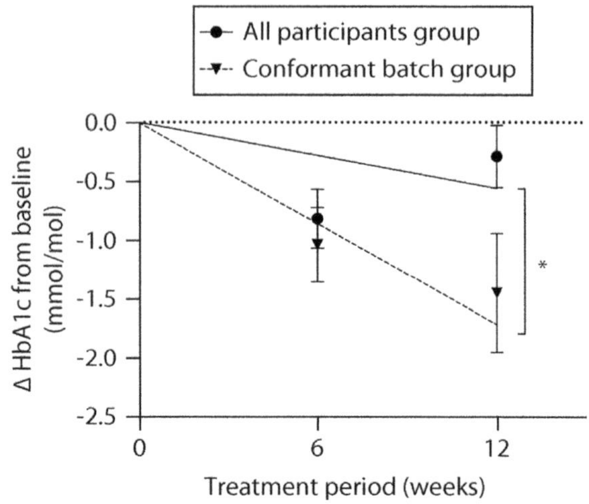

Figure 8.5 Effect of SiPore15 treatment on hemoglobin A1c level (Baek et al., 2022).

and release of antidiabetic drugs (Sarkar et al., 2023). Rinde et al. proposed to use MSNs to treat obesity in mice. The data showed that the nanocomposites reduced adipose tissue formation and leptin levels compared to control (Rinde et al., 2020). Baek et al. synthesized mesoporous silica nanocomposites—SiPore15 to control blood glucose and regulate metabolic disorders in patients with type 2 diabetes. As result, treatment by SiPore15 reduced Hemoglobin A1c significantly and had minimal adverse effects (Fig. 8.5) (Baek et al., 2022).

Jiang et al. loaded MSNs by insulin to provide its pH-dependent release. In vivo study in rats showed the high efficiency of nanocomposites for prolonged hypoglycemic response (Jiang et al., 2018).

8.4.3 Mesoporous silica nanocomposites for cancer treatment

Choi et al. composed mesoporous organosilica nanocomposites (MOS) (Fig. 8.6) loaded with glutathione (GSH, as antioxidant), hyaluronidase (HAase, as enzyme), and doxorubicin (DOX) for chemo-, photothermal, and photodynamic cancer treatment. The data on CD44-overexpressing melanoma cells showed the proposed nanocomposites' enhanced release and high therapeutics efficiency upon laser irradiation (Choi et al., 2022).

Croissant et al. created gold-mesoporous organosilica nanocomposite structures for two-photon photodynamic therapy in cancer cells. As a result, the created nanocomposites induced the cell death of 63% of MCF-7 breast cancer cells (Croissant et al., 2016). Zhan et al. synthesized the biodegradable mesoporous silica nanocomposites with doxorubicin and anticancer drug IR780 to treat breast cancer. The results in vivo in mice showed that the proposed nanocomposites accumulated mainly in tumor tissues and showed an enhanced antitumor effect in breast cancer mouse models (Zhan et al., 2021). Yang et al. conducted a study where they fabricated dendritic mesoporous silica nanocomposites (DMSN) modified with polyethyleneimine (PEI) and

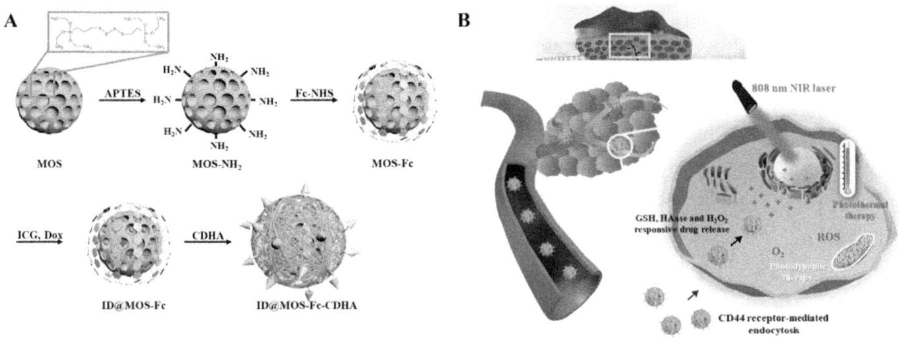

Figure 8.6 (a) The process of synthesis of the multitriggered melanoma targeted ICG-Dox co-loaded MOS-Fc-CDHA nanocomposite and (b) Demonstration of GSH, HAase, and H_2O_2-sensitive drug release, along with the combined effect of chemo-, photothermal, and photodynamic therapies in melanoma cells (Choi et al., 2022).

microRNA-125a (DMSN-PEI@125a). This modification aimed to stimulate an immune response and reverse immunosuppression in cancer cells. The data in vitro showed that the nanocomposites induced the repolarization of tumor-associated macrophages to noncancer M1 type and promoted the immunogenic cell death of the cervical cancer cell line TC-1. The experiments conducted in vivo in mice showed that suppression of tumor growth in TC-1 mice (Fig. 8.7) (Yang et al., 2021).

Yang et al. created magnetic MSNs loaded with photosensitizer chlorin e6 (Ce6) and doxorubicin for photodynamic therapy of cancer. The in vivo studies in tumor-bearing Balb/c mice showed the synergistic antitumor effects both e6 and doxorubicin in mice (Yang et al., 2017). Zhang et al. synthesized and evaluated gold nanobipyramidal-mesoporous silica dioxide nanocomposites loaded with a photodynamic therapy agent IR780, producing nanozyme carriers with photoenhanced peroxidase activity. The results showed that the nanocomposites had high ROS productivity, inducing antitumor immune responses to destroy tumor tissues. Zhang et al. synthesized MSNs loaded with doxorubicine and select phytoestrogen tanshinone IIA to treat breast cancer. The data in vivo showed the perfect targeting and antitumor effects for breast cancer mouse models and the low toxicity in mice normal organs (Zhang et al., 2023).

8.4.4 Mesoporous silica nanocomposites for wound healing

Zhou et al. proposed to use mesoporous silica nanocomposites with cerium oxide (IV) and microRNA miR129 to promote wound healing after radiation-induced skin injury (Fig. 8.7). The main mechanism of action is inhibition of radiation-induced ROS and HIF-1α activation. The mice models were effectively treated with nanocomposites, which facilitated the delivery of miR129 to the blood capillaries of the wounds. This treatment enhanced the radioresistance and improved the efficiency of anti-reactive oxygen species (ROS) measures (Zhou et al., 2022), Fig. 8.8. A schematic

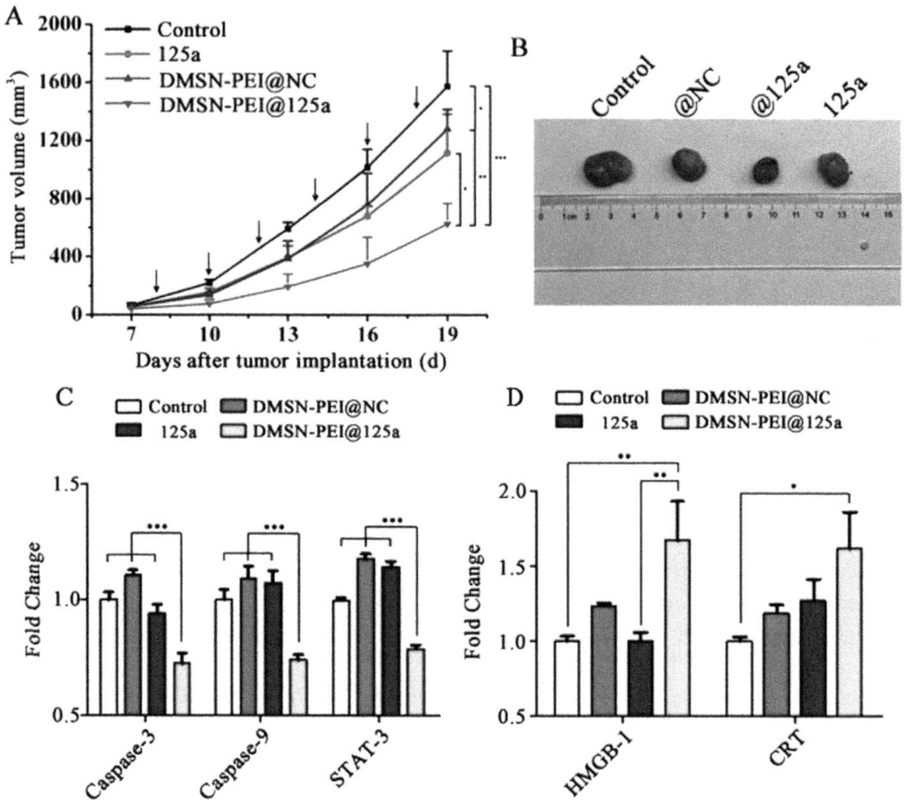

Figure 8.7 Antitumor impact of DMSN-PEI@125a in vivo. The TC-1 tumor models (n = 5) were i.t. injected with ddH_2O, DMSN-PEI@NC, DMSN-PEI@125a, and miR-125a every 2 days for six times from day 8 after tumor inoculation. Growth curves of tumors exhibit different patterns under various treatment regimens. The day of administration is marked by the presence of *arrows* (a). On the 20th day following implantation of the tumors, photographic images to depict representative tumor samples (b). mRNA expression levels of apoptosis-related genes (*caspase-3*, *caspase-9*, *STAT-3*) (c), and damage-associated molecular patterns (*HMGB-1*, *CRT*) in tumors (d) (Yang et al., 2021).

representation showing the process of creating MS-CeO_2-miR 129 and its implementation for radiation-induced skin injury healing Fig. 8.8.

Rasool et al. synthesized nanocomposites based on MSNs with cerium oxide to target biofilms in chronic wounds. The proposed nanocomposites had a strong antibacterial effect and are applicable for chronic wound healing (Rasool et al., 2022). Mahmoudabadi et al. created silver/kaolinite nanocomposites with chitosan and oak extract modification. Studies in vivo showed improved wound healing of the nanocomposites by decrease of inflammation and enhance of the proliferative phase (Mahmoudabadi et al., 2021). Zhang et al. developed Zn containing mesoporous silica nanospheres loaded with ciprofloxacin for wound healing. The data in vivo indicated that the

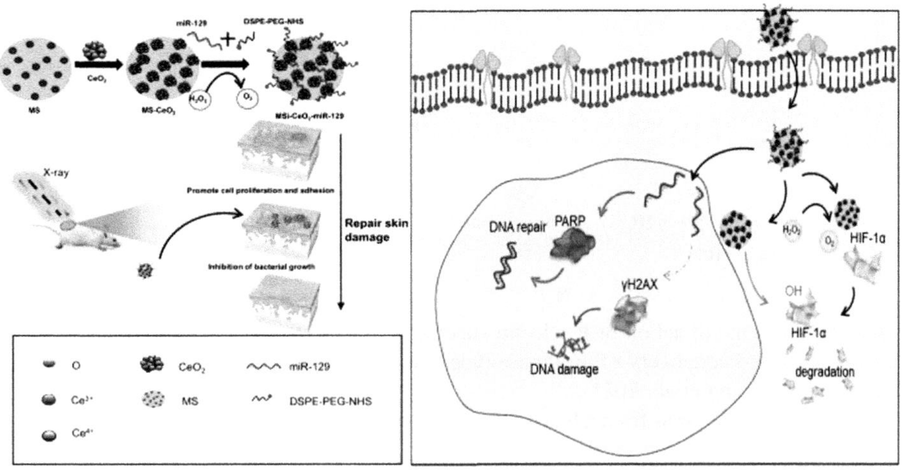

Figure 8.8 A schematic representation showing the process of creating MS-CeO_2-miR 129 and its implementation for radiation-induced skin injury healing (Zhou et al., 2022).

nanocomposites promoted angiogenesis and skin regeneration by Si release, as well as hair follicle regeneration (Zhang et al., 2019). Wu et al. created ceria nanocrystals modified with mesoporous silica nanocomposites and evaluated them in vivo for skin regeneration. As result, the nanocomposites improved the wound healing with enhanced skin appendage morphogenesis and limited scar formation (Wu et al., 2018).

8.4.5 Mesoporous silica nanocomposites for cardiovascular diseases

Chang et al. synthesized P-selectin-specific, mesoporous silica nanocomposite for thrombus-targeted codelivery of urokinase-type plasminogen activator, indocyanine green (probe for near-infrared fluorescence), and fucoidan (an algae-derived anticoagulant) (Fig. 8.7). The data in vivo indicated that the nanocomposite provided a fast release of urokinase-type plasminogen activator for clearing of blood clots and a slow release of fucoidan for long lasting prevention of thrombosis. At this, indocyanine green provided the near-infrared imaging of the nanocomposite redistribution in a body (Chang et al., 2021). Fig. 8.9 Scheme of action of P-selectin-specific, mesoporous silica nanocomposite for thrombus-targeted codelivery of urokinase-type plasminogen activator, indocyanine green, and fucoidan.

Wei et al. synthesized MSNs loaded with honokiol to prevent vascular restenosis after percutaneous coronary intervention. They found that a small-molecule polyphenol honokiol may suppress neointimal hyperplasia after balloon injury, but its solubility is low. The results in vivo in a rat model of balloon injury showed that the nanocomposite inhibited the proliferation and migration of vascular smooth muscle cells and induced regeneration after injury (Wei et al., 2020). Wang et al. created MSNs loaded with microRNA-10b and covered by hybrid neutrophil membranes

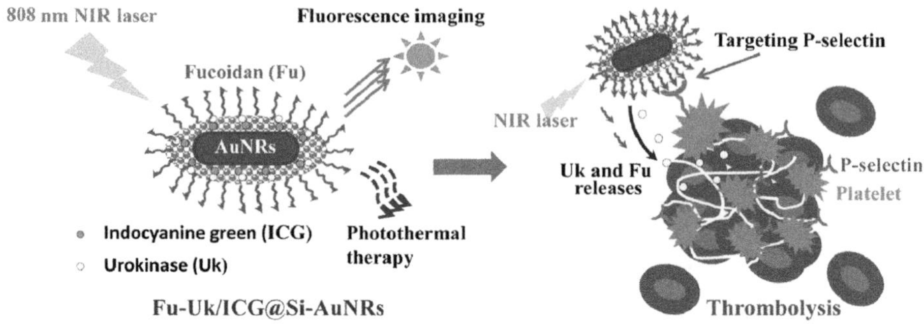

Figure 8.9 Scheme of action of P-selectin-specific, mesoporous silica nanocomposite for thrombus-targeted codelivery of urokinase-type plasminogen activator, indocyanine green, and fucoidan (Chang et al., 2021).

for regeneration therapy after cardiac ischemic injury. The developed nanocomposites were able to penetrate into the injured myocardium, restore the local immunity, and efficiently deliver microRNA-10b to cardiomyocytes. As a result, cardiac function improvement and mitigated ventricular remodeling were observed in vivo (Wang et al., 2023a). Xia et al. developed mesoporous silica nanocomposites with polymer poly(*N*-*n*-propylacrylamide-*co*-*N*-*tert*-butyl acrylamide) for hydrogen sulfide release triggered by hypothermia. As result, they protected cardiomyocytes from ischemic and reperfusion injury (Xia et al., 2022). Liu et al. proposed the use of mesoporous silica nanocomposites loaded with quercetin to treat myocardial ischemia-reperfusion injury. The data in vivo in rats showed that the nanocomposites decreased myocardial infarction size, improved ventricular remodeling, and provided cardiac blood flow recovery through activation of the JAK2/STAT3 pathways (Liu et al., 2021). Yadav et al. estimated the effect of MSNs loaded with curcumin on doxorubicin-induced myocardial toxicity. The data in vivo indicated that nanocomposites protected the myocardium from the toxic effects of doxorubicin by a decreasing ROS and increasing antioxidant enzyme levels (Yadav et al., 2019).

8.4.6 Mesoporous silica nanocomposites as contrast agents for effective imaging

Fu et al. synthesized radiopaque iodinated mesoporous silica nanocomposites covered by liposomes as contrast agents for computer tomography (CT) imaging. As result, they combined the advantages of liposomes and nanocomposites and provided a high and contrast imaging signal (Fu et al., 2024). Zanocco et al. (Fig. 8.10) created mesoporous silica nanocomposites modified with naphthoxazole fluorescent unit and a singlet-oxygen-sensitive furan derivative (NanoFN10) for two-photon imaging. The probe internalized in macrophages and its activity enhanced in presence of singlet oxygen in 180 times, allowing imaging of intracellular singlet oxygen under multiphoton excitation (Zanocco et al., 2023).

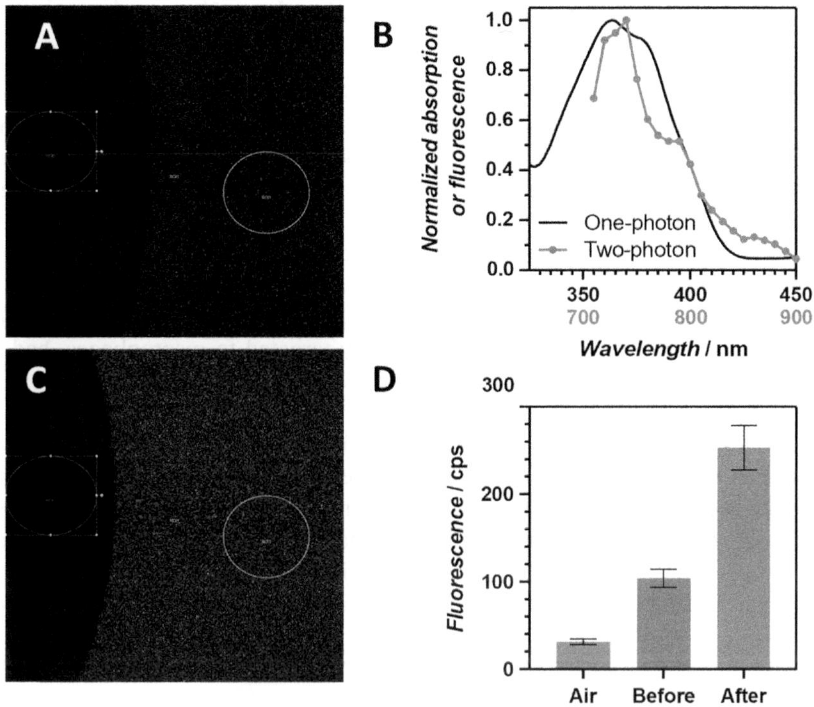

Figure 8.10 Fluorescence images acquired through the utilization of two-photon absorption FN10 fluorescence images ($\lambda_{exc} = 720$ nm; $\lambda_{obs} = 450–540$ nm) before (a) and after (c) reacting with $^1O_2$. FN10 single-photon absorption normalized (black line [black in print version]) and two-photon excitation of fluorescence (*red* line [*gray* in print version]) spectra (b). The fluorescence intensity of the air bubble can be attributed to the phenomenon of two-photon excitation, FN10 and FN10-ox (d) (before and after reacting with $^1O_2$, respectively) (Zanocco et al., 2023).

Wang et al. developed a europium oxide mesoporous silica-coated core–shell structure for magnetic resonance imaging (MRI). The data in vivo in rats showed the nanocomposites had a strong contrast signal and showed excellent efficiency for liver tissue MRI (Wang et al., 2023b). Li et al. synthesized two Janus nanoparticles combining Au and a pirfenidone-loaded mesoporous silica shell modified with thioketal grafted methoxy poly(ethylene glycol) or with 1,2-distearoyl-*sn-glycero*-3-phosphoethanolamine. As a result, Janus nanoparticles internalized into mesenchymal stem cells and provided long-term CT imaging of them in mice with idiopathic pulmonary fibrosis (Li et al., 2023). Sha et al. proposed using manganese-doped gold core mesoporous silica nanocomposites as contrast agents. Using both Au in the core and manganese ions released from the shell allowed for practical dual-modality imaging—both CT and MRI (Sha et al., 2021).

8.4.7 Mesoporous silica nanocomposites for atherosclerosis diagnostics and treatment

Zhang et al. synthesized mesoporous silica nanocomposites loaded with MnO nanoparticles and PPARα and LXRα agonists. Then, these nanocomposites were coated by cRGD-chelated platelet membranes. The proposed nanoconstruction had a dual effect: provided MRI images of atherosclerosis lesions and simultaneously treated atherosclerosis by facilitating the movement of cholesterol through regulating ABCA1 transporters (Zhang et al., 2022a). Song et al. created mesoporous silica nanocomposites loaded with a lipid-lowering drug simvastatin and hyaluronic acid (SIM@HA-MSN) to treat atherosclerosis (Fig. 8.11). The data in vitro and in vivo in mice demonstrated that the nanocomposites efficiently targeted to macrophages, had anti-inflammatory and antifoaming effects, a long plasma-retention time and good biocompatibility (Song et al., 2022).

In similar paper, Liu et al. proposed hyaluronic acid covered dendritic MSNs capped with copper sulfide and heparin to treat atherosclerosis. Results in vitro showed that the nanocomposites provide the controlled release of components in atherosclerotic inflammation locations (Liu et al., 2022). Pham et al. developed hyaluronic acid–coated mesoporous silica nanocomposites modified with CD9 antibody and rosuvastatin to aggravate atherosclerotic plaque formation. As results, the nanocomposites

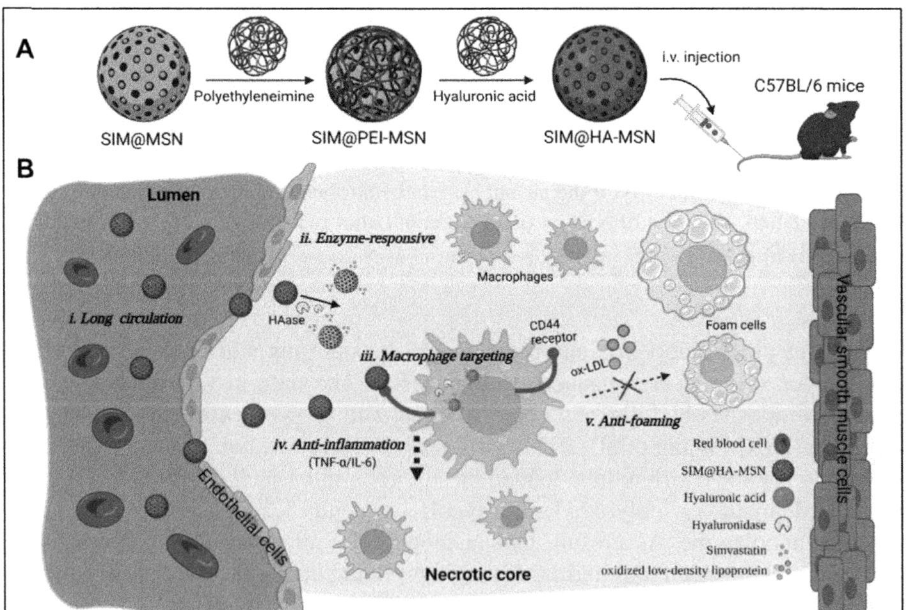

Figure 8.11 Schematization of the construction process of SIM@HA-MSN in the management of atherosclerosis. Production process and application route of SIM@HA-MSN (a). Demonstration of SIM@HA-MSN therapy which may potentially reduce atherosclerosis via long-circulating, enzyme-responsive drug release, macrophage targeting, and anti-inflammatory, and antifoaming impacts (b) (Song et al., 2022).

targeted the lesions, delivered rosuvastatin to the senescent atherosclerotic plaques, and stopped atherosclerosis in $ApoE^{-/-}$ mice (Pham et al., 2021).

8.4.8 Mesoporous silica nanocomposites for tuberculosis diagnostics and treatment

Yahia et al., using a 3D printer, printed a vertebral body scaffold filled with hydrogel composed of gelatine and loaded with mesoporous silica nanocomposites containing two antibiotics to treat tuberculosis. The proposed nanoconstructs showed a high anti-tuberculosis activity in vitro (Yahia et al., 2023). Zhu et al. developed nanoemulsion consisting of MSNs α-linolenic acid and antibacterial drug rifampicin (Fig. 8.12). The data in vitro showed that nanoemulsion had the excellent effect against *Staphylococcus aureus* and drug-sensitive and drug-resistant *Mycobacterium tuberculosis* (Zhu et al., 2022).

Zhu et al. synthesized mesoporous silica nanocomposites loaded with gold for diagnosis and treatment of tuberculosis. The data in vitro showed an excellent antimicrobial property of the proposed nanocomposites against *Mycobacterium tuberculosis*. Also, they can be used in rapid diagnostics of *Mycobacterium tuberculosis* in vitro (Zhu et al., 2022).

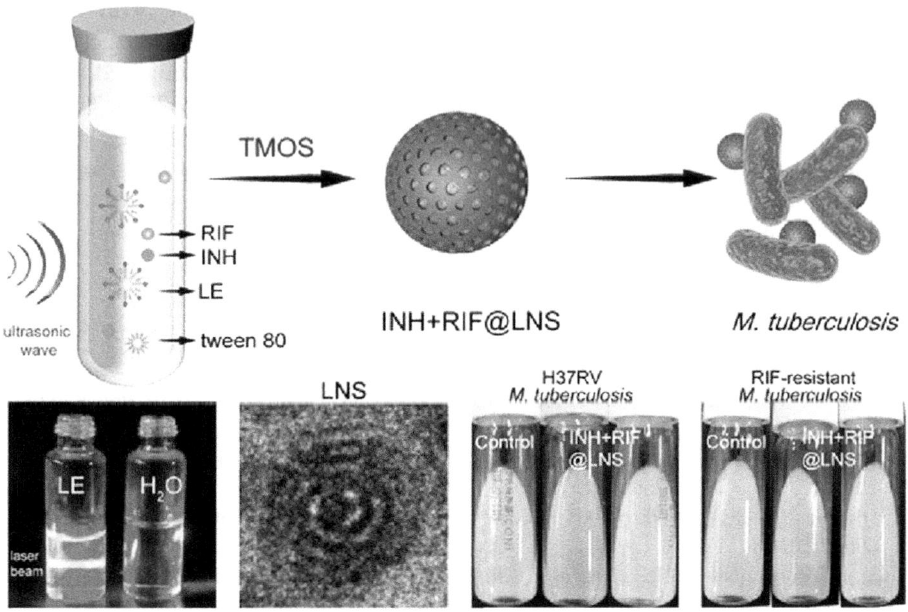

Figure 8.12 Schematic illustration of antituberculosis action of the nanoemulsion consisting of MSNs α-linolenic acid and antibacterial drug rifampicin (Zhu et al., 2022).

8.4.9 Mesoporous silica nanocomposites against alcoholic hepatitis

As it is known, puerarin is a bioactive flavonoid isolated from *Pueraria (Wild.)*. It provides hepatoprotection but is practically insoluble in water. Zhang et al. proposed the use of mesoporous silica nanocomposites to improve the bioavailability of puerarin. The analysis of the administration of the nanocomposites in mice in Gao-binge model induced alcoholic hepatitis pathology showed a high hepatoprotection efficiency of them. The application of nanocomposites decreased the expression of autophagy-related proteins (Atg3, Atg7, LC3, and p62) and alleviated fatty droplets infiltration in damaged liver of mouse (Zhang et al., 2022b).

8.4.10 Future perspectives

Multifunctional silica-based nanoparticles play a crucial role in the nanomaterials library, particularly in the biomedical sector, owing to their easily modifiable structure and composition, high biocompatibility, and versatile surface functionalization properties. Multifunctional silica-based nanoparticles are considered chemically and biologically inert compared to other inorganic nanomaterials. Limited research has been conducted on the biomedical applications leveraging the intrinsic properties of these nanoparticles. These nanoparticles have many applications, such as acting as carriers for delivering therapeutic agents, serving as matrices for constructing nanocomposites tailored to specific needs, or functioning as efficient agents with inherent therapeutic effects or bioactivity. As nanotechnology continues to advance in the biomedical field, the global healthcare nanotechnology market is projected to grow steadily from US$ 17,245 million in 2019 to US$ 252,400 million by 2024. The escalating research endeavors focused on silica-based nanostructures in conjunction with the biomedical field suggest a promising trajectory toward clinical trials for silica-based nanoformulations. It is reasonable to anticipate that this progression will pave the way for successful clinical translation (Xu et al., 2023).

References

Acet, Ö., Önal, B., Sanz, R., Sanz-Pérez, E.S., Erdönmez, D., Odabaşi, M., 2019. Preparation of a new chromatographic media and assessment of some kinetic and interaction parameters for lysozyme. J. Mol. Liq. 276, 480–487. https://doi.org/10.1016/j.molliq.2018.12.037.

Anselmo, A.C., Mitragotri, S., 2021. Nanoparticles in the clinic: an update post <scp>COVID</scp> -19 vaccines. Bioeng Transl. Med. 6. https://doi.org/10.1002/btm2.10246.

Baek, J., Robert-Nicoud, G., Herrera Hidalgo, C., Borg, M.L., Iqbal, M.N., Berlin, R., Lindgren, M., Waara, E., Uddén, A., Pietiläinen, K., Bengtsson, T., 2022. Engineered mesoporous silica reduces long-term blood glucose, HbA1c, and improves metabolic parameters in prediabetics. Nanomedicine 17, 9–22. https://doi.org/10.2217/nnm-2021-0235.

Barenholz, Y.(C.), 2012. Doxil® — the first FDA-approved nano-drug: lessons learned. J. Contr. Release 160, 117–134. https://doi.org/10.1016/j.jconrel.2012.03.020.

Beck, J.S., Vartuli, J.C., Roth, W.J., Leonowicz, M.E., Kresge, C.T., Schmitt, K.D., Chu, C.T.W., Olson, D.H., Sheppard, E.W., McCullen, S.B., Higgins, J.B., Schlenker, J.L., 1992. A new family of mesoporous molecular sieves prepared with liquid crystal templates. J. Am. Chem. Soc. 114, 10834–10843. https://doi.org/10.1021/ja00053a020.

Boissière, C., Martines, M.A.U., Tokumoto, M., Larbot, A., Prouzet, E., 2003. Mechanisms of pore size control in MSU-X mesoporous silica. Chem. Mater. 15, 509–515. https://doi.org/10.1021/cm0211195.

Botella, P., Abasolo, I., Fernández, Y., Muniesa, C., Miranda, S., Quesada, M., Ruiz, J., Schwartz, S., Corma, A., 2011. Surface-modified silica nanoparticles for tumor-targeted delivery of camptothecin and its biological evaluation. J. Contr. Release 156, 246–257. https://doi.org/10.1016/j.jconrel.2011.06.039.

Burns, A., Ow, H., Wiesner, U., 2006. Fluorescent core–shell silica nanoparticles: towards "lab on a particle" architectures for nanobiotechnology. Chem. Soc. Rev. 35, 1028–1042. https://doi.org/10.1039/B600562B.

Carmona, D., Balas, F., Santamaria, J., 2014. Pore ordering and surface properties of FDU-12 and SBA-15 mesoporous materials and their relation to drug loading and release in aqueous environments. Mater. Res. Bull. 59, 311–322. https://doi.org/10.1016/j.materresbull.2014.07.039.

Casals, E., Pfaller, T., Duschl, A., Oostingh, G.J., Puntes, V., 2010. Time evolution of the nanoparticle protein corona. ACS Nano 4, 3623–3632. https://doi.org/10.1021/nn901372t.

Castillo, R.R., Vallet-Regí, M., 2019. Functional mesoporous silica nanocomposites: biomedical applications and biosafety. Int. J. Mol. Sci. 20, 929. https://doi.org/10.3390/ijms20040929.

Cedervall, T., Lynch, I., Lindman, S., Berggård, T., Thulin, E., Nilsson, H., Dawson, K.A., Linse, S., 2007. Understanding the nanoparticle–protein corona using methods to quantify exchange rates and affinities of proteins for nanoparticles. Proc. Natl. Acad. Sci. USA 104, 2050–2055. https://doi.org/10.1073/pnas.0608582104.

Chang, L.-H., Chuang, E.-Y., Cheng, T.-M., Lin, C., Shih, C.-M., Wu, A.T., Jheng, P.-R., Lu, H.-Y., Shih, C.-C., Mi, F.-L., 2021. Thrombus-specific theranostic nanocomposite for codelivery of thrombolytic drug, algae-derived anticoagulant and NIR fluorescent contrast agent. Acta Biomater. 134, 686–701. https://doi.org/10.1016/j.actbio.2021.07.072.

Choi, H.W., Lim, J.H., Kang, T., Chung, B.G., 2022. Antioxidant, enzyme, and H_2O_2-triggered melanoma targeted mesoporous organo-silica nanocomposites for synergistic cancer therapy. Antioxidants 11, 2137. https://doi.org/10.3390/antiox11112137.

Costa, J.A.S., Paranhos, C.M., 2020. Mitigation of silica-rich wastes: an alternative to the synthesis eco-friendly silica-based mesoporous materials. Micropor. Mesopor. Mater. 309, 110570. https://doi.org/10.1016/j.micromeso.2020.110570.

Croissant, J.G., Butler, K.S., Zink, J.I., Brinker, C.J., 2020. Synthetic amorphous silica nanoparticles: toxicity, biomedical and environmental implications. Nat. Rev. Mater. 5, 886–909. https://doi.org/10.1038/s41578-020-0230-0.

Croissant, J.G., Fatieiev, Y., Almalik, A., Khashab, N.M., 2018. Mesoporous silica and organosilica nanoparticles: physical chemistry, biosafety, delivery strategies, and biomedical applications. Adv. Healthc. Mater. 7. https://doi.org/10.1002/adhm.201700831.

Croissant, J.G., Qi, C., Maynadier, M., Cattoën, X., Wong Chi Man, M., Raehm, L., Mongin, O., Blanchard-Desce, M., Garcia, M., Gary-Bobo, M., Durand, J.-O., 2016. Multifunctional gold-mesoporous silica nanocomposites for enhanced two-photon imaging and therapy of cancer cells. Front. Mol. Biosci. 3. https://doi.org/10.3389/fmolb.2016.00001.

Davidson, M., Ji, Y., Leong, G.J., Kovach, N.C., Trewyn, B.G., Richards, R.M., 2018. Hybrid mesoporous silica/noble-metal nanoparticle materials—synthesis and catalytic applications. ACS Appl. Nano Mater. 1, 4386–4400. https://doi.org/10.1021/acsanm.8b00967.

Dobrovolskaia, M.A., Patri, A.K., Zheng, J., Clogston, J.D., Ayub, N., Aggarwal, P., Neun, B.W., Hall, J.B., McNeil, S.E., 2009. Interaction of colloidal gold nanoparticles with human blood: effects on particle size and analysis of plasma protein binding profiles. Nanomedicine 5, 106–117. https://doi.org/10.1016/j.nano.2008.08.001.

Downing, M.A., Jain, P.K., 2020. Mesoporous silica nanoparticles: synthesis, properties, and biomedical applications. In: Nanoparticles for Biomedical Applications. Elsevier, pp. 267–281. https://doi.org/10.1016/B978-0-12-816662-8.00016-3.

Fan, J., Yu, C., Gao, F., Lei, J., Tian, B., Wang, L., Luo, Q., Tu, B., Zhou, W., Zhao, D., 2003. Cubic mesoporous silica with large controllable entrance sizes and advanced adsorption properties. Angew. Chem. 115, 3254–3258. https://doi.org/10.1002/ange.200351027.

Fu, N., Li, A., Zhang, J., Zhang, P., Zhang, H., Yang, S., Zhang, J., 2024. Liposome-camouflaged iodinated mesoporous silica nanoparticles with high loading capacity, high hemodynamic stability, high biocompatibility and high radiopacity. Int. J. Pharm. 650, 123700. https://doi.org/10.1016/j.ijpharm.2023.123700.

Furlan, P.Y., Furlan, A.Y., Kisslinger, K., Melcer, M.E., 2021. Templated mesoporous silica outer shell for controlled silver release of a magnetically recoverable and reusable nanocomposite for water disinfection. ACS Appl. Mater. Interfaces 13, 47972–47986. https://doi.org/10.1021/acsami.1c14669.

Huang, H., Feng, W., Chen, Y., Shi, J., 2020. Inorganic nanoparticles in clinical trials and translations. Nano Today 35, 100972. https://doi.org/10.1016/j.nantod.2020.100972.

Jiang, F., Liu, Y., Wang, X., Yin, Z., 2018. pH-sensitive release of insulin-loaded mesoporous silica particles and its coordination mechanism. Eur. J. Pharmaceut. Sci. 119, 1–12. https://doi.org/10.1016/j.ejps.2018.04.007.

Jiang, K., Zhu, B., Liu, Y., Chen, H., Yuan, M., Qin, Y., Brennan, M., Brennan, C., 2023. Effects of antimicrobial nanocomposite films packaging on the postharvest quality and spoilage bacterial communities of mushrooms (Chanterelles). Food Chem. X 20, 100996. https://doi.org/10.1016/j.fochx.2023.100996.

Kresge, C.T., Leonowicz, M.E., Roth, W.J., Vartuli, J.C., Beck, J.S., 1992. Ordered mesoporous molecular sieves synthesized by a liquid-crystal template mechanism. Nature 359, 710–712. https://doi.org/10.1038/359710a0.

Kreuter, J., Shamenkov, D., Petrov, V., Ramge, P., Cychutek, K., Koch-Brandt, C., Alyautdin, R., 2002. Apolipoprotein-mediated transport of nanoparticle-bound drugs across the blood-brain barrier. J. Drug Target. 10, 317–325. https://doi.org/10.1080/10611860290031877.

Lee, J.E., Lee, N., Kim, T., Kim, J., Hyeon, T., 2011. Multifunctional mesoporous silica nanocomposite nanoparticles for theranostic applications. Acc. Chem. Res. 44, 893–902. https://doi.org/10.1021/ar2000259.

Lesniak, A., Fenaroli, F., Monopoli, M.P., Åberg, C., Dawson, K.A., Salvati, A., 2012. Effects of the presence or absence of a protein corona on silica nanoparticle uptake and impact on cells. ACS Nano 6, 5845–5857. https://doi.org/10.1021/nn300223w.

Li, X., Li, Y., Yu, C., Bao, H., Cheng, S., Huang, J., Zhang, Z., 2023. ROS-responsive Janus Au/mesoporous silica core/shell nanoparticles for drug delivery and long-term CT imaging tracking of MSCs in pulmonary fibrosis treatment. ACS Nano 17, 6387–6399. https://doi.org/10.1021/acsnano.2c11112.

Li, X., Zhang, J., Gu, H., 2011. Adsorption and desorption behaviors of DNA with magnetic mesoporous silica nanoparticles. Langmuir 27, 6099–6106. https://doi.org/10.1021/la104653s.

Liu, C.-J., Yao, L., Hu, Y.-M., Zhao, B.-T., 2021. Effect of quercetin-loaded mesoporous silica nanoparticles on myocardial ischemia-reperfusion injury in rats and its mechanism. Int J Nanomed. 16, 741–752. https://doi.org/10.2147/IJN.S277377.

Liu, S., Zhao, Y., Shen, M., Hao, Y., Wu, X., Yao, Y., Li, Y., Yang, Q., 2022. Hyaluronic acid targeted and pH-responsive multifunctional nanoparticles for chemo-photothermal synergistic therapy of atherosclerosis. J. Mater. Chem. B 10, 562–570. https://doi.org/10.1039/D1TB02000E.

Liu, X., Che, S., 2015. Enhanced release of the poorly soluble drug itraconazole loaded in ordered mesoporous silica. Sci. China Chem. 58, 400–410. https://doi.org/10.1007/s11426-015-5333-x.

Lundqvist, M., Stigler, J., Elia, G., Lynch, I., Cedervall, T., Dawson, K.A., 2008. Nanoparticle size and surface properties determine the protein corona with possible implications for biological impacts. Proc. Natl. Acad. Sci. USA 105, 14265–14270. https://doi.org/10.1073/pnas.0805135105.

Mahmoudabadi, S., Farahpour, M.R., Jafarirad, S., 2021. Effectiveness of green synthesis of silver/kaolinite nanocomposite using *Quercus infectoria* galls aqueous extract and its chitosan-capped derivative on the healing of infected wound. IEEE Trans. NanoBiosci. 20, 530–542. https://doi.org/10.1109/TNB.2021.3105356.

Martinez, D.S.T., Paula, A.J., Fonseca, L.C., Luna, L.A.V., Silveira, C.P., Durán, N., Alves, O.L., 2015. Monitoring the hemolytic effect of mesoporous silica nanoparticles after human blood protein corona formation. Eur. J. Inorg. Chem. 2015, 4595–4602. https://doi.org/10.1002/ejic.201500573.

Mitchell, M.J., Billingsley, M.M., Haley, R.M., Wechsler, M.E., Peppas, N.A., Langer, R., 2021. Engineering precision nanoparticles for drug delivery. Nat. Rev. Drug Discov. 20, 101–124. https://doi.org/10.1038/s41573-020-0090-8.

Monopoli, M.P., Bombelli, F.B., Dawson, K.A., 2011a. Nanobiotechnology: nanoparticle coronas take shape. Nat. Nanotechnol. 6, 11–12. https://doi.org/10.1038/nnano.2011.267.

Monopoli, M.P., Walczyk, D., Campbell, A., Elia, G., Lynch, I., Baldelli Bombelli, F., Dawson, K.A., 2011b. Physical–chemical aspects of protein corona: relevance to *in vitro* and *in vivo* biological impacts of nanoparticles. J. Am. Chem. Soc. 133, 2525–2534. https://doi.org/10.1021/ja107583h.

Motealleh, A., Kart, D., Czieborowski, M., Kehr, N.S., 2021. Functional nanomaterials and 3D-printable nanocomposite hydrogels for enhanced cell proliferation and for the reduction of bacterial biofilm formation. ACS Appl. Mater. Interfaces 13, 43755–43768. https://doi.org/10.1021/acsami.1c13392.

Narayan, R., Nayak, U., Raichur, A., Garg, S., 2018. Mesoporous silica nanoparticles: a comprehensive review on synthesis and recent advances. Pharmaceutics 10, 118. https://doi.org/10.3390/pharmaceutics10030118.

Noma, S.A.A., Ulu, A., Acet, Ö., Sanz, R., Sanz-Pérez, E.S., Odabaşı, M., Ateş, B., 2020. Comparative study of ASNase immobilization on tannic acid-modified magnetic Fe_3O_4/SBA-15 nanoparticles to enhance stability and reusability. New J. Chem. 44, 4440–4451. https://doi.org/10.1039/D0NJ00127A.

Noreen, S., Maqbool, A., Maqbool, I., Shafique, A., Khan, M.M., Junejo, Y., Ahmed, B., Anwar, M., Majeed, A., Abbas, M., Naveed, M., Madni, A., 2022. Multifunctional mesoporous silica-based nanocomposites: synthesis and biomedical applications. Mater. Chem. Phys. 285, 126132. https://doi.org/10.1016/j.matchemphys.2022.126132.

Önal, B., Acet, Ö., Sanz, R., Sanz-Pérez, E.S., Erdönmez, D., Odabaşı, M., 2019. Co-evaluation of interaction parameters of genomic and plasmid DNA for a new chromatographic medium. Int. J. Biol. Macromol. 141, 1183–1190. https://doi.org/10.1016/j.ijbiomac.2019.09.068.

Pajchel, L., Kolodziejski, W., 2018. Synthesis and characterization of MCM-48/hydroxyapatite composites for drug delivery: ibuprofen incorporation, location and release studies. Mater. Sci. Eng. C 91, 734–742. https://doi.org/10.1016/j.msec.2018.06.028.

Paula, A.J., Martinez, D.S.T., Araujo Júnior, R.T., Souza Filho, A.G., Alves, O.L., 2012. Suppression of the hemolytic effect of mesoporous silica nanoparticles after protein corona interaction: independence of the surface microchemical environment. J. Braz. Chem. Soc. 23, 1807–1814. https://doi.org/10.1590/S0103-50532012005000048.

Pham, L.M., Kim, E.-C., Ou, W., Phung, C.D., Nguyen, T.T., Pham, T.T., Poudel, K., Gautam, M., Nguyen, H.T., Jeong, J.-H., Yong, C.S., Park, S.-Y., Kim, J.-R., Kim, J.O., 2021. Targeting and clearance of senescent foamy macrophages and senescent endothelial cells by antibody-functionalized mesoporous silica nanoparticles for alleviating aorta atherosclerosis. Biomaterials 269, 120677. https://doi.org/10.1016/j.biomaterials.2021.120677.

Pino, P. del, Pelaz, B., Zhang, Q., Maffre, P., Nienhaus, G.U., Parak, W.J., 2014. Protein corona formation around nanoparticles – from the past to the future. Mater. Horiz. 1, 301–313. https://doi.org/10.1039/C3MH00106G.

Pisani, C., Gaillard, J.C., Dorandeu, C., Charnay, C., Guari, Y., Chopineau, J., Devoisselle, J.M., Armengaud, J., Prat, O., 2017a. Experimental separation steps influence the protein content of corona around mesoporous silica nanoparticles. Nanoscale 9, 5769–5772. https://doi.org/10.1039/C7NR01654A.

Pisani, C., Gaillard, J.-C., Odorico, M., Nyalosaso, J.L., Charnay, C., Guari, Y., Chopineau, J., Devoisselle, J.-M., Armengaud, J., Prat, O., 2017b. The timeline of corona formation around silica nanocarriers highlights the role of the protein interactome. Nanoscale 9, 1840–1851. https://doi.org/10.1039/C6NR04765C.

Rasool, N., Srivastava, R., Singh, Y., 2022. Cationized silica ceria nanocomposites to target biofilms in chronic wounds. Biomater. Adv. 138, 212939. https://doi.org/10.1016/j.bioadv.2022.212939.

Raza, A., Alavi, S.E., Sime, F.B., Han, F.Y., Roberts, J.A., Popat, A., Falconer, J.R., Kumeria, T., 2021. Microfluidic assembly of pomegranate-like hierarchical microspheres for efflux regulation in oral drug delivery. Acta Biomater. 126, 277–290. https://doi.org/10.1016/j.actbio.2021.03.042.

Rinde, M., Kupferschmidt, N., Iqbal, M.N., Robert-Nicoud, G., Johnston, E.V., Lindgren, M., Bengtsson, T., 2020. Mesoporous silica with precisely controlled pores reduces food efficiency and suppresses weight gain in mice. Nanomedicine 15, 131–144. https://doi.org/10.2217/nnm-2019-0262.

Ryoo, R., Kim, J.M., Ko, C.H., Shin, C.H., 1996. Disordered molecular sieve with branched mesoporous channel network. J Phys Chem 100, 17718–17721. https://doi.org/10.1021/jp9620835.

Sánchez-Salcedo, S., García, A., González-Jiménez, A., Vallet-Regí, M., 2023. Antibacterial effect of 3D printed mesoporous bioactive glass scaffolds doped with metallic silver nanoparticles. Acta Biomater. 155, 654–666. https://doi.org/10.1016/j.actbio.2022.10.045.

Sarkar, S., Ekbal Kabir, M., Kalita, J., Manna, P., 2023. Mesoporous silica nanoparticles: drug delivery vehicles for antidiabetic molecules. Chembiochem 24. https://doi.org/10.1002/cbic.202200672.

Schumacher, K., Grün, M., Unger, K.K., 1999. Novel synthesis of spherical MCM-48. Micropor. Mesopor. Mater. 27, 201–206. https://doi.org/10.1016/j.micromeso.2020.110570.

Şen Karaman, D., Kietz, C., Govardhanam, P., Slita, A., Manea, A., Pamukçu, A., Meinander, A., Rosenholm, J.M., 2022. Core@shell structured ceria@mesoporous silica nanoantibiotics restrain bacterial growth in vitro and in vivo. Biomater. Adv. 133, 112607. https://doi.org/10.1016/j.msec.2021.112607.

Sha, Z., Yang, S., Fu, L., Geng, M., Gu, J., Liu, X., Li, S., Zhou, X., He, C., 2021. Manganese-doped gold core mesoporous silica particles as a nanoplatform for dual-modality imaging and chemo-chemodynamic combination osteosarcoma therapy. Nanoscale 13, 5077–5093. https://doi.org/10.1039/D0NR09220G.

Shishmakova, E.M., Ivchenko, A.V., Bolshakova, A.V., Staltsov, M.S., Urodkova, E.K., Grammatikova, N.E., Rudoy, V.M., Dement'eva, O.V., 2023. Antibacterial bio-nanocomposites based on drug-templated bifunctional mesoporous silica nanocontainers. Pharmaceutics 15, 2675. https://doi.org/10.3390/pharmaceutics15122675.

Song, K., Tang, Z., Song, Z., Meng, S., Yang, X., Guo, H., Zhu, Y., Wang, X., 2022. Hyaluronic acid-functionalized mesoporous silica nanoparticles loading simvastatin for targeted therapy of atherosclerosis. Pharmaceutics 14, 1265. https://doi.org/10.3390/pharmaceutics14061265.

Su, G., Zhong, X., Qiu, S., Fan, J., Zhou, H., Zhou, X., 2022. Preparation of mesoporous silica-based nanocomposites with synergistically antibacterial performance from nano-metal (oxide) and polydopamine. Nanotechnology 33, 155702. https://doi.org/10.1088/1361-6528/ac467a.

Tabriz, P.T., Maghsodi, P.H., Afarin, A.J., Heydari, H., Amiry, F., Sazegar, M.R., 2023. Antibacterial composite: polymeric mesoporous silica nanoparticles and combination of imipenem/meropenem. J. Mater. Chem. B 11, 1971–1977. https://doi.org/10.1039/D2TB02442J.

Tanev, P.T., Pinnavaia, T.J., 1998. Biomimetic assembly of porous lamellar silica molecular sieves with a vesicular particle architecture. Supramol. Sci. 5, 399–404. https://doi.org/10.1016/S0968-5677(98)00039-X.

Tenzer, S., Docter, D., Kuharev, J., Musyanovych, A., Fetz, V., Hecht, R., Schlenk, F., Fischer, D., Kiouptsi, K., Reinhardt, C., Landfester, K., Schild, H., Maskos, M., Knauer, S.K., Stauber, R.H., 2013. Rapid formation of plasma protein corona critically affects nanoparticle pathophysiology. Nat. Nanotechnol. 8, 772–781. https://doi.org/10.1038/nnano.2013.181.

Vallet-Regi, M., Rámila, A., del Real, R.P., Pérez-Pariente, J., 2001. A new property of MCM-41: drug delivery system. Chem. Mater. 13, 308–311. https://doi.org/10.1021/cm0011559.

Wan, Y., Zhao, 2007. On the controllable soft-templating approach to mesoporous silicates. Chem. Rev. 107, 2821–2860. https://doi.org/10.1021/cr068020s.

Wang, G., Zhang, P., Dou, H., Li, W., Sun, K., He, X., Han, J., Xiao, H., Li, Y., 2012. Efficient incorporation of quantum dots into porous microspheres through a solvent-evaporation approach. Langmuir 28, 6141–6150. https://doi.org/10.1021/la300066c.

Wang, Q., Song, Y., Gao, J., Li, Q., Chen, J., Xie, Y., Wang, Z., Tan, H., Yang, H., Zhang, N., Qian, J., Pang, Z., Huang, Z., Ge, J., 2023a. Hippo pathway-manipulating neutrophil-mimic hybrid nanoparticles for cardiac ischemic injury via modulation of local immunity and cardiac regeneration. Acta Pharm. Sin. B 13, 4999–5015. https://doi.org/10.1016/j.apsb.2023.08.021.

Wang, W., Song, S., Liu, W., Xia, T., Du, G., Zhai, X., Jin, B., 2023b. Two-photon excited luminescence of structural light enhancement in subwavelength SiO_2 coating europium ion-doped paramagnetic gadolinium oxide nanoparticle and application for magnetic resonance imaging. Disc. Nano 18, 85. https://doi.org/10.1186/s11671-023-03864-y.

Wei, X., Fang, Z., Sheng, J., Wang, Y., Lu, P., 2020. <p>Honokiol-mesoporous silica nanoparticles inhibit vascular restenosis via the suppression of TGF-β signaling pathway</p>. Int J Nanomed. 15, 5239–5252. https://doi.org/10.2147/IJN.S250911.

Wu, H., Li, F., Wang, S., Lu, J., Li, J., Du, Y., Sun, X., Chen, X., Gao, J., Ling, D., 2018. Ceria nanocrystals decorated mesoporous silica nanoparticle based ROS-scavenging tissue adhesive for highly efficient regenerative wound healing. Biomaterials 151, 66–77. https://doi.org/10.1016/j.biomaterials.2017.10.018.

Xia, W., Yan, T., Wen, L., Zhu, S., Yin, W., Zhu, M., Lang, M., Wang, C., Guo, C., 2022. Hypothermia-triggered mesoporous silica particles for controlled release of hydrogen sulfide to reduce the I/R injury of the myocardium. ACS Biomater. Sci. Eng. 8, 2970–2978. https://doi.org/10.1021/acsbiomaterials.2c00266.

Xu, B., Li, S., Shi, R., Liu, H., 2023. Multifunctional mesoporous silica nanoparticles for biomedical applications. Signal Transduct. Target Ther. 8, 435. https://doi.org/10.1038/s41392-023-01654-7.

Yadav, Y.C., Pattnaik, S., Swain, K., 2019. Curcumin loaded mesoporous silica nanoparticles: assessment of bioavailability and cardioprotective effect. Drug Dev. Ind. Pharm. 45, 1889–1895. https://doi.org/10.1080/03639045.2019.1672717.

Yahia, S., Khalil, I.A., Ghoniem, M.G., El-Sherbiny, I.M., 2023. 3D-bioimplants mimicking the structure and function of spine units for the treatment of spinal tuberculosis. RSC Adv. 13, 17340–17353. https://doi.org/10.1039/D3RA02351F.

Yang, H., Chen, Y., Chen, Z., Geng, Y., Xie, X., Shen, X., Li, T., Li, S., Wu, C., Liu, Y., 2017. Chemo-photodynamic combined gene therapy and dual-modal cancer imaging achieved by pH-responsive alginate/chitosan multilayer-modified magnetic mesoporous silica nanocomposites. Biomater. Sci. 5, 1001–1013. https://doi.org/10.1039/C7BM00043J.

Yang, L., Li, F., Cao, Y., Liu, Q., Jing, G., Niu, J., Sun, F., Qian, Y., Wang, S., Li, A., 2021. Multifunctional silica nanocomposites prime tumoricidal immunity for efficient cancer immunotherapy. J. Nanobiotechnol. 19, 328. https://doi.org/10.1186/s12951-021-01073-2.

Yang, P., Gai, S., Lin, J., 2012. Functionalized mesoporous silica materials for controlled drug delivery. Chem. Soc. Rev. 41, 3679. https://doi.org/10.1039/c2cs15308d.

Yang, Q., Wang, S., Fan, P., Wang, L., Di, Y., Lin, K., Xiao, F.-S., 2005. pH-responsive carrier system based on carboxylic acid modified mesoporous silica and polyelectrolyte for drug delivery. Chem. Mater. 17, 5999–6003. https://doi.org/10.1021/cm051198v.

Yu, C., Yu, Y., Zhao, D., 2000. Highly ordered large caged cubic mesoporous silica structures templated by triblock PEO–PBO–PEO copolymer. Chem. Commun. 575–576. https://doi.org/10.1039/b000603n.

Zanocco, R.P., Bresolí-Obach, R., Nájera, F., Pérez-Inestrosa, E., Zanocco, A.L., Lemp, E., Nonell, S., 2023. NanoFN10: a high-contrast turn-on fluorescence nanoprobe for multiphoton singlet oxygen imaging. Sensors 23, 4603. https://doi.org/10.3390/s23104603.

Zeleňák, V., Halamová, D., Almáši, M., Žid, L., Zeleňáková, A., Kapusta, O., 2018. Ordered cubic nanoporous silica support MCM-48 for delivery of poorly soluble drug indomethacin. Appl. Surf. Sci. 443, 525–534. https://doi.org/10.1016/j.apsusc.2018.02.260.

Zhan, W., Li, H., Guo, Y., Yang, L., Pang, L., Zhang, C., 2021. Hyaluronic acid functionalized biodegradable mesoporous silica nanocomposites for efficient photothermal and chemotherapy in breast cancer. Nanotechnology 32, 165703. https://doi.org/10.1088/1361-6528/abda74.

Zhang, A., Gao, A., Zhou, C., Xue, C., Zhang, Q., Fuente, J.M.D. La, Cui, D., 2023. Confining prepared ultrasmall nanozymes loading ATO for lung cancer catalytic therapy/immunotherapy. Adv. Mater. 35. https://doi.org/10.1002/adma.202303722.

Zhang, W., Lv, Z., Zhang, Y., Gopinath, S.C.B., Yuan, Y., Huang, D., Miao, L., 2022a. Targeted diagnosis, therapeutic monitoring, and assessment of atherosclerosis based on mesoporous silica nanoparticles coated with cRGD-platelets. Oxid. Med. Cell. Longev. 2022, 1–19. https://doi.org/10.1155/2022/6006601.

Zhang, X., Lang, Y., Li, X., Li, Z., Xu, Y., Chu, H., 2022b. The protective effect of puerarin-loaded mesoporous silicon nanoparticles on alcoholic hepatitis through mTOR-mediated autophagy pathway. Biomed. Microdev. 24, 37. https://doi.org/10.1007/s10544-022-00622-2.

Zhang, Y., Chang, M., Bao, F., Xing, M., Wang, E., Xu, Q., Huan, Z., Guo, F., Chang, J., 2019. Multifunctional Zn doped hollow mesoporous silica/polycaprolactone electrospun membranes with enhanced hair follicle regeneration and antibacterial activity for wound healing. Nanoscale 11, 6315–6333. https://doi.org/10.1039/C8NR09818B.

Zhao, D., Feng, J., Huo, Q., Melosh, N., Fredrickson, G.H., Chmelka, B.F., Stucky, G.D., 1998. Triblock copolymer syntheses of mesoporous silica with periodic 50 to 300 angstrom pores. Science (1979) 279, 548–552. https://doi.org/10.1126/science.279.5350.548.

Zhou, D., Du, M., Luo, H., Ran, F., Zhao, X., Dong, Y., Zhang, T., Hao, J., Li, D., Li, J., 2022. Multifunctional mesoporous silica-cerium oxide nanozymes facilitate miR129 delivery for high-quality healing of radiation-induced skin injury. J. Nanobiotechnol. 20, 409. https://doi.org/10.1186/s12951-022-01620-5.

Zhu, P., Cai, L., Liu, Q., Feng, S., Ruan, H., Zhang, L., Zhou, L., Jiang, H., Wang, H., Wang, J., Chen, J., 2022. One-pot synthesis of α-Linolenic acid nanoemulsion-templated drug-loaded silica mesocomposites as efficient bactericide against drug-resistant *Mycobacterium tuberculosis*. Eur. J. Pharmaceut. Sci. 176, 106261. https://doi.org/10.1016/j.ejps.2022.106261.

Biomedical applications of multifunctional polymer based nanostructured coatings

9

Burcu Önal Acet[1], *Dzmitry Shcharbin*[2], *Inessa Halets-Bui*[2], *Mehmet Odabaşı*[1] *and Ömür Acet*[3]
[1]Faculty of Arts and Science, Chemistry Department, Aksaray University, Aksaray, Turkey; [2]Institute of Biophysics and Cell Engineering of the National Academy of Sciences of Belarus, Minsk, Belarus; [3]Vocational School of Health Science, Pharmacy Services Program, Tarsus University, Tarsus, Turkey

9.1 Introduction

Recently, nanomaterials (NMs) are of great importance for the medical field. Especially, these NMs have found applications in the diagnosis of diseases through the administration of contrast agents for imaging purposes. Additionally, NMs have also been employed in the prevention of diseases through the development of vaccines, as well as in the treatment of various ailments through the use of drug delivery systems (DDSs) (Karabasz et al., 2020).

Polymer-based nanostructures (PNMs) have been acting a progressively significant role to make a revolution in the diagnoses and therapy of diseases. Many properties of PNMs (their shape, size, stability, surface charge, mechanical strength, surface chemistry, porosity, etc.) can be specifically functionalized in line with the needs in biomedical applications. For biomedical practices, PNMs should have significant water solubility/dispersibility and good controllable nanoparticle size to prevent rapid clearance and ensure preferentially biodistribution and biodegradability to diminish some side effects. They should have functionality for some important tasks (prodrug studies, targeting or imaging element, etc.) and also have the sensitivity to release the therapeutic agent under stimulated conditions (Tang et al., 2016). In recent times, PNMs have gained prominence as one of the most promising and practical instruments for precise and regulated DDSs (Yıldırım et al., 2022). Producing biopharmaceutical agent/s with developed characteristic is the significant element of the "Quality by Design" paradigm. It should be noted that this includes not only the clinical condition that the new product treats but also its pharmaco-kinetics/dynamics and potential toxicities (Acet, 2023; Brumano et al., 2019; Colombo et al., 2018; Eon-Duval et al., 2012). Thus, effective and nontoxic designs are extremely important for the biomedical field.

In particular, polymer coatings on solid materials are extremely important biomedical science. Thiol and siloxane are frequently utilized to replace noble metal and hydroxylated surfaces, respectively (Wei and Haag, 2015). And also, the extensively

Multifunctional Nanostructured Coatings. https://doi.org/10.1016/B978-0-443-23683-9.00009-X

utilized self-assembled monolayer and chemical surface immobilizations that are stimulated by these and other anchor group(s), Langmuir–Blodgett deposition (Ariga et al., 2013), layer-by-layer assembly (Borges and Mano, 2014), irradiation (He et al., 2009), and electrostatic/hydrophobic adsorption (Wei et al., 2014) are good designed. Nevertheless, many of these approaches also need some specific chemical/physical substrate characteristics. Therefore, they have not become universal coatings (Wei and Haag, 2015).

Universal coatings can replace material surfaces on very large scales. It must also be extremely durable in the applied situations. These kinds of coatings are substrate independent, without taking into consideration both the chemical structure and physical properties of the substrates. In order to create coatings like this, the connections and interactions between polymer structures and various substrates ought to be well adjusted. Chemical functionalities for specifically covalent binding between polymer structure coatings and surfaces should be refrained to alter different kinds of anchor structures that may not be effective on all diverse surface compositions. Even though various irradiation approaches make the surface workable for many different species, it is also stated in the literature that activity and density of active sites are at lower levels on various surfaces. From this perspective, they may need to be strengthened with the help of some compensatory techniques such as polymerization in order to obtain dense surface coatings (Rodriguez-Emmenegger et al., 2011; Teare et al., 2003). Additionally, it should be noted that noncovalent interactions can happen in almost all kinds of interfaces. Multi-noncovalent interactions may be considered as driving forces to form polymer coatings on various kinds of surfaces. There is no doubt that most noncovalent interactions occurring are not strong enough to bond polymer coatings for applications at interfaces. Therefore, some extra physical and chemical crosslinking agents may be needed to increase stabilities of coatings (Wei and Haag, 2015).

There is a constant need for biodegradable and biocompatible NMs that are highly stable in the body at the right time in the disease, new, advanced NMs with optimal pharmacokinetics and biodistribution that will protect healthy tissues in the future of medicine, and multifunctional NMs that will accumulate the drug tissue at the desired level. It is necessary for them to effectively communicate with various chemical entities that can accommodate small pharmaceuticals, nucleic acids, peptides, proteins, and even magnetic nanoparticles, considering the chemical standpoint. Convenient synthesis techniques need to let the biological activities of the transported drug(s) to be maintained. It is particularly significant in providing active therapeutic proteins and antigens, which are necessary constituents of vaccines (Karabasz et al., 2020).

There exist three conventional methods for preparing polymeric nanocarriers, namely layer-by-layer, nanodeposition, and emulsion template technique. Emulsion templating techniques primarily encompass emulsification diffusion, evaporation, coacervation, multiple emulsifications, and emulsion/polymer coating (Couvreur et al., 2002; Deng et al., 2020; Karabasz et al., 2020; Kothamasu et al., 2012; Mora-Huertas et al., 2010). Moreover, other methods such as preparation of polymer covered liposomes, polymersome structures, dendrimers/self-assembling polymeric micelles, inorganic nanoparticles modified with polymeric materials, protein, or peptide self-assembling nanocarrier structures have also been utilized (Fan et al., 2018;

Fu et al., 2020; Karabasz et al., 2020; Kumari et al., 2014; Pattni et al., 2015; Pinto Reis et al., 2006; Rahnfeld and Luciani, 2020; Rostamizadeh and Torchilin, 2020; Soppimath et al., 2001; Sun et al., 2020). Methods for organizing polymeric nanocarrier structures can mainly be examined under two approaches: Dispersion of preformed polymer structures and polymerization of monomers. Synthetic polymers have the advantage of higher purity and better reproducibility compared to natural polymers, and natural polymers have better biocompatibility (Karabasz et al., 2020). Protein-/peptide-based polymeric nanoparticles form another category of nanoparticles. These structures are particularly noteworthy for biomedical applications. Protein-/peptide-based polymeric nanoparticles are composed of amino acid structures and are considered as biologically derived macromolecules, exhibiting excellent biocompatibility and bioavailability. Protein or peptide nanostructures are widely employed to enhance the bioavailability of pharmaceutical drugs (Sun et al., 2020).

In the following sections for this book chapter, we will discuss some important multifunctional PNMs and their coatings and their applications along with the points to be considered during their design. In particular, we touched upon important studies in recent years for applications.

9.2 Important points for multifunctional nanocarriers

Nanocarriers can differ as nanoparticles, nanospheres, and nanocapsules depending on their general structure and nominal morphology. Nanostructures can be defined as homogeneous matrix structures where drug/drugs are dispersed within the material, formed by them, or adsorbed on their pores/surfaces, or formed by conjugation of the drug to them. Another point is that nanocapsule structures are divided into two structures: core and shell (Kothamasu et al., 2012). The core material can exist in various forms depending on the applications. Typically, the shell is composed of polymeric materials. The core contains the drug due to its overall structure, while the shell serves the purpose of safeguarding the core against external influences. Furthermore, it is also feasible to incorporate the drug into the shell material or have it adsorbed on the surfaces of the shell. This feature allows for the simultaneous administration of one or more drugs in different regions of the nanocapsule.

Essentially, the preparation of conventional polymeric nanocarrier systems are as follows: layer-by-layer technique, nanoprecipitation technique, and emulsion template technique. Layer-by-layer technique is an important technique for producing multilayer thin-film structures based on the sequential adsorption of nanostructures with opposite charges (Decher, 1997; Decher et al., 1992). Nanoprecipitation technique is also known in the literature as solvent displacement, interfacial deposition method, or solvent shift technique. Although nanoemulsion structures are thermodynamically unstable, they are kinetically stable. The shell structure can also be permeable, semi-permeable, or impermeable, according as the implementation situation(s). For controlled release applications, permeable or semi-permeable shells are used. The ratio of release from capsules relies mainly on the thickness of the shell, pore size, and material of the core (Karabasz et al., 2020). The properties are created in most technical emulsified systems in the preparation of nanostructures. This involves two steps. One of these involves the creation of an emulsified system, while the second one entails the production of nanostructures through either precipitation/gelation of a polymer or polymerization of monomers (Vauthier and Bouchemal, 2009). An important point to remember is that nanostructures can be produced through intermolecular or intramolecular cross-linking (Harth et al., 2002; van der Ende et al., 2010). In addition, emulsification and emulsion polymerization methods are among the most used techniques, especially for the synthesis of polymer nanostructures (Amato et al., 2015; Teekamp et al., 2015).

9.3 Some multifunctional nanocarriers

Micelles are a group of polymeric nanostructured materials that are structures dispersed in liquid and are a nonmicroscopic structure of polymeric molecules consisting of a colloidal system. Typically, it possesses a core—shell architecture, wherein the internal core comprises the hydrophobic segment of the amphiphiles, while the core region encircled by the corona structure comprises the hydrophilic segment of the amphiphiles (Letchford and Burt, 2007). It is a process in which both attractive and repulsive interactions play a role in micelle formation. There are also different attractive forces in the core section that coordinate its separation from the aqueous phase. Hydrophobic and electrostatic interactions, as well as complexation, give rise to these phenomena (Tang et al., 2016). Various factors such as electrostatic repulsions, hydration, and steric interaction may cause micelles to inhibit their unlimited growth (Gao and Eisenberg, 1993; Jones and Leroux, 1999). Amphiphilic copolymers, similar to low molecular weight surfactants, exhibit a critical micelle concentration necessary for the creation of micelles. At this particular concentration, copolymers are present as individual molecules in a solution. However, when the concentration exceeds this threshold, amphiphilic copolymer chains have the ability to aggregate and create micelles. This aggregation occurs in a manner where the hydrophobic portion of the copolymer avoids direct contact with the aqueous surroundings, as stated by Jones and Leroux (1999). Micellization has an endothermic entropy and is also entropy dependent (Alexandridis et al., 1994; Ma et al., 2007).

Polymersomes exhibit remarkable versatility as polymeric structures, particularly in the field of DDSs. The advantages of polymersomes have been exemplified through the spontaneous arrangement of polymeric chains within vesicular compartments, enabling the targeted delivery of therapeutic agents to specific tissues and organs. Extensive investigations have been carried out in recent times to explore the advancements in versatile polymersomes as targeted DDSs for therapeutic and theranostic applications. These polymersomes offer the unique ability to combine therapeutic delivery through passive and/or active targeting with diagnostic capabilities. Here, they can be controlled by changing some properties of the solution, such as size and morphological structure, chemical structure and size of the copolymer, preparation techniques and initial copolymer concentration, pH, temperature, and solvent kind. There are several methods available to prepare multifunctional polymersomes. One approach involves attaching functional ligands to preformed vesicles. Another method involves the self-assembly of block copolymers with end group functionalization. Additionally, polymers with functionalized hydrophilic blocks can also be utilized (Pawar et al., 2013).

Liposomes represent significant potential as a DDS, thanks to both their physico-chemical and biophysical properties. By encapsulating therapeutic agents, it has aided in mitigating the systemic toxicity and enhancing the in vivo biodistribution, thereby overcoming the challenges associated with conventional free drug therapy. Liposomes, which consist of a lipid bilayer structure, are also comprised of micelles, which are lipid monolayer structures. A liposomal bilayer is composed of cholesterol and phospholipids, forming a dual-layered structure. This unique composition allows liposomes to effectively transport both hydrophilic and lipophilic drugs, owing to their distinct properties (Kim and Jeong, 2021).

The biomedical field has a keen interest in the creation and advancement of molecular nanostructures that possess precise particle sizes and shapes. This is particularly significant in specialized areas like the formulation of active pharmaceuticals, imaging agents, and gene transfection. It is desired that the structures that serve as carriers in drug distribution be nanometer-sized. It is crucial to ensure that the particles are of equal size in order to enhance their ability to traverse cell membranes and minimize the likelihood of undesired elimination through the liver or spleen (Svenson and Tomalia, 2012).

In the year 1978, Vögtle et al. introduced the initial dendritic molecules, which were alternatively referred to as "cascade molecules," in the scientific literature (Buhleier et al., 1978). Tomalia et al. (1985) successfully synthesized poly(amidoamine) (PAMAM) dendrimers, which formed the first complete dendrimer family. Dendrimers are well-defined radial structures with many terminal functions. They are particularly promising for biomedical applications such as drug delivery, diagnosis, and treatment. An important point here is that their translation into clinical use is severely limited due to difficult synthesis steps and laborious purification (Lyu et al., 2020). Structurally, dendrimers basically consist of three different compartments. These can be listed as follows:

(1) the core at the center, where dendrimer maturation begins;
(2) repeated branching units that allow successive layers to form in a radially organized manner;
(3) a large number of terminals that bear a large and diverse number of functional groups and occur on the surface of the molecule (Lyu et al., 2020).

Inorganic nanostructures play a crucial role in various domains such as chemistry, physics, biology, and medicine. In the field of biomedicine, it is very important for nanostructures to demonstrate stability in hydrophilic surroundings and prevent early deterioration or elimination by the immune system. To achieve this, the surface of the nanoparticles is engineered to guarantee colloidal stability and biocompatibility and enable biomedical functionality (Komsthöft et al., 2022).

Despite its possible use for some clinical applications, the absence of colloidal stabilization in biological fluids like blood serum could potentially restrict clinical translation (Wang et al., 2021). The uncontrolled buildup of nanoparticles has the potential to amplify the interaction with proteins, leading to their identification by the immune system or a decrease in their functional abilities (Moore et al., 2015; Sanchez-Cano and Carril, 2020). Even though inorganic nanoparticles are predominantly formed in hydrophobic environments, they must remain stable in hydrophilic and aqueous environments for biomedical applications (Wang and Mattoussi, 2020). Surface modification is an important design parameter that determines the colloidal stability, biocompatibility, and functionality of inorganic nanoparticles for biomedical applications (Heuer-Jungemann et al., 2019). The degree of hydrophilicity of the surface is extremely important in biomedical applications of inorganic nanoparticles. In biological environments, protein fouling and particle aggregation are made worse by the surfaces of hydrophobic nanostructures (Dadfar et al., 2019). Polyethylene glycol (PEG) structure is highly demanded in the surface coating of biomedical nanoparticles due to its important features such as biocompatibility, ease of application to surfaces, and resistance to contamination (Maan et al., 2020; Zheng et al., 2017).

In order to enhance the capabilities of polymeric surfaces, it is possible to incorporate stimuli-responsive agents onto the surface. This additional feature allows for a greater level of functionality and responsiveness in the material. For tunable polymer coatings, multiple coordinated polymeric ligands can also be utilized. Polymer ligands have been reported to offer multisite coordination by introducing various anchoring agents into the backbone. By incorporating secondary polymers, the nanoparticles possess the ability to demonstrate enhanced stability and customized attributes, rendering them appropriate for a diverse array of applications. This method enables researchers to fine-tune the properties of nanoparticles to meet the requirements of specific applications, thereby expanding their potential in various fields (Komsthöft et al., 2022).

Peptide and protein nanostructures offer important advantages such as improving bioavailability, better adjustment of the formulation, a positive effect on drug release, and a decrease in toxicity. They also have the capacity to perform many biochemical and biological tasks. These structures are used in many targeted treatments such as tumor and cancer therapy, pulmonary administration, and vaccines. Protein nanostructures can also be integrated into biodegradable polymers as microspheres for sustained and precise release. In addition to being highly functional, these structures have been accepted as ideal biomaterials for the preparation of nanostructures, thanks to their remarkable amphiphilicity with solvents and drugs. These nanostructures can be easily metabolized and biodegraded. It is also known that they easily adapt to surface modifications in the binding of targeting ligands and drugs (Awuchi et al., 2022) (Fig. 9.1).

9.4 Usability of multifunctional polymer-based nanostructured coatings some diseases

The creation of advanced and multifunctional innovative coatings is extremely important for a more sustainable future (Svinterikos et al., 2023). Functional NMs are

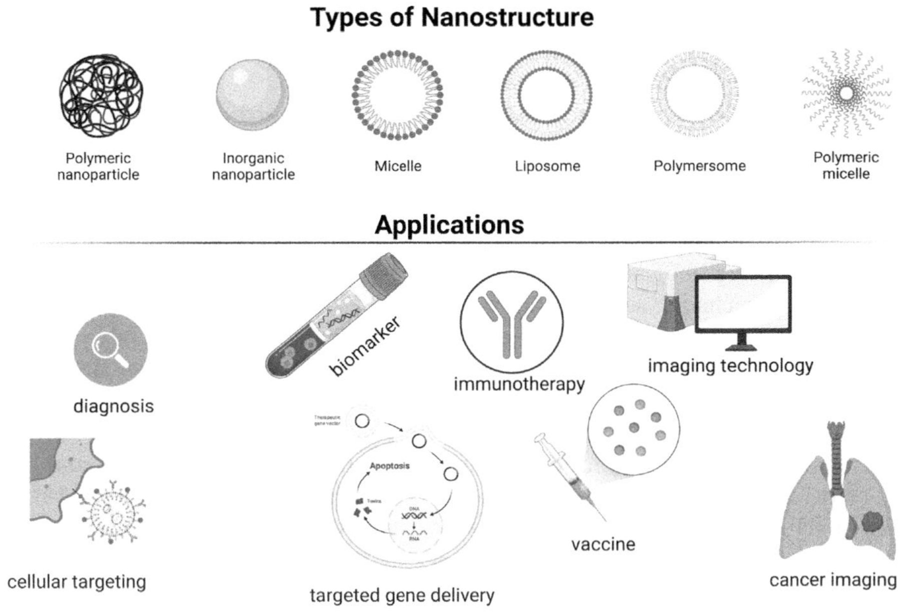

Figure 9.1 Some multifunctional nanocarriers.
Created with BioRender.com.

especially needed in the biomedical field for important fields such as drug delivery, oncology, diagnostics, tissue engineering, and antimicrobial and antifouling biomaterials (Ramburrun et al., 2022).

Despite being considered a crucial treatment for various forms of cancer, chemotherapy's efficacy often falls short of initial expectations. One potential approach to enhance this circumstance could involve focusing on the development of more sophisticated formulations, such as nanoformulations, for currently approved medications. The nanoformulations developed by the clinic exhibit reduced toxicity compared to conventional treatment methods. Furthermore, the therapeutic efficacy of these treatments may remain unaltered. Hence, continuous research is being conducted to explore improved nanoformulations for anticancer medications. Recently, various polymeric nanocarriers have been effectively obtained and analyzed in vivo, showing promising potential as carriers for oral or intravenous transportation (Karabasz et al., 2020). On the other hand, it is also known that actively targeted nanostructures affect cancer more effectively. However, it is significant to bear in mind that the presence of modified nanostructures within tumor structures is contingent upon the enhanced permeability and retention (EPR) effect (Lee et al., 2010). Given that the majority of investigations regarding the EPR effect have been conducted on rodents, the occurrence of this phenomenon in humans and its reliance on the cancer type remain subjects of contention (Nakamura et al., 2016; Prabhakar et al., 2013).

Polymeric nanocarriers have demonstrated their potential in cancer immunotherapy, expanding the scope of NMs' application beyond the delivery of chemotherapeutic agents. Numerous studies have provided evidence for the utilization of these nanocarriers in combating cancer, highlighting their versatility and effectiveness in anticancer treatments (Karabasz et al., 2020).

Nanostructures offer numerous advantages as contemporary therapeutic and prophylactic vaccines for administering tumor, viral, and bacterial antigens (including recombinant proteins and nucleic acid). They also have the potential to be used to increase vaccine effectiveness as a result of increasing antigen integrity and effective protection and immunogenicity. The significance of antigen-containing nanostructures in mimicking the pathogen should not be underestimated, considering their size, shape, density, surface charge, and chemical complexity. Nanostructured vaccines stimulate immune cells (B and T cells) more effectively, resulting in effective antibody synthesis and long-term immunity. Among the recently published articles, it also shows that nanocarriers with polymeric structure can be used for oral delivery of drugs, especially biological drugs such as peptides and proteins, in the future. This necessitates the development of formulation for oral administration that can protect them from degradation in the gastrointestinal tract and without losing their biological activity and aid in their effective absorption. Nanostructures are extensively utilized in the field of regenerative medicine, encompassing a broad range of applications. The primary objective of regenerative medicine is to advance therapeutic approaches that facilitate the recovery and regeneration of damaged tissues caused by either diseases or mechanical trauma. Some lifestyle diseases, such as diet, physical inactivity, and stimulant consumption, become more prevalent each year. Polymeric nanostructures have been extensively discussed in scientific literature as a potential DDS for various medical conditions such as venous thromboembolism, hypertension, peripheral arterial disease, and inflammatory bowel disease. However, it is worth noting that diabetes, in particular, necessitates the development of a novel drug formulation to enhance patient safety and efficacy. Numerous investigations have been conducted with the aim of developing an oral insulin formulation. Polymer-structured nanocarriers are also used for the treatment of various infectious diseases as carriers of antibacterial/fungal/parasitic compounds (Karabasz et al., 2020).

9.5 Applications

9.5.1 Nanocoatings for surface modification of cardiovascular devices

TiCu-nanostructured coatings, which have the ability to release copper in a controlled manner, can serve as a valuable resource in the management of cardiovascular ailments (Yue et al., 2024). Controlled copper release from nanostructures showed decreased platelet adhesion and suppressed smooth muscle cell behavior. They were safe for the human body and had hemolysis less than 5%. Surface modification of cardiovascular devices can be achieved through their application (Yue et al., 2024) (Fig. 9.2).

At the same time, Gallingani et al. synthesized the silver nanoparticles, which were embedded in an organic coating using hexamethyldisiloxane plasma–assisted

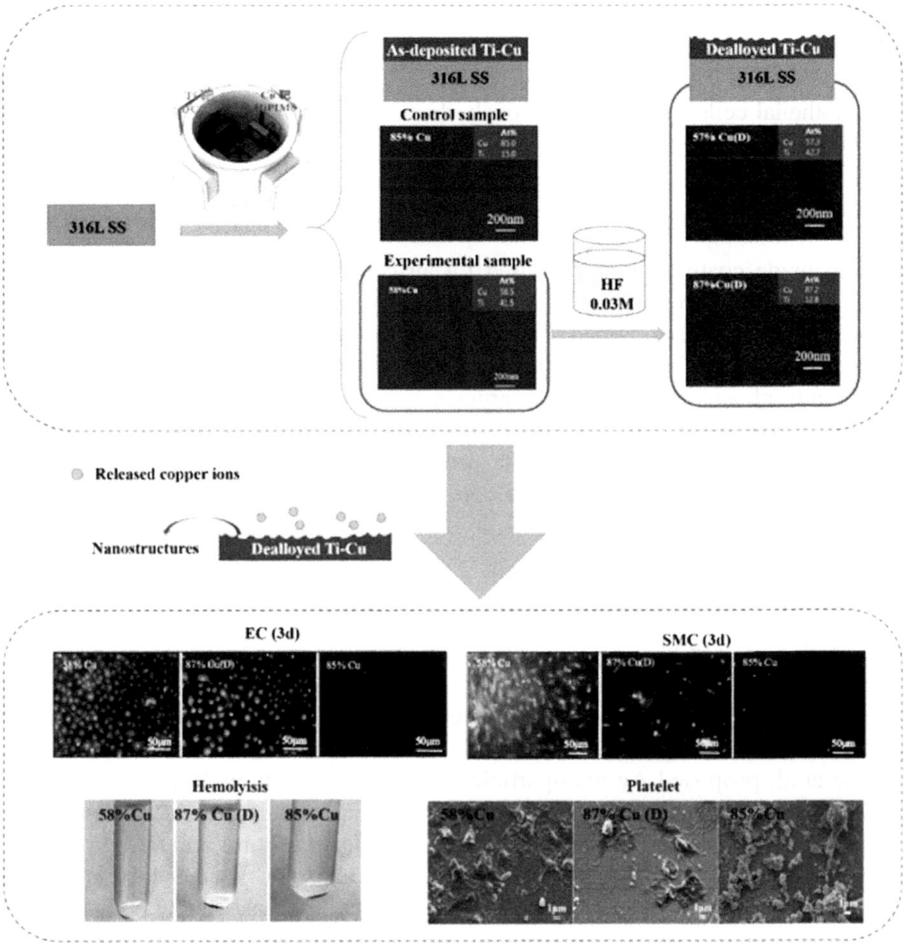

Figure 9.2 TiCu-nanostructured coatings with controlled copper release (Yue et al., 2024).

polymerization. The aim of this study is to provide antibacterial effects of argentum for cardiovascular medical devices and central venous catheters implants (Gallingani et al., 2023). Lakalayeh et al. analyzed the efficiency of bioabsorbable magnesium alloys for cardiovascular stent technology. They proposed magnesium-based alloy coated with dexamethasone-loaded polymeric nanoparticles to improve the antibacterial efficiency of stents. The data showed that nanocoatings inhibited smooth muscle cells proliferation but provide effective adhesion and proliferation for endothelial cells on stents (Lakalayeh et al., 2022). Hataminia et al. synthesized Fe_3O_4 nanoparticles with thermoplastic gel coating for the cardiovascular stents. The data on HUVEC and L929 cell lines as well as on blood cells showed a high biocompatibility of nanocoating (Hataminia et al., 2022). Hwang et al. proposed the utilization of a nanocoating that mimics the endothelium to facilitate the delivery of nitric oxide, thereby promoting the process of cell endothelialization. It was uniformly coated on the left atrial appendage closure device membranes and provided sustained release of nitric oxide. In vitro and in vivo findings demonstrated its ability to enhance the proliferation of endothelial cells while simultaneously decreasing platelet adhesion (Hwang et al., 2021).

9.5.2 Nanocoatings for air filtering and to prevent pneumonia

Cerium oxide coatings were proposed for air-conditioning filters (Abramova et al., 2023). They had a pronounced antibacterial activity and low toxicity. The antibacterial activity of the ceria-modified filters had about 4 by authors' scale. Ventilator-associated pneumonia is often infection for hospitals. Azmi et al. proposed laccase-gadolinium phosphate hybrid nanoparticles combined with enzyme mediator coated on the surface of the endotracheal tube. They were nontoxic toward human lung epithelial cells L929 and A549. The proposed nanocoatings significantly reduced the surface adhesion of *Escherichia coli*, *Pseudomonas aeruginosa*, and *Staphylococcus aureus* (Azmi et al., 2024).

9.5.3 Nanocoatings for tooth and osteo implants

Gao et al. proposed to use silver-embedded sparsely titania nanotube arrays for implants. They demonstrated an enhanced antibactericidal activity against *Staphylococcus aureus*, biocompatibility, and osteogenic activity in vitro and in vivo in rats (Gao et al., 2023) (Fig. 9.3).

Nie et al. proposed Ag nanoparticle-coated polytetrafluorethylene nanoparticle for tooth implants. They demonstrated that nanocoatings had an enchanced (80%–90%) antibacterial effect against *Streptococcus mutans* and *Porphyromonas gingivalis* (Nie et al., 2023). Li et al. synthesized an ascorbic acid-decorated nanostructured titanium surface and studied their effect on breast cancer cells and osteoblasts. The data showed that titanium and ascorbic acid synergistically inhibited the migration and proliferation of breast cancer cells as well as enhanced the proliferation and osteogenic differentiation of osteoblasts (Li et al., 2024). De Santis et al. proposed to use a cerium oxide coating based on pristine and nanotubular-structured Ti and Ti6Al4V surfaces for osseointegration and to prevent bacterial infection. The data showed that cerium oxide on the coatings induced osseointegration. Also, this coating improved oxidative stress resistance and had antibacterial effect (De Santis et al., 2023).

9.5.4 Nanocoatings for skin repair and to prevent adhesion

Skin repair is an important question of modern medicine. Luo et al. proposed extracellular vesicle-derived silk fibroin nanoparticles loaded with milk fat globule EGF factor 8 to improve the state of skin blood vessels. The data showed that nanocoatings decreased mitochondrial autophagy and ferroptosis caused by hypoxia and accelerated wound healing in rats (Luo et al., 2023) (Fig. 9.4).

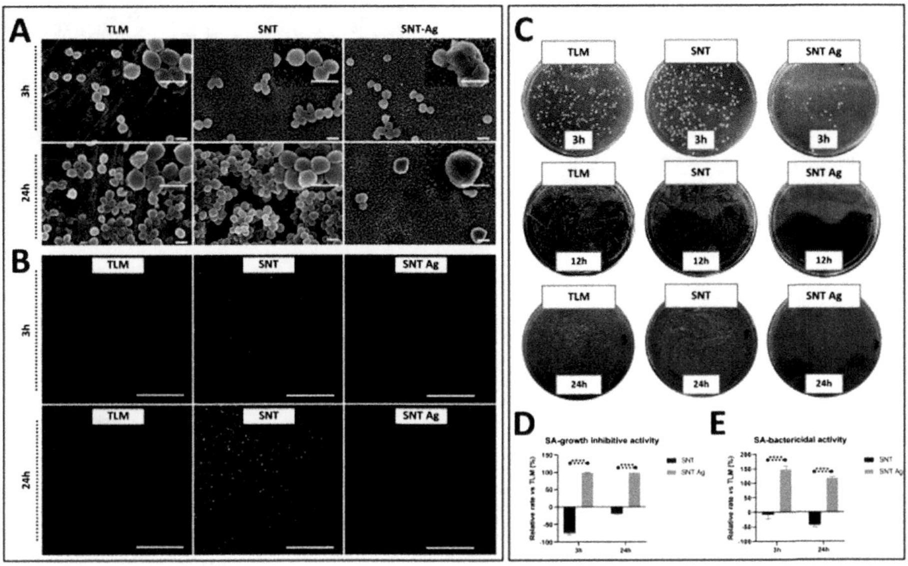

Figure 9.3 Investigation of the antibacterial effectiveness of silver-embedded sparse titanium nanotube arrays for use in implants. (a) Different representative images of *S. aureus* (SA) growth on TLM, SNT, and SNT Ag surfaces are depicted in the FE-SEM images, scale bar = 1 μm; (b) different representative confocal images of live/dead *Staphylococcus aureus* (SA) on transmission light microscopy (TLM), scanning near-field optical microscopy (SNT), and SNT Ag surfaces are presented, scale bar = 100 μm; (c) the development of colonies of SA on TLM, SNT, and SNT Ag surfaces; (d and e) growth inhibitor (d) and bactericidal (e) TLM was standardized at 100% based on confocal images of the activity of SNT and SNT Ag surfaces (Gao et al., 2023).

Hosseinpour-Moghadam et al. synthesized a quercetin-loaded poly(caprolactone)/phosphatidylcholine—silver coatings to stop the formation of postoperative adhesions. An investigation conducted on in vivo studies using animal model of sidewall defect and cecum abrasion demonstrated that the presence of silver in the nanocoating resulted in an increased antibacterial effectiveness. Additionally, the nanocoating exhibited favorable antiadhesion properties (Hosseinpour-Moghadam et al., 2021).

9.5.5 Nanocoatings for drug and RNA delivery and drug screening

The delivery of small RNA (siRNA and microRNA) to treat different diseases is an important question of many studies. Van der Van et al. synthesized a complex nanosystem including miR-mimics, PEG-peptide gold nanoparticles, and an injectable, shear thinning, self-assembling hydrogel for delivery of microRNA toward human model of calcific aortic valve disease. As result, in vivo in calcific aortic valve disease model miR-214 was efficiently delivered into the cytoplasm of human aortic valve interstitial cells. Also, 11 days after subcutaneous injection, miR-67 was cleared

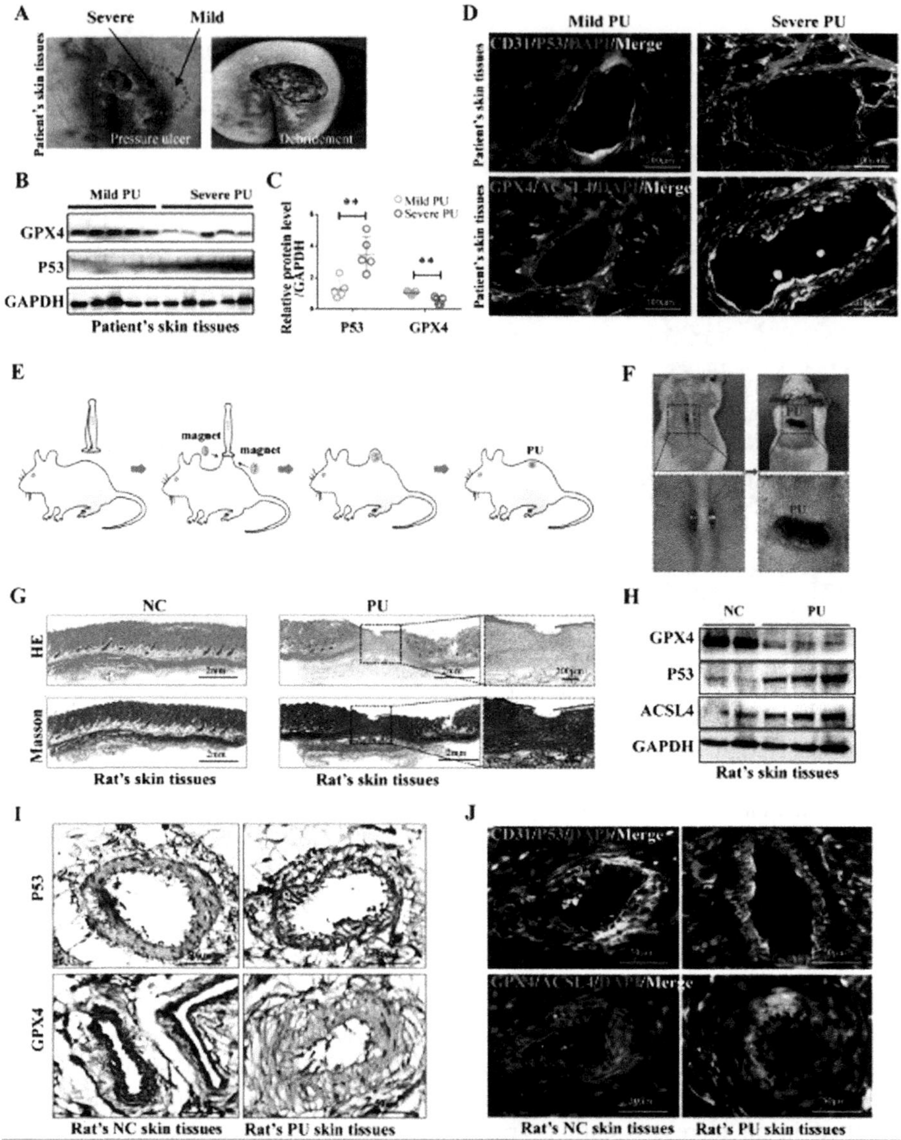

Figure 9.4 Detection and analysis of ferroptosis-related proteins in rats' and patients' PU tissues. (a) Patients who underwent debridement provided both mild and severe PU tissues for collection. (b, c) Quantitative levels of GPX4 and P53 were assessed through representative western blotting in patients exhibiting mild and/or severe PU. (d) Immunofluorescence staining of CD31/P53 and GPX4/ACSL4 on VECs in both mild and severe PU tissues obtained from patients. (e) A technique for constructing a model of PU in rats involved the following steps. Firstly, the rats were administered anesthesia to ensure their comfort during the procedure. Subsequently, the hair on the skin at the back of the rats was carefully removed. Then, the skin was gently lifted and subjected to pressure using cylindrical magnets. This pressure was consistently applied for a duration of 6–8 h each day until the formation of PU occurred. (f) Rats were subjected to the application of cylindrical magnets measuring 8 mm in diameter and 3 mm in height on their back skin in order to create PU. (g) H&E staining and Masson staining were performed on both normal and PU skin tissues in rats. (h) A representative western blot analysis of GPX4, P53, and ACSL4 in both normal and PU skin tissues of rats. (i, j) In rats, normal and PU skin tissues were subjected to representative immunohistochemical staining of P53 and GPX4, along with immunofluorescence staining of CD31/P53 and GPX4/ACSL4 in blood vessels. NC: Normal control, PU: Pressure ulcer (Luo et al., 2023).

from a body via the liver and kidneys (van der Ven et al., 2021). Vasudevan et al. created the liver-specific nanofiber scaffolds with polylactic acid and a decellularized liver extracellular matrix. The nanocoatings that were showcased exhibited higher hydrophilicity and superior swelling characteristics in comparison to the polylactic acid scaffolds. The data on activity of hepatocytes cultured on these scaffolds toward several drugs showed an excellent properties of these scaffolds for ex-vivo drug screening (Vasudevan et al., 2023) (Fig. 9.5).

Ahmed et al. developed niosomal nanogels based on 5-fluorouracil-loaded niosomal nanoparticles coated with chitosan and crosslinked by tripolyphosphate. These nanogels had a continuous 5-FU release within 72 hours of incubation. In contrast to pure 5-FU, the nanogels showed higher anticancer activity against cancer cell line B16F10 and lower cytotoxicity toward normal cell line NIH3T3 (Ahmed et al., 2023). Loskutova et al. proposed to use cellulose nanofiber-shelled perfluoropentane scaffolds with or without paclitaxel to treat cancer cells. They found that this nanoformulation had an enchanced anticancer effect on 4T1 breast cancer cells and did not affect erythrocytes (Loskutova et al., 2023).

9.5.6 Nanocoatings for tissue repair

Different kinds of coatings are actively used in tissue repair. Chen et al. synthesized gelatin methacryloyl (GelMA)-coated, 3D-expanded nanofiber scaffolds for tissue repair. They represented 3D-ordered tissue constructs with clinically appropriate sizes. Different kinds of cells can be used for these microstructures including dermal fibroblasts, bone marrow-derived mesenchymal stem cells, and human neural stem/precursor cells promoting a different tissue construct (Chen et al., 2021). Kittana et al.

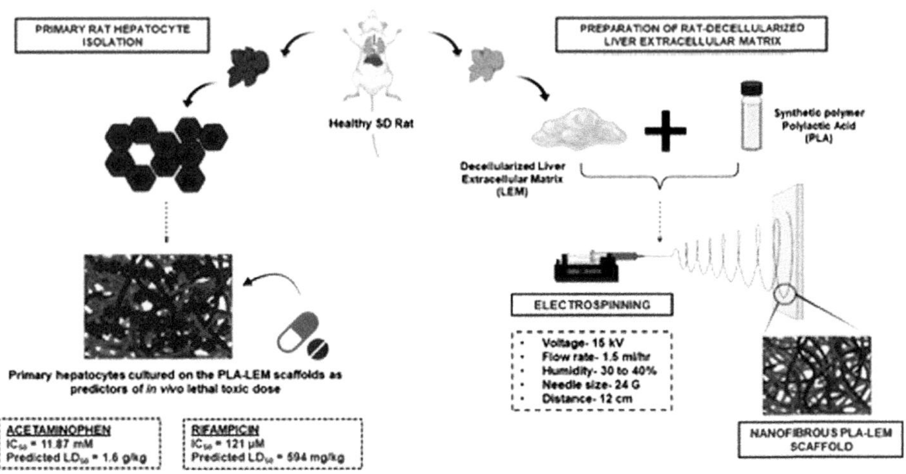

Figure 9.5 Liver-specific nanofiber scaffolds including polylactic acid and decellularized liver extracellular matrix (Vasudevan et al., 2023).
Reprinted with permission from ACS, Copyright (2023).

proposed to use multiwall carbon nanotubes with chitosan to produce nanocoatings for tissue repair. Based on them, they created engineered connective tissues effective for wide range of applications (Kittana et al., 2021). Chen et al. developed polylactic acid nanogels loaded with hyaluronic acid and verteporfin to expedite wound re-epithelialization and regulate scar formation. The effectiveness of these nanocoatings in promoting scarless wound healing was validated through in vitro experiments using skin fibroblasts, as well as in vivo studies conducted on rats and rabbits. Illustration of scarless healing of wounds through the use of hyaluronic acid and verteporfin-loaded polylactic acid nanogels. These nanogels play a crucial role in promoting re-epithelialization of wounds and effectively controlling scar formation. Through schematic representations, the accelerated wound healing process can be visualized, highlighting the potential of this approach to achieve scar-free results (Chen et al., 2023) (Fig. 9.6).

A study conducted by Ailincai et al. focused on the development of chitosan-based nanofibers with the aim of facilitating the healing process of burn wounds. The nanocoatings demonstrated a notable improvement in their ability to combat various strains of bacteria, including Gram-positive, Gram-negative, and fungal strains. Experiments with rats showed that these nanocoatings induced the restoration of dermis and epidermis (Ailincai et al., 2023). Peripheral demyelinating diseases result in the impairment of axons and Schwann cells within the peripheral nervous system. Nagarajan et al. developed a nanofibrous scaffold using polycaprolactone and graphene oxide and coated this bioscaffold with Schwann cell acellular matrix. The main aim for this scaffold was to induce remyelination and to

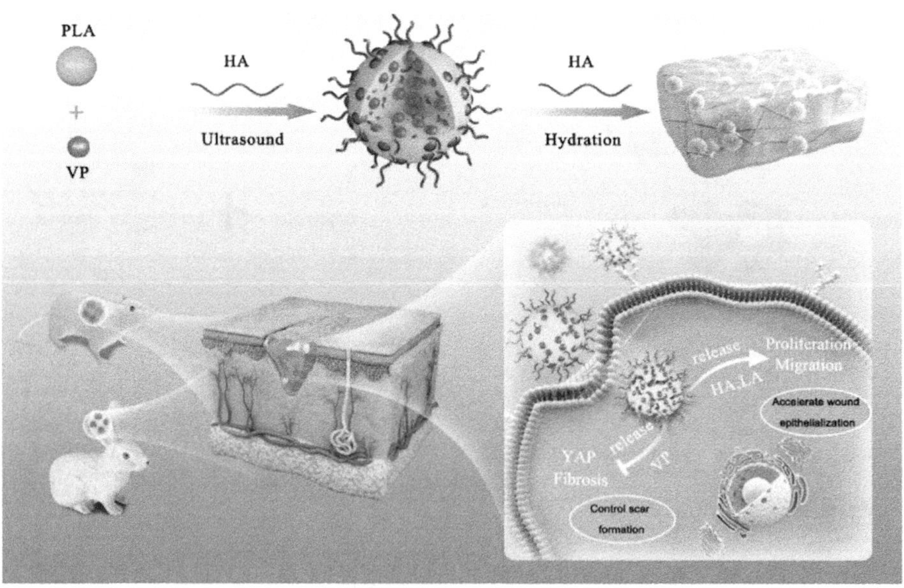

Figure 9.6 Illustration of facilitating scarless healing of wounds through the use of hyaluronic acid and verteporfin-loaded polylactic acid nanogels (Chen et al., 2023).

prevent oxidative stress during on the neurons in the peripheral nervous system. The developed scaffold can be used to treat demyelinating diseases (Nagarajan et al., 2023). Lin et al. synthesized the neural cell–derived membrane coatings with DNA nanogels. The efficiency of these scaffolds was checked for cells representing central nervous system: astrocytes, microglia, cortical neurons, and oligodendrocyte progenitor cells. The data showed high efficiency of scaffolds in vitro and in vivo (Lin et al., 2023). Liu et al. proposed nanofibrous scaffolds decorated with nanoscale SiO_2 protrusions and galectin-1. The main applications of these coatings are neurite outgrowth and neural stem cell migration to repair neural tissues (Liu et al., 2023). Zhu et al. synthesized $Ti_3 C_2 T_x$ MXene with poly(L-lactide) nanofibers coating to enhance neuronal stem cells neurogenesis. As results, this coating provided neuronal stem cells differentiation into both neurons and astrocytes. Also, this coating affected synergistically to provide the growth and maturation of neurites (Aydeger et al., 2023; Zhu et al., 2023). In the similar work, Aydeger et al. created micro-/nano-channeled polycaprolactone/poly-d,l-lactic-co-glycolic acid hybrid film scaffolds modified by gold nanoparticles. This coating induced neurite outgrowth and neuronal differentiation compared with the classic polypyrrole coating. Zhang et al. created dual-functional smart polymer foil coating to prevent septic and aseptic failures. It showed excellent antibacterial properties by killing of several pathogens without direct effect on mammalian cells (Zhang et al., 2023). Salehi et al. synthesized the nanocoating from polylactic acid-Baghdadite scaffold coated with chitosan-vascular endothelial growth factor and checked its osteogenic potential using rat bone marrow–derived mesenchymal stem cells. The data showed that the present coating is an excellent candidate for bone healing applications (Salehi et al., 2023).

9.5.7 Nanocoatings for diagnostics

A colorimetric catalytic nanocomplex was synthesized by Song et al. In this study, dopamine flexible scaffold-coated multishell porous zeolitic imidazolate framework-8 was utilized as a scaffold to encapsulate horseradish peroxidase. The purpose of this nanocomplex was to enhance the detection of cardiac troponin I through lateral flow immunoassay. The proposed nanosystem allowed detection of protein with sensitivity of 0.5 ng mL^{-1} and 0.01 ng mL^{-1}, which were 4/2- and 200/100-fold higher than classic gold nanoparticles based lateral flow immunoassay (Song et al., 2023) (Fig. 9.7).

9.5.8 Nanocoatings for veterinary medicine

Ziabka et al. proposed to use Ag, Cu, and Zn nanoparticle coatings applied on medical implants for veterinary medicine. The antibacterial activity of nanocoatings was accessed using several Gram+ and Gram-bacteria and pathogenic yeasts. The activity in vivo was analyzed on canine patients, and the authors found an enhanced antibacterial activity of nanocoatings. Additionally, silver nanoparticles in coatings accelerated the bone healing (Ziąbka et al., 2023).

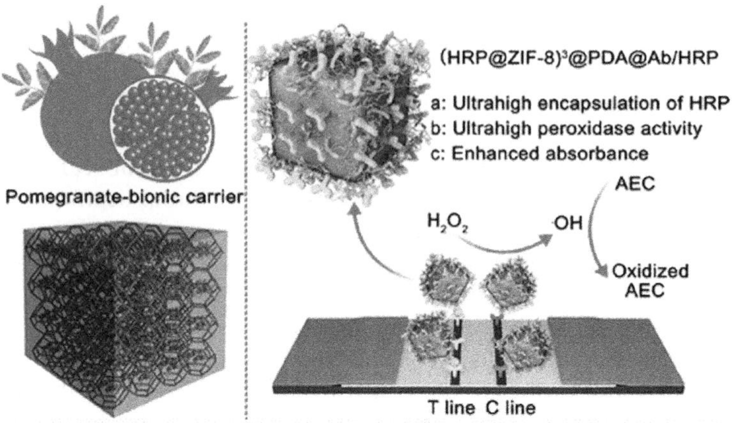

Figure 9.7 A colorimetric catalytic nanocomplex, using dopamine flexible scaffold-coated multishell porous zeolitic imidazolate framework-8 and horseradish peroxidase for lateral flow immunoassay of cardiac troponin I (Song et al., 2023).

9.6 Conclusions and some challenges

In summary, it can be seen in this section that efforts to develop polymeric nanostructured materials have attracted great attention due to their increasing importance in the field of biomedical applications. PNMs have the potential to offer significant benefits in terms of precise diagnosis and efficient treatment of various ailments. PNMs have demonstrated significant potential in transforming the landscape of disease diagnosis and treatment. However, the translation of such promising research into clinical applications is not without its hurdles. One major challenge lies in the efficient large-scale production of PNMs, particularly those that involve complex multistep processes. Additionally, the potential toxicity of PNMs toward cells, tissues, and organs, especially those containing nonbiodegradable or inorganic components, poses a critical concern. Nevertheless, despite these obstacles, PNMs present novel opportunities to enhance or even supplant current systems. It is firmly believed that PNMs will emerge as a pivotal diagnostic and therapeutic tool in the realm of biomedical applications in the foreseeable future (Tang et al., 2016). Additionally, it is crucial to note that within the extensive range of surface modification systems, only a limited number of universal coatings remain viable for effective utilization in practical scenarios. Further development is necessary for the mechanical understanding of universal coatings stabilization and theoretical guidelines for the improvement of such coatings. Hence, forthcoming studies ought to concentrate on investigating the adsorption mechanisms of polymers on various surfaces, quantitatively assessing the role of cross-linking in enhancing the stability of coatings, and formulating a framework of principles to steer the advancement of versatile coatings. The majority of coatings have been studied at the laboratory level. There is a notable lack of comprehensive, large-scale quantitative analyzes of the uniformity of coatings, a crucial consideration for real-world applications. As a result, although the studies still have some deficiencies at this point, universal polymer coatings have introduced a novel approach to enhancing material surface modification (Wei and Haag, 2015).

References

Abramova, A.V., Kozlov, D.A., Veselova, V.O., Kozlova, T.O., Ivanova, O.S., Mikhalev, E.S., Voytov, Y.I., Baranchikov, A.E., Ivanov, V.K., Cravotto, G., 2023. Coating of filter materials with CeO2 nanoparticles using a combination of aerodynamic spraying and suction. Nanomaterials 13, 3168. https://doi.org/10.3390/nano13243168.

Acet, Ö., 2023. Design of enhanced smart delivery systems for therapeutic enzymes: kinetic and release performance of dual effected enzyme-loaded nanopolymers. Catal. Lett. 153, 3174–3184. https://doi.org/10.1007/s10562-023-04418-8.

Ahmed, M.M., Ameen, M.S.M., Abazari, M., Badeleh, S.M., Rostamizadeh, K., Mohammed, S.S., 2023. Chitosan-decorated and tripolyphosphate-crosslinked pH-sensitive niosomal nanogels for Controlled release of fluoropyrimidine 5-fluorouracil. Biomed. Pharmacother. 164, 114943. https://doi.org/10.1016/j.biopha.2023.114943.

Ailincai, D., Cibotaru, S., Anisiei, A., Coman, C.G., Pasca, A.S., Rosca, I., Sandu, A.-I., Mititelu-Tartau, L., Marin, L., 2023. Mesoporous chitosan nanofibers loaded with norfloxacin and coated with phenylboronic acid perform as bioabsorbable active dressings to accelerate the healing of burn wounds. Carbohydr. Polym. 318, 121135. https://doi.org/10.1016/j.carbpol.2023.121135.

Alexandridis, P., Holzwarth, J.F., Hatton, T.A., 1994. Micellization of poly(ethylene oxide)-poly(propylene oxide)-poly(ethylene oxide) triblock copolymers in aqueous solutions: thermodynamics of copolymer association. Macromolecules 27, 2414–2425. https://doi.org/10.1021/ma00087a009.

Amato, D.V., Amato, D.N., Flynt, A.S., Patton, D.L., 2015. Functional, sub-100 nm polymer nanoparticles *via* thiol–ene miniemulsion photopolymerization. Polym. Chem. 6, 5625–5632. https://doi.org/10.1039/C4PY01449A.

Ariga, K., Yamauchi, Y., Mori, T., Hill, J.P., 2013. 25th anniversary article: what can Be done with the Langmuir-blodgett method? recent developments and its critical role in materials science. Adv. Mater. 25, 6477–6512. https://doi.org/10.1002/adma.201302283.

Awuchi, C.G., Akram, M., Owuamanam, I.C., Ogueke, C.C., Awuchi, C.G., Twinomhwezi, H., 2022. Nanotechnological application of peptide- and protein-based therapeutics. In: Applications of Nanotechnology in Drug Discovery and Delivery. Elsevier, pp. 205–238. https://doi.org/10.1016/B978-0-12-824408-1.00005-3.

Aydeger, A., Aysit, N., Baydas, G., Cakici, C., Erim, U.C., Arpa, M.D., Ozcicek, I., 2023. Design of IKVAV peptide/gold nanoparticle decorated, micro/nano-channeled PCL/PLGA film scaffolds for neuronal differentiation and neurite outgrowth. Biomater. Adv. 152, 213472. https://doi.org/10.1016/j.bioadv.2023.213472.

Azmi, A., Mojtabavi, S., Fakhrmousavi, S.A.A., Faizi, M., Forootanfar, H., Samadi, N., Faramarzi, M.A., 2024. Surface functionalization of endotracheal tubes coated with laccase–gadolinium phosphate hybrid nanoparticles for antibiofilm activity and contrasting properties. Biomater. Sci. 12 (3), 674–690. https://doi.org/10.1039/D3BM01406A.

Borges, J., Mano, J.F., 2014. Molecular interactions driving the layer-by-layer assembly of multilayers. Chem. Rev. 114, 8883–8942. https://doi.org/10.1021/cr400531v.

Brumano, L.P., da Silva, F.V.S., Costa-Silva, T.A., Apolinário, A.C., Santos, J.H.P.M., Kleingesinds, E.K., Monteiro, G., Rangel-Yagui, C. de O., Benyahia, B., Junior, A.P., 2019. Development of L-asparaginase biobetters: current research status and review of the desirable quality profiles. Front. Bioeng. Biotechnol. 6. https://doi.org/10.3389/fbioe.2018.00212.

Buhleier, E., Wehner, W., Vögtle, F., 1978. "Cascade"- and "nonskid-chain-like" syntheses of molecular cavity topologies. Synthesis 1978, 155–158. https://doi.org/10.1055/s-1978-24702.

Chen, K., Liu, Y., Liu, X., Guo, Y., Liu, J., Ding, J., Zhang, Z., Ni, X., Chen, Y., 2023. Hyaluronic acid-modified and verteporfin-loaded polylactic acid nanogels promote scarless wound healing by accelerating wound re-epithelialization and controlling scar formation. J. Nanobiotechnol. 21, 241. https://doi.org/10.1186/s12951-023-02014-x.

Chen, S., Carlson, M.A., Li, X., Siddique, A., Zhu, W., Xie, J., 2021. Minimally invasive delivery of 3D shape recoverable constructs with ordered structures for tissue repair. ACS Biomater. Sci. Eng. 7, 2204–2211. https://doi.org/10.1021/acsbiomaterials.1c00344.

Colombo, S., Beck-Broichsitter, M., Bøtker, J.P., Malmsten, M., Rantanen, J., Bohr, A., 2018. Transforming nanomedicine manufacturing toward quality by design and microfluidics. Adv. Drug Deliv. Rev. 128, 115–131. https://doi.org/10.1016/j.addr.2018.04.004.

Couvreur, P., Barratt, G., Fattal, E., Vauthier, C., 2002. Nanocapsule technology: a review. Crit. Rev. Ther. Drug Carrier Syst. 19, 99–134.

Dadfar, S.M., Roemhild, K., Drude, N.I., von Stillfried, S., Knüchel, R., Kiessling, F., Lammers, T., 2019. Iron oxide nanoparticles: diagnostic, therapeutic and theranostic applications. Adv. Drug Deliv. Rev. 138, 302–325. https://doi.org/10.1016/j.addr.2019.01.005.

De Santis, S., Varricchio, R., Ceccucci, A., Sotgiu, G., Di Masi, A., Magna, G., Sennato, S., Orsini, M., 2023. Cerium coatings on pristine and nanostructured Ti and Ti6Al4V surfaces: bioactivity, resistance in simulated inflammatory conditions, and antibacterial performance. ACS Biomater. Sci. Eng. 9, 3262–3272. https://doi.org/10.1021/acsbiomaterials.2c01461.

Decher, G., 1997. Fuzzy nanoassemblies: toward layered polymeric multicomposites. Science 277, 1232–1237. https://doi.org/10.1126/science.277.5330.1232.

Decher, G., Hong, J.D., Schmitt, J., 1992. Buildup of ultrathin multilayer films by a self-assembly process: III. Consecutively alternating adsorption of anionic and cationic polyelectrolytes on charged surfaces. Thin Solid Films 210–211, 831–835. https://doi.org/10.1016/0040-6090(92)90417-A.

Deng, S., Gigliobianco, M.R., Censi, R., Di Martino, P., 2020. Polymeric nanocapsules as nanotechnological alternative for drug delivery system: current status, challenges and opportunities. Nanomaterials 10, 847. https://doi.org/10.3390/nano10050847.

Eon-Duval, A., Broly, H., Gleixner, R., 2012. Quality attributes of recombinant therapeutic proteins: an assessment of impact on safety and efficacy as part of a quality by design development approach. Biotechnol. Prog. 28, 608–622. https://doi.org/10.1002/btpr.1548.

Fan, Z., Chang, Y., Cui, C., Sun, L., Wang, D.H., Pan, Z., Zhang, M., 2018. Near infrared fluorescent peptide nanoparticles for enhancing esophageal cancer therapeutic efficacy. Nat. Commun. 9, 2605. https://doi.org/10.1038/s41467-018-04763-y.

Fu, D., Liu, D., Zhang, L., Sun, L., 2020. Self-assembled fluorescent tripeptide nanoparticles for bioimaging and drug delivery applications. Chin. Chem. Lett. 31, 3195–3199. https://doi.org/10.1016/j.cclet.2020.07.011.

Gallingani, T., Resca, E., Dominici, M., Gavioli, G., Laurita, R., Liguori, A., Mari, G., Ortolani, L., Pericolini, E., Sala, A., Laghi, G., Petrachi, T., Arnauld, G.F., Accorsi, L., Rizzoli, R., Colombo, V., Gherardi, M., Veronesi, E., 2023. A new strategy to prevent biofilm and clot formation in medical devices: the use of atmospheric non-thermal plasma assisted deposition of silver-based nanostructured coatings. PLoS One 18, e0282059. https://doi.org/10.1371/journal.pone.0282059.

Gao, H., Jiang, N., Niu, Q., Mei, S., Haugen, H., Ma, Q., 2023. Biocompatible nanostructured silver-incorporated implant surfaces show effective antibacterial, osteogenic, and anti-inflammatory effects in vitro and in rat model. Int. J. Nanomed. 18, 7359–7378. https://doi.org/10.2147/IJN.S435415.

Gao, Z., Eisenberg, A., 1993. A model of micellization for block copolymers in solutions. Macromolecules 26, 7353–7360. https://doi.org/10.1021/ma00078a035.

Harth, E., Horn, B. Van, Lee, V.Y., Germack, D.S., Gonzales, C.P., Miller, R.D., Hawker, C.J., 2002. A facile approach to architecturally defined nanoparticles via intramolecular chain collapse. J. Am. Chem. Soc. 124, 8653–8660. https://doi.org/10.1021/ja026208x.

Hataminia, F., Majidi, R.F., Najafi Tireh Shabankareh, A., Ghanbari, H., 2022. Green synthesis of oxidized starch with a novel catalyst based on Fe3O4 nanoparticles and H2O2 reagent to form thermoplastic as a stable gel coating on the cardiovascular stents. Int. J. Biol. Macromol. 219, 290–303. https://doi.org/10.1016/j.ijbiomac.2022.07.119.

He, D., Susanto, H., Ulbricht, M., 2009. Photo-irradiation for preparation, modification and stimulation of polymeric membranes. Prog. Polym. Sci. 34, 62–98. https://doi.org/10.1016/j.progpolymsci.2008.08.004.

Heuer-Jungemann, A., Feliu, N., Bakaimi, I., Hamaly, M., Alkilany, A., Chakraborty, I., Masood, A., Casula, M.F., Kostopoulou, A., Oh, E., Susumu, K., Stewart, M.H., Medintz, I.L., Stratakis, E., Parak, W.J., Kanaras, A.G., 2019. The role of ligands in the chemical synthesis and applications of inorganic nanoparticles. Chem. Rev. 119, 4819–4880. https://doi.org/10.1021/acs.chemrev.8b00733.

Hosseinpour-Moghadam, R., Rabbani, S., Mahboubi, A., Tabatabai, S.A., Haeri, A., 2021. Prevention of abdominal adhesion by a polycaprolactone/phospholipid hybrid film containing quercetin and silver nanoparticles. Nanomedicine 16, 2449–2464. https://doi.org/10.2217/nnm-2021-0209.

Hwang, P.T.J., Sherwood, J.A., Millican, R.C., Bobba, P.S., Lynd, T.O., Garner, J.N., Brott, B.C., Hou, D., Jun, H.-W., 2021. Endothelium-mimicking nanomatrix coating to enhance endothelialization after left atrial appendage closure device implantation. ACS Appl. Bio Mater. 4, 4917–4924. https://doi.org/10.1021/acsabm.1c00202.

Jones, M.-C., Leroux, J.-C., 1999. Polymeric micelles—a new generation of colloidal drug carriers. Eur. J. Pharm. Biopharm. 48, 101–111. https://doi.org/10.1016/S0939-6411(99)00039-9.

Karabasz, A., Bzowska, M., Szczepanowicz, K., 2020. Biomedical applications of multifunctional polymeric nanocarriers: a review of current literature. Int. J. Nanomed. 15, 8673–8696. https://doi.org/10.2147/IJN.S231477.

Kim, E.-M., Jeong, H.-J., 2021. Liposomes: biomedical applications. Chonnam Med. J. 57, 27. https://doi.org/10.4068/cmj.2021.57.1.27.

Kittana, N., Assali, M., Zimmermann, W.-H., Liaw, N., Santos, G.L., Rehman, A., Lutz, S., 2021. Modulating the biomechanical properties of engineered connective tissues by chitosan-coated multiwall carbon nanotubes. Int. J. Nanomed. 16, 989–1000. https://doi.org/10.2147/IJN.S289107.

Komsthöft, T., Bovone, G., Bernhard, S., Tibbitt, M.W., 2022. Polymer functionalization of inorganic nanoparticles for biomedical applications. Curr. Opin. Chem. Eng. 37, 100849. https://doi.org/10.1016/j.coche.2022.100849.

Kothamasu, P., Kanumur, H., Ravur, N., Maddu, C., Parasuramrajam, R., Thangavel, S., 2012. Nanocapsules: the weapons for novel drug delivery systems. Bioimpacts 2, 71–81. https://doi.org/10.5681/bi.2012.011.

Kumari, A., Singla, R., Guliani, A., Yadav, S.K., 2014. Nanoencapsulation for drug delivery. EXCLI J 13, 265–286.

Lakalayeh, G.A., Rahvar, M., Nazeri, N., Ghanbari, H., 2022. Evaluation of drug-eluting nanoparticle coating on magnesium alloy for development of next generation bioabsorbable cardiovascular stents. Med. Eng. Phys. 108, 103878. https://doi.org/10.1016/j.medengphy.2022.103878.

Lee, H., Hoang, B., Fonge, H., Reilly, R.M., Allen, C., 2010. In vivo distribution of polymeric nanoparticles at the whole-body, tumor, and cellular levels. Pharm. Res. (N. Y.) 27, 2343–2355. https://doi.org/10.1007/s11095-010-0068-z.

Letchford, K., Burt, H., 2007. A review of the formation and classification of amphiphilic block copolymer nanoparticulate structures: micelles, nanospheres, nanocapsules and polymersomes. Eur. J. Pharm. Biopharm. 65, 259–269. https://doi.org/10.1016/j.ejpb.2006.11.009.

Li, R., Liu, H., Shi, Q., Zhang, G., Pang, G., Xu, Y., Song, J., Lu, Y., 2024. An ascorbic acid-decorated nanostructured surface on titanium inhibits breast cancer development and promotes osteogenesis. Biomed. Mater. 19, 015006. https://doi.org/10.1088/1748-605X/ad0fa2.

Lin, P.H., Huang, C., Hu, Y., Ramanujam, V.S., Lee, E.-S., Singh, R., Milbreta, U., Cheung, C., Ying, J.Y., Chew, S.Y., 2023. Neural cell membrane-coated DNA nanogels as a potential target-specific drug delivery tool for the central nervous system. Biomaterials 302, 122325. https://doi.org/10.1016/j.biomaterials.2023.122325.

Liu, Y., Zhang, X., Wang, Y., Guo, M., Sheng, J., Wang, Y., Wu, T., 2023. Promoting neurite outgrowth and neural stem cell migration using aligned nanofibers decorated with protrusions and galectin-1 coating. Chem. Commun. 59, 10753–10756. https://doi.org/10.1039/D3CC02869K.

Loskutova, K., Torras, M., Zhao, Y., Svagan, A.J., Grishenkov, D., 2023. Cellulose nanofiber-coated perfluoropentane droplets: fabrication and biocompatibility study. Int. J. Nanomed. 18, 1835–1847. https://doi.org/10.2147/IJN.S397626.

Luo, L., Zhang, H., Zhang, S., Luo, C., Kan, X., Lv, J., Zhao, P., Tian, Z., Li, C., 2023. Extracellular vesicle-derived silk fibroin nanoparticles loaded with MFGE8 accelerate skin ulcer healing by targeting the vascular endothelial cells. J. Nanobiotechnol. 21, 455. https://doi.org/10.1186/s12951-023-02185-7.

Lyu, Z., Ding, L., Tintaru, A., Peng, L., 2020. Self-assembling supramolecular dendrimers for biomedical applications: lessons learned from poly(amidoamine) dendrimers. Acc. Chem. Res. 53, 2936–2949. https://doi.org/10.1021/acs.accounts.0c00589.

Ma, J., Guo, C., Tang, Y., Xiang, J., Chen, S., Wang, J., Liu, H., 2007. Micellization in aqueous solution of an ethylene oxide–propylene oxide triblock copolymer, investigated with 1H NMR spectroscopy, pulsed-field gradient NMR, and NMR relaxation. J. Colloid Interface Sci. 312, 390–396. https://doi.org/10.1016/j.jcis.2007.03.013.

Maan, A.M.C., Hofman, A.H., de Vos, W.M., Kamperman, M., 2020. Recent developments and practical feasibility of polymer-based antifouling coatings. Adv. Funct. Mater. 30. https://doi.org/10.1002/adfm.202000936.

Moore, T.L., Rodriguez-Lorenzo, L., Hirsch, V., Balog, S., Urban, D., Jud, C., Rothen-Rutishauser, B., Lattuada, M., Petri-Fink, A., 2015. Nanoparticle colloidal stability in cell culture media and impact on cellular interactions. Chem. Soc. Rev. 44, 6287–6305. https://doi.org/10.1039/C4CS00487F.

Mora-Huertas, C.E., Fessi, H., Elaissari, A., 2010. Polymer-based nanocapsules for drug delivery. Int. J. Pharm. 385, 113–142. https://doi.org/10.1016/j.ijpharm.2009.10.018.

Nagarajan, A., Rizwana, N., Abraham, M., Bhat, M., Vetekar, A., Thakur, G., Chakraborty, U., Agarwal, V., Nune, M., 2023. Polycaprolactone/graphene oxide/acellular matrix nanofibrous scaffolds with antioxidant and promyelinating features for the treatment of peripheral demyelinating diseases. J. Mater. Sci. Mater. Med. 34, 49. https://doi.org/10.1007/s10856-023-06750-2.

Nakamura, Y., Mochida, A., Choyke, P.L., Kobayashi, H., 2016. Nanodrug delivery: is the enhanced permeability and retention effect sufficient for curing cancer? Bioconjugate Chem. 27, 2225–2238. https://doi.org/10.1021/acs.bioconjchem.6b00437.

Nie, H., Wu, Q., Wu, J., Cai, K., Shen, Y., Tang, C., 2023. Bioinspired synthesis of Ag nanoparticles onto polytetrafluoroethylene with enhanced antibacterial activity for dental implant application. Discov. Med. 35, 1104. https://doi.org/10.24976/Discov.Med.202335179.107.

Pattni, B.S., Chupin, V.V., Torchilin, V.P., 2015. New developments in liposomal drug delivery. Chem. Rev. 115, 10938–10966. https://doi.org/10.1021/acs.chemrev.5b00046.

Pawar, P.V., Gohil, S.V., Jain, J.P., Kumar, N., 2013. Functionalized polymersomes for biomedical applications. Polym. Chem. 4, 3160. https://doi.org/10.1039/c3py00023k.

Pinto Reis, C., Neufeld, R.J., Ribeiro, A.J., Veiga, F., 2006. Nanoencapsulation I. Methods for preparation of drug-loaded polymeric nanoparticles. Nanomedicine 2, 8–21. https://doi.org/10.1016/j.nano.2005.12.003.

Prabhakar, U., Maeda, H., Jain, R.K., Sevick-Muraca, E.M., Zamboni, W., Farokhzad, O.C., Barry, S.T., Gabizon, A., Grodzinski, P., Blakey, D.C., 2013. Challenges and key considerations of the enhanced permeability and retention effect for nanomedicine drug delivery in oncology. Cancer Res. 73, 2412–2417. https://doi.org/10.1158/0008-5472.CAN-12-4561.

Rahnfeld, L., Luciani, P., 2020. Injectable lipid-based depot formulations: where do we stand? Pharmaceutics 12, 567. https://doi.org/10.3390/pharmaceutics12060567.

Ramburrun, P., Khan, R.A., Choonara, Y.E., 2022. Design, preparation, and functionalization of nanobiomaterials for enhanced efficacy in current and future biomedical applications. Nanotechnol. Rev. 11, 1802–1826. https://doi.org/10.1515/ntrev-2022-0106.

Rodriguez-Emmenegger, C., Kylián, O., Houska, M., Brynda, E., Artemenko, A., Kousal, J., Alles, A.B., Biederman, H., 2011. Substrate-independent approach for the generation of functional protein resistant surfaces. Biomacromolecules 12, 1058–1066. https://doi.org/10.1021/bm101406m.

Rostamizadeh, K., Torchilin, V.P., 2020. Polymeric nanomicelles as versatile tool for multidrug delivery in chemotherapy. In: Nanopharmaceuticals. Elsevier, pp. 45–72. https://doi.org/10.1016/B978-0-12-817778-5.00003-8.

Salehi, S., Tavakoli, M., Mirhaj, M., Varshosaz, J., Labbaf, S., Karbasi, S., Jafarpour, F., Kazemi, N., Salehi, S., Mehrjoo, M., Emami, E., 2023. A 3D printed polylactic acid-Baghdadite nanocomposite scaffold coated with microporous chitosan-VEGF for bone regeneration applications. Carbohydr. Polym. 312, 120787. https://doi.org/10.1016/j.carbpol.2023.120787.

Sanchez-Cano, C., Carril, M., 2020. Recent developments in the design of non-biofouling coatings for nanoparticles and surfaces. Int. J. Mol. Sci. 21, 1007. https://doi.org/10.3390/ijms21031007.

Song, M., Xing, J., Cai, H., Gao, X., Li, C., Liu, C., Li, X., Fu, X., Ding, S., Cheng, W., Chen, R., 2023. Pomegranate-bionic encapsulating horseradish peroxidase using dopamine flexible scaffold-coated multishell porous ZIF-8 to enhance immunochromatographic diagnosis. ACS Nano 17, 10748−10759. https://doi.org/10.1021/acsnano.3c02164.

Soppimath, K.S., Aminabhavi, T.M., Kulkarni, A.R., Rudzinski, W.E., 2001. Biodegradable polymeric nanoparticles as drug delivery devices. J. Contr. Release 70, 1−20. https://doi.org/10.1016/S0168-3659(00)00339-4.

Sun, M., Hu, H., Sun, L., Fan, Z., 2020. The application of biomacromolecules to improve oral absorption by enhanced intestinal permeability: a mini-review. Chin. Chem. Lett. 31, 1729−1736. https://doi.org/10.1016/j.cclet.2020.02.035.

Svenson, S., Tomalia, D.A., 2012. Dendrimers in biomedical applications—reflections on the field. Adv. Drug Deliv. Rev. 64, 102−115. https://doi.org/10.1016/j.addr.2012.09.030.

Svinterikos, E., Zuburtikudis, I., Abu Khalifeh, H., Farvin Akbar Ali, S., 2023. Multifunctional polymer-based coatings for outdoor glass surfaces: a state of the art. Adv. Ind. Eng. Polym. Res. 6, 310−332. https://doi.org/10.1016/j.aiepr.2023.04.001.

Tang, Z., He, C., Tian, H., Ding, J., Hsiao, B.S., Chu, B., Chen, X., 2016. Polymeric nanostructured materials for biomedical applications. Prog. Polym. Sci. 60, 86−128. https://doi.org/10.1016/j.progpolymsci.2016.05.005.

Teare, D.O.H., Schofield, W.C.E., Roucoules, V., Badyal, J.P.S., 2003. Substrate-independent growth of micropatterned polymer brushes. Langmuir 19, 2398−2403. https://doi.org/10.1021/la020279s.

Teekamp, N., Duque, L.F., Frijlink, H.W., Hinrichs, W.L., Olinga, P., 2015. Production methods and stabilization strategies for polymer-based nanoparticles and microparticles for parenteral delivery of peptides and proteins. Expert Opin. Drug Deliv. 12, 1311−1331. https://doi.org/10.1517/17425247.2015.1003807.

Tomalia, D.A., Baker, H., Dewald, J., Hall, M., Kallos, G., Martin, S., Roeck, J., Ryder, J., Smith, P., 1985. A new class of polymers: starburst-dendritic macromolecules. Polym. J. 17, 117−132. https://doi.org/10.1295/polymj.17.117.

van der Ende, A.E., Harrell, J., Sathiyakumar, V., Meschievitz, M., Katz, J., Adcock, K., Harth, E., 2010. "Click" reactions: novel chemistries for forming well-defined polyester nanoparticles. Macromolecules 43, 5665−5671. https://doi.org/10.1021/ma100711b.

van der Ven, C.F.T., Tibbitt, M.W., Conde, J., van Mil, A., Hjortnaes, J., Doevendans, P.A., Sluijter, J.P.G., Aikawa, E., Langer, R.S., 2021. Controlled delivery of gold nanoparticle-coupled miRNA therapeutics *via* an injectable self-healing hydrogel. Nanoscale 13, 20451−20461. https://doi.org/10.1039/D1NR04973A.

Vasudevan, A., Majumder, N., Sharma, I., Kaur, I., Sundarrajan, S., Venugopal, J.R., Vijayaraghavan, P., Singh, N., Ramakrishna, S., Ghosh, S., M Tripathi, D., Kaur, S., 2023. Liver extracellular matrix-based nanofiber scaffolds for the culture of primary hepatocytes and drug screening. ACS Biomater. Sci. Eng. 9, 6357−6368. https://doi.org/10.1021/acsbiomaterials.3c01216.

Vauthier, C., Bouchemal, K., 2009. Methods for the preparation and manufacture of polymeric nanoparticles. Pharm. Res. (N. Y.) 26, 1025−1058. https://doi.org/10.1007/s11095-008-9800-3.

Wang, W., Mattoussi, H., 2020. Engineering the bio−nano interface using a multifunctional coordinating polymer coating. Acc. Chem. Res. 53, 1124−1138. https://doi.org/10.1021/acs.accounts.9b00641.

Wang, X., Zhong, X., Li, J., Liu, Z., Cheng, L., 2021. Inorganic nanomaterials with rapid clearance for biomedical applications. Chem. Soc. Rev. 50, 8669−8742. https://doi.org/10.1039/D0CS00461H.

Wei, Q., Becherer, T., Angioletti-Uberti, S., Dzubiella, J., Wischke, C., Neffe, A.T., Lendlein, A., Ballauff, M., Haag, R., 2014. Protein interactions with polymer coatings and biomaterials. Angew. Chem. Int. Ed. 53, 8004–8031. https://doi.org/10.1002/anie.201400546.

Wei, Q., Haag, R., 2015. Universal polymer coatings and their representative biomedical applications. Mater. Horiz. 2, 567–577. https://doi.org/10.1039/C5MH00089K.

Yıldırım, M., Acet, Ö., Yetkin, D., Acet, B.Ö., Karakoc, V., Odabası, M., 2022. Anti-cancer activity of naringenin loaded smart polymeric nanoparticles in breast cancer. J. Drug Deliv. Sci. Technol. 74, 103552. https://doi.org/10.1016/j.jddst.2022.103552.

Yue, F., Ayaz, Z., Jiang, Y., Xiang, L., Huang, N., Leng, Y., Akhavan, B., Jing, F., 2024. Dealloyed nano-porous Ti Cu coatings with controlled copper release for cardiovascular devices. Biomater. Adv. 157, 213728. https://doi.org/10.1016/j.bioadv.2023.213728.

Zhang, Y., Cui, J., Chen, K.-Y., Kuo, S.H., Sharma, J., Bhatta, R., Liu, Z., Ellis-Mohr, A., An, F., Li, J., Chen, Q., Foss, K.D., Wang, H., Li, Y., McCoy, A.M., Lau, G.W., Cao, Q., 2023. A smart coating with integrated physical antimicrobial and strain-mapping functionalities for orthopedic implants. Sci. Adv. 9. https://doi.org/10.1126/sciadv.adg7397.

Zheng, L., Sundaram, H.S., Wei, Z., Li, C., Yuan, Z., 2017. Applications of zwitterionic polymers. React. Funct. Polym. 118, 51–61. https://doi.org/10.1016/j.reactfunctpolym.2017.07.006.

Zhu, Y., Ma, X., Li, L., Yang, Q., Jin, F., Chen, Z., Wu, C., Shi, H., Feng, Z., Yin, S., Li, C., 2023. Surface functional modification by $Ti_3C_2T_x$MXene on PLLA nanofibers for optimizing neural stem cell engineering. Adv. Healthcare Mater. 12. https://doi.org/10.1002/adhm.202300731.

Ziąbka, M., Matysiak, K., Cholewa-Kowalska, K., Kyzioł, A., Królicka, A., Sapierzyński, R., Januchta-Kurmin, M., Bissenik, I., 2023. In vitro and in vivo studies of antibacterial coatings on titanium alloy implants for veterinary application. Int. J. Mol. Sci. 24, 8114. https://doi.org/10.3390/ijms24098114.

Nanostructured coatings for biosensing applications in surface plasmon–coupled emission (SPCE) interface

10

Kalathur Mohan Ganesh [1], Seemesh Bhaskar [2] and Sai Sathish Ramamurthy [3,4]
[1]STAR Laboratory, Department of Chemistry, Sri Sathya Sai Institute of Higher Learning, Prasanthi Nilayam, Anantapur, Andhra Pradesh, India; [2]University of Illinois at Urbana-Champaign, Carl R. Woese Institute for Genomic Biology, Department of Electrical and Computer Engineering, Nick Holonyak Jr. Micro and Nanotechnology Laboratory (HMNTL), Urbana, IL, United States; [3]Sri Sathya Sai Institute of Higher Learning, Faculty, Department of Chemistry, Principle Investigator, STAR Lab, CRIF, Puttaparthi, India; [4]University of Maryland Baltimore County, Center for Advanced Sensor Technology, Department of Chemical, Biochemical and Environmental Engineering, Baltimore, MD, United States

10.1 Introduction

Various types of biosensing modalities such as colorimetric, electrochemical, thermal, gravimetric, piezoelectric, acoustic, electro-optic, lateral flow, and optical biosensors to name a few have been greatly refined with the advent of nanostructured material–based coatings and nanoscale thin-film coatings (Barbillon, 2017; Abdulhalim et al., 2008; Mejía-Salazar and Oliveira, 2018). Nanomaterials are used as labels in bioassays and imaging and as transducers in biosensors due to their unique characteristic to respond drastically to local changes in the environment. The absorbance and scattering profiles that are dependent on material properties, charge, size, shape, and refractive index of the local environment characterize the functional properties of any nanomaterial. Covalent, electrostatic, and adsorption (chemisorption and physisorption) of molecules to the plasmonic nanomaterials results in shift (batho-, hypso-, hypo-, and hyperchromic) of their spectral signatures enabling the researchers to detect the changes made in real time (Abdulhalim et al., 2008; Bauch et al., 2014). The use of nanomaterials has even enabled reducing the steps needed in biofunctionalization with concomitant reduction in the analysis time and cost. Moreover, the escalating advancements in the nano- and microfabrication protocols have tremendously encouraged the development of metal-based, dielectric-based, graphene-based, silicon-based plasmonic coating, arrays and interfaces for the biosensing applications including food safety analysis, environmental, and biological health monitoring (Bhaskar et al., 2021a,b,c; Cheerala et al., 2023; Chen et al., 2023; Ganesh et al., 2007; Alharbi et al., 2019).

Multifunctional Nanostructured Coatings. https://doi.org/10.1016/B978-0-443-23683-9.00010-6

Plasmonic biosensors have been developed for the detection of several analytes including enzymes, antibodies, aptamers, DNA, RNA, lectins, exosomes, ctDNA, nucleic acids, and cells (bacterial and cancer) (Xiong et al., 2023; Barya et al., 2023; Chauhan et al., 2023; Bhaskar and Ramamurthy, 2019; Thacharakkal et al., 2024). Two different methods have been incorporated in successfully scrutinizing the qualitative and quantitative information, while plasmonic nanomaterials are used as transducers. Firstly, different microscopic configurations are utilized to image the functional binding events, and necessary information to analyze the results is hence developed. Secondly, spectroscopic analysis is performed where the field enhancement property of the plasmonic nanomaterials is used to amplify the signal intensity of the binding event. It is important to note that plasmonic nanomaterial–based coatings and thin films have been extensively utilized for imaging and spectroscopy-based biosensing applications (Barbillon, 2017; Boriskina et al., 2017; Lakowicz, 2006; Stockman et al., 2018).

Specifically, nanostructured coatings have attracted the attention of researchers in physics, chemistry, biology, and materials science due to their high surface area and remarkable functional properties. The broad domain of photo-plasmonics benefits from such explorations as nanoparticle-based coatings as well as metal and dielectric thin film coatings is extensively explored for myriad applications. Under optical illumination, the plasmonic materials (such as Ag, Au, Pd, Pt, etc.) support charge density oscillations or electron density perturbations generating the quasiparticle called plasmon. These plasmons, while generated on metal nanoparticles, are typically localized in space, and the effect observed is hence termed as localized surface plasmon resonance (LSPR) (Li et al., 2015; Lakowicz et al., 2008). On the other hand, while the plasmons are propagating at the surface as observed in the case of semi-infinite metal-dielectric nanointerfaces, they are referred to as propagating surface plasmon polaritons (SPPs) (where the polariton is the coupled oscillation of the plasmon and the incident light wave) (Li et al., 2015; Jiang et al., 2018). Optical excitation, the field intensity near the nanoparticle-based coatings, as well as the metal thin film coatings are significantly enhanced beyond the diffraction limit. With this as the background, plasmonic as well as high refractive index dielectric nanomaterials as well as their coatings are extensively researched to manipulate the local environment of the bioanalytes of interest that needs to be quantified (Bhaskar et al., 2021a,b,c; Sudha Maria Lis et al., 2023; Rai et al., 2022b). Very minute changes in the nanomaterials' surrounding environment and their coatings result in a significant change in the system's optical response, in terms of absorbance shift or intensity change. Hence, nanostructured coatings based on nanomaterials and thin films are considered ideal candidates for biosensing applications. While the nanostructured coatings can be as such used as the nanoantenna coatings to record changes in the biomolecular events at the interface, the nanostructured thin film coatings are often integrated with coupler technologies: (i) prism and (ii) grating to achieve high sensitivity of the device (Lakowicz, 2006; Li et al., 2015; Anker et al., 2008).

Among the several available approaches adopted for development of biosensing technologies, the prism-based technique has gained major attraction in the past few decades. Following the first works by Otto, Kretchmann, and Reather, numerous biosensing frameworks have been developed for detection of ions, molecules, DNA, RNA, proteins, lipids, and exosomes, to name a few at metal-dielectric interface (Kurihara and Suzuki, 2002). Although numerous optical methods including absorbance-based, fluorescence-based, and photoelectrochemical-based have been established for the sensing of such entities, the luminescence-based methods (fluorescence, chemiluminescence, and phosphorescence) based methods stand out on account of the high sensitivity. Plasmonic materials have been investigated with several fluorescent materials to demonstrate reliable and robust biosensing platforms (Bhaskar, 2023; Badugu et al., 2015; Bhaskar and Ramamurthy, 2021a,b; Rai et al., 2022a). Among several such methodologies, surface plasmon-coupled emission (SPCE) technology developed by Lakowicz and co-workers in the early 2000s based on metal-dielectric coatings has gained tremendous utility in the development of biosensor modalities (Lakowicz, 2005, 2004; Badugu et al., 2014; Dutta Choudhury et al., 2015). This chapter focuses on presenting the latest developments in the domain of SPCE technology where we present case studies elaborating the utility of nanostructured coatings in establishing the desired biosensing platform. We attempt to introduce effective strategies to augment the performance of the existing SPCE-based sensing technology. In this perspective, the readers are introduced to the fundamental fabrication methods of nanostructured coatings in SPCE including the spacer, cavity, and ext. cavity interfacial nanoengineering, fluorescence enhancements and their biosensing applications.

10.2 Metal-enhanced fluorescence (MEF)

The property of the materials at the nanoscale is significantly different from that in the macroscopic scale. The incidence of certain light frequencies over the plasmonic NPs results in the generation of resonant charge density oscillations called plasmons. The characteristics of the plasmons depend on the shape, size, chemical composition, and the properties of the local environment. These characteristics result in strong absorption and scattering of light (which is called overall extinction), thereby enabling scientists to image the nanomaterials in the dark field optical microscopes. It is also important to note that the scattering cross-section of plasmonic NPs is million-fold higher than that of typical fluorescence molecules (Anker et al., 2008). Consequently, plasmonic NPs are utilized as powerful labels for the development of biosensing techniques that are dependent on surface-enhanced spectroscopy and imaging. Moreover, such properties enable them to be useful candidates for fabrication of optical switches, waveguides, and novel lithographic platforms. While this is the functionality of the plasmonic NPs by itself, numerous bioassay approaches have been developed by combining the plasmonic NPs with fluorescence entities such as conventional fluorescent molecules, fluorescent nanoclusters, quantum dots, and nanodiamonds, to name a few (Bauch et al., 2014; Strianese et al., 2012; Lakowicz, 2006). Such attempts have resulted in generation of numerous highly sensitive biosensing platforms including plasmon-enhanced imaging microscopy, ultra-high time-resolved fluorescence spectroscopy, time-correlated single-photon counting, tip-enhanced photoluminescence spectroscopy, surface plasmon-coupled Raman and fluorescence technology, shell-isolated NP enhanced fluorescence, tip-enhanced near-field optical microscopy, and advanced multiphoton fluorescence (Anker et al., 2008; Jahani and Jacob, 2016; Barnes et al., 2003; Chikkaraddy et al., 2016). By and large, it has been observed that placing fluorescent moieties in the close vicinity of plasmonic NPs results in enhanced excitation and enhanced emission (Bhaskar et al., 2021a,b,c; Che et al., 2022; Ganesh et al., 2023a,b). Moreover, sharp nanoarchitectures in plasmonic nanomaterials are demonstrated to present high electric field intensity due to the lightning rod effect (Bhaskar et al., 2020a,b,c,d; Niu et al., 2015). Such high field enhancements result in high fluorescence quantum yield with concomitant reduction in the lifetime of the molecules (Lakowicz et al., 2002; Bhaskar et al., 2023a,b). In other words, plasmon-enhanced fluorescence or metal-enhanced fluorescence platforms result in high photostability, high radiative decay rate, and lower lifetime (Gryczynski et al., 2004). A conceptual schematic of the MEF occurring at the interface between the metal NPs and the radiating dipole is shown as part of Fig. 10.1a.

10.3 SPR and SPCE

Surface plasmon resonance is the phenomenon that occurs when the incoming light resonates with the surface plasmons of the metal thin film. However, for SPR to occur at metal thin film interfaces, the phase matching conditions needs to be satisfied. In other words, for matching the momentum of the light line at free space (FS) to the momentum of the SPPs over the metal thin films, the wave vector of light in FS needs to be increased. This is accomplished by using a number of configurations such as Otto and Kretchmann-Raether (KR) configurations as well as with the use of metallic gratings. Predominantly explored configuration is the KR configuration where the light is incident on the metal film through a prism. The detector is placed on the curved surface of the prism side, and the SPPs are excited by exciting the metal thin film at a particular angle. At the angle that generates the SPPs, a sharp dip in the reflectance is observed at the detector (Lakowicz, 2006, 2004). With the slight changes in the refractive index of the metal-dielectric interface, the angle that generates the sharp dip in the reflectance changes. Concomitantly, the percentage of the dip in the resonance is also observed in certain cases. Using such information, the researchers in the domain of plasmonics have developed promising SPR based sensors for the detection of different analytes. The SPCE technology uses the core principles of SPR itself, however with the incorporation of fluorescent moieties in the dielectric nanointerface (Lakowicz, 2004, 2006; Bhaskar, 2023). Such incorporation results in the coupling of the fluorescence signal with the SPPs of the metal thin film and the emitted photons carry the characteristics of the fluorescent moieties as well as the metal thin film to the far-field. The primary advantage of this methodology (experimental design) is that it allows the experimentalists to perform the sensing in Kretchmann as well as reverse Kretchmann (RK) configurations (Cao et al., 2012).

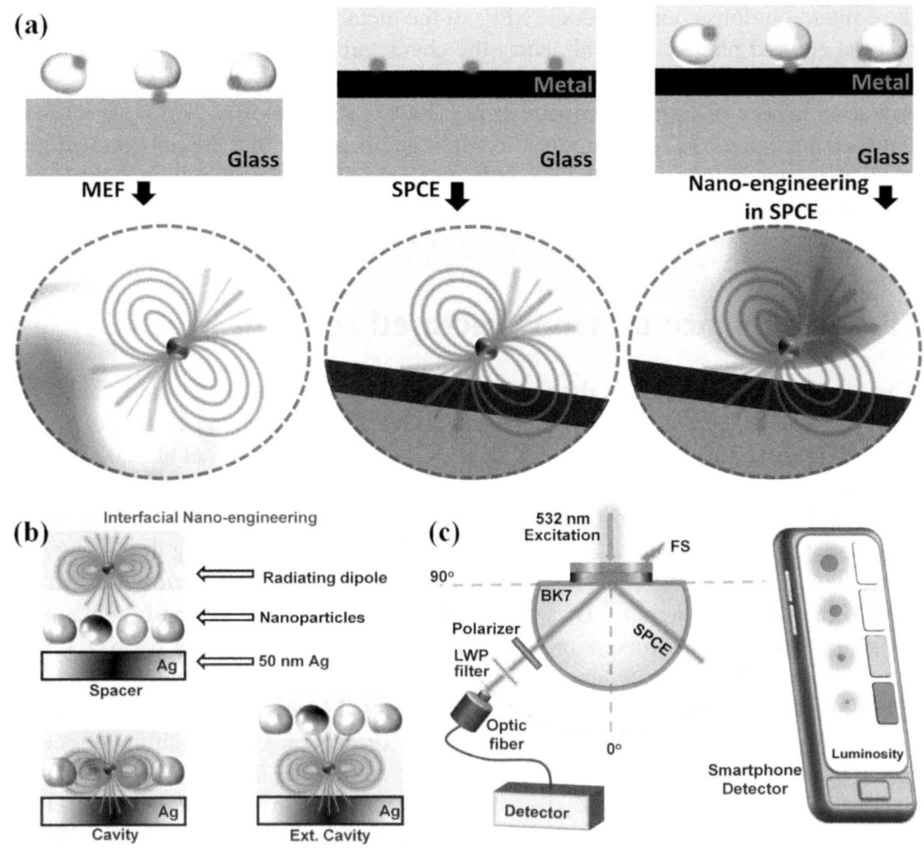

Figure 10.1 Conceptual schematic. (a) Conceptual schematic of metal-enhanced fluorescence (MEF), surface plasmon−coupled emission (SPCE), and nanoengineering in SPCE technology. (b) Spacer, cavity, and ext. Cavity nanointerfaces explored in SPCE technology, and (c) SPCE optical framework.

Reproduced with permission Ganesh, K.M., Rai, A., Battampara, P., Reddy, R., Bhaskar, S., Reddy, N., Ramamurthy, S.S., 2023a. Optical coupling of bio-inspired mustard protein-based bimetallic nanohybrids with propagating surface plasmon polaritons for femtomolar nitrite ion sensing: cellphone-based portable detection device. Nano-Struct. Nano-Objects 35. https://doi.org/10.1016/j.nanoso.2023.101025; Ganesh, K.M., Rai, A., Bhaskar, S., Reddy, N., Ramamurthy, S.S., 2023b. Plasmon-enhanced fluorescence from synergistic engineering of graphene oxide and sharp-edged silver nanorods mediated with castor protein for cellphone-based attomolar sensing. J. Lumin. 260. https://doi.org/10.1016/j.jlumin.2023.119835.

A typical optical configuration of the SPCE setup is shown in Fig. 10.1c. In brief, the laser is incident from the FS side and not the prism side in the RK configuration. A long-wave pass filter is used to cut the laser line reaching the detector. The radiating dipoles that are situated over the metal thin film are excited by the laser radiation, and the emitted photons couple with the SPPs of the metal thin film. It is important to note that the emitted photons not only carry the characteristics of the radiating dipoles (or fluorophores) but also take with them the characteristics of the SPPs, which is the polarization aspect of light (Lakowicz et al., 2008). SPPs selectively couple with the transverse magnetic (TM) component of light and it is seen that the out-coupled radiation on the far-field (prism-side) is highly polarized (TM) as well as highly directional (Lakowicz, 2005).

10.4 Relevance of nanostructured coatings in SPCE

The conceptual schematic of the spacer, cavity, and ext. cavity nanointerfaces typically studied in the SPCE technology development is presented in Fig. 10.1b (Xie et al., 2021a,b; Bhaskar et al., 2020a,b,c,d; Cao et al., 2014; Rai et al., 2021a,b). In the spacer nanointerface, the plasmonic/dielectric NPs function as the active spacer material between the metal thin film and the radiating dipoles. Depending on the functionality of the spacer material, the emission from the radiating dipole is either quenched or enhanced (Bhaskar and Ramamurthy, 2021a,b; Srinivasan et al., 2016). In order to generate cavity interface, the radiating dipoles and the nanomaterials of interest are admixed and spin coated as a single layer over the SPCE platform. This results in the encapsulation of radiating dipole molecules in the infinitesimal nanogap regions between the metal thin film and the nanomaterial. The cavity nanoengineering suffers from the drawback of quenching effects as the NPs and the radiating dipoles are admixed in the first place (Kongsuwan et al., 2020; Rai et al., 2022c; Bhaskar and Ramamurthy, 2023; Bhaskar et al., 2020a,b,c,d). However, this has been overcome with the use of dielectric NPs in place of metal NPs as well as with the use of a similar interface termed ext. cavity nanointerface. This is virtually the opposite of spacer nanointerface, which in turn renders the NPs and the radiating dipoles in two separate nanolayers. However, since the radiating dipoles continue to exist in the extended region of the cavity hotspot region, this interface is called ext. cavity nanointerface (Rai et al., 2022b,c; Rathnakumar et al., 2021). Different groups have investigated the merits and demerits of different nanointerfaces, and it is worth noting that the final performance of particular nanomaterial in SPCE platform depends on several factors including size, shape, porosity, internal nanovoids and nanocavities, magnetic, ferroelectric and piezoelectric properties (Sudha Maria Lis et al., 2023; Rai et al., 2022b; Cao et al., 2014; Venkatesh et al., 2015; Xie et al., 2022; Liu et al., 2015; Cao et al., 2019; Mulpur et al., 2013). The detection in SPCE setup can be performed using the Ocean optics detector and smartphone-based cost-effective detector system. The emission from the SPCE platform is captured using a smartphone camera and processed using a Color Grab app that is freely available in the Google play store.

10.5 Applications of nanostructured coatings for biosensing in SPCE technology

10.5.1 Spacer nanoengineering

As the first case study, we present the functionality of gold nanoparticles (AuNPs) and gold nanostars (AuNSs) in quenching and dequenching the fluorescence signal intensity. The plasmonic metal NPs are largely plagued by parasitic dissipative Ohmic losses due to electron scattering at the lattice of metal, leading to predominant nonradiative decay and quenching (Boriskina et al., 2017; Khurgin, 2015). In comparison to Ag, Al, Cu, and alkali metals, Au exhibits greater chemical stability with excellent performance in visible and near-infrared (NIR) spectral regions. The added advantage of biocompatibility and diverse surface functionalities has rendered them suitable for several biomedical applications.

On irradiation of light, AuNPs like other NPs scatter and absorb. While the incident light reradiates in all directions (at the same frequency) accounts for scattering, the lattice vibrations within the crystal lattice results in dissipative heat losses through absorption (Li et al., 2015; Barnes et al., 2003; Bhaskar et al., 2022a,b). Both these processes work in tandem and reduce the intensity of incident EM-field and are identified as extinction. The enhanced quenching of commonly examined fluorophores such as rhodamine dyes, observed with the usage of AuNPs (at distance <5 nm) is a fundamental caveat for exploiting the utility of AuNPs in SPCE (Kang et al., 2011; Rai et al., 2021a,b; Li et al., 2019). In order to address this limitation, nanostructures that overcome quenching using AuNPs and yield multifold hotspots and EM-field confinement needs to be deployed. NPs with sharp edges, crevices, tips, and voids are known to entrap the incident light energy to subwavelength regions. Although elegant nano-geometries such as nanocubes, nanotriangles, and nanorods assist in this requirement, nanostars (NSs) standout with additional benefits. Variety of NSs such as "multi-branched," "sea urchin-like," "multispiked," "star-shaped" for different interdisciplinary applications has been synthesized (Bhaskar et al., 2020a,b,c,d). Although the "lightning rod effect" supported by nanorods exceptionally enhances the local electric-field, morphologically, similar effects can be obtained from NSs.

The overall dipole moment of the tip plasmons is maximized on account of coupling between the tip and core plasmons (acting as nanoantennas), which in turn increases the excitation cross section and amplifies EM field intensity in their vicinity. Moreover, the NSs present themselves with a rare combination of sharper nanoedges and crevices with pronounced anisotropy making them ideal candidates for achieving template-free dequenching in SPCE even with the usage of gold as the substrate (Bhaskar et al., 2020a,b,c,d). AuNSs have been investigated to understand its efficiency in dequenching the quenched fluorescence signal from gold nanospheres counterparts with intriguing results. The AuNPs synthesized using Turkevich method, displayed the SPR characteristic of AuNPs in the visible region with a λ_{max} at 530 nm (Fig. 10.2a) and AuNSs showed the SPR band with a λ_{max} at $\sim$670 nm

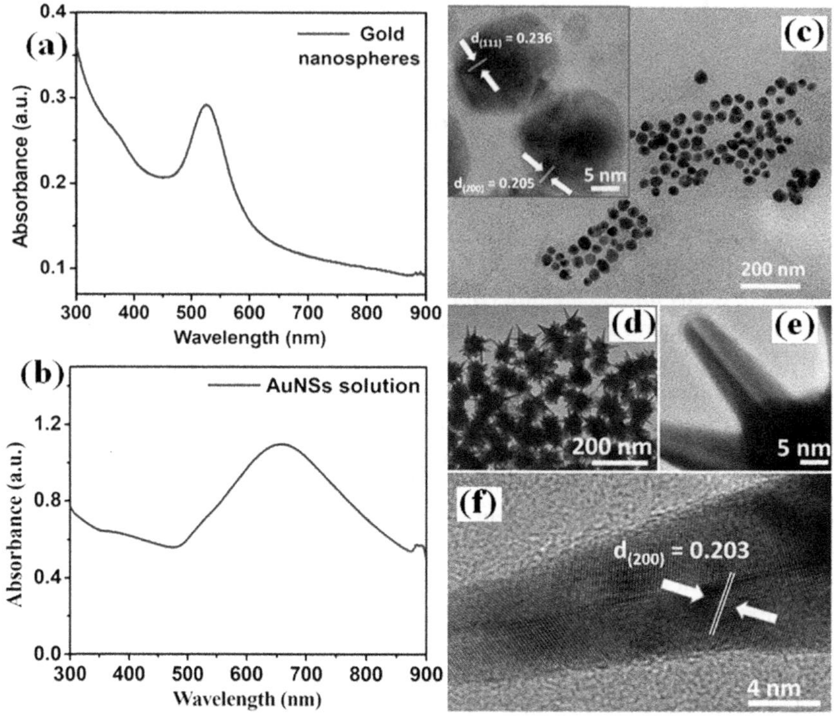

Figure 10.2 UV-vis absorbance, TEM and HRTEM characterization of gold nanospheres and nanostars. (a) UV-vis absorbance profile of AuNPs. (b) UV-vis absorbance profile of AuNSs. (c) TEM image of synthesized AuNPs and HRTEM image of AuNPs indicating the lattice fringes characteristic to gold (inset) (c), (d) and (e): TEM images of synthesized AuNSs. (f) HRTEM image of AuNSs indicating the lattice fringes along with d-spacing and Miller indices characteristic to gold.

Reproduced with permission Bhaskar, S., Kowshik, N.C.S.S., Chandran, S.P., Ramamurthy, S.S., 2020a. Femtomolar detection of spermidine using Au decorated SiO2 nanohybrid on plasmon-coupled extended cavity nanointerface: a smartphone-based fluorescence dequenching approach. Langmuir 36 (11), 2865–2876. https://doi.org/10.1021/acs.langmuir.9b03869; Bhaskar, S., Patra, R., Kowshik, N.C.S.S., Ganesh, K.M., Srinivasan, V., Chandran, S.P., Ramamurthy, S.S., 2020b. Nanostructure effect on quenching and dequenching of quantum emitters on surface plasmon-coupled interface: a comparative analysis using gold nanospheres and nanostars. Phys. E Low-dimens. Syst. Nanostruct. 124. https://doi.org/10.1016/j.physe.2020.114276; Bhaskar, S., Moronshing, M., Srinivasan, V., Badiya, P.K., Subramaniam, C., Ramamurthy, S.S., 2022c. Silver soret nanoparticles for femtomolar sensing of glutathione in a surface plasmon-coupled emission platform. ACS Appl. Nano Mater. 3 (5), 4329–4341. https://doi.org/10.1021/acsanm.0c00470; Bhaskar, S., Singh, A.K., Das, P., Jana, P., Kanvah, S., Bhaktha, S.B.N., Ramamurthy, S.S., 2020d. Superior resonant nanocavities engineering on the photonic crystal-coupled emission platform for the detection of femtomolar iodide and Zeptomolar Cortisol. ACS Appl. Mater. Interf. 12 (30), 34323–34336. https://doi.org/10.1021/acsami.0c07515.

(Fig. 10.2b). The TEM image shown in Fig. 10.2c presents gold nanospheres that are uniform in size with an average size of 16 ± 2 nm. HRTEM imaging further confirms the formation of polycrystalline AuNPs: lattice fringes observed with d-spacings of 2.36 and 2.05 Å correspond to the (111) and (200) Bragg planes of FCC gold. HRTEM images in Figs. 10.2d and e also clearly reveal the star shaped morphology of the reduced gold (AuNSs). Lattice fringes, as seen in the HRTEM image in Fig. 10.2f of a sharp arm of one of the AuNSs, clearly reveal d-spacing corresponding to (200) Bragg planes of FCC gold (2.03 Å), and confirms the formation of AuNSs. The experimentally obtained emission enhancements from AuNSs is ~100-fold and 10 orders of magnitude greater than that realized with AuNPs/gold nanospheres (sixfold). The seed less method as reported by Cheng et al., was adopted for synthesis of NSs of gold resulting in multiple nanospikes (Bhaskar et al., 2020a,b,c,d).

Xie and co-workers utilized the gold nanocages (AuNCs) to generate high field confinement and fluorescence enhancement with factors over 150 and 600 compared to the normal SPCE and FS emission respectively (Xie et al., 2021a,b). This was accomplished by using a thin 10 nm fluorophore layer over the SPCE substrate, thereby presenting an effective approach to dequench the otherwise quenched induced by resonance energy transfer in normal SPCE signal. This enhancement was proven to be triggered by the superior wavelength match, the enhanced electromagnetic field, and new radiation channel and process induced by the AuNC assembly, which provides an opportunity to increase the detection sensitivity and establish an optimal plasmonic enhancement system (Xie et al., 2021a,b).

A schematic diagram of the experiment is shown in Fig. 10.3a. The SPCE signal was measured in the reverse Kretschmann configuration, and the quantum dots (QDs) were excited by incident light (532 nm) to emit the signal through a prism. The emission of original fluorescence from the substrate side is called free space (FS) emission. A typical result for AuNC-enhanced SPCE is shown in Fig. 10.3b, thereby demonstrating the strong localized surface plasmon coupling of AuNCs in the SPCE system. The metal NPs could confine photons to form electron oscillations, which could generate a strong EM field around them. Recently, different simulation tools such as FDTD, Comsol Multiphysics, discrete dipole approximation (DDA) to name a few have been used to evaluate the EM hotspot intensity in the nanogaps generated by the plasmonic nanomaterial. The FDTD simulations (based on Lumerical Solutions) were adopted to understand the near-field EM field patterns of a single AuNC with and without the gold substrate. From Fig. 10.3c–f, we observe that a significant enhancement in the field intensity is observed not only in the outside of the shell but also on the interior nano-gap-based hotspot (Xie et al., 2021a,b). Such coupling of fields from both the plasmonic hotspots within and outside the nanoshell is expected to yield dramatically high fluorescence enhancements due to the mixing or hybridization of modes, resulting in higher local density of states. The high field enhancement thus obtained was used by the Xie and coworkers for development of multiwavelength detection system. This was accomplished by using AuNSs and AuNCs that were mixed with quantum dots solution (Xie et al., 2021a,b). It was observed that without the incorporation of the NPs in to the QDs solution, the multiwavelength detection system was not effective. However, upon mixing the plasmonic NPs of different aspect

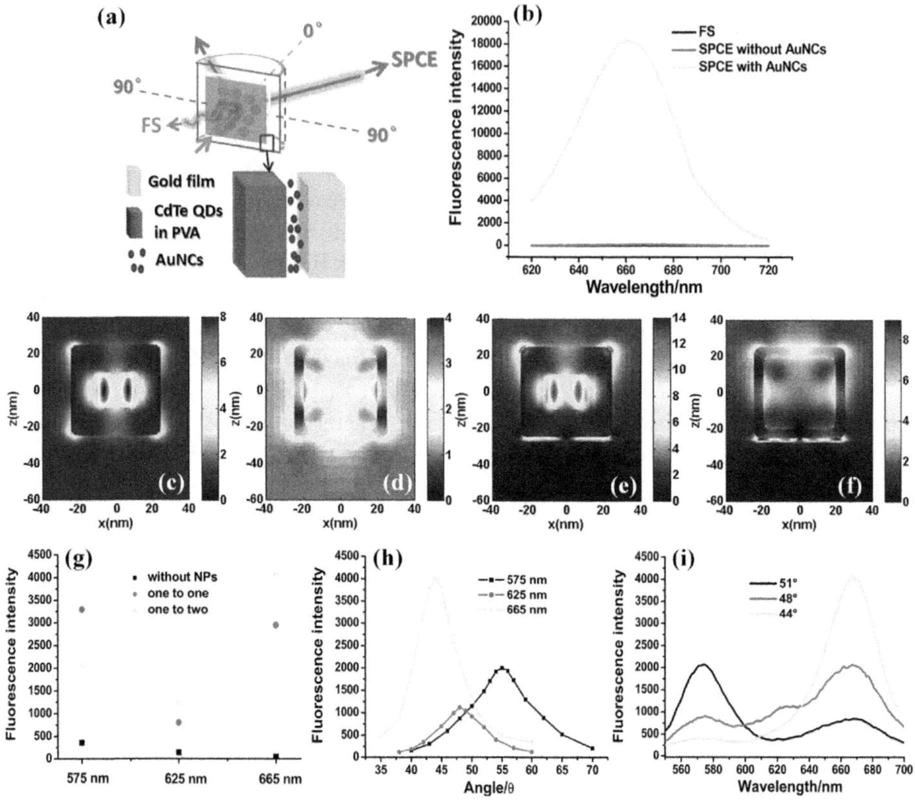

Figure 10.3 Experimental and simulation results for AuNC-enhanced SPCE. (a) Experimental setup for AuNC-enhanced SPCE. The setup is not drawn to scale. (b) The fluorescence spectra for FS and SPCE without and with AuNCs. (c–f) FDTD simulation results of the near-field of a single AuNC and gold substrate with the AuNC at 532 nm (c and e): The outside surface; (d and f): The inside of the AuNC). Length of AuNC: 50 nm; length of hole: 20 nm; shell thickness: 5 nm; distance between AuNC and gold substrate: 2 nm; thickness of gold substrate: 50 nm. (g) SPCE signals of different wavelengths for different mixed ratios, (h) angle distributions for different emission wavelengths, and (i) the fluorescence spectra obtained from different optimal emission angles for different wavelengths (mixed ratio was one to two for AuNSs to AuNCs).

Reproduced with permission from Xie, K.X., Liu, Q., Song, X.L., Huo, R.P., Shi, X.H., Liu, Q.L., 2021a. Amplified fluorescence by hollow-porous plasmonic assembly: a new observation and its application in multiwavelength simultaneous detection . Anal. Chem. 93 (8), 3671–3676. https://doi.org/10.1021/acs.analchem.0c05219; Xie, K.X., Liu, C., Liu, Q., Xiao, X.X., Li, Z., Li, M.F., 2021b. Multiarchitecture-based plasmonic-coupled emission employing gold nanoparticles: an efficient fluorescence modulation and biosensing platform. Langmuir 37 (40), 11880–11886. https://doi.org/10.1021/acs.langmuir.1c01965.

ratios (shells and cubes), the emission intensity at all the desired wavelengths was enhanced. Thus, with the incorporation of NPs, the signals which were otherwise nondetectable were identified with the addition of plasmonic AuNPs (Fig. 10.3g–i). These results present the gold-based nanoengineering with hollow structures and different aspect ratios to be ideal candidates to accomplish multiwavelength simultaneous fluorescence enhancement and detection facilitating multianalyte sensing as well as associated imaging (Xie et al., 2021a,b).

An interesting spacer nanoengineering methodology was accomplished with the use of template-free nano-assemblies. While hybrid nanomaterials comprising more than one component of material is used for augmenting the plasmonic hotspots in SPCE platform, the nanoassemblies enable the elimination of dependence on such template-based materials. Here, we present the snapshot of the highlights from one such study where soret nanoassemblies of AgNPs are interfaced in the SPCE platform to realize high fluorescence enhancements (Bhaskar et al., 2020a,b,c,d). Soret nanoassemblies are synthesized by subjecting the homogeneous nanoparticle solution to temperature gradient, where the word "soret effect" translates to an effect observed upon subjecting a medium to temperature gradient (Bhaskar et al., 2021a,b,c; Moronshing and Subramaniam, 2018; Moronshing et al., 2019). The mechanism of the formation of the silver soret colloids (AgSCs) is presented in Fig. 10.4.

The AgNPs are subjected to adiabatic cooling, triggering the cooling effect to overcome the electrostatic repulsion forces, so as to yield soret nanoassemblies. Thawing the assemblies to room temperature results in phase separation and the nanoassemblies are easily extracted by centrifugation, washing and further concentration. The time interval of adiabatic cooling can be varied to generate tunable number of nanoparticles per nanoassembly. Subjecting the AgNPs to 60-, 90-, and 120-minute results in the nanoassemblies with ~6, ~15, and ~60 number of nanoparticles per assembly. The TEM images of the soret nanoassemblies are shown in Fig. 10.4b–d. From the TEM images, one can clearly observe the formation of nanoassemblies with significant generation of plasmonic hotspots in the voids and crevices. While the pristine AgNPs presented 60-fold SPCE enhancements, the AgSCs yielded 104-fold SPCE enhancements, in spacer nano-interface. This increase in the fluorescence enhancements is attributed to the following interactions occurring at the nanoregime: (i) localized Mie plasmons supported by the entire AgSCs, (ii) delocalized Bragg plasmons supported by the electron cloud density oscillations that partially propagate in and around the nanogaps in the nanoassemblies, and (iii) propagating SPPs supported by the metal thin film (Bhaskar et al., 2020a,b,c,d). The associated publication also discusses about the formation of void and cavity hotspots. While the former is created within the plasmonic nanoassembly, the latter is generated upon interfacing the nanoassemblies over the metal thin film. Such generation of cavity results in nanodomains of high field intensity in the ambient regions. These three effects operate collectively, and the radiating dipoles that are in the immediate vicinity of the AgSCs and the metal thin film demonstrate enhanced excitation and emission (Bhaskar et al., 2020a,b,c,d). In the spacer nanointerface, the laser light irradiates the radiating dipoles in the first place. Following the excitation, the dipoles emit photons, and these emitted photons excite the LSPR of the AgSCs that are in the spacer region between the radiating dipoles and the metal thin film. Consequently, the photons experience a multiplicative effect in the hotspot region generated by the AgSCs and plasmons of metal thin film.

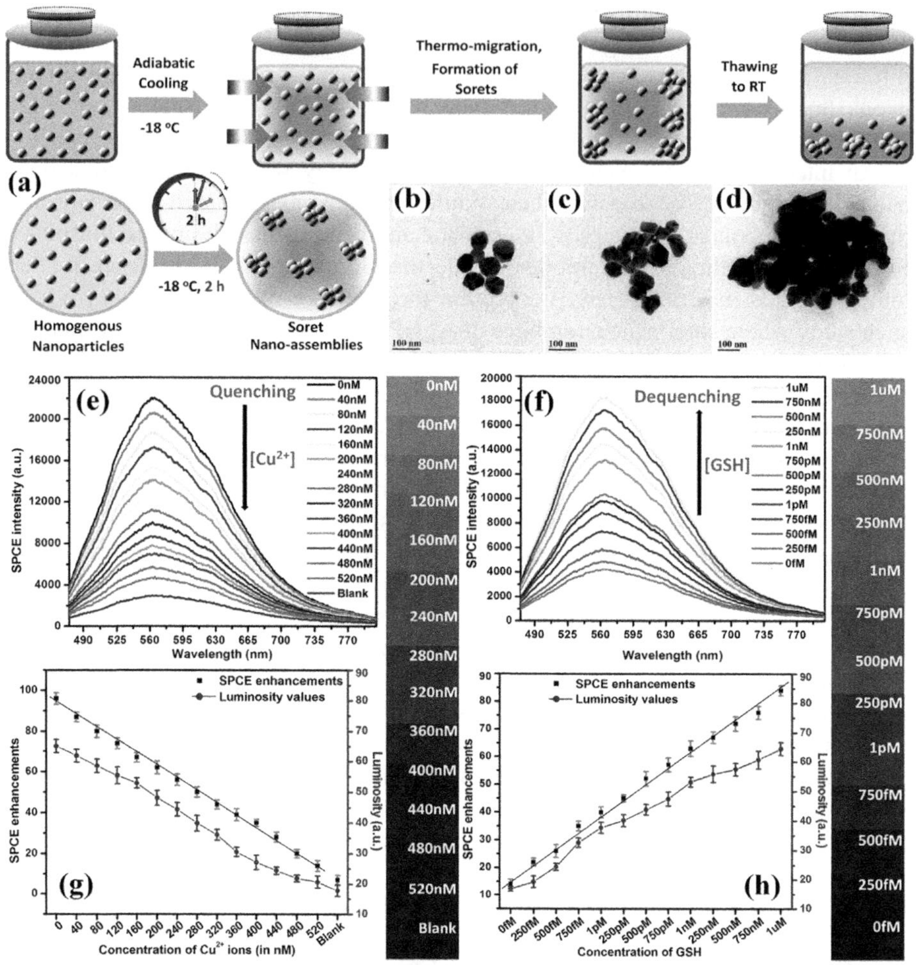

Figure 10.4 Soret colloids nano-engineering for biosensing in SPCE platform. (a) Schematic representation of the formation of SCs utilizing the adiabatic-cooling method. The steps involve adiabatic cooling, thermomigration, and thawing the nanoassembly to room temperature. The reduction in the synthesis time to obtain soret nanoassemblies at different cooling temperatures (−18, −80, −150, and −196°C) is also depicted. (b) TEM images of SC60, SC90, and SC120, respectively, doped in PVA. (e) Quenching in the SPCE spectrum of ARS (0.5 μM) with increasing Cu^{2+} concentration (from 0 to 520 nM with 40 nM increases) using SC60 as the spacer; (f) corresponding SPCE enhancements and luminosity values. (g) Dequenching in the SPCE spectrum of ARS (0.5 μM) + Cu^{2+} (520 nM) with increasing GSH concentration (from 0 fM to 1 μM with 250 fM increases). (h) Corresponding SPCE enhancements and luminosity values.

Reproduced with permission from Bhaskar, S., Kowshik, N.C.S.S., Chandran, S.P., Ramamurthy, S.S., 2020a. Femtomolar detection of spermidine using Au decorated SiO2 nanohybrid on plasmon-coupled extended cavity nanointerface: a smartphone-based fluorescence dequenching approach. Langmuir 36 (11), 2865–2876. https://doi.org/10.1021/

Thus, obtained high fluorescence enhancements are utilized for mobile phone-based detection of glutathione molecule, and the results are captured in Fig. 10.4e–h. A simple alizarin Red S (ARS)–Cu^{2+} ion-GSH ensemble was utilized to perform the experiments. The ARS molecule shows fluorescence in the presence of boric acid (BA) in the buffer solution. Addition of Cu^{2+} ions to the ARS–BA complex results in the quenching of the fluorescence occurring from the ARS. This quenching in the fluorescence signal can be dequenched with the addition of GSH. The resulting fluorescence is captured using the smartphone camera and processed using the freely available Color grab app to analyze the gray-scale values of the luminosity data (Bhaskar et al., 2020a,b,c,d). An excellent correlation between the SPCE enhancements obtained using the conventional ocean optics detector and the smartphone-based detection system is observed. This sensing is performed by incorporating the AgSCs as the functional spacer nanomaterials in the SPCE interface. The gap-based antenna effect featured by the nanoassemblies as discussed here are expected to be further explored with homometallic and heterometallic nanoassemblies of metal-dielectric nanohybrids to achieve further enhancement in the fluorescence signal intensity.

10.5.2 Cavity nanoengineering

The art of cavity nanoengineering technology has been prominently explored ever since the development of laser technology. Cavity-based plasmonic hotspots are generated in several ways. In general, a cavity is produced when two metal-based entities are located adjacent with the nanogap between them (Lee, 2022; Pokhriyal et al., 2013; Ryu and Notomi, 2003). A radiating dipole situated in such a cavity would experience dramatically high electric field intensity in the local environment. Consequently, the emission intensity would significantly enhance, due to high temporal confinement and low spatial confinement. The Purcel factor of a cavity nanoenvironment is directly proportional to the temporal domain and inversely proportional to the spatial domain. That is, the larger the time interval the emitted photon stays in the cavity, the lesser the modal volume of the cavity; such a system would render high fluorescence enhancements due to large photonic density of states (Chikkaraddy et al., 2016; Purcell, 1995; Kolchin et al., 2015). Different types of nanomaterials have been explored in the cavity

acs.langmuir.9b03869; Bhaskar, S., Patra, R., Kowshik, N.C.S.S., Ganesh, K.M., Srinivasan, V., Chandran, S.P., Ramamurthy, S.S., 2020b. Nanostructure effect on quenching and dequenching of quantum emitters on surface plasmon-coupled interface: a comparative analysis using gold nanospheres and nanostars. Phys. E Low-dimens. Syst. Nanostruct. 124. https://doi.org/10.1016/j.physe.2020.114276; Bhaskar, S., Moronshing, M., Srinivasan, V., Badiya, P.K., Subramaniam, C., Ramamurthy, S.S., 2022c. Silver soret nanoparticles for femtomolar sensing of glutathione in a surface plasmon-coupled emission platform. ACS Appl. Nano Mater. 3 (5), 4329–4341. https://doi.org/10.1021/acsanm.0c00470; Bhaskar, S., Singh, A.K., Das, P., Jana, P., Kanvah, S., Bhaktha, S.B.N., Ramamurthy, S.S., 2020d. Superior resonant nanocavities engineering on the photonic crystal-coupled emission platform for the detection of femtomolar iodide and Zeptomolar Cortisol. ACS Appl. Mater. Interf. 12 (30), 34323–34336. https://doi.org/10.1021/acsami.0c07515.

configurations in the SPCE platform. Among them, the lanthanide oxide based high refractive index (HRI) Nd_2O_3—Ag nanohybrids and nanocarbon-florets-Ag nanohybrids presented intriguing results and are discussed in this section (Bhaskar et al., 2022a,b, 2023a,b; Rai et al., 2021a,b). Generally, the plasmonic NPs suffer from parasitic Ohmic losses in them, which adversely affect their performance for hotspot-based plasmonic sensing applications (Khurgin, 2015; Biswas et al., 2015). In order to overcome this bottleneck, recently dielectric nanomaterials, especially with those comprising an HRI, have been investigated on account of their ability to present high field enhancement of both the electric field hotspots as well as the magnetic field hotspots (Jahani and Jacob, 2016; Biswas et al., 2015; Hutter et al., 2013). This is because the HRI dielectric nanomaterials support electric and magnetic field hotspots in the cavity region generated while they are interfaced on metallic thin films. The HRI dielectric Nd_2O_3 nanorods interfaced with the SPCE platform yielded ~118-fold SPCE enhancements (Bhaskar and Ramamurthy, 2019). It is worth noting that this fluorescence enhancement is double than that can be realized with plasmonic AgNPs in SPCE platform. This is accomplished on account of the ability of the Nd_2O_3 NPs to function as effective "Huygen" sources in the rod-shaped morphology, with dominant forward scattering of light as compared to backward scattering (Bhaskar and Ramamurthy, 2019). A frugal nanoengineering methodology was incorporated for generating core-shell Nd_2O_3—Ag (abbreviated as NdAg) nanohybrids. With a simple exposure of the Nd_2O_3 NPs and the Ag metal ion solution, the sunlight helped generate core—shell architectures, and the details are presented with schematic and representative TEM images in Fig. 10.5a (Bhaskar et al., 2023a,b).

The surface interaction between the Nd_2O_3 NPs and the tannic acid solution was used for detecting the latter at picomolar limit of detection (Fig. 10.5b). It is noted that the SPCE enhancements initially increased and further decreased with the addition of tannic acid solution. The initial increase in the SPCE enhancements is initiated by the ability of the tannic acid to interact with the Nd_2O_3 NPs and result in higher number of plasmonic hotspots (Bhaskar et al., 2023a,b). The higher concentrations of tannic acid resulted in a competent interaction with the PVA polymer matrix, which eventually effectuated the formation of globular formations and lessened hotspots. Further, the NdAg nanohybrid was utilized for sensing a food hazard allura red and the attomolar sensitivity is shown in Fig. 10.5c. This was achieved by exploiting the simple interaction between the RhB (SPCE reporter molecule) and the allura red sample. Addition of allura red resulted in quenching of RhB emission presenting long linear range for the indirect detection of the former (Bhaskar et al., 2023a,b).

The latest advancements in the SPCE technology development are captured in Fig. 10.5d. In a conventional fluorescence spectrophotometer, the emission from the molecules placed in the cuvette are collected by placing the detector at 90° with respect to that of the laser line. Consequently, the collection efficiency of the system significantly dwindles. Using plasmonic AgNPs over the glass interface assists in enhanced excitation and enhanced emission intensities with concomitant reduction in the fluorescence lifetimes. This occurs due to the newer channels rendered by the plasmonic NPs toward the radiative decay component of the overall decay (radiative and nonradiative) (Badugu et al., 2015; Lakowicz, 2004). Interfacing such plasmonic Ag over the SPCE

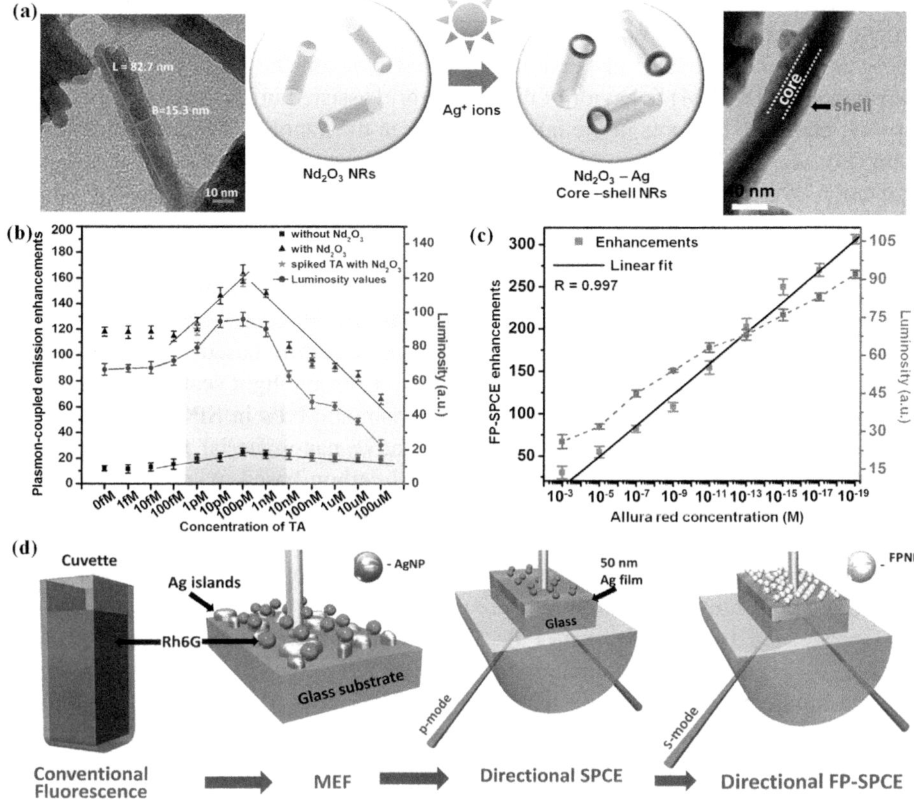

Figure 10.5 Characterization and applications of Nd_2O_3 NPs and Nd_2O_3—Ag nanohybrids for sensing in SPCE. (a) TEM images of Nd_2O_3 NPs and Nd_2O_3—Ag nanohybrids shown along with the conceptual schematic presenting the synthesis of Nd_2O_3—Ag nanohybrids from Nd_2O_3 NPs. (b) Tannic acid sensing using Nd_2O_3 coated SPCE substrates. (c) Allura red sensing using Nd_2O_3—Ag coated SPCE substrates. (d) Conceptual schematic of the developments in the nano-structured coating technologies over the SPCE interface. Reproduced with permission from Bhaskar, S., Thacharakkal, D., Ramamurthy, S.S., Subramaniam, C., 2023a. Metal-dielectric interfacial engineering with Mesoporous nano-carbon florets for 1000-fold fluorescence enhancements: smartphone-enabled visual detection of perindopril erbumine at a single-molecular level. ACS Sustain. Chem. Eng. 11 (1), 78—91. https://doi.org/10.1021/acssuschemeng.2c04064; Bhaskar, S., Srinivasan, V., Ramamurthy, S.S., 2023b. Nd2O3-Ag nanostructures for plasmonic biosensing, antimicrobial, and anticancer applications. ACS Appl. Nano Mater. 6 (2), 1129—1145. https://doi.org/10.1021/acsanm.2c04643.

interface generates high fluorescence emission intensity due to the cavity effect and hence yields 60-folds fluorescence enhancement (Chowdhury et al., 2008). Further, NdAg nanohybrids not only yielded high fluorescence enhancements due to HRI dielectric-plasmonic Ag hybrid coupling, but also resulted in intriguing observations related to the property of the out-coupled signal. Typically, the out-coupled signal in the SPCE interface is TM polarized due to the phase matching conditions involved in the generation of the SPP dispersion diagram. However, the high room temperature ferromagnetic (RTFM) behavior of the nanohybrid assisted in the formation of electric as well magnetic hotspots in the interface between the nanohybrids and the metal thin film (Bhaskar et al., 2023a,b). Consequently, a striking switch in the out-coupled polarization of emission was observed. Hence, the first case study of the cavity nanoengineering presented here demonstrates the functionality of the SPCE technique to selectively couple with different polarizations of light (Bhaskar et al., 2023a,b).

Another striking example of the nanomaterial that has excellently performed in the cavity nanointerface is the nanocarbon florets (NCF) (Bhaskar et al., 2023a,b). The pristine plasmonic NPs suffer from conduction electron based Ohmic losses in them. Additionally, there are issues related to the isotropic light scattering and damping losses that accompany the research with plasmonic NPs in SPCE based sensing applications. NCF serves as an effective alternative nanomaterial for the plasmonic NPs, as NCF presents itself as a nonmetallic, all-carbon-based material with a three-dimensionally arranged dielectric hotspots in their conically graded microcavities (Bhaskar et al., 2023a,b). These structures reminds us of the floral assembly of marigold and therefore render predominantly high regions of dielectric hotspots in the nanocavities sustained by their huge pore volume (0.44 cm^3/g) and a highly accessible surface area (920 m^2/g) (Bhaskar et al., 2023a,b). Low dimensional nanoarchitectures such as 0D, 1D, and 2D nanomaterials have attracted tremendous attention in use for the plasmonic-based biosensing technologies (Bhaskar et al., 2021a,b,c). In this regard, graphene-based structures have been utilized to explore and understand the utility of graphene oxide based plasmon and soliton hybrid coupling in biosensing. It is worth mentioning that the NCF structure integrates the short-range graphitic ordering with long-range domain disorder (Bhaskar et al., 2023a,b). Such nanoarchitectures render unprecedented field confinement and enhancement in the nanoregimes. These structures hence studied in the SPCE framework yielded high fluorescence enhancements in different configurations, the cavity nanointerface being the highest (400-fold). Furthermore, a simple UV-exposure based nanoengineering method was developed to decorate the NCF with the plasmonic AgNPs and the details are presented as part of Fig. 10.6. Fig. 10.6a presents the TEM image of the NCF, and Fig. 10.6b–d presents the TEM images of the NCF-Ag nanohybrids. Fig. 10.6e presents the synthesis methodology used. Such nanohybrids yielded extremely high field enhancement due to the metal-dielectric hybrid coupling at the interfacial nanojunctions in Ag-NCF, presenting >1000-fold SPCE enhancements on the SPCE substrates pre-coated with graphene oxide monolayer (Bhaskar et al., 2023a,b). The high enhancement in the fluorescence was utilized for the detection of biologically and pharmacologically relevant drug, perindopril erbumine (PE) at single-molecular levels (0.1 aM). This has been accomplished using a cost-effective and robust smartphone based SPCE platform, and the results therefrom are captured in Fig. 10.6f–j. The potential of the hybrid photo-plasmonic coupling incorporating the localized SPR of AgNPs, dielectric nanocavities of NCF, and the π-plasmons of graphene oxide with

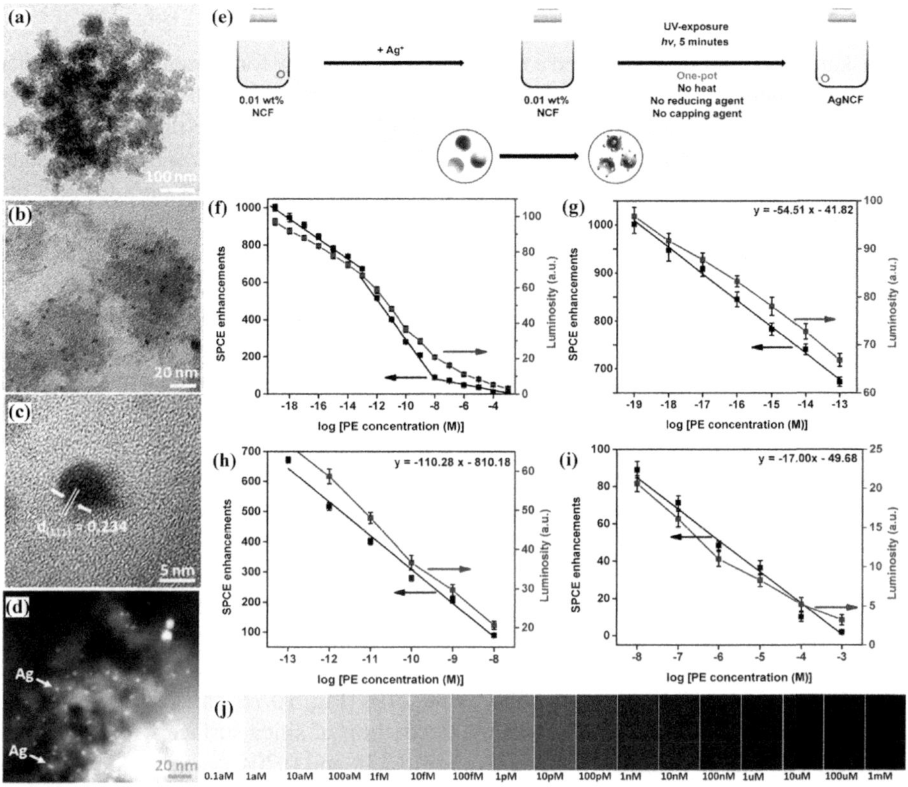

Figure 10.6 Characterization and application of NCF and AgNCF in SPCE-based biosensing application. TEM images of (a) NCF, (b) AgNCF. (c) HRTEM image of AgNCF, and (d) HDAAF image of AgNCF. (e) Schematic representation of the synthesis of AgNCF from NCF via UV-exposure. (f) SPCE enhancements for different concentrations of PE, presenting three linear ranges. These three linear ranges are shown in (g) 0.1 aM.—100 fM, (h) 100 fM to 10 nM, and (i) 10 nM to 1 mM. The respective luminosity values obtained with mobile phone-based detectors are plotted on the right y-axis. (j) Shade card for the entire sensing range from 0.1 aM. to 1 mM.

Reproduced with permission from Bhaskar, S., Thacharakkal, D., Ramamurthy, S.S., Subramaniam, C., 2023a. Metal-dielectric interfacial engineering with Mesoporous nano-carbon florets for 1000-fold fluorescence enhancements: smartphone-enabled visual detection of perindopril erbumine at a single-molecular level. ACS Sustain. Chem. Eng. 11 (1), 78—91. https://doi.org/10.1021/acssuschemeng.2c04064; Bhaskar, S., Srinivasan, V., Ramamurthy, S.S., 2023b. Nd2O3-Ag nanostructures for plasmonic biosensing, antimicrobial, and anticancer applications. ACS Appl. Nano Mater. 6 (2), 1129—1145. https://doi.org/10.1021/acsanm.2c04643.

the propagating SPPs of metal thin film are expected to open newer means of such hybrid engineering, facilitating advanced SPCE-based photonic devices for biosensing applications (Alharbi et al., 2019; Rai et al., 2022a; Bhaskar et al., 2023a,b).

10.5.3 Extended cavity nanoengineering

Bhaskar et al., have demonstrated a rapid and facile strategy for dequenching the otherwise quenched fluorescence emission from radiating dipoles by using Au-decorated SiO_2 NPs (AuSil), a metal (Au)-dielectric (SiO_2) hybrid photonic material to realize 88-fold enhancement in the SPCE platform. Modulation of the local density of states of fluorescent molecules was done using different loading of AuNPs on SiO_2 NPs and was studied in spacer, cavity, and ext. cavity nanointerfaces. They also demonstrated femtomolar detection of spermidine using the AuSil nanohybrids, which showed the best result in ext. cavity interface compared to spacer and cavity architectures (Bhaskar et al., 2020a,b,c,d). They explored the cavity environment extended beyond 30 nm, with a radiating dipole layer (of 30 nm thickness) sandwiched between Ag thin film and AuSil nanohybrid. This architecture completely eliminated the near-field molecular interactions (coulombic interactions) since nanohybrids and dye molecules are situated in two separate nanolayers. From a nanoarchitecture point of view, the ext. cavity is simply a reversal of the spacer interface, where the dipoles are physically separated from AuSil hybrids. Thus, the choice of cavity, spacer, and ext. cavity nano-architecture in SPCE plays a critical role in dequenched augmented emission as far as AuNPs and Au thin films are concerned (Bhaskar et al., 2020a,b,c,d).

The multifold hot-spots rendered by the AuSil nanohybrids assisted in augmented electromagnetic (EM)-field intensity, which was captured by a smartphone-based SPCE platform, presented excellent reliability and reproducibility in spermidine detection. Comparison of TEM images of bare SiO_2 NPs (Fig. 10.7a) and synthesized AuSil hybrids (Fig. 10.7b–d) clearly shows the decoration of silica surface with gold nanospheres. The lattice fringes with d-spacing of 0.235 and 0.204 Å seen in the HRTEM image of AuSil hybrid (Fig. 10.7d) correspond to the Miller indices (111) and (200) of FCC gold. The particle size distribution histogram along with the percentage distribution for AuNPs and SiO_2 NPs presented in Fig. 10.7e and f was used to determine the average size (Bhaskar et al., 2020a,b,c,d). The average size of AuNPs and SiO_2 NPs was determined to be 5.7 ($\pm$0.8) nm and 20.3 ($\pm$4.5) nm, which are in accordance with earlier reports. The property of spermidine to form linkages with AuNPs was utilized to develop a novel sensing method in ext. cavity interface. The 88-fold emission enhancements obtained with AuSil hybrid increased to 207-fold upon addition of 100 μM concentration of spermidine (Fig. 10.7g). This is on account of further aggregation of AuSil hybrid leading to large number of nanogaps between the clusters and individual AuSil. Furthermore, decreasing the concentration of spermidine up to 10 fM allowed for reliable and sensitive detection of this molecule (Bhaskar et al., 2020a,b,c,d). Lowering the concentration to attomolar range did not show reliable output as seen from Fig. 10.7g. The SPCE intensity spectra presenting a clear decrease in emission intensity is in consonance with enhancements (Fig. 10.7h). The modulation in visible color of emission from fluorescent molecules was captured using the smartphone camera, processed using color grab android app, and presented in CIE chromaticity diagram (Fig. 10.7i) along with their respective shade card portion on its right. The results obtained using the smartphone camera showed excellent correlation with emission enhancements obtained using an Ocean Optics spectrometer. The

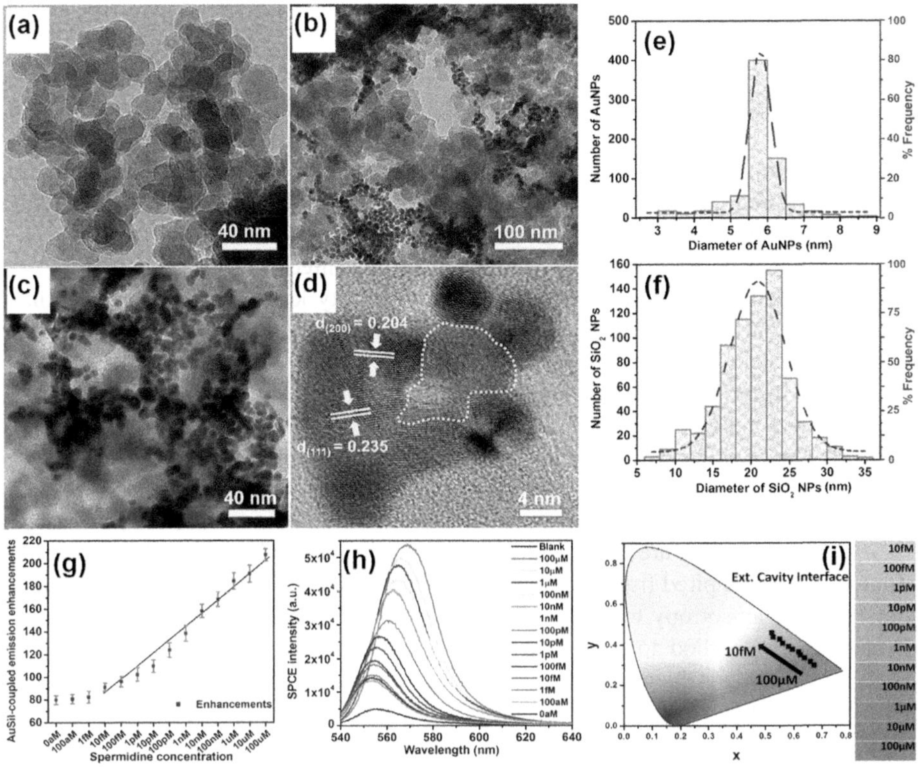

Figure 10.7 Characterization and application of AuSil nanohybrids in SPCE based biosensing application. (a) TEM image of SiO_2 NPs, (b–d) TEM images of synthesized AuSil nanohybrids at different magnifications. In HRTEM image (d) *yellow* dotted curve indicates amorphous SiO_2 decorated with crystalline AuNPs (lattice fringes are shown with corresponding Miller indices for gold), (e and f) particle size distribution histogram of AuNPs and SiO_2 NPs shown along with percentage frequency [right y-axis] obtained from TEM analysis. (g) Plasmon-coupled emission enhancements for increasing concentrations of spermidine taken along with AuSil hybrids in ext. Cavity interface. (h) SPCE spectra for corresponding emission enhancements observed in (g). (i) CIE chromaticity plot and corresponding shade card indicating color change in emission pattern consistent with results in (h). Reproduced with permission Bhaskar, S., Kowshik, N.C.S.S., Chandran, S.P., Ramamurthy, S.S., 2020a. Femtomolar detection of spermidine using Au decorated SiO2 nanohybrid on plasmon-coupled extended cavity nanointerface: a smartphone-based fluorescence dequenching approach. Langmuir 36 (11), 2865–2876. https://doi.org/10.1021/acs.langmuir.9b03869; Bhaskar, S., Patra, R., Kowshik, N.C.S.S., Ganesh, K.M., Srinivasan, V., Chandran, S.P., Ramamurthy, S.S., 2020b. Nanostructure effect on quenching and dequenching of quantum emitters on surface plasmon-coupled interface: a comparative analysis using gold nanospheres and nanostars. Phys. E Low-dimens. Syst. Nanostruct. 124. https://doi.org/10.1016/j.physe.2020.114276; Bhaskar, S., Moronshing, M., Srinivasan, V., Badiya, P.K., Subramaniam, C., Ramamurthy, S.S., 2022c. Silver soret nanoparticles for femtomolar sensing of glutathione in a surface plasmon-coupled emission platform. ACS Appl. Nano Mater. 3 (5), 4329–4341. https://doi.org/10.1021/acsanm.0c00470; Bhaskar, S., Singh, A.K., Das, P., Jana, P., Kanvah, S., Bhaktha, S.B.N., Ramamurthy, S.S., 2020d. Superior resonant nanocavities engineering on the photonic crystal-coupled emission platform for the detection of femtomolar iodide and Zeptomolar Cortisol. ACS Appl. Mater. Interf. 12 (30), 34323–34336. https://doi.org/10.1021/acsami.0c07515.

red shift observed in the emission pattern with increasing spermidine concentration was consistent with the results obtained using the smartphone (Bhaskar et al., 2020a,b,c,d). Hence, the physico–chemical properties and interactions concerning strong coupling of AuSil hybrid nanoresonaters with fluorescent molecules leading to a desired augmented emission and reversal in the quenching behavior of AuNPs in ext. cavity interface was demonstrated.

It is evident by now that SPCE applications are directed toward augmenting the emission enhancements for providing better sensitivity leading to better detection of analytes of concern. In this context, NPs with sharp edges/corners, such as in nanorods (NRs), nanostars (NSs), nanocubes (NCs), and nanoprisms (NPs) are more desirable nanogeometries (Bhaskar and Ramamurthy, 2021a,b; Ganesh et al., 2023a,b; Niu et al., 2015; Rai et al., 2022c; Bhaskar et al., 2020a,b,c,d). This is due to EM hotspots supported by such nanogeometries that are substantially higher compared to isotropic spherical NPs. Although, a number of nanomaterials and nanocomposites are explored by the SPCE technology, little or no attention was being paid toward the processes of their synthesis, which are more often than not toxic and environmentally detrimental. Taking the bio-inspired frugal science approach for synthesizing NPs with defined geometries and anisotropy to be useful in MEF and SPCE, Sriram et al., have demonstrated a new method to synthesize AgNPs using LYCOAT, a biopolymer. They have taken into consideration the key aspects of bio-inspired nanochemistry, disruptive innovation, and frugal science by using a plant-derived product, LYCOAT biopolymer, to synthesize AgNPs in the shapes of nanocubes (NCs) and nanofractals (NFs) (Rathnakumar et al., 2021). Lycoat has the ability to absorb strong high-energy ultraviolet (UV) irradiation, similar to the carbon dots reported earlier. As a result, under UV irradiation, the Ag^{+} ions in the immediate vicinity of LYCOAT in their solution were reduced to Ag^0, upon photo-induced resonance energy electron transfer. They have achieved anisotropic AgNFs and precise AgNCs by regulating the time of UV light exposure.

The SEM images at lower resolution (Fig. 10.8a) and higher resolution (Fig. 10.8b) along with elemental mapping (Fig. 10.8c) for the 60 minutes UV-exposed sample are presented. 60 minutes was the optimized time duration of UV exposure as AgNCs were obtained in this specific duration, and they portrayed intriguing plasmonic couplings and SPCE enhancements in different nanointerfaces. Anisotropic NPs with sharp nanoprotrusions are suitable for MEF and SPCE applications, and AgNCs have shown augmented enhancements in the SPCE platform. The ability of mercury ions to react with AgNPs was adopted in the UV–vis absorbance technique to detect the Hg^{2+} ions at the nanomolar limit of detection. Similarly, the AgNPs synthesized in this report were mixed with different concentrations of Hg^{2+}, and the samples were studied in the SPCE platform (Rathnakumar et al., 2021). On account of the enhanced sensitivity of the SPCE technique and the nanoengineered AgNC morphology, they obtained the attomolar sensitivity of Hg^{2+} ions. They observed that upon the addition of Hg^{2+} ions to AgNP solutions, the color of the solution drastically changed to colorless, indicating the disintegration of stable AgNPs (in the presence of LYCOAT) into Ag^{+} ions. This is schematically illustrated in Fig. 10.8d. The change in the UV–vis absorbance profile (Fig. 10.8e) confirmed this understanding, as the LSPR mode of

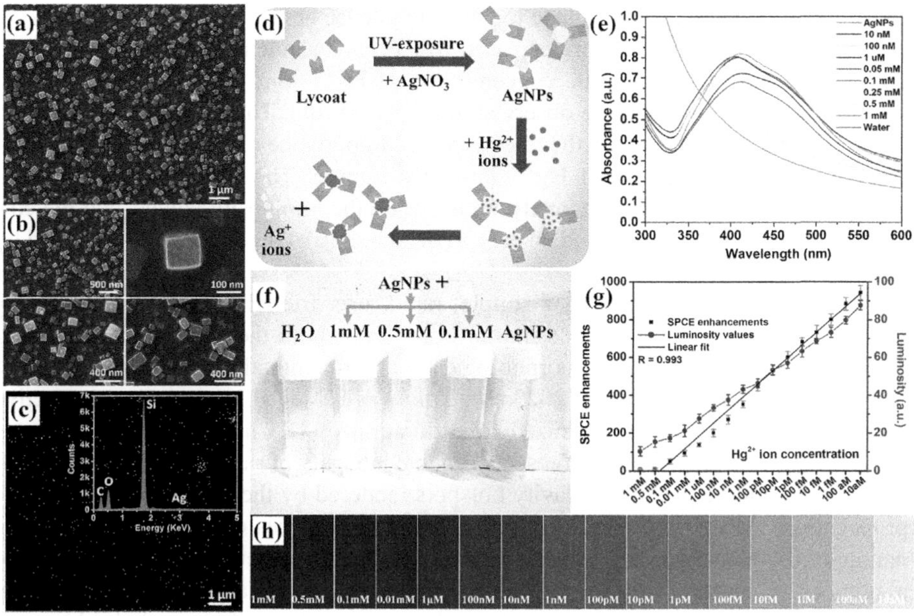

Figure 10.8 Characterization and application of AgNCs for Hg^{2+} ions sensing in SPCE platform. (a and b) Low- and high-resolution SEM images of AgNCs. (c) Corresponding EDAX spectrum of AgNCs, along with the insets showing chemical mapping (d) schematic illustration of the effect of addition of Hg^{2+} ions to AgNPs capped with LYCOAT. (e) UV–vis absorbance pattern obtained upon the addition of different concentrations of Hg^{2+} to AgNPs. (f) Visual change in the color of the solution with and without Hg^{2+} ions. (g) SPCE enhancements and luminosity values obtained using different concentrations of Hg^{2+} ions in the SPCE platform. (h) Shade cards corresponding to the luminosity values at different Hg^{2+} ion concentrations.

Reproduced with permission from Rathnakumar, S., Bhaskar, S., Rai, A., Saikumar, D.V.V., Kambhampati, N.S.V., Sivaramakrishnan, V., Ramamurthy, S.S., 2021. Plasmon-coupled silver nanoparticles for mobile phone-based attomolar sensing of mercury ions. ACS Appl. Nano Mater. 4 (8), 8066–8080. https://doi.org/10.1021/acsanm.1c01347.

AgNPs was absent upon Hg^{2+} ion addition (Rathnakumar et al., 2021). Due to the redox chemical reaction between the zero valent Ag (part of AgNPs) and divalent Hg^{2+} ions, Hg^{2+} ions were reduced to Hg atoms and the AgNPs were oxidized to Ag^{+}, as observed in the color change shown in Fig. 10.8f. These samples were studied in the SPCE platform using a conventional Ocean optics detector and the mobile phone-based technique. The AgNCs were adopted in the ext. cavity nanointerface for all the sensing studies. The SPCE enhancements obtained for all the samples were plotted in the left y-axis, and the luminosity values obtained from the mobile phone-based analysis were plotted in the right y-axis to provide a direct comparison. An excellent correlation was noted between the SPCE enhancements and the trend in the luminosity values obtained using the mobile phone, as seen in Fig. 10.8g. As the AgNPs are oxidized into silver ions, on addition of higher concentration of Hg^{2+}, the ability of AgNPs in the cubic shape to generate intense hotspots in the SPCE platform was compromised, resulting in decreased SPCE enhancements (Rathnakumar et al., 2021). Whereas, decreasing the concentration of Hg^{2+} ions (till 10.0 aM.) restored the enhancements, which were observed to be > 900-fold. The shade cards presented in Fig. 10.8h clearly indicated the relevance and importance of the visual platform for assessing the concentration of Hg^{2+} ions in a sample. The obtained emission was captured using a mobile phone, analyzed to obtain the luminosity values, and the corresponding shade cards provided an easy understanding of the results, not distorted by human bias.

Ganesh et al., have reported a simple, rapid, user-friendly, and green chemistry approach by using biocompatible mustard protein as both the reducing and capping agent for the nanosynthesis of morphologically and functionally tunable plasmophore (Li et al., 2015; Ganesh et al., 2023a,b). This was achieved by exposing respective metal ions and protein in optimized time duration of UV irradiation. More than a 950-fold improvement in the fluorescence signal intensity of rhodamine B was demonstrated utilizing the plasmonic cavity hotspots rendered by the hybrid coupling of the propagating SPPs of Ag with localized plasmon resonances of the bimetallic AgAu nanohybrids (Ganesh et al., 2023a,b). Such high enhancements can be attributed to a number of factors working in tandem at the photo -plasmonic nanointerface: (i) high EM field intensity sustained by the anisotropic AgAu nanohybrids at the nano-edges and nano-curvatures, (ii) the large gradient of integrated plasmonic hotspots supported by the bimetallic nanosystem that aid in the generation of loss less plasmon passages in the AuAgAu trimers that are inevitably formed in the nondirected synthesis protocol, (iii) interplasmonic coupling between the proximally located AgAu nanohybrids in the discontinuously embedded dielectric interface of the metal–dielectric waveguide, and (iv) hybrid photon cascading effect from the radiating dipoles sandwiched between the bimetallic hybrid and the thin film, on account of new pathways of radiative emission and enhanced photostability (Ganesh et al., 2023a,b).

The schematic representation of the one-pot frugal synthesis of AgAu nanohybrids with the use of MPS is shown in Fig. 10.9a. The samples were investigated for their performance as SPCE enhancers in spacer, cavity, and ext. cavity interfaces. It was observed that for all the samples, the SPCE enhancements increased from cavity to spacer to ext. cavity nanointerface (Fig. 10.9b). In spite of the use of the cavity nanointerface, where large quenching effects were envisaged, the SPCE enhancements presented dequenched as well as augmented enhancements (Ganesh et al., 2023a,b). The dispersion diagram presented an overlap of the experimental (shaded region with color) and theoretical data (dotted region), depicted in Fig. 10.9c, for the variant which presented the highest SPCE enhancements. The highly directional characteristic of the spectra obtained is shown in the radiation pattern presented in Fig. 10.9d. The SEM experimentation was performed with the EDAX elemental analysis (Fig. 10.9e–g). It was observed that Au had a higher percentage compared to Ag in terms of the contribution of the elements toward the entire composition of the AgAu nanohybrid, thereby corroborating the possible formation of less loss generating plasmonic trimers, semi-quantitatively. The loss generated in Ag is drastically minimized in the plasmonic

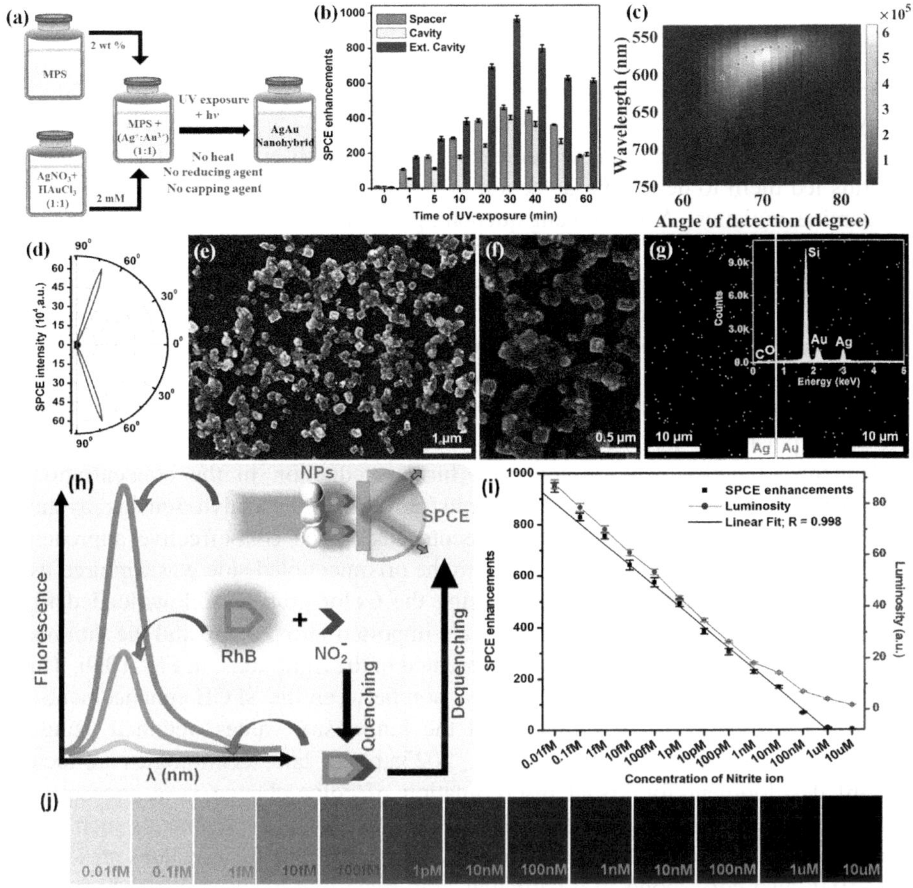

Figure 10.9 Characterization and sensing applications of AgAu nanohybrids obtained via bioinspired methodology. (a) Schematic representation of the AgAu nanohybrids synthesis using mustard protein solution (MPS) through a single step process. (b) SPCE enhancements obtained for differently UV-irradiated samples studied in spacer, cavity, and ext. Cavity nanointerfaces. (c) Overlap of the experimentally (shaded region) and theoretically (dots) obtained dispersion data and (d) sharply directional radiation pattern for the 30 minutes UV-irradiated sample, showing augmented enhancements. (e) and (f) are SEM images of the 30 minutes UV-irradiated samples at lower and higher magnification, respectively. (g) Elemental mapping obtained from EDAX for the Ag−Au nanohybrids, along with insets showing the chemical mapping. (h) Schematic representation of quenching produced with addition of the nitrite ion to RhB and subsequent dequenching generated with the use of Ag−Au nanohybrids. (i) Overlap of the SPCE enhancements (left y-axis) obtained from an ocean optics detector and the luminosity values (right y-axis) extracted from mobile phone-based detector. (j) The gray-scale shade cards correspond to the luminosity values. Reproduced with permission from Ganesh, K.M., Rai, A., Battampara, P., Reddy, R., Bhaskar, S., Reddy, N., Ramamurthy, S.S., 2023a. Optical coupling of bio-inspired mustard protein-based bimetallic nanohybrids with propagating surface plasmon polaritons for femtomolar nitrite ion sensing: cellphone-based portable detection device. Nano-Struct. Nano-Objects 35. https://doi.org/10.1016/j.nanoso.2023.101025; Ganesh, K.M., Rai, A., Bhaskar, S., Reddy, N., Ramamurthy, S.S., 2023b. Plasmon-enhanced fluorescence from synergistic engineering of graphene oxide and sharp-edged silver nanorods mediated with castor protein for cellphone-based attomolar sensing. J. Lumin. 260. https://doi.org/10.1016/j.jlumin.2023.119835.

hybrid in addition to the formation of highly stabilized systems on account of the stability conferred by the AuNPs (Ganesh et al., 2023a,b).

The substantial increase in the fluorescence intensity observed with the AgAu nanohybrids led them to develop a detection system of analytes that closely interact with rhodamine moieties. A significant quenching in the SPCE intensity was observed upon the addition of nitrite ions, and hence, a sensor was developed using this quenching-dequenching as the key aspect, which is schematically presented in Fig. 10.9h. Since the angle of plasmonic resonance is the angle of collection of the out-coupled fluorescence signal, the SPCE framework is analogous to SPR spectroscopic methods (Ganesh et al., 2023a,b). While the normal fluorescence of RhB is quenched upon the addition of NO_2^- ions (1 M), the utility of AgAu nanohybrids as the active plasmonic material in the ext. cavity nanointerface results in realization of a dequenched emission. Consequently, further reduction in the concentration of NO_2^- ions assisted in gathering a linear trend (range R) in the analyte sensing, as shown in Fig. 10.9i. The sensing was also executed using the cost-effective approach of adopting a smartphone. The emission from the prism coupled side was captured using the smartphone camera and analyzed using the Color grab app, downloaded freely from the Google Play Store. The images are imported into the app and the luminosity values for each image is extracted and presented in the right -axis, in Fig. 10.9i. This is performed so as to make a reliable comparison between the SPCE enhancements obtained using the spectrophotometer and the luminosity values obtained using the smartphone sensing device (Ganesh et al., 2023a,b). Of late, technologies are focused toward development for visual discrimination of color change as a consequential change in the value of a particular parameter in an experiment (parameters such as concentration, doping and temperature). In this context, the luminosity values presented in Fig. 10.9i also correspond to a particular gray-scale luminosity value, which is presented in Fig. 10.9j for all the concentrations, thus enabling semi-quantitative analysis of the concentration prior to decision-making through extended gold standard technologies (Ganesh et al., 2023a,b).

10.6 Futuristic scope and perspectives

This chapter discusses different types of nanoengineering methodologies incorporated in the SPCE platform for the development of biosensing technologies. In this regard, the utility and performance of three major interfaces, spacer, cavity, and ext. cavity are elaborated in detail with specific case studies. It is important to discuss the futuristic scope and perspectives from the viewpoint of nanostructured thin films and coatings. In general, the fundamental aim of the different nanoengineering strategies in SPCE platform is to realize augmented fluorescence enhancements (Lakowicz, 2006). While different types of nanohybrids of metal-dielectric core—shell and decorated architectures are yet to be explored in different nanointerfaces, the coating studies can also be investigated with low-dimensional nanomaterials. For instance, MoS_2, WS_2, WSe_2, and similar other low-dimensional materials can be explored in different nanointerfaces to understand the photo-plasmonic coupling efficiency with the other plasmonic and dielectric nanomaterials. Additionally, different types of light emitting entities are yet to be studied, where the coupling behavior of quantum dots, nanodiamonds, and f-block element-based nanomaterials is to be comprehensively understood for biosensing applications. Furthermore, the SPCE platform (50 nm Ag thin film) can be further structured at the nanoregime by using advanced lithographic techniques such as focused ion beam and nonlinear optical setups to generate nanotrenches and nanotriangles with sharp tips. Generation of such platforms with nano-patterned interfaces is expected to yield unprecedented fluorescence enhancements. Also, these interfaces may be used to develop fluorescence, Raman, and analogous biosensing platforms. Further, sputtering methods need to be considered to effectively generate nanofilms of hybrid nanomaterials. Judicious utilization of understanding of functional plasmonic and dielectric materials (such as their dielectric properties) as well as the strength of technologies including atomic layer deposition, chemical layer deposition, arc vapor deposition, pulsed laser deposition, dip coating, and molecular beam epitaxy spin coating to name a few is expected to result in optically active platforms that has immense potential for development of biosensing technologies with high sensitivity and selectivity (Barbillon, 2017; Schaefer, 2010; Bayda et al., 2020).

Further, it is worth emphasizing that currently there is an urgent need to explore biocompatible and biodegradable materials in the field of nanoscience and nanotechnology. This is extremely important as the nanomaterials that are being synthesized have direct access to natural resources such as air and water, especially when not segregated and separated carefully. The toxicity of nanoengineered materials has been an issue of major concern in the past decade, and there is a growing demand toward incorporating the 12 principles of green chemistry in nanotechnology research. Considering these aspects, in this chapter, we also discuss the utility of go-green approaches in the development of bioinspired nanomaterials for the generation of effective nanointerfaces in SPCE-based biosensing frameworks (Duan et al., 2015; Albrecht et al., 2006). As the resources available from the natural world are technically many, it is recommended to study biocompatible materials to function as reducing and capping agents in the synthesis of nanomaterials. Additionally, it is important to effectively optimize the synthesis conditions by exploring parameters such as pH, temperature, and surfactant effects to name a few. In light of these observations, the versatility and performance of the SPCE based nanoengineering technologies are expected to significantly advance the current biosensing modalities (Balzani, 2005; Nalwa, 2004; Binns, 2021; Hyeon et al., 2015).

10.7 Conclusions

The resonance properties of metal thin films and metallic nanomaterials have gathered tremendous attention of researchers from various walks of scientific domains. In the past decade, SPCE technology has been used to understand the fundamental physicochemical interactions at the metal–dielectric interfaces and to augment the performance of the conventional fluorescence-based biosensing platforms (Xie et al., 2021a,b). SPCE based sensing frameworks show great potential as a new class of subwavelength photonic devices for ameliorating the existing biosensing platforms (Lakowicz, 2006). Nanoscale engineering at the metal–dielectric interface in SPCE has enabled the researchers to comprehensively understand the underlying principles of Purcell enhancement, Casimir force, Fabry-Perot mode coupling along with the proof-of-principle demonstrations of the functional waveguides and steering emission platforms to name a few. It is expected that the effective nanoengineering of spacer, cavity, and ext. cavity interfaces would notably find way into the real-time world of miniaturized photonics research as independent discrete research fields. The key highlights from recent exciting research in this domain are succinctly captured in this chapter. Developments such as these are anticipated to serve as guiding principles for advancing the field of SPCE and the fundamental concepts and applications discussed in this chapter from the perspective of nanostructured coatings would be of benefit to chemists, physicists, and materials scientists.

References

Abdulhalim, I., Zourob, M., Lakhtakia, A., 2008. Surface plasmon resonance for biosensing: a mini-review. Electromagnetics 28 (3), 214–242. https://doi.org/10.1080/02726340801921650.

Albrecht, M.A., Evans, C.W., Raston, C.L., 2006. Green chemistry and the health implications of nanoparticles. Green Chem. 8 (5), 417–432. https://doi.org/10.1039/b517131h.

Alharbi, R., Irannejad, M., Yavuz, M., 2019. A short review on the role of the metal-graphene hybrid nanostructure in promoting the localized surface plasmon resonance sensor performance. Sensors 19 (4), 862. https://doi.org/10.3390/s19040862.

Anker, J.N., Hall, W.P., Lyandres, O., Shah, N.C., Zhao, J., Van Duyne, R.P., 2008. Biosensing with plasmonic nanosensors. Nat. Mater. 7 (6), 442–453. https://doi.org/10.1038/nmat2162.

Badugu, R., Szmacinski, H., Ray, K., Descrovi, E., Ricciardi, S., Zhang, D., Chen, J., Huo, Y., Lakowicz, J.R., 2015. Metal-dielectric waveguides for high-efficiency coupled emission. ACS Photon. 2 (7), 810–815. https://doi.org/10.1021/acsphotonics.5b00219.

Badugu, R., Descrovi, E., Lakowicz, J.R., 2014. Radiative decay engineering 7: Tamm state-coupled emission using a hybrid plasmonic-photonic structure. Anal. Biochem. 445 (1), 1–13. https://doi.org/10.1016/j.ab.2013.10.009.

Balzani, V., 2005. Nanoscience and nanotechnology: a personal view of a chemist. Small 1 (3), 278–283. https://doi.org/10.1002/smll.200400010.

Barbillon, G., 2017. Nanoplasmonics: Fundamentals and Applications, vol. 978. BoD – Books on Demand, 2017.

Barnes, W.L., Dereux, A., Ebbesen, T.W., 2003. Surface plasmon subwavelength optics. Nature. 424 (6950), 824–830. https://doi.org/10.1038/nature01937.

Barya, P., Xiong, Y., Shepherd, S., Gupta, R., Akin, L.D., Tibbs, J., Lee, H., Singamaneni, S., Cunningham, B.T., 2023. Photonic-plasmonic coupling enhanced fluorescence enabling digital-resolution ultrasensitive protein detection. Small 19 (44). https://doi.org/10.1002/smll.202207239.

Bauch, M., Toma, K., Toma, M., Zhang, Q., Dostalek, J., 2014. Plasmon-enhanced fluorescence biosensors: a review. Plasmonics 9 (4), 781–799. https://doi.org/10.1007/s11468-013-9660-5.

Bayda, S., Adeel, M., Tuccinardi, T., Cordani, M., Rizzolio, F., 2020. The history of nanoscience and nanotechnology: from chemical–physical applications to nanomedicine. Molecules 25 (1), 112. https://doi.org/10.3390/molecules25010112.

Bhaskar, S., Kowshik, N.C.S.S., Chandran, S.P., Ramamurthy, S.S., 2020a. Femtomolar detection of spermidine using Au decorated SiO_2 nanohybrid on plasmon-coupled extended cavity nanointerface: a smartphone-based fluorescence dequenching approach. Langmuir 36 (11), 2865–2876. https://doi.org/10.1021/acs.langmuir.9b03869.

Bhaskar, S., Thacharakkal, D., Ramamurthy, S.S., Subramaniam, C., 2023a. Metal-dielectric interfacial engineering with Mesoporous nano-carbon florets for 1000-fold fluorescence enhancements: smartphone-enabled visual detection of perindopril erbumine at a single-molecular level. ACS Sustain. Chem. Eng. 11 (1), 78–91. https://doi.org/10.1021/acssuschemeng.2c04064.

Bhaskar, S., Visweswar Kambhampati, N.S., Ganesh, K.M., Pa, M.S., Srinivasan, V., Ramamurthy, S.S., 2021a. Metal-free, graphene oxide-based tunable soliton and plasmon engineering for biosensing applications. ACS Appl. Mater. Interf. 13 (14), 17046–17061. https://doi.org/10.1021/acsami.1c01024.

Bhaskar, S., Patra, R., Kowshik, N.C.S.S., Ganesh, K.M., Srinivasan, V., Chandran S, P., Ramamurthy, S.S., 2020b. Nanostructure effect on quenching and dequenching of quantum emitters on surface plasmon-coupled interface: a comparative analysis using gold nanospheres and nanostars. Phys. E Low-dimens. Syst. Nanostruct. 124. https://doi.org/10.1016/j.physe.2020.114276.

Bhaskar, S., Das, P., Moronshing, M., Rai, A., Subramaniam, C., Bhaktha, S.B.N., Ramamurthy, S.S., 2021b. Photoplasmonic assembly of dielectric-metal, Nd_2O_3-Gold soret nanointerfaces for dequenching the luminophore emission. Nanophotonics 10 (13), 3417–3431. https://doi.org/10.1515/nanoph-2021-0124.

Bhaskar, S., Das, P., Srinivasan, V., Bhaktha, S.B.N., Ramamurthy, S.S., 2022a. Plasmonic-silver sorets and dielectric-Nd_2O_3 nanorods for ultrasensitive photonic crystal-coupled emission. Mater. Res. Bull. 145. https://doi.org/10.1016/j.materresbull.2021.111558.

Bhaskar, S., Rai, A., Ganesh, K.M., Reddy, R., Reddy, N., Ramamurthy, S.S., 2022b. Sericin-based bio-inspired nano-engineering of Heterometallic AgAu nanocubes for attomolar Mefenamic acid sensing in the mobile phone-based surface plasmon-coupled interface. Langmuir 38 (39), 12035–12049. https://doi.org/10.1021/acs.langmuir.2c01894.

Bhaskar, S., Moronshing, M., Srinivasan, V., Badiya, P.K., Subramaniam, C., Ramamurthy, S.S., 2020c. Silver soret nanoparticles for femtomolar sensing of glutathione in a surface plasmon-coupled emission platform. ACS Appl. Nano Mater. 3 (5), 4329–4341. https://doi.org/10.1021/acsanm.0c00470.

Bhaskar, S., Singh, A.K., Das, P., Jana, P., Kanvah, S., Bhaktha B N, S., Ramamurthy, S.S., 2020d. Superior resonant nanocavities engineering on the photonic crystal-coupled emission platform for the detection of femtomolar iodide and Zeptomolar Cortisol. ACS Appl. Mater. Interf. 12 (30), 34323–34336. https://doi.org/10.1021/acsami.0c07515.

Bhaskar, S., Ramamurthy, S., 2023. 161 performance enhancement of light emitting radiating dipoles (LERDs) using surface plasmon-coupled and photonic crystal-coupled emission platforms. In: Organic and Inorganic Light Emitting Diodes: Reliability Issues and Performance Enhancement. CRC Press, pp. 161–184.

Bhaskar, S., Ramamurthy, S.S., 2021a. High refractive index dielectric TiO2and graphene oxide as salient spacers for > 300-fold enhancements, 9781665423045. In: Proceedings of 2021

IEEE International Conference on Nanoelectronics, Nanophotonics, Nanomaterials, Nanobioscience and Nanotechnology, 5NANO 2021. Institute of Electrical and Electronics Engineers Inc., India https://doi.org/10.1109/5NANO51638.2021.9491131, 4 29 2021/04/29. http://ieeexplore.ieee.org/xpl/mostRecentIssue.jsp?punumber=9491110.

Bhaskar, S., Ramamurthy, S.S., 2019. Mobile phone-based picomolar detection of tannic acid on Nd2O3 nanorod-metal thin-film interfaces. ACS Appl. Nano Mater. 2 (7), 4613−4625. https://doi.org/10.1021/acsanm.9b00987.

Bhaskar, S., Srinivasan, V., Ramamurthy, S.S., 2023b. Nd_2O_3-Ag nanostructures for plasmonic biosensing, antimicrobial, and anticancer applications. ACS Appl. Nano Mater. 6 (2), 1129−1145. https://doi.org/10.1021/acsanm.2c04643.

Bhaskar, S., 2023. Biosensing technologies: a focus review on recent advancements in surface plasmon coupled emission. Micromachines 14 (3), 574. https://doi.org/10.3390/mi14030574.

Bhaskar, S., Jha, P., Subramaniam, C., Sathish Ramamurthy, S., 2021c. Multifunctional hybrid soret nanoarchitectures for mobile phone-based picomolar Cu2+ ion sensing and dye degradation applications. Phys. E Low-dimens. Syst. Nanostruct. 132, 114764. https://doi.org/10.1016/j.physe.2021.114764.

Bhaskar, S., Ramamurthy, S.S., 2021b. Synergistic coupling of titanium carbonitride nanocubes and graphene oxide for 800-fold fluorescence enhancements on smartphone based surface plasmon-coupled emission platform. Mater. Lett. 298, 130008. https://doi.org/10.1016/j.matlet.2021.130008.

Binns, C., 2021. Introduction to Nanoscience and Nanotechnology. John Wiley & Sons, 978−1, 2021.

Biswas, S., Kar, G.P., Arora, D., Bose, S., 2015. A unique strategy towards high dielectric constant and low loss with multiwall carbon nanotubes anchored onto graphene oxide sheets. RSC Adv. 5 (31), 24132−24138. https://doi.org/10.1039/c4ra15885g.

Boriskina, S.V., Cooper, T.A., Zeng, L., Ni, G., Tong, J.K.K., Tsurimaki, Y., Huang, Y., Meroueh, L., Mahan, G., Chen, G., 2017. Losses in plasmonics: from mitigating energy dissipation to embracing loss-enabled functionalities. Adv. Opt Photon 9 (4), 775−827. https://doi.org/10.1364/AOP.9.000775.

Cao, S.H., Cai, W.P., Liu, Q., Xie, K.X., Weng, Y.H., Huo, S.X., Tian, Z.Q., Li, Y.Q., 2014. Label-free aptasensor based on ultrathin-linker-mediated hot-spot assembly to induce strong directional fluorescence. J. Am. Chem. Soc. 136 (19), 6802−6805. https://doi.org/10.1021/ja500976a.

Cao, S.H., Weng, Y.H., Xie, K.X., Wang, Z.C., Pan, X.H., Chen, M., Zhai, Y.Y., Xu, L.T., Li, Y.Q., 2019. Surface plasmon coupled fluorescence-enhanced interfacial "molecular beacon" to probe biorecognition switching: an efficient, versatile, and facile signaling biochip. ACS Appl. Bio Mater. 2 (2), 625−629. https://doi.org/10.1021/acsabm.8b00751.

Cao, S.H., Cai, W.P., Liu, Q., Li, Y.Q., 2012. Surface plasmonCoupled emission: what can directional fluorescence bring to the analytical sciences? Annu. Rev. Anal. Chem. 5, 317−336. https://doi.org/10.1146/annurev-anchem-062011-143208.

Chauhan, N., Xiong, Y., Ren, S., Dwivedy, A., Magazine, N., Zhou, L., Jin, X., Zhang, T., Cunningham, B.T., Yao, S., Huang, W., Wang, X., 2023. Net-shaped DNA nanostructures designed for rapid/sensitive detection and potential inhibition of the SARS-CoV-2 virus. J. Am. Chem. Soc. 145 (37), 20214−20228. https://doi.org/10.1021/jacs.2c04835.

Che, C., Xue, R., Li, N., Gupta, P., Wang, X., Zhao, B., Singamaneni, S., Nie, S., Cunningham, B.T., 2022. Accelerated digital biodetection using Magneto-plasmonic nanoparticle-coupled photonic resonator absorption microscopy. ACS Nano 16 (2), 2345−2354. https://doi.org/10.1021/acsnano.1c08569.

Cheerala, V.S.K., Ganesh, K.M., Bhaskar, S., Ramamurthy, S.S., Neelakantan, S.C., 2023. Smartphone-based attomolar Cyanide ion sensing using Au-graphene oxide Cryosoret nanoassembly and Benzoxazolium-based fluorophore in a surface plasmon-coupled enhanced fluorescence interface. Langmuir 39 (22), 7939–7957. https://doi.org/10.1021/acs.langmuir.3c00801.

Chen, Y.L., Hu, Y.H., Yang, X., Gu, Y.L., Wang, X.Y., Xia, Y.H., Zhang, X.Y., Zhang, Y.S., 2023. Plasmon-enhanced fluorescence of gold nanoparticle/graphene quantum dots for detection of Cr3+ ions. Photon. Res. 11 (10), 1781–1790. https://doi.org/10.1364/PRJ.495683.

Chikkaraddy, R., De Nijs, B., Benz, F., Barrow, S.J., Scherman, O.A., Rosta, E., Demetriadou, A., Fox, P., Hess, O., Baumberg, J.J., 2016. Single-molecule strong coupling at room temperature in plasmonic nanocavities. Nature 535 (7610), 127–130. https://doi.org/10.1038/nature17974.

Chowdhury, M.H., Ray, K., Geddes, C.D., Lakowicz, J.R., 2008. Use of silver nanoparticles to enhance surface plasmon-coupled emission (SPCE). Chem. Phys. Lett. 452 (1–3), 162–167. https://doi.org/10.1016/j.cplett.2007.12.047.

Duan, H., Wang, D., Li, Y., 2015. Green chemistry for nanoparticle synthesis. Chem. Soc. Rev. 44 (16), 5778–5792. https://doi.org/10.1039/C4CS00363B.

Dutta Choudhury, S., Badugu, R., Lakowicz, J.R., 2015. Directing fluorescence with plasmonic and photonic structures. Acc. Chem. Res. 48 (8), 2171–2180. https://doi.org/10.1021/acs.accounts.5b00100.

Ganesh, K.M., Rai, A., Battampara, P., Reddy, R., Bhaskar, S., Reddy, N., Ramamurthy, S.S., 2023a. Optical coupling of bio-inspired mustard protein-based bimetallic nanohybrids with propagating surface plasmon polaritons for femtomolar nitrite ion sensing: cellphone-based portable detection device. Nano-Struct. Nano-Objects 35. https://doi.org/10.1016/j.nanoso.2023.101025.

Ganesh, K.M., Rai, A., Bhaskar, S., Reddy, N., Ramamurthy, S.S., 2023b. Plasmon-enhanced fluorescence from synergistic engineering of graphene oxide and sharp-edged silver nanorods mediated with castor protein for cellphone-based attomolar sensing. J. Lumin. 260. https://doi.org/10.1016/j.jlumin.2023.119835.

Ganesh, N., Zhang, W., Mathias, P.C., Chow, E., Soares, J.A.N.T., Malyarchuk, V., Smith, A.D., Cunningham, B.T., 2007. Enhanced fluorescence emission from quantum dots on a photonic crystal surface. Nat. Nanotechnol. 2 (8), 515–520. https://doi.org/10.1038/nnano.2007.216.

Gryczynski, I., Malicka, J., Gryczynski, Z., Lakowicz, J.R., 2004. Radiative decay engineering 4. Experimental studies of surface plasmon-coupled directional emission. Anal. Biochem. 324 (2), 170–182. https://doi.org/10.1016/j.ab.2003.09.036.

Hutter, T., Huang, F.M., Elliott, S.R., Mahajan, S., 2013. Near-field plasmonics of an individual dielectric nanoparticle above a metallic substrate. J. Phys. Chem. C 117 (15), 7784–7790. https://doi.org/10.1021/jp400963f.

Hyeon, T., Manna, L., Wong, S.S., 2015. Sustainable nanotechnology . Royal society of chemistry. South Korea Chem. Soc. Rev. 44 (16), 5755–5757. https://doi.org/10.1039/c5cs90072g.

Jahani, S., Jacob, Z., 2016. All-dielectric metamaterials. Nat. Nanotechnol 11 (1), 23–36. https://doi.org/10.1038/nnano.2015.304.

Jiang, N., Zhuo, X., Wang, J., 2018. Active plasmonics: principles, structures, and applications. Chem. Rev. 118 (6), 3054–3099. https://doi.org/10.1021/acs.chemrev.7b00252.

Kang, K.A., Wang, J., Jasinski, J.B., Achilefu, S., 2011. Fluorescence manipulation by gold nanoparticles: from complete quenching to extensive enhancement. J. Nanobiotechnol. 9. https://doi.org/10.1186/1477-3155-9-16.

Khurgin, J.B., 2015. How to deal with the loss in plasmonics and metamaterials. Nat. Nanotechnol. 10 (1), 2–6. https://doi.org/10.1038/nnano.2014.310.

Kolchin, P., Pholchai, N., Mikkelsen, M.H., Oh, J., Ota, S., Islam, M.S., Yin, X., Zhang, X., 2015. High purcell factor due to coupling of a single emitter to a dielectric slot waveguide. Nano Lett. 15 (1), 464–468. https://doi.org/10.1021/nl5037808.

Kongsuwan, N., Demetriadou, A., Horton, M., Chikkaraddy, R., Baumberg, J.J., Hess, O., 2020. Plasmonic nanocavity modes: from near-field to far-field radiation. ACS Photonics 7 (2), 463–471. https://doi.org/10.1021/acsphotonics.9b01445.

Kurihara, K., Suzuki, K., 2002. Theoretical understanding of an absorption-based surface plasmon resonance sensor based on Kretchmann's theory. Anal. Chem. 74 (3), 696–701. https://doi.org/10.1021/ac010820+.

Lakowicz, J.R., Ray, K., Chowdhury, M., Szmacinski, H., Fu, Y., Zhang, J., Nowaczyk, K., 2008. Plasmon-controlled fluorescence: a new paradigm in fluorescence spectroscopy. Analyst 133 (10), 1308–1346. https://doi.org/10.1039/b802918k.

Lakowicz, J.R., 2006. Plasmonics in biology and plasmon-controlled fluorescence. Plasmonics 1 (1), 5–33. https://doi.org/10.1007/s11468-005-9002-3.

Lakowicz, J.R., Shen, Y., D'Auria, S., Malicka, J., Fang, J., Gryczynski, Z., Gryczynski, I., 2002. Radiative decay engineering: 2. Effects of silver island films on fluorescence intensity, lifetimes, and resonance energy transfer. Anal. Biochem. 301 (2), 261–277. https://doi.org/10.1006/abio.2001.5503.

Lakowicz, J.R., 2004. Radiative decay engineering 3. Surface plasmon-coupled directional emission. Anal. Biochem. 324 (2), 153–169. https://doi.org/10.1016/j.ab.2003.09.039.

Lakowicz, J.R., 2005. Radiative decay engineering 5: metal-enhanced fluorescence and plasmon emission. Anal. Biochem. 337 (2), 171–194. https://doi.org/10.1016/j.ab.2004.11.026.

Lakowicz (Ed.), 2006. Principles of Fluorescence Spectroscopy. Springer, US, p. 2006.

Lee, S., 2022. Nanoparticle-on-mirror cavity: a historical view across nanophotonics and nanochemistry. J. Kor. Phys. Soc. 81 (6), 502–509. https://doi.org/10.1007/s40042-022-00407-z.

Li, M., Cushing, S.K., Wu, N., 2015. Plasmon-enhanced optical sensors: a review. Analyst 140 (2), 386–406. https://doi.org/10.1039/c4an01079e.

Li, M., An, X., Jiang, M., Li, S., Liu, S., Chen, Z., Xiao, H., 2019. "Cellulose spacer" strategy: anti-aggregation-caused quenching membrane for mercury ion detection and removal. ACS Sustain. Chem. Eng. 7 (18), 15182–15189. https://doi.org/10.1021/acssuschemeng.9b01928.

Liu, Q., Cao, S.H., Cai, W.P., Liu, X.Q., Weng, Y.H., Xie, K.X., Huo, S.X., Li, Y.Q., 2015. Surface plasmon coupled emission in micrometer-scale cells: a leap from interface to bulk targets. J. Phys. Chem. B 119 (7), 2921–2927. https://doi.org/10.1021/jp512031r.

Mejía-Salazar, J.R., Oliveira, O.N., 2018. Plasmonic biosensing. Chem. Rev. 118 (20), 10617–10625. https://doi.org/10.1021/acs.chemrev.8b00359.

Moronshing, M., Bhaskar, S., Mondal, S., Sathish Ramamurthy, S., Subramaniam, C., 2019. Surface-enhanced Raman scattering platform operating over wide pH range with minimal chemical enhancement effects: test case of tyrosine. J. Raman Spectrosc. 50 (6), 826–836. https://doi.org/10.1002/jrs.5587.

Moronshing, M., Subramaniam, C., 2018. Room temperature, multiphasic detection of explosives, and volatile organic compounds using thermodiffusion driven soret colloids. ACS Sustain. Chem. Eng. 6 (7), 9470–9479. https://doi.org/10.1021/acssuschemeng.8b02050.

Mulpur, P., Podila, R., Lingam, K., Vemula, S.K., Ramamurthy, S.S., Kamisetti, V., Rao, A.M., 2013. Amplification of surface plasmon coupled emission from graphene-ag hybrid films. J. Phys. Chem. C 117 (33), 17205−17210. https://doi.org/10.1021/jp406122s.

Nalwa, H.S., 2004. Encyclopedia of Nanoscience and Nanotechnology (V. 8. Ne-P). American Scientific Publishers, p. 2004.

Niu, W., Chua, Y.A.A., Zhang, W., Huang, H., Lu, X., 2015. Highly symmetric gold nanostars: crystallographic control and surface-enhanced Raman scattering property. J. Am. Chem. Soc. 137 (33), 10460−10463. https://doi.org/10.1021/jacs.5b05321.

Pokhriyal, A., Lu, M., Chaudhery, V., George, S., Cunningham, B.T., 2013. Enhanced fluorescence emission using a photonic crystal coupled to an optical cavity. Appl. Phys. Lett. 102 (22). https://doi.org/10.1063/1.4809513.

Purcell, E.M., 1995. Spontaneous Emission Probabilities at Radio Frequencies. Springer Science and Business Media LLC. https://doi.org/10.1007/978-1-4615-1963-8_40, 839−839.

Rai, A., Bhaskar, S., Ganesh, K.M., Ramamurthy, S.S., 2022a. Biocompatible Gellucire® inspired bimetallic nanohybrids for augmented fluorescence emission based on graphene oxide interfacial plasmonic architectures. ECS Trans. 107 (1), 4527−4535. https://doi.org/10.1149/10701.4527ecst.

Rai, A., Bhaskar, S., Reddy, N., Ramamurthy, S.S., 2021a. Cellphone-Aided attomolar zinc ion detection using Silkworm protein-based nanointerface engineering in a plasmon-coupled dequenched emission platform. ACS Sustain. Chem. Eng. 9 (44), 14959−14974. https://doi.org/10.1021/acssuschemeng.1c05437.

Rai, A., Bhaskar, S., Ganesh, K.M., Ramamurthy, S.S., 2021b. Engineering of coherent plasmon resonances from silver soret colloids, graphene oxide and Nd_2O_3 nanohybrid architectures studied in mobile phone-based surface plasmon-coupled emission platform. Mater. Lett. 304. https://doi.org/10.1016/j.matlet.2021.130632.

Rai, A., Bhaskar, S., Ganesh, K.M., Ramamurthy, S.S., 2022b. Hottest hotspots from the coldest cold: welcome to nano 4.0. ACS Appl. Nano Mater. 5 (9), 12245−12264. https://doi.org/10.1021/acsanm.2c02556.

Rai, A., Bhaskar, S., Battampara, P., Reddy, N., Ramamurthy, S.S., 2022c. Integrated photo-plasmonic coupling of bioinspired sharp-edged silver nano-particles with nano-films in extended cavity functional interface for cellphone-aided femtomolar sensing. Mater. Lett. 316, 132025. https://doi.org/10.1016/j.matlet.2022.132025.

Rathnakumar, S., Bhaskar, S., Rai, A., Saikumar, D.V.V., Kambhampati, N.S.V., Sivaramakrishnan, V., Ramamurthy, S.S., 2021. Plasmon-coupled silver nanoparticles for mobile phone-based attomolar sensing of mercury ions. ACS Appl. Nano Mater. 4 (8), 8066−8080. https://doi.org/10.1021/acsanm.1c01347.

Ryu, H.Y., Notomi, M., 2003. Enhancement of spontaneous emission from the resonant modes of a photonic crystal slab single-defect cavity. Opt Lett. 28 (23), 2390−2392. https://doi.org/10.1364/OL.28.002390.

Schaefer, H.E., 2010. Nanoscience: The Science of the Small in Physics, Engineering, Chemistry, Biology and Medicine. Springer Berlin Heidelberg, Germany, pp. 1−772. https://doi.org/10.1007/978-3-642-10559-3.

Srinivasan, V., Vernekar, D., Jaiswal, G., Jagadeesan, D., Ramamurthy, S.S., 2016. Earth abundant iron-rich N-doped graphene based spacer and cavity materials for surface plasmon-coupled emission enhancements. ACS Appl. Mater. Interf. 8 (19), 12324−12329. https://doi.org/10.1021/acsami.5b12038.

Stockman, M.I., Kneipp, K., Bozhevolnyi, S.I., Saha, S., Dutta, A., Ndukaife, J., Kinsey, N., Reddy, H., Guler, U., Shalaev, V.M., Boltasseva, A., Gholipour, B., Krishnamoorthy, H.N.S., Macdonald, K.F., Soci, C., Zheludev, N.I., Savinov, V.,

Singh, R., Groß, P., Lienau, C., Vadai, M., Solomon, M.L., Barton, D.R., Lawrence, M., Dionne, J.A., Boriskina, S.V., Esteban, R., Aizpurua, J., Zhang, X., Yang, S., Wang, D., Wang, W., Odom, T.W., Accanto, N., De Roque, P.M., Hancu, I.M., Piatkowski, L., Van Hulst, N.F., Kling, M.F., 2018. Roadmap on plasmonics. J. Opt. 20 (4). https://doi.org/10.1088/2040-8986/aaa114.

Strianese, M., Staiano, M., Ruggiero, G., Labella, T., Pellecchia, C., D'Auria, S., 2012. Fluorescence-based biosensors. Methods Mol. Biol. 875, 193–216. https://doi.org/10.1007/978-1-61779-806-1_9.

Sudha Maria Lis, S., Bhaskar, S., Dahiwadkar, R., Kanvah, S., Ramamurthy, S.S., Bhaktha B. N, S., 2023. Plasmon-rich BCZT nanoparticles in the photonic crystal-coupled emission platform for cavity hotspot-driven attomolar sensing. ACS Appl. Nano Mater. 6 (20), 19312–19326. https://doi.org/10.1021/acsanm.3c03731.

Thacharakkal, D., Bhaskar, S., Sharma, T., Rajaraman, G., Sathish Ramamurthy, S., Subramaniam, C., 2024. Plasmonic synergism in tailored metal–carbon interfaces for real-time single molecular level sniffing of PFOS and PFOA. Chem. Eng. J. 480, 148166. https://doi.org/10.1016/j.cej.2023.148166.

Venkatesh, S., Badiya, P.K., Ramamurthy, S.S., 2015. Low-dimensional carbon spacers in surface plasmon-coupled emission with femtomolar sensitivity and 1000-fold fluorescence enhancements. Chem. Commun. 51 (37), 7809–7811. https://doi.org/10.1039/C5CC01040C.

Xie, K.X., Liu, Q., Song, X.L., Huo, R.P., Shi, X.H., Liu, Q.L., 2021a. Amplified fluorescence by hollow-porous plasmonic assembly: a new observation and its application in multi-wavelength simultaneous detection. Anal. Chem. 93 (8), 3671–3676. https://doi.org/10.1021/acs.analchem.0c05219.

Xie, K.X., Li, Z., Fang, J.H., Cao, S.H., Li, Y.Q., 2022. Au-Ag Alloy nanoshuttle mediated surface plasmon coupling for enhanced fluorescence imaging. Biosensors 12 (11). https://doi.org/10.3390/bios12111014.

Xie, K.X., Liu, C., Liu, Q., Xiao, X.X., Li, Z., Li, M.F., 2021b. Multiarchitecture-based plasmonic-coupled emission employing gold nanoparticles: an efficient fluorescence modulation and biosensing platform. Langmuir 37 (40), 11880–11886. https://doi.org/10.1021/acs.langmuir.1c01965.

Xiong, Y., Shepherd, S., Tibbs, J., Bacon, A., Liu, W., Akin, L.D., Ayupova, T., Bhaskar, S., Cunningham, B.T., 2023. Photonic crystal enhanced fluorescence: a review on design strategies and applications. Micromachines 14 (3). https://doi.org/10.3390/mi14030668.

Section IV

Multifunctional nanostructured coatings for environmental applications

Recent advances in bio-inspired multifunctional coatings for corrosion protection

11

Subodh Barthwal[1] *and Sumit Barthwal*[2]
[1]Department of Mechanical Engineering, Amity University, Greater Noida, Uttar Pradesh, India; [2]Nanomechatronics Lab, Kookmin University, Seoul, South Korea

11.1 Introduction

Corrosion refers to the degradation of materials resulting from their interaction with the surrounding environment. While, in theory, all materials can corrode based on this definition, the term is commonly linked to the deterioration of metals (Palanisamy, 2019). Metals are favored materials for their notable characteristics, including high electrical and thermal conductivity, ductility, malleability, and strength. It is a universal issue that affects numerous industries, infrastructure, and everyday objects, leading to significant economic losses and safety hazards.

The corrosion process can be broadly categorized into several types, including uniform corrosion, localized corrosion (such as pitting and crevice corrosion), galvanic corrosion, and stress corrosion cracking, among others. Each type involves distinct mechanisms and factors influencing its occurrence, making corrosion a multifaceted challenge to address. Several environmental factors contribute to the initiation and progression of corrosion. Exposure to moisture, oxygen, aggressive chemicals, and varying temperatures accelerates corrosion reactions (Foorginezhad et al., 2021). In addition, the presence of impurities, contaminants, or pollutants in the environment can intensify the corrosion process. Corrosion imposes substantial economic burdens, resulting in material degradation, increased maintenance costs, and potential safety hazards. Industries such as transportation, infrastructure, oil and gas, and manufacturing face significant challenges in mitigating corrosion effects on their assets. Moreover, corrosion-related failures can compromise the structural integrity of critical components, leading to catastrophic consequences. A global study conducted by NACE International on corrosion costs and preventative strategies in 2013, which was utilized in the World Bank economic sector and global Gross Domestic Product (GDP) data to relate the cost of corrosion. This comprehensive analysis revealed that the estimated global cost of corrosion amounted to approximately 3.4% of the GDP (Mazumder, 2020).

The corrosion rate of metals can be effectively managed through various measures, with common methods including cathodic protection, galvanizing, painting, alloying element addition, and coatings (Guo et al., 2014; Deshpande et al., 2014; Gandel et al., 2014).

Multifunctional Nanostructured Coatings. https://doi.org/10.1016/B978-0-443-23683-9.00011-8

Chrome-based coatings have demonstrated high efficacy in safeguarding metal surfaces against corrosion. However, the use of chromium VI is restricted in several countries due to its toxicity and carcinogenic nature.

Coatings emerge as a convenient and widely employed approach for preventing metal corrosion, owing to their low cost, accessibility of raw materials, flexibility, and simplicity. Despite the various advantages, coatings are susceptible to issues like cracking and degradation, necessitating repair or replacement. Prolonged exposure to corrosive environments can weaken the mechanical properties of coatings, leading to the formation and propagation of microcracks. To enhance coating properties, various methods have been proposed, with one notable approach involving the incorporation of fillers into a polymer matrix to increase both the mechanical properties and corrosion inhibition performance of coatings. Fig. 11.1 depicts the decade-long trend in publications related to "anticorrosion coatings" and "anticorrosion" surfaces, sourced from the Web of Science. The noticeable surge in the number of papers in recent years underscores the significant concern about corrosion in our society. Researchers are actively engaged in the exploration and advancement of anticorrosion coatings and surfaces, often incorporating features inspired by organisms with natural anti-icing properties.

In recent years, a new class of protective coatings, termed bio-inspired coatings, has emerged. These coatings possess the ability to repair damage inflicted by corrosive media and thereby enhancing the lifespan of the coating. Despite being in the early stages of development, it is crucial to broadly understand various aspects of these coatings, encompassing their properties, mechanisms, and performance. The internal structure of bioinspired coating systems allows for the tuning of potential properties, expanding their application possibilities. Integrating multiple functionalities into coatings has become increasingly appealing for various applications, showcasing remarkable performance benefits. Examples include coatings with self-healing coupled with corrosion resistance, self-cleaning combined with antifouling, and antifouling integrated

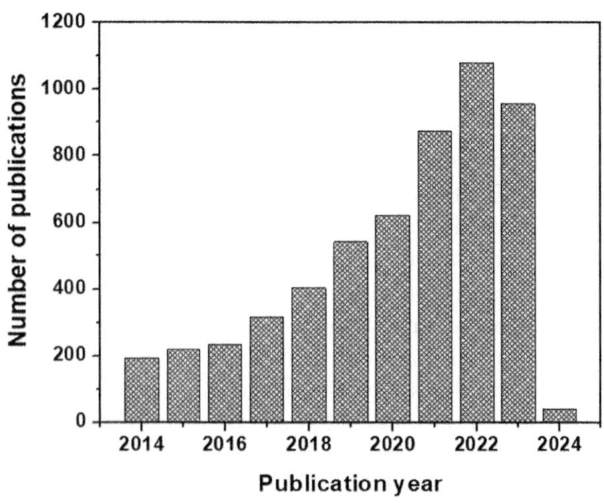

Figure 11.1 Statistics from the web of Science indicate the total number of publications between 2014 and 2024 when using "anticorrosion" and "anticorrosion coating" as keywords in the search.

with anticorrosion measures. These multifunctional bio-inspired smart coatings represent a novel approach to corrosion protection strategies, contributing to extended service life and reduced maintenance costs.

11.2 Functional surfaces in nature

Nature is an abundant source of inspiration for developing innovative materials and technologies with bio-inspired properties. The diverse and intricate structures observed in the natural world have evolved over millions of years, resulting in remarkable variations that address various challenges. In the context of materials science, bio-inspired design principles have become a focal point for researchers aiming to replicate or adapt nature's solutions to enhance the performance of synthetic materials. These bio-inspired properties span a wide range of applications, including but not limited to materials with self-healing capabilities, antifouling surfaces, superhydrophobic coatings, and adaptive functionalities. For example, the lotus leaf's superhydrophobic surface, with its hierarchical micro/nanostructures and hydrophobic epicuticular wax, inspired the development of self-cleaning coatings that repel water and dirt, mimicking nature's ability to stay clean and pristine.

Water contact angle measurement serves as a crucial parameter in quantifying surface wettability, offering insights into the degree of hydrophobicity. Higher contact angles indicate greater water resistance and lower surface energy. Materials are commonly categorized as hydrophobic or hydrophilic based on wetting behavior. Derived from the Greek roots "hydro" meaning "water" and "phobos" or "philia" representing "fear" or "love" respectively, these terms categorize materials as hydrophobic if they repel water and are difficult to wet, and hydrophilic if they readily interact with water.

Static contact angle analysis, measuring the angle formed by a static liquid droplet at the interphase of a solid surface and surrounding vapor, offers insights into water-surface interactions. Utilizing a contact angle goniometer, the water contact angle (WCA) is determined, with materials classified as hydrophilic if their WCA is below 90 degrees, and hydrophobic if it exceeds 90 degrees. Meanwhile, surfaces exhibiting WCAs ranging between 150 and 180 degrees are termed superhydrophobic, while those with minimal contact angles leading to droplet spreading and complete wetting are labeled as superhydrophilic, as depicted in Fig. 11.2a.

Superhydrophobic surfaces, characterized by maximal contact angles and nearly spherical droplet shapes, boast remarkable self-cleaning abilities (Josyula et al., 2024; Prakash and Prasanth, 2021). Their high water repellency and appropriate surface roughness enable easy droplet rolling, effectively carrying away contaminants and dust particles. This inherent self-cleaning property holds significant implications across diverse applications, from antifouling surfaces in marine environments to low-maintenance coatings in architectural settings.

Drawing inspiration from nature, researchers have developed bio-inspired slippery surfaces (SLIPS), with one notable example being inspired by the carnivorous

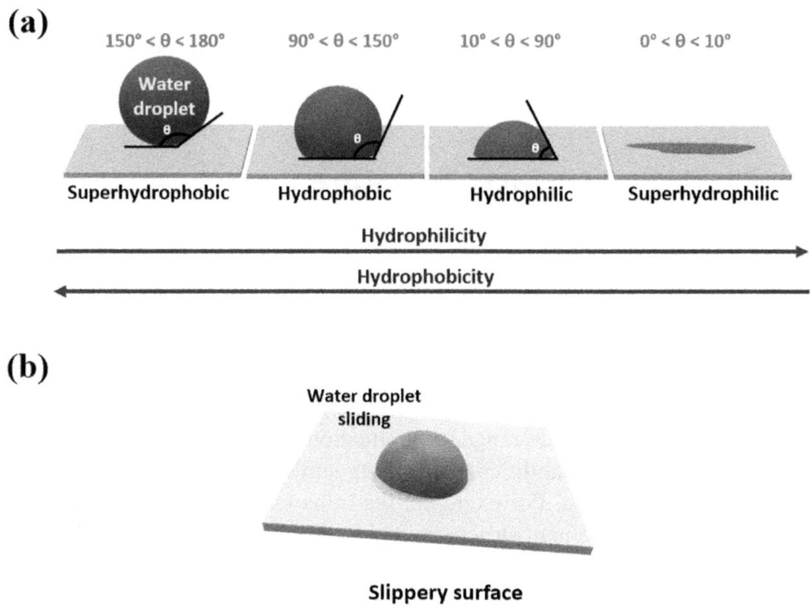

Figure 11.2 Shows water droplets on surfaces exhibiting varying degrees of wetting properties.

Nepenthes pitcher plant found in tropical regions, as shown in Fig. 11.2b (Wong et al., 2011). Unlike the lotus leaf, the pitcher plant possesses a unique structure characterized by a moisture-absorbing nectar component and a specialized rough structure on the inner side of the rim (peristome). This configuration creates a homogenous liquid film in a humid environment, allowing the plant to capture and trap insects, which are attracted to its nectar but ultimately slip and fall into the pitcher, where they are digested by the plant. The microstructures on the pitcher plant's surface generate a lubricating layer of water, reducing friction between the insect's feet and the surface, and facilitating their slip.

By extrapolating the concept of infiltrating porous surfaces with a liquid to achieve a wet and slippery surface, researchers are innovating slippery surfaces for diverse applications. These applications include fluid manipulation, antibiofouling, anti-icing/frosting, self-healing, drag reduction, corrosion prevention, and self-cleaning. Such surfaces offer exceptionally low liquid friction and adhesion, promising advancements across various fields.

Understanding the intricate interplay between surface roughness and chemistry is crucial for advancing the design and application of functional surfaces with tailored wetting characteristics. By unraveling these relationships, researchers can develop advanced materials, technologies, and strategies that benefit human society and contribute to the preservation of the environment.

Nature offers numerous examples of plants, insects, and birds that exhibit remarkable properties when in contact with water (Bixler and Bhushan, 2014; Gao et al., 2007; Han et al., 2017; Stark et al., 2016; Hurchalla and Drelich, 2019; Barthlott and Neinhuis,

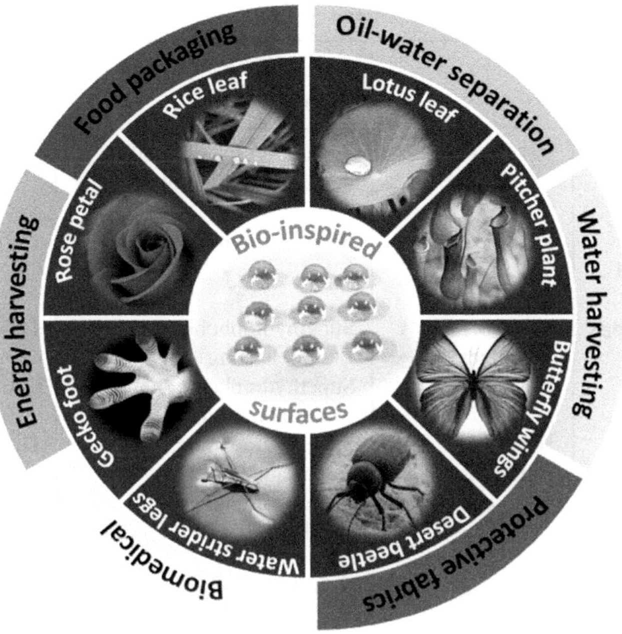

Figure 11.3 Superhydrophobic surfaces available in nature.

1997; Otten and Herminghaus, 2004; Liu and Li, 2012; Kostal et al., 2018; Feng et al., 2008; Chen et al., 2016a). Some of these surfaces are summarized below, as shown in Fig. 11.3 and Table 11.1.

11.2.1 Lotus leaf

The lotus leaf, particularly from the *Nelumbo nucifera* plant, serves as a natural model with its ability to achieve superhydrophobicity, evident by a high contact angle (>150 degrees) that repels water droplets (Mozumder et al., 2011; Latthe et al., 2014). The superhydrophobicity of the lotus leaf is a fascinating natural phenomenon that has inspired numerous scientific studies and technological advancements. The surface of the lotus leaf is characterized by a hierarchical structure consisting of micro- and nano-scale protrusions covered with a layer of epicuticular wax. This unique structure, often referred to as the "lotus effect," enables the leaf to repel water droplets with exceptional efficiency (Zorba et al., 2008; Xu et al., 2016).

When water comes into contact with the surface of a lotus leaf, it forms spherical droplets that roll off easily, carrying dirt and contaminants away with them. This self-cleaning property ensures that the leaf remains clean and dry, even in environments where dirt and debris are prevalent. In addition, the low adhesion between water droplets and the lotus leaf surface prevents the formation of ice, making it resistant to icing even at low temperatures.

Table 11.1 Superhydrophobic surfaces present in nature.

Natural surface	WCA	Properties	References
Rice leaf	164 degrees	Superhydrophobic, self-cleaning, antifouling, low drag	Bixler and Bhushan (2014)
Mosquito compound eyes	155 degrees	Superhydrophobic, anti-fog	Gao et al. (2007)
Butterfly wings	152 ± 1.7 degrees	Superhydrophobic, self-cleaning	Han et al., (2017)
Gecko foot	150 degrees	Superhydrophobic, anti-adhesion	Stark et al., (2016)
Water strider legs	167.6 ± 4.4 degrees	Superhydrophobic, anti-adhesion	Hurchalla and Drelich (2019)
Lotus leaf	>150 degrees	Superhydrophobic, self-cleaning	Barthlott and Neinhuis (1997)
Indian cress	180 degrees	Superhydrophobic, self-cleaning	Otten and Herminghaus (2004)
Shark skin	160 degrees	Superhydrophobic, self-cleaning, antifouling	Liu and Li (2012)
Desert beetle	>150 degrees	Superhydrophobic, fog-collection behavior	Kostal et al. (2018)
Rose petal	154.6 degrees	Superhydrophobic, high surface adhesion	Feng et al. (2008)
Pitcher plant	~120 degrees−145 degrees	Slippery, antiadhesive, trapping mechanism	Chen et al. (2016)

Scientists and engineers have been inspired by the superhydrophobic properties of the lotus leaf to develop a wide range of applications, including self-cleaning coatings, anti-icing surfaces, and water-repellent materials (Koch et al., 2009; Gu et al., 2017). By mimicking the hierarchical structure and hydrophobic chemistry of the lotus leaf surface, researchers have created artificial materials that exhibit similar properties, offering solutions to various challenges in industries such as aerospace, automotive, and healthcare. Overall, the superhydrophobic lotus leaf serves as a model for bio-inspired design, demonstrating the remarkable capabilities that can be achieved through the imitation of natural systems.

11.2.2 Rose petal

In contrast to the self-cleaning characteristics of lotus leaves, red rose petals exhibit a distinct behavior with small water droplets adhering to their surfaces. These smaller droplets tend to pin on the surface, maintaining a fresh appearance, while larger droplets, like raindrops, roll off. The petal surface comprises an array of micropapillae, each featuring nanofoldings or cuticles at their tops. Despite water droplets not spreading on the rose petal surface, they adhere to it, even when the surface is inverted, a phenomenon known as the "petal effect" (Wang et al., 2017; Dou et al., 2015; Liu et al., 2017).

This superhydrophobic surface, characterized by high solid-liquid hysteresis, adopts a Cassie impregnating wetting state, allowing water to penetrate the micropapillae while retaining air gaps in the nanoscale folds (Bharat and Michael, 2010). This property of the red rose petals contributes to the fabrication of artificial superhydrophobic surfaces with high adhesive properties. These surfaces offer an effective solution for transporting small liquid volumes in microfluidic devices.

Beyond superhydrophobicity and high water adhesion, red rose petals also exhibit structural color resulting from tiny nanorods with a periodic pattern on the nanofolders (Feng et al., 2010). The collaboration between chemical color from pigments and structural color from nanostructures imparts vibrant colors to flowers, providing unique visual functions for both human and animal visual systems.

11.2.3 Rice leaf

The lotus leaf surface is renowned for enabling water droplets to freely roll in all directions, demonstrating isotropic superwettability. In contrast, the rice leaf (*Oryza sativa*) exhibits a distinct anisotropic wettability (Feng et al., 2002). Guo and Liu (2007) discovered that rice leaves possess remarkable superhydrophobicity, preventing water droplets from wetting their surface. The surface exhibits a binary micro- and nanostructure similar to lotus leaves, featuring papillae with an average diameter of 5–8 mm in a one-dimensional order. The sublayer beneath the surface contains uniformly distributed nanometer-scale pins, enhancing air-trapping capability. This surface achieves superhydrophobicity with a WCA of 157.28 degrees, facilitating the effortless rolling off of water droplets (Bixler et al., 2014).

11.2.4 Water striders

Water striders are common insects found in ponds and brooks, displaying remarkable abilities to stand, walk, and leap effortlessly on water surfaces (Dickinson, 2003; Hu et al., 2007; Gao and Feng, 2011). An insightful study by Gao and Jiang (2004) highlighted the remarkable superhydrophobic properties exhibited by water striders, focusing on their legs. The water striders' ability to stand and glide swiftly on water is attributed to the complex combination of hierarchical micro- and nanostructures. These structures include numerous oriented tiny hairs known as microsetae, which possess fine nanogrooves. These needle-shaped setae exhibit diameters ranging from 3 microns down to several hundred nanometers, with an average length of about 50 μm and an inclined angle of approximately 20 degrees from the leg's surface. This unique adaptation allows water striders to stand and move swiftly on the water's surface. Water striders serve as a unique example, showcasing resilient and long-lasting superhydrophobicity even during extended contact and significant pressure disturbances with water. This presents an opportunity for the innovative design of aquatic devices.

Replicating the structure of the water strider's leg has garnered significant attention in research, driven by its potential applications. The design of microdevices capable of standing and walking on water, inspired by the water strider, presents opportunities for activities like exploration and monitoring on water surfaces (Lepore et al., 2013; Yao et al., 2011). This inventive approach holds the potential for developing practical microdevices with applications in diverse fields.

11.2.5 Butterfly wings

The wings of the Butterfly are fascinating natural structures that have inspired numerous innovations in the field of biomimicry. These wings exhibit remarkable properties that have sparked interest and research in various scientific disciplines and engineering applications. Zheng et al. (2007) observed that butterfly wings exhibit overlapping quadrate scales (150 mm in length and 70 mm in width), forming a periodic hierarchy in one direction. Each scale's surface features separate ridging stripes, consisting of fine nano-stripes with multilayers of cuticle lamellae of different lengths and tilted slightly upward with emerging nano-tips. These distinctive structures create air pockets that significantly reduce the surface contact area with water on the wings, showcasing superhydrophobicity with a water contact angle (WCA) of 152 degrees.

Butterfly wings exhibit remarkable optical features such as the "Christmas-tree structure" providing brilliant structural color in Morpho butterflies (Butt et al., 2016) and nipple arrays enhancing the luminosity of glasswing butterfly wings (Narasimhan et al., 2018). These unique optical features result from the intricate hierarchical structures on wing scales. These hierarchical structures such as consists of scales and nano-ridges, interact with light in a way that gives rise to vibrant and iridescent colors.

In addition to their optical and self-cleaning features, butterfly wings are known for their lightweight yet strong structure. Mimicking the structural composition of butterfly wings has influenced the development of lightweight materials with enhanced strength and durability, impacting fields such as aerospace and materials science. The inspiration drawn from the properties of butterfly wings remains a driving force in pioneering research, providing solutions for challenges in diverse areas, including optics, materials engineering, and surface technology.

11.2.6 Gecko foot

The gecko possesses the remarkable capability to navigate and climb on diverse natural surfaces, due to its finely layered adhesion system on its feet. The gecko foot, with its remarkable bio-inspired properties, has been a subject of extensive research and exploration in various scientific fields. The gecko foot comprises precisely aligned microhairs known as setae, measuring approximately 5 μm in diameter and 110 μm in length, which further branch into smaller structures termed spatulae. This intricate microstructure enables the gecko to achieve exceptional adhesion through van der

Waals forces, creating molecular interactions between the setae and the surface (Autumn et al., 2000; Autumn and Gravish, 2008; Boesel et al., 2010).

Beyond its renowned reversible adhesive properties, gecko feet, featuring multi-scale structures, also demonstrate self-cleaning capabilities and superhydrophobicity, exhibiting robust adhesive forces to water. Autumn et al. in 2002 revealed the superhydrophobic nature of gecko setae, with a water contact angle (CA) of approximately 160 degrees (Autumn et al., 2002).

11.2.7 Pitcher plants

Nepenthe's pitcher plants, native to tropical regions, are famous for their carnivorous tendencies. These plants feature specialized leaves that form tall pitchers, acting as traps for insects and other small prey. When these creatures land on the rounded rim, known as the peristome, of the pitcher, they often slip and fall into a pool of digestive juices, meeting their demise (Huang and Guo, 2019; Li et al., 2024).

Researchers have recognized two primary factors contributing to the peristome's low-friction surface. First, the rim exhibits micron-sized ridges comprised of overlapping epidermal cells, resembling roof tiles, which provide directional texture. This texture facilitates easier sliding toward the interior of the pitcher, making escape more challenging for trapped insects. Second, the presence of liquid, whether from humidity, rainfall, or the plant's own nectar, creates a thin film on the microtextured surface. This moisture significantly reduces friction between the plant and insect feet, particularly when movement occurs toward the pitcher's interior. Unlike many plants' leaf surfaces that repel water, the peristome of the pitcher plant readily wets, enhancing its effectiveness as a slippery surface.

These remarkable properties of pitcher plants have served as inspiration for the development of slippery liquid-infused porous surface (SLIPS) across diverse applications. These include the design of antifouling and anticorrosion surfaces, self-healing, drag reduction, and manipulation of fluids.

11.3 Bio-inspired multifunctional coatings

Traditional coatings typically serve singular or dual purposes, such as providing decoration or protecting the substrate. In contrast, multifunctional coatings, also known as smart or intelligent coatings, represent a paradigm shift in materials science. These coatings, whether single- or multilayered thin films are engineered to offer a diverse array of functionalities within a single coating system (Moghadam et al., 2021). Beyond their conventional roles, these coatings integrate advanced features like self-healing, corrosion resistance, antifouling, anti-icing, antimicrobial properties, sensing capabilities, and even energy harvesting or storage (Wang et al., 2023; Barthwal and Lim, 2020; Lv et al., 2014; Qu et al., 2023; Barthwal et al., 2024; Ren et al., 2021). These coatings are designed to showcase a blend of properties that significantly improve performance, durability, and functionality across diverse applications. By responding to external triggers like temperature, pH, moisture, or mechanical damage, they provide on-demand protection against chemical and physical damage. This adaptability allows for the timely release of stored or embedded corrosion-inhibitor agents, safeguarding the coated material effectively.

Bio-inspired structures have garnered significant attention due to their remarkable hierarchical surfaces and exceptional functionalities. These structures serve as the foundation for bio-inspired multifunctional coatings, which mimic the properties observed in living organisms. By drawing inspiration from nature, these coatings adopt design principles found in various organisms to achieve features such as self-cleaning, antifouling, anti-icing, and enhanced mechanical strength. By harnessing nature's ingenuity, bio-inspired multifunctional coatings offer a promising solution to complex engineering challenges, providing superior performance and durability across diverse applications. Although integrating multiple bioinspired functionalities into a coating presents scientific challenges, ongoing research into biological systems continues to unveil new possibilities. The application potential of multifunctional bioinspired anticorrosive coatings is particularly appealing in areas requiring exceptional water repellence, self-cleaning, anti-icing, self-healing, antifouling, and drag reduction. Numerous multifunctional coatings have been reported in the literature, including those combining various characteristics with corrosion resistance properties as shown below.

11.3.1 Nature-inspired corrosion-resistant coating featuring self-cleaning characteristics

In the natural world, numerous surfaces exhibit superhydrophobic characteristics, with the lotus leaf serving as a well-known example due to its inherent self-cleaning ability. The lotus leaf's structure comprises hierarchical micro- and nanostructures combined with epicuticular wax. This distinctive arrangement enables water droplets to effortlessly roll off, carrying away any adhered dirt and leaving the surface clean, a phenomenon known as "self-cleaning." Inspired by the lotus leaf, researchers have achieved the creation of artificial superhydrophobic surfaces with self-cleaning properties over the past few decades (Foorginezhad and Zerafat, 2019; Kamegawa et al., 2017; Wang et al., 2016; Zheng et al., 2015; Barthwal et al., 2020; Li et al., 2022). These engineered surfaces hold great potential for diverse applications in daily life, spanning industries, agriculture, and military contexts. In architectural applications, they help maintain the esthetic appeal of buildings by preventing the accumulation of dust, pollutants, and biological growth. In automotive and aerospace industries, self-cleaning surfaces contribute to improved visibility and aerodynamic efficiency. Furthermore, the use of self-cleaning coatings in solar panels ensures sustained energy production by preventing dust and dirt accumulation. The self-cleaning functionality reduces maintenance costs, conserves water (as less cleaning is required), and contributes to a sustainable and eco-friendly approach.

Bio-inspired anticorrosion coatings with self-cleaning properties have emerged as innovative solutions in materials science, drawing inspiration from nature's design principles to address corrosion-related challenges. These coatings leverage biomimetic approaches to replicate unique features observed in living organisms, combining anti-corrosion capabilities with self-cleaning functionalities. The incorporation of superhydrophobicity is a common strategy in developing bio-inspired anticorrosion coatings with self-cleaning properties. Surfaces that exhibit superhydrophobicity repel water droplets, carrying away contaminants and preventing their adhesion, as shown in Fig. 11.4. The self-cleaning aspect is particularly valuable in corrosive environments where pollutants or debris can accumulate and compromise the integrity of traditional coatings. The development of bio-inspired anticorrosion coatings with self-cleaning properties not only addresses the challenges of corrosion protection but also aligns with sustainable and eco-friendly practices.

Zhang et al. (2024) recently employed an organic-inorganic hybrid strategy and a spray-coating technique to create a superamphiphobic coating on Q235 carbon steel, utilizing aluminum oxide nanoparticles (Al_2O_3 NPs), organosilanes, and waterborne

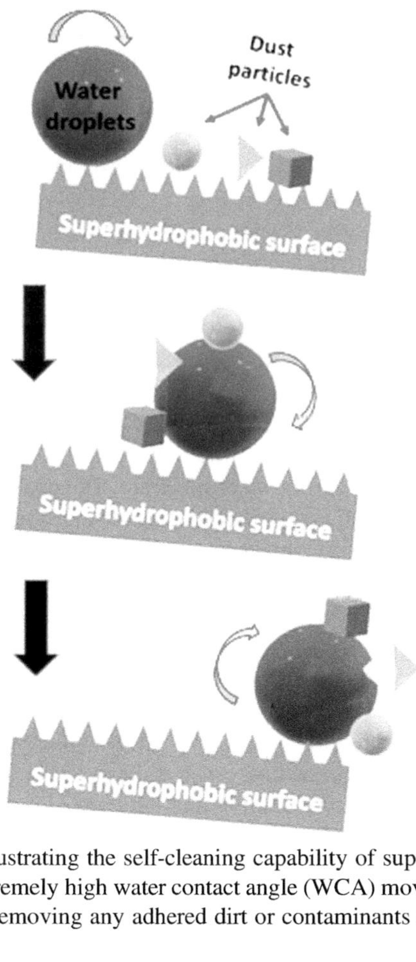

Figure 11.4 Diagram illustrating the self-cleaning capability of superhydrophobic surfaces. A water droplet with an extremely high water contact angle (WCA) moves effortlessly across the surface, picking up and removing any adhered dirt or contaminants along the way.

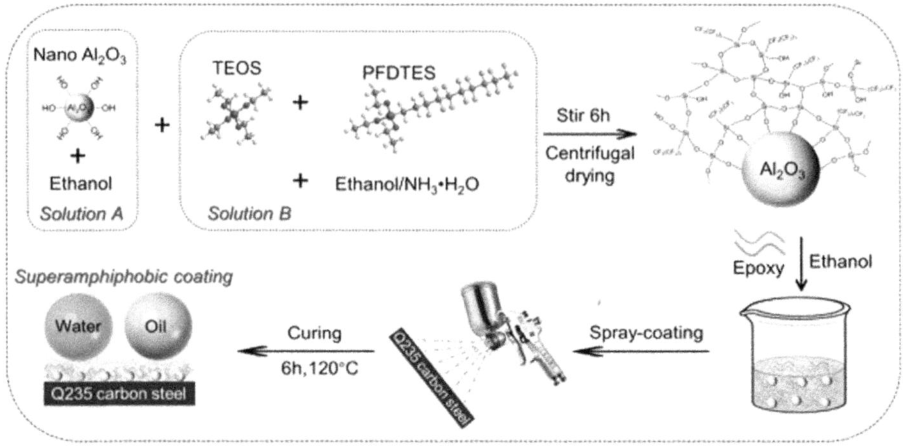

Figure 11.5 Presents a schematic depiction of the fabrication process for POS@Al_2O_3 and POS@Al_2O_3@Epoxy superamphiphobic coating.
From Zhang, B., Yang, G., Xu, W., Duan, J., Hou, B., 2024. Hybrid superamphiphobic anticorrosion coating with integrated functionalities of liquid repellency, self-cleaning, and anti-icing. J. Mater. Sci. Technol. 184, 256–268.

epoxy resin. The fabrication process of the POS@Al_2O_3 and POS@Al_2O_3@Epoxy superamphiphobic coating is shown in Fig. 11.5. The self-cleaning efficacy of the superamphiphobic coatings was assessed with liquids of varying surface tensions. Solid sand particles were initially scattered on the coating, and droplets with different surface tensions, including water, glycerol, ethylene glycol, and peanut oil, were introduced to observe the cleaning process, as displayed in Fig. 11.6. Results showed that droplets effortlessly rolled on the coating's surface, removing contaminants during the process. Moreover, the electrochemical tests revealed a significant improvement in corrosion resistance, with the coating exhibiting a positive shift in corrosion potential by 590 mV and a four-order-of-magnitude decrease in corrosion current density compared with the bare Q235 carbon steel substrate. The coating's enhanced corrosion resistance is attributed to its Cassie–Baxter nonwetting contact and low surface energy, reducing the solid-liquid contact area while increasing the liquid-gas contact area through trapped air layers in the surface structure gaps. This configuration results in excellent corrosion insulation and barrier effects.

Chu et al. (2021) developed a one-step hydrothermal method to create a cerium-based superhydrophobic coating on Mg alloy, incorporating CeO_2 and cerium stearate ($Ce(CH_3(CH_2)16COO)_3$). This resulted in an Allium giganteum-like structure with micro-nano hierarchical roughness and low surface energy, providing the coating with a high WCA of 164 degrees and a low rolling angle (RA) of 4.1 degree. The coating exhibits outstanding self-cleaning and anticorrosion properties, demonstrated through the gradual removal of contaminants like acetylene black and Al_2O_3 powder with the dropwise addition of water. The superhydrophobic surface effectively resists

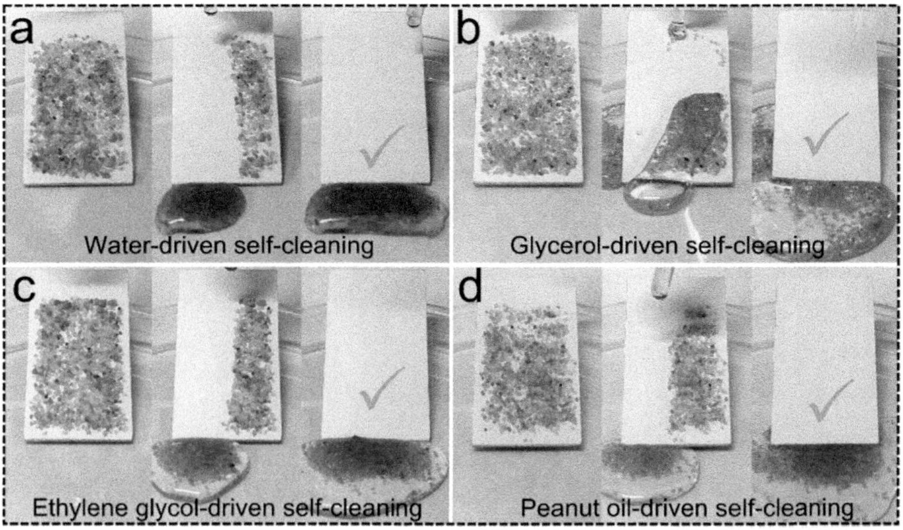

Figure 11.6 The self-cleaning process on the engineered POS@Al_2O_3@Epoxy superamphiphobic coating is induced by (a) water, (b) glycerol, (c) ethylene glycol, and (d) peanut oil.
From Zhang, B., Yang, G., Xu, W., Duan, J., Hou, B., 2024. Hybrid superamphiphobic anticorrosion coating with integrated functionalities of liquid repellency, self-cleaning, and anti-icing. J. Mater. Sci. Technol. 184, 256–268.

residual contaminants due to its low surface adhesion, highlighting excellent self-cleaning characteristics. Moreover, the corrosion rate in a 3.5 wt% NaCl aqueous solution is approximately two orders of magnitude lower than that of the bare Mg alloy, showcasing remarkable corrosion resistance attributed to the coating's water-repellent nature, reducing the contact area with the corrosive solution.

Inspired by the pinecone structure, Bai and Zhang (2020) utilized an electrodeposition method to create a reduced-graphene oxide (rGO)/Ni composite coating on stainless steel, featuring pinecone-like micro/nanostructures. The self-cleaning ability was evaluated using graphite powders as contaminants, revealing that water droplets efficiently carried away the contaminants due to the lower surface adhesion, enabling easy rolling on the superhydrophobic coating surface without tilting. Remarkably, the coating maintained robust self-cleaning and superhydrophobicity with a WCA of 155.8 degrees $\pm$ 1.2 degrees and an SA of 5.9 degrees $\pm$ 1.2 degrees even after 100 cycles of sandpaper abrasion. In addition, the coating exhibited excellent anticorrosion performance in a 3.5 wt% NaCl solution, with 99.98% inhibition efficiency.

Zhang et al. (2023) developed a carbon steel slippery coating by spraying silicone oil-soaked SiO_2 on acrylic polyurethane, using irregular coral cluster-structured SiO_2 particles prepared via a biphasic method. A sol-gel-based spherical SiO_2 coating was also prepared for comparison. The irregular coral cluster SiO_2 exhibited superior oil-locking ability, leading to reduced mass loss after the wear test and enhanced self-cleaning and

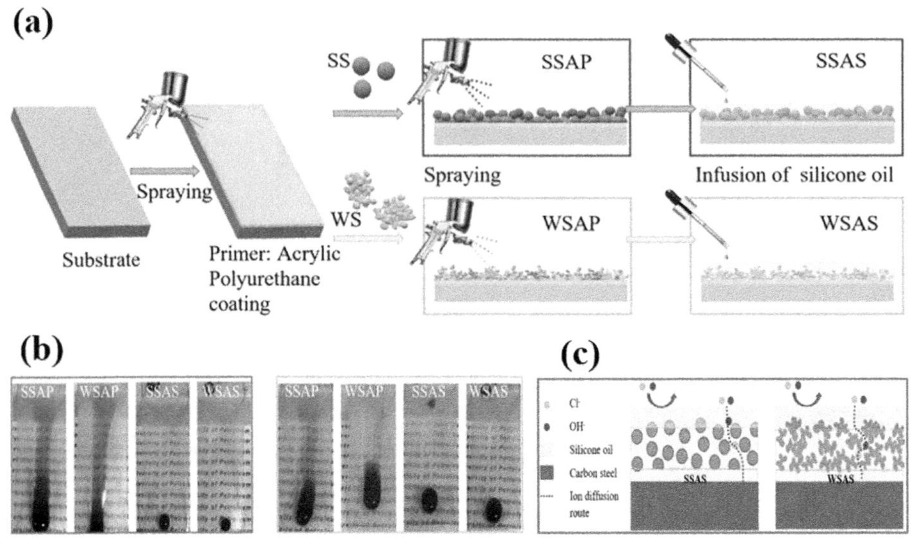

Figure 11.7 Illustrates (a) the schematic image of the preparation process, (b) the self-cleaning performance of the coatings before and after 40 abrasion cycles, and (c) the schematic of the corrosion protection mechanism of coatings on carbon steel.
From Zhang, K., Xu, S., Tan, J., Li, X., Du, L., Ma, J., Ping, Q., 2023. Fabrication of liquid-infused porous slippery coating with stability, self-cleaning and corrosion resistance properties for carbon steel. Colloids Surf. A, 669. (Original work published 2023).

corrosion resistance properties. The self-cleaning performance was evaluated through a sliding test, depicted in Fig. 11.7a and b. Coffee droplets smoothly flowed on both the SSAS (Dimethyl silicone oil injected coating mixed with SiO_2 prepared by the sol-gel method) and WSAS (Dimethyl silicone oil injected coating with SiO_2 prepared by biphasic method) surfaces before the wear test. However, postwear test, a small amount of coffee residue was observed on the SSAS, while no visible droplets remained on the WSAS. This indicates that only the WSAS coating exhibits superior self-cleaning ability both before and after the wear test. Anticorrosion results revealed a charge transfer resistance three orders of magnitude higher than conventional SiO_2 coating after a 20-day immersion test, emphasizing the WSAS coating's superior barrier against corrosive ions. The anticorrosion mechanism involves silicone oil acting as an insulator, replacing air in the base layer pores and enhancing the coating's shielding performance. With a higher silicon oil content, the WSAS coating outperforms the SSAS, providing an improved barrier against corrosive ions. In addition, the coral cluster SiO_2 filler in WSAS forms a complex labyrinth structure in acrylic polyurethane resin, prolonging the diffusion path for corrosive ions and resulting in superior corrosion resistance, as shown in Fig. 11.7c.

Ma et al. (2019) utilized a nanosecond laser treatment process to fabricate a biomimetic liquid-infused slippery surface on carbon steel. Laser-processing induced a porous multilevel micro/nanostructure on the steel. After FAS-17 modification and lubricant oil infusion, the resulting slippery surface exhibited low contact angle hysteresis. Self-cleaning performance was evaluated by tilting the samples at 15 degrees and pouring water containing Fe_3O_4 powder as a contaminant. On the base steel surface water droplets stick due to its hydrophilic nature, leaving contaminant particles after the self-cleaning test. Conversely, on the slippery surface, water droplets swiftly slid off, effectively carrying away solid contaminants, and showcasing superior self-cleaning. In addition, the anticorrosion results revealed the slippery steel's highest polarization resistance (*Rp*) value of 1778.38 kΩ cm^2 compared with the base and superhydrophobic steel surfaces, with *Rp* values of 19.18 kΩ cm^2 and 173.69 kΩ cm_2, respectively. The slippery surface demonstrated exceptional self-cleaning and anticorrosion performances, attributed to its antiwetting and insulating effects on liquid mediums.

11.3.2 Nature-inspired corrosion-resistant coating with self-healing properties

The selection of an optimal corrosion prevention method depends on several factors, including cost, efficacy, metal type, and corrosive environment. Among various methods, coatings stand out as a convenient option due to their affordability, availability of materials, versatility, and ease of application. Despite their advantages, coatings are susceptible to issues such as cracking and degradation over time, necessitating repair or replacement. Prolonged exposure to corrosive environments weakens the mechanical properties of coatings, leading to the formation and propagation of microcracks. Consequently, coating failure becomes predictable, requiring repair or replacement.

In recent years, a new class of protective coatings known as self-healing coatings has emerged. These coatings have the remarkable ability to repair damage caused by corrosive agents, thereby extending the lifespan of the coating. The development of self-healing coatings has sparked significant interest in the field of coatings research, with investigations focusing on their properties, mechanisms, synthesis techniques, and applications.

Bio-inspired self-healing multifunctional coatings represent a cutting-edge paradigm in materials science, drawing inspiration from the regenerative abilities observed in biological systems. These coatings are designed to mimic nature's ability to repair and adapt, offering a revolutionary approach to enhance the performance and durability of various surfaces. The bio inspired aspect often involves incorporating self-healing mechanisms inspired by living organisms, such as plants or animals, into the coating matrix.

To safeguard metal surfaces from corrosion, corrosion-resistant and self-healing agents are incorporated into polymeric matrices. These self-healing species exhibit the ability to autonomously repair damages or cracks, often triggered by environmental stimuli (Liu et al., 2021). Generally, self-healing coatings can be categorized as either extrinsic or intrinsic (Cui et al., 2021). One of the most effective strategies for designing extrinsic self-healing coatings involves the controlled release of healing agents. This is achieved by incorporating one or more healing agents, such as monomers, dyes, catalysts, and hardeners, into containers (microcapsules, hollow fibers, sheets, and microvascular structures). These containers are subsequently integrated into a polymeric matrix deposited on the metal substrate. When damage occurs, the self-healing agents are released to halt the corrosion reaction or autonomously repair coating defects (Dieleman et al., 2018; Ahmad et al., 2023). This biomimetic approach enables autonomous and rapid healing, akin to the way blood vessels in living organisms transport healing agents to wounded tissues. Fig. 11.8 illustrates a schematic representation of the healing process, showcasing the release of healing agents from these containers. In intrinsic self-healing protective coatings, the healing effect is primarily achieved through the material's molecular capabilities, allowing it to restore reversible internal features in response to external stimuli, thereby regaining its inherent structure. The integration of nano/microcontainers with self-healing species and corrosion inhibitors, reinforced within a polymer matrix, enhances the self-healing capability of metals for corrosion prevention. Numerous researchers are actively contributing to the advancement of smart self-healing coatings within this field (Shchukin et al., 2006; Barthwal and Barthwal, 2023; Li et al., 2015; Qian et al., 2019; Chen et al., 2016, 2020; Igwe et al., 2020; Barthwal et al., 2019).

Despite the promising advancements, the development of self-healing coatings is still in its initial stages, highlighting the importance of comprehensive exploration into their various aspects to unlock their full potential. As research in this field progresses, scientists are exploring new bio-inspired strategies, sustainable materials, and scalable manufacturing processes. Bio-inspired self-healing multifunctional coatings hold tremendous potential for revolutionizing material engineering, offering solutions for enhanced sustainability, reduced maintenance costs, and improved performance across diverse applications.

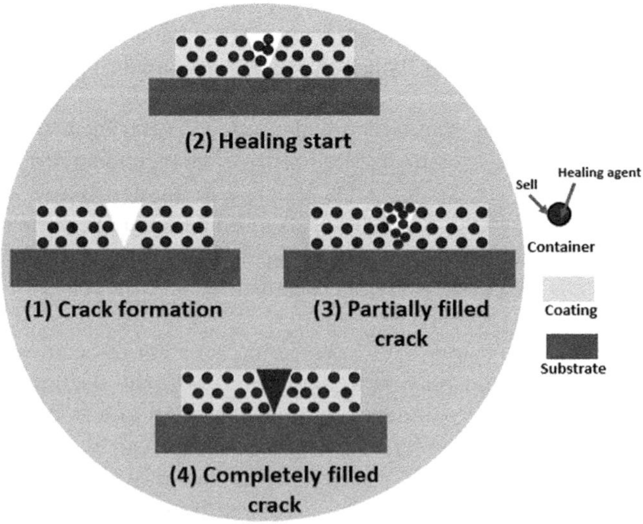

Figure 11.8 A schematic of the healing process using a release of healing agents from containers.

Ma et al. (2021) introduced a nanocomposite coating featuring dual-action self-healing corrosion protection activated through the photothermal response of plasmonic titanium nitride nanoparticles (TiN NPs). The design involved the development of TiN@mesoporous SiO_2 core-shell nanocontainers, serving as reservoirs for benzotriazole (BTA) corrosion inhibitors, which were then incorporated into the matrix of a shape memory epoxy coating. Leveraging their plasmonic property, TiN NPs efficiently absorbed light energy and converted it into heat. Upon near-infrared (NIR) irradiation, the release rate of BTA corrosion inhibitors from TiN-BTA@SiO_2 nanoparticles significantly increased. Simultaneously, the photothermal effect of TiN rapidly elevated the surface temperature of the coating, initiating the shape memory effect to restore coating scratches and impeding the penetration of corrosive solutions. Electrochemical measurements validated that the nanocomposite coating, enriched with 2 wt% of TiN-BTA@SiO_2, exhibited optimal corrosion protection along with an exceptional self-healing performance that could be activated within 30 s of NIR illumination.

Layered double hydroxides (LDHs) find extensive use in providing corrosion protection for metals. LDHs are versatile materials with a layered structure, typically consisting of positively charged metal hydroxide layers and exchangeable interlayer anions. This unique structure makes LDHs effective in hindering the corrosion process (Asl et al., 2021). When applied as coatings on metal surfaces, LDHs act as a barrier, preventing corrosive agents from directly interacting with the underlying metal. The ability of LDHs to exchange interlayer anions enhances their corrosion inhibition properties (Guo et al., 2018). The widespread use of LDHs in corrosion protection underscores their importance in enhancing the durability and longevity of metal structures in various applications. Huang et al. (2023) developed a multifunctional coating on Mg alloy by sequential surface etching, in-situ growth of layered double hydroxides (LDHs), and octadecyl-trimethoxy-silane (OTMS) modification, as shown in Fig. 11.9. The coating exhibited a WCA (>150 degrees) and WSA (<10 degrees) when exposed to various aqueous solutions. The LDHs' layered structure and the lubricating property of OTMS contributed to exceptional wear resistance. The dense silane network structure and the LDHs intermediate ion layer effectively blocked corrosive ion penetration, resulting in prolonged anticorrosion performance during a 6-week immersion in a 3.5 wt% NaCl solution. In addition, the self-healing property of the superhydrophobic coating was evident, as scratches could be spontaneously repaired through immersion in water, driven by the swelling and hydrolytic polycondensation of the silane coating, along with the attachment of LDHs corrosion products to the defects. This dual mechanism endowed the composite coating with remarkable self-healing capabilities.

Tian et al. (2023) recently introduced a novel epoxy-modified polyurethane composite coating (FD- EPU), inspired by animal cartilage tissue. This innovative coating incorporated functional reduced graphene oxide (FrGO) and 2-octyl-4,5-dichloroisothiazolinone (DCOIT), providing exceptional resistance to cavitation, corrosion, and fouling. The introduction of FrGO, characterized by multiple hydrogen bonds, not only increased the tensile strength of FD-EPU but also provided a barrier effect, endowing the coating with exceptional anticorrosion properties. Immersed in a 3.5 wt% NaCl solution for 15 days, FD-EPU

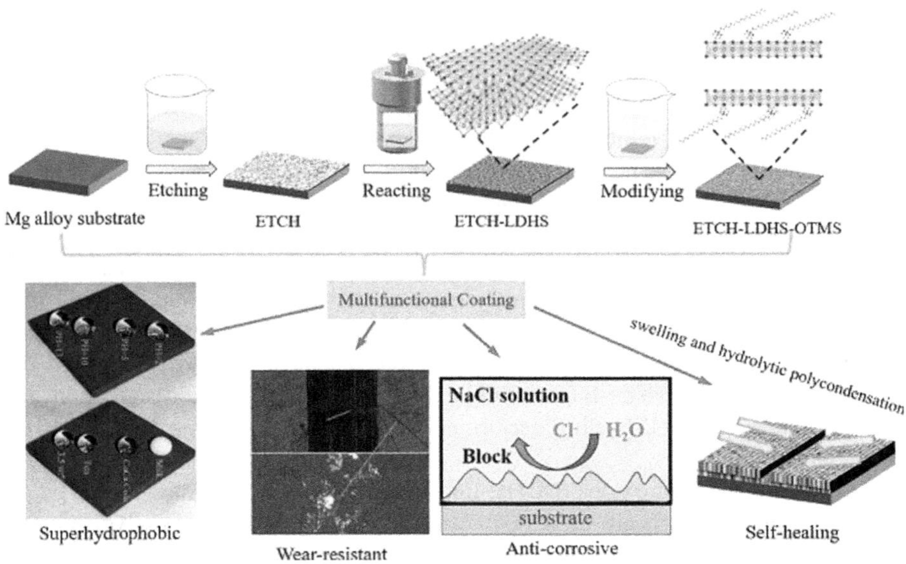

Figure 11.9 Schematic illustration of the preparation process of the composite coating on Mg alloy with multifunctional properties.
From Huang, N., Wang, Y., Zhang, Y., Liu, L., Yuan, N., Ding, J., 2023. Multifunctional coating on magnesium alloy: superhydrophobic, self-healing, anti-corrosion and wear-resistant. Surf. Coat. Technol. 463, 129539.

exhibited a significantly higher resistance to corrosion compared with EPU1. Moreover, FD-EPU demonstrated notable photothermal self-healing capabilities, leveraging graphene's ability to absorb near-infrared light. Upon exposure to 0.5 w/cm^2 near-infrared light for 30 s, FD-EPU achieved a remarkable self-healing efficiency of up to 61%. This process involves the absorption of light energy by FrGO, generating heat that breaks DA bonds to form furan rings and maleimides. Upon removal of the NIR light, the irradiated area may cool, allowing reorganization of free furan rings and maleimides, contributing to the coating's self-healing capacity.

Shi et al. (2017) proposed a method that combines galvanic dealloying and chemical vapor deposition to create a slippery liquid-infused porous surface (SLIPS) on CuZn alloy. Initially, CuZn undergoes dealloying with a CuSO4 aqueous solution, resulting in a porous Cu_2O surface through a dealloying-deposition process. The surface is then modified with dodecanethiol to achieve superhydrophobicity. The infusion of perfluoropolyether lubricant to form SLIPS is facilitated using a spin coating technique to precisely control lubricant thickness. The SLIPS exhibit remarkable corrosion inhibition capabilities, reducing the corrosion current density to 1/500 of bare CuZn. Notably, SLIPS demonstrate effective self-healing capabilities against external mechanical damage, such as cutting and erosion in seawater. Even after multiple instances of cutting or drilling, the corrosion inhibition is maintained, highlighting the coating's high tolerance to external mechanical damage facilitated by the self-healing fluid.

Jiang et al. (2021a) proposed a liquid-infused surface (LR), with self-healing properties achieved through laser ablation and regional hydrothermal growth of layered double hydroxide (LDH) films on Al alloys. Initially, laser ablation created a porous structure with a certain depth, facilitating the storage of lubricant oil in nanostructures, as shown in Fig. 11.10a. Subsequently, regional growth of LDH film via the hydrothermal method (using $Zn(NO_3)_2$ and $C_6H_{12}N_4$ at 95°C) effectively prevented the loss of lubricant oil. Finally, the prepared surface was modified to be superhydrophobic and slippery by immersion in 0.5% FAS (fluoroalkylsilane) and krytox100, respectively.

The anticorrosion and self-healing properties of sample LR were assessed through electrochemical tests after immersion in a 3.5 wt.% NaCl solution for various durations. Regarding anticorrosion ability, the corrosion current density of sample LR was recorded as 0.02 $\mu A/cm^2$ after 48 h of immersion, approximately one order of magnitude lower than that of untreated AA1060 alloy (0.25 $\mu A/cm^2$), as revealed in Fig. 11.10b. As for self-healing properties, the lubricant oil enhanced the barrier effect of LDH films, reducing susceptibility to water uptake. In addition, the porous structure

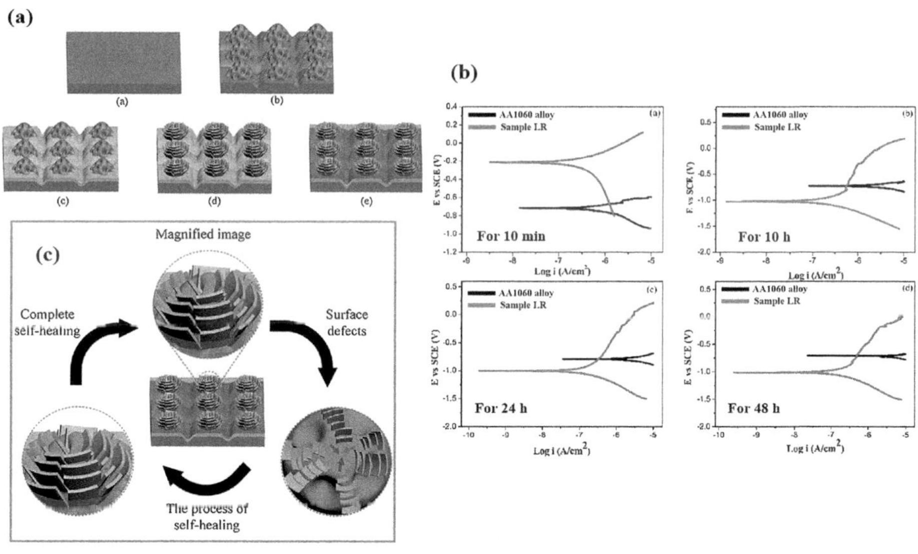

Figure 11.10 (a) Schematic diagram illustrating the preparation of a liquid-infused surface (LR), (b) Tafel curves comparing the untreated AA1060 alloy and the LR sample immersed in 3.5% NaCl solution over varying time periods, and (c) schematic diagram depicting the self-healing mechanism of the LR sample.
From Jiang, S., Zhang, H., Song, K., Liu, X., 2021. Corrosion protection application of liquid-infused surface with self-healing via regional growth of layered double hydroxide films on aluminum alloy. Colloids Surf. A 612, 125996.

created by laser ablation effectively stored the lubricant oil in gaps between nanostructures, ensuring the achievement of self-healing ability, Fig. 11.10c. Moreover, the regional growth of LDH films effectively prevented the loss of lubricant oil, ensuring the reliability of anticorrosion and self-healing abilities.

11.3.3 Nature-inspired corrosion-resistant coating with antifouling properties

Fouling is a universal issue across various industries and applications, involving the unwanted accumulation of substances such as biological organisms, sediments, and other debris on surfaces. This phenomenon is encountered in diverse fields such as marine transportation, water treatment, oil and gas production, power generation, and biomedical applications. The consequences of fouling can be severe, ranging from reduced efficiency and performance degradation to increased maintenance costs.

Fouling materials encompass both living organisms, referred to as biofouling, and inorganic/organic substances. Biofouling involves the accumulation of microorganisms, such as bacteria, algae, and barnacles, on solid surfaces, predominantly affecting the marine industry. This issue is particularly pronounced in maritime transportation, which facilitates 90% of global trade. The marine biofouling problem leads to significant economic losses, with billions of dollars annually attributed to the shipping industry. Global implications arise as fouling organisms on ship hulls contribute to increased surface roughness, elevating drag resistance and fuel consumption. Studies indicate that a micro-thick biofilm can cause a 70% rise in surface drag (Schultz et al., 2015), and heavy calcareous fouling can lead to an 86% reduction in cruising speed (Schultz, 2007). The resulting escalated fuel consumption not only contributes to financial burdens but also produces greenhouse gas emissions, posing risks of global warming (Selim et al., 2017). In addition, fouling organisms traveling with ships can introduce ecological threats by transporting invasive species across diverse marine environments (Davidson et al., 2020).

Biofouling is commonly represented as a sequential process comprising three key stages, as depicted in Fig. 11.11. Upon immersion in seawater, polysaccharides, proteins, and organic/inorganic macromolecules swiftly adhere to the surface, creating a conditioning film (Jin et al., 2022). Within minutes or hours, bacteria and diatoms initiate settlement on this conditioning film. Once settled, these microorganisms release chemicals that attract protozoa and invertebrates, culminating in the formation

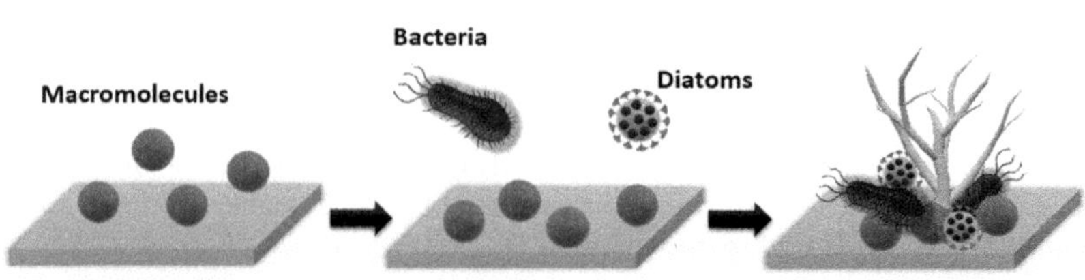

Figure 11.11 Schematic illustration of biofouling stages.

of microscopic communities. Simultaneously, bacteria and diatoms produce extracellular polymeric substances (EPS) to facilitate proliferation and counteract harmful substances, leading to the gradual development of a dense biofilm on the surface. Subsequently, macroscopic fouling organisms, including macroalgae, barnacles, shellfish, and other larger entities, adhere to the biofilm, resulting in the establishment of macroscopic fouling communities over the coming days or weeks.

The industrial fouling problem is multifaceted, presenting varied challenges across sectors including energy production, water treatment, chemical processing, shipping, and food production. For example, in the energy sector, fouling is a general issue in power plants, oil refineries, and heat exchangers. In power generation, fouling on the surfaces of boilers, condensers, and heat exchangers reduces thermal efficiency, leading to increased fuel consumption and operational costs. In oil refineries, fouling deposits on heat exchangers and piping can hinder the flow of fluids and decrease heat transfer efficiency. The impact extends to water treatment processes, affecting membrane-based technologies, desalination plants, and water distribution systems, leading to reduced filtration efficiency and increased maintenance. In the medical field, the fouling of medical devices can compromise their functionality and pose risks to patients.

Biofouling not only poses challenges to surfaces but can also accelerate corrosion, known as microbially influenced corrosion (MIC) or biocorrosion (Usher et al., 2014). This intricate phenomenon involves interactions between microorganisms and metallic structures, leading to an accelerated corrosion process. Approximately 20% of corrosion in aqueous systems is estimated to be MIC (Scholz, 1997). The global marine industry bears a significant financial burden, with biofouling costs exceeding $15 billion annually. Marine corrosion induces small cracks on surfaces, which can progressively develop into extensive corrosion, compromising surface strength and presenting serious security concerns. In MIC, microbial activity significantly impacts the corrosion behavior of metals, alloys, or materials in diverse environments, from natural settings to industrial systems.

Conventional antifouling or anticorrosion coatings, designed with a singular purpose, fall short of meeting contemporary demands. Advances in materials science and synthesis techniques have introduced both opportunities and challenges to the development of integrated coatings that simultaneously address antifouling and anticorrosion requirements.

To address fouling challenges, there is a growing focus on developing environmentally friendly antifouling coatings. Inspired form the nature, researchers exploring eco-friendly formulations or designing surfaces that resist the adherence of contaminants, thereby minimizing fouling and its associated complications. For instance, marine organisms like sharks and dolphins have evolved skin textures that inhibit the attachment of marine organisms, showcasing the potential for biomimicry in antifouling strategies. Sharks showcase inherent antifouling and self-cleaning characteristics crucial for their survival. The necessity for low-drag skin, essential for swift navigation to capture prey, underscores the importance of antifouling to maintain optimal low drag. The skin of sharks features microstructured riblets on dermal denticles, distinguishing it from other aquatic species. This unique microstructure serves a pivotal role in minimizing drag and preventing biofouling. The presence of these microstructured surfaces significantly inhibits the aggregation of barnacles, algae, and bacteria on shark skin, contributing to its antifouling and self-cleaning properties (Vincent, 2009; Sullivan and Callaghan, 2020).

Liu et al. (2022) developed a superhydrophobic coating by in-situ growing SiO_2 particles on the surface of carbon nanotubes (CNTs) using tetraethyl orthosilicate (TEOS). The resulting CNTs@SiO_2 composite particles were further modified with perfluorodecyltrimethoxysilane and dispersed within a thermoplastic polyurethane (TPU) matrix, yielding the TPU/CNTs@SiO_2 superhydrophobic composite coating. The coating, featuring multilevel micro/nanostructures of CNTs@SiO_2 and robust bonding with TPU, demonstrates remarkable mechanical durability, retaining superhydrophobicity even after sandpaper abrasion or tape peeling. The micro/nanostructures on the coating surface create air-trapping pores, enhancing resistance to corrosive media like acids, alkalis, and salt solutions. Moreover, the coating exhibits excellent chemical stability in various extreme environments and demonstrates notable antifouling and self-cleaning properties.

Li et al. (2021) developed a one-step electrodeposition method to create a superhydrophobic (SHP) coating on Cu substrates using Ca-myristic acid complex. The resulting SHP surfaces exhibit a high water contact angle of 159.30 degrees, low surface adhesive force, and excellent self-cleaning properties. Biofouling resistance was evaluated with diatoms *Navicula minima*, revealing minimal attachment on the superhydrophobic Cu surface compared with bare Cu after 7 days of immersion (Fig. 11.12a). Statistical analysis (Fig. 11.12b) indicates two orders of magnitude reduction in diatom presence on the superhydrophobic Cu surface (ca. 5.60×10^8 cells/cm^2) compared with bare Cu (ca. 6.27×10^{10} cells/cm^2). The self-cleaning

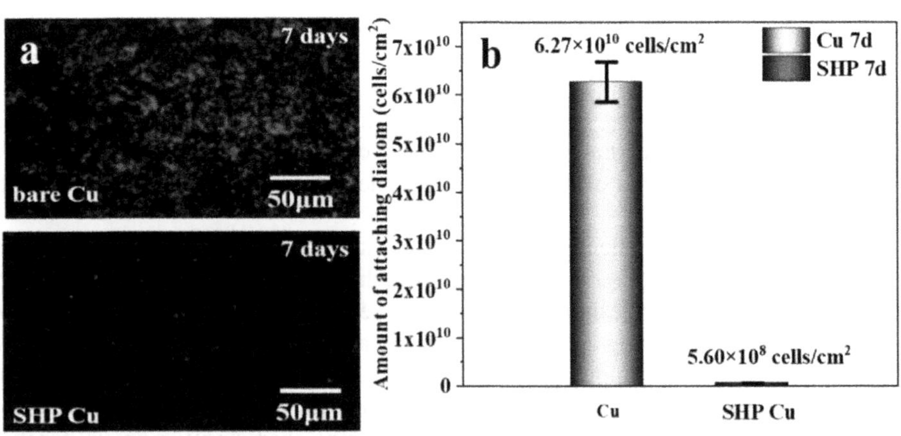

Figure 11.12 Depicts (a) the fouling resistance test of bare Cu and SHP Cu after immersion in diatom suspension for 7 days and (b) the statistical results of diatom settlement on bare Cu and SHP Cu after 7 days.
From Li, B., Miller, R.H.B., Zhang, H., Ouyang, Y., Qiu, R., Hu, S., Niu, H., 2021. A facile one-step route to fabricate bio-inspired superhydrophobic matrix with high corrosion inhibition to Cu metal. Surf. Interfaces, 25, 101189.

efficiency in seawater with diatoms is attributed to the composite nature of the superhydrophobic coating, limiting bio-organism attachment to hydrophobic protrusions. In addition, anticorrosion results reveal that, even after 120 h of immersion in 3.5wt% NaCl solution, the corrosion current density of superhydrophobic Cu remains significantly lower (1/27.5) than that of bare Cu, demonstrating its potential for effective corrosion protection in harsh environments.

Recent studies have focused on understanding interactions between synthetic materials and biological components across various scales. Self-assembled monolayers (SAMs) (Gillich et al., 2011; Chelmowski et al., 2008), and grafted polymers (Barbey et al., 2009) have been extensively explored to resist protein and cell adsorption. Despite initial success, challenges remain in maintaining long-term stability, with issues such as degradation and detachment of SAMs and grafted polymer chains. Therefore, Kuroki et al. (2013) introduced an innovative stimuli-responsive material with an antifouling surface by 3D polymer grafting of pH-responsive poly(2-vinylpyridine) chains both onto the network surface and inside the host network material (poly(ethylene oxide)). Compared with the material featuring only surface-grafted polymer chains, the proposed material with 3D polymer grafting exhibited a four-fold increase in antifouling properties. This enhanced antifouling capability is maintained through the spontaneous replacement of detached or damaged polymeric chains by the grafted PEO chains within the network.

More recently, Ma et al. (2024) proposed a fluorine-free, robust, and large-area functional slippery lubricated coating (FSLC) based on a simple blend of polysilazane (PSZ), silicone oil, and nano-silica. This straightforward, and cost-effective protocol not only promotes a hierarchical micro-nano structure but also favorable surface chemistry, facilitating robust silicone oil adhesion and excellent slippery properties (sliding angle: ~1.6 degrees) on various solid materials without additional processing or redundant treatments, as shown in Fig. 11.13a and b. The prepared surface displayed multifunctional properties, including antifouling, self-cleaning, fog collection, anticorrosion, and anti-icing functionalities.

To test the fouling resistance of FSLCs, the prepared sample was immersed in cold water (~0°C), hot water (~85°C), a 1 M HCl solution (pH = 0), a 3.5 wt% NaCl solution, and a 1 M NaOH solution (pH = 14). After a few seconds, the immersed specimens were removed, and the surfaces of all the samples exhibited no water stains, indicating good antifouling behavior, Fig. 11.13c. Corrosion tests indicated that the FSLC exhibited a lower corrosion current ($\sim 3.0 \times 10^{-12}$ A) than untreated Al ($\sim 4.0 \times 10^{-8}$ A). Furthermore, the impedance of FSLC was $\sim 6.0 \times 10^{7}\ \Omega$, significantly higher than that of Al ($\sim 1.4 \times 10^{6}$), as shown in Fig. 11.13d and e. These results demonstrate that FSLCs offer excellent corrosion resistance, crucial for protecting metal substrates.

11.3.4 Nature-inspired corrosion-resistant coating with anti-icing properties

The icing problem is a pervasive challenge in various industries and environmental conditions, posing threats to safety, efficiency, and functionality. Ice formation on surfaces can lead to severe consequences, particularly in transportation, energy, and

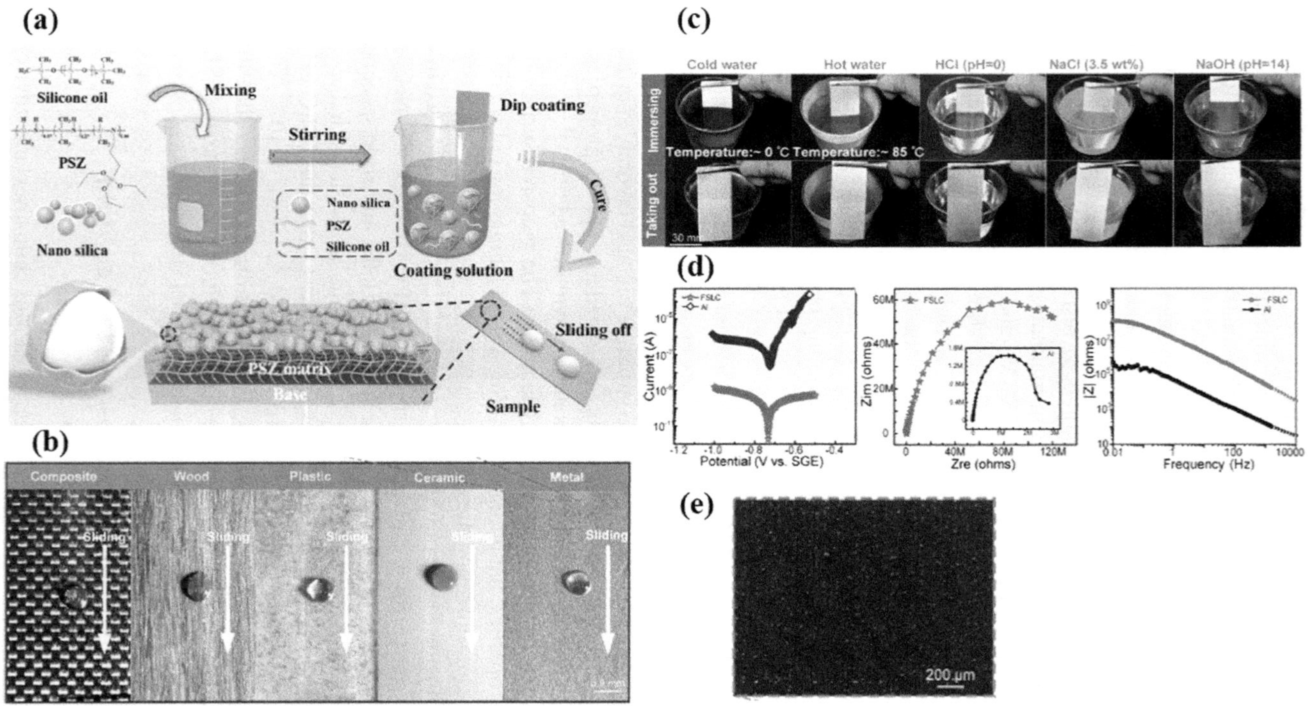

Figure 11.13 (a) Schematic of the fabrication process of FSLC surfaces, (b) FSLC applied on various substrates, (c) antifouling properties of FSLC surfaces, (d) anticorrosion properties of the prepared surface analyzed using Tafel curves, Nyquist plots, and Bode modulus plots, and (e) SEM image of the FSLC surfaceS.

From Ma, J., Zhang, C., Zhang, P., Song, J., 2024. One-step synthesis of functional slippery lubricated coating with substrate independence, antifouling property, fog collection, corrosion resistance, and icephobicity. J. Colloid Interface Sci. 664, 228–237.

infrastructure sectors. The accumulation of ice on aircraft wings, power lines, wind turbines, and other surfaces hampers aerodynamics increases energy consumption and compromises structural integrity.

To combat the challenges posed by ice formation on solid surfaces, there is a critical need for efficient strategies that either prevent ice buildup or facilitate its swift removal. Conventional active anti-icing methods, such as mechanical, heating, or chemical approaches commonly used in various applications, are often associated with drawbacks including time-consuming procedures, labor intensiveness, high energy consumption, and environmental unfriendliness. Given the significance of anti-icing surfaces in daily life, transportation, and industrial processes, the demand for an anti-icing solution that is effective, stable, environmentally friendly, and cost-efficient has become increasingly urgent.

Anti-icing or icephobicity may be characterized as the properties of the surface that prevent ice/frost formation, delay the crystallization of condensed water and reduce the ice adhesion strength (between ice and surface) for easy removal. To address this issue, researchers have turned to bio-inspired solutions for developing anti-icing coatings. Nature provides many examples where organisms have evolved mechanisms to prevent ice formation. For instance, certain plants and insects exhibit supercooling abilities, allowing them to resist ice nucleation even at subzero temperatures. Mimicking these natural strategies, bio-inspired anti-icing coatings aim to create surfaces that actively deter ice formation or facilitate easy ice removal. Presently, there are two primary approaches for designing anti-icing surfaces: employing structured surfaces modified with low-energy molecular chains, creating superhydrophobic surfaces, and utilizing slippery liquid-infused porous surfaces (SLIPSs), comprising a porous or structured substrate infused with a lubricating liquid, as shown in Fig. 11.14.

Superhydrophobic surfaces, drawing inspiration from lotus leaves and characterized by a high water contact angle (>150°C) and low contact angle hysteresis, offer numerous advantages in enhancing anti-icing properties (Barthwal and Lim, 2019; Parvate et al., 2020; Drelich and Marmur, 2014; Shen et al., 2018; Wenxuan et al., 2018; Liao et al., 2014; Alamri et al., 2020). They effectively restrain ice/frost formation, improve the anti-icing performance of impacting water droplets, and reduce ice adhesion strength. The anti-icing effectiveness of superhydrophobic surfaces is attributed to the significant air trapped between the structured surfaces (nanostructures and/

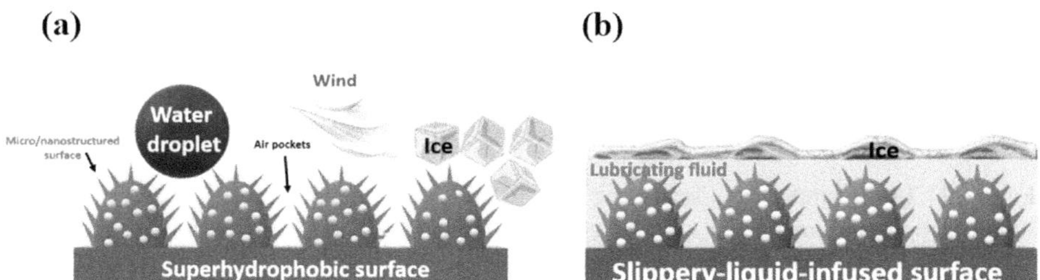

Figure 11.14 Schematic illustration of anti-icing surfaces (a) superhydrophobic and (b) SLIPs surfaces.

or microstructures). These air cushions serve as an efficient thermal barrier between the surface and supercooled water droplets, maintaining the droplets in the Cassie–Baxter state. Consequently, condensed water droplets bounce off the surface before freezing, preventing ice/frost formation. However, it is noteworthy that certain reports suggest superhydrophobic surfaces are not consistently icephobic. Under specific conditions, such as low temperatures and high humidity, a transition from the Cassie–Baxter mode to the Wenzel mode may occur, leading to a decrease in superhydrophobicity and anti-icing performance of superhydrophobic surfaces (Chen et al., 2012; Nosonovsky and Hejazi, 2012).

Cui et al. (2023) engineered hierarchical micro/nanostructures on the surface of Zr-based metallic glasses (MG) through nanosecond laser ablation, followed by heat treatment in the air. The resulting structured surface exhibited remarkable water repellency, with a water contact angle (CA) of 168.2 $\pm$ 1.5 degrees and a low sliding angle (SA) of 4 $\pm$ 0.7 degrees. The introduction of a superhydrophobic surface on the MG substrate not only enhanced its corrosion resistance by approximately 81.78% but also increased the charge-transfer impedance from 2.251×10^3 to $1.157 \times 10^4\ \Omega\ cm^2$. This improved resistance to icing was evident in the extended freezing time from 179 to 285 s, the reduced freezing temperature from -1.8 to -17.8°C, and the decreased adhesion force from 19.33 to 8.92 N.

Similarly, Lan et al. (2021) introduced a superhydrophobic surface on aluminum alloy substrates by employing ammonia etching followed by modification with 1,1,2H,2H-perfluorodecyltriethoxysilane (PFDTES), as shown in Fig. 11.15a. The resulting surface featured both hierarchical micro-nano structures and low surface energy, contributing to its superhydrophobic characteristics. The anti-icing performance

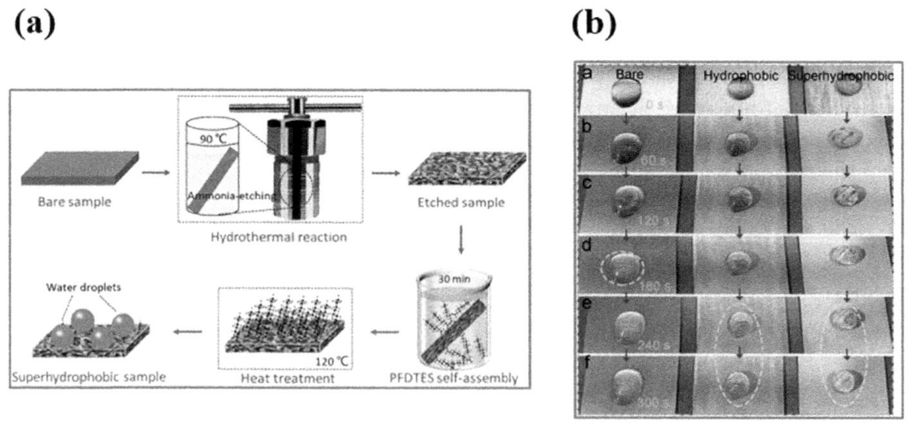

Figure 11.15 Illustrates (a) a schematic depiction of the fabrication process of the superhydrophobic samples on AA5083 sheet and (b) the static anti-icing performance of bare, hydrophobic, and superhydrophobic surfaces.
From Lan, X., Zhang, B., Wang, J., Fan, X., Zhang, J., 2021. Hydrothermally structured superhydrophobic surface with superior anti-corrosion, anti-bacterial and anti-icing behaviors. Colloids Surf. A 624, 126820.

was evaluated at −15°C, revealing that the untreated sample started freezing at 60 s and was fully iced at 180 s. In contrast, the superhydrophobic surface exhibited improved ice delay performance, with droplets initiating freezing at 120 s and complete icing occurring at 300 s, as displayed in Fig. 11.15b. In addition, the anticorrosion assessment demonstrated a four-order increase in charge transfer resistance (Rct) for the superhydrophobic surface compared with the untreated sample.

Similarly, taking cues from natural systems, the development of anti-icing slippery surfaces aims to minimize ice formation on various surfaces. Inspired by the Nepenthes pitcher plants, renowned for their slippery peristome that facilitates insect sliding into the pitcher, slippery liquid-infused porous surface (SLIPS) utilize a structured or porous substrate infused with a lubricating liquid (Wong et al., 2011; Kim et al., 2012; Kreder et al., 2016; Anand et al., 2012; Yu et al., 2024; Zhang et al., 2019; Anttonen et al., 2018; Xiang et al., 2023). This integration creates a slippery interface where the lubricating liquid, often a low-viscosity oil or a suitable substance, forms a continuous layer with exceptional anti-icing properties. These surfaces demonstrate remarkably low ice adhesion strength at low temperatures, thanks to their ultraslippery and chemically homogeneous interface. However, challenges related to the longevity and durability of SLIPSs persist, as the ice can wick away the lubricant over time, or the infused liquid may deplete during icing/deicing cycles or mechanical abrasion.

Zhang et al. (2017) developed an anti-icing slippery liquid-infused porous surface (SLIPS) on AZ31 Mg alloy using a hydrothermal process to create a double-layered structure. The porous structure was modified with a low surface-energy material to form superhydrophobic coatings, followed by the infusion of perfluoropolyether (PFPE) liquid. The results showed that the ice adhesion strength on the superhydrophobic and SLIPS coatings was measured at 62 ± 3 and 40 ± 3 MPa, respectively. The SLIPS coating exhibited lower ice adhesion strength than the superhydrophobic coating due to the reduced ice–substrate interface resulting from the fully infused lubricant. In addition, the SLIPS coating, infused with a lubricant featuring a low freezing point, provided an ultrasmooth solid–liquid interface with fewer defects, heterogeneities, and pinning points compared with the solid–solid interface of the superhydrophobic coating. Moreover, the SLIPS coating demonstrated superior anticorrosion behavior, the impedance of the SLIPS coating is four orders of magnitude larger than the LDH–carbonate coating.

11.3.5 Nature-inspired corrosion-resistant coating with oil–water separation properties

Corrosion of metal surfaces and the contamination of water bodies by oil spills represent critical challenges with significant environmental and economic implications (Aurell and Gullett, 2010; Peterson et al., 2003). Traditional approaches to address these issues often involve separate solutions for corrosion protection and oil–water separation (Naghdi and Schenk, 2016; Al-Shamrani and Xiao, 2002; Song et al., 2015).

Oil contamination of water bodies represents a pervasive environmental concern, arising from various sources such as industrial discharges, transportation accidents, and offshore drilling mishaps. The consequences of oil spills are manifold, ranging

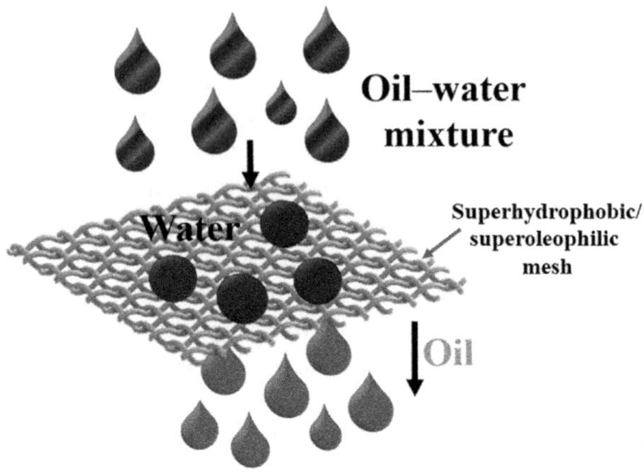

Figure 11.16 Separation mechanism of superhydrophobic surfaces.

from ecological damage to economic losses and public health hazards. Traditional cleanup methods, while employed extensively, often fall short in terms of efficiency, speed, and environmental impact, prompting the exploration of alternative approaches.

However, emerging research in biomimicry has sparked a paradigm shift toward integrated solutions inspired by nature's ingenious designs. Inspired by these natural designs, researchers have developed innovative coatings that not only shield metal substrates from corrosion but also facilitate the effective separation of oil and water. These nature-inspired coatings aim to replicate the hierarchical surface morphologies and chemical functionalities found in nature, harnessing the principles of biomimicry to create multifunctional materials with superior performance and sustainability.

Superhydrophobic surfaces utilize the disparity in surface wettability between oil and water to produce selective permeability and achieve better oil−water separation efficiency, as shown in Fig. 11.16 When a mixture of oil and water is introduced onto the surface, the water droplets are repelled and roll off, leaving behind the oil droplets, which adhere to the surface. This process effectively separates the oil from the water, allowing for the collection and removal of the oil phase while retaining the water phase. As a result, various functional materials exhibiting superhydrophobic and superoleophilic properties have been extensively investigated for efficient oil−water separation, including meshes, membranes, sponges, foams, and fabrics (Barthwal and Lim, 2021; Crick et al., 2013; Wei et al., 2020; Zhang et al., 2019; Wang and Uyama, 2016; Barthwal et al., 2022; Gu et al., 2017).

Sow et al. (2021) reported the development of an economical, fluorine-free superhydrophobic coating featuring reentrant surface topology, utilizing fly ash (FA) and room-temperature-vulcanizing silicone (RTVS). Fly ash, a by-product of coal combustion in thermal power plants and abundantly available, was used to create superhydrophobic coatings endowed with corrosion inhibition, self-cleaning, and oil−water separation properties. The microstructural roughness crucial for superhydrophobicity was achieved through FA particles, while RTVS acted as the fluorine-free low-surface energy compound and substrate-binder in the formulation of the superhydrophobic (FA-SHO) coating, as depicted in Fig. 11.17a and b. These coatings demonstrated compatibility

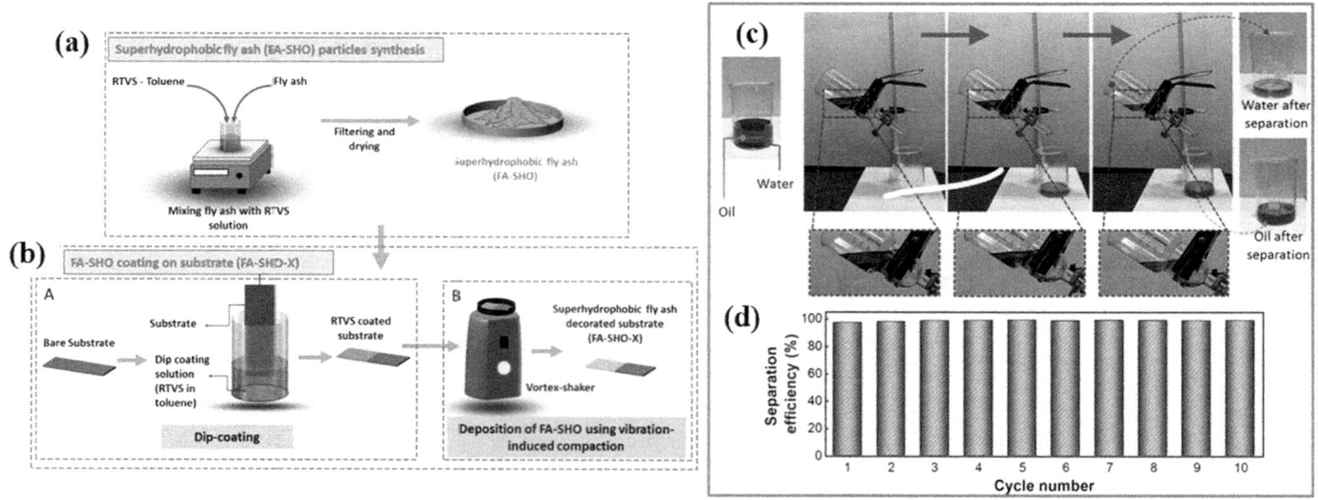

Figure 11.17 (a) Schematic of the preparation process for superhydrophobic FA (FA-SHO) particles, (b) application of the FA-SHO coating on a substrate, (c) Photographs showcasing the oil-water separation process using FA–SHO–cloth as the membrane in a gravity-driven filtration assembly, and (d) separation efficiency of oil (diesel) and water as a function of cycle number.
From Sow, P.K., Singhal, R., Sahoo, P., Radhakanth, S., 2021. Fabricating low-cost, robust superhydrophobic coatings with re-entrant topology for self-cleaning, corrosion inhibition, and oil-water separation. J. Colloid Interface Sci. 600, 358–372.

with diverse base substrates and maintained their superhydrophobic characteristics even after undergoing multiple tape-peeling cycles and exposure to various environmental conditions such as temperature, pH, and UV radiation.

Remarkably, the synthesized coating exhibited outstanding corrosion resistance when subjected to a 3.5 wt% NaCl solution, displaying an exceptionally low corrosion rate of 0.328 nm yr^{-1}, a low corrosion current density of 2.80×10^{-11} A cm^{-2}, a high impedance of 8.44×10^{9} Ω cm^{-2} at a frequency of 0.01 Hz, and an impressive corrosion inhibition efficiency of 99.999%, as shown in Fig. 11.17c and d. In addition, the FA-SHO coating, when applied to a cotton cloth, demonstrated remarkable efficacy in oil–water separation, achieving separation efficiencies exceeding 99%. Therefore, the fabricated superhydrophobic coatings hold tremendous promise for a wide range of practical and industrial applications, while simultaneously offering innovative idea for the utilization of fly ash.

Similarly, Jiang et al. (2021b) developed a hydrophobic/underwater lipophilic carbonized bacterial cellulose-polyvinylidene fluoride (PVDF) foam, with PVDF and bacterial cellulose (BC) by a simple environmentally friendly synthesis step. Initially, BC undergoes carbonization to yield carbonized bacterial cellulose (CBC), characterized by a three-dimensional porosity with larger surface roughness. PVDF, known for its low surface energy properties, is then incorporated, resulting in a PVDF-CBC foam that combines the benefits of CBC and PVDF, showcasing hydrophobicity and underwater lipophilicity.

The resulting porous PVDF-CBC foam demonstrates a WCA of 142 degrees $\pm$ 3.2. Moreover, the foam exhibits exceptional adsorption capabilities, with a cooking oil adsorption capacity exceeding 600% and an organic solvent adsorption capacity reaching up to 923%. The PVDF-CBC foam can be efficiently reused by releasing the absorbed solvent using ethanol, retaining its adsorption capacity even after 10 cycles with minimal degradation. Furthermore, the foam displays impressive corrosion resistance, even in harsh environmental conditions. Immersion tests in HCl, NaOH, and saturated NaCl solutions for 24 h reveal that the foam maintains its ability to separate more than 98% of added cooking oil, underscoring its robust performance in challenging scenarios. Gao et al. (2023) utilized a straightforward constant-potential electrodeposition technique to produce a superhydrophobic Ni-plated stainless-steel mesh (SNSSM) characterized by exceptional corrosion resistance and chemical stability, ideal for oil–water separation applications. The Ni plating featured a mountain-like micro/nanostructure, resulting in a rough surface that was further modified with stearic acid to enhance superhydrophobicity. Under optimal deposition conditions of a potential of -2.0 V and a deposition time of 60 min, the SNSSM achieved the highest water contact angle of 156.5 degrees.

Comparative analysis with the untreated SSM revealed a significant reduction in the corrosion current density of the SNSSM, decreasing from 1.728×10^{-6} mA/cm^{2} to 1.737×10^{-8} mA/cm^{2}. This reduction was accompanied by a notable increase in the charge transfer resistance and a remarkable corrosion protection efficiency of 98.99%. An air film layer within the superhydrophobic micro/nanostructure acts as a cushion and hinders the contact between the corrosive fluid and the plating, enhancing the corrosion protection efficiency. Furthermore, the SNSSM demonstrated superior corrosion protection performance compared with the SSM substrate when immersed in a 3.5 wt% NaCl solution for 192 h, retaining its superhydrophobicity even across a pH range of 2–12. In addition, the SNSSM exhibited high separation efficiency and flux for mixtures of oil or organic solvents with water, maintaining a separation efficiency exceeding 94.0% and stable reusability even after 30 cycles of separation.

Recently, Gritsevich et al. (2024) developed an SLIPS based on thin silicone films featuring grafted flexible side chains of monovinyl-terminated PDMS. These uniform coatings were synthesized using an environmentally friendly pressurized (subcritical) CO_2 process, both on smooth silicon wafers and rough porous substrates like carbon aerogels. In addition, the coatings could be achieved through traditional dip-coating methods, as evidenced on titanium slides. The resulting coatings exhibited low sliding angles toward the water and a broad spectrum of alcohol solution droplets. In the case of aerogels, the film deposition covered the entire surface and internal structure. Notably, initially, hydrophilic aerogels underwent a transformation into highly hydrophobic entities after coating, selectively absorbing oil from water/oil mixtures. Modified aerogels demonstrated the ability to absorb up to 6.6 times their dry weight of nonpolar liquids such as hexane, pump oil, and chloroform, as depicted in Fig. 11.18a and b. Furthermore, anticorrosion tests conducted on coated titanium substrates showcased significantly enhanced corrosion resistance. The anodic currents for coated titanium slides decreased by 1.5 orders of magnitude, while the corrosion potentials exhibited approximately 0.9 V more positive values compared with bare titanium slides, Fig. 11.18c and d. This innovative technique of creating slippery, hydrophobic, and anticorrosion PDMS films while preserving the initial material structure holds promise for various applications.

11.4 Challenges and future scopes

Recent advances in bio-inspired multifunctional coatings for corrosion protection have shown substantial promise in tackling corrosion-related issues while incorporating additional functionalities derived from nature. By emulating natural structures and mechanisms, researchers have created innovative coatings that offer not only excellent corrosion resistance but also features such as self-healing, self-cleaning, and environmental responsiveness. Applying nature-inspired coatings with customized functionalities can significantly benefit industries facing surface protection and maintenance challenges, reducing maintenance costs, and extending the lifespan of various substrates. Here, we have discussed several corrosion-resistant coatings inspired by lotus leaves, and slippery surfaces, highlighting their multifunctional properties.

This review aims to present the latest developments and key parameters affecting the durability and stability of bio-inspired coatings. Factors such as surface roughness and surface chemistry are crucial for the fabrication of these advanced coatings. The field of bio-inspired coatings is poised to revolutionize corrosion protection strategies across various industries, fostering the development of more durable and sustainable materials.

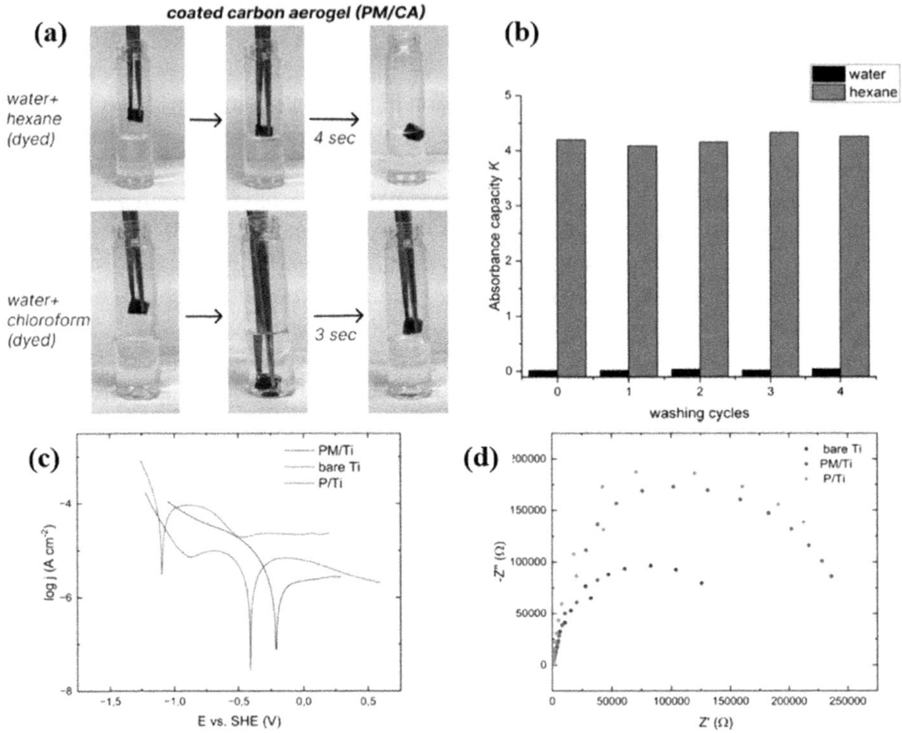

Figure 11.18 (a) Demonstration of selective oil absorption using coated CA: first row—absorbing hexane from the surface of water, second row—absorbing chloroform from beneath the water, (b) hexane and water absorbance capacity of PM/CA samples after multiple drying cycles, (c) polarization curves, and (d) phase plot of bare Ti compared with PM/Ti and P/Ti-coated surfaces.
From Gritsevich, D.K., Stamer, K.S., Zefirov, V.V., Pestrikova, A.A., Kondratenko, M.S., Gallyamov, M.O., Frank, I.V., Gulin, A.A., Kazaryan, P.S., 2024. One step synthesis of durable slippery, oil/water selective, corrosion-resistant silicone coatings with grafted flexible sidechains. Prog. Org. Coat. 192, 108496.

Despite the promising advancements in bio-inspired coatings for corrosion protection, several challenges remain to be addressed. A significant hurdle is the transition from laboratory-scale research to scalable and cost-effective industrial applications. This involves scaling up production processes, optimizing coating application methods, and ensuring compatibility with existing manufacturing infrastructure. In addition, the long-term performance and durability of bio-inspired coatings under real-world conditions require further investigation to validate their efficacy and reliability. Integrating multiple functionalities into a single coating without compromising its primary protective role is another technical challenge.

Bio-inspired coatings aim to be more sustainable, but their production processes and materials must be evaluated for environmental impact. It is crucial to ensure these coatings do not introduce new environmental hazards. Furthermore, establishing regulatory requirements and standards for bio-inspired coatings is essential to facilitate their widespread adoption and commercialization.

Looking ahead, the future of bio-inspired multifunctional coatings for corrosion protection holds tremendous promise, with several avenues for further exploration and development. Future research efforts will likely focus on enhancing the durability, sustainability, and multifunctionality of bio-inspired coatings through innovative materials design, surface engineering techniques, and advanced characterization methods. Tailoring coatings to specific application requirements, such as in marine environments or aerospace structures, will be a key area of focus. In addition, integrating bio-inspired coatings with smart materials and sensing technologies could enable real-time monitoring of corrosion processes and early detection of degradation, enhancing the overall performance and lifespan of coated materials.

References

Ahmad, S., Habib, S., Nawaz, M., Shakoor, R.A., Kahraman, R., Al Tahtamouni, T.M., 2023. The role of polymeric matrices on the performance of smart self-healing coatings: a review. J. Ind. Eng. Chem. 124, 40−67.

Alamri, S., Vercillo, V., Aguilar-Morales, A.I., Schell, F., Wetterwald, M., Lasagni, A.F., Bonaccurso, E., Kunze, T., 2020. Self-limited ice formation and efficient de-icing on superhydrophobic micro-structured airfoils through direct laser interference patterning. Adv. Mater. Interfac. 7, 2001231.

Al-Shamrani, A.J.A.A., Xiao, H., 2002. Destabilisation of oil-water emulsions and separation by dissolved air flotation. Water Res. 36, 1503−1512.

Anand, S., Paxson, A.T., Dhiman, R., Smith, J.D., Varanasi, K.K., 2012. Enhanced condensation on lubricant-impregnated Nanotextured surfaces. ACS Nano 6 (11), 10122−10129.

Anttonen, H.N., Koivuluoto, H., Tuominen, M., Teisala, H., Juuti, P., Haapanen, J., Harra, J., Stenroos, C., Lahti, J., Kuusipalo, J., Mäkelä, J.M., Vuoristo, P., 2018. Icephobicity of slippery liquid infused porous surfaces under multiple freeze−thaw and ice accretion-detachment cycles. Adv. Mater. Interfac. 5 (20), 1800828.

Asl, V.Z., Chini, S.F., Zhao, J., Palizdar, Y., Shaker, M., Sadeghi, A., 2021. Corrosion properties and surface free energy of the Zn-Al LDH/rGO coating on MAO pretreated AZ31 magnesium alloy. Surf. Coat. Technol. 426, 127764.

Aurell, J., Gullett, B.K., 2010. Aerostat sampling of PCDD/PCDF emissions from the gulf oil spill in situ burns. Environ. Sci. Technol. 44, 9431−9437.

Autumn, K., Gravish, N., 2008. Gecko adhesion: evolutionary nanotechnology. Philos. Trans. R. Soc. A 366 (1870), 1575−1590.

Autumn, K., Liang, Y.A., Hsieh, S.T., Zesch, W., Chan, W.P., Kenny, T.W., Fearing, R., Full, R.J., 2000. Adhesive force of a single gecko foot-hair. Nature 405 (6787), 681−685.

Autumn, K., Sitti, M., Liang, Y.C.A., Peattie, A.M., Hansen, W.R., Sponberg, S., Kenny, T.W., Fearing, R., Israelachvili, J.N., Full, R.J., 2002. Evidence for van der Waals adhesion in gecko setae. Proc. Natl. Acad. Sci. USA 99 (19), 12252−12256.

Bai, Z., Zhang, B., 2020. Fabrication of superhydrophobic reduced-graphene oxide/nickel coating with mechanical durability, self-cleaning and anticorrosion performance. Nano Mater. Sci. 2, 151−158.

Barbey, R., Lavanant, L., Paripovic, D., Schuewer, N., Sugnaux, C., Tugulu, S., Klok, H.A., 2009. Polymer brushes via surface-initiated controlled radical polymerization: synthesis, characterization, properties, and applications. Chem. Rev. 109, 5437.

Barthlott, W., Neinhuis, C., 1997. Purity of the sacred lotus, or escape from contamination in biological surfaces. Planta 202, 1–8.

Barthwal, S., Barthwal, S., 2023. A highly robust, non-fluorinated, and economical PDMS-based superhydrophobic flexible surface with repairable and flame-retardant properties. J. Appl. Polym. Sci. 140 (17), e53766.

Barthwal, S., Lim, S.H., 2019. Rapid fabrication of a dual-scale micro-nanostructured superhydrophobic aluminum surface with delayed condensation and ice formation properties. Soft Matter 15 (39), 7945–7955.

Barthwal, S., Lim, S.H., 2020. Robust and chemically stable superhydrophobic aluminum-alloy surface with enhanced corrosion-resistance properties. Int. J. Precis. Eng. Manuf.-Green Tech. 7, 481–492.

Barthwal, S., Lim, S.H., 2021. Durable, fluorine-free and repairable superhydrophobic aluminum surface with hierarchical micro/nanostructures and its application for continuous oil-water separation. J. Membr. Sci. 618, 118716.

Barthwal, S., Lee, B., Lim, S.H., 2019. Fabrication of robust and durable slippery anti-icing coating on textured superhydrophobic aluminum surfaces with infused silicone oil. Appl. Surf. Sci. 496, 143677.

Barthwal, S., Barthwal, S., Singh, B., Singh, N.B., 2020. Multifunctional and fluorine-free superhydrophobic composite based on PDMS modified MWCNTs/ZnO coating with oil-water separation, self-cleaning and flame retardant Properties. Colloids Surf., A 597, 124776.

Barthwal, S., Jeon, Y., Lim, S.H., 2022. Superhydrophobic sponge decorated with hydrophobic MOF-5 nanocoating for efficient oil-water separation and antibacterial applications. Sustain. Mater. Technol. 33, e00492.

Barthwal, S., Barthwal, S., Robust, E.A., 2024. Multifunctional superhydrophobic/oleophobic microporous aluminum surface via a two-step chemical etching process. Surface. Interfac. 46, 103933.

Bharat, B., Michael, N., 2010. The rose petal effect and the modes of superhydrophobicity. Phil. Trans. R. Soc. A. 368, 4713–4728.

Bixler, G.D., Bhushan, B., 2014. Rice-and butterfly-wing effect inspired self-cleaning and low drag micro/nanopatterned surfaces in water, oil, and air flow. Nanoscale 6, 76–96.

Bixler, G.D., Theiss, A., Bhushan, B., Lee, S.C., 2014. Anti-fouling properties of micro-structured surfaces bio-inspired by rice leaves and butterfly wings. J. Colloid Interface Sci. 419, 114–133.

Boesel, L.F., Greiner, C., Arzt, E., Campo, A., 2010. Gecko-inspired surfaces: a path to strong and reversible dry adhesives. Adv. Mater. 22 (19), 2125–2137.

Butt, H., Yetisen, A.K., Mistry, D., Khan, S.A., Hassan, M.U., Yun, S.H., 2016. Morpho butterfly-inspired nanostructures. Adv. Opt. Mater. 4, 497–504.

Chelmowski, R., Koster, S.D., Kerstan, A., Prekelt, A., Grunwald, C., Winkler, T., Metzler-Nolte, N., Terfort, A., Woll, C., 2008. J. Am. Chem. Soc. 130, 14952.

Chen, J., Liu, J., He, M., Li, K., Cui, D., Zhang, Q., Zeng, X., Zhang, Y., Wang, J., Song, Y., 2012. Superhydrophobic surfaces cannot reduce ice adhesion. Appl. Phys. Lett. 101, 111603.

Chen, H., Zhang, P., Zhang, L., Liu, H., Jiang, Y., Zhang, D., Han, Z., Jiang, L., 2016a. Continuous directional water transport on the peristome surface of Nepenthes alata. Nature 532, 85–89.

Chen, K., Zhou, S., Wu, L., 2016b. self-healing underwater superoleophobic and anti-biofouling coatings based on the assembly of hierarchical microgel spheres. ACS Nano 10, 1386–1394.

Chen, C., Xiao, G., He, Y., Zhong, F., Li, H., Wu, Y., Chen, J., 2020. Bio-inspired superior barrier self-healing coating: self-assemble of graphene oxide and polydopamine coated halloysite nanotubes for enhancing corrosion resistance of waterborne epoxy coating. Prog. Org. Coating 139, 105402.

Chu, J.H., Sun, G.X., Tong, L.B., Jiang, Z.H., 2021. Facile one-step hydrothermal fabrication of Allium giganteum-like superhydrophobic coating on Mg alloy with self-cleaning and anti-corrosion properties. Colloids Surf. A Physicochem. Eng. 617, 126370.

Crick, C.R., Gibbins, J.A., Parkin, I.P., 2013. Superhydrophobic polymer-coated copper-mesh; membranes for highly efficient oil-water separation. J. Mater. Chem. A 1, 5943–5948.

Cui, G., Zhang, C., Wang, A., Zhou, X., Xing, X., Liu, J., Li, Z., Chen, Q., Lu, Q., 2021. Research progress on self-healing polymer/graphene anticorrosion coatings. Prog. Org. Coating 155, 106231.

Cui, M., Huang, H., Wu, H., Zhang, L., Yan, J., 2023. Achieving superhydrophobicity of Zr-based metallic glass surface with anti-corrosion and anti-icing properties by nanosecond laser ablation and subsequent heat treatment. Surf. Coat. Technol. 475, 130159.

Davidson, I.C., Smith, G., Ashton, G.V., Ruiz, G.M., Scianni, C., 2020. An experimental test of stationary lay-up periods and simulated transit on biofouling accumulation and transfer on ships. Biofouling 36 (4), 455–466.

Deshpande, P.P., Jadhav, N.G., Gelling, V.J., Sazou, D., 2014. Conducting polymers for corrosion protection: a review. J. Coating Technol. Res. 11, 473–494.

Dickinson, M., 2003. How to walk on water. Nature 424, 621–622.

Dieleman, C.D., Denissen, P.J., Garcia, S.J., 2018. Long-term active corrosion protection of damaged coated-AA2024-T3 by embedded electrospun inhibiting nanonetworks. Adv. Mater. Interfac. 5 (12), 1800176.

Dou, X.Q., Zhang, D., Feng, C.L., Jiang, L., 2015. Bioinspired hierarchical surface structures with tunable wettability for regulating bacteria adhesion. ACS Nano 9, 10664–10672.

Drelich, J., Marmur, A., 2014. Physics and applications of superhydrophobic and super-hydrophilic surfaces and coatings. Surf. Innovations 2, 211–227.

Feng, L., Li, S.H., Li, Y.S., Li, H.J., Zhang, L.J., Zhai, J., Song, Y.L., Liu, B.Q., Jiang, L., Zhu, D.B., 2002. Super-hydrophobic surfaces: from natural to artificial. Adv. Mater. 14 (24), 1857–1860.

Feng, L., Zhang, Y., Xi, J., Zhu, Y., Wang, N., Xia, F., Jiang, L., 2008. Petal effect: a super-hydrophobic state with high adhesive force. Langmuir 24, 4114–4119.

Feng, L., Zhang, Y.A., Li, M.Z., Zheng, Y.M., Shen, W.Z., Jiang, L., 2010. The structural color of red rose petals and their duplicates. Langmuir 26 (18), 14885–14888.

Foorginezhad, S., Zerafat, M.M., 2019. Fabrication of superhydrophobic coatings with self-cleaning properties on cotton fabric based on Octa vinyl polyhedral oligomeric silsesquioxane/polydimethylsiloxane (OV-POSS/PDMS) nanocomposite. J. Colloid Interface Sci. 540, 78–87.

Foorginezhad, S., Mohseni-Dargah, M., Firoozirad, K., Aryai, V., Razmjou, A., Abbassi, R., Garaniya, V., Beheshti, A., Asadnia, M., 2021. Recent advances in sensing and assessment of corrosion in sewage pipelines. Process Saf. Environ. 147, 192–213.

Gandel, D.S., Easton, M.A., Gibson, M.A., Abbott, T., Birbilis, N., 2014. The influence of zirconium additions on the corrosion of magnesium. Corrosion Sci. 81, 27–35.

Gao, P., Feng, J.J., 2011. A numerical investigation of the propulsion of water walkers. J. Fluid Mech. 668, 363–383.

Gao, X., Jiang, L., 2004. Water-repellent legs of water striders. Nature 432 (7013), 36.

Gao, X., Yan, X., Yao, X., Xu, L., Zhang, K., Zhang, J., Yang, B., Jiang, L., 2007. The dry-style antifogging properties of mosquito compound eyes and artificial analogues prepared by soft lithography. Adv. Mater. 19, 2213–2217.

Gao, Q., Li, J., Ma, R., Fan, Y., Du, A., Yang, M., Zhao, X., Bian, Y., 2023. Facile electrodeposition of corrosion-resistant superhydrophobic Ni-plated stainless-steel mesh for oil-water separation. Mater. Today Commun. 37, 107032.

Gillich, T., Benetti, E.M., Rakhmatullina, E., Konradi, R., Li, W., Zhang, A., Schluter, A.D., Textor, M., 2011. Self-assembly of focal point oligo-catechol ethylene glycol dendrons on titanium oxide surfaces: adsorption kinetics, surface characterization, and nonfouling properties. J. Am. Chem. Soc. 133, 10940.

Gritsevich, D.K., Stamer, K.S., Zefirov, V.V., Pestrikova, A.A., Kondratenko, M.S., Gallyamov, M.O., Frank, I.V., Gulin, A.A., Kazaryan, P.S., 2024. One step synthesis of durable slippery, oil/water selective, corrosion-resistant silicone coatings with grafted flexible sidechains. Prog. Org. Coating 192, 108496.

Gu, J.C., Xiao, P., Chen, P., Zhang, L., Wang, H.L., Dai, L.W., Song, L.P., 2017. Functionalization of biodegradable PLA nonwoven fabric as superoleophilic and superhydrophobic material for efficient oil absorption and oil/water separation. ACS Appl. Mater. Interfaces 9, 5968–5973.

Gu, Y., Zhang, W., Mou, J., Zheng, S., Jiang, L., Sun, Z., Wang, E., 2017. Research progress of biomimetic superhydrophobic surface characteristics, fabrication, and application. Adv. Mech. Eng. 9, 1–13.

Guo, Z., Liu, W., 2007. Biomimic from the superhydrophobic plant leaves in nature: binary structure and unitary structure. Plant Sci. 172, 1103–1112.

Guo, W.X., Li, X.Y., Chen, M.X., Xu, L., Dong, L., Cao, X., Tang, W., Zhu, J., Lin, C., Pan, C., Wang, Z.L., 2014. Electrochemical cathodic protection powered by triboelectric nanogenerator. Adv. Funct. Mater. 24 (42), 6691–6699.

Guo, L., Wu, W., Zhou, Y.F., Zhang, F., Zeng, R.C., Zeng, J.M., 2018. Layered double hydroxide coatings on magnesium alloys: a review. J. Mater. Sci. Technol. 34, 1455–1466.

Han, Z., Fu, J., Wang, Z., Wang, Y., Li, B., Mu, Z., Zhang, J., Niu, S., 2017. Long-term durability of superhydrophobic properties of butterfly wing scales after continuous contact with water. Colloids Surf. A A518, 139–144.

Hu, D.L., Prakash, M., Chan, B., Bush, J.W.M., 2007. Water-walking devices. Exp. Fluid 43, 769–778.

Huang, C., Guo, Z., 2019. Fabrications and applications of slippery liquid-infused porous surfaces inspired from nature: a review. J. Bionic Eng. 16, 769–793.

Huang, N., Wang, Y., Zhang, Y., Liu, L., Yuan, N., Ding, J., 2023. Multifunctional coating on magnesium alloy: superhydrophobic, self-healing, anti-corrosion and wear-resistant. Surf. Coat. Technol. 463, 129539.

Hurchalla, G., Drelich, J.W., 2019. Water-repellency of hierarchical-structured legs of water-walking striders and fire ants. Surf. Innovations 7 (3–4), 184–193.

Igwe, C., May, C., Emmanuel, E.O., Hassan, A., 2020. Recently emerging nanotechnological advancements in polymer nanocomposite coatings for anticorrosion, anti-fouling and self-healing. Surface. Interfac. 21, 10074.

Jiang, R., Bian, T., Zheng, X., Cui, Y., Zhang, Y., Li, M., Li, Z., 2021b. Corrosion-resistant porous hydrophobic PVDF-CBC foam for the treatment of oil-water separation. Mater. Chem. Phys. 273, 125080.

Jiang, S., Zhang, H., Song, K., Liu, X., 2021a. Corrosion protection application of liquid-infused surface with self-healing via regional growth of layered double hydroxide films on aluminum alloy. Colloids Surf. A 612, 125996.

Jin, H., Wang, J., Tian, L., Gao, M., Zhao, J., Ren, L., 2022. Recent advances in emerging integrated antifouling and anticorrosion coatings. Mater. Des. 213, 110307.

Josyula, T., Malla, L.K., Thomas, T.M., Kalichetty, S.S., Mahapatra, P.S., Pattamatta, A., 2024. Fundamentals and applications of surface wetting. Langmuir 40 (16), 8293–8326.

Kamegawa, T., Irikawa, K., Yamashita, H., 2017. Multifunctional surface designed by nanocomposite coating of polytetrafluoroethylene and TiO_2 photocatalyst: self-cleaning and superhydrophobicity. Sci. Rep. 7 (1), 1–8.

Kim, P., Wong, T.S., Alvarenga, J., Kreder, M.J., Adorno-Martinez, W.E., Aizenberg, J., 2012. Liquid-infused nanostructured surfaces with extreme anti-ice and anti-frost performance. ACS Nano 6, 6569–6577.

Koch, K., Bhushan, B., Jung, Y.C., Barthlott, W., 2009. Fabrication of artificial lotus leaves and significance of hierarchical structure for superhydrophobicity and low adhesion. Soft Matter 5, 1386–1393.

Kostal, E., Stroj, S., Kasemann, S., Matylitsky, V., Domke, M., 2018. Fabrication of biomimetic fog-collecting superhydrophilic–superhydrophobic surface micropatterns using femtosecond lasers. Langmuir 34, 2933–2941.

Kreder, M.J., Alvarenga, J., Kim, P., Aizenberg, J., 2016. Design of Anti-icing Surfaces: smooth, textured or slippery? Nat. Rev. Mater. 1, 15003.

Kuroki, H., Tokarev, I., Nykypanchuk, D., Zhulina, E., Minko, S., 2013. Stimuli-responsive materials with self-healing antifouling surface via 3D polymer grafting. Adv. Funct. Mater. 23, 4593–4600.

Lan, X., Zhang, B., Wang, J., Fan, X., Zhang, J., 2021. Hydrothermally structured superhydrophobic surface with superior anti-corrosion, anti-bacterial and anti-icing behaviors. Colloids Surf. A 624, 126820.

Latthe, S.S., Terashima, C., Nakata, K., Fujishima, A., 2014. Superhydrophobic surfaces developed by mimicking hierarchical surface morphology of lotus leaf. Molecules 19, 4256–4283.

Lepore, E., Giorcelli, M., Saggese, C., Tagliaferro, A., Pugno, N., 2013. Mimicking water striders' legs superhydrophobicity and buoyancy with cabbage leaves and nanotube carpets. J. Mater. Res. 28, 976–983.

Li, Z., Qin, B., Zhang, X., Wang, K., Wei, Y., Ji, Y., 2015. Self-healing anti-corrosion coatings based on polymers of intrinsic microporosity for the protection of aluminum alloy. RSC Adv. 5 (126), 104451–104457.

Li, B., Miller, R.H.B., Zhang, H., Ouyang, Y., Qiu, R., Hu, S., Niu, H., 2021. A facile one-step route to fabricate bio-inspired superhydrophobic matrix with high corrosion inhibition to cu metal. Surf. Interface Anal. 25, 101189.

Li, H., Lai, J., Yao, M., Leng, Y., Wu, Z., Zhang, J., Peng, H., Qi, Z., 2022. Development of high performance superhydrophobic coating with excellent corrosion resistance, durability, and self-cleaning properties using M-TiO_2@EP composites. Appl. Surf. Sci. 601, 154109.

Li, J., Lu, B., Cheng, Z., Cao, H., An, X., 2024. Designs and recent progress of "pitcher plant effect" inspired ultra-slippery surfaces: a review. Prog. Org. Coating 191, 108460.

Liao, R., Zuo, Z., Guo, C., Yuan, Y., Zhuang, A., 2014. Fabrication of superhydrophobic surface on aluminum by continuous chemical etching and its anti-icing property. Appl. Surf. Sci. 317, 701709.

Liu, Y., Li, G., 2012. A new method for producing "Lotus Effect" on a biomimetic shark skin. J. Colloid Interface Sci. 388, 235–242.

Liu, Y., Li, X.L., Jin, J.F., Liu, J.A., Yan, Y.Y., Han, Z.W., Ren, L.Q., 2017. Anti-icing property of bio-inspired micro-structure superhydrophobic surfaces and heat transfer model. Appl. Surf. Sci. 400, 498–505.

Liu, C., Wu, H., Qiang, Y., Zhao, H., Wang, L., 2021. Design of smart protective coatings with autonomous self-healing and early corrosion reporting properties. Corrosion Sci. 184, 109355.

Liu, Y., Cao, X., Shi, J., Shen, B., Huang, J., Hu, J., Chen, Z., Lai, Y., 2022. A superhydrophobic TPU/CNTs@SiO_2 coating with excellent mechanical durability and chemical stability for sustainable anti-fouling and anti-corrosion. Chem. Eng. J. 434, 134605.

Lv, J., Song, Y., Jiang, L., Wang, J., 2014. Bio-inspired strategies for anti-icing. ACS Nano 8, 3152–3169.

Ma, Q., Wang, W., Dong, G., 2019. Facile fabrication of biomimetic liquid-infused slippery surface on carbon steel and its self-cleaning, anti-corrosion, anti-frosting and tribological properties. Colloids Surf. A 577, 17–26.

Ma, L., Wang, J., Zhang, D., Huang, Y., Huang, L., Wang, P., Qian, H., Li, X., Terryn, H.A., Mol, J.M.C., 2021. Dual-action self-healing protective coatings with photothermal responsive corrosion inhibitor nanocontainers. Chem. Eng. J. 404, 127118.

Ma, J., Zhang, C., Zhang, P., Song, J., 2024. One-step synthesis of functional slippery lubricated coating with substrate independence, anti-fouling property, fog collection, corrosion resistance, and icephobicity. J. Colloid Interface Sci. 664, 228–237.

Mazumder, M.A.J., 2020. Global impact of corrosion: occurrence, cost and mitigation. Glob. J. Eng. Sci. 5 (4), 1–5.

Moghadam, S.G., Parsimehr, H., Ehsani, A., 2021. Multifunctional superhydrophobic surfaces. Adv. Colloid Interface Sci. 290, 102397.

Mozumder, M.S., Zhang, H., Zhu, J., 2011. Mimicking lotus leaf: development of micro-nanostructured biomimetic superhydrophobic polymeric surfaces by ultrafine powder coating technology. Macromol. Mater. Eng. 296, 929–936.

Naghdi, F.G., Schenk, P.M., 2016. Dissolved air flotation and centrifugation as methods for oil recovery from ruptured microalgal cells. Bioresour. Technol. 218, 428–435.

Narasimhan, V., Siddique, R.H., Lee, J.O., Kumar, S., Ndjamen, B., Du, J., Hong, N., Sretavan, D., Choo, H., 2018. Multifunctional biophotonic nanostructures inspired by the longtail glasswing butterfly for medical devices. Nat. Nanotechnol. 13, 512–519.

Nosonovsky, M., Hejazi, V., 2012. Why Superhydrophobic surfaces are not always Icephobic. ACS Nano 6, 8488–8491.

Otten, A., Herminghaus, S., 2004. How plants keep dry: a physicist's point of view. Langmuir 20, 2405–2408.

Palanisamy, G., 2019. Corrosion Inhibitors. IntechOpen. https://doi.org/10.5772/intechopen.80542.

Parvate, S., Dixit, P., Chattopadhyay, S., 2020. Superhydrophobic surfaces: insights from theory and experiment. J. Phys. Chem. B 124, 1323–1360.

Peterson, C.H., Rice, S.D., Short, J.W., Esler, D., Bodkin, J.L., Ballachey, B.E., Irons, D.B., 2003. Long-term ecosystem response to the exxon valdez oil spill. Science 302, 2082.

Prakash, C.G.J., Prasanth, R., 2021. Approaches to design a surface with tunable wettability: a review on surface properties. J. Mater. Sci. 56, 108–135.

Qian, B., Zheng, Z., Michailids, M., Fleck, N., Bilton, M., Song, Y., Li, G., Shchukin, D., 2019. Mussel-inspired self-healing coatings based on polydopamine-coated nanocontainers for corrosion protection. ACS Appl. Mater. Interfaces 11, 10283–10291.

Qu, C., Rozsa, J.L., Jung, H.J., Williams, A.R., Markin, E.K., Running, M.P., McNamara, S., Walsh, K.M., 2023. Bio-inspired antimicrobial surfaces fabricated by glancing angle deposition. Sci. Rep. 13, 207.

Ren, J., Liu, Q., Pei, Y., Wang, Y., Yang, S., Lin, S., Chen, W., Ling, S., Kaplan, D.L., 2021. Bioinspired energy storage and harvesting devices. Adv. Mater. Technol. 6, 2001301.

Scholz, F., 1997. Microbially influenced corrosion of materials: scientific and engineering aspects. J. Solid State Electrochem. 1 (2).

Schultz, M.P., Walker, J.M., Steppe, C.N., Flack, K.A., 2015. Impact of diatomaceous biofilms on the frictional drag of fouling-release coatings. Biofouling 31 (9—10), 759—773.

Schultz, M.P., 2007. Effects of coating roughness and biofouling on ship resistance and powering. Biofouling 23 (5), 331—341.

Selim, M.S., Shenashen, M.A., El-Safty, S.A., Higazy, S.A., Selim, M.M., Isago, H., Elmarakbi, A., 2017. Recent progress in marine foul-release polymeric nanocomposite coatings. Prog. Mater. Sci. 87, 1—32.

Shchukin, D.G., Zheludkevich, M., Yasakau, K., Lamaka, S., Ferreira, M.G.S., Moehwald, H., 2006. Layer-by-layer assembled nanocontainers for self-healing corrosion protection. Adv. Mater. 18 (13), 1672—1678.

Shen, Y., Wu, Y., Tao, J., Zhu, C., Chen, H., Wu, Z., Xie, Y., 2018. Spraying fabrication of durable and transparent coatings for anti-icing application: dynamic water repellency, icing delay, and ice adhesion. ACS Appl. Mater. Interfaces 11 (3), 3590—3598.

Shi, Z., Xiao, Y., Qiu, R., Niu, S., Wang, P., 2017. A facile and mild route for fabricating slippery liquid-infused porous surface (SLIPS) on CuZn with corrosion resistance and self-healing properties. Surf. Coat. Technol. 330, 102—112.

Song, J., Lu, Y., Luo, J., Huang, S., Wang, L., Xu, W., Parkin, I.P., 2015. Barrel-shaped oil skimmer designed for collection of oil from spills. Adv. Mater. Interfac. 2, 1500350.

Sow, P.K., Singhal, R., Sahoo, P., Radhakanth, S., 2021. Fabricating low-cost, robust superhydrophobic coatings with re-entrant topology for self-cleaning, corrosion inhibition, and oil-water separation. J. Colloid Interface Sci. 600, 358—372.

Stark, A.Y., Subarajan, S., Jain, D., Niewiarowski, P.H., Dhinojwala, A., 2016. Superhydrophobicity of the gecko toe pad: biological optimization versus laboratory maximization. Philos. Trans. Royal Soc. A 374, 20160184.

Sullivan, T., Callaghan, I.O., 2020. Recent developments in biomimetic antifouling materials: a review. Biomimetics 5, 1—16.

Tian, W., Wang, S., Wang, C., Xu, H., Zhao, Y., Jin, H., Tian, L., 2023. An epoxy-modified polyurethane composite coating with repetitive self-healing function for anti-cavitation, anticorrosion, and antifouling applications. Chem. Eng. J. 477, 146849.

Usher, K.M., Kaksonen, A.H., Cole, I., Marney, D., 2014. Critical review: microbially influenced corrosion of buried carbon steel pipes. Int. Biodeterior. Biodegrad. 93, 84—106.

Vincent, J.F.V., 2009. Biomimetics—a review. Proc. IMechE 223, 919—939.

Wang, G., Uyama, H., 2016. Facile synthesis of flexible macroporous polypropylene sponges for separation of oil and water. Sci. Rep. 6, 21265.

Wang, H., Zhu, Y., Hu, Z., Zhang, X., Wu, S., Wang, R., Zhu, Y., 2016. A novel electrodeposition route for fabrication of the superhydrophobic surface with unique self-cleaning, mechanical abrasion and corrosion resistance properties. Chem. Eng. J. 303, 37—47.

Wang, P.W., Zhao, T.Y., Bian, R.X., Wang, G.Y., Liu, H., 2017. Robust superhydrophobic carbon nanotube film with Lotus leaf mimetic multiscale hierarchical structures. ACS Nano 11, 12385—12391.

Wang, L., Wang, X., Liu, T., Sun, F., Li, S., Geng, Y., Yao, B., Xu, J., Fu, J., 2023. Bio-inspired self-healing and anti-corrosion waterborne polyurethane coatings based on highly oriented graphene oxide. Npj Mater. Degrad. 7, 96.

Wei, Y., Xie, Z., Qi, H., 2020. Superhydrophobic-superoleophilic SiC membranes with micro-nano hierarchical structures for high-efficient water-in-oil emulsion separation. J. Membr. Sci. 601, 117842.

Wenxuan, Y., Yuan, Y., Guoyong, L., Bing, Z., Rongkai, Y., 2018. The anti-icing/frosting aluminum surface with hydrangea-like micro/nanostructure prepared by chemical etching. Mater. Lett. 226, 4–7.

Wong, T.S., Kang, S.H., Tang, S.K.Y., Smythe, E.J., Hatton, B.D., Grinthal, A., Aizenberg, J., 2011. Bioinspired self-repairing slippery surfaces with pressure-stable Omniphobicity. Nature 477, 443–447.

Xiang, H., Yuan, Y., Zhu, T., Dai, X., Zhang, C., Gai, Y., Liao, R., 2023. A novel durable anti-icing slippery surfaces with dendritic porous structure. Mater. Today Phys. 35, 101137.

Xu, Q., Zhang, W., Dong, C., Sreeprasad, T.S., Xia, Z., 2016. Biomimetic self-cleaning surfaces: synthesis, mechanism and applications. J. R. Soc. Interface 13, 20160300.

Yao, X., Song, Y.L., Jiang, L., 2011. Applications of bio-inspired special wettable surfaces. Adv. Mater. 23, 719–734.

Yu, A., He, S., Fu, F., Guo, Z., 2024. Solid-like slippery surface for anti-icing and efficient fog collection. Mater. Today Sustain. 26, 100754.

Zhang, J., Gu, C., Tu, J., 2017. Robust slippery coating with superior corrosion resistance and anti-icing performance for AZ31B mg alloy protection. ACS Appl. Mater. Interfaces 9, 11247–11257.

Zhang, M., Yu, J., Chen, R., Liu, Q., Liu, J., Song, D., Liu, P., Gao, L., Wang, J., 2019. Highly transparent and robust slippery lubricant-infused porous surfaces with anti-icing and anti-fouling performances. J. Alloys Compd. 803, 51–60.

Zhang, M., Ma, W., Wu, S., Tang, G., Cui, J., Zhang, Q., Chen, F., Xiong, R., Huang, C., 2019. Electrospun frogspawn structured membrane for gravity-driven oil-water separation. J. Colloid Interface Sci. 547, 136–144.

Zhang, K., Xu, S., Tan, J., Li, X., Du, L., Ma, J., Qiu, P., 2023. Fabrication of liquid-infused porous slippery coating with stability, self-cleaning and corrosion resistance properties for carbon steel. Colloids Surf. A 669, 131495.

Zhang, B., Yang, G., Xu, W., Duan, J., Hou, B., 2024. Hybrid superamphiphobic anti-corrosion coating with integrated functionalities of liquid repellency, self-cleaning, and anti-icing. J. Mater. Sci. Technol. 184, 256–268.

Zheng, Y., Gao, X., Jiang, L., 2007. Directional adhesion of superhydrophobic butterfly wings. Soft Matter 3 (2), 178–182.

Zheng, S., Li, C., Fu, Q., Li, M., Hu, W., Wang, Q., Du, M., Liu, X., Chen, Z., 2015. Fabrication of self-cleaning superhydrophobic surface on aluminum alloys with excellent corrosion resistance. Surf. Coat. Technol. 276, 341–348.

Zorba, V., Stratakis, E., Barberoglou, M., Spanakis, E., Tzanetakis, P., Anastasiadis, S.H., Fotakis, C., 2008. Biomimetic artificial surfaces quantitatively reproduce the water repellency of a lotus leaf. Adv. Mater. 20, 4049–4054.

Recent trends of micro and nanostructured conducting polymers for environmental applications

12

Isabela Jasper, Vanessa Klobukoski, Andrei Elias Deller, Raquel Anastácio Amaral, Patricio Peralta Zamora and Marcio Vidotti
Universidade Federal do Paraná (UFPR), Departamento de Química, Curitiba, Brazil

12.1 Introduction

Conductive polymers, also known as conjugated polymers, belong to a special class of polymers due to their electrical and optical properties similar to inorganic semiconductors. These characteristics are attributed to the structure of their carbon chain, which features alternating single and double bonds, with highly delocalized, polarized, and electron-dense π bonds (Namsheer and Rout, 2021; Le et al., 2017). This imparts significant electrical conductivity to conductive polymers, along with intriguing optical properties such as light absorption and emission.

Among the most common conductive polymers, we can mention polyaniline (Pani), polypyrrole (Ppy), poly(3,4-ethylenedioxythiophene) (Pedot), and polyacetylene (Pa). These materials exhibit a unique combination of properties, making them attractive for a variety of applications. The crystallinity and partially amorphous nature of conductive polymers directly influence their solubility and processability. This means that the incorporation of side groups into the polymer chain can affect their physical and chemical properties, allowing for the adaptation of conductive polymers to different needs.

In addition, the electrical and optical properties of conductive polymers are controlled through doping, which involves the addition of dopant ions (Namsheer and Rout, 2021). Dopant ions have a significant impact on the electronic structure of the polymer, influencing the mobility of charge carriers and the efficiency of electron transport. These characteristics are crucial for applications in electronic devices, such as organic transistors, solar cells, and sensors.

The transition of conductive polymers from an insulating state to a metallic state is closely associated with the generation of charge carriers, such as polarons, bipolarons, and solitons. These localized and delocalized states play a crucial role in determining the electrical properties of conductive polymers, being intimately connected to their electronic structure. Under photoexcitation or doping, conjugated polymers undergo a remarkable transformation. The π bonds within the polymer become self-

Multifunctional Nanostructured Coatings. https://doi.org/10.1016/B978-0-443-23683-9.00012-X

localized, causing the material to transition from a nonlinear transmission state to a metallic state. Conversely, in the undoped state, the polymer conductor exhibits anisotropic behavior in its electronic structure, displaying both the electrical and optical characteristics of semiconductors, alongside the mechanical properties typical of polymers (Namsheer and Rout, 2021).

Lastly, electrochemical synthesis is an important approach for the fabrication of conductive polymers. It allows for the control of doping and the polymer's structure, resulting in materials with specific properties. Detailed characterization of conductive polymers through techniques such as spectroscopy, electrochemical microscopy, and electrochemical analyses is essential to understand their properties and optimize their application in various fields, including electronics, optoelectronics, and energy devices (Stenger-Smith, 1998).

In this chapter, we have comprehensively discussed the possibilities of electrochemical synthesis of conductive polymers, their key characterizations, and various environmental applications. Fig. 12.1 schematically illustrates the concepts covered, providing an overview of these materials and their applications. Understanding and exploring the properties of conductive polymers is a promising field for the development of advanced and sustainable technologies.

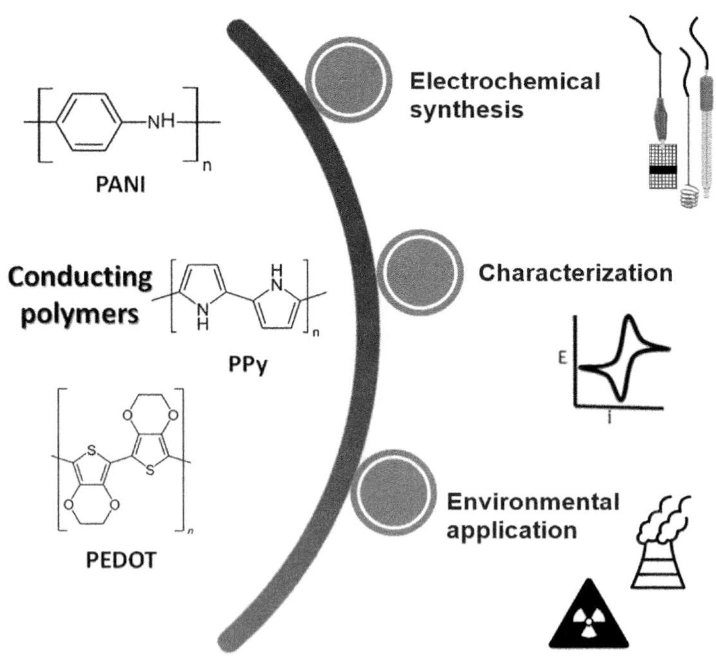

Figure 12.1 Summary scheme of the topics covered in the chapter.

12.2 Electrochemical synthesis of conducting polymers

The electrochemical synthesis of conducting polymers has been widely studied and applied due to its relevance in several areas such as electronics, sensors, electrochromic devices, and environmental applications (Namsheer and Rout, 2021; Ramanavicius et al., 2020; Inzelt, 2008).

Conducting polymers have many applications due to their electrical properties related to high theoretical capacitance, good wave absorption, high redox activity, and excellent electrochemical behavior. They are available in different morphologies such as sheets, particles, hydrogels, and rod-like structures (Namsheer and Rout, 2021).

In recent studies, much of it focuses on typical applications of conducting polymers (2016–23) as active electrode materials for energy storage applications, electrochemical sensing, and conversion fields such as electrochemical supercapacitors, lithium-ion batteries, fuel cells, and solar cells (Ramanavicius et al., 2020; Tajik et al., 2020).

12.2.1 Advantages and disadvantages over chemical synthesis

The electrochemical synthesis of conducting polymers offers several advantages when compared to conventional chemical methods. The first and most notable advantage relates to the simplicity of the process. This approach eliminates the need for extreme reaction conditions, allowing synthesis under milder conditions.

This methodology avoids the use of oxidants as the conductive polymeric material is obtained at the anode after the application of a positive potential, leading to a higher purity. Moreover, as the polymer is deposited on the electrode, subsequent electrochemical characterization is facilitated (Namsheer and Rout, 2021; Gvozdenovic et al., 2014).

Another important advantage is the higher time efficiency of the electrochemical synthesis, The formation of the conducting polymer occurs in situ during the application of an electrical potential, resulting in reduced synthesis times compared with chemical methods that require long reaction times and subsequent purification steps (Namsheer and Rout, 2021; Singh and Riaz, 2022).

This synthesis method also offers greater control over the surface morphology of the material and the chemical structure of the polymer obtained. From adjustments of electrochemical parameters, depending on the technique, it is possible to control the way in which the deposition occurs, the electrolyte used, the potential applied, and the rate at which the monomer is oxidized, so that the final shape of the material can change as a function of the modification of these variables. One of the main control factors is the thickness of the deposited film, and electrochemical synthesis allows this control easily.

Despite numerous advantages, the electrochemical technique for the synthesis of conducting polymers can offer some disadvantages. Among them, the main one could be pointed out as a limited possibility in front of the production scale. In chemical synthesis, materials can be synthesized on a large scale, which becomes a great advantage in industrial productions, for example. The electrochemical method, on the contrary, becomes less advantageous in this case, since it is a technique much more employed on a laboratory scale and the amount of product obtained depends on many limiting factors, such as the area of the electrode on which the material is deposited (anode) and the direct use of electrical energy in carrying out the synthesis (Piaskowski and Bourret, 2022).

In addition, some chemical compounds may exhibit low electrochemical reactivity, making electrochemical synthesis unfeasible or challenging for certain materials, and the technique is more sensitive to impurities and contamination than conventional chemical synthesis. The need for the use of some solvents and electrolytes may involve the need for specific and expensive components, which is a major drawback against the need for low cost (Watts et al., 2014; Guimard et al., 2007).

Another limiting factor lies in the adhesion of the conducting polymer to the substrate surface. In electrochemical synthesis, the polymer is deposited directly on the metal substrate, also called the current collector. This adhesion to the electrode can be considered a disadvantage when the synthesized polymer is to be used in other applications independent of the substrate. However, this same characteristic is an advantage when the aim is to employ the conducting polymer in the form it was synthesized, as an electrode, which is common in devices such as electrochemical sensors and supercapacitors (Namsheer and Rout, 2021; Ramanavicius et al., 2020; Gvozdenovic et al., 2014).

12.2.2 Deposition mechanisms

Electrochemical synthesis of conducting polymers is a process that utilizes electrochemical techniques for the deposition of conducting polymers onto a substrate. The most common electrochemical techniques are electropolymerization techniques such as potentiodynamic, potentiostatic, or galvanostatic (Gvozdenovic et al., 2014).

Electrochemical synthesis of conducting polymers is usually carried out in a three electrode cell arrangement, although galvanostatic synthesis can be carried out in a two electrode system. A three electrode electrochemical cell arrangement is composed of a working electrode (WE), a reference electrode (RE), and an auxiliary electrode, also called a counter electrode (CE). The working electrode is where the reaction of interest takes place, such as the polymerization and deposition of the conducting polymer. The reference electrode is used to measure the potential of the working electrode and maintain a constant potential during the reaction. The auxiliary electrode is used to complete the electrical circuit, allowing current to flow through the cell. It also helps to balance the charge generated at the working electrode.

In this type of electrochemical cell, the precursor material is used, which is the monomer that will be oxidized to promote polymerization, as well as a supporting electrolyte to allow ion transport and electron exchange. In some cases, it is necessary to use a solvent or catalysts (Chien et al., 2022; da Silva et al., 2009).

In these electrochemical methods, the electrodes are connected to the potentiostat, which is an electronic instrument used to control the potential of a working electrode relative to a reference electrode in an electrochemical cell. It does this by adjusting the current flowing to the auxiliary electrode to maintain the desired potential at the working electrode. This equipment allows the control of several parameters through the computer, which allows customizing the procedures used, making it very versatile. A three electrode cell connected to the potentiostat is shown in Fig. 12.2.

The technique is defined from the control of these parameters, since it allows to control the application of potential and current, being able to make them fixed or variable, in addition to controlling the amplitude of these quantities, application time, etc. The galvanostatic technique refers to the formation of electroconductive polymers at a constant current rate. The potentiostatic technique involves the application of a constant positive potential for oxidative polymerization. The potentiodynamic technique refers to the application of a cyclic voltammetry technique for electrochemical polymerization (Sharma, 2021).

In the galvanostatic technique, polymerization is carried out under a constant electric current, which is applied during the process. In this technique, the polymerization takes place under a constant electric current, which is applied during the process, which configures an electrolysis process. In general, the application of a positive and constant electric current result in an increasing response of the working electrode potential, which eventually promotes the desired polymerization, also known as galvanostatic oxidation (Chen, 2011; Kobayashi et al., 1984; Lv et al., 2020; Zhang et al., 2010).

The main advantage of the galvanostatic technique is the possibility to synthesize large surface area films continuously or in batches as required. This technique is especially suitable for the synthesis of polypyrrole. The synthesized films can be easily removed from the electrode substrate and stand alone. However, the galvanostatic technique has some limitations. It provides little information about the oxidative processes occurring at the electrode interface, in other words, it does not allow a detailed analysis of the reaction mechanisms involved in the polymerization process (Eftekhari and Jafarkhani, 2014; Unsworth et al., 1992).

In the potentiostatic technique, a constant and positive electrical potential is applied to promote the oxidative polymerization of the monomers present in the electrolyte. This technique is suitable for obtaining conductive polymers through the direct oxidation of the monomers, leading to the formation of conjugated bonds that give the polymer its conductive properties. Potentiostatics is useful to control the polymerization

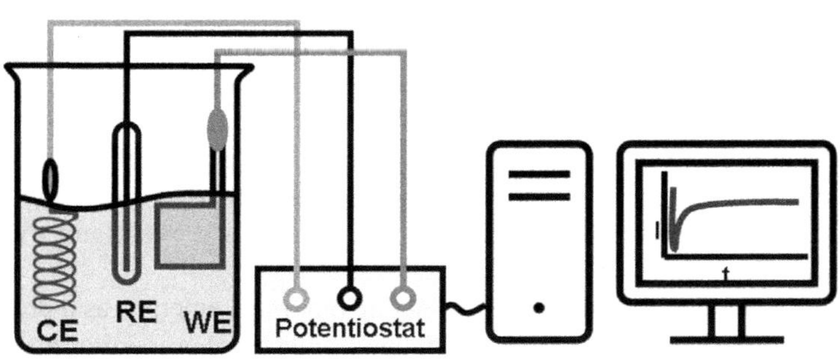

Figure 12.2 Three electrode electrochemical cell connected to potentiostat, whose parameters are set on the computer.

speed and the quality of the final product, and is commonly employed in the synthesis of conductive polymers with high purity and homogeneity (Fomo et al., 2019; Kirti et al., 2022; Tang et al., 2000; Zhou et al., 2007; Molina et al., 2014).

Among the most common potentiostatic techniques, amperometry stands out. In amperometry, a constant potential is applied to the electrode and the variation of the electric current is measured as a function of time. This technique is valuable for evaluating the kinetics of the polymerization reaction and enables the determination of important parameters such as the efficiency of electropolymerization and the rate of formation of the conducting polymer (Garcia-Cabezon et al., 2020).

One of the notable advantages of the potentiostatic technique is the ability to perform selective monomer oxidation at a specific relative potential. This avoids problems arising from mixed oxidation mechanisms, which can arise in galvanostatic techniques, and provides greater control and precision in the synthesis of the conducting polymer. The potentiostatic technique presents a significant advantage in that it allows selective oxidation of the monomer at a specific relative potential, thus avoiding the problems of mixed oxidation mechanisms that are inherent in galvanostatic techniques (Unsworth et al., 1992).

The potentiodynamic technique employs cyclic voltammetry, a potential sweep technique, to perform the electrochemical polymerization of the monomers. During cyclic voltammetry, the applied electrical potential varies cyclically between predefined limits, which allows controlling the electrochemical reactivity of the monomers and the rate of formation of the conducting polymer. This technique is useful for the synthesis of polymers with more complex structures and can be tailored to obtain materials with specific properties (Namsheer and Rout, 2021).

Cyclic voltammetry is used to synthesize conducting polymers such as polypyrrole, polyaniline, and poly(3,4-ethylenedioxythiophene) (Pedot). The synthesis of conducting polymers by cyclic voltammetry involves the repeated oxidation and reduction in a monomer in an electrolyte containing a counter ion. Oxidation of the monomer forms a cation radical that polymerizes to form the conducting polymer. Reduction in the conducting polymer forms an anionic radical that depolymerizes to form the monomer. The process is repeated several times to increase the film thickness of the conducting polymer (Lo et al., 2020; Han et al., 2020).

The cyclic potential sweep creates favorable conditions for the formation of covalent bonds between the monomers, resulting in the electrochemical polymerization of the conductive material. The electric current peaks obtained in the voltammogram during the sweep represent the redox processes of oxidation and reduction in the monomers and the conductive polymer formed. The electrochemical variables can be easily controlled and monitored, such as potential, electric charge, and electric current1. Cyclic voltammetry is a relatively simple and rapid technique that can be used to study a wide variety of chemical systems (Krische and Zagorska, 1989; Ramya et al., 2013; Shabani Shayeh et al., 2015).

The electrochemical synthesis of conducting polymers emerges as a promising and versatile technique to obtain materials with unique electrical properties, highly relevant in several application areas. Through the application of potentiostatic techniques, such as potentiostatic and cyclic voltammetry, it is possible to control the oxidative polymerization of monomers, resulting in materials with high purity, homogeneity and specific morphologies.

Although it presents some limitations, such as the limited possibility in front of the industrial production scale and the sensitivity to impurities and contamination, the electrochemical synthesis of conducting polymers has proven to be a valuable approach to meet the demands of advanced materials in technological and environmental applications (Tajik et al., 2020; Gvozdenovic et al., 2014; Casado and Mecerreyes, 2020).

Thus, electrochemical techniques play a key role in the synthesis of conducting polymers, allowing the production of materials with customized and optimized properties for various purposes. The constant advancement in this field of electrochemistry promises new discoveries and applications for these conducting polymers, driving the development of more efficient, sustainable, and innovative technologies for the future.

12.3 Conducting polymers characterization

12.3.1 Electrochemical characterization

The electrochemical characterization of conductive polymers (CPs) is performed mainly with the techniques of cyclic voltammetry and electrochemical impedance spectroscopy. They are used to investigate oxy reduction and diffusion processes that occur in CPs.

Cyclic voltammetry (CV) is an electrochemical technique generally used to acquire qualitative information on electrochemical processes under different conditions (Pacheco et al., 2013). It has wide applicability for quickly providing information on the thermodynamics of redox reactions, the kinetics of heterogeneous electron transfer reactions, and chemical reactions coupled to adsorptive processes. Generally, CV is the first experiment to be performed to study the electrochemical behavior of a compound or the surface of an electrode. It allows the quick study of the redox behavior of a system in a large range of potentials. In this technique, the potential is scanned in the direct and inverse directions in several successive cycles observing the cathodic and anodic peaks of the electroactive species (Rountree et al., 2014).

Cyclic voltammetry is one of the most popular electrochemical techniques for investigating the reduction and oxidation processes of conducting polymers and com posites. CV is also widely used in electron transfer initiated chemical reactions study, including electrocatalysis (Elgrishi et al., 2018). Each conducting polymer has a typical voltammogram with characteristic oxidation and reduction peaks that will be discussed separately.

Among the main CPs of interest, polyaniline (Pani) has a voltammetric profile with more redox peaks. Pani presents three oxidation states: the reduced state of Pani is leucoemeraldine, the redox species that provides electrical conductivity is protonated emeraldine, and pernigraniline is the fully oxidized state. During the CV scan, redox processes can be observed at less positive potentials, attributed to the reversible oxidation of leucoemeraldine to emeraldine. The second peak has been assigned to the reversible transformation of emeraldine into pernigraniline. The intermediate peak around 0.7 V is usually associated with a cross linking between polymer chains or the presence of the hydrolysis products (Huerta et al., 2021).

The electrochemical behavior of CPs is related to the change in the electrical properties of the film from an insulator to an electrical conductor, involving both electron and ion transport within the film. For the polypyrrole (Ppy), the reduction and oxidation processes involve changes in mass and resistance, as well as electronic transitions. During the reduction process, the polymer maintains its electroneutrality by expelling anions or incorporating cations (dedoping). Conversely, when a more positive potential is applied, the polymer is oxidized, and electroneutrality is maintained by the insertion of anions (doping). The Ppy properties and characteristics can be modified according to the nature of the dopant and counter ion, which significantly modifies the voltammetric profile (Ansari Khalkhali et al., 2003). Ppy synthesized in solutions with small counter ions, such as Cl^-, NO_3^-, and SO_4^{2-}, exhibits preferably anionic exchange behavior due to the high mobility of these ions in the polymer matrix (Tamm et al., 1997). Cationic exchange occurs mainly in Ppy films doped with large anions such as dodecyl sulfonate (DS), dodecyl benzene sulfonate (DBS), and polystyrene sulfonate (Pss), due to the immobility of these ions in the Ppy chain (Weidlich et al., 2001).

Regarding Pedot, the voltammetric profile exhibits a characteristic capacitive current followed by reduction and oxidation processes that are associated with the doping/dedoping process of the polymer. These processes are influenced by the type of dopant used, as compared to Ppy.

12.3.2 Electrochemical impedance spectroscopy

Another electrochemical technique of great interest in the characterization of CPs is electrochemical impedance spectroscopy (EIS). This technique allows the investigation of the electrical properties of modified surface electrodes and the charge conduction mechanism in the electrode/electrolyte interface processes. The technique involves applying a sinusoidal potential with low amplitude, causing only a minor disturbance in the system. While the potential is applied, a frequency scan is performed (Zhang and Pan, 2015). By varying the excitation frequency of the applied potential over a range of frequencies, it is possible to calculate the complex impedance. The complex impedance represents the sum of the real and imaginary impedance components of the system as a function of the frequency (angular frequency ω). Impedance results can be represented using Nyquist or Bode plots. In the case of Nyquist plots, the complex impedance (Z) can be presented as a sum of the real impedance ($Z'(\Omega)$) and the imaginary impedance ($Z''(\Omega)$). These components primarily arise from the resistance and capacitance of the system. The typical electrochemical impedance spectrum can be displayed in the form of Nyquist plots, where Z'' is plotted versus Z', with the data recorded as a function of frequency (Laschuk et al., 2021).

The typical EIS response of a conducting polymer exhibits the formation of a semicircle in the high frequency region, which is attributed to the faradaic processes occurring at the electrode/electrolyte interface. At the low frequency, an inclined line is observed, which is related to the amount of charge stored within the polymer film after the oxidation process. This charge is responsible for maintaining charge neutrality (Wolfart et al., 2017). To fit the data, an equivalent circuit consisting of several elements can be used to model the system's electrical response over a wide frequency range (Ferloni et al., 1996). Impedance data from conductive polymers can be analyzed using a modified Randles circuit. This circuit consists of a series resistance (R_s), which accounts for the resistance of the electrolyte, electrical connections, and the resistance of the film deposited on the electrode. It also includes a constant phase element associated with the capacitance of the double electric layer (Q_{dl}); the charge transfer resistance (R_{ct}), which represents the resistance of the faradaic interfacial process from the redox reactions, and another constant phase element related to the low frequency capacitance (Q_{lf}). The values of n_{dl} and n_{lf} vary from zero to one, and when n assumes a value of one, it indicates a perfectly flat electrode (Ferloni et al., 1996; Marchesi et al., 2015).

Another form of representing EIS data is through a Bode plot. This representation describes the frequency dependent behavior of electrical systems and contains crucial information about the sensitivity to changes in capacitance, resistance, and ion diffusion corresponding to the applied frequency. The Bode plot is constructed with a logarithmic frequency scale on the x-axis and both the logarithm of the absolute value of the impedance ($|Z|$) and the phase angle (θ) on the y-axis (Laschuk et al., 2021). A typical representation of Nyquist and Bode plots for a conducting polymer is shown in Fig. 12.3.

12.3.3 Electrical conductivity

One of the most intriguing properties of CPs is their intrinsic conductivity and the ability to modulate this conductivity by an order of magnitude of 10 times. The conductivity is not present in the reduced form of the polymer, which corresponds to the

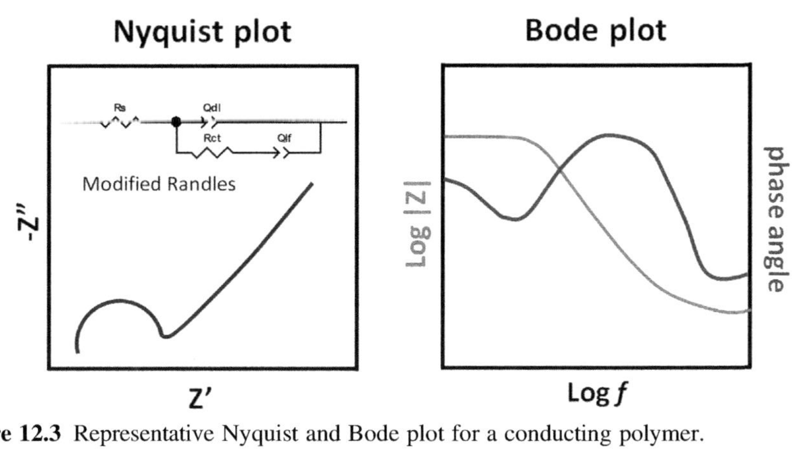

Figure 12.3 Representative Nyquist and Bode plot for a conducting polymer.

undoped state. However, the conductivity arises from the formation of charge carriers upon oxidizing (p-doping) or reduction (n-doping) of the polymer's conjugated backbone. This process of charge formation is accompanied by changes in optical and electrical properties and is counterbalanced by the introduction of dopants (counter ions) into the polymer chain (Tajik et al., 2020).

The electrical conductivity of CPs can be measured using either the two-point or four-point technique. In the two-point technique, the CP is deposited between two electrodes with a small gap between them. A small direct current probe is applied between the two electrodes and the resulting current or voltage is measured. The resistance measured by this method includes the polymer's resistance and the contact between the polymer and substrate. In the four-probe method, the measurement of conductance includes only the bulk polymer, without considering the potential drop at the polymer/substrate interface. In this method, a constant current is applied between two external electrodes, and the potential difference between the internal electrodes is measured (Lange and Mirsky, 2011).

12.3.4 Morphological characterization

The morphology and structure of CPs can be examined using scanning electron microscopy (SEM) and transmission electron microscopy (TEM), while the topography is investigated by atomic force microscopy (AFM). These techniques allow for detailed visualization and analysis of the CP's surface and internal structure at different magnifications.

The scanning electron microscope is used to obtain detailed topographic, morphological, and compositional information from a wide variety of materials, including metals, alloys, polymers, minerals, and biological samples, among others. The image can be obtained using secondary electrons (for topographic contrast) and backscattered electrons (for topographic and/or compositional contrast) with resolution capacity at the 1 nm scale. In addition, it is also possible to gather information about the chemical composition of the sample through the detection of X-rays and energy dispersion (EDS). Although SEM can provide rich morphological and compositional data for a wide range of materials, it is often necessary to employ additional tools to achieve a complete characterization of materials. For instance, if the objective is to examine the true microstructure of a material (rather than just surface topography), transmission electron microscopy (TEM) is employed (Ul-Hamid, 2018). TEM utilizes a transmitted electron beam to capture images of the sample, offering a resolution of up to 0.2 nm. This microscopic technique has been extensively employed in nanomaterial research and can reveal the relationship between nanoparticles and conductive polymers and investigate other characteristics of CPs nanostructures (Malatesta, 2021).

The morphological structure of CPs plays a crucial role in determining their performance, making morphology control a valuable tool for enhancing material properties. Consequently, the structural characterization of CPs is essential, in addition to the electrochemical characterization. It is worth noting that the morphology of CPs, especially in the nanoscale, can significantly impact their electrochemical properties.

It is well-known that the morphology of CPs is influenced by various synthetic conditions, such as temperature, pH, oxidant species, template, oxidant/monomer ratio, and polymerization type, among others. Typically, CP film morphology is characterized by the formation of globules that are stacked together in an irregular shape. This morphology arises primarily from the growth process occurring in multiple layers. However, it is possible to prepare CPs with various nanostructures including nanospheres, nanorods, nanowires, nanotubes, and nanofibers (Xue et al., 2020). Fig. 12.4 illustrates some examples of CP morphology and structures.

For an example of how CP morphology can be easily modulated by the synthesis conditions, in the work of Iurchenkova and collaborators (Iurchenkova et al., 2021), SEM and High-Resolution Transmission Microscopy (HRTEM) were used to characterize the morphology of Ppy synthesized in different media (acetonitrile, ethanol, and water). The microscopy investigation revealed that all samples had an amorphous microstructure, forming a dense structure. The nature of the medium only affected the particle size, while the shape and microstructure of the particles remained consistent across all samples.

One of the most reliable techniques for investigating changes in polymer surface topography is Atomic Force Microscopy (AFM), which enables the examination of surface architecture at the nanoscale. This technique has been widely employed in the morphological characterization of insulating materials, semiconductors, and conducting polymers (Alessandrini and Facci, 2005). The significant advantage of using AFM for morphological characterization, in comparison with SEM and TEM is that this technique does not require sample preparation and can be performed in an ambient atmosphere. As a result, scanning is faster and introduces fewer artifacts. However, image retrieval in AFM is slower than in SEM (Eaton and West, 2010).

Atomic force microscopy is a technique that enables the visualization and precise measurement of surface structures with high resolution. AFM allows the acquisition of images that depict the individual arrangement of atoms in a sample and the observation of the individual structure of molecules (Eaton and West, 2010). The sample's topography is scanned by a probe composed of a sharp tip near the end of the cantilever. This probe detects the forces between itself and the sample, with a sensitivity of piconewton and sub-nanoscale resolution (Krieg et al., 2019). In addition to imaging the topography at the nanometer scale, AFM's phase imaging mode is utilized to map the heterogeneity of composites, blends, and hybrid materials. This is achieved due to the cantilever phase response being sensitive to surface properties such as adhesiveness, electrical forces, capacitance, magnetic forces, conductivity, viscoelasticity, and resistance. Moreover, AFM has demonstrated its capability to provide qualitative and semiquantitative information about material degradation and aging processes (Eaton and West, 2010).

AFM has been employed in various applications to study alterations in the topography of conducting polymer films and nanostructures. These investigations encompass changes resulting from different electrosynthesis parameters, the composites formation, and variations in film volume due to expansion and contraction processes, among other factors (Smela and Gadegaard, 2001; Ahmed et al., 2011; Patois et al., 2011).

Figure 12.4 (a) SEM image of Ppy film; (b) TEM image of Pedot nanotube decorated AuNps; (c) SEM image of Ppy nanospheres; (d) SEM image of Pani nanorods; (e) SEM image of polypyrrole nanofiber arrays obtained by biphasic electropolymerization on a preroughened Pt electrode; (f) low-magnification SEM image of polypyrrole nanowires synthesized by cathodic electropolymerization different morphologies of conductive polymers, figures taken from different articles.
From Soares, A.L., Hryniewicz, B.M., Deller, A.E., Volpe, J., Marchesi, L.F., Souto, D.E.P., Vidotti, M., 2021. Electrodes based on PEDOT nanotubes decorated with gold nanoparticles for biosensing and energy storage. ACS Appl. Nano Mater. 4(9), 9945−9956. https://doi.org/10.1021/acsanm.1c02677. (a) Reprinted with permission from reference Iurchenkova, A.A., Kallio, T., Fedorovskaya, E.O., 2021, Relationships between polypyrrole synthesis conditions, its morphology and electronic structure with supercapacitor properties measured in electrolytes with different ions and pH values. Electrochim. Acta. 391, 138892, 00134686. https://doi.org/10.1016/j.electacta.2021.138892; (b) Reprinted with permission from Soares, A.L., Hryniewicz, B.M., Deller, A.E., Volpe, J., Marchesi, L.F., Souto, D.E.P., Vidotti, M., 2021. Electrodes based on PEDOT nanotubes decorated with gold nanoparticles for biosensing and energy storage. ACS Appl. Nano Mater. 4(9), 9945−9956. https://doi.org/10.1021/acsanm.1c02677; (c) Reprinted with permission from reference Wu, W., Yang, L., Chen, S., Shao, Y., Jing, L., Zhao, G., Wei, H., 2015. Core-shell nanospherical polypyrrole/graphene oxide composites for high performance supercapacitors. Royal Society of Chemistry, China RSC Adv. 20462069. 5 (111), 91645−91653. https://doi.org/10.1039/c5ra17036b; (d) Reprinted with permission from reference Hu, Q., Sun, D., Ma, Y., Qiu, B., Guo, Z., 2017. Conductive polyaniline nanorods enhanced methane production from anaerobic wastewater treatment. Elsevier Ltd, China Polymer. 00323861. 120, 236−243. https://doi.org/10.1016/j.polymer.2017.05.073; (e) Reprinted with permission from reference Li, M., Wei, Z., Jiang, L., 2008. Polypyrrole nanofiber arrays synthesized by a biphasic electrochemical strategy. J. Mater. Chem. 13645501, 18 (19), 2276−2280, https://doi.org/10.1039/b800289d; (f) Reprinted with permission from reference Nam, D.H., Kim, M.J., Lim, S.J., Song, I.S., Kwon, H.S., 2013. Single-step synthesis of polypyrrole nanowires by cathodic electropolymerization. J. Mater. Chem. A. 20507496. 1 (27), 8061−8068, https://doi.org/10.1039/c3ta11227f.

12.4 Conducting polymers applied to degradation of environmental contaminants

The growing concern for sustainability has sparked extensive research on the development and responsible usage of chemical products across various sectors, including energy, health, and the environment. Ensuring a sustainable trajectory for societal development has become a critical and pressing priority. Industrial practices, unrestricted chemical use, and inadequate waste management have led to the spread of pollutants in soil, water, and air, detrimentally affecting both the quality of life and ecological equilibrium. Proper disposal and treatment of these products postuse are essential to mitigate adverse environmental impacts that may arise if they are not managed responsibly (Petrie et al., 2015; Bolan et al., 2014).

Early detection and efficient degradation of environmental contaminants are crucial to mitigate their harmful effects. The ability to detect even low levels of pollutants is highly sought after and paramount, as it enables more effective strategies for combating the presence of contaminants, even at high concentrations. By implementing proactive monitoring and remediation techniques, we can address environmental issues at their onset and prevent further escalation of detrimental impacts (Gavrilescu et al., 2015). The competent authorities have established limits for pesticide concentrations in the environment, and these values may vary depending on local legislation. Due to this difference, many works use local legislation as a parameter but also use the limits consumed by the World Health Organization (WHO) (World Health Organization, WHO recommended classification of pesticides by hazard and guidelines to classification, 2019).

In scientific research, achieving low detection limits is always a desirable goal. However, for practical real-world applications, the reliability of the device within a range that aligns with local legislation or WHO limits becomes the determining factor. In this context, conductive polymers emerge as a promising and viable solution for environmental monitoring and pollutant remediation. Their unique properties make them well-suited for sensitive detection and efficient removal of pollutants, aligning with the stringent regulatory requirements set forth by environmental authorities. Harnessing the potential of conductive polymers holds great promise for addressing environmental challenges and ensuring a safer and healthier future for our planet (Gavrilescu et al., 2015; Fang et al., 2018).

Conductive polymers have properties that enhance the systems into which they are incorporated. Thus, it is possible to develop more robust and high-performance devices, which can be applied for environmental monitoring and combating environmental contaminants. This makes conductive polymers ideal for the development of sensitive and selective sensors, capable of detecting even low concentrations of contaminants in the environment. With these sensors, it is possible to continuously monitor the presence of pollutants, allowing quick and efficient responses, thus helping in environmental monitoring (Fang et al., 2018; Facure et al., 2017; Sall et al., 2020).

In addition to monitoring, conductive polymers can also be applied for the degradation of environmental contaminants. Conductive polymers can act as catalysts in the degradation processes of organic and inorganic pollutants, making the process more efficient and economically viable. Therefore, the next topics will be dedicated to addressing the applications of conductive polymers both in monitoring and combating environmental contaminants.

12.4.1 Polluting and harmful gases

The detection and degradation of polluting and harmful gases is of great importance to ensure human and environmental health. In this context, conductive polymers are considered suitable materials to be used, as they offer advantages of operating at room temperature, ease of chemical modification, mechanical stability under external stress application (Wong et al., 2020; Bag and Lee, 2021).

Different types of gas sensors can be used together with conductive polymers, including those that use quartz-crystal microbalance (Torad et al., 2021), field effect transistors (Lee et al., 2005), and chemiresistors (Lange and Mirsky, 2011). Of all the sensors, chemoresistive sensors are the most widely used due to the easy and cheap synthesis process, easy operation, and small size (Shakeel et al., 2022). This sensor can determine the change in resistance of the detection material after target gas exposure. When gas adsorption occurs, polymers chemically interact and react with gaseous molecules through weak intermolecular forces and can change the doping concentration of conductive polymers. When a reducing gas is adsorbed onto the surface of a p-type conducting polymer, the conductivity of the polymer decreases as the gas molecules donate electrons to the polymer. In turn, the opposite phenomenon happens with an oxidizing gas. Thus, if the conductivity of the polymer is detected, the type and concentration of adsorbed gas analytes can be predicted (Bag and Lee, 2021).

Another area of broad application for conductive polymers is in photocatalysts and photoelectrocatalysts. These materials have spatially extended π-conjugated electron systems, which have sparked increasing interest in research due to their wide absorption capacity in the visible light range, high charge carrier mobility, and excellent stability (Dai et al., 2015). The properties of commonly used conductive polymers for photocatalytic applications are shown in Table 12.1.

Recent studies have demonstrated that conductive polymers, when used in hybrid structures modified with materials such as TiO_2(Gao et al., 2016), Cu_2O (de la Cruz et al., 2021) and ZnO (Liu et al., 2019), for example, exhibit enhanced photocatalytic performance for the degradation of polluting gases. Polymer-inorganic material compounds increase active sites, potential conductivity, selectivity, and detection and degradation capacity (Shakeel et al., 2022). Furthermore, when these materials are on the nanoscale, it is possible to synthesize smaller devices that can be easily portable. The excellent conductivity of conductive polymers contributes to accelerating the transfer of photo-generated electrons, thereby reducing electron-hole recombination during the photocatalytic process. This characteristic is crucial for the efficiency and effectiveness of these materials in atmospheric pollutant degradation.

Table 12.1 Properties of conductive polymers commonly used in photocatalytic processes.

Conducting polymer	Conductivity (S cm^{-1})	Bandgap (eV)
Polypyrrole (Ppy)	10^2–7.5×10^3	3.1
Polyaniline (Pani)	30–200	3.2
Poly(3,4-ethylenedioxythiophene) (Pedot)	300	1.1
Polyacetylene (Pa)	10–10^3	1.5
Polythiophene (Pth)	10^2–1.7×10^5	2.1
Poly(phenylene vinylene)s (Ppv)	3–5×10^3	2.5

Conductive polymers, such as polyaniline (Pani), play a multifunctional role in enhancing the transfer of photo-generated carriers and positively influencing gas chemisorption. A notable example of this situation is the chemisorption of carbon dioxide (CO_2) by the nitrogen-containing functional groups present in the structure of Pani. Recent studies reveal that the presence of CO_2 significantly accelerates the reduction in water (H_2O) to hydrogen (H_2) for catalysts containing Pani (Xie et al., 2016).

In this context, a relevant aspect is the ability of Pani to become highly conductive when exposed to an acid. The addition of an acid, known as acid doping, has the power to dramatically increase the electrical conductivity of Pani. Thus, the interaction with CO_2, an acidic molecule, may similarly promote an increase in the conductivity of catalysts containing Pani. This resulting phenomenon has significant implications for electron transfer and electron-hole separation, which are fundamental in photocatalytic and photoelectrocatalytic processes (Xie et al., 2016).

Furthermore, the CO_2 chemisorption capacity by the nitrogen-functional groups of Pani paves the way for the potential application of these materials in CO_2 capture and conversion technologies. The selective capture of CO_2, followed by its conversion into other useful chemical products, could significantly contribute to mitigating the greenhouse effect and producing high-value-added products such as ethanol and methanol.

These findings highlight the versatility and promise of conductive polymers, especially polyaniline, in various technological applications, ranging from enhancing the performance of photocatalysts and photoelectrocatalysts to the potential detection of gases by sensors. However, it is important to emphasize that further research and developments are needed to fully comprehend the potential of these materials and optimize their properties for practical applications.

12.4.2 Organic molecules

The sensitivity of conductive polymers to organic molecules is based on the affinity of the polymer backbone, the affinity of side groups, and the binding of immobilized receptors. For this, biological receptors (such as nucleic acids, antibodies, enzymes) or synthetic receptors (such as cyclodextrins, calixarenes, phenylboronic acid) can be used (Lange et al., 2008). Immobilization of these receptors is based on noncovalent interactions or covalent binding of a receptor to a matrix. As an example of these interactions, we can mention the incorporation of β-cyclodextrin in polypyrrole for the detection of dopamine, L-dopa, ascorbic acid, chlorpromazine and neurotransmitters derived from pyrogallol and catechol (IZAOUMEN et al., 2005) and also the incorporation of the enzyme horseradish peroxidase in a thin Pani film for the detection of glyphosate (Roman et al., 2020).

The application of conductive polymers in fuel cells also stands out, making it possible to convert organic waste into electricity through electrochemical reactions. These materials are used as a support material for electrocatalysts, thus enabling the electrooxidation of small molecules such as hydrogen, methanol, and formic acid using hybrids based on conductive polymers with metals or metal oxides deposited on the polymer's surface (Ghosh et al., 2020).

The advantages of using conductive polymers in these cells are related to the following items (Dutta and Kundu, 2014).

- Conductive polymers can be used as a catalyst support matrix for electrodes, independently and in combination with carbonaceous materials, primarily due to their electronic conductivity;
- The presence of a conjugated backbone in conductive polymers leads to a more efficient delocalization of electrons compared with the commercial carbon matrix, which conducts electric charge through an extensive hopping mechanism;
- The broad surface area provided by conductive polymers promotes more efficient dispersion and distribution of catalyst particles. Consequently, the particles exhibit a larger active surface area, which significantly benefits electrochemical reactions;
- The incorporation of heteroatoms, especially nitrogen, forms a strong interaction with the metal particles of the catalyst, ensuring a robust anchoring of the catalytic particles dispersed in the polymeric matrix, thereby resulting in greater stability and efficiency of the catalyst;
- Conductive polymers are a cost-effective option, leading to a significant reduction in the usage and consumption of precious metal catalysts. This advantage results in an effective decrease in the overall cost involved in the manufacturing of electrodes;
- Conductive polymers are lightweight, which contributes to the generation of high power and high current per unit of weight;
- The synthesis of these conductive polymers involves simple reaction routes and, therefore, represents an advantage in reducing the complexity associated with fuel cell manufacturing;
- Conductive polymers demonstrate high thermal stability within the temperature range in which fuel cells operate, providing stability to the overall system;

Due to all these characteristics, the use of conductive polymers holds promising potential for applications in fuel cells. In addition, the degradation of toxic organic pollutants through advanced oxidation processes (AOPs) using conductive polymers is also possible (Ait El Fakir et al., 2021). For AOPs, conductive polymers containing nitrogen can facilitate electron transfer due to their conjugated chains. As nitrogen atoms are located within the conjugated chains, they are more stable than their counterparts in the defective sites of carbocatalysts. These heteroatoms can activate the inert carbon structure by increasing the charge density of adjacent carbon atoms, thus promoting interactions occurring within the catalyst.

Among conductive polymers, polypyrrole stands out in the preparation of conductive membranes (Karimi et al., 2014; Liu et al., 2012; Bhattacharya et al., 2008), due to its relatively high conductivity and good environmental stability. Its compact and rigid structure, with weakly basic anionic groups that are interchangeable, allows for easy polymerization through chemical or electrochemical oxidation. However, due to the vulnerability of these polymers to electrochemical oxidation corrosion, conductive polymeric membranes are often used as cathodes (Pan et al., 2019).

Thus, it is evident that conductive polymers exhibit remarkable and promising characteristics in various technological applications, both for the degradation and detection of organic molecules. Their potential is noteworthy and can significantly contribute to progress in various fields, including fuel cells, advanced oxidation processes, receptors, and other applications.

12.4.3 Pesticides

The application of pesticide controllers in the development of pesticide sensors has shown a promising approach to developing devices that can detect pesticide concentrations in agricultural environments and the environment as a whole.

According to IUPAC, pesticides belong to the class of agrochemicals, which in turn are chemicals used in the production of crops and food. The IUPAC provides a complete glossary, not only with a definition of a pesticide but also with various terms and definitions of concepts related to pesticide studies. This glossary contains two definitions for the term pesticide, the main one being:

> *"Strictly, a substance intended to kill pests: in common usage, any substance used for controlling, preventing, or destroying animal, microbiological, or plant pests."*

Still, the IUPAC suggests 20 different types of classifications for pesticides, where we can highlight here some of the best known: herbicides (fight unwanted weeds), insecticides (fight insects that harm crops), fungicides (control fungi that are pathogenic to plants), and nematicides (combat of worms), among others (Stephenson et al., 2006).

In general, pesticides can also be classified according to their chemical composition, where the most common are as follows: organophosphates, carbamates, pyrethroids, and neonicotinoids (Sarath Chandran et al., 2019). Pesticides contribute significantly to increasing agricultural productivity, but they are substances that can generate environmental impacts, especially if they are used indiscriminately and inappropriately. In these cases, they can impact the environment, becoming harmful to nontarget organisms, accumulating in the environment and thus contaminating soils, water bodies, and even the local air. In addition, the World Health Organization (WHO) classifies pesticides according to their toxicity to humans. By categorizing pesticides based on their levels of toxicity, the WHO provides essential information to guide the regulation, proper handling, and responsible use of these substances to minimize risks to health and the environment (World Health Organization, WHO recommended classification of pesticides by hazard and guidelines to classification, 2019).

There are great increases and advances in research for the combat and environmental monitoring of pesticides, where increasingly efficient materials have been developed for these applications. In this context, conductive polymers can be used due to their properties, like electroactivity and high surficial area. Conductive polymers have shown promising potential in the application for the degradation of pesticides. The development of detection and degradation systems for pesticides based on conductive polymers is still a growing research area (Senthil et al., 2022).

For pesticide degradation, conductive polymers can act as catalysts or as support for anchoring a catalyst group, increasing the efficiency of the process due to its electroactivity and large surface area (Bano et al., 2021; Lin et al., 2021). When these conductive polymers are exposed to matrices containing pesticides, some physical-chemical properties of the polymer may be disturbed. The interaction between polymer and pesticide occurs primarily through an adsorption process, which already disturbs the system. After this interaction, the polymer may act as a catalyst for the interfacial reaction of the pesticide. Even if the polymer does not act as a catalyst, and the pesticide does not react, the adsorption process that took place earlier disturbs the system (Stejskal, 2020).

This generates a disturbance in the signal of electric current, electric potential, electric resistance, capacitance, or some other parameter. Conductive polymers are very sensitive to changes mainly electrochemical. This sensitivity becomes very interesting for developing sensors, since even at low concentrations of contaminants, such as pesticides; there will be sensitivity to the imposed disturbance (Lange et al., 2008). The variation in the signal can then be correlated with the concentration of the pesticide, thus being a quantitative analysis. Less robust sensors may not provide a quantitative response, but a qualitative one, where the result may provide something like "positive" or "false" for the presence or absence of the pesticide in that sample (Lange et al., 2008; Park et al., 2014; Rehman and Zeng, 2020).

Because they have excellent electrochemical properties, conductive polymers are mainly applied for the development of electrochemical sensors. In these sensors made conductive polymers can be present in the form of thin film, particle or composite, deposited on a solid substrate. The material obtained is what we know as modified electrode, where there is a substrate responsible for supporting the presence of modifiers, in this case, the conductive polymer (Wang et al., 2020c; Adarakatti and Kempahanumakkagari, 2019; March et al., 2015).

12.4.4 Toxic metals

Several studies have long shown that certain metals are potentially toxic to the environment (Gumpu et al., 2015). These metals have long been called "Heavy Metals," which is a very inconsistent definition and one that is falling out of use. As the IUPAC itself already discussed in 2002, it is possible to find different attempts to conceptualize the term "heavy metal," but all inconsistent. The purpose of this topic is not to discuss the term "heavy metals," but to inform the reader that this term is inconsistent, and here, therefore, studies that detect and combat some metals that are proven to be toxic to the environment will be discussed and cited (Duffus, 2002).

Conductive polymers have been explored for the development of sensors for detecting toxic metals due to their physicochemical properties that may be sensitive to the presence of these metals (March et al., 2015; Garg et al., 2022). As previously mentioned, conductive polymers improve the transduction system of the devices, helping to detect specific metals in different matrices (Wang et al., 2020b; Deshmukh et al., 2018; Kour et al., 2021).

Sensors using conductive polymers generally study the electrochemical properties of the system, such as electrical resistance, electrical current, capacitance, or changes in the system's redox process. The variation in these parameters can be correlated with the presence of toxic metals, in addition to being possible to obtain a quantitative relationship between variation in signal and concentration of the analyte (Wang et al., 2020b; Kour et al., 2021; She et al., 2017; Zegebreal et al., 2023).

The remarkable versatility of conductive polymers extends to the functionalization of their structures, resulting in further enhanced sensor performance. Beyond serving as an improved transduction system, these polymers also act as a solid foundation for additional modifications, allowing for even more tailored sensor capabilities. One prominent application lies in the use of conductive polymers as adsorbents for heavy metals. The presence of multiple functional groups in their structure facilitates robust interactions between the metal ions and the polymer matrix, enabling effective and selective adsorption of heavy metals from various environments. This dual functionality makes conductive polymers a highly promising choice for both advanced sensor technologies and eco-friendly approaches to mitigate heavy metal contamination. The functional groups of conducting polymers can act as chelating sites for binding metal ions, such as poly(1,5-diaminonaphthalene) (Pdan), which has an amine group capable of chelating metal ions through its amine group (Stejskal, 2020; Deshmukh et al., 2018; Lo et al., 2020; Jadoun et al., 2023).

In addition, they also allow the functionalization of their structure with other functional groups of interest, improving the adsorption performance. The presence of nitrogen (N) and sulfur (S) atoms in the structure of some conductive polymers is vital in the adsorption process of toxic metals. These elements provide high attraction properties for electrostatic interactions, π bonds, chelation, and ion exchange, thus contributing to improved adsorption. The diverse interaction mechanisms exhibited by conductive polymers render them highly efficient in the removal of various contaminants, making them a promising choice for purification and environmental remediation applications. With a deeper understanding of these adsorption processes, continuous advancements in conductive polymers are being pursued, aiming to develop increasingly effective and sustainable solutions for treating pollutants present in the environment. Leveraging the inherent properties of these polymers, coupled with ongoing research and innovation, holds the key to tackling environmental challenges and fostering a cleaner and healthier ecosystem for generations to come (Stejskal, 2020; Lo et al., 2020; Jadoun et al., 2023).

There are several types of conductive polymers, but some are more studied and stand out for their wide range of applications. As in the case of polypyrrole, probably the most studied was used by Zhang et al. (2020) to form a composite also containing phytic acid and a ZIF-type metal-organic structure (Pa/Ppy/ZIF-8@ZIF-67), and this composite was used to modify a vitreous carbon electrode. The modified electrode was used to individually and simultaneously detect Pb(II) and Cu(II) ions in solution, where the detection limits were 2.9 nM for Pb(II) and 14.8 nM for Cu(II) (Zhang et al.,

2020). The use of polypyrrole has also been reported to develop sensors to detect arsenic (Salunke et al., 2018; Salunke et al., 2017, 2021; Sadrolhosseini et al., 2017), lead (Xu et al., 2020; Sadrolhosseini et al., 2014; Zhang et al., 2023), mercury (Sadrolhosseini et al., 2014; Sabri et al., 2011; Palanisamy et al., 2017; Arulraj et al., 2016), and cadmium (Song et al., 2019; Anwar et al., 2020; Verma et al., 2020; Wang et al., 2017).

Pedot also has several studies reporting its use to detect toxic metals, as in the study by Giannetto et al. who modified glassy carbon electrode with a Pedot/gold composite, where through the anodic stripping technique, they detected mercury in concentrations as low as 0.83 ng/mL (Giannetto et al., 2011). Other studies also report the use of Pedot to detect mercury (Ding et al., 2013; Ramdzan et al., 2021), cadmium (Machhindra and Yen, 2022; Anandhakumar et al., 2011; Ponrathnam et al., 2012), and lead (Anandhakumar et al., 2011; Yasri et al., 2011; Sayyad et al., 2020, 2021). Other conductive polymers are also applied to detect various toxic metals, such as polyaniline (Yang et al., 2015; Gui et al., 2017; Wang et al., 2011a,b; Ong et al., 2021; Wang et al., 2013; Hashemi et al., 2020; Wang et al., 2020a), polyacetylene (Gui et al., 2017), and polythiophene (Iranzad et al., 2018; Alireza Hashemi et al., 2022).

12.5 Conclusion

Conductive polymers have been widely employed in various environmental applications due to their unique characteristics, such as conductivity, compatibility, ease of processing, and low cost. The synthesis of these polymers can be achieved through electrochemical polymerization, which is a simpler approach compared with traditional chemical synthesis. This method requires fewer reagents and exhibits faster reaction rates. Several types of depositions can be employed in this process, including cyclic voltammetry, amperometry, and potentiometry.

Furthermore, in this chapter, various characterizations of these materials have been explored, covering aspects such as conductivity (using techniques like two and four-point methods), electrochemical characterizations (via cyclic voltammetry and EIS), and morphology (utilizing techniques like SEM, TEM, and AFM). Finally, it is important to emphasize the significant versatility of these polymers, as evidenced by their diverse applications, including sensing, electrocatalysis, fuel cells, advanced oxidation processes, and many other areas.

References

Adarakatti, P.S., Kempahanumakkagari, S.K., 2019. Modified electrodes for sensing. Royal Society of Chemistry, India SPR Electrochem. 15, 58–95. https://doi.org/10.1039/9781788013895-00058.

Ahmed, F., Kumar, S., Arshi, N., Anwar, M.S., Su-Yeon, L., Kil, G.S., Park, D.W., Koo, B.H., Lee, C.G., 2011. South Korea Preparation and characterizations of polyaniline (PANI)/ZnO nanocomposites film using solution casting method, 9 30 2011/09/30 Thin Solid Films 519 (23), 8375–8378. https://doi.org/10.1016/j.tsf.2011.03.090.

Alessandrini, A., Facci, P., 2005. AFM: a versatile tool in biophysics. Institute of Physics Publishing, Italy Meas. Sci. Technol. 16 (6), R65. https://doi.org/10.1088/0957-0233/16/6/R01.

Alireza Hashemi, S., Bahrani, S., Mousavi, S.M., Mojoudi, F., Omidifar, N., Lankarani, K.B., Arjmand, M., Ramakrishna, S., 2022. Development of sulfurized Polythiophene-Silver Iodide-Diethyldithiocarbamate nanoflakes toward Record-High and selective absorption and detection of mercury derivatives in aquatic substrates. Chem. Eng. J. 440, 135896. https://doi.org/10.1016/j.cej.2022.135896.

Anandhakumar, S., Mathiyarasu, J., Phani, K.L.N., Yegnaraman, V., 2011. Simultaneous determination of cadmium and lead using PEDOT/PSS modified glassy carbon electrode. Am. J. Anal. Chem. 02 (04), 470–474. https://doi.org/10.4236/ajac.2011.24056.

Ansari Khalkhali, R., Price, W.E., Wallace, G.G., 2003. Quartz crystal microbalance studies of the effect of solution temperature on the ion-exchange properties of polypyrrole conducting electroactive polymers. React. Funct. Polym. 56 (3), 141–146. https://doi.org/10.1016/S1381-5148(03)00055-5.

Anwar, N., Asif, M., Shakoor, A., Niaz, N.A., Qasim, M., Irfan, M., Ali, G., Mahmood, A., 2020. Electrical properties and characteristics of polypyrrole cadmium oxide (PPy-CdO) nanocomposite Schottky diodes. Pleiades journals, Pakistan Polym. Sci. 62 (5), 543–549. https://doi.org/10.1134/S0965545X20050028.

Arulraj, A.D., Devasenathipathy, R., Chen, S.M., Vasantha, V.S., Wang, S.F., 2016. Femtomolar detection of mercuric ions using polypyrrole, pectin and graphene nanocomposites modified electrode. Academic Press Inc., India J. Colloid Interface Sci. 483, 268–274. https://doi.org/10.1016/j.jcis.2016.08.026.

Bag, A., Lee, N.-E., 2021. Recent advancements in development of wearable gas sensors. Adv. Mater. Technol. 6 (3), 2000883. https://doi.org/10.1002/admt.202000883.

Bano, K., Kaushal, S., Singh, P.P., 2021. A review on photocatalytic degradation of hazardous pesticides using heterojunctions. Polyhedron 209, 115465. https://doi.org/10.1016/j.poly.2021.115465.

Bhattacharya, A., Mukherjee, D.C., Gohil, J.M., Kumar, Y., Kundu, S., 2008. Preparation, characterization and performance of conducting polypyrrole composites based on polysulfone. Desalination 225 (1–3), 366–372. https://doi.org/10.1016/j.desal.2006.07.018.

Bolan, N., Kunhikrishnan, A., Thangarajan, R., Kumpiene, J., Park, J., Makino, T., Kirkham, M.B., Scheckel, K., 2014. Remediation of heavy metal(loid)s contaminated soils—to mobilize or to immobilize? Elsevier, Australia J. Hazard Mater. 266, 141–166. https://doi.org/10.1016/j.jhazmat.2013.12.018.

Casado, N., Mecerreyes, D., 2020. Introduction to Redox Polymers: Classification, Character ization Methods and Main Applications, vol 2021. Royal Society of Chemistry (RSC), pp. 1–26. https://doi.org/10.1039/9781788019743-00001.

Chen, W., 2011. Electrochemical polymerization of benzoxazines. In: Handbook of Benzoxazine Resins. Elsevier, China, pp. 175–181. https://doi.org/10.1016/B978-0-444-53790-4.00053-9. Available from: http://www.sciencedirect.com/science/book/9780444537904.

Chien, Y.C., Brandell, D., Lacey, M.J., 2022. Towards reliable three-electrode cells for lithium-sulfur batteries. Royal Society of Chemistry, Sweden Chem. Commun. 58 (5), 705–708. https://doi.org/10.1039/d1cc04553a. Available from: http://pubs.rsc.org/en/journals/journal/cc.

da Silva, R.A.B., de Almeida, E.G.N., Rabelo, A.C., da Silva, A.T.C., Ferreira, L.F., Richter, E.M., 2009. Three electrode electrochemical microfluidic cell: construction and characterization. Sociedade Brasileira de Quimica, Brazil J. Braz. Chem. Soc. 20 (7), 1235–1241. https://doi.org/10.1590/s0103-50532009000700006. Available from: http://jbcs.sbq.org.br/online/2009/vol20_n7/05-08755AR.pdf.

Dai, W., Xu, H., Yu, J., Hu, X., Luo, X., Tu, X., Yang, L., 2015. Photocatalytic reduction of CO2 into methanol and ethanol over conducting polymers modified Bi2WO6 microspheres under visible light. Elsevier B.V., China Appl. Surf. Sci. 356, 173–180. https://doi.org/10.1016/j.apsusc.2015.08.059. Available from: http://www.journals.elsevier.com/applied-surface-science/.

de la Cruz, P.T.S., Irikura, K., Lachgar, A., Cardoso, J.C., Cavero, H.A., Zanoni, M.V.B., 2021. Preparation of FTO/CU2O electrode protected by PEDOT:PSS and its better performance in the photoelectrocatalytic reduction of CO2 to methanol. Springer, Peru Electrocatalysis 11 (5), 546–554. https://doi.org/10.1007/s12678-020-00612-z. Available from: http://www.springer.com/chemistry/electrochemistry/journal/12678.

Deshmukh, M.A., Shirsat, M.D., Ramanaviciene, A., Ramanavicius, A., 2018. Composites based on conducting polymers and carbon nanomaterials for heavy metal ion sensing (review). Taylor and Francis Ltd., India Crit. Rev. Anal. Chem. 48 (4), 293–304. https://doi.org/10.1080/10408347.2017.1422966.

Ding, L., Zhai, J., Bond, A.M., Zhang, J., 2013. Polystyrenesulfonate doped poly(-Hydroxymethyl 3,4-Ethylenedioxythiophene) stabilized Au nanoparticle modified glassy carbon electrode as a reusable sensor for mercury(II) detection in chloride media. J. Electroanal. Chem. 704, 96–101. https://doi.org/10.1016/j.jelechem.2013.06.014.

Duffus, J.H., 2002. "Heavy metals"—a meaningless term? (IUPAC technical report). Walter de Gruyter GmbH, United Kingdom Pure Appl. Chem. 74 (5), 793–807. https://doi.org/10.1351/pac200274050793. Available from: http://www.degruyter.com/view/j/pac.

Dutta, K., Kundu, P.P., 2014. A review on aromatic conducting polymers-based catalyst supporting matrices for application in microbial fuel cells. Polym. Rev. 54 (3), 401–435. https://doi.org/10.1080/15583724.2014.881372.

Eaton, P., West, P., 2010. Atomic Force Microscopy. Oxford University Press, Portugal, pp. 1–256. https://doi.org/10.1093/acprof:oso/9780199570454.001.0001. Available from: http://www.oxfordscholarship.com/view/10.1093/acprof:oso/9780199570454.001.0001/acprof-9780199570454.

Eftekhari, A., Jafarkhani, P., 2014. Galvanodynamic synthesis of polyaniline: a flexible method for the deposition of electroactive materials. J. Electroanal. Chem. 717–718, 110–118. https://doi.org/10.1016/j.jelechem.2014.01.018.

El Fakir, A.A., Anfar, Z., Amedlous, A., Mohamed, Z., Hafidi, Z., El Achouri, M., Jada, A., El Alem, N., 2021. Engineering of new hydrogel beads based conducting polymers: metal-free catalysis for highly organic pollutants degradation. Appl. Catal. B Environ. 286, 119948. https://doi.org/10.1016/j.apcatb.2021.119948.

Elgrishi, N., Rountree, K.J., McCarthy, B.D., Rountree, E.S., Eisenhart, T.T., Dempsey, J.L., 2018. A practical beginner's guide to cyclic voltammetry. American Chemical Society, United States J. Chem. Educ. 95 (2), 197–206. https://doi.org/10.1021/acs.jchemed.7b00361. Available from: http://pubs.acs.org/loi/jceda8.

Facure, M.H.M., Mercante, L.A., Mattoso, L.H.C., Correa, D.S., 2017. Detection of trace levels of organophosphate pesticides using an electronic tongue based on graphene hybrid nanocomposites. Elsevier B.V., Brazil Talanta 167, 59–66. https://doi.org/10.1016/j.talanta.2017.02.005. Available from: https://www.journals.elsevier.com/talanta.

Fang, X., Zong, B., Mao, S., 2018. Metal–organic framework-based sensors for environmental contaminant sensing. SpringerOpen, China Nano-Micro Lett. 10 (4). https://doi.org/

10.1007/s40820-018-0218-0. Available from: http://www.springer.com/engineering/journal/40820.

Ferloni, P., Mastragostino, M., Meneghello, L., 1996. Impedance analysis of electronically conducting polymers. Electrochim. Acta 41 (1), 27−33. https://doi.org/10.1016/0013-4686(95)00290-u.

Fomo, G., Waryo, T., Feleni, U., Baker, P., Iwuoha, E., 2019. Electrochemical Polymerization. Springer Science and Business Media LLC, pp. 105−131. https://doi.org/10.1007/978-3-319-95987-0_3.

Gao, F., Hou, X., Wang, A., Chu, G., Wu, W., Chen, J., Zou, H., 2016. Preparation of polypyrrole/TiO2 nanocomposites with enhanced photocatalytic performance. Elsevier B.V., China Particuology 26, 73−78. https://doi.org/10.1016/j.partic.2015.07.003. Available from: http://nonsolus/IGT/Audit/DocsST/China%20Particuology_rename_IGT.doc.

Garcia-Cabezon, C., Garcia-Hernandez, C., Rodriguez-Mendez, M.L., Martin-Pedrosa, F., 2020. A new strategy for corrosion protection of porous stainless steel using polypyrrole films. J. Mater. Sci. Technol. 37, 85−95. https://doi.org/10.1016/j.jmst.2019.05.071.

Garg, N., Deep, A., Sharma, A.L., 2022. Recent trends and advances in porous metal-organic framework nanostructures for the electrochemical and optical sensing of heavy metals in water. Taylor and Francis Ltd., India Crit. Rev. Anal. Chem.. https://doi.org/10.1080/10408347.2022.2106543. Available from: www.tandf.co.uk/journals/titles/10408347.asp.

Gavrilescu, M., Demnerová, K., Aamand, J., Agathos, S., Fava, F., 2015. Emerging pollutants in the environment: present and future challenges in biomonitoring, ecological risks and bioremediation. Elsevier, Romania N. Biotech. 32 (1), 147−156. https://doi.org/10.1016/j.nbt.2014.01.001. Available from: http://www.elsevier.com.

Ghosh, S., Das, S., Mosquera, M.E.G., 2020. Conducting polymer-based nanohybrids for fuel cell application. MDPI AG, Spain Polymers 12 (12), 1−19. https://doi.org/10.3390/polym12122993. Available from: https://www.mdpi.com/2073-4360/12/12/2993/pdf.

Giannetto, M., Mori, G., Terzi, F., Zanardi, C., Seeber, R., 2011. Composite PEDOT/Au nanoparticles modified electrodes for determination of mercury at trace levels by anodic stripping voltammetry. Wiley-VCH Verlag, Italy Electroanalysis 23 (2), 456−462. https://doi.org/10.1002/elan.201000469. Available from: http://onlinelibrary.wiley.com/journal/10.1002/(ISSN)1521-4109.

Gui, M., Jiang, J., Wang, X., Yan, Y., Li, S., Xiao, X., Liu, T., Liu, T., Feng, Y., 2017. Copper ion-mediated glyphosate detection with N-heterocycle based polyacetylene as a sensing platform. Sensor. Actuator. B Chem. 243, 696−703. https://doi.org/10.1016/j.snb.2016.12.037.

Guimard, N.K., Gomez, N., Schmidt, C.E., 2007. Conducting polymers in biomedical engineering. Prog. Polym. Sci. 32 (8−9), 876−921. https://doi.org/10.1016/j.progpolymsci.2007.05.012.

Gumpu, M.B., Sethuraman, S., Krishnan, U.M., Rayappan, J.B.B., 2015. A review on detection of heavy metal ions in water—an electrochemical approach. Sens. Actuators, B 213, 515−533. https://doi.org/10.1016/j.snb.2015.02.122.

Gvozdenovic, M., Jugovic, B., Stevanovic, J., Grgur, B., 2014. Electrochemical synthesis of electroconducting polymers. Chem. Ind. 68 (6), 673−684. https://doi.org/10.2298/hemind131122008g.

Han, J.J., Zhang, N., Liu, D.L., Ma, H., Han, T., Sun, D.D., 2020. Cyclic voltammetry for the determination of the selectivity of PANI-HClO4 sensor to different acids. Springer, China Ionics 26 (2), 1029−1038. https://doi.org/10.1007/s11581-019-03238-4. Available from: http://www.springerlink.com/content/120106/.

Hashemi, S.A., Mousavi, S.M., Bahrani, S., Ramakrishna, S., Hashemi, S.H., 2020. Picomolar-level detection of mercury within non-biological/biological aqueous media using ultra-sensitive polyaniline-Fe3O4-silver diethyldithiocarbamate nanostructure. Springer, Singapore Anal. Bioanal. Chem. 412 (22), 5353–5365. https://doi.org/10.1007/s00216-020-02750-1. Available from: https://link.springer.com/journal/216/volumes-and-issues.

Hu, Q., Sun, D., Ma, Y., Qiu, B., Guo, Z., 2017. Conductive polyaniline nanorods enhanced methane production from anaerobic wastewater treatment. Elsevier Ltd, China Polymer 120, 236–243. https://doi.org/10.1016/j.polymer.2017.05.073.

Huerta, F., Quijada, C., Montilla, F., Morallón, E., 2021. Revisiting the redox transitions of polyaniline. Semiquantitative interpretation of electrochemically induced IR bands. J. Electroanal. Chem. 897, 115593. https://doi.org/10.1016/j.jelechem.2021.115593.

Inzelt, G., 2008. Springer chemical and electrochemical syntheses of conducting polymers. In: Conducting Polymers: A New Era in Electrochemistry, pp. 123–148. https://doi.org/10.1007/9783540759300_4.

Iranzad, F., Gheibi, M., Eftekhari, M., 2018. Synthesis and application of polythiophene-coated Fe3O4 nanoparticles for preconcentration of ultra-trace levels of cadmium in different real samples followed by electrothermal atomic absorption spectrometry. Taylor and Francis Ltd., Iran Int. J. Environ. Anal. Chem. 98 (1), 16–30. https://doi.org/10.1080/03067319.2018.1426757.

Iurchenkova, A.A., Kallio, T., Fedorovskaya, E.O., 2021. Relationships between polypyrrole synthesis conditions, its morphology and electronic structure with supercapacitor properties measured in electrolytes with different ions and pH values. Electrochim. Acta 391, 138892. https://doi.org/10.1016/j.electacta.2021.138892.

Izaoumen, N., Bouchta, D., Zejli, H., Kaoutit, M., Stalcup, A., Temsamani, K., 2005. Electrosynthesis and analytical performances of functionalized poly (pyrrole/β-cyclodextrin) films. Talanta 66 (1), 111–117. https://doi.org/10.1016/j.talanta.2004.10.003.

Jadoun, S., Fuentes, J.P., Urbano, B.F., Yáñez, J., 2023. A review on adsorption of heavy metals from wastewater using conducting polymer-based materials. Elsevier Ltd, Chile J. Environ. Chem. Eng. 11 (1). https://doi.org/10.1016/j.jece.2022.109226. Available from: http://www.journals.elsevier.com/journal-of-environmental-chemical-engineering/.

Karimi, M., Mohsen-Nia, M., Akbari, A., 2014. Electro-separation of synthetic azo dyes from a simulated wastewater using polypyrrole/polyacrylonitrile conductive membranes. J. Water Proc. Eng. 4 (C), 6–11. https://doi.org/10.1016/j.jwpe.2014.08.008.

Kirti, Gupta, R., Srivastava, D.N., 2022. A biodegradable polymer-based plastic chip electrode as a current collector in supercapacitor application. Multidisciplinary Digital Publishing Institute (MDPI), India Electrochem 3 (3), 379–396. https://doi.org/10.3390/electrochem3030026.

Kobayashi, T., Yoneyama, H., Tamura, H., 1984. Polyaniline film-coated electrodes as electrochromic display devices. J. Electroanal. Chem. 161 (2), 419–423. https://doi.org/10.1016/S0022-0728(84)80201-6.

Kour, G., Kothari, R., Azam, R., Majhi, P.K., Dhar, S., Pathania, D., Tyagi, V.V., 2021. Conducting Polymer Based Nanoadsorbents for Removal of Heavy Metal Ions/Dyes from Wastewater Engineering Materials. Springer Science and Business Media B.V., India, pp. 135–157. https://doi.org/10.1007/978-3-030-62090-5_7. Available from: springer.com/series/4288.

Krieg, M., Fläschner, G., Alsteens, D., Gaub, B.M., Roos, W.H., Wuite, G.J.L., Gaub, H.E., Gerber, C., Dufrêne, Y.F., Müller, D.J., 2019. Atomic force microscopy-based mechanobiology. Springer Nature, Spain Nature Rev. Phys. 1 (1), 41–57. https://doi.org/10.1038/s42254-018-0001-7. Available from: nature.com/natrevphys/.

Krische, B., Zagorska, M., 1989. Overoxidation in conducting polymers. Synth. Met. 28 (1−2), 257−262. https://doi.org/10.1016/0379-6779(89)90530-4.

Lange, U., Mirsky, V.M., 2011. Chemiresistors based on conducting polymers: a review on measurement techniques. Anal. Chim. Acta 687 (2), 105−113. https://doi.org/10.1016/j.aca.2010.11.030.

Lange, U., Roznyatovskaya, N.V., Mirsky, V.M., 2008. Conducting polymers in chemical sensors and arrays. Anal. Chim. Acta 614 (1), 1−26. https://doi.org/10.1016/j.aca.2008.02.068.

Laschuk, N.O., Easton, E.B., Zenkina, O.V., 2021. Reducing the resistance for the use of electrochemical impedance spectroscopy analysis in materials chemistry. Royal Society of Chemistry, Canada RSC Adv. 11 (45), 27925−27936. https://doi.org/10.1039/d1ra03785d. Available from: http://pubs.rsc.org/en/journals/journal/ra.

Le, T.H., Kim, Y., Yoon, H., 2017. Electrical and electrochemical properties of conducting polymers. MDPI AG, South Korea Polymers 9 (4). https://doi.org/10.3390/polym9040150. Available from: http://www.mdpi.com/2073-4360/9/4/150/pdf.

Lee, S.M., Uhm, S.J., Bang, J.I., Song, K.D., Joo, B.S., Lee, Y.S., Lee, D.D., 2005. A field effect transistor type gas sensor based on polyaniline, 11 2005/11. In: Digest of Technical Papers—International Conference on Solid State Sensors and Actuators and Microsystems, TRANSDUCERS '05, vol. 2, pp. 1935−1938. South Korea.

Li, M., Wei, Z., Jiang, L., 2008. Polypyrrole nanofiber arrays synthesized by a biphasic electrochemical strategy. J. Mater. Chem. 18 (19), 2276−2280. https://doi.org/10.1039/b800289d.

Lin, C.H., Lin, J.H., Chen, C.F., Ito, Y., Luo, S.C., 2021. Conducting polymer-based sensors for food and drug analysis. Taiwan Food and Drug Administration, Taiwan J. Food Drug Anal. 29 (4), 544−558. https://doi.org/10.38212/2224-6614.3374. Available from: https://www.jfda-online.com/cgi/viewcontent.cgi?article=3374&context=journal.

Liu, L., Liu, J., Gao, B., Yang, F., Chellam, S., 2012. Fouling reductions in a membrane bioreactor using an intermittent electric field and cathodic membrane modified by vapor phase polymerized pyrrole. J. Membr. Sci. 394−395, 202−208. https://doi.org/10.1016/j.memsci.2011.12.042.

Liu, J., Zhang, Y.H., Bai, Z.M., Huang, Z.A., Gao, Y.K., 2019. Photoelectrocatalytic oxidation of methane into methanol and formic acid over ZnO/graphene/polyaniline catalyst. Institute of Physics Publishing, China Chin. Phys. B 28 (4). https://doi.org/10.1088/1674-1056/28/4/048101. Available from: https://iopscience.iop.org/article/10.1088/1674-1056/28/4/048101/pdf.

Lo, M., Ktari, N., Gningue-Sall, D., Madani, A., Aaron, S.E., Aaron, J.J., Mekhalif, Z., Delhalle, J., Chehimi, M.M., 2020. Polypyrrole: a reactive and functional conductive polymer for the selective electrochemical detection of heavy metals in water. Springer Nature, Senegal Emergent Mater. 3 (6), 815−839. https://doi.org/10.1007/s42247-020-00119-9.

Lv, H., Pan, Q., Song, Y., Liu, X.X., Liu, T., 2020. A review on nano-/microstructured materials constructed by electrochemical technologies for supercapacitors. Springer, China Nano-Micro Lett. 12 (1). https://doi.org/10.1007/s40820-020-00451-z. Available from: http://www.springer.com/engineering/journal/40820.

Machhindra, L.A., Yen, Y.K., 2022. A highly sensitive electrochemical sensor for Cd2+ detection based on Prussian Blue-PEDOT-loaded laser-scribed graphene-modified glassy carbon electrode. MDPI, Taiwan Chemosensors 10 (6). https://doi.org/10.3390/chemosensors10060209. Available from: https://www.mdpi.com/2227-9040/10/6/209/pdf?version=1654156231.

Malatesta, M., 2021. Transmission electron microscopy as a powerful tool to investigate the interaction of nanoparticles with subcellular structures. MDPI, Italy Int. J. Mol. Sci. 22 (23). https://doi.org/10.3390/ijms222312789. Available from: https://www.mdpi.com/1422-0067/22/23/12789/pdf.

March, G., Nguyen, T.D., Piro, B., 2015. Modified electrodes used for electrochemical detection of metal ions in environmental analysis. MDPI, France Biosensors 5 (2), 241–275. https://doi.org/10.3390/bios5020241. Available from: http://www.mdpi.com/2079-6374/5/2/241/pdf.

Marchesi, L.F., Jacumasso, S.C., Quintanilha, R.C., Winnischofer, H., Vidotti, M., 2015. The electrochemical impedance spectroscopy behavior of poly(aniline) nanocomposite electrodes modified by layer-by-layer deposition. Elsevier Ltd, Brazil Electrochim. Acta 174, 864–870. https://doi.org/10.1016/j.electacta.2015.05.077.

Molina, J., Fernández, J., del Río, A.I., Bonastre, J., Cases, F., 2014. Synthesis of Pt nanoparticles on electrochemically reduced graphene oxide by potentiostatic and alternate current methods. Mater. Char. 89, 56–68. https://doi.org/10.1016/j.matchar.2014.01.003.

Nam, D.H., Kim, M.J., Lim, S.J., Song, I.S., Kwon, H.S., 2013. Single-step synthesis of polypyrrole nanowires by cathodic electropolymerization. J. Mater. Chem. A 1 (27), 8061–8068. https://doi.org/10.1039/c3ta11227f.

Namsheer, K., Rout, C.S., 2021. Conducting polymers: a comprehensive review on recent advances in synthesis, properties and applications. Royal Society of Chemistry, India RSC Adv. 11 (10), 5659–5697. https://doi.org/10.1039/d0ra07800j. Available from: http://pubs.rsc.org/en/journals/journal/ra.

Ong, J.-Y., Law, Z.J., Pua, C.H., Phang, S.W., 2021. Effect of acid dopants toward polyaniline based optical sensor for lead detection. Pleiades journals, Malaysia Polym. Sci. 63 (5), 485–492. https://doi.org/10.1134/S0965545X21050102. Available from: https://rd.springer.com/journal/11498.

Pacheco, W.F., Semaan, F.S., De Almeida, V.G.K., Ritta, A.G.S.L., Aucélio, R.Q., 2013. Voltammetry: a brief review about concepts. Sociedade Brasileira de Quimica, Brazil Rev. Virt. Quimica 5 (4), 516–537. https://doi.org/10.5935/1984-6835.20130040. Available from: http://www.uff.br/RVQ/index.php/rvq/article/view/380/345.

Palanisamy, S., Thangavelu, K., Chen, S.M., Velusamy, V., Chang, M.H., Chen, T.W., Al-Hemaid, F.M.A., Ali, M.A., Ramaraj, S.K., 2017. Synthesis and characterization of polypyrrole decorated graphene/β-cyclodextrin composite for low level electrochemical detection of mercury (II) in water. Sens. Actuators, B 243, 888–894. https://doi.org/10.1016/j.snb.2016.12.068.

Pan, Z., Song, C., Lin, L., Wang, H., Pan, Y., Wang, C., Li, J., Wang, T., Feng, X., 2019. Membrane technology coupled with electrochemical advanced oxidation processes for organic wastewater treatment: recent advances and future prospects. Chem. Eng. J. 376, 120909. https://doi.org/10.1016/j.cej.2019.01.188.

Park, S., Kwon, O., Lee, J., Jang, J., Yoon, H., 2014. Conducting polymer-based nanohybrid transducers: a potential route to high sensitivity and selectivity sensors. Sensors 14 (2), 3604–3630. https://doi.org/10.3390/s140203604.

Patois, T., Lakard, B., Monney, S., Roizard, X., Fievet, P., 2011. Characterization of the surface properties of polypyrrole films: influence of electrodeposition parameters. Synth. Met. 161 (21–22), 2498–2505. https://doi.org/10.1016/j.synthmet.2011.10.003.

Petrie, B., Barden, R., Kasprzyk-Hordern, B., 2015. A review on emerging contaminants in wastewaters and the environment: current knowledge, understudied areas and recommendations for future monitoring. Elsevier Ltd, United Kingdom Water Res. 72, 3–27. https://doi.org/10.1016/j.watres.2014.08.053.

Piaskowski, J., Bourret, G.R., 2022. Electrochemical synthesis of plasmonic nanostructures. MDPI, Austria Molecules 27 (8). https://doi.org/10.3390/molecules27082485. Available from: https://www.mdpi.com/1420-3049/27/8/2485/pdf.

Ponrathnam, T., Robertson, S., Cho, J., Kurup, P., Nagarajan, R., 2012. Detection of cadmium using conjugated polymer modified electrodes. MRS Proceedings 1436 (January). https://doi.org/10.1557/opl.2012.1510.

Ramanavicius, A., Tereshchenko, A., Plikusiene, I., Ratautaite, V., Deshmukh, M.A., Smyntyna, V., Oztekin, Y., Bubniene, U., Ramanaviciene, A., 2020. Electrochemical formation of 'synthetic receptors' based on conducting polymers, 1 2020/01. In: Springer Proceedings in Physics, vol 244. Springer Science and Business Media Deutschland GmbH, Lithuania, pp. 239–245. https://doi.org/10.1007/978-981-15-3996-1_24. http://www.springer.com/series/361.

Ramdzan, N.S.M., Fen, Y.W., Omar, N.A.S., Anas, N.A.A., Liew, J.Y.C., Daniyal, W.M.E.M.M., Hashim, H.S., 2021. Detection of mercury ion using surface plasmon resonance spectroscopy based on nanocrystalline cellulose/poly(3,4-ethylenedioxythiophene) thin film. Elsevier B.V., Malaysia Measurement: Journal of the International Measurement Confederation 182. https://doi.org/10.1016/j.measurement.2021.109728. Available from: https://www.journals.elsevier.com/measurement.

Ramya, R., Sivasubramanian, R., Sangaranarayanan, M.V., 2013. Conducting polymers-based electrochemical supercapacitors—progress and prospects. Electrochim. Acta 101, 109–129. https://doi.org/10.1016/j.electacta.2012.09.116.

Rehman, A., Zeng, X., 2020. Interfacial composition, structure, and properties of ionic liquids and conductive polymers for the construction of chemical sensors and biosensors: a perspective. Elsevier B.V., Saudi Arabia Curr. Opin. Electrochem. 23, 47–56. https://doi.org/10.1016/j.coelec.2020.03.010.

Roman, R.L., Nagi, L., Silva, L.L., Fernandes, S.C., de Mello, J.M.M., Magro, J.D., Fiori, M.A., 2020. Monocrystalline silicon/polyaniline/horseradish peroxidase enzyme electrode obtained by the electrodeposition method for the electrochemical detection of glyphosate. Springer, Brazil J. Mater. Sci. Mater. Electron. 31 (12), 9443–9456. https://doi.org/10.1007/s10854-020-03484-7.

Rountree, E.S., McCarthy, B.D., Eisenhart, T.T., Dempsey, J.L., 2014. Evaluation of homogeneous electrocatalysts by cyclic voltammetry. American Chemical Society, United States Inorg. Chem. 53 (19), 9983–10002. https://doi.org/10.1021/ic500658x.

Sabri, Y.M., Kojima, R., Ippolito, S.J., Wlodarski, W., Kalantar-zadeh, K., Kaner, R.B., Bhargava, S.K., 2011. QCM based mercury vapor sensor modified with polypyrrole supported palladium. Sensor. Actuator. B Chem. 160 (1), 616–622. https://doi.org/10.1016/j.snb.2011.08.038.

Sadrolhosseini, A.R., Noor, A.S.M., Bahrami, A., Lim, H.N., Talib, Z.A., Mahdi, M.A., 2014. Application of polypyrrole multi-walled carbon nanotube composite layer for detection of mercury, lead and iron ions using surface plasmon resonance technique. Public Library of Science, Malaysia PLoS One 9 (4). https://doi.org/10.1371/journal.pone.0093962. Available from. http://www.plosone.org/article/fetchObject.action?uri=info%3Adoi%2F10.1371%2Fjournal.pone.0093962&representation=PDF.

Sadrolhosseini, A.R., Naseri, M., Kamari, H.M., 2017. Surface plasmon resonance sensor for detecting of arsenic in aqueous solution using polypyrrole-chitosan-cobalt ferrite nanoparticles composite layer. Opt Commun. 383, 132–137. https://doi.org/10.1016/j.optcom.2016.08.065.

Sall, M.L., Diaw, A.K.D., Gningue-Sall, D., Efremova Aaron, S., Aaron, J.J., 2020. Toxic heavy metals: impact on the environment and human health, and treatment with conducting organic polymers, a review. Springer, Senegal Environ. Sci. Pollut. Control Ser. 27 (24), 29927–29942. https://doi.org/10.1007/s11356-020-09354-3. Available from: https://link.springer.com/journal/11356.

Salunke, R.S., Kasar, C.K., Bangar, M.A., Chavan, P.G., Shirale, D.J., 2017. Electrodeposition of gold nanoparticles decorated single polypyrrole nanowire for arsenic detection in potable water: a chemiresistive sensor device. J. Mater. Sci. Mater. Electron. 28 (19), 14672–14677. https://doi.org/10.1007/s10854-017-7332-5.

Salunke, R.S., Chavan, P.G., Shirale, D.J., 2018. Anodic stripping voltammetry studies of electrochemically engineered silver nanoparticles over single polypyrrole nanowire device for tracing of arsenic(III): an environmental perspective. Springer Science and Business Media Deutschland GmbH, India Nanotechnology for Environmental Engineering 3 (1). https://doi.org/10.1007/s41204-018-0042-2. Available from: springer.com/earth+sciences+and+geography/engineering+geology/journal/41204.

Salunke, R.S., Nakate, Y.T., Umar, A., Nakate, U.T., Ahmad, R., Shirale, D.J., 2021. Anodic stripping voltammetry analysis of gold nanoparticles functionalized one-dimensional single polypyrrole nanowire for arsenic sensing. Elsevier B.V., India Surface. Interfac. 23. https://doi.org/10.1016/j.surfin.2020.100895. Available from. http://www.journals.elsevier.com/surfaces-and-interfaces.

Sarath Chandran, C., Thomas, S., Unni, M.R., 2019. Pesticides: Classification, Detection, and Degradation. Springer Science and Business Media LLC, pp. 71–87. https://doi.org/10.1007/978-3-030-04657-6_5.

Sayyad, P.W., Ingle, N.N., Al-Gahouari, T., Mahadik, M.M., Bodkhe, G.A., Shirsat, S.M., Shirsat, M.D., 2020. Sensitive and selective detection of Cu2+ and Pb2+ ions using Field Effect Transistor (FET) based on L-Cysteine anchored PEDOT:PSS/rGO composite. Elsevier B.V., India Chem. Phys. Lett. 761. https://doi.org/10.1016/j.cplett.2020.138056. Available from: http://www.elsevier.com/locate/cplonline.

Sayyad, P.W., Ansari, T.R., Ingle, N.N., Al-Gahouari, T., Bodkhe, G.A., Mahadik, M.M., Shirsat, S.M., Shirsat, M.D., 2021. L-Cysteine peptide-functionalized PEDOT-PSS/rGO nanocomposite for selective electrochemical detection of lead Pb(II) ions. Springer Science and Business Media Deutschland GmbH, India Appl. Phys. Mater. Sci. Process 127 (5). https://doi.org/10.1007/s00339-021-04511-y. Available from. http://www.springer.com/materials/journal/339.

Senthil, T., Dutta, K., Dhamodharan, D., Divakaran, N., Wu, L., 2022. Conjugated polymers-based sensors for detection of water pollutants Conjugated Polymers for Next-Generation Applications. In: Synthesis, Properties and Optoelectrochemical Devices, vol. 1. Elsevier, India, pp. 273–323. https://doi.org/10.1016/B978-0-12-823442-6.00014-3. Available from: https://www.sciencedirect.com/book/9780128234426.

Shabani Shayeh, J., Ehsani, A., Ganjali, M.R., Norouzi, P., Jaleh, B., 2015. Conductive polymer/reduced graphene oxide/Au nano particles as efficient composite materials in electrochemical supercapacitors. Appl. Surf. Sci. 353, 594–599. https://doi.org/10.1016/j.apsusc.2015.06.066.

Shakeel, A., Rizwan, K., Farooq, U., Iqbal, S., Ali Altaf, A., 2022. Advanced polymeric/inorganic nanohybrids: an integrated platform for gas sensing applications. Chemosphere 294, 133772. https://doi.org/10.1016/j.chemosphere.2022.133772.

Sharma, M.K., 2021. Synthesis of Advanced Materials by Electrochemical Methods. Springer Science and Business Media LLC, pp. 435–466. https://doi.org/10.1007/978-981-16-1807-9_14.

She, Z.W., Kibsgaard, J., Dickens, C.F., Chorkendorff, I., Nørskov, J.K., Jaramillo, T.F., 2017. Combining theory and experiment in electrocatalysis: insights into materials design.

American Association for the Advancement of Science, United States Science 355 (6321). https://doi.org/10.1126/science.aad4998. Available from: http://science.sciencemag.org/content/sci/355/6321/eaad4998.full.pdf.

Singh, N., Riaz, U., 2022. Recent trends on synthetic approaches and application studies of conducting polymers and copolymers: a review. Springer Science and Business Media Deutschland GmbH, India Polym. Bull. 79 (12), 10377–10408. https://doi.org/10.1007/s00289-021-03987-1. Available from: https://www.springer.com/journal/289.

Smela, E., Gadegaard, N., 2001. Volume change in polypyrrole studied by atomic force microscopy. J. Phys. Chem. B 105 (39), 9395–9405. https://doi.org/10.1021/jp004126u.

Soares, A.L., Hryniewicz, B.M., Deller, A.E., Volpe, J., Marchesi, L.F., Souto, D.E.P., Vidotti, M., 2021. Electrodes based on PEDOT nanotubes decorated with gold nanoparticles for biosensing and energy storage. American Chemical Society, Brazil ACS Appl. Nano Mater. 4 (9), 9945–9956. https://doi.org/10.1021/acsanm.1c02677. Available from: https://pubs.acs.org/journal/aanmf6.

Song, Y., Bian, C., Hu, J., Li, Y., Tong, J., Sun, J., Gao, G., Xia, S., 2019. Porous polypyrrole/graphene oxide functionalized with carboxyl composite for electrochemical sensor of trace cadmium (II). Electrochemical Society Inc., China J. Electrochem. Soc. 166 (2), B95. https://doi.org/10.1149/2.0801902jes.

Stejskal, J., 2020. Conducting polymers are not just conducting: a perspective for emerging technology. John Wiley and Sons Ltd, Czech Republic Polymer International 69 (8), 662–664. https://doi.org/10.1002/pi.5947. Available from: http://onlinelibrary.wiley.com/journal/10.1002/(ISSN)1097-0126.

Stenger-Smith, J.D., 1998. Intrinsically electrically conducting polymers. Synthesis, characterization, and their applications. Elsevier Ltd, United States Prog. Polym. Sci. 23 (1), 57–79. https://doi.org/10.1016/S0079-6700(97)00024-5.

Stephenson, G.R., Ferris, I.G., Holland, P.T., Nordberg, M., 2006. Glossary of terms relating to pesticides (IUPAC Recommendations 2006). Pure Appl. Chem. 78 (11), 2075–2154. https://doi.org/10.1351/pac200678112075.

Tajik, S., Beitollahi, H., Nejad, F.G., Shoaie, I.S., Khalilzadeh, M.A., Asl, M.S., Van Le, Q., Zhang, K., Jang, H.W., Shokouhimehr, M., 2020. Recent developments in conducting polymers: applications for electrochemistry. Royal Society of Chemistry, Iran RSC Adv. 10 (62), 37834–37856. https://doi.org/10.1039/d0ra06160c. Available from: http://pubs.rsc.org/en/journals/journal/ra.

Tamm, J., Hallik, A., Alumaa, A., Sammelselg, V., 1997. Electrochemical properties of polypyrrole/sulphate films. Electrochim. Acta 42 (19), 2929–2934. https://doi.org/10.1016/s0013-4686(97)00114-x.

Tang, Z., Liu, S., Wang, Z., Dong, S., Wang, E., 2000. Electrochemical synthesis of polyaniline nanoparticles. Electrochem. Commun. 2 (1), 32–35. https://doi.org/10.1016/S1388-2481(99)00133-2.

Torad, N.L., Minisy, I.M., Sharaf, H.M., Stejskal, J., Yamauchi, Y., Ayad, M.M., 2021. Gas sensing properties of polypyrrole/poly(N-vinylpyrrolidone) nanorods/nanotubes-coated quartz-crystal microbalance sensor. Elsevier Ltd, Egypt Synth. Met. 282. https://doi.org/10.1016/j.synthmet.2021.116935. Available from: http://www.journals.elsevier.com/synthetic-metals/.

Ul-Hamid, A., 2018. A Beginners' Guide to Scanning Electron Microscopy. https://doi.org/10.1007/978-3-319-98482-72018.

Unsworth, J., Lunn, B.A., Innis, P.C., Jin, Z., Kaynak, A., Booth, N.G., 1992. Technical review : conducting polymer electronics. J. Intell. Mater. Syst. Struct. 3 (3), 380–395. https://doi.org/10.1177/1045389x9200300301.

Verma, A., Choudhary, R.B., Nayak, D., Mandal, G., 2020. Structural analysis with augmented optoelectronic performance of polyindole–polypyrrole–cadmium sulfide nanohybrid: PIN/PPY/CdS. Springer, India J. Inorg. Organomet. Polym. Mater. 30 (7), 2683–2691. https://doi.org/10.1007/s10904-020-01525-6.

Wang, Z., Liu, E., Gu, D., Wang, Y., 2011a. Singapore glassy carbon electrode coated with polyaniline-functionalized carbon nanotubes for detection of trace lead in acetate solution. Thin Solid Films 519 (15), 5280–5284. https://doi.org/10.1016/j.tsf.2011.01.175, 5 31 2011/05/31.

Wang, Z., Liu, E., Zhao, X., 2011b. Singapore Glassy carbon electrode modified by conductive polyaniline coating for determination of trace lead and cadmium ions in acetate buffer solution, 5 31 2011/05/31 Thin Solid Films 519 (15), 5285–5289. https://doi.org/10.1016/j.tsf.2011.01.176.

Wang, Z., Li, L., Liu, E., 2013. Graphene ultrathin film electrodes modified with bismuth nanoparticles and polyaniline porous layers for detection of lead and cadmium ions in acetate buffer solutions, 10 1 2013/10/01 Thin Solid Films 544, 362–367. https://doi.org/10.1016/j.tsf.2013.02.098. Elsevier B.V. Singapore. http://www.journals.elsevier.com/journal-of-the-energy-institute.

Wang, Y., Wang, L., Huang, W., Zhang, T., Hu, X., Perman, J.A., Ma, S., 2017. A metal-organic framework and conducting polymer based electrochemical sensor for high performance cadmium ion detection. Royal Society of Chemistry, China J. Mater. Chem. A 5 (18), 8385–8393. https://doi.org/10.1039/c7ta01066d.

Wang, C.M., Tu, H.L., Li, Y., Xie, J.F., Chang, C.C., 2020. Highly sensitive detection of mercury ions using zincophosphite framework nanoparticle–polyaniline composites. American Chemical Society, Taiwan ACS Appl. Nano Mater. 3 (10), 9724–9730. https://doi.org/10.1021/acsanm.0c01837.

Wang, W., Wang, X., Cheng, N., Luo, Y., Lin, Y., Xu, W., Du, D., 2020. Recent advances in nanomaterials-based electrochemical (bio)sensors for pesticides detection. TrAC, Trends Anal. Chem. 132, 116041. https://doi.org/10.1016/j.trac.2020.116041.

Wang, Y., Liu, A., Han, Y., Li, T., 2020. Sensors based on conductive polymers and their composites: a review. John Wiley and Sons Ltd, China Polym. Int. 69 (1), 7–17. https://doi.org/10.1002/pi.5907. Available from: http://onlinelibrary.wiley.com/journal/10.1002/(ISSN)1097-0126.

Watts, K., Baker, A., Wirth, T., 2014. Electrochemical synthesis in microreactors. Akademiai Kiado Rt., United Kingdom J. Flow Chem. 4 (1), 2–11. https://doi.org/10.1556/JFC-D-13-00030. Available from: http://www.metapress.com/content/122156/.

Weidlich, C., Mangold, K.-M., Jüttner, K., 2001. Conducting polymers as ion-exchangers for water purification. Electrochim. Acta 47 (5), 741–745. https://doi.org/10.1016/s0013-4686(01)00754-x.

Wolfart, F., Hryniewicz, B.M., Marchesi, L.F., Orth, E.S., Dubal, D.P., Gómez-Romero, P., Vidotti, M., 2017. Direct electrodeposition of imidazole modified poly(pyrrole) copolymers: synthesis, characterization and supercapacitive properties. Elsevier Ltd, Brazil Electrochim. Acta 243, 260–269. https://doi.org/10.1016/j.electacta.2017.05.082.

Wong, Y.C., Ang, B.C., Haseeb, A.S.M.A., Baharuddin, A.A., Wong, Y.H., 2020. Review—conducting polymers as chemiresistive gas sensing materials: a review. Institute of Physics Publishing, Malaysia J. Electrochem. Soc. 167 (3). https://doi.org/10.1149/2.0032003JES. Available from: https://iopscience.iop.org/article/10.1149/2.0032003JES/pdf.

World Health Organization, 2019. WHO Recommended Classification of Pesticides by Hazard and Guidelines to Classification, p. 2019.

Wu, W., Yang, L., Chen, S., Shao, Y., Jing, L., Zhao, G., Wei, H., 2015. Core-shell nano-spherical polypyrrole/graphene oxide composites for high performance supercapacitors. Royal Society of Chemistry, China RSC Adv. 5 (111), 91645–91653. https://doi.org/10.1039/c5ra17036b.

Xie, S., Zhang, Q., Liu, G., Wang, Y., 2016. Photocatalytic and photoelectrocatalytic reduction of CO2 using heterogeneous catalysts with controlled nanostructures. Royal Society of Chemistry, China Chem. Commun. 52 (1), 35–59. https://doi.org/10.1039/c5cc07613g.

Xu, T., Dai, H., Jin, Y., 2020. Electrochemical sensing of lead(II) by differential pulse voltammetry using conductive polypyrrole nanoparticles. Springer, China Microchim. Acta 187 (1). https://doi.org/10.1007/s00604-019-4027-z. Available from: http://www.springer/at/mca.

Xue, Y., Chen, S., Yu, J., Bunes, B.R., Xue, Z., Xu, J., Lu, B., Zang, L., 2020. Nanostructured conducting polymers and their composites: synthesis methodologies, morphologies and applications. Royal Society of Chemistry, China J. Mater. Chem. C 8 (30), 10136–10159. https://doi.org/10.1039/d0tc02152k. Available from: http://pubs.rsc.org/en/journals/journal/tc.

Yang, Y., Kang, M., Fang, S., Wang, M., He, L., Zhao, J., Zhang, H., Zhang, Z., 2015. Electrochemical biosensor based on three-dimensional reduced graphene oxide and polyaniline nanocomposite for selective detection of mercury ions. Sens. Actuators, B 214, 63–69. https://doi.org/10.1016/j.snb.2015.02.127.

Yasri, N.G., Halabi, A.J., Istamboulie, G., Noguer, T., 2011. Chronoamperometric determination of lead ions using PEDOT:PSS modified carbon electrodes. Talanta 85 (5), 2528–2533. https://doi.org/10.1016/j.talanta.2011.08.013.

Zegebreal, L.T., Tegegne, N.A., Hone, F.G., 2023. Recent progress in hybrid conducting polymers and metal oxide nanocomposite for room-temperature gas sensor applications: a review. Elsevier B.V., Ethiopia Sensor Actuator Phys. 359. https://doi.org/10.1016/j.sna.2023.114472. Available from: https://www.journals.elsevier.com/sensors-and-actuators-a-physical.

Zhang, S., Pan, N., 2015. Supercapacitors performance evaluation. Wiley-VCH Verlag, United States Adv. Energy Mater. 5 (6). https://doi.org/10.1002/aenm.201401401. Available from: http://onlinelibrary.wiley.com/journal/10.1002/(ISSN)1614-6840.

Zhang, J., Kong, L.B., Li, H., Luo, Y.C., Kang, L., 2010. Synthesis of polypyrrole film by pulse galvanostatic method and its application as supercapacitor electrode materials. J. Mater. Sci. 45 (7), 1947–1954. https://doi.org/10.1007/s10853-009-4186-0.

Zhang, W., Fan, S., Li, X., Liu, S., Duan, D., Leng, L., Cui, C., Zhang, Y., Qu, L., 2020. Electrochemical determination of lead(II) and copper(II) by using phytic acid and polypyrrole functionalized metal-organic frameworks. Springer, China Microchim. Acta 187 (1). https://doi.org/10.1007/s00604-019-4044-y. Available from: http://www.springer/at/mca.

Zhang, H., Li, Y., Zhang, Y., Wu, J., Li, S., Li, L., 2023. A disposable electrochemical sensor for lead ion detection based on in situ polymerization of conductive polypyrrole coating. Springer, China J. Electron. Mater. 52 (3), 1819–1828. https://doi.org/10.1007/s11664-022-10175-y. Available from: https://www.springer.com/journal/11664.

Zhou, H., Wen, J., Ning, X., Fu, C., Chen, J., Kuang, Y., 2007. Electrosynthesis of polyaniline films on titanium by pulse potentiostatic method. Synth. Met. 157 (2–3), 98–103. https://doi.org/10.1016/j.synthmet.2006.12.013.

Recent developments in multifunctional nanofibrous membranes for oily wastewater treatment

13

Fayez Uddin Ahmed and Debarun Dhar Purkayastha
Department of Physics, National Institute of Technology Nagaland, Chumukedima, Nagaland, India

13.1 Introduction

In today's industrial landscape, the generation of oily wastewater poses a ubiquitous challenge across sectors such as petrochemicals, manufacturing, and food processing (Saravanan et al., 2021; Chen et al., 2024a,b; Xu et al., 2024; Aier and Dhar Purkayastha, 2024). These effluents, characterized by intricate compositions of hydrocarbons, oils, and diverse contaminants, present substantial environmental and regulatory challenges (Saravanan et al., 2021; Xu et al., 2024; Liu et al., 2024). The importance of effective wastewater treatment methods cannot be overstated, as untreated or inadequately treated oily wastewater can result in detrimental consequences like water pollution, ecological damage, and health hazards (Mazumder et al., 2023; Chowdhary et al., 2018). Traditional wastewater treatment methods, widely employed, often struggle to efficiently address the multifaceted nature of oily wastewater (Singh et al., 2023; Zhang et al., 2024). Limitations in techniques such as gravity separation, electrochemical treatment, biological media utilization, coalescence, and centrifugation necessitate a paradigm shift toward more advanced and sustainable solutions (Singh et al., 2023; Zhang et al., 2024; Abuhasel et al., 2021). It is within this context that multifunctional nanofibrous membranes emerge as a promising and innovative solution for the treatment of complex oily wastewater (Saleem et al., 2020). The incorporation of metal oxides, growing special morphologies of metal oxides, and integrating different nanomaterials such as MXenes, graphene oxide, carbon nanotubes, and metal-organic frameworks are key strategies to develop multifunctional nanofibrous membranes (Deshmukh et al., 2024; Zeng et al., 2024; Imsong and Dhar Purkayastha, 2024b). These modifications enhance the membrane's capabilities in various ways. Metal oxides, for example, can impart photocatalytic activity to the membrane, enabling the degradation of organic pollutants (Xing and Hong, 2024). Special morphologies of metal oxides, such as nanorods or nanosheets, can enhance the membrane's surface area which enhances the wettability, reactivity and improve its overall performance (Zhou et al., 2024). In addition, the integration of

Multifunctional Nanostructured Coatings. https://doi.org/10.1016/B978-0-443-23683-9.00013-1

nanomaterials can further enhance the membrane's properties, such as its oil—water separation efficiency, stimuli-responsiveness, chemical stability, mechanical stability, antibacterial activity, and antifouling properties (Huang et al., 2024). Thus, nanofibrous membranes with these advanced functionalities offer a transformative approach to wastewater treatment by combining unique structural properties and advanced filtration capabilities (Saleem et al., 2020; Nasir et al., 2021; Barambu et al., 2021; Yan et al., 2023; Jiang et al., 2017). They can selectively separate various types of oily wastewater, including oil-in-water and water-in-oil emulsions, while effectively removing heavy metal ions and suspended particles (Yan et al., 2023; Meng et al., 2023). Furthermore, they demonstrate the capability to degrade organic pollutants, including pharmaceutical compounds, through photocatalytic activity. Nanofibrous membranes also exhibit antibacterial properties, reducing the risk of bacterial contamination in treated water, and their antifouling properties enhance performance by preventing the accumulation of contaminants. This multifunctionality makes nanofibrous membranes an ideal candidate for tackling the challenges associated with complex industrial effluents, highlighting their potential for enhancing treatment efficiency and sustainability in oily wastewater treatment applications (Cao et al., 2023; Wang et al., 2024; Moatmed et al., 2024).

This chapter provides an overview of recent developments in multifunctional nanofibrous membranes for oily wastewater treatment. It begins with a background on oil/water separation methods, highlighting the challenges and limitations of traditional approaches. The chapter then discusses the basics of electrospun nanofibrous membranes and their utilization in separating oil and water. The concept of multifunctionality in nanofibrous membrane design is introduced, focusing on stimulus-responsive membranes that can undergo reversible changes in their properties when exposed to external stimuli. These membranes, such as photo-responsive, thermo-responsive, pH-responsive, and gas-responsive membranes, offer enhanced flexibility and efficiency in oil/water separation processes. Furthermore, the chapter explores the incorporation of photocatalytic and fouling-resistant materials into nanofibrous membranes to improve their performance in treating oily wastewater. In addition, it discusses the importance of chemical and mechanical durability in ensuring the long-term effectiveness of these membranes in harsh industrial environments. Recent breakthroughs in nanofibrous membrane technology for treating oily wastewater are presented, including the enhancement of nanostructured membranes for efficient oil/water separation, functionalization of membranes for improved performance, and the use of novel materials in membrane design. The economic and environmental sustainability of nanofibrous membrane technology is also discussed, highlighting its capacity to decrease energy usage and waste generation in comparison with conventional treatment approaches. The chapter concludes with a discussion on the real-world impact of nanofibrous membranes in oily wastewater treatment, showcasing their applications in industries such as petrochemicals and textiles. The concluding remark outlines the future standpoints and concerns in the field, emphasizing the need for continued research and innovation to address the evolving needs of oily wastewater management.

13.2 Background of oil/water separation

Oil/water separation is a vital technique employed in various industries and environmental applications to address the challenge of separating oil and water mixtures. This process is essential for wastewater treatment, oil spill cleanup, enabling industrial operations, and ensuring the production of potable water. The need for effective oil/water separation arises from the widespread occurrence of oil contamination in water sources, whether from industrial discharges, accidental spills, or natural seepage. This wastewater may include fats, hydrocarbons, and petroleum components like diesel oil and gasoline (Kumar et al., 2022). The increasing volume of such wastewater, estimated at 10 to 15 billion m^3 worldwide in 2013, underscores the urgency of addressing this environmental concern (Mohshim et al., 2023; Aghaei et al., 2023). These oily wastewater pose serious environmental risks, threatening aquatic ecosystems, human health, and economic activities dependent on clean water sources (Moatmed et al., 2024). Numerous research institutions, both domestic and international, have extensively explored and deliberated on methods for treating oily wastewater. Various treatment approaches have been developed to minimize or eliminate the detrimental effects of oil impurities in wastewater. Several technologies have been developed to achieve this, ranging from conventional methods like gravity separation and skimming to advanced processes such as electrochemical treatment, biological media utilization, demulsification, air flotation, coalescence, adsorption, centrifugation, coagulation flocculation, and membrane filtration are among the methods devised for oily wastewater treatment (Aghaei et al., 2023; Subramanian et al., 2022). Each method has its advantages, limitations, and suitability for specific applications depending on factors like the type and volume of oil, the characteristics of the water matrix, and the desired purity levels. Understanding the principles underlying oil/water separation processes is crucial for designing efficient separation systems, optimizing performance, and addressing emerging challenges in environmental management and resource recovery. In the sections that follow, we will look at the numerous approaches and technologies employed in oil/water separation, their mechanisms of action, and recent advancements.

13.2.1 Gravity separation

Gravity separation is a physical separation process that exploits the density differences between components in a mixture (Seader et al., 2006). It is commonly employed for separating oil and water in the treatment oily wastewater. The basic principle is to allow the mixture to settle in a tank or basin, relying on gravity to separate the phases based on their density. However, its effectiveness is notably reduced when dealing with stable emulsions, where oil and water are finely dispersed and exhibit resistance to easy separation. Particularly with emulsified oils, the method's efficiency is constrained, and its application often requires the use of large settling tanks.

13.2.2 Electrochemical treatment

Electrochemical treatment stands as an advanced approach for addressing oily wastewater remediation, utilizing electrochemical processes to eliminate contaminants, including oil and various pollutants, from water (Ghaffarian Khorram et al., 2023). The fundamental concept involves applying an electric current to initiate chemical reactions that result in the breakdown or alteration of pollutants. A commonly employed electrochemical method for treating oily wastewater is electrocoagulation, where metal electrodes, typically aluminum or iron, undergo electrolysis to produce coagulating agents. These agents play a role in destabilizing and aggregating oil droplets and suspended particles, facilitating their separation from water. Despite its efficacy in contaminant removal, electrochemical treatment may present challenges, including elevated energy consumption, potential corrosion of electrode materials, and the generation of sludge.

13.2.3 Biological media utilization

Biological media utilization is a method employed for the treatment of oily wastewater, relying on biological processes and microorganisms to break down and degrade contaminants (Sanghamitra et al., 2021). In this approach, microorganisms, such as bacteria and fungi, are introduced into the wastewater to consume and metabolize the organic components, including oil and grease. The microorganisms utilize the contaminants as a food source, converting them into simpler and less harmful byproducts through biological reactions. While biological media utilization can be effective in reducing oil content in wastewater, its efficiency can be impacted by variables such as microbial activity, ambient circumstances, and the specific properties of the oily wastewater. In addition, the process may require longer retention times, and the effectiveness can be hindered by the presence of certain toxic pollutants or chemicals.

13.2.4 Demulsification

Demulsification is a technique employed for separating the emulsified oil and water phases in oily effluent. In this method, demulsifying agents or chemicals are introduced to disrupt the stable emulsion and promote the amalgamation of oil droplets (Zolfaghari et al., 2016). The demulsifying agents typically work by decreasing the interfacial tension within water and oil, facilitating the oil droplets to merge and form larger aggregates. Once the emulsion is destabilized, gravity separation or other separation techniques can be applied to facilitate the removal of the coalesced oil phase. While demulsification can be effective in breaking down emulsions and separating oil from water, the choice of demulsifying agents, dosage, and the efficiency of the process can be influenced by the specific characteristics of the emulsified oil and wastewater composition. In addition, the use of demulsifiers may introduce chemical additives to the wastewater, requiring careful consideration of environmental impacts.

13.2.5 Air flotation

Air flotation represents a specialized approach for treating oily wastewater by introducing fine air bubbles into the water, which attach themselves to oil droplets and suspended particles (Yu et al., 2017). The buoyant force generated by the air bubbles causes the contaminants to rise to the water's surface, forming a froth or foam layer that can be easily removed. Although air flotation is effective, its effectiveness is influenced by variables such as the size and concentration of oil droplets, the chemical composition of the wastewater, and the flotation system's design. Despite its advantages, air flotation may face challenges in treating highly concentrated or stable emulsions and might require the addition of coagulants or flocculants to enhance contaminant removal.

13.2.6 Coalescence

Coalescence is a method employed in the remediation of oily wastewater, specifically targeting the separation of oil droplets from the water phase. In this technique, coalescing agents or chemicals are introduced to promote the merging or coalescence of small oil droplets into larger aggregates (Zhao et al., 2021). These coalescing agents function by reducing the interfacial tension between oil droplets, facilitating their combination into larger, more easily separable entities. Once coalescence occurs, gravity separation or other separation methods can be applied to effectively remove the coalesced oil phase from the water. While coalescence is efficient in breaking down small oil droplets and enhancing their removal, the choice of coalescing agents and their dosage can be influenced by the specific characteristics of the emulsified oil and the composition of the wastewater. In addition, environmental considerations are crucial to evaluate potential impacts associated with the introduction of coalescing agents into the wastewater treatment process.

13.2.7 Adsorption

Adsorption is a frequently employed technique for treating oily wastewater, involves the removal of oil and contaminants by utilizing porous materials with a substantial surface area, such as activated carbon, silica gel, zeolites, sponges, etc. as adsorbents (Sabir, 2015). Through physical or chemical forces, the contaminants adhere to the surface of these materials, efficiently capturing and eliminating oil from the water. Although adsorption is effective in reducing oil content, its efficiency depends on factors such as the choice of adsorbent, contact time, and the specific characteristics of the oily wastewater. However, a limitation is the potential saturation of adsorbents over time, resulting in reduced effectiveness. In addition, the regeneration or disposal of spent adsorbents can pose environmental challenges.

13.2.8 Centrifugation

Centrifugation is a mechanical separation method widely employed in the treatment of oily wastewater (Cambiella et al., 2006). This technique employs centrifugal force to differentiate and segregate different components by their varying densities. In the context of oily wastewater, centrifugation is applied to isolate the oil phase from water and other impurities. The wastewater is introduced into a rapidly spinning centrifuge, creating a centrifugal force that drives the heavier oil droplets outward, allowing their separation from the water. Centrifugation is particularly effective for removing larger oil droplets and solid particles from the wastewater. However, its efficiency may be limited when dealing with stable emulsions or smaller oil droplets. In addition, the equipment and energy requirements for centrifugation can contribute to higher operational costs.

13.2.9 Coagulation and flocculation

Coagulation and flocculation are widely employed in the treatment of oily wastewater, serving as chemical processes that facilitate the removal of contaminants, including oil and suspended particles (de Medeiros et al., 2022). In coagulation, chemicals known as coagulants are added to the wastewater, promoting the destabilization of colloidal particles and the formation of larger aggregates. Following coagulation, flocculation involves the addition of flocculants, which induce the agglomeration of the destabilized particles into larger, easily removable flocs. These flocs can be separated from the water through sedimentation or filtration processes. While coagulation and flocculation are effective in reducing the concentration of oil and suspended particles, their effectiveness can be affected by factors such as the selection of coagulants and flocculants, pH levels, and the specific characteristics of the oily wastewater. In addition, the generation of sludge during the process requires careful management and disposal considerations.

13.2.10 Membrane filtration

Membrane filtration serves as a sophisticated and effective method for the remediation of oily effluent, relying on the application of selective permeable membranes to achieve separation of contaminants from water (Barambu et al., 2021; Cao et al., 2023; Moatmed et al., 2024; Ahmad et al., 2020). This method exploits the varying sizes of particles and molecules, allowing water molecules to pass through while obstructing the passage of oil droplets and other impurities. The membrane filtration process encompasses various techniques, each tailored to specific pore sizes and separation requirements. Microfiltration (MF) employs membranes with larger pores, typically in the range of 0.1−10 μm, making it suitable for removing suspended solids and larger particles (Bardhan et al., 2022). Ultrafiltration (UF) takes the separation a step further with smaller pore sizes in the range of 0.001−0.1 μm, effectively removing colloidal particles and smaller oil droplets (Bardhan et al., 2022). Nanofiltration (NF) and reverse osmosis (RO) utilize even smaller pore sizes, typically in the range of 0.0001–0.001 μm, allowing for the removal of dissolved salts, organic compounds, and finer contaminants (Joseph et al., 2023). While these advanced filtration methods significantly contribute to achieving high-quality treated water, they come with certain challenges. One major challenge is membrane fouling, where contaminants accumulate on the membrane surface, reducing its effectiveness over time (Zheng et al., 2024). Maintenance of membranes to address fouling, as well as considerations related to energy consumption and operational costs, are crucial factors in determining the feasibility and efficiency of membrane filtration for oily wastewater treatment. Despite these challenges, membrane filtration stands as a valuable and versatile technique in the arsenal of methods for addressing the complex issue of oily wastewater (Manouchehri, 2024). While membrane filtration serves as a sophisticated method for the treatment of oily wastewater, emerging technologies like electrospun nanofibrous membranes offer promising advancements in this field. Electrospun nanofibrous membranes represent a more refined and efficient approach to oil/water separation, leveraging innovative fabrication techniques and unique structural properties to enhance performance and address specific challenges associated with membrane filtration (Jiang et al., 2017; Moatmed et al., 2024; Tian et al., 2023; Hu et al., 2024; Cui et al., 2023). Unlike conventional membrane filtration, which relies on selective permeable membranes with predefined pore sizes, electrospun nanofibrous membranes offer greater flexibility in tailoring pore sizes, surface properties, and structural characteristics to suit diverse separation requirements (Wang and Hsiao, 2016). This versatility enables precise control over the separation process, allowing for more effective removal of oil droplets, colloidal particles, and dissolved contaminants from wastewater streams.

13.3 Background of electrospun nanofibrous membranes

13.3.1 Historical evolution of nanofiber technology

The historical context and evolution of nanofiber technology trace its roots back to the early 19th century. Pioneered by Rayleigh in 1897 and further explored by Zeleny, Formhals marked a significant milestone in the 1930s by patenting electrospinning as a feasible fiber-spinning technology (Tucker et al., 1995). This groundwork laid the foundation for subsequent researchers such as Taylor, Saville, and Denn, who delved into the intricacies of electrically powered jets, paving the way for the surge in electrospinning research in the late 1960s (Ray et al., 2019). As we delve deeper into the realm of nanotechnology, the electrospinning process has undergone extensive research and development (R&D), experiencing a notable resurgence in attention between 2005 and 2015 (Tucker et al., 1995; Ray et al., 2019; Wang and Hsiao, 2016). This surge can be attributed to the growing attention toward nanoengineering and nanotechnology, as seen by the growing number of publications in this context.

13.3.2 Electrospinning process

The exploration of nanofibrous membranes centers around the electrospinning process, a cornerstone mechanism that encompasses various approaches, each contributing to the versatile role these membranes play in treating oily wastewater (Ge et al., 2018; Yalcinkaya et al., 2022). The traditional electrospinning setup, as depicted in Fig. 13.1, comprises a spinneret fitted with a needle, an injection pump, a high-voltage direct current (DC) power source, and connected collector (Figen, 2020). Primarily, the process involves introducing a mixture, composite, sol-gel, polymeric solution/melt derived from either natural or synthetic source into a syringe. This solution is then forced toward the needle's tip using a syringe pump, generating a spherical droplet. When a voltage ranging between 5 and 50 kV is applied to the blend on the needle, the droplet expands into a conical shape known as the Taylor cone (schematic representation in Fig. 13.1) (Xiong et al., 2021). An electrified jet develops and travels to the oppositely charged metal collector depending on the viscosity of the fluid, sufficient to sustain elongation and flailing without generating beads. During this travel, the jet solidifies and forms microfiber webs on the collector due to solvent evaporation. Surprisingly, changing the solution's physicochemical properties allows for modification of the shape and structure of the produced nanofibers.

13.3.3 Impact of various factors on electrospinning process

Numerous well-organized papers have exhaustively studied the impact of various factors on the electrospinning process (Ray et al., 2019). Furthermore, the solvent and cosolvent used in the synthesis have a substantial impact on the geometry and structure of the nanofibers. The adaptability and robustness of the electrospinning process are evident, allowing for the transformation of a broad range of materials into nanofibrous structures (Wang and Hsiao, 2016; Ge et al., 2018; Yalcinkaya et al., 2022; Xiong et al., 2021). Its exceptional adaptability and efficiency surpass traditional technologies, yet the resulting nanofiber formation and structure are significantly influenced by specific parameters. These parameters fall into three main categories: intrinsic

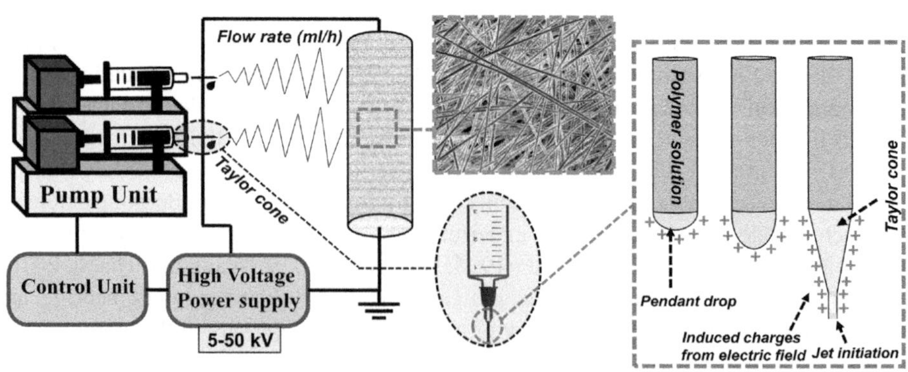

Figure 13.1 Schematical depiction of electrospinning set up and Taylor cone formation.

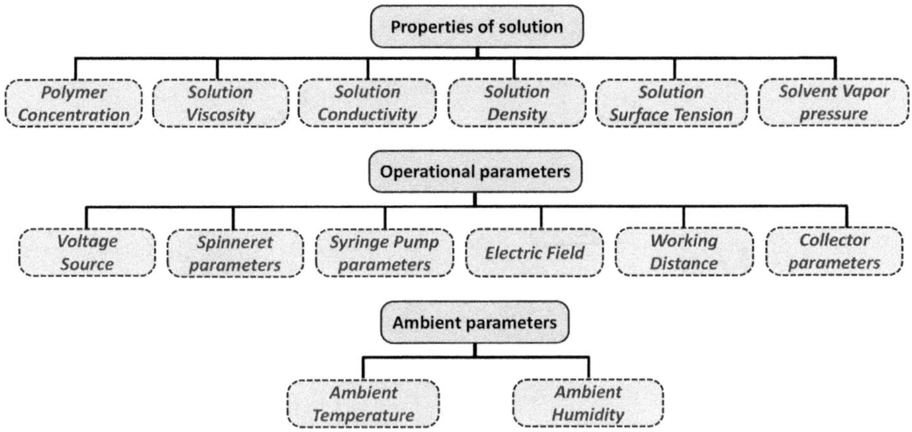

Figure 13.2 Electrospinning dependent parameters.

properties of the polymeric solution, operational parameters, and surrounding parameters (Figen, 2020; Theron et al., 2004). Fig. 13.2 provides a concise overview of these parameters, highlighting the optimal conditions crucial for fabricating high-performance electrospun nanofibrous membranes.

13.3.4 Properties and applications of nanofibrous membranes

Solution electrospinning, in particular, stands out as an exceptionally versatile process capable of producing nanofibers from a wide range of polymeric materials. Recent advancements in scalability for mass production have significantly improved the practicality of electrospinning, resulting in increased production rates and reduced material costs. Nanofibrous membranes exhibit distinctive properties that make them stand out in various applications. Their nanoscale dimensions provide an extraordinary surface area, enhancing reactivity and adsorption capacity (El-Aswar et al., 2022). Despite their diminutive size, nanofibers demonstrate remarkable mechanical strength, ensuring the durability and robustness of the membranes across diverse applications (El-Aswar et al., 2022; Han et al., 2020). In addition, the high degree of flexibility in nanofibrous structures allows them to conform to different shapes and surfaces, providing adaptability crucial for applications involving varied forms or deformations (Han et al., 2020). Notably, the high aspect ratio, huge surface area, and interwoven porous structure further contributes to the unique electrical, mechanical, and thermal properties of nanofibrous membranes (El-Aswar et al., 2022; Han et al., 2020; Ibrahim and Klingner, 2020; Wang et al., 2019a,b,c). This combination of properties positions nanofibrous membranes as versatile materials with significant potential in fields such as filtration, catalysis, and beyond. Numerous composite membrane configurations, featuring various nanofiber arrangements, have been successfully illustrated for a spectrum of water treatment applications, including nanofiltration, ultrafiltration, microfiltration, reverse osmosis, and adsorption processes, as well as membrane distillation

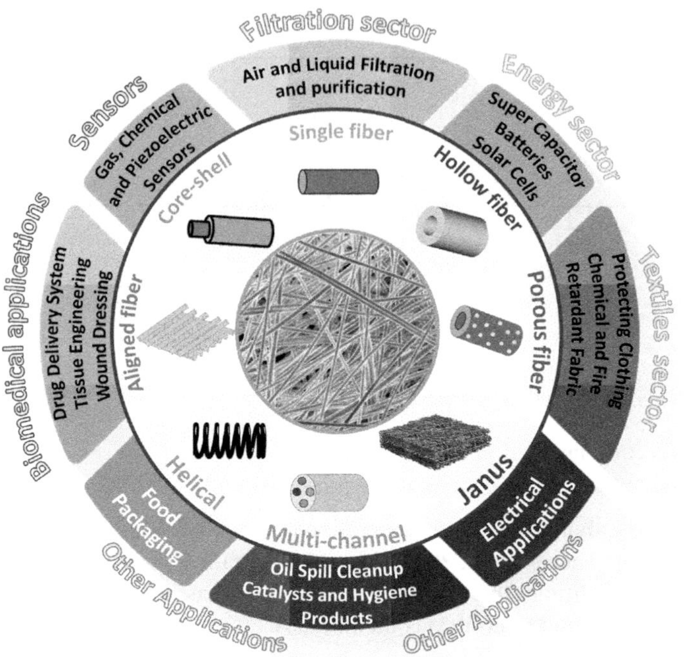

Figure 13.3 Schematical presentation of different types of nanofibers and their applications.

(Ge et al., 2018; Yalcinkaya et al., 2022; Wang et al., 2019a,b,c). The nanofibers, characterized by an extensive surface-to-volume ratio, represent an exciting material class applicable to numerous domains, including environmental engineering, healthcare, biotechnology, energy storage, and information technology (Ibrahim and Klingner, 2020). Understanding these fundamental aspects of nanofibrous membranes lays the groundwork for exploring their diverse applications, including but not limited to filtration. The huge potential in various applications offered by different types of nanofibers based on electrospinning is summarized schematically in Fig. 13.3.

13.4 Oil/water separation by membrane technology

The remarkable properties of electrospun nanofibrous membrane make them highly attractive for an array of applications, including environmental cleanup and industrial wastewater treatment. One particularly promising application is their use in oil/water separation, where these membranes demonstrate exceptional efficiency and versatility. In this section, we delve into the mechanisms and strategies employed by nanofibrous membranes to achieve effective oil/water separation, highlighting their pivotal role in addressing environmental challenges and advancing sustainable solutions. A significant progress has been made in researching the importance of surface wetting in oil/

water separation, offering potential solutions for real-world oil spills and the purification of intricate oily wastewater. The concept of wettability, characterized by contact angles (CAs) exceeding 150 degrees or below 10 degrees, has garnered attention for its applications in self-cleaning, desalination, water harvesting, anti-icing corrosion resistance, liquid transfer, gas separation, and oil/water separation (Wang et al., 2016). In the realm of oily wastewater treatment, materials demonstrating superhydrophobic/superoleophilic traits, coined as "oil-removing" materials, are intended to specifically separate oil from oil-in-water mixtures or emulsions (Wang et al., 2016; Yue et al., 2019). Conversely, materials demonstrating superhydrophilic/superoleophobic abilities are termed "water-removing" materials (Wang et al., 2016; Yue et al., 2019; Rasouli et al., 2021). The materials with in-air superhydrophobicity and in-air superoleophilicity, known as "oil-removing," and those with superhydrophilicity and underwater superoleophobicity, known as "water-removing," are the top choices for addressing oil spill scenario and treating intricate oily wastewater, as depicted in Fig. 13.4a and b (Rasouli et al., 2021). The wettability-regulated separation technology, with its ultrahigh separation performance and exceptional reusability, surpasses conventional separation methods, making it a preferred alternative for managing oil spills and filtering oily wastewater (Wang et al., 2016). The pioneering work by Jiang et al. in 2004 introduced the first "oil-removing" material, showcasing gravity-driven oil/water separation (Feng et al., 2004). Subsequently, numerous superwetting materials have been developed, encompassing those with singular wettability as well as Janus materials exhibiting opposite wettability. This overview encompasses the mechanisms involved in oil/water separation, as well as nanofibrous membranes designed for oil removal, water removal, and Janus characteristics.

13.4.1 Mechanism of oil/water separation

The effective separation of oil/water is considerably influenced by the distinct interfacial interaction between oil and water, with three crucial physical attributes: porosity, pore size, and intrusion pressure (Zhang et al., 2023). These factors are essential in shaping the overall performance. Porosity and pore size, in particular, are influential determinants that impact the permeation rate of either water or oil through the membrane, adding complexity to the dynamics of the separation process. The model depicted in Fig. 13.4c–f can be adopted to elaborate on the oil and water separation mechanism. The intrusion pressure (ΔP) refers to the highest static pressure that the membrane can endure and is mathematically represented by the Eq. 13.1 (Ahmed and Dhar Purkayastha, 2023a):

$$\Delta P = \frac{2\gamma}{R} = -\frac{C\gamma \, Cos\theta}{d} \tag{13.1}$$

where γ represents the liquid (water or oil) surface tension, θ is the advancing contact angle between the fluid and the membrane, R is the radius of the meniscus, C denotes the circumference of the pore, and d is the pore size of the membrane. This

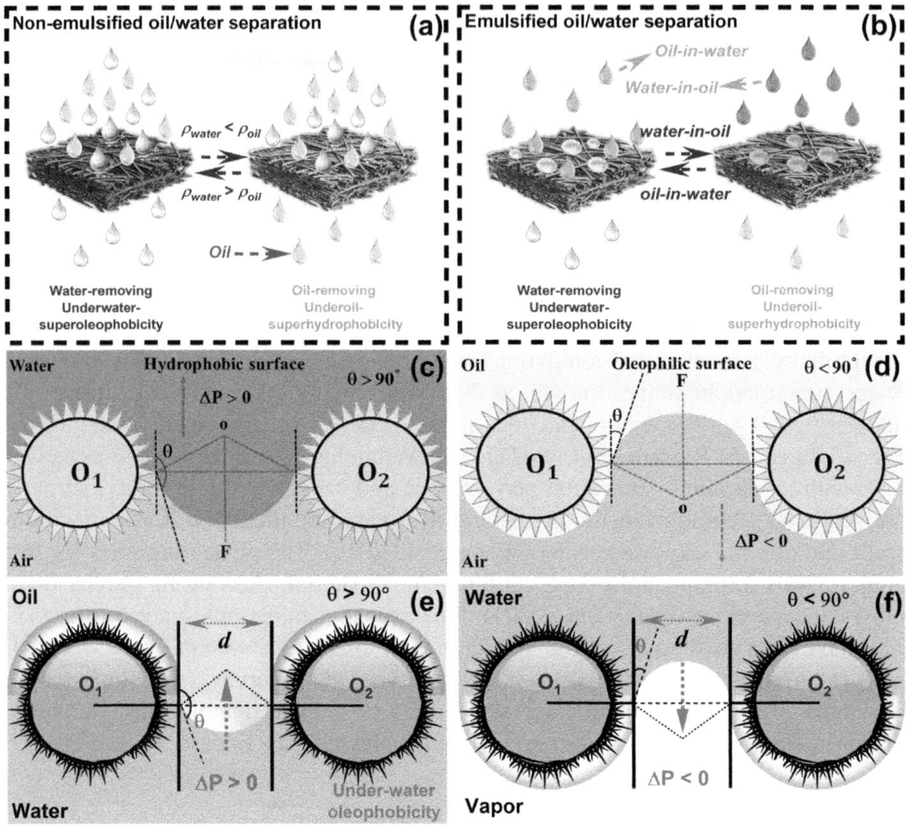

Figure 13.4 Schematic illustration of water/oil-removing type membrane (a) nonemulsified oil/water separation and (b) emulsified oil/water separation; Schematic diagram of oil/water separation mechanism (c) water phase (CA, $\theta > 90$ degrees) is unable to permeate the membrane as$\Delta P > 0$, (d) oil phase (OCA, $\theta < 90$ degrees) can permeate the membrane as $\Delta P < 0$, (e) oil phase ($\theta > 90$ degrees) is unable to penetrate the pre-wetted membrane as $\Delta P > 0$, and (f) water phase ($\theta < 90$ degrees) easily penetrates the pre-wetted membrane as $\Delta P < 0$.

equation highlights the substantial impact of interface wettability on the intrusion pressure. In specific conditions, where $\theta > 90$ degrees, the ΔP becomes positive (Fig. 13.3c), allowing it to surpass external pressure. On the contrary, when $\theta < 90$ degrees, $\Delta P < 0$ (Fig. 13.3d), it is easier for the liquid to pass through the pores and entirely wet the lyophilic interface. Hence, surface exhibiting superhydrophobic/superoleophilic properties is inherently "oil-removing" facilitating rapid oil penetration while obstructing water passage (Feng et al., 2004). Similarly, when oil droplets encounter a water-prewetted surface, they cannot infiltrate the surface due to its underwater superoleophobicity, where the oil contact angle exceeds 90 degrees, and $\Delta P > 0$ (Fig. 13.3e). Typically, when a surface is in-air hydrophilic, it exhibits underwater superoleophobicity owing to the high surface tension of water in comparison with oil. As a result, surface exhibiting superhydrophilic/underwater superoleophobic properties can act as "water-removing" particularly segregating immiscible oil/water mixtures by permitting the water phase to flow through and forming a water film that repels oil completely (Ahmed and Dhar Purkayastha, 2023b). In the same way, emulsions of oil/water may also be segregated through either demulsification or size-sieving. In the size-sieving approach, it is crucial that the size of the pores of the superwetting membrane be either less or equal to the size of surfactant-stabilized droplets (Rachel Imsong and Dhar Purkayastha, 2024c). Typically, surfaces exhibiting superhydrophobicity and superoleophilicity, referred to as "oil-removing" surfaces, are utilized for the separation of water-in-oil emulsions. On the contrary, materials with "water-removing" characteristics are commonly applied for the remediation of oil-in-water emulsions. It is noteworthy that a smaller pore size, while effective for size-sieving, may lead to a reduced separation flux when dealing with emulsions as compared to immiscible oil/water mixtures. In contrast, demulsification provides a swift method for emulsion separation, involving two key steps (Saravanan et al., 2021): upon contacting the membrane surface, emulsified droplets in oil/water emulsions are captured, and (Chen et al., 2024a,b) these droplets amalgamate on the surface, causing the emulsions to transform into immiscible mixture of oil and water (Zhang et al., 2023). Subsequently, by utilizing either superhydrophobic or superhydrophilic materials to separate these immiscible mixtures, the process achieves selective separation, thereby improving the efficiency of oil/water emulsion separation.

13.4.2 Oil-removing membrane

Membranes categorized as "oil-removing" possess the capability of selectively eliminating oil from oil/water mixtures, a characteristic owing to its remarkable superhydrophobic/superoleophilic properties (Ahmed and Dhar Purkayastha, 2023a). The superwetting nature of a surface is intricately tied to its hierarchical micro/nanostructures and suitable surface energy. For achieving simultaneous superhydrophobicity and superoleophilicity, the membrane should have surface energy within the oil and water. In addition, augmenting the roughness of the surface emerges as a crucial strategy to escalate the superhydrophobic/superoleophilic property of the membrane (Ma et al., 2018). Over the years, numerous strategies have been devised and implemented for designing and fabricating "oil-removing" materials. For instance, of an "oil-removing" membrane where T. Zhang et al. developed a superoleophilic and underoil superhydrophobic PVDF/graphene nanofibrous membrane (GPNM) through one-step electrospinning (Zhang et al., 2021a,b). The membrane, with tunable properties, efficiently separated water-in-oil emulsions, exhibiting water contact angle of 140.1 $\pm$ 1.1 degree, nearly 0° oil contact angle, and excellent ultralow water adhesion under oil. Its morphology and performance were adjusted by varying graphene concentration. The GPNM demonstrated exceptional oil absorption, effectively collecting floating oils from the water surface. Furthermore, T. Zhang et al. developed an affordable polyurethane sponge hollow tube with graphite coating for spontaneous and effective oil removal (Zhang et al., 2020). The sponge, created by a simple dip-coating process, exhibited high selectivity in absorbing various oils. The fabricated membrane demonstrated excellent performance in collecting different oils from water surfaces, boasting a high flux, outstanding separation efficiency for toluene, and notable reusability. The development of this strategy opened up practical opportunities for attaining effective execution in the separation of oil/water mixtures. Subsequent efforts have led to the design and synthesis of growing array of materials for oil removal. Researchers typically utilize diverse materials such as metallic mesh, polymeric membranes, fabrics, sponges, etc., as substrates. Combining these materials with arrayed micro/nanostructures and low surface energy to design materials with oil removal properties (Rasouli et al., 2021). Despite the success of introducing super wetting properties into the realm of oily wastewater management, these materials have revealed certain drawbacks in the course of real-world scenarios. The bottleneck of these materials lies in their susceptibility toward oil fouling as a consequence of their inherent oleophilic nature (Zheng et al., 2021). Moreover, these materials are ill-suited for continuously separating light oil under the assistance of gravity. This is because a hydrated film forms on the water surface beneath the oil, impeding direct contact with the oil. This inherent limitation significantly hinders the practical performance of "oil-removing" materials. Consequently, there is a critical research focus on engineering novel membranes with oil-removal, antifouling, and anti-adhesion capabilities in recent years.

13.4.3 Water-removing membrane

In contrast to "oil-removing" membranes, "water-removing" membranes with superhydrophilic/underwater superoleophobic properties particularly segregate water from oil/water mixtures. In contrast to oil removal membranes, the latter have favorable benefits. The superoleophobic nature of the membrane protects against oil fouling, overcoming limitations of oil contamination and enhancing the service life, oil, and membrane recyclability. In addition to this, they excel in separation processes driven by gravity, particularly with less dense oils; and they demonstrate unique advantages in oil spills by enabling oil enrichment and recovery (Zhang et al., 2023; Ahmed and Dhar Purkayastha, 2023b; Rachel Imsong and Dhar Purkayastha, 2024c; Ahmed et al., 2022). With these benefits, the engineering of "water-removing" membranes have garnered considerable attention. Inspired by the advancements in underwater superoleophobic materials, F. Sun et al. presented a straight-forward approach to fabricate a hierarchical membrane that exhibits efficient photocatalysis and oil/water separation (Sun et al., 2021). The membrane's complex structure involves electrospinning hydrophilic polyacrylonitrile (PAN) nanofibers and electrospraying PAN/TiO_2 spheres, with subsequent in-situ growth of polydopamine (PDA) to enhance the membrane. This arrangement confers superhydrophilic and underwater superoleophobic properties in membrane. The resulting hierarchical nanofibrous membrane offers intriguing applications in oil—water separation and photocatalysis. In summary, water removal membranes effectively prevent the membrane from being tainted by oil. Regardless, these membranes are ill-suited for heavy oil, since the water settle beneath and adhere to the membrane surface because of its low density, creating an obstacle that prevents water from permeating (Yue et al., 2019). In addition, in comparison with superhydrophobic/superoleophilic membranes, achieving superhydrophilic/superoleophobic membranes poses greater challenge due to the difficulty in finding a substance having surface energy more than water and less than oil. Therefore, there is a pressing requirement to develop improved membranes that can switch their wettability between oil and water, enabling the effective and timely separation of oil/water mixtures and emulsions as needed.

13.4.4 Janus membrane

Single-wettability membranes face substantial limitations in practical applications due to various environmental conditions. Therefore, membranes having switchable wettability of water and oil-removing properties are of utmost importance. Janus membranes, featuring asymmetric surface wettability on two opposing surfaces, have garnered significant attention recently. The opposite wettability of these surfaces enables them to serve as a switchable blockade in the separation process of oil and water, as depicted in Fig. 13.5 Moreover, the distinctive capability of Janus materials to enable unidirectional liquid transport makes them well-suited for applications in intelligent control, an area that has attracted considerable attention in recent years. Typically, Janus membrane with asymmetric wettability can be created through different methods: asymmetric fabrication and modification on the membranes. The asymmetric fabrication method is the most straightforward approach in creating Janus membranes. It entails separately fabricating each side of the Janus material and then combining them to form the resultant membrane. For instance, C. Cheng et al. engineered a well-thought-out design for a Janus nanofibrous membrane using deacetylated cellulose acetate/polyvinylidene fluoride (Cheng et al., 2022). This membrane exhibits asymmetric wetting properties in air, water, or oil environments, achieved through a sequential electrospinning process involving PVDF and CA nanofiber membranes

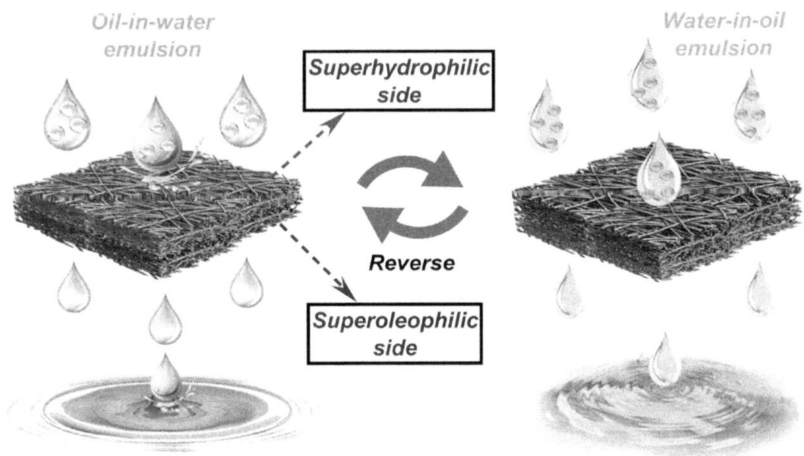

Figure 13.5 Illustration of the Janus membrane in the separation of oil/water.

followed by alkaline remediation. The resultant membrane demonstrates enhanced efficiency, long-term separation performance, and resistance to acids and alkalis, positioning it as a promising candidate for segregating emulsified water-in-diesel mixture. In a study by X. Huang et al., they propose a simple method involving the decoration of carbon nanotubes on nanofibrous membrane to create a durable Janus membrane (Huang et al., 2022). The membrane featuring a superhydrophilic and hydrophobic nanofiber composite layer, allows for high efficiency and flux for separating of various intricate oily wastewater. The asymmetry wettability of the resultant membrane is adjusted by modifying the hydrophobic surface. It exhibits long-term stability and mechanical robustness, making it promising for purifying diverse oil-contaminated wastewater. In a separate study, J.-H. Zuo et al. designed a durable Janus membrane with polyvinylidene fluoride (PVDF) using thermal-induced phase separation (TIPS) combined with a spray-coating technique (Zuo et al., 2020). The TIPS-fabricated PVDF membranes had one surface coated with polydopamine/polyethyleneimine and the other with micro/nanoscaled silica particles interlinked with diphenyldichlorosilane. The spray-coating imparted superhydrophilic and superhydrophobic properties to the opposite surface, creating a modified Janus membrane with asymmetric surface wettability. This membrane exhibited excellent separation performance, including high permeation rates for toluene-in-water emulsion and oil rejection, showcasing its potential in practical applications. Nonetheless, the primary limitation of Janus membranes lies in their inability to achieve continuous oil/water separation without the constant need for mechanical rotation or artificial inversion based on different separation targets. In addition, Janus membranes exhibit other drawbacks, including a complex synthesis procedure and poor separation performance, which hampers their widespread application. Consequently, a more efficient approach to continually address oil spills and purify oily wastewater could involve in achieving an uninterrupted separation of oil/water mixtures or emulsions using a single superwetting surface without the requirement for mechanical or artificial rotation.

13.5 Multifunctionality in nanofibrous membrane design

Multifunctionality in nanofibrous membranes design refers to the integration of diverse features and capabilities within a single membrane structure (Cao et al., 2023). Unlike conventional membranes designed for a singular purpose, multifunctional nanofibrous membranes are engineered to execute numerous tasks simultaneously (Cao et al., 2022). This innovation is achieved through the strategic incorporation of various materials, functionalities, or treatments into the nanofiber matrix, with the goal to enhance overall performance and versatility for a diverse spectrum of applications. The multifunctional aspects of these membranes open avenues for addressing complex challenges across different fields. By integrating functionalities such as superhydrophilicity/superhydrophobicity (Ma et al., 2018, 2021), stimulus-responsiveness (Su et al., 2022), photocatalytic activity, biofouling resistance, chemical and mechanical durability these membranes offer unique solutions for treating complex oily wastewater (Ahmed and Dhar Purkayastha, 2023a,b; Ahmed et al., 2022). The integration of such diverse features into a single membrane structure demonstrates the advancement of nanofiber technology. Superhydrophilicity ensures enhanced water affinity, facilitating efficient water transport and minimizing fouling. Stimuli-responsiveness adds another level of sophistication, allowing the membrane to triggered by external stimuli, such as light, pH, temperature, and gas for controlled release or other desired reactions and also allow them to adapt to changing conditions, ensuring optimal performance under varying circumstances (Cao et al., 2023). The incorporation of photocatalytic activity is particularly significant, as it enables the membranes to degrade organic pollutants under light irradiation, further enhancing their efficiency and helps to irradiate fouling issues (Zhang et al., 2024; Sun et al., 2021). In addition, the membranes' inherent biofouling resistance combats the accumulation of unwanted biological substances, ensuring sustained performance over time. Moreover, the chemical and mechanical durability further enhances their longevity and effectiveness, making them reliable solutions for complex challenges in wastewater treatment (Han et al., 2020; Huang et al., 2022). This multifunctional approach is particularly relevant in the context of oily wastewater treatment, where conventional methods may fall short in addressing the complexity of contaminants. The ability of nanofibrous membranes to simultaneously tackle issues like oil—water separation, biofouling, and response to external stimuli makes them valuable assets in environmental remediation efforts. The design considerations for multifunctional nanofibrous membranes involve a careful selection of materials, incorporation of functional additives, and optimization of fabrication processes. Understanding the synergies between different functionalities is crucial to achieving a balance that maximizes the membrane's overall performance. As the field continues to advance, the exploration of new materials and innovative combinations becomes paramount. Researchers in nanofiber technology are at the forefront of this exploration, pushing the boundaries of what multifunctional membranes can achieve. This strategic combination enhances overall membrane performance and versatility, making them suitable for a wide range of applications. The following key aspects highlight the significance of multifunctional nanofibrous membranes:

13.5.1 Stimulus-responsive nanofibrous membrane

Stimulus-responsive membranes have significantly advanced the domain of oily wastewater treatment, offering more sophisticated methods for treating complex pollutants. These membranes are designed to triggered by external stimuli, such as light, pH, temperature or specific chemicals, triggering specific changes in their structure or properties. In the context of oily sewage treatment, stimulus-responsive membranes can exhibit several key benefits such as enhanced selectivity and improved efficiency.

13.5.1.1 Photo-responsive

Nanofibrous photo-responsive membranes for oil–water separation signify a notable advancement in membrane technology. These membranes, engineered to regulate wettability, could be reversibly altered by switching between UV and visible light exposure. This capability provides effective and on-demand oil/water separation, allowing for precise tuning of surface properties based on specific separation requirements. The membranes' reversible nature enhances their flexibility and adaptability, making them applicable for a wide range of oil–water separation applications. Achieving this photo-responsive behavior often involves incorporating photoactive materials, like photocatalysts or photoresponsive polymers, into the membrane structure (Imsong and Dhar Purkayastha, 2024b). For instance, Z. Chen et al. developed a membrane with regulated surface wettability by grafting azobenzene groups onto a SiO_2 roughened polypropylene (PP) membrane (Chen et al., 2021). This membrane can rapidly switch between hydrophobic and hydrophilic states upon exposure to UV and visible light, respectively. These photosensitive membranes exhibit excellent physical and chemical stability, enduring harsh environmental conditions, mechanical deformation, and multiple rinsing cycles. The remarkable photosensitivity allows for the creation of various removable and reproducible patterns with distinct wetting properties upon light exposure. These promising results suggest that photosensitive grafted PP membranes with switchable wettability hold great potential as materials for oil/water separation membranes. In another work, T. Huo et al. created a UV-driven selective super-wettable electrospun poly(vinylidene fluoride-co-hexafluoropropylene) membrane blended with fluorinated TiO_2 nanoparticles (Huo et al., 2022). This membrane's wetting properties can tuned from superhydrophobic/superoleophilic to superhydrophilic/underwater superoleophobic through light irradiation and heating. During these processes, the dynamic WCA instantly changes from 170 to 0° and retrieves to 160 degrees. This regulated wettability was demonstrated through a 15-cycle retrieval test, with the electrospun membrane exhibiting superhydrophobic property. This has reliable, effective, and flexible solution to treat complex oil–water mixtures in wastewater environment. S.-y. Chen et al. developed a composite membrane by combining polysulfone (PSF) with titanium dioxide (TiO_2) nanoparticles using a hydrothermal process (Chen et al., 2024a,b). This membrane FPT-A-UV, demonstrated unique superwetting properties when activated by UV light. In air, the membrane displayed strong attraction to both water and oil, with contact angles close to 0°. When submerged in water, it repelled oil, and when in oil, it repelled water, with contact angles exceeding 150 degrees. The membrane showed excellent ability to remove methylene blue (MB), achieving a removal rate of 90.8% under complex conditions involving MB-containing emulsions, with a water permeance of 780.3 $L/m^2/hour$. Furthermore, the FPT-A-UV membrane efficiently separated various emulsified solutions. These findings suggest promising applications for this composite membrane in wastewater treatment, particularly in challenging environments. Z. Li et al. created a membrane by incorporating zinc oxide (ZnO) nanorods onto polysulfone (PSF) nano-fibers through electrospinning and hydrothermal methods (Li et al., 2023). This process produced a hierarchical micro-/nano-structure on the PSF nanofibers, increasing their surface area significantly. The ZnO nanorods, being photosensitive, rendered the membrane hydrophilic, enabling effective oil—water separation under ultraviolet (UV) light. In addition, the membrane exhibited self-cleaning properties, facilitating the breakdown of contaminants under UV light exposure. This innovative membrane shows promise for applications requiring intelligent separation processes.

13.5.1.2 Thermo-responsive

Thermo-responsive nanofibrous membranes for oil—water separation are another innovative class of materials that are attracting significant interest in membrane technology. These membranes can alter their wettability or permeability in response to temperature changes, offering a unique approach for separation processes. The ability to alter their wettability between hydrophilic and hydrophobic states in response to temperature changes makes thermo-responsive membranes valuable for applications requiring precise control over separation processes. For example, these membranes can selectively separate oil and water based on temperature variations, which is particularly useful in scenarios where temperature manipulation is feasible, such as in industrial processes or environmental remediation. One approach to creating thermo-responsive membranes is by incorporating temperature-sensitive polymers or nanoparticles into the membrane matrix. When the temperature changes, these components undergo conformational changes, altering the membrane's properties. For instance, poly(N-isopropylacrylamide) (PNIPAM) is a commonly used temperature-responsive polymer having a low critical solution temperature (LCST) of roughly 32°C. Below this temperature, PNIPAM is hydrophilic, but above this temperature, it becomes hydrophobic, allowing for the preferential separation of oil and water. R. Ou et al. designed a durable, thermoresponsive polymer membrane by combining elastic polyurethane (TPU) microfibers with PNIPAM (Liu et al., 2018). They coated the TPU microfibers' surface with PNIPAM hydrogel, leveraging the membrane's hierarchical structure with enhancement of surface roughness to improve wettability. This TPU-PNIPAM membrane can alter between superhydrophilic and superhydrophobic properties depending on the temperature (25—45°C). It efficiently separates water-in-oil and oil-in-water emulsions at 45 and 25°C respectively, achieving a high separation performance. Moreover, the resultant membrane exhibits outstanding mechanical and flexible properties. These membranes hold significant promise for practical and efficient oil—water separation applications, suggesting room for further development. In another work, Y. Wang et al. produced a nanofibrous membrane from regenerated cellulose (RC) using electrospinning (Ou et al., 2016). They then grafted PNIPAAM onto the RC nanofibers through surface-initiated atom transfer radical polymerization. The resultant membrane demonstrated temperature-responsive surface wettability, transitioning between hydrophilic and hydrophobic states. It also showed tuneable super-lyophilic/super-lyophobic behavior at the water-oil-solid interface, making it favorable for regulated oil/water separation in wastewater treatment and oil purification. Using temperature changes as a trigger is considered a potent method for controlling oil/water separation without needing additional chemical inputs. However, maintaining the membranes' hydrophobicity during separation is difficult. Also, it requires significant energy consumption to maintain the liquid temperature above the LCST. In addition, research has predominantly focused on PNIPAM for desirable oil/water separation, but there is potential for more studies on polymers having temperature-responsive behavior in future.

13.5.1.3 pH-responsive

pH-responsive membranes for oil—water separation can particularly segregate oil from water based on the pH of the solution. These membranes are designed to change their behaviors, such as wettability or pore size, in response to pH variations, allowing for controlled oil—water separation. One approach is to incorporate pH-responsive polymers into the membrane matrix or as a coating on the membrane surface. These polymers can undergo conformational changes or solubility transitions in response to pH changes, leading to alterations in membrane properties. For example, a membrane may be designed with pH-responsive polymer brushes that change from a collapsed to a swollen state in response to pH changes. At low pH (acidic conditions), the polymer brushes collapse, reducing membrane pore size and increasing hydrophobicity, which can enhance oil—water separation efficiency. Conversely, at high pH (basic conditions), the polymer brushes swell, increasing pore size and hydrophilicity, which can facilitate water permeation while repelling oil. For instance, B. Cheng et al. designed a pH-responsive electrospun PVDF-g-PAA, consisting of poly(vinylidene fluoride) membrane grafted with poly(acrylic acid) (Cheng et al., 2017). This membrane can switch its wettability and structure in response to the pH of the surrounding aqueous solution. By undergoing protonation and deprotonation, the membrane can change its surface properties, enabling switchable oil/water wettability. Remarkably, this membrane achieved desirable oil/water separation using only gravity, with the pH of the medium serving as the trigger. The membrane's tree-like structure contributed to its excellent separation efficiency and flux in both scenarios. This smart membrane can be utilized in diverse applications, including water purification, oil recovery, and other separation processes, given its exceptional controllable permeation properties. In another work, Z. Zhao et al. used solution blow spinning (SBS) to create polyimide membrane (PI-NFMs) with fluffy structures (Zhao et al., 2023). The membrane quickly switches between ultra hydrophobic and superhydrophilic states by simple treatment of acid and alkali, showing excellent pH responsiveness. They can also switch between modes for removing oil or water depending on the oil—water mixture. The membranes demonstrate decent separation efficiency, considerable permeate flux for water and oil, and good reusability. The study offers a viable method for removing industrial oil effluents and oil spills efficiently and effectively. R. Su et al. fabricated a nanofibrous membrane having pH-responsive behavior by modifying a polyvinylidene fluoride (PVDF) electrospun nanofibrous membrane (Su et al., 2022). They used self-polymerization of polydopamine (PDA), silver nanoparticle (AgNP) immobilization, and thiol self-assembly. This membrane can switch between superhydrophobic and superhydrophilic states at pH 7 and 13, respectively. It efficiently separated heavy and light oil/water mixtures with efficiency over 96.5% and 99.2% respectively, while maintaining a stable flux. It also separated diverse surfactant-stabilized oil-in-water mixtures with over 99% efficiency and had good structural stability against mechanical abrasion. This membrane has potential for energy-efficient oil/water separation in wastewater treatment. Nevertheless, existing pH-controlled nanofibrous membranes can only alter their surface wettability under specific pH conditions, making it challenging to separate intricate multiphase mixtures of oil and water. Hence, there is dire need to design nanofibrous membranes for separating oil and water with a broader pH stimulus range. Furthermore, chemical stability is crucial for their use in real-world scenarios, as these membranes inevitably encounter acidic or alkaline solutions while altering their surface properties.

13.5.1.4 Gas-responsive

Nanofibrous membranes with gas responsiveness offer a revolutionary approach in oil—water separation. Unlike other stimuli, such as pH or temperature, gas can be easily operated and is not restricted by sample volume, making it highly suitable for large-scale applications. These membranes can be designed to respond to specific gases by incorporating gas-responsive materials or functional groups. When exposed to the target gas, these materials undergo a change in properties, such as pore size or surface energy, which selectively affects the passage of oil or water molecules through the membrane. Y. Wang et al. created CO_2-responsive membrane for efficient oil/water separation. This method uses capillary force-driven confinement self-assembly to attach a CO_2-responsive copolymer uniformly to the membrane surface, creating a large membrane area up to 3600 cm^2. The membrane can switch between high hydrophobicity/underwater superoleophilicity and superhydrophilicity/underwater superoleophobicity when exposed to CO_2/N_2. It could separate diverse oil/water combinations with over 99.9% efficiency, including immiscible mixtures, emulsions, and pollutant-containing mixtures. The membrane is also recyclable and self-cleaning. This innovative membrane, with its strong separation properties and scalability, is promising for smart liquid separation applications. In another work, H. Che et al. created nanostructured electrospun polymer membranes that can change their oil/water wettability using CO_2 (Che et al., 2015). These membranes can act as an oil/water switch due to their integration of CO_2 reactance and porous nanostructure. This system offers a new method for controlling surface wettability with gas and provides advantages for precise oil/water separation and CO_2 applications. In conclusion, gas-responsive nanofibrous membranes offer a versatile and efficient approach to selective oil—water separation. Their responsiveness to gases like CO_2 allows for precise control over surface wettability, enabling the membranes to switch between hydrophobic and hydrophilic states. This capability, coupled with their scalability and ease of operation, makes them highly promising for large-scale applications. In addition, the combination of gas responsiveness and nanostructured design opens up new possibilities for innovative separation processes and applications in CO_2 capture and other areas.

13.5.1.5 Multiple-responsive

We have previously discussed materials with smart surfaces that trigger to external factors like light, temperature, pH, and gas. However, there are also materials with smart surfaces that respond to combinations of these external environments. For example, a membrane could be designed to respond to both pH and temperature, changing its wettability in response to specific pH levels and temperatures. This multiple-responsive behavior can enhance the membrane's versatility and efficiency in separating complex oil−water mixtures. This section covers recent advancements in dual or multiple-responsive behavior in materials and their potential applications in separating oil/water systems. S. Zhang et al. created a composite nanofiber membrane that responds to both light and temperature, with switchable superwettability (Yin et al., 2022). They achieved this by first attaching carbon nanotubes (CNTs) to nanofibrous polyurethane membrane and then modifying them with PNIPAM. The membrane exhibited excellent photothermal conversion, allowing for rapid and reversible switching between superhydrophilicity and good hydrophobicity. It effectively separated oil/water mixtures with varying oil densities and oil-in-water emulsions at ambient temperature, as well as water-in-oil emulsions under infrared light, achieving superior liquid permeation flux and separation efficiency. This multifunctional resultant membrane has promising application and controllable separation of oil-contaminated wastewater. S. Li et al. developed a responsive composite membrane for oil−water separation by incorporating MXene/PPy onto polypropylene (PP) fibers using in-situ polymerization (Li et al., 2024). They improved adhesion by modifying with dopamine (PDA) and enhanced durability with a pH-responsive polymer (PDM). The membrane showed excellent pH responsiveness and photothermal conversion. It had a water contact angle (WCA) of 52 degrees at pH = 2 and 125 degrees at pH = 12, and could reach 104.6°C in 10 minutes under 2 kW/m^2 irradiation. It separated oil/water mixtures for 12 hours with a maximum flux of 3.87 kg/m^2/hour, and showed corrosion resistance, stability, and recyclability. The study offers a novel approach for fabricating intelligent and responsive fiber composite membranes for oil−water separation. In another work, Y. Cao et al. created a dual-responsive oil/water separation material by coating a mesh with PDMAEMA hydrogel (Cao et al., 2014). The material exhibits superhydrophilicity and underwater superoleophobicity under specific pH and temperature conditions. It separates water from oil/water mixtures by selectively allowing water to penetrate the mesh below 55°C (pH 7) or at pH less than 13 (T = 25°C), while retaining oil. Above 55°C or at pH greater than 13, the material allows oil to pass through the mesh and can be collected. This material shows promise for practical applications and introduces new possibilities for intelligent water/oil separation processes. In conclusion, the development of materials with multiple-responsive surfaces has advanced the domain of oil−water separation by allowing for intelligent and efficient separation processes. These materials can respond to combinations of external stimuli, including light, pH, temperature, and more, enhancing their versatility and effectiveness in separating complex oil−water mixtures. These membranes have considerable promise for practical applications in the separation of oil-contaminated wastewater, with benefits such as high separation efficiency, recyclability, and stability. Overall, these advancements open up new possibilities for intelligent and controllable oil–water separation processes.

13.5.2 Photocatalytic and fouling resistance

Photocatalytic nanofibrous membranes are designed to utilize photocatalytic materials, such as TiO_2, ZnO, ZnS, SnO_2, Fe_2O_3, CdS, ZrO_2, etc., to degrade organic contaminants in water, including oil, through photocatalysis. These membranes combine the advantages of nanofibrous structures, which offer high surface area and porosity, with the photocatalytic properties of the embedded materials. In the context of oil–water separation, these membranes can be effective in breaking down organic pollutants in oil, improving the efficiency of the separation process. When exposed to ultraviolet (UV) light or visible light, the photocatalytic material generates reactive oxygen species (ROS), which can degrade organic compounds in the oil phase, making it easier to separate from water. The metal oxide-modified photocatalytic nanofibrous membranes can also exhibit superhydrophilic or superhydrophobic properties, depending on the surface modification, which can further enhance their separation efficiency. The superhydrophilic surface allows the water to flow through while repelling oil, and vice versa for the superhydrophobic surface. F. U. Ahmed et al. utilized a straightforward and low-temperature hydrothermal method to intricately decorate the electrospun PAN nanofibrous membrane's surface with a micro–nano structure composed of ZnO nano-needles (Ahmed and Dhar Purkayastha, 2023b). The developed membrane combines membrane filtration and photodegradation processes, showcasing enhanced permeability, pollutant removal, antifouling properties, and self-cleaning capabilities. Furthermore, integrating materials with biofouling resistance properties can help combat the accumulation of unwanted biological substances on the membrane surface, ensuring sustained performance over time. This property is particularly important for maintaining the efficiency of the membrane in real-world applications where biofouling is a common challenge. X. You et al. created a metal-free Cl/S double-doped graphitic carbon nitride photocatalyst embedded in nanofibrous polyacrylonitrile membrane via electrospinning (You et al., 2023). The resulting porous membrane showed superior superhydrophilicity and superoleophobicity underwater, enabling high (>98.0%) oil-in-water emulsion separation efficiency. Its porous structure increased contact between bacteria/contaminant and reactive species, enhancing antibacterial and photodegradation properties, and providing high barrier to contaminants. This multifunctional, porous, antifouling nanofibrous membrane holds promise for practical oily wastewater treatment. N.H. Alias et al. developed a photocatalytic nanofiber-coated fibrous membrane for treating water produced from oilfield (Alias et al., 2019). The membrane, consisting of polyacrylonitrile nanofibers embedded with graphitic carbon nitride, was coated onto an alumina hollow fiber. The porous membrane allowed water permeation while capturing oil droplets, achieving high oil rejection efficiency. Its sparse mesh morphology prevented fouling and maintained high permeate flux. The membrane showed a permeate flux of 640 $L/m^2/hour$ and oil rejection of 99% after 180 minutes of separation. It also demonstrated excellent cleaning and fouling resistance performance, making it suitable for long-term filtration operations in oilfield water treatment. In another work, H. Shan et al. introduced a method for creating flexible nanofibrous membranes incorporating CuO nanoparticles anchored to ZrO_2 (Shan et al., 2023). This involved employing successive ionic layer adsorption and reaction assembly, followed by thermal oxidation processes. The resulting membranes exhibited stable hierarchical structures with mechanical robust performance, enabling exceptional reusability and straightforward recycling. The effective fabrication of these ceramic-centered nanofibrous membranes, characterized by efficient catalytic performance and superior flexibility, holds promise for inspiring the development of other synergistic membrane catalysts for wastewater treatment. This approach is particularly effective for separating water from oil or oil from water with significant oil content while nanoparticles contribute unique catalytic or antibacterial properties. Q.-h. Hu et al. innovated a new hybrid nanofibrous membrane by incorporating a molecularly imprinted metal—organic framework (MOF) into polyacrylonitrile (PAN) (Hu et al., 2024). This hybrid membrane, displaying effective selectivity, prolonged cyclic adsorption capability, and antibacterial fouling resistance, holds promise as a versatile filtration membrane for practical applications in wastewater treatment. Z. Yin et al. developed a straightforward technique to create a versatile cellulose-centered membrane with superhydrophobic properties with a water contact angle of 163 degrees (Yin et al., 2022). This membrane exhibits self-cleaning abilities, oil—water separation, photocatalytic degradation, and antibiofouling capabilities. The high separation efficiency, considerable flux, and recyclability in oil/water separation are attributed to its robust superhydrophobicity, achieved through a combination of stearic acid modification over CuO nanostructure coating. The membrane exhibits superior antibiofouling properties, resisting algae adhesion for over 20 days due to its self-cleaning ability and the antimicrobial properties of CuO nanoparticles. Overall, the integration of materials with tailored properties into nanofibrous membranes offers a promising approach to achieving high-performance membranes for oil/water separation with enhanced durability and efficiency. Overall, photocatalytic nanofibrous membranes offer a promising technique for oil—water separation, combining efficient separation properties with the ability to degrade organic pollutants and antifouling capability, making them suitable for various environmental remediation applications.

13.5.3 Chemical and mechanical durability

Chemical and mechanical durability are crucial factors in the performance and longevity of nanofibrous membranes used in oil—water separation applications. Chemical durability refers to the membrane's ability to withstand exposure to various chemicals present in the oil—water mixture, cleaning agents, or environmental factors without significant degradation. Nanofibrous membranes can be chemically durable by choosing materials that are resistant to degradation, such as polymeric materials like polytetrafluorethylene (PTFE), polyethylene terephthalate (PET), and polyvinylidene fluoride (PVDF). These materials are known for superior chemical stability and resistance to solvents, acids, and bases. For example, F.U. Ahmed et al. introduce a method to create an electrospun PVDF membrane adorned with ZnO nanorods for separating oil—water emulsions and mixtures. The surface of the PVDF@ZnO nanorods is superhydrophilic and features hierarchical micro—nano structures that trap water molecules, displaying underwater superoleophobicity. The resultant membrane shows high pure water flux and separation efficiency of approximately 99%. Moreover, it maintains its superhydrophilic-underwater superoleophobic properties in adverse environmental conditions. In another work, S. K. Hong et al. developed a nanofibrous cellulosic membrane (NFC membrane) for efficiently separating large quantities of oil/water mixtures (Hong et al., 2018). The membrane was created using electrospinning, a simple method for layering nanofibrous membranes with uniform porosity on a substrate. Because of its cellulose composition, the membrane exhibited strong resistance to oils and high stability against chemicals, achieving oil/water separation efficiencies of over 99%.

Mechanical durability refers to the membrane's ability to maintain its structural integrity and mechanical strength under operational conditions, including pressure, flow rates, and physical stresses during cleaning or handling. Nanofibrous membranes are inherently fragile due to their small fiber diameters and porous structures. To improve mechanical durability, nanofibrous membranes can be reinforced with other materials or fabricated using techniques that enhance mechanical strength, such as electrospinning with high polymer concentrations or crosslinking the polymer chains. S. Sahu et al. developed a PVDF/Ti_3C_2/ZnO membrane by incorporating 1D ZnO nanoneedles onto a PVDF electrospun membrane using electrospinning, vacuum deposition, and hydrothermal methods. This membrane showed superhydrophilicity and stable underwater superoleophobicity, achieving significant oil/water separation performance and can be reused (Sahu and Dhar Purkayastha, 2023). It also showed significant thermal, chemical, and mechanical stability under adverse conditions and maintained its morphology after multiple operation cycles. This study presents a scalable approach for fabricating multifunctional membranes for water purification applications. F.U. Ahmed et al. achieved the surface modification of an electrospun PVDF nanofibrous membrane by depositing a TiO_2 nano cauliflower-like structure with photocatalytic activity using an easy and low-cost hydrothermal method. Leveraging the membrane's micro/nanostructure, the TiO_2@PVDF membrane exhibited excellent surface wetting property, capability of photocatalytic degradation, and resilience to chemical and mechanical stress tests (Ahmed and Dhar Purkayastha, 2023a). In summary, chemical and mechanical durability are critical considerations in the design and synthesis of nanofibrous membranes for oil—water separation to ensure stability in long-term and performance under harsh operating conditions.

13.6 Recent advances in nanofibrous membranes for oily wastewater treatment

Recent advances in nanofibrous membranes for oily wastewater treatment have significantly enhanced the efficiency and sustainability of oil/water separation processes.

These advancements have been driven by the need for effective solutions to address environmental challenges posed by industrial activities, particularly in the petrochemical and manufacturing sectors. By incorporating innovative materials and fabrication techniques, researchers have developed nanofibrous membranes with improved performance characteristics, making them highly suitable for treating complex oily wastewater streams.

13.6.1 Nanostructured membranes for enhanced oil/water separation

Nanostructured membranes are at the forefront of enhancing oil/water separation efficiency, offering improved selectivity and flux rates. These membranes are designed with nanoscale features that exploit the unique properties of materials at this scale. Various nanostructured membranes, such as nanoporous membranes, nanocomposite membranes, and those with surface nanostructures, offer distinct benefits and serve different purposes. A nanoporous membranes consist of a dense layer with nanoscale pores that allow water molecules to penetrate through while blocking larger oil molecules. The nanopores are designed to be smaller than the size of oil droplets, ensuring efficient separation. Nanoporous membranes are particularly effective for ultrafiltration and nanofiltration applications. For instance, H. M. Hegab et al. developed a nanoporous graphene membrane with tripartite nanochannels, resembling aquaporin, using a 3D hierarchical multifunctionalization approach (Hegab et al., 2024). Graphene oxide (GO) nanosheets, known for their atomic thinness and amphiphilic properties, were used to create this membrane, which exhibited exceptional permeability. The prepared membrane demonstrated ultra-high water permeability and outstanding selectivity. As a result, the fabricated membrane achieved higher flux compared with the GO membrane, while maintaining a remarkable oil rejection rate of approximately 97%. In another work, Y. Cai et al. created a highly flexible and durable membrane using polydivinylbenzene (PDVB) coated on porous wood, resulting in a super-hydrophobic surface with hydrophobic nanopores (Cai et al., 2022). This PDVB-wood membrane exhibits exceptional durability, flexibility, and chemical resistance. It achieves over 99.98% efficiency in separating water-in-oil emulsions. Moreover, the membrane shows superior recyclability, maintaining a separation efficiency of up to 99.98% even after 20 cycles. The combination of nanopores and super-hydrophobicity contributes to its high performance. This functional membrane, derived from biomass, offers durability, easy scalability, low cost, and environmentally friendly regeneration, making it a promising alternative to conventional filter membranes for environmental applications. The advancements in nanostructured membranes, exemplified by the innovative works of H. M. Hegab et al. and Y. Cai et al., underscore the potential of nanotechnology in revolutionizing oil/water separation processes(Hegab et al., 2024; Cai et al., 2022). Moving beyond nanoporous membranes, the exploration of nanocomposite membranes represents another exciting avenue for enhancing membrane performance. Nanocomposite membranes incorporate nanoparticles or nanofillers into the membrane matrix to enhance its performance. These nanoparticles can improve membrane selectivity, mechanical strength, and antifouling properties. For example, incorporating graphene oxide or carbon nanotubes can improve membrane properties due to their large surface area and mechanical strength. X. Wang et al. developed a hydrophobic-oleophilic nanofibrous membrane by blending cellulose nanocrystals (CNC) with PVDF using electrospinning (Wang et al., 2019a,b,c). The incorporation of CNC enhanced the emulsion separation flux of the membranes by optimizing their microstructure and porosity. For water-in-toluene emulsions, membranes containing 4 wt% CNC showed a significantly higher flux and a separation efficiency, which was 4 times greater than that of bare PVDF membranes. Importantly, the membranes maintained good mechanical properties as CNC loading increased and could be easily recycled, making them flexible, environmentally friendly, and sustainable. A. M. Pornea et al. developed a simple and effective method to create a Janus nanocomposite membrane with asymmetric oil and water separation capabilities (Pornea et al., 2020). They achieved this by incorporating organosilane functionalized carbon nanotubes (f-CNT) onto an inter-linked hydrophilic electrospun polyvinyl alcohol (c-PVA) membrane using vacuum filtration. The functionalization of fluorine-containing 1,1,2H,2H-perfluoroctyltriethoxysilane on CNT which enhance the hydrophobicity of the interface, enabling the CNT-decorated surface to efficiently separate water-in-oil emulsions. This technique not only contribute interfacial hydrophobicity but also improved the compatibility within CNT and the c-PVA membrane, resulting in a mechanically stable Janus membrane. The resultant membrane showed excellent selectivity for oil and water from various emulsions, making it promising for various applications. The innovative approaches demonstrated by X. Wang et al. and A. M. Pornea et al. highlight the significant progress in membrane technology for oil and water separation (Wang et al., 2019a,b,c; Pornea et al., 2020). While X. Wang et al. focused on enhancing membrane performance through the incorporation of cellulose nanocrystals, A. M. Pornea et al. explored the development of a Janus nanocomposite membrane for asymmetric oil and water separation. Moving forward, exploring membranes with surface nanostructures could further enhance the efficiency and selectivity of membrane processes. These nanostructured surfaces can be tailored to exhibit superhydrophobic or superhydrophilic properties, offering exciting opportunities for advanced membrane applications. R. Imsong et al. created a nanofibrous PVDF membrane having a 2D nanostructured composite of MXene nanosheets and TiO_2 nanofrost to treat complex oily wastewater (Imsong and Dhar Purkayastha, 2024a). This membrane demonstrated excellent separation performance for surfactant-stabilized emulsions and high water permeability. It also showed outstanding antifouling and self-cleaning properties, along with efficient removal of metal cations from oily wastewater, likely due to TiO_2 nanofrost intercalation. The combined effect of MXene nanosheets and TiO_2 nanofrost in the TiO2@MX-PVDF also exhibited superior photocatalytic activity to degrade organic pollutants under visible light. This work presents an advanced and effective method for developing a multifunctional 2D MXene composite membrane for the treatment of oily wastewater. In another work, F. U. Ahmed et al. improved an electrospun PVDF membrane by coating it with a photocatalytic TiO_2 nano cauliflower-like structure using a simple hydrothermal method (Ahmed and Dhar Purkayastha, 2023a). This modification enhanced the membrane's surface wettability, photocatalytic activity, and durability.

The TiO_2@PVDF nano cauliflower-like membrane effectively separates water-in-oil mixtures and surfactant-stabilized emulsions due to its wetting behavior. In addition, the photocatalytic properties of TiO_2 enable the membrane to degrade organic and pharmaceutical pollutants in treatment of wasterwater. This modified NF PVDF membrane demonstrates superior multifunctionality, suggesting potential for treating complex oily wastewater. S. Sahu et al. fabricated a PVDF/Ti_3C_2/$WO_3 \cdot H_2O$ membrane by growing $WO_3 \cdot H_2O$ micro-flowers hydrothermally onto porous PVDF/Ti_3C_2 electrospun mats(Sahu and Dhar Purkayastha, 2024). The $WO_3 \cdot H_2O$ micro-flowers grew across Ti_3C_2 MXene sheets, enhancing interfacial contact, controlling wettability, and boosting photo-reactivity. The fibrous membrane exhibited superhydrophilicity, sustained underwater superoleophobic characteristics which leads to minimal oil adhesion. It achieved outstanding separation effectiveness and reusability, successfully separating various oil/water emulsions with high permeation flux. The membrane was more superior than PVDF/Ti_3C_2 membrane in degrading RhB ($\sim$91%) and showed good recyclability under visible light radiation. Such dual-functional materials with exceptional photocatalytic degradation and oil/water emulsion separation properties hold promise for treating complex oily wastewater. Thus, the development of nanostructured membranes has significantly advanced the field of oil/water separation, offering higher efficiency and improved performance compared with traditional membranes. These membranes are being increasingly adopted in various industries, including wastewater treatment, oil and gas due to their superior properties and effectiveness in separating oil and water mixtures.

13.6.2 Functionalization of nanofibrous membranes for improved performance

Functionalization of nanofibrous membranes involves modifying their surface properties to enhance performance. This can include altering their hydrophilicity, hydrophobicity, and antifouling characteristics, all of which contribute to more efficient oil/water separation processes. One common approach in functionalization is the introduction of specific chemical groups onto the membrane surface. For example, hydrophilic groups like −OH or −COOH can be added to enhance water affinity, promoting the passage of water molecules while repelling oil. Conversely, hydrophobic groups can be introduced to improve oil affinity, aiding in the selective exclusion of oil from water. For example, Y. Wang et al. developed superamphiphilic SiO_2-TiO_2 porous nanofibrous membranes modified with amino-silanization (ASTPNM) to eliminate water-soluble heavy-metal ions and oil contaminants from wastewater (Wang et al., 2019a,b,c). These membranes show high efficiency in separating water and oils, including emulsions, due to their underwater superoleophobicity. The optimal modification time for maintaining under-oil superhydrophilicity is 15 minutes (ASTPNM-15), allowing easy cleaning of oil contaminants with water alone. ASTPNM-15 also demonstrates excellent adsorption capacity for Pb^{2+} and stable separation efficiency for various heavy-metal ions and oil spills over multiple cycles, showcasing its recyclability and reusability in practical applications. In another work, L. Liu et al. developed a biodegradable electrospun nanofibrous poly(lactic acid) membrane with a hierarchical surface modification for efficient oil/water separation (Liu and Yuan, 2018). The membrane was coated with polydopamine, silver nanoparticles, and fluorinated thiol to achieve superhydrophobicity and superoleophilicity. This resultant membrane effectively separated water-in-oil emulsions, exhibited low water adhesion, and demonstrated antibacterial and antifouling properties. Overall, this biodegradable membrane shows promise for treating oily wastewater. Thus, the functionalization of nanofibrous membranes as discussed above provides a foundation for enhancing their performance in oil/water separation. Moving forward, surface modification can also involve the incorporation of nanoparticles or nanomaterials with specific properties. For instance, adding nanoparticles with photocatalytic properties can enable the degradation of organic pollutants in the water phase, enhancing the overall efficiency of the membrane. In addition, incorporating nanoparticles with antimicrobial properties can prevent biofouling, maintaining the membrane's performance over time. An example of this approach is demonstrated by L. Zhang et al., utilized membrane-pores as immobilization foundation to anchor highly dispersed β-FeOOH through an in-situ mineralization technique (Zhang et al., 2021a,b). This not only subjected to highly active sites for the degradation of refractory pollutants but also conferred hydrophilic/underwater superoleophobic properties to the membrane. The resulting membrane exhibited effective separation capacity and demonstrated impressive catalytic oxidation performance through the activation of peroxymonosulfate. This dual functionality is particularly advantageous as it not only enhances the efficiency of the treatment process by effectively removing oil-based pollutants but also contributes significantly to the environmental sustainability of the overall process. By addressing both the physical separation and chemical breakdown of contaminants, nanotechnology-enabled nanofibrous membranes present a holistic and synergistic approach to tackling the challenges associated with oily wastewater. In a parallel effort, S. Sahu et al. achieved uniform decoration of 1D ZnO nanoneedles decorated over Ti_3C_2 incorporated in a PVDF nanofibrous membrane (Sahu and Dhar Purkayastha, 2023). This resulted in a membrane exhibiting superhydrophilicity and stable underwater superoleophobicity, showcasing remarkable separation performance and reusability. Notably, the composite membrane demonstrated exceptional photocatalytic degradation performance and maintained robust thermal, chemical, and mechanical stability in adverse environmental conditions. Even after numerous cycles of multifunctional operation, the membrane retained its morphological durability. Collectively, these innovations represent a promising frontier in the realm of oily wastewater treatment, underscoring ongoing efforts to enhance efficiency and address environmental challenges. Overall, functionalization of nanofibrous membranes offers a versatile approach to tailoring their properties for specific applications, leading to improved performance in oil/water separation processes.

13.6.3 Novel materials in nanofibrous membranes for oil–water separation

Researchers are increasingly investigating novel materials to enhance the performance of nanofibrous membranes in oil/water separation applications. Graphene oxide (GO) is known for its high mechanical strength, chemical stability, and large surface area. These properties enable GO to improve membrane selectivity by effectively blocking oil molecules while allowing water to pass through. In addition, GO can enhance membrane durability, making it more resistant to degradation over time. For instance, Y. Zhan et al. developed stable composite membranes with halloysite nanotubes and graphene oxide on a porous poly(arylene ether nitrile) nanofibrous mat (Zhan et al., 2018). The membranes, prepared with a mussel-inspired polydopamine coating, showed super-hydrophilic and underwater super-oleophobic behavior. The resultant membrane shows high rejection ratios and excellent antifouling against various oil/water emulsions, including high-temperature and corrosive ones. These membranes offer promising applications in water cleaning and resource recovery. In addition to graphene oxide, carbon nanotubes (CNTs) have also been widely studied for their ability to enhance membrane durability and performance. CNTs possess high aspect ratios and unique structural properties, such as high tensile strength and thermal conductivity. These properties can enhance water permeability through the membrane while also providing resistance to fouling. CNTs can thus improve the overall efficiency and lifespan of nanofibrous membranes. For example, X. Huang et al. developed an easy method to create a durable Janus membrane (JM) using carbon nanotubes to decorate the nanofibrous membrane (Huang et al., 2022). This JM consists of a composite layer of superhydrophilic and hydrophobic, both with a core-shell structure for flexibility and stretchability. The strong interfacial compatibility within the layers confirms the durability and stability even after several stretches. The JM effectively separates various types of oily wastewater, including different oil densities, water-in-oil and oil-in-water emulsions with high separation flux and efficiency. This mechanically robust and stretchable JM shows promise for purifying oil-contaminated wastewater. Similarly, MXenes have also emerged as a promising material for enhancing membrane properties, including durability and selectivity. MXenes, a class of two-dimensional materials, are being studied for their hydrophilic nature and potential for improving membrane performance. For instance, R. Imsong et al. developed a simple method to create a superhydrophilic/underwater superoleophobic membrane by incorporating two-dimensional structures of $Ti_3C_2T_X$ MXene onto a nanofibrous PAN membrane by vacuum-assisted filtration (Imsong and Dhar Purkayastha, 2023). This membrane exhibited excellent separation capabilities for various oil—water emulsions due to its inherent hydrophilicity, low oil droplet adhesion, and ordered 2D $Ti_3C_2T_X$ MXene stack nanosheets. The membrane also demonstrated exceptional chemical stability when exposed to harsh environmental conditions and can effectively separated emulsions in such environments. In addition, the MXene-PAN membrane could efficiently adsorb organic contaminants from oily effluents. The membrane's dual functionality, combining excellent organic dye adsorption and separation performance, presents novel possibilities for treatment of oily wastewater. In addition to MXenes, researchers are also exploring metal-organic frameworks (MOFs) for enhancing nanofibrous membrane performance in oil/water separation. These porous materials offer high surface areas and tunable structures, making them promising for selective adsorption and separation processes. For instance, F. Zhou et al. produced superhydrophobic nanofibrous ZIF-71/PVDF-HFP membranes using electrospinning and an etching-mediated in-situ growth technique for separating water-in-oil emulsions(Zhou et al., 2022). They etched ZnO to develop adsorption sites, enabling strong anchoring of ZIF-71 nanoparticles on the nanofibers. These membranes displayed superhydrophobicity under oil, with a contact angle of 162.1 degree. They efficiently separated various water-in-oil emulsions under gravity, achieving over 99% separation efficiency for all emulsions. The water-in-chloroform flux in the absence of surfactant attained up to 6577.68 L/m^2/hour. Moreover, the membranes showed superior cyclic stability and resistance to fouling. Hence, the etching-assisted MOF/nanofibrous membranes hold great promise for effective oil/water separation. In conclusion, the integration of novel materials such as graphene oxide, carbon nanotubes, MXenes, and MOFs has significantly enhanced the performance of nanofibrous membranes in oil/water separation applications. These materials offer unique properties such as high mechanical strength, chemical stability, and tunable structures, improving membrane selectivity, durability, and antifouling properties. Researchers have demonstrated the effectiveness of these membranes in separating a wide range of oil/water emulsions, including high-temperature and corrosive ones, with high separation efficiency and flux. Moreover, the membranes show excellent cyclic stability and resistance to fouling, making them promising for practical applications in water cleaning and resource recovery. Overall, the development of these advanced membranes represents a significant advancement in the domain of oily wastewater treatment, offering efficient and sustainable solutions to environmental challenges.

13.6.4 Economic and environmental sustainability of nanofibrous membrane technology

Researchers are increasingly focused on enhancing the economic and environmental sustainability of nanofibrous membrane technology. One key area of development involves the creation of cost-effective fabrication techniques. These methods, such as electrospinning, can be scaled up for mass production while minimizing energy consumption and material waste. Another approach is the use of renewable and eco-friendly materials in membrane fabrication. Biopolymers like cellulose and chitosan, sourced from natural origins, are being explored as alternatives to synthetic polymers. These materials are biodegradable and sustainable, offering environmental benefits. Optimizing membrane performance is also crucial for sustainability. Improved selectivity and durability can reduce energy consumption and waste generation. Membranes that operate more efficiently require less energy input and have a lower environmental impact. Recycling and reusing membranes play a significant role in sustainability efforts. Researchers are investigating methods to regenerate and refurbish membranes, extending their lifespan and reducing the need for new membrane production. Conducting life cycle assessments (LCAs) of nanofibrous membrane technology is essential for understanding its overall environmental impact. LCAs consider the environmental effects of membrane production, operation, and disposal, providing insights for improving sustainability. In conclusion, efforts to enhance the economic and environmental sustainability of nanofibrous membrane technology include cost-effective fabrication techniques, the use of renewable materials, performance optimization, and recycling initiatives. These endeavors aim to reduce energy consumption, waste generation, and environmental impact, making nanofibrous membranes a more sustainable solution for various applications.

13.7 Application of nanofibrous membranes in oily wastewater treatment: Real-world impact

Nanofibrous membranes have revolutionized oily wastewater treatment across industries, offering a paradigm shift in environmental management. In sectors such as petrochemicals and textiles, these membranes have emerged as indispensable tools, efficiently and cost-effectively mitigating environmental challenges. The petrochemical industry faces the complex task of treating oily wastewater containing hydrocarbons, oils, and various pollutants. Nanofibrous membranes, with their unique structure created through electrospinning techniques, excel in capturing and separating oil droplets and contaminants. Their high surface area and porosity allow for selective filtration based on size and hydrophobicity, meeting stringent environmental regulations. These membranes also reduce the burden on downstream treatment processes, leading to overall cost savings and operational efficiency. Their longevity and durability in challenging industrial environments minimize the need for frequent replacements and maintenance, further enhancing their environmental compliance and operational efficiency. A key advancement in nanofibrous membrane technology is the ability to modify the membrane's surface morphology to change its wettability. By introducing special structures like hierarchical roughness or micro/nanostructures, the membrane's interaction with water and oil can be altered. This approach has been successfully applied to enhance oil/water separation efficiency. Moreover, the integration of nanomaterials, such as metal oxides or carbon-based materials, has significantly improved membrane filtration efficiency, particularly in treating textile wastewater. These nanoparticles enhance the membranes' ability to capture and immobilize pollutants, including dyes and oils, aligning with the industry's emphasis on environmental sustainability. Stimuli-responsive nanofibrous membranes represent another advancement in oily wastewater treatment. These membranes can undergo reversible transformation in their properties or structure in response to external stimuli, offering enhanced flexibility and efficiency. For instance, pH-responsive membranes can selectively switch between hydrophilic and hydrophobic states, allowing for controlled oil/water separation. Similarly, temperature-responsive membranes can change their permeability or selectivity based on temperature variations. Chemical and mechanical durability are critical for nanofibrous membranes used in oily wastewater treatment. These membranes must withstand harsh industrial environments and frequent cleaning processes without compromising their effectiveness. Strategies such as incorporating durable polymers, cross-linking agents, or reinforcing materials like carbon nanotubes can enhance their stability. Surface modification techniques, such as plasma treatment or functionalization with hydrophobic coatings, improve resistance to fouling and degradation, prolonging membrane lifespan and performance. In conclusion, nanofibrous membranes offer a versatile, efficient, and sustainable solution for separating oil and contaminants from water. By incorporating nanomaterials, modifying surface morphology, and designing stimuli-responsive membranes, researchers continue to enhance their performance and versatility, paving the way for more effective oily wastewater treatment solutions across industries.

13.8 Future perspectives and challenges

The future of nanofibrous membranes for oily wastewater treatment holds exciting possibilities, marked by emerging trends and avenues for further research. One prospective direction lies in enhancing the multifunctionality of these membranes, where researchers may delve into incorporating advanced features like self-cleaning capabilities and selective pollutant adsorption. Exploration of novel nanostructured materials, such as graphene-based membranes, is another promising trend that could elevate the efficiency of oil—water separation. The field is also witnessing a shift toward sustainability and green manufacturing practices, emphasizing the exploration of bio-based polymers and environmentally friendly synthesis methods. Advanced fabrication techniques, including template-assisted methods, offer potential for better control over membrane properties. Real-time monitoring systems and sensor integration within membranes represent an innovative approach, providing opportunities for on-site performance evaluation. However, as researchers chart this promising course, several challenges loom. Issues of durability, ensuring long-term stability in harsh industrial conditions, scalability for large-scale deployment, and achieving high selectivity in oil—water separation, particularly in the presence of diverse contaminants, pose significant hurdles. In addition, understanding and mitigating any potential environmental impact of nanofibrous membrane materials is crucial for their responsible integration into wastewater treatment processes. Successfully navigating these challenges will be pivotal in realizing the full potential of nanofibrous membranes, making them transformative tools in the ongoing pursuit of sustainable and efficient solutions for oily wastewater treatment on both domestic and industrial fronts.

13.9 Conclusion: The path forward in oily wastewater treatment

In conclusion, our exploration of nanofibrous membranes for oily wastewater treatment has illuminated the transformative potential of these innovative solutions across various industries. From delving into fabrication techniques, challenges, multifunctional design, recent advances, applications, and future perspectives, we have gained insights into the dynamic landscape of membrane technology. Our examination of challenges in oily wastewater treatment, spanning traditional methods like gravity separation to advanced techniques like membrane filtration, sets the stage for understanding the limitations of conventional approaches. The concept of multifunctionality in membrane design emerged as a crucial theme, emphasizing the integration of materials, versatility across applications, and the capability to address complex challenges. The breakthrough technologies in electrospinning techniques, smart nanofibers, surface modification, scalability, hybrid structures, biofabrication, and in situ monitoring underscore the continuous evolution of nanofiber technology. Recent advances in nanofibrous membranes for oily wastewater treatment showcased innovations in membrane design, novel materials, breakthrough technologies, responsive and smart membranes, and the integration of nanotechnology for catalysis. These advancements have significantly progressed the field. Nanofibrous membranes integrate diverse materials and functionalities to address oily wastewater treatment challenges. Breakthroughs in electrospinning and hybrid structures enable scalable, efficient membrane production. Materials like graphene oxide, carbon nanotubes, MXenes, and MOFs enhance membrane performance, offering high selectivity and durability. Stimuli-responsive membranes demonstrate adaptive properties, enabling controlled oil/water separation based on external stimuli. Their multifunctionality enables them to selectively capture oil droplets and contaminants while allowing water to pass through, meeting stringent environmental regulations. Advanced fabrication techniques offer better control over membrane properties, enhancing efficiency and durability. The integration of novel materials like graphene oxide and carbon nanotubes further improves membrane performance, making them more resistant to fouling and degradation. Smart membranes, with their stimuli-responsive properties, represent a significant advancement in oily wastewater treatment. They can change properties in response to external stimuli, such as pH or temperature, improving efficiency and selectivity. For example, pH-responsive membranes can selectively switch between hydrophilic and hydrophobic states, allowing for controlled oil/water separation. Looking ahead, the future of nanofibrous membranes in oily wastewater treatment is promising. Continued research aims to enhance multifunctionality, exploring features like self-cleaning capabilities and selective pollutant adsorption. Further exploration of novel materials holds promise for elevating separation efficiency. Emphasis on sustainability and green practices is expected to increase, focusing on bio-based polymers and environmentally friendly synthesis methods for membrane production. The integration of real-time monitoring systems and sensors within membranes represents a promising avenue for future research. As researchers innovate, nanofibrous membranes have the potential to play a vital role in shaping the future of oily wastewater treatment, offering hope for a more sustainable and resilient world.

References

Abuhasel, K., Kchaou, M., Alquraish, M., Munusamy, Y., Jeng, Y.T., 2021. Oilywastewater treatment: overview of conventional and modern methods, challenges, and future opportunities. MDPI AG, Saudi Arab. Water (Switzerland) 13 (7). https://doi.org/10.3390/w13070980.

Aghaei, A., Suresh, K., Firouzjaei, M.D., Elliott, M., Rahimpour, A., Sadrzadeh, M., 2023. Hybrid/Integrated Treatment Technologies for Oily Wastewater Treatment Advanced Technologies in Wastewater Treatment: Oily Wastewaters. Elsevier, Canada Elsevier, Canada, pp. 377–419. https://doi.org/10.1016/B978-0-323-99916-8.00002-X.

Ahmad, T., Guria, C., Mandal, A., 2020. A review of oily wastewater treatment using ultrafiltration membrane: a parametric study to enhance the membrane performance. J. Water Proc. Eng. 36. https://doi.org/10.1016/j.jwpe.2020.101289.

Ahmed, F.U., Dhar Purkayastha, D., 2023a. Decoration of cauliflower-like TiO2 on nanofibrous PVDF membranes: a strategy for wastewater treatment. Am. Chem. Soc., India ACS Appl. Polym. Mater. 5 (2), 1241–1253. https://doi.org/10.1021/acsapm.2c01753.

Ahmed, F.U., Dhar Purkayastha, D., 2023b. Superhydrophilic ZnO nano-needle decorated over nanofibrous PAN membrane and its application towards oil/water separation. J. Environ. Chem. Eng. 11 (6). https://doi.org/10.1016/j.jece.2023.111166.

Ahmed, F.U., Upadhaya, D., Dhar Purkayastha, D., Krishna, M.G., 2022. Stable hydrophilic and underwater superoleophobic ZnO nanorod decorated nanofibrous membrane and its application in wastewater treatment. J. Membr. Sci. 659. https://doi.org/10.1016/j.memsci.2022.120803.

Aier, I., Dhar Purkayastha, D., 2024. Hierarchical 0D CuO wrapped by petal-like 2D ZnO: a strategic approach of superhydrophobic melamine sponge toward wastewater treatment. Langmuir. https://doi.org/10.1021/acs.langmuir.4c00651.

Alias, N.H., Jaafar, J., Samitsu, S., Matsuura, T., Ismail, A.F., Othman, M.H.D., Rahman, M.A., Othman, N.H., Abdullah, N., Paiman, S.H., Yusof, N., Aziz, F., 2019. Photocatalytic nanofiber-coated alumina hollow fiber membranes for highly efficient oilfield produced water treatment. Chem. Eng. J. 360, 1437–1446. https://doi.org/10.1016/j.cej.2018.10.217.

Barambu, N.U., Bilad, M.R., Bustam, M.A., Kurnia, K.A., Othman, M.H.D., Nordin, N.A.H.M., 2021. Development of membrane material for oily wastewater treatment: a review. Ain Shams Eng. J. 12 (2), 1361–1374. https://doi.org/10.1016/j.asej.2020.08.027.

Bardhan, A., Akhtar, A., Subbiah, S., 2022. Microfiltration and Ultrafiltration Membrane Technologies Advancement in Polymer-Based Membranes for Water Remediation. Elsevier, India Elsevier, India, pp. 3–42. https://doi.org/10.1016/B978-0-323-88514-0.00001-2.

Cai, Y., Yu, Y., Wu, J., Wang, K., Dong, Y., Qu, J., Hu, J., Zhang, L., Fu, Q., Li, J., Zhang, Q., Tian, D., 2022. Durable, flexible, and super-hydrophobic wood membrane with nanopore by molecular cross-linking for efficient separation of stabilized water/oil emulsions. EcoMat 4 (6). https://doi.org/10.1002/eom2.12255.

Cambiella, A., Benito, J.M., Pazos, C., Coca, J., 2006. Centrifugal separation efficiency in the treatment of waste emulsified oils. Chem. Eng. Res. Des. 84 (1), 69–76. https://doi.org/10.1205/cherd.05130.

Cao, Y., Liu, N., Fu, C., Li, K., Tao, L., Feng, L., Wei, Y., 2014. Thermo and pH dual-responsive materials for controllable oil/water separation. ACS Appl. Mater. Interfaces 6 (3), 2026–2030. https://doi.org/10.1021/am405089m.

Cao, W., Ma, W., Lu, T., Jiang, Z., Xiong, R., Huang, C., 2022. Multifunctional nanofibrous membranes with sunlight-driven self-cleaning performance for complex oily wastewater remediation. J. Colloid Interface Sci. 608, 164–174. https://doi.org/10.1016/j.jcis.2021.09.194.

Cao, W., Zhang, M., Ma, W., Huang, C., 2023. Multifunctional electrospun nanofibrous membrane: an effective method for water purification. Separ. Purif. Technol. 327. https://doi.org/10.1016/j.seppur.2023.124952.

Che, H., Huo, M., Peng, L., Fang, T., Liu, N., Feng, L., Wei, Y., Yuan, J., 2015. CO_2 Responsive nanofibrous membranes with switchable oil/water wettability. Angew. Chem. Int. Ed. 54 (31), 8934–8938. https://doi.org/10.1002/anie.201501034.

Chen, Z., Xie, H.Y., Li, Y.J., Chen, G.E., Xu, S.J., Xu, Z.L., 2021. Smart light responsive polypropylene membrane switching reversibly between hydrophobicity and hydrophilicity for oily water separation. J. Membr. Sci. 638. https://doi.org/10.1016/j.memsci.2021.119704.

Chen, L., Zhou, J., Li, K., Wu, X., Chen, Z., Chen, S., Li, X., Xie, H., Su, X., Lin, P., Wu, W., 2024a. Robust and stable organic sulfonate sodium salt-coated superhydrophilic mesh for ultra-fast gravity-drive oil-water separation. Surface. Interfac. 49, 104331. https://doi.org/10.1016/j.surfin.2024.104331.

Chen, S.Y., Li, L., Feng, M., Huang, T., Zhang, N., Wang, Y., 2024b. UV-activated superwetting ability of electrospun polysulfone/titanium dioxide membranes toward highly efficient methylene blue removal and oil/water separation. J. Membr. Sci. 695, 122450. https://doi.org/10.1016/j.memsci.2024.122450.

Cheng, B., Li, Z., Li, Q., Ju, J., Kang, W., Naebe, M., 2017. Development of smart poly(vinylidene fluoride)-graft-poly(acrylic acid) tree-like nanofiber membrane for pH-responsive oil/water separation. J. Membr. Sci. 534, 1–8. https://doi.org/10.1016/j.memsci.2017.03.053.

Cheng, C., Wei, Z., Gu, J., Wu, Z., Zhao, Y., 2022. Rational design of Janus nanofibrous membranes with novel under-oil superhydrophilic/superhydrophobic asymmetric wettability for water-in-diesel emulsion separation. J. Colloid Interface Sci. 606, 1563–1571. https://doi.org/10.1016/j.jcis.2021.08.164.

Chowdhary, P., Raj, A., Bharagava, R.N., 2018. Environmental pollution and health hazards from distillery wastewater and treatment approaches to combat the environmental threats: a review. Chemosphere 194, 229–246. https://doi.org/10.1016/j.chemosphere.2017.11.163.

Cui, J., Wan, M., Wang, Z., Zhao, Y., Sun, L., 2023. Preparation of PAN/SiO2/CTAB electrospun nanofibrous membranes for highly efficient air filtration and sterilization. Separ. Purif. Technol. 321. https://doi.org/10.1016/j.seppur.2023.124270.

de Medeiros, A.D.M., da Silva Junior, C.J.G., de Amorim, J.D.P., Durval, I.J.B., de Santana Costa, A.F., Sarubbo, L.A., 2022. Oily wastewater treatment: methods, challenges, and trends. Processes 10 (4). https://doi.org/10.3390/pr10040743.

Deshmukh, K., Pandey, M., Hussain, C.M., Sivan, A., Pandey, M., Deshmukh, K., Appukuttan, S., 2024. MXene-Reinforced Polymer Composite Membranes for Water Desalination and Wastewater Treatment MXene Reinforced Polymer Composites: Fabrication, Characterization and Applications. Wiley, pp. 459–500. https://doi.org/10.1002/9781119901280.ch14. Wiley.

El-Aswar, E.I., Ramadan, H., Elkik, H., Taha, A.G., 2022. A comprehensive review on preparation, functionalization and recent applications of nanofiber membranes in wastewater treatment. J. Environ. Manag. 301. https://doi.org/10.1016/j.jenvman.2021.113908.

Feng, L., Zhang, Z., Mai, Z., Ma, Y., Liu, B., Jiang, L., Zhu, D., 2004. A super-hydrophobic and super-oleophilic coating mesh film for the separation of oil and water. Angew. Chem. Int. Ed. 43 (15), 2012–2014. https://doi.org/10.1002/anie.200353381.

Figen, A.K., 2020. History, Basics, and Parameters of Electrospinning Technique Electrospun Materials and Their Allied Applications. wiley, Turkey wiley, Turkey, pp. 53–69. https://doi.org/10.1002/9781119655039.ch2.

Ge, J., Fu, Q., Yu, J., Ding, B., 2018. Electrospun Nanofibers for Oil-Water Separation Electrospinning: Nanofabrication and Applications. Elsevier, China Elsevier, China, pp. 391–417. https://doi.org/10.1016/B978-0-323-51270-1.00013-3.

Ghaffarian Khorram, A., Fallah, N., Nasernejad, B., Afsham, N., Esmaelzadeh, M., Vatanpour, V., 2023. Electrochemical-based processes for produced water and oily wastewater treatment: a review. Chemosphere 338. https://doi.org/10.1016/j.chemosphere.2023.139565.

Han, Y., Xu, Y., Zhang, S., Li, T., Ramakrishna, S., Liu, Y., 2020. Progress of improving mechanical strength of electrospun nanofibrous membranes. Macromol. Mater. Eng. 305 (11). https://doi.org/10.1002/mame.202000230.

Hegab, H.M., Elmekawy, A., Aubry, C., Kallem, P., Wadi, V.S., Banat, F., Hasan, S.W., 2024. 3D hierarchical aquaporin-like nanoporous graphene membrane with engineered tripartite nanochannels for efficient oil/water separation. Clean. Water(Washingt.) 7 (1). https://doi.org/10.1038/s41545-024-00303-y.

Hong, S.K., Bae, S., Jeon, H., Kim, M., Cho, S.J., Lim, G., 2018. An underwater superoleophobic nanofibrous cellulosic membrane for oil/water separation with high separation flux and high chemical stability. Nanoscale 10 (6), 3037–3045. https://doi.org/10.1039/c7nr08199e.

Hu, Q.h., Tang, D.y., Wang, X.h., Yan, L.l., Deng, L.l., Zhao, M.q., Deng, E.n., Zhou, Q.h., 2024. Molecularly imprinted MOF/PAN hybrid nanofibrous membranes for selective bisphenol A adsorption and antibacterial fouling in water treatment. Separ. Purif. Technol. 328. https://doi.org/10.1016/j.seppur.2023.124984.

Huang, X., Wu, Z., Zhang, S., Xiao, W., Zhang, L., Wang, L., Xue, H., Gao, J., 2022. Mechanically robust Janus nanofibrous membrane with asymmetric wettability for high efficiency emulsion separation. J. Hazard Mater. 429. https://doi.org/10.1016/j.jhazmat.2022.128250.

Huang, J., Ran, X., Sun, L., Bi, H., Wu, X., 2024. Recent advances in membrane technologies applied in oil–water separation. Discov. Nano 19 (1). https://doi.org/10.1186/s11671-024-04012-w.

Huo, T., Li, F., Jiang, K., Kong, W., Zhao, X., Zhuang, H., Pan, Y., 2022. Fluorocarbon-based selective-superwetting nanofibrous membranes with ultraviolet-driven switchable wettability for oil–water separation. ACS Appl. Nano Mater. 5 (9), 13018 13026. https://doi.org/10.1021/acsanm.2c02809.

Ibrahim, H.M., Klingner, A., 2020. A review on electrospun polymeric nanofibers: production parameters and potential applications. Polym. Test. 90. https://doi.org/10.1016/j.polymertesting.2020.106647.

Imsong, R., Dhar Purkayastha, D., 2023. Dual-functional superhydrophilic/underwater superoleophobic 2D $Ti_3C_2T_X$ MXene-PAN membrane for efficient oil-water separation and adsorption of organic dyes in wastewater. India Separation and Purification Technology 306. https://doi.org/10.1016/j.seppur.2022.122636.

Imsong, R., Dhar Purkayastha, D., 2024a. A synergistically enhanced photocatalytic 2D MXene nanofibrous composite membrane for wastewater treatment. Appl. Surf. Sci. 654. https://doi.org/10.1016/j.apsusc.2024.159402.

Imsong, R., Dhar Purkayastha, D., 2024b. Superhydrophilic photothermal-responsive CuO@MXene nanofibrous membrane with inherent biofouling resistance for treating complex oily wastewater. ACS Appl. Mater. Interfaces 16 (15), 19537–19550. https://doi.org/10.1021/acsami.4c00646.

Imsong, R., Dhar Purkayastha, D., 2024c. A synergistically enhanced photocatalytic 2D MXene nanofibrous composite membrane for wastewater treatment. Appl. Surf. Sci. 654, 159402. https://doi.org/10.1016/j.apsusc.2024.159402.

Jiang, Y., Hou, J., Xu, J., Shan, B., 2017. Switchable oil/water separation with efficient and robust Janus nanofiber membranes. Carbon 115, 477–485. https://doi.org/10.1016/j.carbon.2017.01.053.

Joseph, T.M., Al-Hazmi, H.E., Śniatała, B., Esmaeili, A., Habibzadeh, S., 2023. Nanoparticles and nanofiltration for wastewater treatment: from polluted to fresh water. Poland Environ. Res. 238. https://doi.org/10.1016/j.envres.2023.117114.

Kumar, L., Chugh, M., Kumar, S., Kumar, K., Sharma, J., Bharadvaja, N., 2022. Remediation of petrorefinery wastewater contaminants: a review on physicochemical and bioremediation strategies. Proc. Saf. Environ. Protect. 159, 362–375. https://doi.org/10.1016/j.psep.2022.01.009.

Li, Z., Tio, W., Yang, J.C.E., Sun, D.D., 2023. A photo-responsive micro/nanomembrane for smart separation and self-cleaning. J. Chem. 47 (44), 20272–20278. https://doi.org/10.1039/d3nj02982d.

Li, S., Lin, J.H., Ren, H.T., Shiu, B.C., Sun, F., Lou, C.W., Li, T.T., 2024. Photothermal-driven pH-responsive fiber composite membrane for oil-wastewater purification. J. Environ. Chem. Eng. 12 (1). https://doi.org/10.1016/j.jece.2023.111757.

Liu, L., Yuan, W., 2018. A hierarchical functionalized biodegradable PLA electrospun nanofibrous membrane with superhydrophobicity and antibacterial properties for oil/water separation. New J. Chem. 42 (21), 17615–17624. https://doi.org/10.1039/C8NJ03112F.

Liu, Y., Tas, S., Zhang, K., De Vos, W.M., Ma, J., Vancso, G.J., 2018. Thermoresponsive membranes from electrospun mats with switchable wettability for efficient oil/water separations. Macromolecules 51 (21), 8435–8442. https://doi.org/10.1021/acs.macromol.8b01853.

Liu, H., Xie, J., Zhao, J., Wang, R., Qi, Y., Sun, S., 2024. Temperature and photo sensitive PVDF-g-PNIPAAm/BN@PDA-Ag nanocomposite membranes with superior wasterwater separation and light-cleaning capabilities. Separ. Purif. Technol. 331. https://doi.org/10.1016/j.seppur.2023.125692.

Ma, W., Zhao, J., Oderinde, O., Han, J., Liu, Z., Gao, B., Xiong, R., Zhang, Q., Jiang, S., Huang, C., 2018. Durable superhydrophobic and superoleophilic electrospun nanofibrous membrane for oil-water emulsion separation. J. Colloid Interface Sci. 532, 12–23. https://doi.org/10.1016/j.jcis.2018.06.067.

Ma, W., Ding, Y., Li, Y., Gao, S., Jiang, Z., Cui, J., Huang, C., Fu, G., 2021. Durable, self-healing superhydrophobic nanofibrous membrane with self-cleaning ability for highly-efficient oily wastewater purification. J. Membr. Sci. 634. https://doi.org/10.1016/j.memsci.2021.119402.

Manouchehri, M., 2024. A comprehensive review on state-of-the-art antifouling super(wetting and anti-wetting) membranes for oily wastewater treatment. Adv. Colloid Interface Sci. 323, 103073. https://doi.org/10.1016/j.cis.2023.103073.

Mazumder, A., Sarkar, S., Sen, D., Bhattacharjee, C., 2023. Environmental Effects and Human Health Challenges Originated from Oily Wastewater Advanced Technologies in Wastewater Treatment: Oily Wastewaters. Elsevier, India. Elsevier, India, pp. 29–47. https://doi.org/10.1016/B978-0-323-99916-8.00007-9.

Meng, L., Shi, W., Li, Y., Li, X., Tong, X., Wang, Z., 2023. Janus membranes at the water-energy nexus: a critical review. Adv. Colloid Interface Sci. 318. https://doi.org/10.1016/j.cis.2023.102937.

Moatmed, S.M., Khedr, M.H., El-dek, S.I., El-Deen, A.G., 2024. Multifunctional nanofiber membrane for high removal efficiency of biological/organic contaminations and oil-in-water emulsion under gravity-driven separation. J. Mol. Liq. 394. https://doi.org/10.1016/j.molliq.2023.123748.

Mohshim, D.F., Nasir, R., Mannan, H.A., Abdulrahman, A., Mukhtar, H., 2023. Challenges of Industries in Dealing with Oily Wastewater Release and Treatments Advanced Technologies in Wastewater Treatment: Oily Wastewaters. Elsevier, Malaysia. Elsevier, Malaysia, pp. 49–70. https://doi.org/10.1016/B978-0-323-99916-8.00009-2.

Nasir, A.M., Awang, N., Jaafar, J., Ismail, A.F., Othman, M.H.D., Rahman, M.A., Aziz, F., Mat Yajid, M.A., 2021. Recent progress on fabrication and application of electrospun nanofibrous photocatalytic membranes for wastewater treatment: a review. J. Water Proc. Eng. 40. https://doi.org/10.1016/j.jwpe.2020.101878.

Ou, R., Wei, J., Jiang, L., Simon, G.P., Wang, H., 2016. Robust thermoresponsive polymer composite membrane with switchable superhydrophilicity and superhydrophobicity for efficient oil-water separation. Am. Chem. Soc., Aust. Environ. Sci. Technol. 50 (2), 906–914. https://doi.org/10.1021/acs.est.5b03418.

Pornea, A.M., Puguan, J.M.C., Deonikar, V.G., Kim, H., 2020. Robust Janus nanocomposite membrane with opposing surface wettability for selective oil-water separation. Separ. Purif. Technol. 236. https://doi.org/10.1016/j.seppur.2019.116297.

Rasouli, S., Rezaei, N., Hamedi, H., Zendehboudi, S., Duan, X., 2021. Superhydrophobic and superoleophilic membranes for oil-water separation application: a comprehensive review. Mater. Des. 204. https://doi.org/10.1016/j.matdes.2021.109599.

Ray, S.S., Chen, S.S., Nguyen, N.C., Nguyen, H.T., 2019. Electrospinning: A Versatile Fabrication Technique for Nanofibrous Membranes for use in Desalination Nanoscale Materials in Water Purification. Elsevier, Taiwan. Elsevier, Taiwan, pp. 247–273. https://doi.org/10.1016/B978-0-12-813926-4.00014-8.

Sabir, S., 2015. Approach of cost-effective adsorbents for oil removal from oily water. Crit. Rev. Environ. Sci. Technol. 45 (17), 1916–1945. https://doi.org/10.1080/10643389.2014.1001143.

Sahu, S., Dhar Purkayastha, D., 2023. 1D/2D ZnO nanoneedles/Ti_3C_2 MXene enrobed PVDF electrospun membrane for effective water purification. Appl. Surf. Sci. 622. https://doi.org/10.1016/j.apsusc.2023.156905.

Sahu, S., Dhar Purkayastha, D., 2024. $WO_3 \cdot H_2O$ micro-flowers decorated PVDF/Ti_3C_2 MXene membrane for oily wastewater treatment. Separ. Purif. Technol. 330. https://doi.org/10.1016/j.seppur.2023.125486.

Saleem, H., Trabzon, L., Kilic, A., Zaidi, S.J., 2020. Recent advances in nanofibrous membranes: production and applications in water treatment and desalination. Desalination 478. https://doi.org/10.1016/j.desal.2019.114178.

Sanghamitra, P., Mazumder, D., Mukherjee, S., 2021. Treatment of wastewater containing oil and grease by biological method- a review. J. Environ. Sci. Health, Part A 56 (4), 394–412. https://doi.org/10.1080/10934529.2021.1884468.

Saravanan, A., Senthil Kumar, P., Jeevanantham, S., Karishma, S., Tajsabreen, B., Yaashikaa, P.R., Reshma, B., 2021. Effective water/wastewater treatment methodologies for toxic pollutants removal: processes and applications towards sustainable development. Chemosphere 280, 130595. https://doi.org/10.1016/j.chemosphere.2021.130595.

Seader, J.D., Henley, E.J., Wiley, J., 2006. Separation Process Principles, second ed., p. 2006

Shan, H., Wang, W., Wang, Z., Ge, J., Liu, Q., Zhang, W., Fu, Q., 2023. General synthesis of flexible CuO nanoparticles-anchored ZrO_2 nanofibrous membranes for catalytic oxidation of tetracycline. Chem. Eng. J. 466. https://doi.org/10.1016/j.cej.2023.143063.

Singh, B.J., Chakraborty, A., Sehgal, R., 2023. A systematic review of industrial wastewater management: evaluating challenges and enablers. J. Environ. Manag. 348. https://doi.org/10.1016/j.jenvman.2023.119230.

Su, R., Li, L., Kang, J., Ma, X., Chen, D., Fan, X., Yu, Y., 2022. AgNPs-thiols modified PVDF electrospun nanofiber membrane with a highly rough and pH-responsive surface for controllable oil/water separation. J. Environ. Chem. Eng. 10 (5). https://doi.org/10.1016/j.jece.2022.108235.

Subramanian, E.L., Muthu, S., Khadir, A., 2022. Environmental Footprints and Eco-Design of Products and Processes Inorganic-Organic Composites for Water and Wastewater Treatment Volume.

Sun, F., Li, T.T., Ren, H.T., Shiu, B.C., Peng, H.K., Lin, J.H., Lou, C.W., 2021. Multi-scaled, hierarchical nanofibrous membrane for oil/water separation and photocatalysis: preparation, characterization and properties evaluation. Prog. Org. Coating 152. https://doi.org/10.1016/j.porgcoat.2020.106125.

Theron, S.A., Zussman, E., Yarin, A.L., 2004. Experimental investigation of the governing parameters in the electrospinning of polymer solutions. Polymer 45 (6), 2017–2030. https://doi.org/10.1016/j.polymer.2004.01.024.

Tian, S., He, Y., Zhang, L., Li, S., Bai, Y., Wang, Y., Wu, J., Yu, J., Guo, X., 2023. CNTs/TiO_2 loaded carbonized nanofibrous membrane with two-type self-cleaning performance for high efficiency oily wastewater remediation. Colloids Surf. A Physicochem. Eng. Asp. 656. https://doi.org/10.1016/j.colsurfa.2022.130306.

Tucker, N., Stanger, J.J., Staiger, M.P., Razzaq, H., Hofman, K., 1995. The History of the Science and Technology of Electrospinning from 1600 to, p. 1995.

Wang, X., Hsiao, B.S., 2016. Electrospun nanofiber membranes. Curr. Opin. Chem. Engin. 12, 62–81. https://doi.org/10.1016/j.coche.2016.03.001.

Wang, Z., Elimelech, M., Lin, S., 2016. Environmental applications of interfacial materials with special wettability. Environ. Sci. Technol. 50 (5), 2132–2150. https://doi.org/10.1021/acs.est.5b04351.

Wang, C., Wang, J., Zeng, L., Qiao, Z., Liu, X., Liu, H., Zhang, J., Ding, J., 2019a. Fabrication of electrospun polymer nanofibers with diverse morphologies. Molecules 24 (5). https://doi.org/10.3390/molecules24050834.

Wang, X., Cheng, W., Wang, D., Ni, X., Han, G., 2019b. Electrospun polyvinylidene fluoride-based fibrous nanocomposite membranes reinforced by cellulose nanocrystals for efficient separation of water-in-oil emulsions. J. Membr. Sci. 575, 71–79. https://doi.org/10.1016/j.memsci.2018.12.057.

Wang, Y., Wang, B., Wang, Q., Di, J., Miao, S., Yu, J., 2019c. Amino-functionalized porous nanofibrous membranes for simultaneous removal of oil and heavy-metal ions from wastewater. ACS Appl. Mater. Interfaces 11 (1), 1672–1679. https://doi.org/10.1021/acsami.8b18066.

Wang, C., Wang, F., Zhang, H., Zhang, Y., Zhang, C., Wen, Z., Peng, M., Yang, Y., Wang, S., Xu, C., Wu, A., Zhang, Y., 2024. Multifunctional polyaniline modified calcium alginate aerogel membrane with antibacterial, oil-water separation, dye and heavy metal ions removal properties for complex water purification. Sci. Total Environ. 927, 172058. https://doi.org/10.1016/j.scitotenv.2024.172058.

Xing, X., Hong, M., 2024. Oil/water separation copper membranes modified by laser induced ZnO nanowires growth with enhanced catalytic function. Colloids Surf. A Physicochem. Eng. Asp. 682, 132939. https://doi.org/10.1016/j.colsurfa.2023.132939.

Xiong, J., Liu, Y., Li, A., Wei, L., Wang, L., Qin, X., Yu, J., 2021. Mass Production of High-Quality Nanofibers via Constructing Pre-Taylor Cones with High Curvature on Needleless

Electrospinning. Elsevier Ltd, China Materials and Design, p. 197. https://doi.org/10.1016/j.matdes.2020.109247.

Xu, L., Zhang, J., Wang, T., Cai, Y., Feng, F., He, F., Yi, L., 2024. Demulsification and separation of crude oil-in-water emulsion using superhydrophilic silica-based nanocomposite membranes. Colloids Surf. A Physicochem. Eng. Asp. 691. https://doi.org/10.1016/j.colsurfa.2024.133863.

Yalcinkaya, F., Gunduz, A., Boyraz, E., Bryjak, M., 2022. Nanofibers for Oil-Water Separation and Coalescing Filtration Electrospun and Nanofibrous Membranes: Principles and Applications. Elsevier, Czech Republic Elsevier, Czech Republic, pp. 409–432. https://doi.org/10.1016/B978-0-12-823032-9.00017-9.

Yan, X., Wang, Y., Huang, Z., Gao, Z., Mao, X., Kipper, M.J., Huang, L., Tang, J., 2023. Janus polyacrylonitrile/carbon nanotube nanofiber membranes for oil/water separation. ACS Appl. Nano Mater. 6 (6), 4511–4521. https://doi.org/10.1021/acsanm.3c00006.

Yin, Z., Yuan, F., Xue, M., Xue, Y., Xie, Y., Ou, J., Luo, Y., Hong, Z., Xie, C., 2022. A multifunctional and environmentally safe superhydrophobic membrane with superior oil/water separation, photocatalytic degradation and anti-biofouling performance. J. Colloid Interface Sci. 611, 93–104. https://doi.org/10.1016/j.jcis.2021.12.070.

You, X., Wang, M., Jiang, G., Zhao, X., Wang, Z., Liu, F., Zhao, C., Qiu, Z., Zhao, R., 2023. Multifunctional porous nanofibrous membranes with superior antifouling properties for oil-water separation and photocatalytic degradation. J. Membr. Sci. 668. https://doi.org/10.1016/j.memsci.2022.121245.

Yu, L., Han, M., He, F., 2017. A review of treating oily wastewater. Arab. J. Chem. 10, S1913. https://doi.org/10.1016/j.arabjc.2013.07.020.

Yue, X., Li, Z., Zhang, T., Yang, D., Qiu, F., 2019. Design and fabrication of superwetting fiber-based membranes for oil/water separation applications. Chem. Eng. J. 364, 292–309. https://doi.org/10.1016/j.cej.2019.01.149.

Zeng, J., Su, J., Sun, N., Han, L., Yuan, M., Deng, H., Zhuang, Y., Wang, J., Zhang, Y., 2024. A hierarchical cerium metal organic framework-graphene oxide heterostructure composite for rapid and efficient scavenging of fluoride. Separ. Purif. Technol. 345, 127327. https://doi.org/10.1016/j.seppur.2024.127327.

Zhan, Y., He, S., Wan, X., Zhao, S., Bai, Y., 2018. Thermally and chemically stable poly(arylene ether nitrile)/halloysite nanotubes intercalated graphene oxide nanofibrous composite membranes for highly efficient oil/water emulsion separation in harsh environment. J. Membr. Sci. 567, 76–88. https://doi.org/10.1016/j.memsci.2018.09.037.

Zhang, T., Xiao, C., Zhao, J., Liu, X., Huang, Y., Ji, D., 2020. Graphite powder coated polyurethane sponge hollow tube as a high-efficiency and cost-effective oil-removal materials for continuous oil collection from water surface. J. Appl. Polym. Sci. 137 (31). https://doi.org/10.1002/app.48921.

Zhang, L., Yang, N., Han, Y., Wang, X., Zhang, L., Sun, Y., Jiang, B., 2021a. Highly dispersed β-FeOOH nanocatalysts anchored in confined membrane pores for simultaneously improving catalytic and separation performance. Separ. Purif. Technol. 279. https://doi.org/10.1016/j.seppur.2021.119684.

Zhang, T., Xiao, C., Zhao, J., Liu, X., Ji, D., Zhang, H., 2021. One-step facile fabrication of PVDF/graphene composite nanofibrous membrane with enhanced oil affinity for highly efficient gravity-driven emulsified oil/water separation and selective oil absorption. Separ. Purif. Technol. 254. https://doi.org/10.1016/j.seppur.2020.117576.

Zhang, J., Peng, K., Xu, Z.K., Xiong, Y., Liu, J., Cai, C., Huang, X., 2023. A comprehensive review on the behavior and evolution of oil droplets during oil/water separation by membranes. Adv. Colloid Interface Sci. 319. https://doi.org/10.1016/j.cis.2023.102971.

Zhang, J., Wu, H., Shi, L., Wu, Z., Zhang, S., Wang, S., Sun, H., 2024. Photocatalysis coupling with membrane technology for sustainable and continuous purification of wastewater. Separ. Purif. Technol. 329. https://doi.org/10.1016/j.seppur.2023.125225.

Zhao, C., Zhou, J., Yan, Y., Yang, L., Xing, G., Li, H., Wu, P., Wang, M., Zheng, H., 2021. Application of coagulation/flocculation in oily wastewater treatment: a review. Sci. Total Environ. 765. https://doi.org/10.1016/j.scitotenv.2020.142795.

Zhao, Z., Dong, J., Dong, W., Wang, L., Wang, P., Zhou, Y., Li, H., Liu, Q., Deng, B., Li, D., 2023. PI nanofiber membranes with pH-responsive wettability fabricated through solution blow spinning for efficient on-demand oil-water separation. J. Environ. Chem. Eng. 11 (6). https://doi.org/10.1016/j.jece.2023.111565.

Zheng, W., Huang, J., Li, S., Ge, M., Teng, L., Chen, Z., Lai, Y., 2021. Advanced materials with special wettability toward intelligent oily wastewater remediation. ACS Appl. Mater. Interfaces 13 (1), 67–87. https://doi.org/10.1021/acsami.0c18794.

Zheng, W., Chen, Y., Xu, X., Xing, P., Niu, Y., Xu, P., Li, T., 2024. Research on the factors influencing nanofiltration membrane fouling and the prediction of membrane fouling. J. Water Proc. Eng. 59, 104876. https://doi.org/10.1016/j.jwpe.2024.104876.

Zhou, F., Wang, Y., Dai, L., Xu, F., Qu, K., Xu, Z., 2022. Anchoring metal organic frameworks on nanofibers via etching-assisted strategy: toward water-in-oil emulsion separation membranes. Separ. Purif. Technol. 281. https://doi.org/10.1016/j.seppur.2021.119812.

Zhou, B., Zhang, X., Wang, P., Zhang, X., Wei, C., Wang, Y., Wen, G., 2024. Application of metal oxide catalysts for water treatment – a review. J. Mol. Liq. 401, 124644. https://doi.org/10.1016/j.molliq.2024.124644.

Zolfaghari, R., Fakhru'l-Razi, A., Abdullah, L.C., Elnashaie, S.S.E.H., Pendashteh, A., 2016. Demulsification techniques of water-in-oil and oil-in-water emulsions in petroleum industry. Elsevier B.V., Malaysia Separat. Purificat. Technol. 170, 377–407. https://doi.org/10.1016/j.seppur.2016.06.026.

Zuo, J.H., Gu, Y.H., Wei, C., Yan, X., Chen, Y., Lang, W.Z., 2020. Janus polyvinylidene fluoride membranes fabricated with thermally induced phase separation and spray-coating technique for the separations of both W/O and O/W emulsions. J. Membr. Sci. 595. https://doi.org/10.1016/j.memsci.2019.117475.

Section V

Multifunctional nanostructured coatings for energy management

Multifunctional self-cleaning nanostructured coatings for PV panels, CSP mirrors, and related solar devices

14

Bandana Bera[1,2], *Sudipta Senapati*[2] *and Yitzhak Mastai*[1,2]
[1]Institute for Nanotechnology and Advanced Materials, Bar-Ilan University, Ramat-Gan, Israel; [2]Department of Chemistry, Bar-Ilan University, Ramat-Gan, Israel

14.1 Introduction

Solar energy is one of the best renewable and sustainable power sources, harnessing the sun's radiation. Improving power conversion efficiency is an essential focus in the development of photovoltaic (PV) devices and concentrated solar power (CSP) mirrors. Boosting efficiency not only helps promote environmental sustainability but also makes solar energy systems competitive alternatives to traditional energy sources. These systems are often deployed in semiarid or desert regions where sunlight is abundant, but the presence of sand and dust particles poses a significant challenge. Solar systems utilized in both rural and urban areas encounter diverse forms of dust and pollutants originating from agricultural, construction, and industrial activities. In addition, airborne contaminants such as pollen, plant debris, soot, and particulate matter can settle on solar panels, along with bird droppings and salt deposits in coastal areas. Insects, common in agricultural regions, may leave residues that create a layer diminishing sunlight absorption.

The accumulation of dust and environmental contaminants, known as "soiling" or "fouling," is a growing concern for the efficiency and reliability of PV modules (Bessa et al., 2021; Adak et al., 2022). These accumulations reduce the amount of incident light reaching the solar cells, leading to a reduction in electricity conversion efficiency. The accumulation of dust on PV panels has the potential to substantially diminish the efficiency and power output of the solar energy system, with efficiency reductions reaching up to 80% (Kazem et al., 2020). Numerous studies have been conducted in different regions, including the Thar desert in India, which reveal that solar modules tend to lose their efficiency over time due to soiling, with potential losses of up to 40% (Sayyah et al., 2014). The impact of contaminants on solar panel efficiency varies from region to region, with areas prone to dust storms being affected more severely.

Regions with ample solar resources can generate electricity by converting solar irradiance into energy. These regions typically exhibit PV capacity factors (CFs) ranging between 0.2 and 0.25 (Fig. 14.1a). (Li et al., 2020) CF is the actual annual electricity

Multifunctional Nanostructured Coatings. https://doi.org/10.1016/B978-0-443-23683-9.00014-3

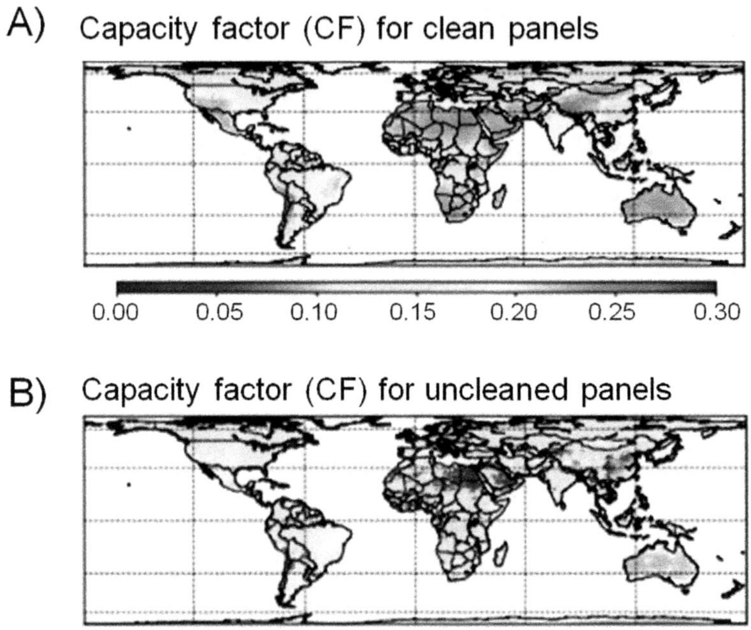

Figure 14.1 The capacity factor (CF) is a measure that reflects the actual energy output of a solar PV system relative to its maximum potential output if it were operating at full capacity continuously. (a) Photovoltaic CFs are depicted for fixed-panel PV systems with no aerosol deposition. (b) Panels covered by aerosol deposition, without undergoing cleaning, exhibit reduced Plane of Array irradiance (POAI) compared with clean panels, primarily due to soiling effects. The significant decline in PV efficiency attributed to soiling is evident in both (a) and (b).
Adopted from Li, X., Mauzerall, D.L., Bergin, M.H., (2020). Global reduction of solar power generation efficiency due to aerosols and panel soiling. Nat. Sustain. 3(9), 720–727.

generation divided by the total generation that would occur if the PV panels run at their nameplate capacity throughout the year. Some regions like western China and the west coast of South America showed high CFs, ranging from 0.25 to 0.3, thereby representing some of the world's highest efficiency levels in solar energy production. In contrast, PV CFs in various regions hover between 0.1 and 0.2. However, the impact of soiling over an entire annual cycle factored in, assuming removal only by precipitation, PV CFs experiences a significant decrease. In the most resource-abundant regions (Fig. 14.1b), CFs decrease by over 30%. Regions like northern Africa and the Middle East, affected by substantial dust deposition, exhibit CFs lower than 0.1, which is more than 60% lower than those of clean panels. Other resource-rich areas, including the western United States, southern Africa, and Central Asia, witness CF reductions exceeding 25%, dropping from 0.2 to 0.25 to 0.15–0.2, due to soiling effects. Even some low-resource regions, like eastern China and northern India, experience CF reductions in approximately 0.1 due to soiling impacts.

The cover glass is a crucial component in maintaining solar power conversion efficiency because it serves as the initial interface for PV modules to interact with incident photons. Regular cleaning of PV module surfaces is essential to mitigate soiling. However, accessing solar module installations for maintenance can be inconvenient. While solar panels have a lot of potential, their performance faces challenges, primarily the decline in efficiency due to the accumulation of dust, dirt, and debris on the panel surfaces. Manual cleaning is a common maintenance practice to counter efficiency drops, but it introduces potential problems such as thermal stress, cracks, scratches, and abrasions. Manual cleaning also involves risks of electrical hazards, water damage to electrical connections, and the threat of falls, especially during rooftop cleaning. The manual cleaning process is not only difficult and costly for large-scale installations but may also lead to damages, resulting in substantial resource wastage in terms of money, water, labor, and energy. As solar energy continues to play a crucial role in sustainable power generation, addressing the impact of soiling becomes increasingly important to ensure optimal performance and longevity of solar panel installations.

Traditional methods of cleaning solar panels are costly, require a lot of resources and labor, and pose safety risks. Fortunately, new technologies like antisoiling coatings and robotic cleaning systems are being developed to improve efficiency, minimize water usage, and reduce the need for manual intervention. These advancements aim to make solar panel cleaning safer, more affordable, and more eco-friendly.

In the quest for solutions beyond conventional cleaning methods, engineers are exploring innovative self-cleaning approaches to minimize surface soiling and optimize PV module performance. Drawing inspiration from nature, researchers have embraced a "biomimetic" strategy, replicating micro-/nanomorphologies found in lotus plant leaves, rice leaves, butterfly wings, and springtail cuticles (Si et al., 2018; Barthlott and Neinhuis, 1997; Sun et al., 2020; Bixler and Bhushan, 2014). Among the most effective solutions to combat dust accumulation on solar panels is the incorporation of nanomaterials onto their surfaces (Das et al., 2017; Aljdaeh et al., 2021). Nanostructured materials, leveraging unique properties like high surface area, have found application in diverse fields, spanning electronics and optoelectronics (semiconductors, quantum dots, dye-sensitized solar cells) (Maurya et al., 2018, 2019), battery technology (Hur, 2022; Zhan et al., 2013), catalysis (Ashik et al., 2018; Qian et al., 2021), textiles (Jadoun et al., 2020; Andra et al., 2021), and biological domains such as drug delivery (Senapati and Maiti, 2020; Senapati et al., 2016, 2018a,b; Kumar et al., 2019), gene delivery (Senapati et al., 2019a, 2021) as well as therapeutics for Alzheimer's disease (Senapati et al., 2023, 2024; Habashi et al., 2022, 2023) and diagnostics applications (Singh et al., 2018; Senapati et al., 2019b). Utilizing nanomaterial coatings for self-cleaning of solar PV panels has emerged as a promising strategy to enhance their efficiency. Inspired by biomimicry, where micro-/nanomorphologies from natural surfaces are replicated employing nanotechnology to develop innovative solutions. These coatings leverage the unique properties of nanomaterials, such as their high surface area and specific functionalities, to create a self-cleaning mechanism. The application of nanocoatings on solar panels helps prevent the accumulation of dust, dirt, and other contaminants on the surface. This not only ensures the optimal exposure of solar cells to sunlight but also minimizes the need for frequent manual cleaning. Several types of nanomaterials such as titania, silica, graphene oxide, and polymer-based have been explored for coating solar PV panels to impart self-cleaning properties.

In this chapter, we provide an overview of the preparation, structure, versatility, and applications of nanostructured self-cleaning materials, highlighting their attractive and useful properties. We will also delve into discussions on effectiveness, challenges, and future perspectives in this evolving field.

14.2 Wetting ability: The science underlying self-cleaning mechanism

Wetting ability refers to the ability of a liquid to spread across a solid surface. It is determined by the balance between adhesive forces (attraction between the liquid and solid) and cohesive forces (attraction between liquid molecules). The degree to which a liquid wets a surface is measured by the contact angle (θ) formed between the liquid, solid, and air at the point where they meet. Young's equation is used to determine wetting behaviors of a solid surface. It takes into account the interfacial energies between the three phases involved: the liquid, solid, and air (Fig. 14.2). The equation describes the relationship between the contact angle (θ), surface tensions, and the interfacial energies of the three-phase system (Senapati et al., 2017). This formula is essential for understanding and predicting the behavior of liquids on solid surfaces and is expressed as given below.

$$\gamma_{SA} = \gamma_{SL} + \gamma_{LA}\cos\theta$$

where: θ is the contact angle, γ_{SA} is the surface tension between the solid (S) and air (A) phases, γ_{SL} is the surface tension between the solid (S) and liquid (L) phases, γ_{LA} is the surface tension between the liquid (L) and air (A) phases. A contact angle between 0 degree and 10 degrees indicates superhydrophilicity, 10 degrees−90 degrees signifies hydrophilicity, and 90 degrees−150 degrees represents hydrophobicity.

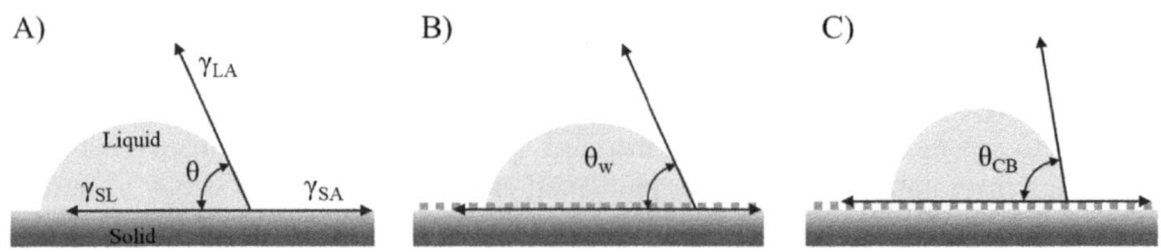

Figure 14.2 Schematic depicting the wetting states of liquid droplets on diverse solid surfaces in air, along with the associated contact angle (CA) models: (a) Young's model, (b) Wenzel's model, and (c) Cassie-Baxter's model.

(Danish, 2022) Surfaces exhibiting a contact angle exceeding 150 degrees are categorized as superhydrophobic.

Although Young's model provides a foundational understanding of wetting on smooth surfaces, the Wenzel and Cassie-Baxter models become crucial when dealing with textured, rough, or patterned surfaces (Huang et al., 2023). They offer insights into how surface morphology impacts wetting behavior, influencing practical applications in fields such as materials science, biomimicry, and the development of advanced coatings or surfaces with tailored wetting properties. After introducing roughness factor r, Wenzel's model follows the following equation.

$$\cos \theta_w = r . \cos \theta$$

This model helps explain how surface roughness influences the wetting behavior of a liquid on a rough surface. The roughness ratio (r) is a metric that indicates how much the surface roughness affects the wetting behavior. If r is more than 1, the roughness amplifies wetting (increases apparent contact angle). On the contrary, if r is less than 1, the roughness inhibits wetting (decreases apparent contact angle).

The Cassie-Baxter model is used to describe the wetting behavior of a liquid droplet on a surface that has wettable and nonwettable regions, which are often characterized by microscale or nanoscale roughness. This model explains the nonwetting behavior of liquid droplets on superhydrophobic surfaces due to the presence of air pockets trapped beneath the droplet because of the microscale or nanoscale structures on the surface.

The Cassie-Baxter equation relates the apparent contact angle (θ_{CB}) on the composite surface to the intrinsic contact angle (θ), the fraction of the wettable surface (f_w), and the intrinsic contact angle on the wettable surface (θ_w). The equation is as follows:

$$\cos(\theta_{CB}) = f_w \cdot \cos(\theta_w) + (1 - f_w) \cdot \cos(\theta)$$

Where: θ_{CB} is the apparent contact angle on the composite surface. θ is the intrinsic contact angle on a smooth, nonwettable surface. f_w is the fraction of the surface that is wettable. θ_w is the intrinsic contact angle on the wettable surface. Understanding the Cassie-Baxter equation helps explain how surface roughness influences the wetting behavior of a liquid on a rough surface.

In the context of self-cleaning coatings, wetting ability is crucial. The Wenzel and Cassie-Baxter models serve pivotal roles in shaping surface design for enhanced wetting behavior. The Wenzel model proves invaluable in optimizing surface roughness to either amplify or diminish wetting, providing insights into the durability and prevention of liquid adhesion. This model informs the customization of coatings for specific applications where the prevention of dirt and pollutants is crucial. On the contrary, the Cassie-Baxter model is instrumental in designing superhydrophobic coatings that capitalize on air-trapping beneath liquid droplets. This model finds applications in scenarios demanding advanced self-cleaning properties, anti-icing capabilities, and resistance to fogging. By leveraging these models, researchers and engineers can tailor self-cleaning coatings to meet diverse environmental and functional requirements, spanning from outdoor surfaces to specialized optical devices. Self-cleaning coatings are designed to repel or shed liquids and particles, preventing them from adhering to the surface. When a surface has high wetting ability, water or other liquids spread out evenly, forming a thin film. This can help in the removal of contaminants like dust, dirt, and pollutants when the surface is exposed to rain or other cleaning mechanisms.

Surface texture and roughness are carefully designed to enhance hydrophobicity and facilitate the easy removal of dirt and debris. Drawing inspiration from nature, particularly the lotus effect, materials are tailored to mimic micro- and nano-scale structures combined with hydrophobic coatings, promoting the formation of water droplets that roll off easily. Lowering the surface energy of materials further contributes to enhanced hydrophobicity, achieved through the incorporation of low-energy coatings or chemical modifications. Understanding wetting transitions guides the design process, allowing for the strategic manipulation of surface properties to optimize self-cleaning efficiency in various conditions.

The self-cleaning mechanism often involves creating superhydrophobic or superoleophobic surfaces, where water or oil-based liquids bead up and roll off the surface, carrying away contaminants. This minimizes the need for manual cleaning and helps maintain the efficiency of surfaces, such as solar panels or other coated materials.

14.3 Self-cleaning coating types

Self-cleaning coatings come in diverse types, each tailored to specific properties and applications. **Hydrophilic coatings**, featuring materials like titanium dioxide (TiO_2) and zinc oxide (ZnO) nanoparticles, enhance water spreading and pollutant degradation, contributing to the self-cleaning capabilities of PV panels. On the contrary, **hydrophobic coatings**, often employing fluoropolymers like polytetrafluoroethylene (PTFE) and perfluoroalkoxy (PFA), create water-repelling surfaces on CSP mirrors, preventing water droplets from adhering and facilitating easy cleaning. For superior water repellency, **superhydrophobic coatings** utilize materials such as silica nanoparticles and hierarchical structures, mimicking the self-cleaning properties observed in nature, and are beneficial for preventing dust accumulation on solar devices. **Oleophobic coatings**, featuring fluorinated polymers, provide protection against oil-based contaminants, ensuring sustained efficiency. **Antireflective coatings**, utilizing nanostructured materials like nanoporous silica, minimize reflection on PV panels, optimizing light absorption and energy conversion. **Photocatalytic coatings** with materials like TiO_2 and ZnO nanoparticles utilize sunlight to break down organic pollutants, contributing to self-cleaning mechanisms. In addition, **antibacterial coatings**, incorporating agents like silver nanoparticles, prevent microbial growth on solar devices, ensuring hygiene in diverse environments. The selection of materials for these coatings is tailored to the specific environmental conditions and contaminants expected, reflecting a dynamic and evolving landscape in solar energy technology where ongoing research aims to optimize durability and performance. In the following section, we will delve into a more detailed exploration of each type of material used in these applications.

14.4 Materials used for self-cleaning coating

14.4.1 *Titanium dioxide nanoparticles*

Titanium dioxide (TiO_2) stands out as an exceptionally promising material for self-cleaning coatings, owing to its extraordinary photocatalytic properties. These properties stem from the creation of photogenerated hole and electron charge carriers, initiated by the absorption of ultraviolet (UV) light within the band gap of TiO_2 (Prawira and Ariyanti, 2021; Afzal et al., 2021). The photo-generated hole, found in the valence band, diffuses to the TiO_2 surface and interacts with absorbed water molecules, forming hydroxyl radicals ($\cdot OH$). Concurrently, electrons in the conduction band partake in the reduction process, reacting with molecular oxygen to produce superoxide radical anions ($O2\cdot^-$). This dynamic interplay results in the effective breakdown of organic compounds, dirt, and pollutants on the coated surface. Beyond its cleansing capabilities, TiO_2 introduces hydrophilicity to surfaces, facilitating the effortless removal of debris through rain or rinsing. Its ability to thwart mold and algae growth enhances its appeal across applications, from exterior building surfaces to automotive coatings. The enduring nature of TiO_2's photocatalytic activity ensures a sustained self-cleaning effect, making it a valuable solution for surfaces prioritizing cleanliness and low maintenance. However, ongoing research is crucial to scrutinize potential environmental and health implications, particularly concerning the use of nanoscale TiO_2 particles in these coatings. Despite its numerous advantages, challenges persist, such as the dependence of TiO_2's photocatalytic activity on UV light, limiting effectiveness in shaded or overcast conditions.

There are various ways to make self-cleaning coatings using TiO_2. Each method has its own properties. The sol-gel method is a process that allows you to control the composition of the film by hydrolyzing and condensing metal alkoxides (Guldin et al., 2013; Huang and Kunitake, 2003). Chemical vapor deposition (CVD) is a way to create thin, pure TiO_2 films by reacting to volatile precursors. Spray pyrolysis is a more affordable method that involves spraying a solution made with a titanium precursor onto a hot surface to form TiO_2 films. Hydrothermal synthesis is a technique that yields crystalline TiO_2 nanoparticles or films with a specific structure and morphology by conducting the process under high temperature and pressure conditions. Sputtering is a physical process that allows you to control the thickness and adhesion of the film. Electrochemical deposition is a process that allows you to control the thickness and composition of the film by electrodeposition. Template-assisted methods use templates like polymers to create nanoscale structures in TiO_2 coatings. Each method has its own advantages depending on the application requirements, desired properties of the film, and scalability.

14.4.2 *Zinc oxide nanoparticles*

Zinc oxide (ZnO) nanoparticles are another promising hydrophilic self-cleaning coating material, capitalizing on its intrinsic photocatalytic properties and versatility. Functioning as a semiconductor, ZnO exhibits remarkable photocatalytic activity under

UV light, generating reactive oxygen species that efficiently decompose organic pollutants on the coated surface. The hydrophilic nature of ZnO fosters water droplet spreading, facilitating the removal of dirt and contaminants, resulting in an effective self-cleaning effect upon exposure to rain or water. Furthermore, ZnO coatings showcase antibacterial properties, impeding microbial growth and preventing biofilm accumulation. Offering UV protection, ZnO coatings prove valuable in outdoor applications, enhancing the durability and longevity of coated materials. Synthesized through methods like spray coating and sol-gel processes, ZnO emerges as a versatile material for self-cleaning applications, particularly in architectural coatings, automotive finishes, and diverse outdoor surfaces. In a recent report, hydrophilic and superhydrophilic ZnO, with varying morphologies, were developed for self-cleaning applications in PV systems (Fig. 14.3). (Nundy et al., 2020) Three distinct ZnO microstructures—nanorods (R−ZnO), microflowers (F−ZnO), and microspheres (M-ZnO)—were created, each with unique crystallite sizes and band gap values. Notably, M-ZnO exhibited superhydrophilic behavior with a water contact angle (WCA) of 2.8 degrees, showcasing its potential in advanced self-cleaning technologies.

14.4.3 Silicones

Silicones serve as highly effective self-cleaning coating materials, leveraging their distinctive attributes such as hydrophobicity, durability, and ease of application (Bogdanowicz et al., 2022). Their inherent hydrophobic nature creates a water-repellent surface, preventing water droplet adhesion and facilitating effortless self-cleaning as rain or liquids wash away dirt and contaminants. Silicones, such as polydimethylsiloxane (PDMS), exhibit exceptional chemical and UV resistance, ensuring durability and prolonged performance. Their flexibility makes them adaptable to diverse substrates like glass, metal, and plastics. Commonly applied in architectural surfaces, automotive coatings, and electronic devices, silicones play a crucial role in maintaining cleanliness and protecting against environmental factors. Notably, specific formulations of silicone-based coatings may vary, and ongoing advancements in silicone technology continuously enhance their self-cleaning capabilities. PDMS, with its ultralow surface energy, stands out as a representative silicone material, offering advantages such as thermal stability, nontoxicity, and biocompatibility. Techniques like femtosecond laser etching enable the creation of three-dimensional patterned superhydrophobic surfaces on PDMS. In addition, PDMS can be combined with materials like SiO_2 nanoparticles to achieve superhydrophobicity, showcasing its versatility (Gong and He, 2020). Beyond PDMS, silicones like heptadecafluoro-1,1,2,2-tetrahydrophilictrimethoxysilane (FAS-17) can also be employed for superhydrophobic surfaces, accompanied by properties such as oil−water separation (Liu et al., 2015).

14.4.4 Polymeric nanomaterials

Hydrophobic polymers, including well-known examples like polymethylmethacrylate (PMMA), polytetrafluoroethylene (PTFE), and polyurethanes, play a pivotal role in advancing self-cleaning coatings. At the forefront of material science innovation,

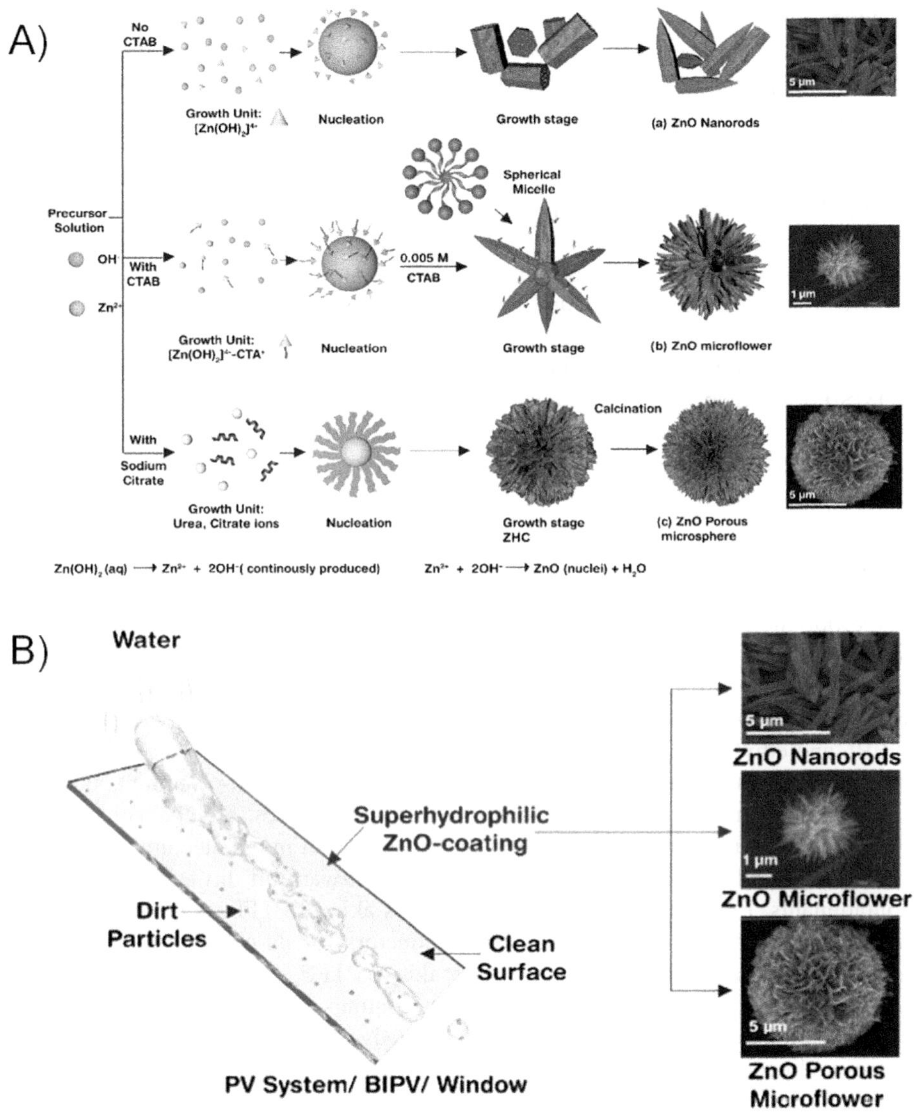

Figure 14.3 (a) Schematic illustrating a facile synthesis process for producing morphologically diverse hydrophilic and superhydrophilic ZnO microstructures, including nanorods, flowerlike structures, and microspheres. (b) Schematic representation highlighting the morphologically varied ZnO microstructures designed to enhance hydrophilicity in photovoltaic and glazing systems.
From Nundy, S., Ghosh, A., Mallick, T.K., (2020). Hydrophilic and superhydrophilic self-cleaning coatings by morphologically varying ZnO microstructures for photovoltaic and glazing applications. ACS Omega 5 (2), 1033–39. https://doi.org/10.1021/acsomega.9b02758.

polymer-based self-cleaning coatings leverage the distinctive properties of these polymers to create surfaces with exceptional self-cleaning capabilities. Engineered to be highly hydrophobic, these coatings actively repel water and prevent contaminant adhesion. Inspired by nature, some coatings mimic the Lotus Effect, incorporating microscopic structures akin to lotus leaves for effortless water and particle shedding. Certain formulations may integrate photocatalytic materials like titanium dioxide, activated by light to break down organic substances on the coated surface. This synergy of hydrophobicity, surface texture, and photocatalytic activity ensures easy dirt removal and coating durability, making it a promising solution for diverse applications, from automotive windshields to architectural glass. Comparative studies, such as those by the Sutar group (Sutar et al., 2020), highlight the superiority of coatings like PDMS over PMMA, emphasizing factors like abrasion resistance. In other developments, PDMS/Polyurethane coatings have achieved 99% light transmission, showcasing omniphobic properties that enable the slip of concentrated oil and hexadecane with impressive hysteresis angles of 22 degrees and 12 degrees, respectively (Khan et al., 2020).

14.4.5 Fluoropolymers

Fluoropolymers, renowned for their distinct attributes like low surface energy, chemical resistance, and nonstick qualities, are extensively employed as self-cleaning coating materials. Notable examples encompass polytetrafluoroethylene (PTFE), polyvinylidene fluoride (PVDF), perfluoroalkoxy (PFA), fluorinated ethylene propylene (FEP), and ethylene tetrafluoroethylene (ETFE). PTFE, commonly known as Teflon, boasts exceptional nonstick properties, rendering it ideal for self-cleaning applications seen in cookware, bakeware, and various industrial surfaces. PVDF, with its robust chemical resistance and weatherability, finds application in architectural coatings for structures, benefitting from its nonstick features that ward off dirt and environmental contaminants. PFA, a copolymer with properties akin to PTFE, is utilized in critical industries like pharmaceuticals and semiconductors for its chemical resistance and nonstick attributes. FEP, another copolymer akin to PTFE, delivers nonstick properties and is preferred in transparent or translucent coatings for cookware, bakeware, and chemical processing equipment. ETFE, acknowledged for transparency and durability, graces architectural surfaces with self-cleaning functionalities, notably on building facades. These chosen fluoropolymers, by establishing smooth, low-friction surfaces, actively repel water, oil, and contaminants, curbing dirt adhesion and facilitating easier cleaning—ultimately enhancing the overall self-cleaning functionality across diverse applications and industries.

In recent years, fluorinated polymers have been extensively employed to create superhydrophobic coatings. Various methods, such as roughening the surface of polymers through technologies like cold spraying or plasma treatment have been explored (Qin and Lu, 2023). For instance, Ravi et al. (2019) utilized cold spraying technology to prepare a superhydrophobic perfluoro-alkoxy alkane (PFA) coating on stainless steel, forming a nanorough structure (Ravi et al., 2019). Ryu et al. (2017) achieved excellent superhydrophobic coating using PTFE as a base material with plasma treatment (Ryu et al., 2017). While some long-chain fluorides are expensive, studies focus on synthesizing fluorinated polymers using different reagents, such as copolymerization of methacrylic acid 2, 3, 4, 5, 5-hexafluoro-2, 4-bis (trifluoromethyl) pentyl ester (FMA) and 3-methylpropenyloxypropyltrimethoxysilane (TSMA) to create a fluorinated polymer sol−gel precursor (PFT). This precursor is then introduced into the sol−gel process to prepare a superhydrophobic coating on glass substrates. In addition, Roslizar et al. (2019) induced superhydrophobicity by hot pressing microtexture on fluorinated ethylene propylene (FEP) film (Roslizar et al., 2019), known for high transmittance and good photostability, making it suitable for superhydrophobic coatings. In another study, a simple procedure was devised to create a stable, self-cleaning, highly hydrophobic composite coating (Rudra et al., 2023). Silica nanoparticles were coated with two fluorinated polymers, poly(pentafluorostyrene) (pPFS) and poly(pentafluorophenyl acrylate) (pPFPA), resulting in FPB-g-SiNPs. These nanofillers were applied to form a multiscale roughness coating on polystyrene (PS), yielding a highly hydrophobic surface (FPB-g-SiNP/PS) with a contact angle of ~143 degrees. The coating, featuring an ungrafted silica unit for hydrophilicity, exhibited good wetting toward water and various liquids, as well as excellent self-cleaning properties akin to a superhydrophobic surface. In addition, it demonstrated robust chemical durability against alkali, saline water, and organic solvents.

14.5 Durability

While self-cleaning coatings offer improved performance for solar PV modules and CSP plants, aligning their life expectancy with that of the PV/CSP system remains a challenge. The durability of these coatings is pivotal for their large-scale commercial applications, necessitating an understanding of the challenges posed by external factors during outdoor operation. Notably, the photocatalytic activity of superhydrophilic coatings diminishes over time, particularly in dry and less rainy environments, impacting their sustained efficacy on solar panels (Qin and Lu, 2023). Similarly, the fragile nature of superhydrophobic coatings, characterized by a water contact angle (WCA) exceeding 150 degrees, makes them susceptible to wear and performance deterioration. Addressing these challenges, researchers have pursued innovative strategies such as introducing fluorine atoms, increasing coating layers, and creating protective microstructures to enhance the longevity and robustness of superhydrophobic coatings. The introduction of fluorine atoms aligned with carbon atoms around the resin's main chain significantly improves durability, as demonstrated by various studies. In addition, augmenting the number of coating layers enhances surface adhesion and durability, as evidenced by Sarkın et al (2020) and Latthe et al (2019). The creation of protective "armor" microstructures, as reported by Wang et al. involves a meticulous analysis of nanostructures for waterproofing and micrometer structures for durability, resulting in coatings with improved wear resistance. These advancements contribute to extending the service life of self-cleaning coatings and ensuring their reliable performance in diverse environmental conditions. Researchers underscore the importance of considering factors such as antifog and anti-icing properties in the preparation of self-cleaning coatings for PV modules, emphasizing the need for comprehensive considerations in future studies to optimize durability, antireflection, and self-cleaning performance for sustained efficiency in solar energy applications.

14.6 Current challenges and outlook

The implementation of self-cleaning coatings for solar devices such as PV panels and concentrated solar power (CSP) mirrors faces numerous challenges. These challenges include durability, environmental adaptability, cost-effectiveness, maintenance requirements, integration with other functionalities, long-term stability, and scalability. One of the biggest hurdles is achieving an optimal balance between enhancing efficiency and matching the coating's life expectancy with that of the solar system. Coatings are often subject to wear and tear due to exposure to external mechanical and environmental stresses during outdoor operation, which results in durability issues. This requires innovative solutions. In addition, the effectiveness of coatings can vary depending on different environments, and cost-effective manufacturing methods are needed to enhance adaptability and economic viability. Maintenance requirements, especially for coatings sensitive to wear, necessitate the development of formulations with minimal upkeep demands.

The outlook for self-cleaning coatings in solar applications is promising, driven by ongoing research and technological advancements. Researchers are focusing on developing more durable coatings adaptable to diverse environmental conditions for consistent performance. To ensure economic viability on a large scale, efforts are directed toward optimizing manufacturing processes and identifying cost-effective materials. Future advancements may yield coatings with minimal maintenance requirements and increased wear resistance, addressing longevity concerns. Anticipated developments include multifunctional coatings integrating self-cleaning properties with antireflective or antifog features. Researchers aim to explore scalable application methods for expansive surfaces and enhance compatibility with various substrates. In summary, a holistic approach combining advancements in materials, formulations, and application methods is poised to unlock the full potential of self-cleaning coatings, advancing the efficiency and sustainability of solar energy technologies.

14.7 Conclusions

In summary, the exploration of nanostructured materials for self-cleaning coatings in solar panels, CSP mirrors, and related devices emerges as a dynamic and promising trajectory, contributing significantly to the advancement of efficiency and sustainability in solar energy technologies. This chapter has provided an extensive review encompassing statistics, theoretical foundations, materials, preparation methods, and applications of various self-cleaning coatings employed in solar panels, CSP mirrors, and analogous devices. The discussion delves into the research progress, elucidates the advantages and disadvantages of different materials and methods, and encapsulates an analysis of challenges and potential solutions. Emphasizing the crucial role of achieving surface roughness and low surface energy for superhydrophobic coatings, the chapter categorizes superhydrophilic coating preparation into molecular structure deposition and surface chemical modification. Commonly utilized materials such as TiO_2, fluorinated polymers, and silicones are highlighted, with considerations for applying these coatings to PV modules encompassing self-cleaning, transparency, antireflection, anti-icing, and durability. The identified challenge of addressing the poor durability of superhydrophobic coatings underscores the need for ongoing research and innovation in this domain. Overall, this chapter underscores the transformative impact of nanotechnology on the pursuit of sustainable and efficient solar energy solutions, envisioning a cleaner and more resilient future in the realm of renewable energy.

References

Adak, D., Bhattacharyya, R., Barshilia, H.C., 2022. A state-of-the-art review on the multifunctional self-cleaning nanostructured coatings for PV panels, CSP mirrors and related solar devices. Renew. Sustain. Energy Rev. 159, 112145.

Afzal, A., Habib, A., Ulhasan, I., Shahid, M., Rehman, A., 2021. Antireflective self-cleaning TiO2 coatings for solar energy harvesting applications. Front. Mater. 8, 687059.

Aljdaeh, E., Kamwa, I., Hammad, W., Abuashour, M.I., Sweidan, T.E., Khalid, H.M., Muyeen, S.M., 2021. Performance enhancement of self-cleaning hydrophobic nanocoated photovoltaic panels in a dusty environment. Energies 14 (20), 6800.

Andra, S., Balu, S.K., Jeevanandam, J., Muthalagu, M., 2021. Emerging nanomaterials for antibacterial textile fabrication. N. Schmied. Arch. Pharmacol. 394, 1355–1382.

Ashik, U.P., Viswan, A., Kudo, S., Hayashi, J.I., 2018. Nanomaterials as catalysts. In: Applications of Nanomaterials. Woodhead Publishing, pp. 45–82.

Barthlott, W., Neinhuis, C., 1997. Purity of the sacred lotus, or escape from contamination in biological surfaces. Planta 202, 1–8.

Bessa, J.G., Micheli, L., Almonacid, F., Fernández, E.F., 2021. Monitoring photovoltaic soiling: assessment, challenges, and perspectives of current and potential strategies. iScience 24, 102165.

Bixler, G.D., Bhushan, B., 2014. Rice-and butterfly-wing effect inspired self-cleaning and low drag micro/nanopatterned surfaces in water, oil, and air flow. Nanoscale 6 (1), 76–96.

Bogdanowicz, K.A., Dutkiewicz, M., Maciejewski, H., Nowicki, M., Przybył, W., Plebankiewicz, I., Iwan, A., 2022. Siloxane resins as hydrophobic self-cleaning layers for silicon and dye-sensitized solar cells: material and application aspects. RSC Adv. 12 (30), 19154–19170.

Danish, M., 2022. Contact angle studies of hydrophobic and hydrophilic surfaces. In: Handbook of Magnetic Hybrid Nanoalloys and Their Nanocomposites. Springer International Publishing, Cham, pp. 1–22.

Das, S., Silic, S., Das, B., April 15, 2017. Nanotechnology for water-less cleaning of solar panels. InElectrochem. Soc. Meet. Abstract. 231 (43).

Gong, X., He, S., 2020. Highly durable superhydrophobic polydimethylsiloxane/silica nanocomposite surfaces with good self-cleaning ability. ACS Omega 5 (8), 4100–4108.

Guldin, S., Kohn, P., Stefik, M., Song, J., Divitini, G., Ecarla, F., Ducati, C., Wiesner, U., Steiner, U., 2013. Self-cleaning antireflective optical coatings. Nano Lett. 13 (11), 5329–5335.

Habashi, M., Vutla, S., Tripathi, K., Senapati, S., Chauhan, P.S., Haviv-Chesner, A., Richman, M., Mohand, S.A., Dumulon-Perreault, V., Mulamreddy, R., Okun, E., 2022. Early diagnosis and treatment of Alzheimer's disease by targeting toxic soluble Aβ oligomers. Proc. Natl. Acad. Sci. USA 119 (49).

Habashi, M., Chauhan, P.S., Vutla, S., Senapati, S., Diachkov, M., El-Husseini, A., Guérin, B., Lubell, W.D., Rahimipour, S., 2023. Aza-residue modulation of cyclic d, l-α-Peptide nanotube assembly with enhanced anti-amyloidogenic activity. J. Med. Chem. 66 (4), 3058–3072.

Huang, J., Kunitake, T., 2003. Nano-precision replication of natural cellulosic substances by metal oxides. J. Am. Chem. Soc. 125 (39), 11834–11835.

Huang, F., Motealleh, B., Wang, D., Cornelius, C.J., 2023. Tailoring intrinsic hydrophobicity and surface energy on rough surface via low-T Cassie–Wenzel wetting transition method. AIChE J. 69 (3), e17908.

Hur, J., 2022. Nanomaterials for ion battery applications. Nanomaterials 12 (13), 2293.

Jadoun, S., Verma, A., Arif, R., 2020. Modification of Textiles via Nanomaterials and Their Applications. Frontiers of Textile Materials: Polymers, Nanomaterials, Enzymes, and Advanced Modification Techniques, pp. 135–152.

Kazem, H.A., Chaichan, M.T., Al-Waeli, A.H., Sopian, K., 2020. A review of dust accumulation and cleaning methods for solar photovoltaic systems. J. Clean. Prod. 276, 123187.

Khan, A., Huang, K., Sarwar, M.G., Cheng, K., Li, Z., Tuhin, M.O., Rabnawaz, M., 2020. Self-healing and self-cleaning clear coating. J. Colloid Interface Sci. 577, 311–318.

Kumar, S., Singh, A.P., Senapati, S., Maiti, P., 2019. Controlling drug delivery using nanosheet-embedded electrospun fibers for efficient tumor treatment. ACS Appl. Bio Mater. 2 (2), 884–894.

Latthe, S.S., Sutar, R.S., Kodag, V.S., Bhosale, A.K., Kumar, A.M., Sadasivuni, K.K., Xing, R., Liu, S., 2019. Self–cleaning superhydrophobic coatings: potential industrial applications. Prog. Org. Coating 128, 52–58.

Li, X., Mauzerall, D.L., Bergin, M.H., 2020. Global reduction of solar power generation efficiency due to aerosols and panel soiling. Nat. Sustain. 3 (9), 720–727.

Liu, S., Latthe, S.S., Yang, H., Liu, B., Xing, R., 2015. Raspberry-like superhydrophobic silica coatings with self-cleaning properties. Ceram. Int. 41 (9), 11719–11725.

Maurya, I.C., Senapati, S., Singh, S., Srivastava, P., Maiti, P., Bahadur, L., 2018. Effect of particle size on the performance of TiO2 based dye-sensitized solar cells. ChemistrySelect 3 (34), 9872–9880.

Maurya, I.C., Singh, S., Senapati, S., Srivastava, P., Bahadur, L., 2019. Green synthesis of TiO2 nanoparticles using Bixa orellana seed extract and its application for solar cells. Sol. Energy 194, 952–958.

Nundy, S., Ghosh, A., Mallick, T.K., 2020. Hydrophilic and superhydrophilic self-cleaning coatings by morphologically varying ZnO microstructures for photovoltaic and glazing applications. ACS Omega 5 (2), 1033–1039.

Prawira, R.A., Ariyanti, D., 2021. Recent study on TiO2 based self-cleaning coating. In: IOP Conference Series: Materials Science and Engineering, vol. 1053. IOP Publishing, p. 012061.

Qian, W., Xu, S., Zhang, X., Li, C., Yang, W., Bowen, C.R., Yang, Y., 2021. Differences and similarities of photocatalysis and electrocatalysis in two-dimensional nanomaterials: strategies, traps, applications and challenges. Nano-Micro Lett. 13, 1−38.

Qin, J., Lu, H., 2023. A review of self-cleaning coatings for solar photovoltaic systems: theory, materials, preparation, and applications. Environ. Sci. Pollut. Control Ser. 30 (40), 91591−91616.

Ravi, K., Sulen, W.L., Bernard, C., Ichikawa, Y., Ogawa, K., 2019. Fabrication of micro-/nano-structured super-hydrophobic fluorinated polymer coatings by cold-spray. Surf. Coating. Technol. 373, 17−24.

Roslizar, A., Dottermusch, S., Vüllers, F., Kavalenka, M.N., Guttmann, M., Schneider, M., Paetzold, U.W., Hölscher, H., Richards, B.S., Klampaftis, E., 2019. Self-cleaning performance of superhydrophobic hot-embossed fluoropolymer films for photovoltaic modules. Sol. Energy Mater. Sol. Cell. 189, 188−196.

Rudra, S., Dhara, M., Chakraborty, M., Jana, T., 2023. Fluorinated polymer brush-grafted silica nanoparticles: robust and durable self-cleaning coating materials. ACS Appl. Polym. Mater. 5 (9), 7443−7457.

Ryu, J., Kim, K., Park, J., Hwang, B.G., Ko, Y., Kim, H., Han, J., Seo, E., Park, Y., Lee, S.J., 2017. Nearly perfect durable superhydrophobic surfaces fabricated by a simple one-step plasma treatment. Sci. Rep. 7 (1), 1981.

Sarkın, A.S., Ekren, N., Sağlam, Ş., 2020. A review of anti-reflection and self-cleaning coatings on photovoltaic panels. Sol. Energy 199, 63−73.

Sayyah, A., Horenstein, M.N., Mazumder, M.K., 2014. Energy yield loss caused by dust deposition on photovoltaic panels. Sol. Energy 107, 576−604.

Senapati, S., Maiti, P., 2020. Emerging bio-applications of two-dimensional nanoheterostructure materials. In: 2D Nanoscale Heterostructured Materials. Elsevier, pp. 243−255.

Senapati, S., Thakur, R., Verma, S.P., Duggal, S., Mishra, D.P., Das, P., Shripathi, T., Kumar, M., Rana, D., Maiti, P., 2016. Layered double hydroxides as effective carrier for anticancer drugs and tailoring of release rate through interlayer anions. J. Contr. Release 224, 186−198.

Senapati, S., Rana, D., Maiti, P., 2017. Crystal-melt phase change of food and biopolymers. In: Glass Transition and Phase Transitions in Food and Biological Materials, vol. 24, pp. 119−139. https://doi.org/10.1002/9781118935682.ch4.

Senapati, S., Mahanta, A.K., Kumar, S., Maiti, P., 2018a. Controlled drug delivery vehicles for cancer treatment and their performance. Signal Transduct. Targeted Ther. 3 (1), 7.

Senapati, S., Shukla, R., Tripathi, Y.B., Mahanta, A.K., Rana, D., Maiti, P., 2018b. Engineered cellular uptake and controlled drug delivery using two dimensional nanoparticle and polymer for cancer treatment. Mol. Pharm. 15 (2), 679−694.

Senapati, S., Sarkar, T., Das, P., Maiti, P., 2019a. Layered double hydroxide nanoparticles for efficient gene delivery for Cancer treatment. Bioconjugate Chem. 30 (10), 2544−2554.

Senapati, S., Patel, D.K., Ray, B., Maiti, P., 2019b. Fluorescent-functionalized graphene oxide for selective labeling of tumor cells. J. Biomed. Mater. Res. 107 (9), 1917−1924.

Senapati, S., Upadhyaya, A., Dhruw, S., Giri, D., Maiti, P., 2021. Controlled DNA delivery using poly (lactide) nanoparticles and understanding the binding interactions. J. Phys. Chem. B 125 (35), 10009−10017.

Senapati, S., Secchi, V., Cova, F., Richman, M., Villa, I., Yehuda, R., Shenberger, Y., Campione, M., Rahimipour, S., Monguzzi, A., 2023. Noninvasive treatment of Alzheimer's disease with scintillating nanotubes. Adv. Healthcare Mater. 12 (32), 2301527.

Senapati, S., Tripathi, K., Awad, K., Rahimipour, S., 2024. Multifunctional liposomes targeting amyloid-β oligomers for early diagnosis and therapy of Alzheimer's disease. Small 2311670.

Si, Y., Dong, Z., Jiang, L., 2018. Bioinspired designs of superhydrophobic and superhydrophilic materials. ACS Cent. Sci. 4 (9), 1102−1112.

Singh, S., Mitra, K., Senapati, S., Singh, R., Biswas, Y., Sen Gupta, S.K., Misra, N., Mandal, T.K., Maiti, P., Ray, B., 2018. Water soluble fluorescent graphene nanodots. ChemNanoMat 4 (11), 1177−1188.

Sun, D., Böhringer, K.F., 2020. An active self-cleaning surface system for photovoltaic modules using anisotropic ratchet conveyors and mechanical vibration. Microsystems & Nanoengineering 6 (1), 87.

Sutar, R.S., Gaikwad, S.S., Latthe, S.S., Kodag, V.S., Deshmukh, S.B., Saptal, L.P., Kulal, S.R., Bhosale, A.K., 2020. Superhydrophobic nanocomposite coatings of hydrophobic silica NPs and poly (methyl methacrylate) with notable self-cleaning ability. In: Macromolecular Symposia, vol. 393, p. 2000116.

Zhan, H., Xiao, J., Nie, Z., Li, X., Wang, C., Zhang, J.G., Liu, J., 2013. Nanostructured materials for rechargeable batteries: synthesis, fundamental understanding and limitations. Current Opinion in Chemical Engineering 2 (2), 151−159.

Applications of thermal spray—based coatings for renewable energy

15

Durvesh Gautam and Yogendra K. Gautam
Smart Materials and Sensor Laboratory, Department of Physics, Chaudhary Charan Singh University, Meerut, Uttar Pradesh, India

Thermal spray coating emerges as a promising technology, offering a versatile approach to improve the performance and durability of critical components in various renewable energy applications. Different chemicals are utilized for coating to enhance the performance and longevity of renewable energy technologies, for solar panels, anti-reflective coatings such as silicon dioxide (SiO_2) and titanium dioxide (TiO_2) are employed to improve light transmission and absorption. Additionally, self-cleaning coatings featuring titanium dioxide (TiO_2) utilize photocatalytic properties to break down organic pollutants.Hydrophobic coatings, including fluoropolymers like; polytetrafluoroethylene (PTFE) and perfluorooctyltrichlorosilane (PFOTS), are used to repel water and contaminants, thereby reducing maintenance. In wind turbine applications, epoxy resins and polyurethane coatings offer superior anti-corrosion properties and UV resistance. Thermal barrier coatings, such as zirconia (ZrO_2) and yttria-stabilized zirconia (YSZ), provide thermal insulation for high-temperature components. This chapter provides an inclusive analysis of the application of thermal spray coatings in renewable energy systems, focusing on their impact on key components such as solar panels, wind turbine blades, geothermal, hydro turbine, photocatalytic hydrogen production, and thermoelectric power generator systems. The chapter also highlights recent advancements in thermal spray coating materials, deposition techniques, and their compatibility with different renewable energy technologies.

15.1 Introduction

A new age of technical innovation has been brought about by the global search for sustainable and renewable energy sources. This has accelerated the development and integration of sophisticated materials to improve the durability and efficiency of renewable energy systems. Within this framework, thermal spray coating has become a viable option, providing a flexible toolkit to tackle the difficulties of essential components in different renewable energy applications. Thermal spray coating techniques are extremely flexible techniques that can be used for practically any component and in a wide range of applications. Because of this, it has expanded to become a significant

Multifunctional Nanostructured Coatings. https://doi.org/10.1016/B978-0-443-23683-9.00015-5

global business worth many billions of dollars since coatings were first created in the early 1900s (Siegmann and Abert, 2013). Thermal spray technology consists of a collection of coating processes and techniques that deposit metallic or nonmetallic materials in a molten or semi-molten state to create a coating. The coating material can be powder, ceramic rods, wire, or molten materials that add functionality to surfaces and enhance component performance. The objective of thermal spray methods is to provide materials a certain attribute or to extend their lifespan relative to their structural processes of breakdown (e.g., optical and electrical) (Lombardi et al., 2013).

Dr. Max Ulrich Schoop (Zurich, Switzerland) and collaborators were working in the late 19th and early 20th centuries to develop a range of devices that could melt metal and spray it onto surfaces to obtain a coating (Thermal Spray for Extreme Environments, 2019). Tin and lead were the only low melting point metals utilized in the early stages of this innovation, and the procedure was known as metalizing (Hermanek, 2001; Davis, 2004). Dr. Schoop and collaborators developed the first thermal spray device for metal wires after a few years and several patents later. This device functioned by melting a metal wire in a flame produced by burning fuel and oxygen together. To create a coating, a pressurized gas was used to atomize the molten metal, causing the metal droplets to be driven onto a substrate or component (Thermal Spray for Extreme Environments, 2019). This approach provided the groundwork for the development of more successful methods, such as high-velocity oxygen fuel (HVOF), which initiated the large family of combustion thermal spray methods shown in Fig. 15.1. The electric arc was a breakthrough presented by Dr. Schoop's group, in the metalizing process in response to the demand for new kinds of metals to prevent corrosion (Espallargas, 2015a). The emergence of the second huge group of methods for thermal spraying, which depend on the application of electrical energy, was facilitated by the electric arc, which made it easier to spray higher melting point metals, like steel, stainless steel, zinc, etc (Crawmer, 2013; Amin and Panchal, 2016; Talib et al.,

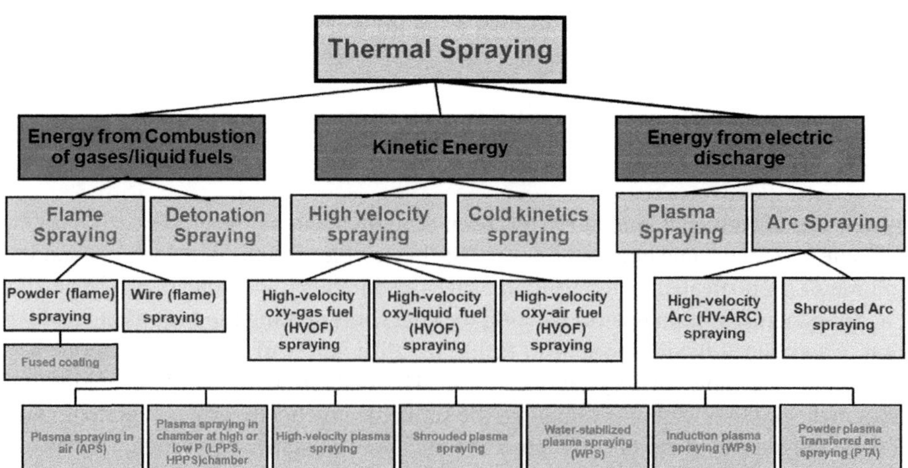

Figure 15.1 Schematic of family tree of different thermal spray technologies.

2003). To produce coatings without combustion or electrical energy, the feedstock (coating material) is sprayed onto the substrate using solid-state spraying, which is the foundation of the third and most recent group of methods for thermal spraying, this process started in the 1980s (Li et al., 2018, 2020). The dynamics behind the development of coating in this most recent generation, referred to as cold spraying, are still incompletely understood (Li et al., 2019). Warm spray and high-velocity air fuel are examples of low-temperature combustion methods that have emerged recently, while the more conventional methods of combustion have been modified around the idea of cold spraying (Yan et al., 2021). All other spraying techniques employ a combusted gas. There is just one solid-state spraying technique that can be categorized: cold gas spraying (i.e., without melting the feedstock material). Metal coatings by cold spraying methods aim to preserve the feedstock material's original structure and attributes together with a minimal degree of oxidation. The methods of thermal spraying are put together like a genealogy as shown in Fig. 15.1 (Espallargas, 2015a). The extreme advancement in thermal spray occurred after WW II due to the demands and advancements of aircraft and spacecraft. This contrasted with the early periods of the first part of the 20th century that saw the use of thermal spray when numerous patents were developed around the metalizing approach using electric (arc) or combustion methods. During this time, methods for developing hotter flames were required to coat materials with high melting points, such as ceramics and refractory metals. The greatest advancements in thermal spray technology were made during this time in terms of tools and machinery (consumables/feedstock) (Li et al., 2023). Thermal spray systems such as the ceramic rod flame spray, the detonation gun (D-gun), and the plasma spray were developed to protect engine components used at high temperatures in the 20th-century mid (Fauchais and Montavon, 2008). The advancement of the feedstock materials, which came in wire and powder form, was largely responsible for these improvements. After the 1950s, the growth of methods, procedures, and research surrounding thermal spray has coincided with an exponential expansion in sales of equipment and supplies as well as the industrial sectors where thermal spraying is employed. Furthermore, thermal spray technology has grown significantly due to the contribution of modeling, nanotechnologies, environmental concerns, and optimization of spraying processes.

The depletion of natural resources and environmental problems has received much interest in renewable energy sources in recent decades. To obtain energy from renewable energy resources at the best cost, a great deal of research has been done in the area of coating technology. Since coating methods are becoming increasingly important in the generation of renewable energy, their expenditure is significant. Advanced coating methods, including physical vapor deposition (PVD) (Selvakumar and Barshilia, 2012), chemical vapor deposition (CVD) (Wang and Dai, 2002), and simple paint and spray methods (Wijewardane and Goswami, 2012) are employed in renewable energy to accomplish a range of goals, including self-cleaning and more effective spectrum selectivity as well as traditional device protection. This is because thermal spray technology can provide coatings that are corrosion and resistant to oxidation, wear prevention, repair and dimensional restoration, lubricity, thermal insulation, and biocompatibility, among other properties. The possibilities for applying innovative uses for thermal spraying and breaking into new marketplaces are being further expanded by the recent advancements in feedstock materials and thermal spray technologies. The application of thermal spray coatings presents an exciting avenue to augment the performance, resilience, and operational lifespan of key components, ultimately contributing to the viability and widespread adoption of renewable energy solutions. Thermal spray coating is used in solar energy industry to solve problems with photovoltaic module deterioration, wear, and corrosion, and energy conversion efficiency. In wind energy, coating reduces turbine blade stress and erosion, and increase the operating life. This chapter delves into the diverse applications of thermal spray coatings within the renewable energy sector, exploring their transformation impact on energy conversion, structural integrity, and resistance to environmental degradation.

15.2 Thermal spray coating techniques

Thermal spray coating techniques are an agile and adaptable technology that can be applied to almost any component and in a wide range of applications. Its flexibility makes this technology suitable against wear, aggressive, corrosion, and high-temperature environments, and for repair and restoration of components (Espallargas, 2015b). The coating techniques known as thermal spraying include the spraying of heated or melted materials onto a surface. Heat is applied to the "feedstock" (coating precursor) via either chemical or electrical methods (plasma or arc) (combustion flame) and heated or melted materials are sprayed on a substrate/component to form a coating as shown in Fig. 15.2 (Mehta et al., 2017).

Since earlier times, coating techniques have been used to safeguard the substrate material from the environment. Modern coating techniques improve the substrate's and its components' performance in addition to protecting them. Multiple effects are made possible by sophisticated coating processes, such as protection against corrosion and erosion, wear reduction, atmospheric inertness, desired hardness, coating the substrate with the appropriate material, and strengthening the substrate material (Davis,

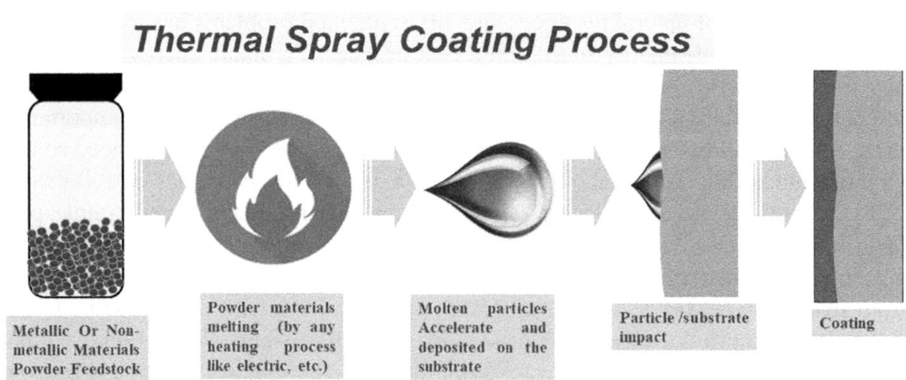

Figure 15.2 Schematic representation of the thermal spray coating process (Mehta et al., 2017).

2004; Talib et al., 2003). Because there are several coating materials in the market, there are numerous procedures that may be used to apply coatings. The thickness and substrate temperature attained during coating, as well as the coating material, determine the variations in all coating properties and functions. These three criteria determine how the coaters' and users' ruling gateways function for their respective purposes. An analogy of the various coating methods in the market based on the temperature of the substrate attained during coating and the thickness of the coating is presented in Fig. 15.3 (Davis, 2004).

Table 15.1, provides the advantages and disadvantages of several coating techniques, including PVD, CVD, Sol-gel, Electroplating, and Thermal Spray.

A layer of the required substance is provided for atmospheric protection or aesthetics in some of the coatings, which are straightforward to create and reasonably priced. It is necessary to coat the material that is subjected to dynamic stress, erosion, wear, and friction so that the coating will withstand the challenging working circumstances. One such coating process that has gained popularity is thermal spray coating, in which high-temperature (up to 1500°C) coating material is sprayed over substrate material, with high-velocity gas or air assisting in the particle flow. Different thermal spray coating processes, such as high-velocity oxyfuel coating (HVOF), plasma spray coating, and arc spray coating, are classified according to the heat source (Dorfman, 2018). Plasma spray coating is the most used method of thermal spray coating, which uses a heat source that heats feedstock materials above 1500°C. The versatility of thermal spray coatings extends to their compatibility with a wide range of materials, including metals, ceramics, polymers, and composites. This adaptability allows engineers and researchers to tailor coatings to meet the specific demands of diverse

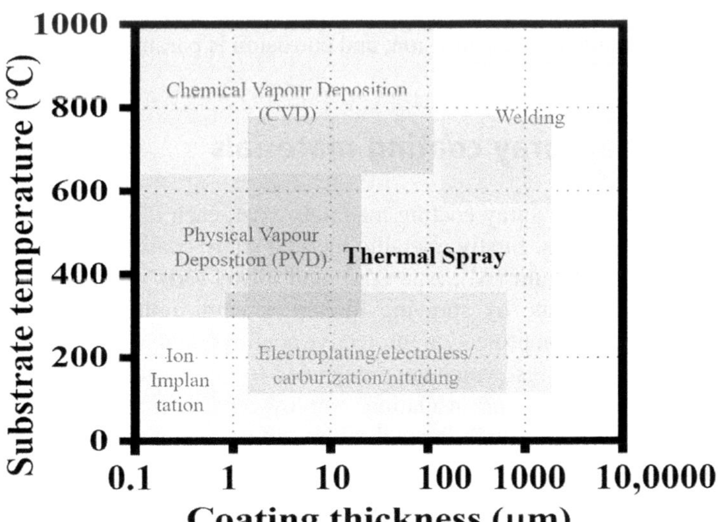

Figure 15.3 Overview of the various market-available coating processes (Davis, 2004).

Table 15.1 Advantages and disadvantages of coating techniques.

Techniques	Advantages	Disadvantages
Thermal spray deposition	• Low price • High rate of deposition • Appropriate to almost all substrate materials • Use several coating materials.	• Coating thicknesses irregularities • The coated specimen has a significant concentration of residual stress • Issues with nanoporous coating.
Electro deposition	• Moderate deposition rate • Coating matrix controllable • Reduced coating porosity	• Caustic waste • Reduced coating strength
Physical vapor deposition (PVD)	• Used to coat complex materials • The dense coating has a thickness at the nanoscale. • Chosen for insulating coating because of poor tribological performance.	• Slow deposition rate • Relatively high price • Poor coating strength
Chemical vapor deposition (CVD)	• The uniform coating on complex shape • Coating complex substances • Coating without porosity • Uniform and controlled thickness	• Restricted to volatile gas use and chemical composition • The extremely low rate of deposition • High initial cost

environments, such as those encountered in renewable energy systems, where resistance to extreme temperatures, abrasion, and corrosion is paramount (Dorfman, 2018).

15.3 Thermal spray coating materials

Several types of thermal spray coating materials exist, each tailored to specific applications and requirements. Firstly, metallic thermal spray coatings commonly utilized metals including aluminum (Al), zinc (Zn), and nickel (Ni), and their alloys. These are applied to a substrate by spraying molten or semi-molten metal particles for wear resistance, corrosion protection, protective coatings, and improved electrical conductivity. Ceramic coatings, which are recognized for their hardness, wear resistance, high-temperature stability, and insulation, employ ceramic materials such as oxides, carbides, and nitrides. The flexibility, adhesion, and corrosion resistance of these coatings are well recognized. Composite coatings are made of a combination of materials, such as metals, ceramics, and polymers, to provide synergistic applications such as increased wear and corrosion resistance and reduced friction. Tungsten carbide and chromium carbide are examples of carbide materials that may be used to coat cutting tools and other industrial items that are subjected to abrasive wear for increased hardness and wear resistance (Moridi et al., 2014).

Coating materials for tribological improvement, such as graphite, Fe, Ni, WS_2, oxides of Al, Fe, and Cr, and various polymeric compounds, have been effectively used (Chen et al., 2023). The majority of tribological testing was conducted in dry regions, with a relatively low number conducted in lubricated conditions. These studies involved the examination of various friction modifiers and antiwear additives. Conventional methods may be used to coat the substrate material; however, the problem with these methods is that the coating may not be stable (comparable to the vapor deposition technique) or the material may experience induced thermal strains that could lead to the substrate material failing mechanically and the hybrid coating method can help with these problems (Tejero-Martin et al.). Coating materials such as iron oxide (Fe_2O_3) and aluminum oxide (Al_2O_3) (Ji et al., 2005), 25 wt% alumina (Al_2O_3) coating (Miguel et al., 2011), ceramic oxides, carbides, and bromides (Torres et al., 2010; Spencer et al., 2009; Wilden et al., 2002) had a distinct impact on the substrate's tribological, chemical, and mechanical properties. Coating materials such as graphite and MoS_2 can be utilized to create antifriction coatings that have improved wear resistance and self-lubricating properties (Domínguez-Meister et al., 2017; Berman et al., 2018). To increase the hardness and wear resistance using chromium oxide, polymeric coating materials such as PAO, PU, and nylon not only improved wear resistance but also significantly decreased the coefficient of friction (Espallargas et al., 2013; Armada et al., 2013; Joshi and Sivakumar, 2015). The effects of different coating techniques and materials on the various substrate material properties is given in Table 15.2.

15.4 Renewable energy technology's big threat

Due to the depletion of fossil fuels and worries about their effects on the environment, renewable energy sources have received a lot of attention in recent decades. The use of coating technologies has emerged as a key participant in the conversion of renewable energy. One feature that makes these technologies so important is that they are reasonably priced. In the renewable energy sector, basic spray and paint techniques, chemical vapor deposition, and physical vapor deposition (PVD) are employed to achieve diverse tasks. Features might range from basic device protection to very sophisticated ones like spectrum selectivity and self-cleaning. The application of protective coatings (using a thermal spray procedure) modifies the surface characteristics of components efficiently and reasonably. In an era of environmental consciousness, renewable energies have taken over as the preferred means of producing energy, even if the gas and oil companies have also effectively utilized thermal spray coatings. While thermal spray coatings have many advanced uses now, their original application in the renewable energy sector was not limited to protecting renewable energy components from corrosion, erosion, and lifespan also used for increasing efficiency. On a more complex level, thermal spray methods can change the characteristics of coatings in applications like plasma and detonation spraying, which causes serious issues. For this reason,

Table 15.2 Effects of different coating techniques and materials/feedstocks on the various substrate material properties.

Researchers	Coating technique	Coating material	Substrate material	Result
(Xie et al., 2024)	Electrodeposited	Ni/Cu/Ni composite	Polyetherimide (PEI) composite substrates	The adhesion force of the Ni/58.6 μm Cu/Ni multilayer on the substrate can be improved by heat treatment at 200°C for 4 hours.
(Wu et al., 2023)	Lasers and micro-arc oxidation	LT/MAO	Titanium alloys	The LT/MAO coating's wear rate and average friction coefficient were lowered by 26% and 28%, respectively, which were 0.262×10^{-6} mm(Siegmann and Abert, 2013)· N^{-1} m^{-1} and 0.23, respectively.
(Varis et al., 2023)	APS, HVOF, and HVAF sprayed	NiCr and NiCrBSi coatings		Not only the stress displacement curve for the coatings, but also the fracture strength of the coatings
(Fefekos et al., 2023)	HVAF sprayed	WC-CoCr and Inconel 625	Aluminum alloy, carbon steel, and Hastelloy-X	The results showed that initial splat characteristics were substantially altered by changing the substrate and the spray angle
(Prasad et al., 2024)	HVOF	WC-12Co/FeNiCrMo composite	Boiler steel substrate	The hardness has increased and the wear resistance is improved compared to the substrate
(Song et al., 2020)	HVOF	Spherical morphology present in WC-CrxCy-Ni and WC-Co-Cr.	410 stainless steel	WC-CrxCy-Ni coating has higher wear concert as compared to WC-Co-Cr coating at high temperatures but inferior wear performance at room temperature at lower loads.
(Abu-warda et al., 2020)	HVOF	Ni20Cr	T24 steel pipes	At 260 μm coating thickness, T24 underwent high-temperature oxidation; the coated sample's microhardness was sluggish at 460 HV, and its adhesive strength was 29 MPa.

(Singh et al., 2019)	HVOF	WC-Co-Cr synergized with Mo2C/Y2O3/ZrO_2	Stainless steel (SDSS) 2507	The substrate material's resistance to erosion and corrosion was improved by the coating.
(Wen et al., 2019)	HVOF	CoNiCrAlY	Mild steel	Compared to the substrate, the coated sample's overall mechanical qualities were superior.
(Pulsford et al., 2019)	HVOAF	WC-10Co-4Cr	416 stainless steels	The material of the substrate became more microhard after coating. Performance of the coated sample in terms of dry tribology. The material of the substrate became more microhard after coating. Performance of the coated sample in terms of dry tribology also improved.
(Rodrigues et al., 2020)	HVOF	Cr_3C_2-25(80Ni20Cr)	316 L SS	In the presence of both water and NaCl, the tribological performance of the coated material improved. Additionally, an increase in wear rate was noted in the presence of NaCl + H_2SO_4 solution.
(Matikainen et al., 2020)	HVOF and HVAF	Cr_3C_2–25NiCr, Cr_3C_2–50NiCrMoNb, Cr_3C_2–37WC–18NiCoCr and WC-10Co-4Cr		HVAF coating has a greater elastic modulus, wear, cavitation, and erosion resistance than HVOF. Elevated Cr and the presence of WC enhanced the resistance to erosion and abrasion.
(Alcover Junior and Pukasiewicz, 2019)	ASP, FS	Babbitt (ASTM B23/2)	Low carbon steel	When comparing coated samples to casted samples, the COF of coated samples was greater in dry circumstances and lower in lubricated conditions. Coated samples have a reduced rate of wear
(Sharma et al., 2020)	APS, HVOF	Cr_3C_2–NiCr; T-800; PS-400	Hastelloy-X and Inconel X-750	Whereas PS400 had the lowest wear rate, T 800 had the highest. That decrease was caused by the existence of a glaze oxide layer. In the instance of PS400, the COF was the lowest

Continued

Table 15.2 Continued

Researchers	Coating technique	Coating material	Substrate material	Result
(Pimentel et al., 2014)	APS, HVOF	Ni60 and WC-17Co	17-4 PH stainless steel	The highest WDE resistance is exhibited by the HVOF-WC17Co coating material, which exhibits 50.6% less shrinkage than that of the substrate material
(Deng et al., 2018)	APS	8YSZ-C/MoS_2	Stainless steel	Friction wear and fragility breakage were the wear mechanisms. Because of their lower porosity, C or MoS_2 blended coatings performed better tribologically than 8YSZ coatings.
(Çelik, 2020)	Sol-gel	B + TiO_2	Ti6Al4V	Because of the enhanced hardness, the COF and wear rate decreased by up to 72% and 68%, respectively

thermal spray is a useful tool in thermal electric power generation, photocatalytic hydrogen production, and thermal photovoltaic systems (Tucker, 1986).

The working environment for components of renewable energy is severe, tigerish, and even hostile. Thus, the renewable energy industry was introduced to thermally spray protective coatings for the lifespan and efficiency improvement of renewable energy components (Wijewardane, 2015).

The protective coatings on wind turbines have a deeper understanding of the construct of thermal sprayed coatings and the renewable energy business. Up to 8000 distinct parts may be found in a wind turbine. Power systems, gearboxes, generators, sensors, electronic control systems, and other components might be included in this. Each of them needs to be able to withstand conditions unique to where it is located, including high temperatures, frequent vibrations, and little opportunity for maintenance (because of its limited accessibility). These parts must be shielded from environmental stressors, but the makers also agreed that no amount of protection could significantly increase weight. However, protective coatings produced by thermal spraying can protect against a variety of environmental factors, including severe winds, seawater, and hot, dry temperatures. Some protective coatings are even resistant to methane and other harmful gases (Gulfam and Zhang, 2019). Thermal spray protective coatings are the lifesaver for components used in renewable energy sources since they are impervious to weather and temperature variations.

15.5 Application of thermal spray coatings for renewable energy sources

15.5.1 Biomass renewable energy

The greatest source of renewable energy at present is bioenergy. It comes to the conversion of biomass into thermal energy, and this tendency is projected to continue. However, the highly reactive species that induce slagging and high-temperature corrosion are responsible for the damage that biomass combustion, when combined with increased recovered fuel (waste) combustion, causes to power plant components (Oksa et al., 2016). Thermal spray coating is a useful technical development that might improve the longevity and efficiency of important parts of biomass systems when used in the context of renewable energy from biomass. The production of biomass energy, which comes from organic resources like wood, agricultural leftovers, and organic waste, requires powerful infrastructure and equipment to endure the demanding conditions involved in energy conversion and combustion. When chemical interactions with the combustion gases and deposits at high temperatures result in the loss of heat exchanger surfaces, such as superheaters, severe fireside corrosion develops. This type of corrosion is particularly common in boilers that burn biomass and recycled fuel. It is frequently brought on by a molten or gaseous assault that is triggered by heavy metal, sulfur, alkali metal (K, Na), bromine, and chlorine compounds (Zn, Pb, Cu, etc.) (Montgomery et al., 2011, 2013; Enestam et al., 2013). Leaks caused by corrosion can force the boiler to shut down, which results in high maintenance and availability costs (Enestam et al., 2013; Oksa et al., 2014; Adamiec, 2009). For corrosion protection with low cost and excellent thermo-mechanical qualities, thermal spray coatings can be applied on low-alloyed tubes instead of higher alloyed steels or alloy tubes based on nickel (Gulfam and Zhang, 2019; Oksa and Metsäjoki, 2015). A protective chromium oxide layer shielded the plasma-sprayed NiCr coating from high-temperature corrosion in deposits that approximated coal-biomass burning circumstances; the effectiveness of the coating was strongly associated with its chromium content (Hussain et al., 2013). Thermal spray coatings provide a layer of defense for parts in biomass combustion chambers, boilers, and gasification units that are subjected to high temperatures, corrosive byproducts, and abrasive conditions. Through the application of thermal spray coatings to critical surfaces, such the walls of combustion chambers or heat exchanger components, engineers can reduce the effects of wear, erosion, and corrosion. This protective layer not only prolongs the lifespan of critical equipment but also contributes to the overall efficiency and reliability of biomass renewable energy systems, making them more resilient and sustainable in the long run.

15.5.2 Solar energy

Solar PV cell efficiency is dependent upon various factors, including module temperature, sunlight spectrum range, installation angle, meteorological conditions, radiation density in that area, buildup of dust or soil, shadow, lifetime/aging, semiconductors, wires, and uncertainties in the manufacturer's rating (Aziz and Ismail, 2015). The capacity of the cover glass to permit radiation absorption or reflection is a significant factor influencing the solar panel's efficiency Also, the accumulation of dust (dust from sources such as pollution, fabric fibers, wind-borne soil particles, smoke particles from moving cars, pedestrian dust, salt stains on water, bird droppings, etc.) reduces the solar system's efficiency since less solar light is absorbed (Rubino et al., 2021). These particles of dust are presented in the environment, and it is difficult to stop them from accumulating. One layer of dust accumulates first, eventually covering the entire surface. Subsequently, the following layer accumulation begins throughout the entire surface, and this process keeps on until accumulations of many layers are formed (Łatka et al., 2020). Because accumulated dust particles scatter and block incoming radiation, it is imperative to clean PV solar panels regularly. Furthermore, the buildup of dust and rain on the glass surface of solar panels results in corrosion, which further reduces the efficiency of the solar cells. When module surfaces are exposed to these materials, solar panel output power and efficiency can decrease by up to 50% (Senthil and Ravi, 2021). Due to their high levels of solar radiation and lack of shadow and cloud cover, hot deserts and semiarid regions are home to most solar power facilities. Nevertheless, dust is a significant issue in these locations as well, so routine cleaning is required. The most popular techniques for cleaning these solar panels involve using detergents and water for washing. Robotic brush cleanings are also utilized in some places. However, these techniques need a lot of work, money, manpower, huge water tanks, pumps, and energy. Solar energy equipment may be shielded with a variety of specialist coatings to shield these elaborate pieces of machinery from harm and dangers. The operational expenses related to cleaning and maintaining solar panels might be reduced by applying coatings to solar arrays and individual panels (Simpson et al., 2015). On the other hand, it has been expected that solar panels covered with superhydrophobic materials can increase their efficiency by up to 91% in comparable conditions. Nanoparticle (NP)-based coatings have been demonstrated to reduce the amount of dust, snow, and ice that may collect and stick to glass surfaces by forming a hydrophobic layer between surfaces (Senthil and Ravi, 2021). Therefore, the application of thermal spray coating in solar energy plays a crucial technological advancement, addressing challenges associated with the durability and performance of solar systems. Thermal spray coatings serve as a protective layer for solar panels, offering defense against corrosive elements and physical wear. These coatings contribute to the resilience of PV modules by enhancing resistance to environmental stressors, thereby extending their operational life. Additionally, thermal spray coatings can be tailored to optimize light absorption and increase energy conversion efficiency, making them instrumental in improving the overall performance of solar energy systems. Whether applied to enhance the durability of solar collectors or improve the efficiency of PV modules, thermal spray coatings play a vital role in advancing the reliability and effectiveness of solar energy technology.

15.5.3 Wind energy

Using massive wind turbines that harness the aerodynamic force of their rotor blades to convert wind energy into electrical power, wind energy is a feasible source of energy. The generation of wind energy is subject to several challenging environmental factors, such as high temperatures, dust, air pollution, hail, and more. The application of thermal spray coating in the domain of wind energy represents a significant advancement in addressing the unique challenges faced by wind turbine components. Thermal spray coating applied through techniques like HVOF spraying or plasma spraying for wind energy assists in the best possible performance from wind turbine parts while protecting against regular environmental problems including wear, corrosion, and high-temperature oxidation (Wind Energy Thermal Spray Coatings, 2023; Mubarok et al., 2013). Also, these coatings enhance the durability and operational lifespan of turbine blades, mitigating the detrimental effects of wind-induced forces. For wind turbines, corrosion-resistant coating is sprayed on the exterior and interior components. It offers several advantages, such as heat management, sealing systems, abrasion reduction, vibration and sound absorption, and component maintenance. For instance, modern thermal spray coatings are used to provide improved protection against assembly damage and mechanical wear and tear on internal bolting fixtures and transmission components in wind energy systems (Momber and Marquardt, 2018). The lifespan of wind turbines and its numerous parts and components is extended by a deep understanding of material technologies and procedures. Additionally, by doing these economical preventive, protective, and restorative actions, maintenance expenses and replacement part costs can be reduced (Valaker et al., 2015). By preserving the aerodynamic integrity of turbine components, the use of thermal spray coatings in wind energy not only protects against wear and tear but also increases the efficiency of energy generation. Therefore, thermal spray coatings play a vital role in ensuring the reliability and longevity of wind turbines, making them indispensable for the sustainable harnessing of wind energy.

15.5.4 Geothermal energy

Geothermal sources are very corrosive and erosive because geothermal energy systems involve components exposed to elevated temperatures, corrosive fluids, and abrasive environments. The reliability of the different components of geothermal power plants is seriously threatened by high temperatures, pressures, and corrosive salts. Due to its low cost and ease of welding, carbon steel—which has poor corrosion resistance—is frequently utilized in geothermal applications (Anand et al., 2024). Geothermal plant components have been coated with Cr_3C_2-based materials in a variety of ways. The investigation was performed in terms of coating characterization to see if the obtained material could be an option for use in a geothermal environment. Thermal spray coatings provide a protective layer on critical infrastructure such as pipes, well casings, and heat exchanger components. This protective coating acts as a barrier against corrosion and erosion, ensuring the integrity and longevity of these components in an aggressive geothermal environment. By applying thermal spray coatings, engineers can enhance the resistance of materials to high temperatures and chemical attacks, ultimately contributing to the overall efficiency and reliability of geothermal energy systems. The versatility of thermal spray coatings makes them a valuable tool in mitigating the deleterious effects of geothermal conditions and ensuring the sustained and efficient operation of geothermal power plants (Barbier, 2002).

15.5.5 Hydroelectric power

Turbine blades in hydroelectric power plants need to be able to withstand the constant, high-speed impact of water (with and without erodent particles), and they also need to have outstanding mechanical qualities and resistance to erosion and corrosion (Aryal et al., 2022a). Chronic exposure to an erosive environment can be detrimental and result in equipment failure because of weight loss from materials being removed off surfaces all the time. One way to decrease the impact of erosion is to apply an erosion-resistant hard surface coating on the turbine's hard surface. Thermal spray coatings offer a protective solution by forming a durable, wear-resistant layer on turbine components such as blades, runners, and casings. These coatings, applied through methods like HVOF or plasma spraying, provide resistance against erosion, corrosion, and cavitation damage. Ceramic pastes, ceramic paints, and strong welding alloys are commonly used in the hydropower industry in Nepal (Mann and Arya, 2001). Tungsten carbide-based ceramics (86WC-10Co-4Cr) have recently been used by hydropower companies in the HVOF process. Such coatings' performance is influenced by spray parameters and particle properties. The HVOF coating was applied to the turbine components of the Kaligandaki "A" Hydropower Plant (KGA HPP). When the plant was first put into service in 2002, the runners (material 13/4 CrNi) were coated with HVOF. However, the face plates and guiding vanes were left uncoated. After a year of operation, sand erosion destroyed the non-coated components (while running time = 7768) (Aryal et al., 2022b). During the same time period, the Cahua power plant's coated turbine increased energy output by 13.1 GWh, about 50% more than the output of the uncoated turbine (Davis, 2004). Due to silt erosion, MB-II (476 MW) turbine components require yearly repair and maintenance costs of about Rs. Eight crore (in Nepalese Rupees). Using a high-velocity oxyfuel technique, a thick coating of tungsten carbide was applied to the turbine components of the MB-II to fight silt erosion (Bhati et al., 2021; Tripathi et al., 2021). Thermal spray coatings contribute to increased operational efficiency and extended service life of hydro turbines. The adaptability of thermal spray coatings to various materials and their capacity to withstand harsh hydraulic conditions make them indispensable for ensuring the reliability and sustainability of hydropower generation.

15.5.6 Photocatalytic hydrogen production

Photocatalytic systems typically involve catalysts that are crucial for the conversion of solar energy into chemical energy, facilitating the splitting of water into hydrogen and oxygen. Improved photocatalytic properties are dependent on any material's ability to absorb light within a certain range, namely the ultraviolet (UV)-visible (Vis)-and near-infrared (NIR) spectrums. The application of thermal spray coating in the realm of photocatalytic hydrogen production represents an innovative approach to enhance the efficiency and durability of key components within the hydrogen generation process. Thermal spray coatings can be applied to these catalyst surfaces to improve their stability, corrosion resistance, and catalytic activity. For instance, the photocatalytic property of TiO_2 (n-type semiconductor coating material) may be activated when exposed to UV light. This characteristic has been the subject of several studies on coatings formed using thermal spray methods. Thermally sprayed photocatalyst semiconducting materials have been developed for a variety of photocatalytic applications; two examples are ZnO and TiO_2 coatings (Faisal et al., 2022). All of these studies have focused more on improving the material's absorption capacity (which may be accomplished by adjusting the thermal spray process's parameters), which may improve sunlight-driven pure water splitting, which separates water into hydrogen and oxygen. Observations of the literature suggest that more research examples focus on using TiO_2 semiconductors as photocatalysts to increase absorption capability (for nonhydrogen production applications, such as self-cleaning surfaces, air, and water purification systems, and sterilization), while there are relatively few examples that directly investigate or characterize water splitting-based hydrogen production. TiO_2 and other semiconductors could be explored and investigated using thermal spray techniques for water-splitting-based hydrogen production, as there has been a lot of work done in the last 10 years to control the phase transformation of feedstock materials during spraying (e.g., APS, HVOF, and cold spray) (Vanka et al., 2022). The adaptability of thermal spray coatings to various catalyst materials and their capacity to withstand the demanding conditions of photocatalysis make them a valuable tool in advancing the reliability and performance of hydrogen production systems.

15.5.7 *Thermoelectric power generation*

Thermal spray coating plays a crucial role in enhancing the performance and efficiency of thermoelectric power generators. Thermoelectric power generators convert heat energy into electrical energy through the Seebeck effect, where a temperature gradient across a thermoelectric material generates a voltage (Longtin et al., 2013; Tregenza et al., 2024).

Thermal spray coatings are utilized in thermoelectric power generators to reduce heat loss and provide insulation, thermoelectric module surfaces can be coated with thermal spray coatings, and temperature differential is improved across the thermoelectric material by insulation, which increases power production efficiency. such as metallic or ceramic coatings, can protect the thermoelectric modules from corrosion caused by exposure to harsh operating environments (Tregenza et al., 2024). Additionally, they provide crucial corrosion protection, ensuring the longevity of generators exposed to challenging environments. By enhancing heat transfer through coatings with high thermal conductivity, the efficiency of power generation is improved. Thermal spray coatings also enable the customization of material properties, allowing for the tailoring of electrical conductivity and other characteristics to optimize overall performance. Furthermore, the application of coatings with wear-resistant properties and the ability to create specific surface textures contribute to increased durability and better control over thermal radiation (Tewolde et al., 2016).

15.6 Conclusion

Applications of thermal spray coating in renewable energy sources have become a significant technical breakthrough, significantly improving the effectiveness, robustness, and overall performance of different components in the renewable sector.

These coatings have the potential to optimize heat transfer, improve thermal insulation, and enhance the overall lifespan of critical components. Thermal spray coatings help maintain the long-term dependability of renewable energy systems by reducing the impacts of wear and corrosion, which eventually supports the sustainability and financial feasibility of clean energy sources. The use of thermal spray coatings in renewable energy applications not only strengthens the resilience of the infrastructure but also highlights the dedication to developing technologies that lead to a cleaner, more sustainable energy landscape as we look toward a future dependent on sustainable energy. Furthermore, as the renewable energy industry grows and expands, there is a growing need for possible solutions to deal with issues like performance optimization and material deterioration. Thermal spray coating research and development will probably continue to be at the forefront of initiatives to increase the durability and efficiency of renewable energy systems, gradually creating a strong and sustainable foundation for the transition to a more sustainable and environmentally friendly future.

Abbreviations

APS	Atmospheric plasma spraying
CVD	Chemical vapor deposition
FS	Flame spray
HVAF	High-velocity air fueling
HVAS	High-velocity-arc-spray
HVOF	High-velocity oxyfuel coating
NIR	Near-infrared
PTFE	Polytetrafluoroethylene
PFOTS	Perfluorooctyltrichlorosilane
PVD	Physical vapor deposition
PEI	Polyetherimide
PTAS	Power Plasma transferred arc spray
PAO	Poly-Alpha-Olefins
PU	Polyurethane
UV-Vis	Ultraviolet visible
YSZ	Yttria-stabilized zirconia

Acknowledgments

The authors thank to Government of Uttar Pradesh for Research grant under Research and Development Scheme [No. 23/2023/680/Sattar-4-2023—21—4(59)/2022] and CCS University, Meerut [DEV/URGS/2022—23/39].

References

Abu-warda, N., López, A.J., López, M.D., Utrilla, M.V., 2020. Ni20Cr coating on T24 steel pipes by HVOF thermal spray for high temperature protection. Surf. Coat. Technol. 381, 125133. https://doi.org/10.1016/j.surfcoat.2019.125133.

Adamiec, J., 2009. High temperature corrosion of power boiler components cladded with nickel alloys. Mater. Char. 60, 1093—1099. https://doi.org/10.1016/j.matchar.2009.03.017.

Alcover Junior, P.R.C., Pukasiewicz, A.G.M., 2019. Evaluation of microstructure, mechanical and tribological properties of a Babbitt alloy deposited by arc and flame spray processes. Tribol. Int. 131, 148—157. https://doi.org/10.1016/j.triboint.2018.10.027.

Amin, S., Panchal, H., 2016. A Review on Thermal Spray Coating Processes. http://www.ijcter.com.

Anand, R.S., Li, A., Huang, W., Chen, J., Li, Z., Ma, Q., Jiang, F., 2024. Super-long gravity heat pipe for geothermal energy exploitation - a comprehensive review. Renew. Sustain. Energy Rev. 193, 114286. https://doi.org/10.1016/j.rser.2024.114286.

Armada, S., Schmid, R., Equey, S., Fagoaga, I., Espallargas, N., 2013. Liquid-solid self-lubricated coatings. J. Therm. Spray Technol. 22, 10—17. https://doi.org/10.1007/s11666-012-9847-x.

Aryal, S., Chitrakar, S., Shrestha, R., kumar Jha, A., 2022a. Coating technology in hydro-turbines for sediment affected power plants: a review. IOP Conf. Ser. Earth Environ. Sci. 1037, 012021. https://doi.org/10.1088/1755-1315/1037/1/012021.

Aryal, S., Chitrakar, S., Shrestha, R., Jha, A.K., 2022b. Coating technology in hydro-Turbines for sediment affected power plants: a review. In: IOP Conf Ser Earth Environ Sci. Institute of Physics. https://doi.org/10.1088/1755-1315/1037/1/012021.

Aziz, F., Ismail, A.F., 2015. Spray coating methods for polymer solar cells fabrication: a review. Mater. Sci. Semicond. Process. 39, 416–425. https://doi.org/10.1016/j.mssp.2015.05.019.

Barbier, E., 2002. Geothermal energy technology and current status: an overview. Renew. Sustain. Energy Rev. 6, 3–65. https://doi.org/10.1016/S1364-0321(02)00002-3.

Berman, D., Narayanan, B., Cherukara, M.J., Sankaranarayanan, S.K.R.S., Erdemir, A., Zinovev, A., Sumant, A.V., 2018. Operando tribochemical formation of onion-like-carbon leads to macroscale superlubricity. Nat. Commun. 9, 1164. https://doi.org/10.1038/s41467-018-03549-6.

Bhati, P., Shrama, V., Jha, R., Gupta, S., Vishnoi, M., Mamatha, T.G., 2021. The coatings developed by thermal spraying technique for hydroturbine applications—a brief review, pp. 559–574. https://doi.org/10.1007/978-981-16-0159-0_50.

Çelik, A., 2020. Improving structural, tribological and electrochemical properties of Ti6Al4V alloy with B-doped TiO_2 thin films. Tribol. Int. 146, 106210. https://doi.org/10.1016/j.triboint.2020.106210.

Chen, Z., Fan, H., Tan, H., Chen, W., Zhu, S., Cheng, J., Zhang, Y., Yang, J., 2023. Effects of graphite contents on the microstructure evolution, mechanical properties and high temperature tribological behavior of Cu–Ni–Al/Gr solid-lubricating composites. Tribol. Int. 179, 108193. https://doi.org/10.1016/j.triboint.2022.108193.

Crawmer, D.E., 2013. Thermal spray processes. In: Thermal Spray Technology. ASM International, pp. 33–53. https://doi.org/10.31399/asm.hb.v05a.a0005718.

Davis, J.R., 2004. & Associates, ASM International Thermal Spray Society Training Committee, Handbook of Thermal Spray Technology. ASM International, Materials Park, OH.

Deng, W., Zhao, X., An, Y., Hao, E., Li, S., Zhou, H., Chen, J., 2018. Improvement of tribological properties of as-sprayed 8YSZ coatings by in-situ synthesis C/MoS2 composite lubricant. Tribol. Int. 128, 260–270. https://doi.org/10.1016/j.triboint.2018.07.038.

Domínguez-Meister, S., Rojas, T.C., Brizuela, M., Sánchez-López, J.C., 2017. Solid lubricant behavior of MoS_2 and WSe_2 -based nanocomposite coatings. Sci. Technol. Adv. Mater. 18, 122–133. https://doi.org/10.1080/14686996.2016.1275784.

Dorfman, M.R., 2018. Thermal spray coatings. In: Handbook of Environmental Degradation of Materials. Elsevier, pp. 469–488. https://doi.org/10.1016/B978-0-323-52472-8.00023-X.

Enestam, S., Backman, R., Mäkelä, K., Hupa, M., 2013. Evaluation of the condensation behavior of lead and zinc in BFB combustion of recovered waste wood. Fuel Process. Technol. 105, 161–169. https://doi.org/10.1016/j.fuproc.2011.09.002.

Espallargas, N., 2015a. Introduction to thermal spray coatings. In: Future Development of Thermal Spray Coatings. Elsevier, pp. 1–13. https://doi.org/10.1016/B978-0-85709-769-9.00001-4.

Espallargas, N., 2015b. Introduction to thermal spray coatings. In: Future Development of Thermal Spray Coatings: Types, Designs, Manufacture and Applications. Elsevier Inc., pp. 1–13. https://doi.org/10.1016/B978-0-85709-769-9.00001-4

Espallargas, N., Vitoux, L., Armada, S., 2013. The wear and lubrication performance of liquid–solid self-lubricated coatings. Surf. Coat. Technol. 235, 342–353. https://doi.org/10.1016/j.surfcoat.2013.07.063.

Faisal, N.H., Prathuru, A., Ahmed, R., Rajendran, V., Hossain, M., Venkatachalapathy, V., Katiyar, N.K., Li, J., Liu, Y., Cai, Q., Horri, B.A., Thanganadar, D., Sodhi, G.S., Patchigolla, K., Fernandez, C., Joshi, S., Govindarajan, S., Kurushina, V., Katikaneni, S., Goel, S., 2022. Application of thermal spray coatings in electrolysers for hydrogen production: advances, challenges, and opportunities. ChemNanoMat 8. https://doi.org/10.1002/cnma.202200384.

Fauchais, P., Montavon, G., 2008. Thermal and cold spray: recent developments. Key Eng. Mater. 384, 1–59. https://doi.org/10.4028/www.scientific.net/KEM.384.1.

Fefekos, A.G., Gupta, M., Mahade, S., Björklund, S., Joshi, S., 2023. Effect of spray angle and substrate material on formation mechanisms and properties of HVAF sprayed coatings. Surf. Coat. Technol. 452, 129115. https://doi.org/10.1016/j.surfcoat.2022.129115.

Gulfam, R., Zhang, P., 2019. Power generation and longevity improvement of renewable energy systems via slippery surfaces – a review. Renew. Energy 143, 922–938. https://doi.org/10.1016/j.renene.2019.05.055.

Hermanek, F.J., 2001. Thermal Spray Society, Thermal Spray Terminology and Company Origins, first ed. ASM International, Materials Park, Ohio.

Hussain, T., Dudziak, T., Simms, N.J., Nicholls, J.R., 2013. Fireside corrosion behavior of HVOF and plasma-sprayed coatings in advanced coal/biomass Co-fired power plants. J. Therm. Spray Technol. 22, 797–807. https://doi.org/10.1007/s11666-013-9887-x.

Ji, G., Elkedim, O., Grosdidier, T., 2005. Deposition and corrosion resistance of HVOF sprayed nanocrystalline iron aluminide coatings. Surf. Coat. Technol. 190, 406–416. https://doi.org/10.1016/j.surfcoat.2004.04.091.

Joshi, S.V., Sivakumar, G., 2015. Hybrid processing with powders and solutions: a novel approach to deposit composite coatings. J. Therm. Spray Technol. 24, 1166–1186. https://doi.org/10.1007/s11666-015-0262-y.

Łatka, L., Pawłowski, L., Winnicki, M., Sokołowski, P., Małachowska, A., Kozerski, S., 2020. Review of functionally graded thermal sprayed coatings. Appl. Sci. 10, 5153. https://doi.org/10.3390/app10155153.

Li, W., Yang, K., Yin, S., Yang, X., Xu, Y., Lupoi, R., 2018. Solid-state additive manufacturing and repairing by cold spraying: a review. J. Mater. Sci. Technol. 34, 440–457. https://doi.org/10.1016/j.jmst.2017.09.015.

Li, W., Assadi, H., Gaertner, F., Yin, S., 2019. A review of advanced composite and nanostructured coatings by solid-state cold spraying process. Crit. Rev. Solid State Mater. Sci. 44, 109–156. https://doi.org/10.1080/10408436.2017.1410778.

Li, W., Cao, C., Yin, S., 2020. Solid-state cold spraying of Ti and its alloys: a literature review. Prog. Mater. Sci. 110, 100633. https://doi.org/10.1016/j.pmatsci.2019.100633.

Li, K., Liu, Z., Xu, J., Liu, H., Xu, S., Wang, C., Qin, J., 2023. Evaluation of high-speed aircraft thermal management system based on spray cooling technology: energy analysis, global cooling, and multi-objective optimization. Appl. Therm. Eng. 229, 120632. https://doi.org/10.1016/j.applthermaleng.2023.120632.

Lombardi, A.N., Casteletti, L.C., Totten, G.E., 2013. Thermal spray technologies: an overview. In: Encyclopedia of Tribology. Springer US, Boston, MA, pp. 3607–3617. https://doi.org/10.1007/978-0-387-92897-5_684.

Longtin, J.P., Zuo, L., Hwang, D., Fu, G., Tewolde, M., Chen, Y., Sampath, S., 2013. Fabrication of thermoelectric devices using thermal spray: application to vehicle exhaust systems. J. Therm. Spray Technol. 22, 577–587. https://doi.org/10.1007/s11666-013-9903-1.

Mann, B.S., Arya, V., 2001. Abrasive and erosive wear characteristics of plasma nitriding and HVOF coatings: their application in hydro turbines.

Matikainen, V., Koivuluoto, H., Vuoristo, P., 2020. A study of Cr3C2-based HVOF- and HVAF-sprayed coatings: abrasion, dry particle erosion and cavitation erosion resistance. Wear 446–447, 203188. https://doi.org/10.1016/j.wear.2020.203188.

Mehta, J., Mittal, V.K., Gupta, P., 2017. Role of thermal spray coatings on wear, erosion and corrosion behavior: a review. J. Appl. Sci. Eng. 20, 445–452. https://doi.org/10.6180/jase.2017.20.4.05.

Miguel, J.M., Vizcaino, S., Lorenzana, C., Cinca, N., Guilemany, J.M., 2011. Tribological behavior of bronze composite coatings obtained by plasma thermal spraying. Tribol. Lett. 42, 263–273. https://doi.org/10.1007/s11249-011-9769-7.

Momber, A.W., Marquardt, T., 2018. Protective coatings for offshore wind energy devices (OWEAs): a review. J Coat Technol Res 15, 13–40. https://doi.org/10.1007/s11998-017-9979-5.

Montgomery, M., Jensen, S.A., Borg, U., Biede, O., Vilhelmsen, T., 2011. Experiences with high temperature corrosion at straw-firing power plants in Denmark. Mater. Corros. 62, 593–605. https://doi.org/10.1002/maco.201005856.

Montgomery, M., Hansson, A.N., Jensen, S.A., Vilhelmsen, T., Nielsen, N.H., 2013. In situ corrosion testing of various nickel alloys at Måbjerg waste incineration plant. Mater. Corros. 64, 14–25. https://doi.org/10.1002/maco.201006019.

Moridi, A., Hassani-Gangaraj, S.M., Guagliano, M., Dao, M., 2014. Cold spray coating: review of material systems and future perspectives. Surf. Eng. 30, 369–395. https://doi.org/10.1179/1743294414Y.0000000270.

Mubarok, F., Armada, S., Fagoaga, I., Espallargas, N., 2013. Thermally sprayed SiC coatings for offshore wind turbine bearing applications. J. Therm. Spray Technol. 22, 1303–1309. https://doi.org/10.1007/s11666-013-9991-y.

Oksa, M., Metsäjoki, J., 2015. Optimizing NiCr and FeCr HVOF coating structures for high temperature corrosion protection applications. J. Therm. Spray Technol. 24, 436–453. https://doi.org/10.1007/s11666-014-0192-0.

Oksa, M., Auerkari, P., Salonen, J., Varis, T., 2014. Nickel-based HVOF coatings promoting high temperature corrosion resistance of biomass-fired power plant boilers. Fuel Process. Technol. 125, 236–245. https://doi.org/10.1016/j.fuproc.2014.04.006.

Oksa, M., Metsäjoki, J., Kärki, J., 2016. Corrosion testing of thermal spray coatings in a biomass Co-firing power plant. Coatings 6, 65. https://doi.org/10.3390/coatings6040065.

Pimentel, J.V., Danek, M., Polcar, T., Cavaleiro, A., 2014. Effect of rough surface patterning on the tribology of W–S–C–Cr self-lubricant coatings. Tribol. Int. 69, 77–83. https://doi.org/10.1016/j.triboint.2013.09.004.

Prasad, C.D., Kollur, S., Aprameya, C.R., Chandramouli, T.V., Jagadeesha, T., Prashanth, B.N., 2024. Investigations on tribological and microstructure characteristics of WC-12Co/FeNiCrMo composite coating by HVOF process. J. Occup. Med. 76, 186–195. https://doi.org/10.1007/s11837-023-06242-2.

Pulsford, J., Venturi, F., Pala, Z., Kamnis, S., Hussain, T., 2019. Application of HVOF WC-Co-Cr coatings on the internal surface of small cylinders: effect of internal diameter on the wear resistance. Wear 432–433, 202965. https://doi.org/10.1016/j.wear.2019.202965.

Rodrigues, I., Nunes, B., Almeida, A., Vilar, R., Figueiredo-Pina, C.G., 2020. Are Cr3C2-25(80Ni20Cr) coatings deposited by HVOF suitable for reciprocating sliding contacts under aqueous media? Tribol. Int. 149, 105825. https://doi.org/10.1016/j.triboint.2019.06.018.

Rubino, F., Poza, P., Pasquino, G., Carlone, P., 2021. Thermal spray processes in concentrating solar power technology. Metals 11, 1377. https://doi.org/10.3390/met11091377.

Selvakumar, N., Barshilia, H.C., 2012. Review of physical vapor deposited (PVD) spectrally selective coatings for mid- and high-temperature solar thermal applications. Sol. Energy Mater. Sol. Cell. 98, 1–23. https://doi.org/10.1016/j.solmat.2011.10.028.

Senthil, S., Ravi, K.R., 2021. A brief review on self-cleaning coatings for photovoltaic systems, pp. 197—234. https://doi.org/10.1007/978-981-16-0594-9_7.

Sharma, A., Sadeghi, F., Sharma, A., 2020. Experimental investigation of fretting wear of coated spring clip and inlet ring in land-based gas turbines at elevated temperature. Wear 446—447, 203200. https://doi.org/10.1016/j.wear.2020.203200.

Siegmann, S., Abert, C., 2013. 100 years of thermal spray: about the inventor Max Ulrich Schoop. Surf. Coating. Technol. 220, 3—13. https://doi.org/10.1016/j.surfcoat.2012.10.034.

Simpson, J.T., Hunter, S.R., Aytug, T., 2015. Superhydrophobic materials and coatings: a review. Rep. Prog. Phys. 78, 086501. https://doi.org/10.1088/0034-4885/78/8/086501.

Singh, J., Kumar, S., Mohapatra, S.K., 2019. An erosion and corrosion study on thermally sprayed WC-Co-Cr powder synergized with Mo2C/Y2O3/ZrO2 feedstock powders. Wear 438—439. https://doi.org/10.1016/j.wear.2019.01.082.

Song, B., Murray, J.W., Wellman, R.G., Pala, Z., Hussain, T., 2020. Dry sliding wear behaviour of HVOF thermal sprayed WC-Co-Cr and WC-CrxCy-Ni coatings. Wear 442—443, 203114. https://doi.org/10.1016/j.wear.2019.203114.

Spencer, K., Fabijanic, D.M., Zhang, M.-X., 2009. The use of Al—Al2O3 cold spray coatings to improve the surface properties of magnesium alloys. Surf. Coat. Technol. 204, 336—344. https://doi.org/10.1016/j.surfcoat.2009.07.032.

Talib, R.J., Saad, S., Toff, M.R.M., Hashim, H., 2003. Thermal spray coating technology-a review.

Tejero-Martin, D., Rad, M.R., Mcdonald, A., Hussain, T., Beyond traditional coatings, a review on thermal sprayed functional and smart coatings.

Tewolde, M., Fu, G., Hwang, D.J., Zuo, L., Sampath, S., Longtin, J.P., 2016. Thermoelectric device fabrication using thermal spray and laser micromachining. J. Therm. Spray Technol. 25, 431—440. https://doi.org/10.1007/s11666-015-0351-y.

Thermal spray for extreme environments. J. Therm. Spray Technol. 28 (7), 2019, 1339—1345. https://doi.org/10.1007/s11666-019-00929-2.

Torres, B., Garrido, M.A., Rico, A., Rodrigo, P., Campo, M., Rams, J., 2010. Wear behaviour of thermal spray Al/SiCp coatings. Wear 268, 828—836. https://doi.org/10.1016/j.wear.2009.12.006.

Tregenza, O., Hutasoit, N., Khalik, M.A., Palanisamy, S., Hulston, C., 2024. An application of cold spray additive manufacturing for thermoelectric generator thermal interfacing. Int. J. Heat Mass Tran. 220, 124996. https://doi.org/10.1016/j.ijheatmasstransfer.2023.124996.

Tripathi, A., Pandey, A., Rathore, A., Singh, A., Vishnoi, M., 2021. Study on tribological behavior of HVOF developed coatings especially for hydroturbine runner application—a concise review, pp. 575—595. https://doi.org/10.1007/978-981-16-0159-0_51.

Tucker, R.C., 1986. Thermal spray coatings: broad and growing applications. Int. J. Powder Metall. 38, 2002.

Valaker, E.A., Armada, S., Wilson, S., 2015. Droplet erosion protection coatings for offshore wind turbine blades. Energy Proc. 80, 263—275. https://doi.org/10.1016/j.egypro.2015.11.430.

Vanka, S., Zeng, G., Deutsch, T.G., Toma, F.M., Mi, Z., 2022. Long-term stability metrics of photoelectrochemical water splitting. Front. Energy Res. 10. https://doi.org/10.3389/fenrg.2022.840140.

Varis, T., Mäkelä, A., Suhonen, T., Laurila, J., Vuoristo, P., 2023. Integrity of APS, HVOF and HVAF sprayed NiCr and NiCrBSi coatings based on the tensile stress-strain response. Surf. Coat. Technol. 452, 129068. https://doi.org/10.1016/j.surfcoat.2022.129068.

Wang, D., Dai, H., 2002. Low-temperature synthesis of single-crystal germanium nanowires by chemical vapor deposition. Angew. Chem. Int. Ed. 41, 4783–4786. https://doi.org/10.1002/anie.200290047.

Wen, W., Jackson, G.A., Li, H., Sun, W., 2019. An experimental and numerical study of a CoNiCrAlY coating using miniature specimen testing techniques. Int. J. Mech. Sci. 157–158, 348–356. https://doi.org/10.1016/j.ijmecsci.2019.04.001.

Wijewardane, S., 2015. Thermal spray coatings in renewable energy applications. In: Future Development of Thermal Spray Coatings: Types, Designs, Manufacture and Applications. Elsevier Inc., pp. 241–257. https://doi.org/10.1016/B978-0-85709-769-9.00010-5

Wijewardane, S., Goswami, D.Y., 2012. A review on surface control of thermal radiation by paints and coatings for new energy applications. Renew. Sustain. Energy Rev. 16, 1863–1873. https://doi.org/10.1016/j.rser.2012.01.046.

Wilden, J., Schnick, T., Wank, A., Schreiber, F., 2002. In: Berndt, C.C., Lugscheider, E. (Eds.), Performance of HVOF Nanostructured Titanium Diboride Composite Coatings, pp. 1033–1037. https://doi.org/10.31399/asm.cp.itsc2002p1033.

Wind Energy Thermal Spray Coatings Common Equipment for Wind Energy Industry Request A Quote HAVE A QUESTION?, 2023. https://www.thermioninc.com/wind-energy.html#:~:text=Wind.

Wu, G., Yin, Y., Zhang, S., Wang, Y., Xiang, Y., Li, L., Yao, J., 2023. Effect of laser texturing on the antiwear properties of micro-arc oxidation coating formed on Ti-6Al-4V. Surf. Coat. Technol. 453, 129114. https://doi.org/10.1016/j.surfcoat.2022.129114.

Xie, D., Wu, W., Huang, J., Wang, X., Zhang, Y., Wang, Z., Jiang, P., Tang, L., Wu, F., Wang, Q., 2024. Effect of electrodeposited Cu interlayer thickness on characterizations and adhesion force of Ni/Cu/Ni coatings on polyetherimide composite substrates. Int. J. Adhesion Adhes. 129, 103580. https://doi.org/10.1016/j.ijadhadh.2023.103580.

Yan, Z., Gainey, B., Gohn, J., Hariharan, D., Saputo, J., Schmidt, C., Caliari, F., Sampath, S., Lawler, B., 2021. A comprehensive experimental investigation of low-temperature combustion with thick thermal barrier coatings. Energy 222, 119954. https://doi.org/10.1016/j.energy.2021.119954.

Ternary metal-based composites for coating in electrochemical reactions

16

Nikhil Thakur[1], *Pawan Kumar*[1], *Sorabh Aggarwal*[2], *Bandna Bharti*[3], *Praveen Vummadisetty Naidu*[4] *and Abhishek Kandwal*[1,5]
[1]School of Physics and Materials Science, Shoolini University, Solan, Himachal Pradesh, India; [2]School of Core Engineering, Shoolini University, Solan, Himachal Pradesh, India; [3]Department of Chemistry, DAV University, Jalandhar, Punjab, India; [4]Interdisciplinary Centre for Security, Reliability and Trust, University of Luxembourg, Luxembourg; [5]School of Chips, Xi'an Jiaotong-Liverpool University, Taicang, Suzhou, China

16.1 Introduction

In the past few years, the world of nanotechnology and nanoscience has experienced a revolutionary transformation, sparking unprecedented advances in diverse fields, from electronics to medicine and renewable energy. The ability to create and manipulate nanostructures has led to a wide range of nanomaterials that are now readily accessible for diverse applications. A crucial approach to improve the efficiency of these nanomaterials involves adjusting their chemical composition, which in turn tunes their intrinsic properties and unlocks novel functionalities (Ben Saber et al., 2020). Nanostructured mono-components, for example, metallic as well as metal oxide nanoparticles, have been widely investigated for their distinctive properties. However, a more intriguing avenue that has gained significant interest among scientists is the development of multi-nanostructure materials, comprising hybrid structures or nanocomposites. These hybrid materials demonstrate an exciting array of additional properties and activities, resulting from the synergistic improvements in the individual nanocomponents (Cortie and McDonagh, 2011). Creating heterogeneous nanomaterials within a single nanocomposite has become an enticing endeavor in recent years, as it offers the potential to harness multifunctional intrinsic properties and generate entirely new characteristics that are absent in the individual nanocomponents (Aljaafari et al., 2019; Ansari et al., 2014; Sajid Ali Ansari and Cho, 2017; Ansari and Cho, 2016; S.A. Ansari and Cho, 2017; Ansari et al., 2015,2019; Sawant et al., 2018). Among these, the noble metal/semiconductor binary nanocomposite system has emerged as a popular choice, primarily owing to its transformative impact on the electronic, magnetic, and optical properties of the constituent materials (Lee et al., 2018). This transformation, in turn, has paved the way for these hybrid materials to be utilized in a wide array of applications, opening new possibilities in various domains (Xu et al., 2010; Elemike et al., 2019). While the noble metal/semiconductor

Multifunctional Nanostructured Coatings. https://doi.org/10.1016/B978-0-443-23683-9.00016-7

binary nanocomposite system holds significant promise, recent progress has led to the development of a novel category of ternary nanostructures, which involves the integration of noble metal/semiconductors/semiconductors or either noble metal/metal/semiconductors (Fang et al., 2013; Kim et al., 2016; Zhuang et al., 2015; Shi et al., 2006; Flomin et al., 2017; S. Li et al., 2018; An et al., 2019; Caudillo-Flores et al., 2020). The incorporation of an additional component in these ternary nanostructures has shown immense potential for further enhancing the performance of nanomaterials, culminating in a surge of newly generated properties that surpass those observed in the binary counterparts. These novel ternary metal-based composites are proving to be highly efficient and showcase unique properties, offering exciting opportunities for various applications. By carefully selecting and combining different metallic elements, researchers can fine-tune properties such as electrical conductivity, catalytic activity, corrosion resistance, and mechanical strength. This makes ternary metal-based composites ideal candidates for coating applications in electrochemical reactions, where coating performance and stability are of paramount importance.

Researchers are extensively studying the electrochemical performance and mechanisms of ternary metal-based coatings to assess their efficiency and suitability for specific electrochemical reactions. Factors such as charge transfer kinetics, catalytic activity, ion diffusion, and stability under various operating conditions are investigated to optimize the design and performance of these composites in coating applications. Considering these developments, the present work aims to delve into the realm of ternary metal-based composites for coating in electrochemical reactions, seeking to unveil their extraordinary potential and uncover the untapped possibilities they hold. Through a comprehensive investigation, we endeavor to shed light on the exciting world of ternary nanomaterials, with the goal of advancing their implementation and furthering the scope of electrochemical applications.

16.2 Design and synthesis of ternary metal-based composites for coating applications

The development of ternary metal-based composites for coating applications in electrochemical reactions requires a multidisciplinary approach, encompassing material selection, fabrication techniques, and desired coating properties. The objective is to create coatings with enhanced functionalities, such as improved electrical conductivity, corrosion resistance, catalytic activity, and mechanical strength, to meet the specific requirements of different electrochemical systems. Choosing the right metallic elements is the first crucial step in the design of ternary metal-based composites. Metals with diverse characteristics, including noble metals, transition metals, and rare earth elements, are selected to achieve a balanced combination of properties. Noble metals like platinum or palladium are incorporated to enhance catalytic activity, while transition metals such as nickel or iron are introduced to improve electrical conductivity. The selection process takes into account factors like cost, availability, and compatibility with the target electrochemical system.

Following the metal component selection, various synthesis methods are employed to fabricate ternary metal-based composite coatings. The hydrothermal synthesis method involves using a high-temperature, high-pressure aqueous environment for creating these composites. Metal precursors are mixed in a solvent and placed in a sealed vessel, allowing for controlled growth of nanoscale materials with specific compositions, sizes, and shapes. Hydrothermal synthesis is extensively employed in electrochemical coatings because of its capacity to generate uniform and highly active materials suitable for a range of electrochemical reactions. For example, Fang et al. (2013) have achieved a remarkable result by ingeniously synthesizing captivating Au@TiO_2−CdS ternary nanostructures (as shown in Fig. 16.1a and b). This pioneering work involved a meticulously crafted hydrothermal process, seamlessly integrating ultrasmall CdS (cadmium sulfide) nanoparticles onto the pure Au@TiO_2 heteronanostructure. They observed that the ternary nanostructures exhibited a significantly high rate of photocatalytic hydrogen production within visible light, as compared to the binary nanostructures (Au@TiO_2 or CdS−TiO_2) alone (Fig. 16.1c). The significant enhancement observed in this study can be attributed to the groundbreaking advancement in ternary nanostructure design. The introduction of this newly developed ternary nanostructure has demonstrated remarkable capabilities in facilitating efficient electron transport from the CdS layer toward the Au core. This seamless electron migration is facilitated through the incorporation of a TiO_2 nanocrystal bridge, which acts as a conduit for the electron flow. As a result of this ingenious configuration, the undesired electron−hole recombination events that typically plague CdS photocatalysts are

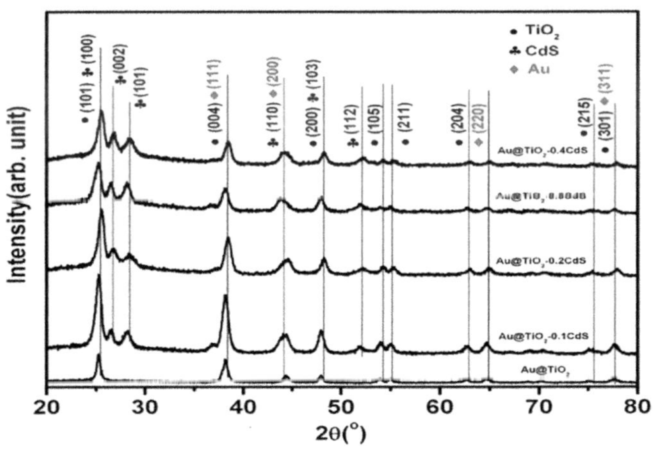

(A)

Figure 16.1 Au@TiO_2−CdS nanocomposites; (a) different XRD patterns; (b) SEM and TEM images; (c) hydrogen evolution results; (d) photocatalytic H_2 production is shown schematically.

Reprinted with permission from J. Fang, L. Xu, Z. Zhang, Y. Yuan, S. Cao, Z. Wang, L. Yin, Y. Liao, C. Xue, Au@TiO2-CdS ternary nanostructures for efficient visible-light-driven hydrogen generation . ACS Appl. Mater. Interfaces. 19448252. 5 (16) (2013), 8088−8092, https://doi.org/10.1021/am4021654 Copyright (2013) by ACS Publications.

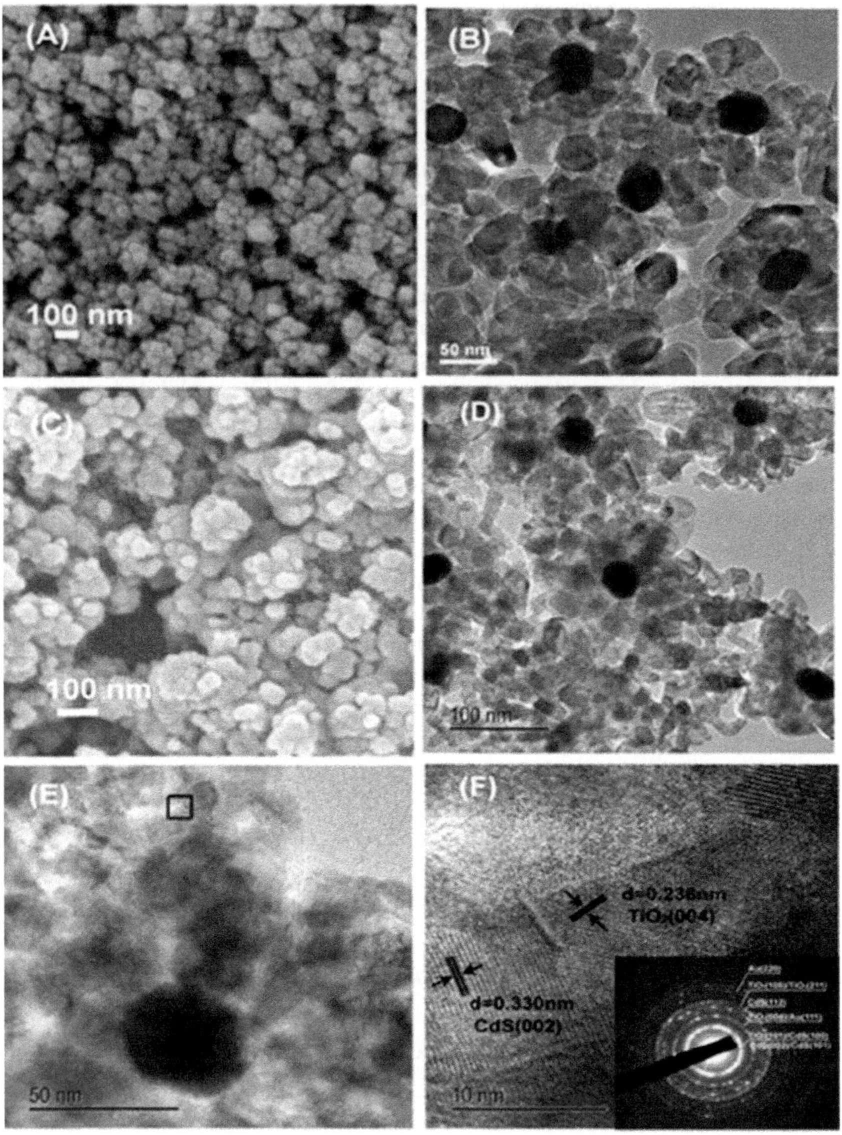

(B)

Figure 16.1 cont'd.

drastically minimized, leading to improved overall performance (Fig. 16.1d). These findings highlight the potential of ternary metal-based composite coatings for enhancing photocatalytic efficiency.

Kaneti et al. (2015) proposed an innovative method for the synthesis of ternary α-Fe_2O_3−ZnO−Au nanocomposites using a simplified hydrothermal process under

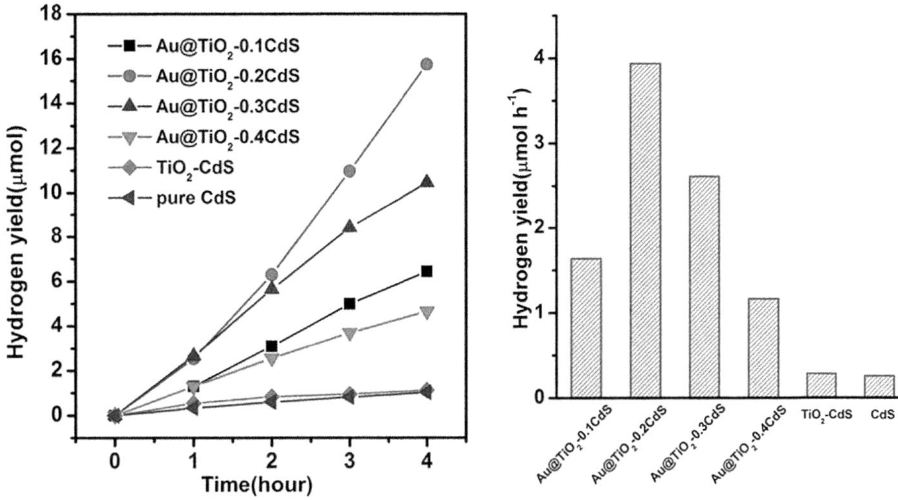

(C)

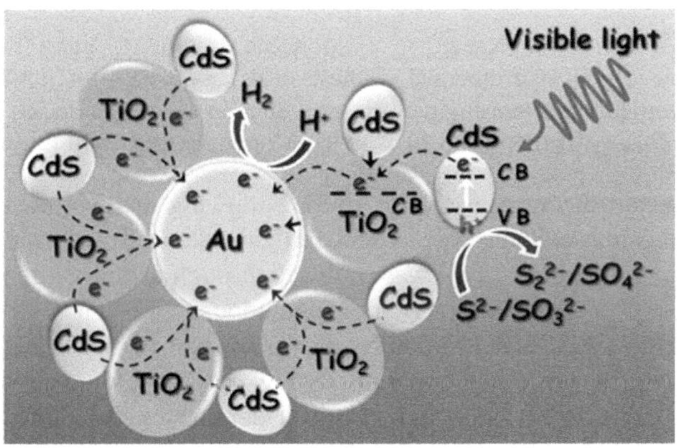

(D)

Figure 16.1 cont'd.

mild circumstances, as illustrated in Fig. 16.2. The ternary α-Fe_2O_3—ZnO—Au system exhibited several noteworthy characteristics: (1) enhanced sensitivity toward acetone and n-butanol in comparison with both single-component α-Fe_2O_3 as well as binary α-Fe_2O_3—ZnO nanomaterials, and (2) a smaller optimal working temperature of approximately 225°C. This study presents a promising advancement in the development of original ternary nanostructures for gas sensing applications, offering enhanced performance and lower operating temperatures.

For gas sensor applications, a novel hybrid ternary nanocomposite comprised of Pt and TiO_2 was presented in a recent study by Luo et al. (Luo and Zhang, 2018) Using a hydrothermal technique, the researchers first created a binary TiO2—MoS_2

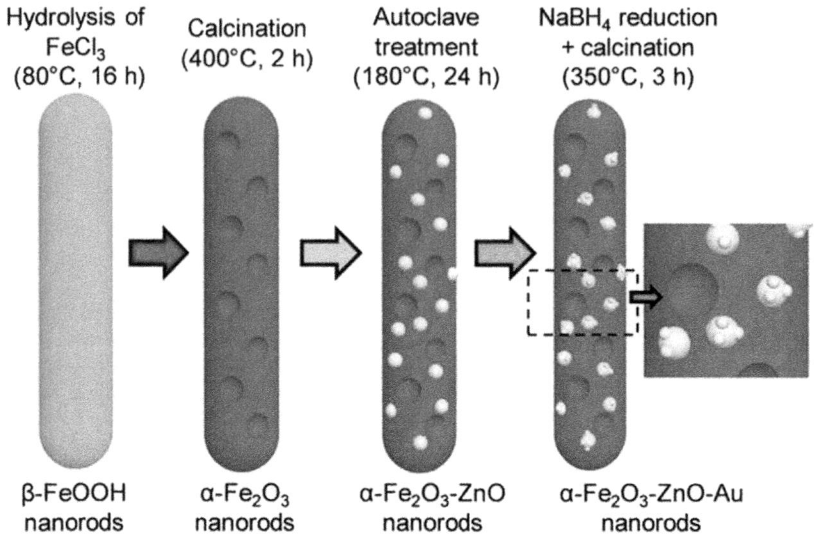

Figure 16.2 Illustration depicting the synthesis process of ternary α-Fe_2O_3−ZnO−Au nanocomposites.
Reprinted with permission from Kaneti, Y. V., Moriceau, J., Liu, M., Yuan, Y., Zakaria, Q., Jiang, X. & Yu, A. (2015). Hydrothermal synthesis of ternary α-Fe_2O_3−ZnO−Au nanocomposites with high gas-sensing performance. Sensors and Actuators, B: Chemical, 209, 889−897. https://doi.org/10.1016/j.snb.2014.12.065 Copyright (2014) by Elsevier.

nanostructure to start the synthesis (Fig. 16.3). Following the synthesis of the binary TiO_2−MoS_2 nanostructure, the researchers further enhanced the nanocomposite by incorporating Pt nanoparticles. This was achieved through a thermal treatment process carried out at 400°C. The resulting ternary nanocomposite, Pt−TiO_2−MoS_2, was then utilized to fabricate a gas sensor. Notably, this gas sensor demonstrated its most significant response to H_2 gas at a comparatively low operating temperature of approximately 100°C. Importantly, the designed sensor demonstrated significant selectivity toward H_2 when compared to other gases. This research presents a promising approach for the development of gas sensors with enhanced performance and selectivity by utilizing ternary nanocomposites.

Zhang et al. (Zhang et al., 2018) developed a highly effective benzene gas sensor utilizing a Pd@TiO_2/MoS_2 nanocomposite, synthesized through a straightforward hydrothermal method (Fig. 16.4). The nanocomposite achieved by them demonstrated exceptional sensing capabilities specifically for benzene detection. It exhibited rapid response times, meaning it quickly reacted to the presence of benzene gas, and fast recovery times, indicating it returned to its initial state promptly after the gas was removed. In addition, the nanocomposite showed excellent repeatability, consistently delivering accurate and reliable results across multiple sensing cycles. The binary TiO_2−MoS_2 system was outperformed by the Pd@TiO_2/MoS_2 nanocomposite in terms of sensing properties. The remarkable sensing capabilities of this ternary nanostructure arise from a harmonious interplay between the Pd@TiO_2/MoS_2 combination

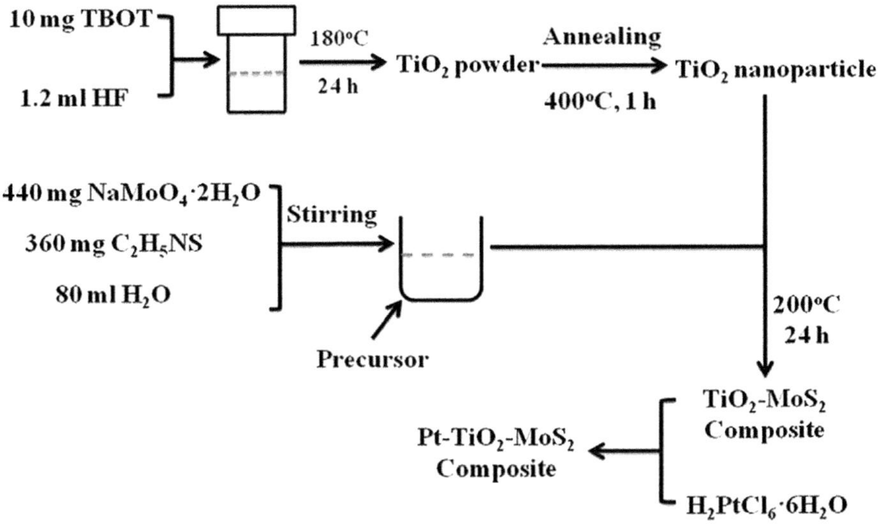

Figure 16.3 Illustration depicting the synthesis process of ternary Pt–TiO_2–MoS_2 nanocomposites.
Reprinted with permission from Luo, Y. & Zhang, C. (2018). Pt-activated TiO_2–MoS_2 nanocomposites for H_2 detection at low temperature. Journal of Alloys and Compounds, 747, 550–557. https://doi.org/10.1016/j.jallcom.2018.03.068 Copyright (2018) by Elsevier.

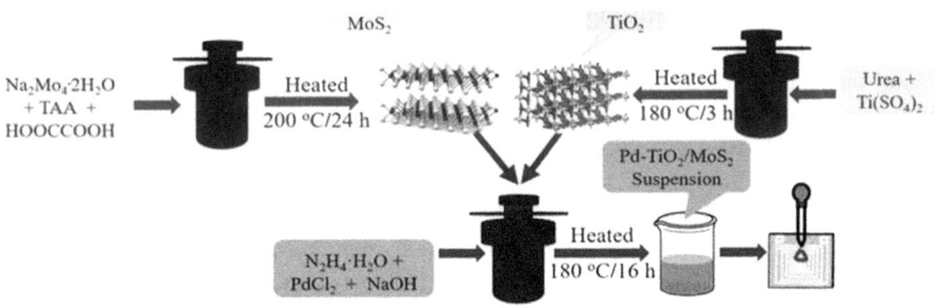

Figure 16.4 Illustration depicting the synthesis process of the Pd–TiO_2/MoS_2 nanocomposite. Reprinted with permission from Zhang, D., Jiang, C. & Zhou, X. (2018). Fabrication of Pd-decorated TiO2/MoS2 ternary nanocomposite for enhanced benzene gas sensing performance at room temperature. Talanta, 182, 324–332. https://doi.org/10.1016/j.talanta.2018.01.064 Copyright (2018) by Elsevier.

and the designed nano-Schottky barrier, both contributing to the enhanced gas sensing capabilities. This research highlights the potential of ternary nanocomposites for advanced gas sensing applications, offering improved performance and selectivity for detecting benzene gas.

Zhang et al. (D. Zhanget al., 2017) successfully engineered a highly efficient hydrogen gas sensor by introducing a novel Pd–SnO_2/MoS_2 ternary nanostructure, synthesized through a precise and reproducible hydrothermal approach (Fig. 16.5).

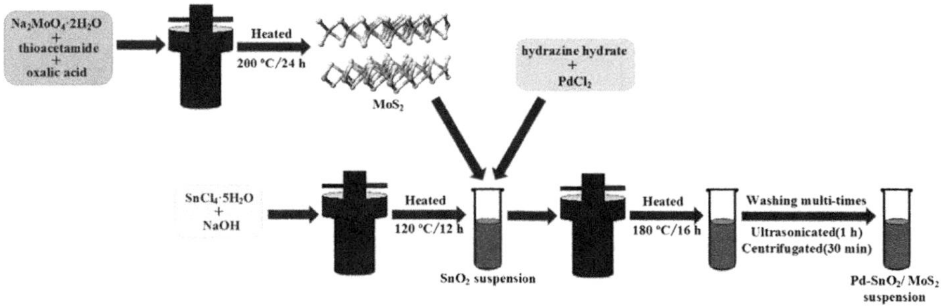

Figure 16.5 Hydrothermal synthesis method for preparing Pd–SnO_2/MoS_2 suspension. Reprinted with permission from Zhang, D., Sun, Y., Jiang, C. & Zhang, Y. (2017). Room temperature hydrogen gas sensor based on palladium decorated tin oxide/molybdenum disulfide ternary hybrid via hydrothermal route. Sensors and Actuators, B: Chemical, 242, 15–24. https://doi.org/10.1016/j.snb.2016.11.005 Copyright (2017) by Elsevier.

The Pd–SnO_2/MoS_2 nanostructure film's gas-sensing characteristics were extensively assessed by subjecting the sensor to various hydrogen gas concentrations at ambient temperature. The preliminary experimental results demonstrated the hydrogen gas sensor's remarkable attributes, revealing its high sensitivity, reactivity, and selectivity toward hydrogen gas. This research effectively highlights the immense potential of the Pd–SnO_2/MoS_2 ternary nanostructure as an efficient sensing material for H_2 gas detection, offering a promising approach for hydrogen sensing applications.

Another technique, the sol–gel process, is a versatile technique for creating ternary metal-based composites. It starts with the conversion of metal alkoxides or metal salts into a sol (colloidal suspension) followed by gelation and drying. This method allows precise control over the composition and thickness of the coatings. The sol–gel method is particularly useful for generating thin films on complex geometries and finds applications in corrosion protection and electrocatalysis. Li et al. (S. Li et al., 2018) effectively fabricated a ternary TiO_2@Pt@CeO_2 nanocomposite through a template/sol–gel approach, which was further subjected to calcination treatment for optimal results (Fig. 16.6). This nanocomposite serves as a multifunctional photocatalyst, simultaneously enhancing both the photo-oxidation of benzyl alcohol as well as the photo-reduction properties of Cr (VI). The incorporation of Pt nanoparticles in the TiO_2@Pt@CeO_2 structure acts as an electron trap site, effectively reducing the recombination of photoexcited electron–hole pairs. Furthermore, the incorporation of Pt nanoparticles within the structure of TiO_2@Pt@CeO_2 hollow nanospheres results in a notable enhancement of the oxygen absorption capacity, addressing and enhancing the main challenge in heterogeneous photocatalysis. This research offers valuable insights into the design and utilization of ternary nanocomposites for enhanced photocatalytic applications.

Hematite (α-Fe_2O_3), a metal oxide frequently employed as a photoanode within semiconductor nanomaterials, possesses an excellent band gap (approximately, 2.10 eV) that makes it possible to absorb visible light effectively. The combination

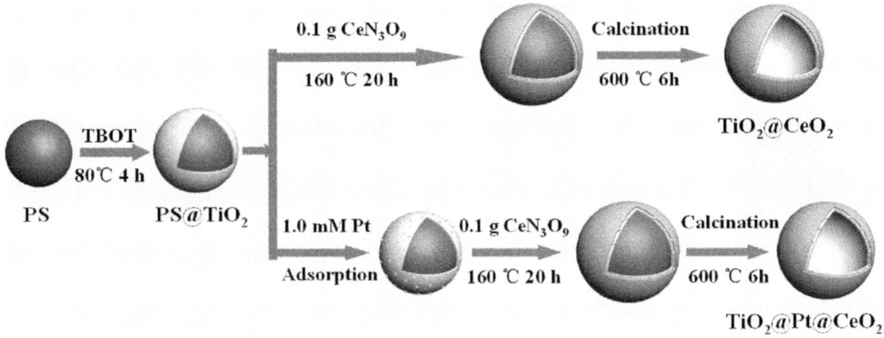

Figure 16.6 Diagram showing the steps involved in creating the ternary TiO_2@Pt@CeO_2 nanocomposite making use of a template/sol−gel technique and calcination treatment. Reprinted with permission from Li, S., Cai, J., Wu, X., Liu, B., Chen, Q., Li, Y. & Zheng, F. (2018). TiO 2 @Pt@CeO 2 nanocomposite as a bifunctional catalyst for enhancing photo-reduction of Cr (VI) and photo-oxidation of benzyl alcohol. Journal of Hazardous Materials, 346, 52−61. https://doi.org/10.1016/j.jhazmat.2017.12.001 Copyright (2018) by Elsevier.

of hematite with noble metals and other metal oxides can lead to the creation of highly efficient hybrid nanocomposites with enhanced photocatalytic performance. By forming a heterojunction, a distinct electric field emerges at the interface between the semiconductor and metal, facilitating an efficient charge separation mechanism and minimizing electron−hole recombination. Notably, Kim et al. (2016) developed a straightforward sol−gel method to fabricate a ternary Au/Fe_2O_3/TiO_2 nanocomposite, where the inclusion of Au or Ag nanoparticles on the Fe_2O_3@TiO_2 surface significantly improves the photocatalytic efficiency by enhancing the charge separation process (Fig. 16.7).

Ternary metal-based composites can also be synthesized using solvothermal methods, which use organic solvents instead of water. Solvothermal synthesis allows for the creation of unique nanocomposite structures, enhancing the electrochemical performance of the coatings. Badhe et al. (Badhe et al., 2018) successfully fabricated ternary Pd@CdS@TiO_2 nanocomposites through the solvothermal method, as illustrated in Fig. 16.8. These nanocomposites exhibited remarkable photocatalytic activity in reducing toxic Cr(VI) to Cr(III) at room temperature, making use of HCOOH as a reducing agent. Photocatalyst performance was further improved by adding Pd nanoparticles to the CdS@TiO_2 binary system. In addition, in comparison with the binary CdS@TiO_2 nanocomposite, there was a decline in electron−hole pair recombination, which raised the photocatalytic reduction in Cr(VI), according to photoluminescence (PL) studies. Notably, HCOOH acted as a scavenging molecule for holes, further promoting the photo-reduction reaction of Cr(VI) to Cr(III). Overall, this study presents a unique and effective approach to address the chromium toxicity issue.

In addition, sonochemical processes, which utilize high-frequency ultrasound, induce chemical reactions in a liquid medium, producing nanomaterials with controlled sizes and shapes. Flores et al. (Caudillo-Flores et al., 2020) employed a

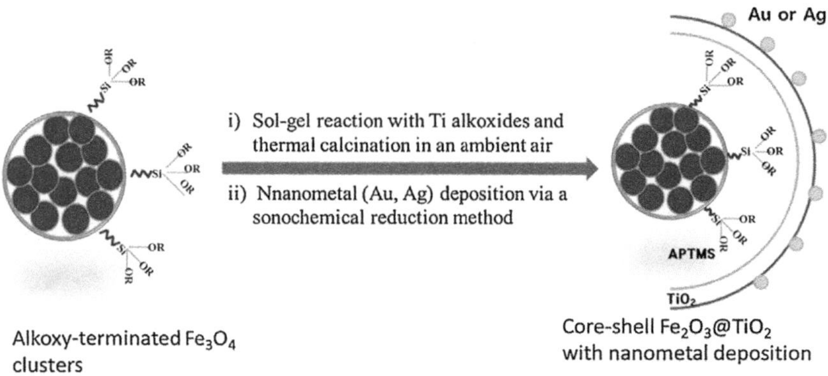

Figure 16.7 Schematic representation of the core-shell nanostructure formation of Fe_2O_3@ TiO_2 with nanometal (Au or Ag) deposition.
Reprinted with permission from Kim, S. E., Woo, J. Y., Kang, S. Y., Min, B. K., Lee, J. K. & Lee, S. W. (2016). A facile general route for ternary Fe2O3@TiO2@nanometal (Au, Ag) composite as a high-performance and recyclable photocatalyst. Journal of Industrial and Engineering Chemistry, 43, 142–149. https://doi.org/10.1016/j.jiec.2016.07.060 Copyright (2016) by Elsevier.

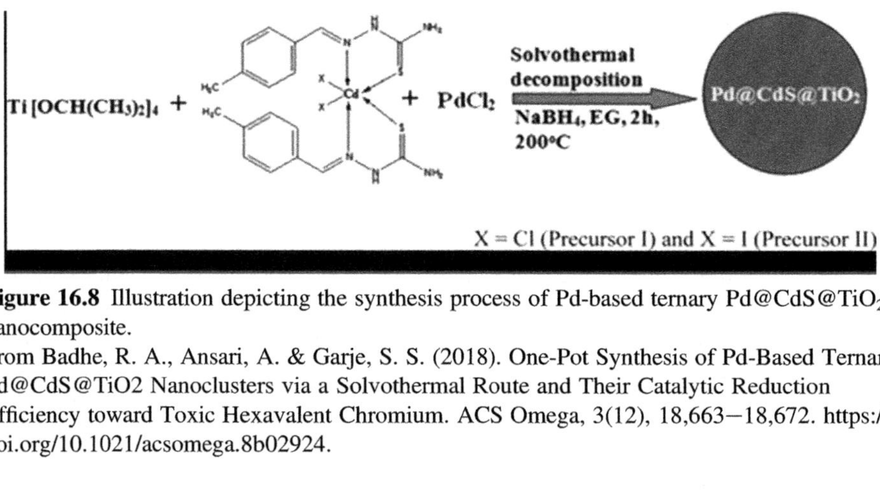

Figure 16.8 Illustration depicting the synthesis process of Pd-based ternary Pd@CdS@TiO_2 nanocomposite.
From Badhe, R. A., Ansari, A. & Garje, S. S. (2018). One-Pot Synthesis of Pd-Based Ternary Pd@CdS@TiO2 Nanoclusters via a Solvothermal Route and Their Catalytic Reduction Efficiency toward Toxic Hexavalent Chromium. ACS Omega, 3(12), 18,663–18,672. https://doi.org/10.1021/acsomega.8b02924.

sonochemical method to successfully synthesize ternary $CuGaS_2$-TiO_2-Pt nanocomposites. The researchers then conducted a comprehensive evaluation of these nanocomposites to assess their capability to generate hydrogen when exposed to solar light excitation. The results obtained from the study of the $CuGaS_2$-TiO_2-Pt nanocomposites were truly remarkable, as they displayed an exceptional level of activity. Specifically, the nanocomposites demonstrated a true quantum activity value of approximately 6.50%. The findings of this study highlight the potential of ternary nanocomposites for efficient hydrogen generation under solar light, opening up new possibilities for renewable energy applications.

Chemical Deposition, also known as chemical vapor deposition (CVD), is a widely used method for depositing thin films or coatings of various materials onto substrates. It is a vapor-phase process that involves the chemical reaction of gaseous precursors to form solid materials on the substrate's surface. A novel ternary core-shell rodlike hybrid nanocomposite, Au-WO_3@TiO_2, was developed by Yang et al. (2020) using the chemical deposition method. This nanocomposite featured gold nanoparticles decorating the surface of WO_3@TiO_2. To assess the photocatalytic performance, they conducted experiments using simulated solar light. They evaluated the nanocomposite's effectiveness by measuring the reduction in methylene blue (MB) concentration over a given period of exposure to the simulated solar light conditions. The experimental findings revealed a substantial enhancement in photocatalytic efficiency for the Au-WO_3@TiO_2 hybrid nanocomposites when compared to the binary WO_3@ TiO_2 counterparts. This original design and synthesis strategy hold promise for enhancing the photocatalytic performance of materials, opening up new opportunities for solar-light-driven applications.

Other techniques such as Layer-by-Layer (LbL) Self-Assembly Process and Facile Colloidal Processes are also employed in the synthesis of ternary metal-based composites. The LbL self-assembly process involves the sequential deposition of alternating positively and negatively charged layers onto a substrate. In the case of ternary metal-based composites, each layer can be composed of different metal precursors or metal oxides. Using the LbL self-assembly method, Zhang et al. (D. Zhanget al., 2017) developed a highly efficient Pt/Co_3O_4/MoS_2 nanocomposites for methane gas sensors (Fig. 16.9). The gas sensing efficiency of the nanocomposite toward methane was evaluated at various operating temperatures, with an optimal temperature of 170°C identified. The experimental results demonstrated that the Pt/Co_3O_4/MoS_2 nanocomposite exhibited remarkable gas sensing performance, surpassing that of the binary Co_3O_4/MoS_2 nanocomposites. The enhanced sensing capability of the designed gas sensor system was attributed to the combination of the large surface area of MoS_2 as well as the enhanced activity provided by the inserted Pt nanoparticles. This research showcases the potential of ternary nanocomposites for developing efficient and selective gas sensors, emphasizing the contribution of both surface area and catalytic activity in achieving high gas sensing efficiency.

Facile colloidal processes encompass various methods that stabilize nanoscale particles or colloids in a solution. These methods can include simple mixing, emulsification, co-precipitation, among others. By combining three different metal colloids or nanoparticles, facile colloidal processes can create ternary metal-based composites. A simple colloidal method was created by Zhuang et al. (2015) to synthesize ultrathin one-dimensional, binary, as well as ternary nanostructures. This method, which involves cation exchange as well as postsynthetic adjustment, presents several benefits over traditional hot-injection chemical processes for preparing hetero-nanostructures. The research findings provide a new pathway for fabricating ternary nanocomposite systems, expanding the possibilities for photoinduced applications. This research significantly advances manufacturing techniques, while simultaneously expanding the horizons of potential applications within the domain of ternary nanocomposites.

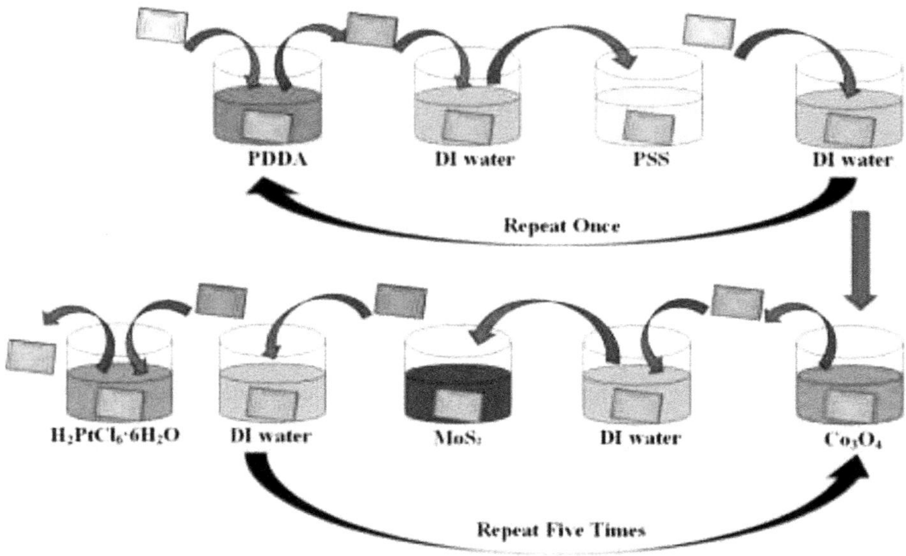

Figure 16.9 Illustration depicting the synthesis process of Pt–Co_3O_4/MoS_2 nanocomposite using LbL self-assembly fabrication.
Reprinted with permission from Zhang, D., Chang, H., Sun, Y., Jiang, C., Yao, Y. & Zhang, Y. (2017). Fabrication of platinum-loaded cobalt oxide/molybdenum disulfide nanocomposite toward methane gas sensing at low temperature. Sensors and Actuators, B: Chemical, 252, 624–632. https://doi.org/10.1016/j.snb.2017.06.063 Copyright (2017) by Elsevier.

The synthesis of ternary metal-based composites for coating applications requires careful characterization and analysis to evaluate their structure, morphology, and properties. Advanced techniques such as scanning electron microscopy (SEM), transmission electron microscopy (TEM), X-ray diffraction (XRD), and energy-dispersive X-ray spectroscopy (EDS) can be employed to assess the microstructure, phase composition, and elemental distribution within the coating.

16.3 Characterization techniques for ternary metal-based composites

Characterization techniques play a crucial role in the evaluation and understanding of the structure, morphology, composition, and properties of ternary metal-based composites used in coating applications. These techniques enable researchers to assess the performance, stability, and functionality of the coatings, providing valuable insights for optimizing their design and fabrication. In this section, some of the key characterization techniques commonly employed for ternary metal-based composites are discussed.

SEM is a widely used technique for studying the microstructure and surface morphology of ternary metal-based composite coatings (Deyab et al., 2022; Ashwini et al., 2021). SEM allows for high-resolution imaging, providing information on the grain size, porosity, and distribution of different components within the coating. By utilizing EDS in conjunction with SEM, elemental mapping and analysis can be performed, enabling the quantification of the composition and distribution of the metallic elements within the composite (Nouseen et al., 2022). In a groundbreaking study by Liu et al. (2018), SEM provided captivating insights into the microstructures of ternary metal-based composites. The examination of pure FeS_2 revealed polyhedral crystals forming agglomerated spheres with smooth surfaces (Fig. 16.10a). Upon introducing Co species, the $Co_{0.33}Fe_{0.67}S_2$ composite showcased a remarkable reduction in polyhedral crystal size (Fig. 16.10b). The graphene/FeS_2 composite displayed irregular particles dispersed within graphene sheets, attributed to graphene's nucleation assistance (Fig. 16.10c). In addition, the GCFS-0.33 composite presented a loose structure with smaller nanoparticles embedded in graphene layers (Fig. Fig. 16.10d). These fascinating results shed light on the intricate relationships between composition and

Figure 16.10 Images obtained with a SEM showing; (a) a pure FeS_2 image; (b) a $Co_{0.33}Fe_{0.67}S_2$ image; (c) a graphene/FeS_2 image; and (d) a GCFS-0.33 composite image.
Reprinted with permission from Liu, W., Niu, H., Yang, J., Cheng, K., Ye, K., Zhu, K., Wang, G., Cao, D. & Yan, J. (2018). Ternary Transition Metal Sulfides Embedded in Graphene Nanosheets as Both the Anode and Cathode for High-Performance Asymmetric Supercapacitors. Chemistry of Materials, 30(3), 1055–1068. https://doi.org/10.1021/acs.chemmater.7b04976 Copyright (2018) by ACS Publications.

morphology, unlocking the potential of these composites for diverse applications in materials science, arousing excitement for future developments in this evolving field.

TEM is another powerful tool for characterizing ternary metal-based composites at the nanoscale (Jo, Sagaya and Selvam, 2003). TEM provides detailed information about the crystal structure, grain boundaries, and defects within the coating. It can also be used to visualize the distribution and size of nanoparticles or other nanostructures embedded in the composite. Electron diffraction techniques, such as selected area electron diffraction (SAED), can be utilized to investigate the crystallographic structure as well as phase composition of the coating. In the fascinating study conducted by Mummoorthi et al. (2022), the remarkable internal morphology and crystal structure of the ternary α-Fe_2O_3/NiO/rGO composite were unveiled using the powerful technique of TEM, depicted vividly in Fig. 16.11. This captivating TEM image revealed the presence of two distinct particles within the composite. The first particle, with an intriguing irregular cube-like morphology, exuded a mysterious darkness and boasted a broader size, indicative of α-Fe_2O_3. The second particle, on the contrary, emitted a striking brightness and showcased a smaller size, characteristic of NiO. The fascinating ternary α-Fe_2O_3/NiO/rGO composite was created by intricately weaving a film of rGO over both components and gracefully enveloping the surfaces. This captivating insight into the composite's internal architecture provides a captivating glimpse into the mesmerizing world of ternary metal-based composites, pushing the boundaries of scientific understanding.

XRD is a commonly employed technique for investigating the phase composition, crystal structure, and crystallinity of ternary metal-based composites (Ramakrishnan

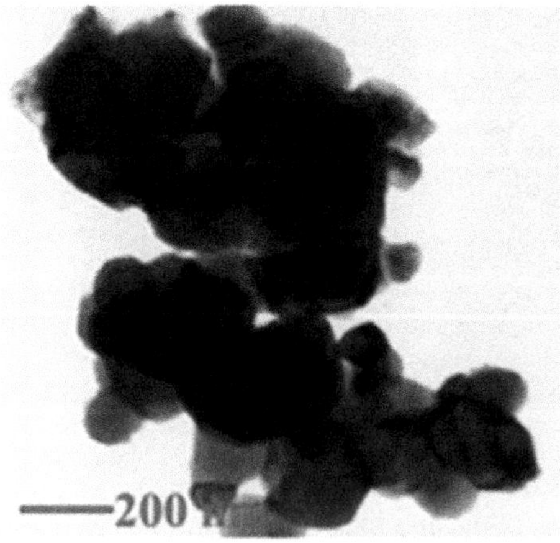

Figure 16.11 TEM image showcasing the synthesized ternary α-Fe_2O_3/NiO/rGO composite sample.

From Mummoorthi, G., Shajahan, S., Abu Haija, M., Mahalingam, U. & Rajendran, R. (2022). Synthesis and Characterization of Ternary α-Fe2O3/NiO/rGO Composite for High-Performance Supercapacitors. ACS Omega, 7(31), 27,390–27,399. https://doi.org/10.1021/acsomega.2c02418

et al., 2021; Liu et al., 2015; Imani et al., 2021; Rahman et al., 2021). By analyzing the diffraction pattern produced by X-rays interacting with the coating, researchers can identify the phases present and determine their relative abundance. XRD is particularly useful for assessing the formation of new phases or identifying any phase transformations that may occur during the synthesis or annealing processes. For example, Nouseen et al. (2022) conducted XRD analysis on ternary microspheres composed of $CuCO_2S_4$, $CuMn_2S_4$, $CuZn_2S_4$, and $CuNi_2S_4$, depicted in Fig. 16.12. The XRD graph for $CuCo_2S_4$ nanoparticles (Fig. 22.12a) exhibited peaks at 2θ values of 31.3degrees, 39.2degrees, 47.3degrees, 50degrees, and 54.9degrees, corresponding to the cubic thiospinel phase, consistent with existing literature (Soni et al., 2019; Chauhan et al., 2019). For $CuZn_2S_4$ microspheres (Fig. 16.12b), the peaks at 2θ values of 28.5degrees, 33.1degree, 47.4degrees, 56.3degrees, and 69.4degrees matched the crystalline planes A(111), A(200), A(220), A(311), and A(400) of face-centered cubic ZnS. In addition, the appearance of peaks at 29.63degrees and 61.57degrees indicated the presence of cubic copper (I) oxide nanocrystals (Srivastava et al., 2020). In the case of $CuMn_2S_4$ microspheres (Fig. 16.12c), the XRD pattern showed peaks at 2θ values of 28.5degrees, 33.1degree, 47.4degrees, and 56.3degrees, which were indexed to the cubic phase α-MnS, along with additional peaks indicating cubic copper (I) oxide nanocrystals (Khan et al., 2016). Similarly, the XRD spectra of $CuNi_2S_4$ (Fig. 16.12d) displayed peaks at 31.4degrees, 38.2degrees, 50.1degree, and

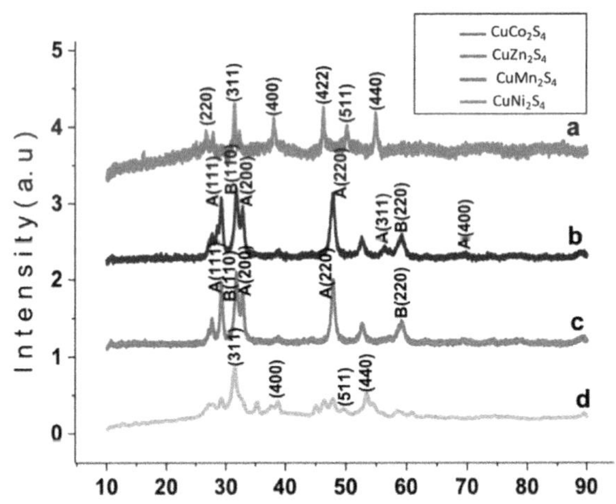

Figure 16.12 (a) $CuCo_2S_4$, (b) $CuZn_2S_4$, (c) $CuMn_2S_4$, and (d) $CuNi_2S_4$ are ternary metal sulfide microspheres that have XRD patterns.

55.1degree, corresponding to the simple cubic phase of $CuNi_2S_4$, confirming the presence of CuS and NiS structures and demonstrating high crystallinity and crystal strength.

Fourier transform infrared spectroscopy (FTIR) provides valuable information about the chemical bonding and functional groups present in ternary metal-based composites (Katowah et al., 2019). By analyzing the absorption and transmission of infrared light, FTIR can identify characteristic vibrational modes associated with different chemical bonds. This technique is useful for confirming the presence of specific coatings or surface modifications, as well as providing insights into the chemical interactions between the coating and the substrate. In the field of ternary-based metal oxide research, Ozkan et al. (2023) utilized ATR(Attenuated Total Reflection) FTIR to demonstrate the dehydrogenation of POA units and formation of C=N bonds through IR heating of PPOA with SWCNTs or MWCNTs, $Co(CH_3CO_2)_2 \cdot 4H_2O$, and $FeCl_3 \cdot 6H_2O$ salts. ATR FTIR spectra of Co-Fe/CNT/PPOA revealed intriguing shifts and broadening in bands at 1587 and 1483 cm^{-1}, which corresponds to νC—C bond stretching in aromatic rings (Fig. 16.13), while absorption bands at 3380 and 3060 cm^{-1} (ν_{N-H} and ν_{C-H} bond stretching in POA units) weakened with increasing synthesis temperature. Similar absorption bands at 869 and 836 cm^{-1} indicated out-of-plane bending vibrations of δ_{C-H} bonds in the 1,2,4-substituted benzene ring, and a new band at 737 cm^{-1} denoted out-of-plane bending vibrations of δ_{C-H} bonds in the 1,2-substituted benzene ring of end groups (Ozkan et al., 2011). Elemental analysis confirmed POA unit dehydrogenation, with CoFe/CNT/PPOA nanocomposites displaying reduced hydrogen content at higher temperatures, resulting in a significant rise in the C/H ratio. The C/H ratio, which once stood at 16.1 for PPOA, now soars to 35.6 and 39.5 for Co-Fe/SWCNT/PPOA and Co-Fe/MWCNT/PPOA, respectively, as marvelously illustrated in Table 16.1. The release of hydrogen facilitated metal

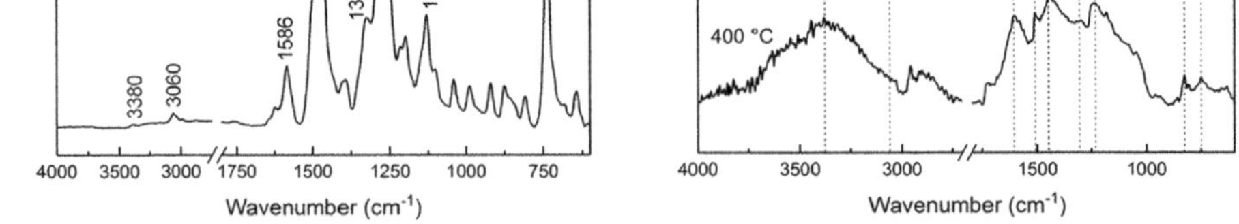

Figure 16.13 ATR FTIR spectra depicting (a) PPOA and (b) Co-Fe/CNT/PPOA at varying synthesis temperatures.
From Ozkan, S., Petrov, V., Vasilev, A., Chernavskii, P., Efimov, M., Muratov, D., Pankina, G. & Karpacheva, G. (2023). Formation Features of Polymer—Metal—Carbon Ternary Electromagnetic Nanocomposites Based on Polyphenoxazine. Polymers, 15.

Table 16.1 ICP-AES data for nanocomposites and information on the PPOA component's elements (Ozkan et al., 2023).

Materials	Cobalt %	Iron%	Carbon%	Nitrogen%	Hydrogen%	Oxygen%	C/N	C/H
PPOA	–	–	78.70	7.70	4.0	8.70	10.20	16.10
Co-Fe/PPOA	11.40	10.30	53.50	5.50	1.80	17.50	9.70	29.70
Co-Fe/SWCNT/PPOA	10.20	7.80	56.90	4.80	1.60	18.70	11.90	35.60
Co-Fe/MWCNT/PPOA	14.50	14.10	55.30	4.30	1.40	10.40	12.90	39.50

From Ozkan, S., Petrov, V., Vasilev, A., Chernavskii, P., Efimov, M., Muratov, D., Pankina, G. & Karpacheva, G. (2023). Formation Features of Polymer–Metal–Carbon Ternary Electromagnetic Nanocomposites Based on Polyphenoxazine. Polymers, 15.

reduction, yielding captivating bimetallic Co-Fe particles, as indicated by reflection peaks in diffraction angles ($2\theta = 69.04$degrees and 106.5degrees).

Raman Spectroscopy is a nondestructive technique that provides information about the molecular structure and composition of ternary metal-based composites (Fechler et al., 2012; Gopal et al., 2022; Abdullah et al., 2020). Raman spectroscopy utilizes the inelastic scattering of laser light to probe the vibrational modes of molecules in the coating. By analyzing the Raman spectrum, researchers can identify the presence of specific compounds, detect structural changes, and assess the level of crystallinity or disorder in the coating. In the realm of ternary-based metal composites, Raman spectroscopy has proven to be a valuable tool for investigating the impact of annealing on the vibrational modes and defect states arising from oxygen-containing functional groups in MoO_2—Ni-graphene composites. Mehmood et al. (2021) conducted an insightful study, as depicted in Fig. 16.14a, wherein they analyzed Raman spectra for various samples, including pristine graphene, Ni-graphene, and composites labeled as S_1, S_2, and S_3. The pristine graphene spectrum revealed three dominant peaks at 1590 cm^{-1}, 1300 cm^{-1}, and 2700 cm^{-1}. The 1590 cm^{-1} peak corresponds to circumferential C—C vibrations (TO phonon) or sp^2 bonding (Quezada-Renteria et al., 2019), while the peaks at 1300 cm^{-1} and 2700 cm^{-1} are attributed to D band as well as 2G bands (Liu et al., 2020), respectively, indicating the presence of defects at the boundary of graphene sheets. The ID/IG ratio, used as a metric for crystal quality, was found to be 0.02 for pristine graphene, implying a low level of defects. However, upon analyzing the Raman modes for S_1, S_2, and S_3 samples, it was observed that the ID/IG ratio increased to 0.94 for S_3, indicating a higher presence of defective sites. In addition, the G-D value decreased to 225 cm^{-1} from 251 cm^{-1} for S_3, further confirming the prevalence of defects on the graphene surface (Fig. 16.14b). These results suggest that the annealing process influences the defect density and crystal quality in

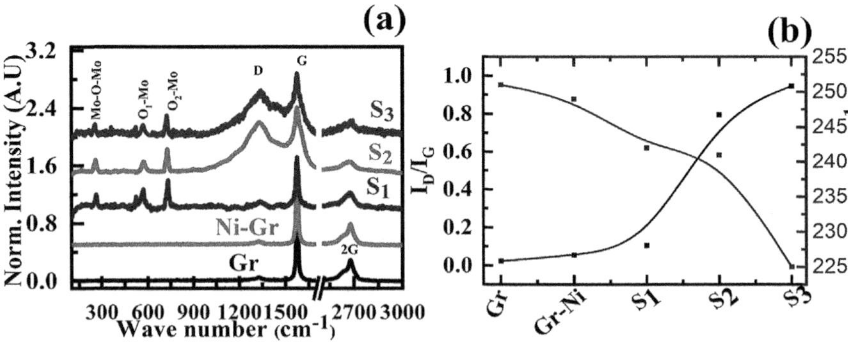

Figure 16.14 Distinct analyses presented: (a) Raman spectra and (b) ID/IG and G-D plot, comparing S_1, S_2, and S_3 samples.

MoO_2–Ni-graphene composites. The insights gained from this investigation shed light on the structural changes occurring in these ternary-based metal composites and provide valuable knowledge for enhancing their performance in various applications. Raman spectroscopy proves to be a crucial tool in characterizing and understanding the intricate properties of such advanced materials.

Electrochemical impedance spectroscopy (EIS) is a powerful electrochemical technique for characterizing the electrochemical behavior and performance of ternary metal-based composite coatings (Lodhi et al., 2022; R. Zhang et al., 2017; Jose et al., 2021). EIS measures the impedance response of the coating to an applied small amplitude AC signal across a range of frequencies. By analyzing the impedance spectrum, researchers can extract information about the coating's charge transfer resistance, capacitance, and other electrochemical properties. EIS is particularly valuable for assessing the corrosion resistance and stability of the coating in corrosive environments.

These are just a few examples of the characterization techniques commonly used for ternary metal-based composites in coating applications. It is important to note that a combination of different techniques is often employed to gain a comprehensive understanding of the coating's structure, morphology, composition, and properties. By utilizing these characterization techniques, researchers can optimize the design and synthesis of ternary metal-based composites and tailor their properties to meet the specific requirements of various electrochemical reactions and coating applications.

16.4 Electrochemical performance and mechanisms of ternary metal-based coatings

The electrochemical performance and mechanisms of ternary metal-based coatings are of critical importance in understanding their behavior and optimizing their functionality in various electrochemical reactions. These coatings are designed to exhibit enhanced performance, such as improved catalytic activity, charge transfer kinetics, and stability, which can significantly impact the efficiency and durability of electrochemical systems. In this section, the electrochemical performance and underlying mechanisms of ternary metal-based coatings were explored.

One key aspect of the electrochemical performance of ternary metal-based coatings is their catalytic activity (Zhou et al., 2021; Shih et al., 2021; Wu et al., 2018). Ternary composites can exhibit synergistic effects due to the combination of different metallic elements, leading to enhanced catalytic properties. For example, the incorporation of noble metals like platinum or palladium in ternary composites can significantly improve the catalytic activity for various electrochemical reactions, including oxygen reduction and hydrogen evolution. The presence of transition metals such as nickel or iron can further enhance the catalytic performance by promoting charge transfer and improving reaction kinetics. The charge transfer kinetics of ternary metal-based coatings are also crucial for their electrochemical performance (Yu et al., 2016; Sarkar et al., 2021; Majumdar, 2021). Coatings with high electrical conductivity and optimized interfacial properties facilitate efficient charge transfer between the electroactive species and the electrode surface. The design of ternary composites allows for tailoring the electrical conductivity by combining metals with different conductivities. In addition, optimizing the microstructure and morphology of the coatings can improve the accessibility of active sites, enhance the electroactive surface area, and reduce charge transfer resistance, thereby enhancing the overall electrochemical performance.

Stability is another essential factor in the electrochemical performance of ternary metal-based coatings (Sun et al., 2020; Wang et al., 2019; Hussain et al., 2022; Li et al., 2014; Gupta et al., 2020). Coatings must exhibit good stability and resistance to degradation during prolonged exposure to harsh electrochemical environments. Ternary composites can provide improved stability by mitigating issues such as corrosion, dissolution, or electrode fouling. The careful selection of metals with enhanced corrosion resistance, along with suitable synthesis techniques to achieve a robust coating structure, contributes to the long-term stability of the coatings. Understanding the underlying mechanisms governing the electrochemical performance of ternary metal-based coatings is crucial for further optimization. Mechanistic studies involve investigating factors such as the reaction pathways, adsorption processes, and surface restructuring occurring at the electrode-electrolyte interface. Techniques such as in-situ electrochemical measurements, surface analysis, and computational modeling play a significant role in unraveling these mechanisms. By elucidating the fundamental processes involved, researchers can develop strategies to enhance the performance of ternary metal-based coatings and overcome limitations associated with their electrochemical behavior.

16.5 Applications of ternary metal-based composites in electrochemical reactions

Ternary metal-based composites have demonstrated significant potential in a wide range of electrochemical reactions, offering enhanced performance, stability, and versatility compared with traditional materials. These composites have found applications in diverse fields, including energy storage, catalysis, corrosion protection, sensors, and electrochemical synthesis. In this section, several prominent applications of ternary metal-based composites in electrochemical reactions were discussed.

One of the primary applications of ternary metal-based composites is in energy storage systems, particularly in rechargeable batteries. These composites can be used as electrode materials, improving the charge storage capacity, cycling stability, and rate capability, offering improvements in energy density and cycle life (Cao et al., 2019; Idrees et al., 2021; L. Li et al., 2018). ZnO is a promising electrode material for lithium-ion batteries (LIBs), with a theoretical capacity of approximately 978 mAh. g^{-1} (Pavitra et al., 2023). It has properties that include a 3.37 eV band gap, high electron mobility, and chemical stability (Zhang et al., 2016). However, ZnO faces challenges like capacity fading and low-rate capability due to volume expansion during cycling (Hsieh et al., 2013). To address this, ternary composites, particularly ZnO-based composites, offer solutions. For example, Ma et al. (2018) fabricated a NiO–ZnO/rGO composite that exhibited a discharge capacity of 1017 mAh. g^{-1} after two hundred cycles and 185 mAh. g^{-1} at a higher current density of 2000 mAh. g^{-1}. Huang et al. (2015) developed a ZnO–NiO/graphene ternary composite with a discharge capacity of 452.7 mAh. g^{-1} after 50 cycles at 300 mAh. g^{-1}. Similarly, He et al. (2019) achieved a discharge capacity of 1000 mAh. g^{-1} with ZnO/GeO_x/C ternary composite nanofibers. These composites enhance the electrochemical performance of ZnO, making it a promising candidate for next-generation batteries. Silicon (SiO_x) is another crucial component for high-capacity anodes in LIBs, with a theoretical capacity of 3579 mAh g^{-1}. However, its widespread use faces challenges such as poor electrical conductivity and volume expansion. Combining SiO_x into ternary composites addresses these issues. Gu et al. (2020) developed SiO_x@SnO_2@C ternary composite microspheres through hydrothermal synthesis, achieving a specific capacity of 796 mAh. g^{-1} at 1000 mAh. g^{-1} after 300 cycles. Hu et al. (2020) synthesized Fe_3O_4@SiO_2@rGO ternary composite anodes, showing a specific capacity of 514 mAh. g^{-1} at 5000 mAh. g^{-1} after 1000 cycles. Wang et al. (2015) created SiO/Ni/graphite ternary composites using a two-step ball-milling method, maintaining a stable capacity of 522 mAh. g^{-1} after 50 cycles at 1000 mAh. g^{-1}. These composites offer solutions to the challenges associated with silicon, making them promising candidates for next-generation batteries.

Fuel cells are another area where ternary metal-based composites have shown promise (Liu et al., 2014; Baruah and Kumar, 2018; Nazal et al., 2018). These composites can serve as catalysts or catalyst supports, enhancing the efficiency of fuel cell reactions, such as oxygen reduction in fuel cell cathodes. By incorporating ternary composites with suitable catalytic properties, fuel cell performance can be significantly improved, leading to higher power output, improved durability, and reduced costs. Ternary metal-based composites have found applications in electrochemical sensors as well (Zhang and Fu, 2019; Arsalan et al., 2021; Turunc, 2022; Lin et al., 2020). The incorporation of specific metal combinations can improve the selectivity, sensitivity, and response time of electrochemical sensors for the detection of various analytes. These sensors have been employed in environmental monitoring, biomedical diagnostics, and industrial process control, enabling rapid and accurate detection of target substances. Electrochemical synthesis is another area where ternary metal-based composites have demonstrated their utility (Li et al., 2022; Iqbal et al., 2021). These composites can act as efficient electrocatalysts, promoting desired electrochemical reactions with high selectivity and yield. By carefully designing the composition and structure of the composites, researchers can achieve enhanced reaction rates and improved product selectivity, facilitating the development of more sustainable and efficient electrochemical synthesis processes.

16.6 Future perspectives and challenges in the advancement of ternary metal-based coatings

The advancement of ternary metal-based coatings for electrochemical applications shows considerable potential, although it also encounters challenges and offers opportunities for future research and progress. In this section, the future perspectives, potential directions, and challenges in the development of ternary metal-based coatings were explored.

One of the future perspectives lies in expanding the range of metal combinations and compositions in ternary composites. While significant progress has been made in exploring various metal combinations, there are still many unexplored possibilities. By considering different metal ratios, alternative metal choices, and novel synthesis techniques, researchers can uncover new ternary composites with tailored properties for specific electrochemical applications. Another future direction involves the integration of ternary metal-based coatings into emerging technologies such as flexible electronics, wearable devices, and energy harvesting systems. These technologies often require coatings with unique properties, including flexibility, stretchability, and self-healing capabilities. By incorporating ternary composites with the desired properties, these coatings can enable the development of advanced devices with enhanced functionality and durability.

The design and optimization of ternary metal-based coatings also require a deeper understanding of the fundamental structure—property relationships. Researchers need to explore the effects of composition, morphology, crystal structure, and interfacial interactions on the coating's performance. Advanced characterization techniques, coupled with theoretical modeling and simulations, can provide insights into the complex phenomena occurring within the coating, enabling the rational design of coatings with improved properties. Another challenge lies in the scale-up and commercialization of ternary metal-based coatings. While promising at the laboratory scale, translating these coatings to industrial applications requires addressing scalability, cost-effectiveness, and manufacturing reproducibility. Researchers need to develop scalable synthesis methods, ensure reliable coating quality, and establish cost-efficient production processes to enable the widespread adoption of ternary metal-based coatings. Furthermore, the long-term stability and durability of ternary metal-based coatings remain critical challenges. Coatings must withstand harsh operating conditions, such as temperature fluctuations, chemical exposure, and mechanical stresses, without degradation. Research efforts should focus on understanding the degradation mechanisms, developing protective strategies, and exploring novel coating architectures or hybrid systems that offer enhanced durability. Environmental sustainability is another important aspect that needs to be considered in the development of ternary metal-based coatings. The choice of materials and fabrication techniques should strive for reduced environmental impact and enhanced recyclability. Green synthesis approaches, eco-friendly materials, and efficient coating processes should be explored to minimize the use of toxic or rare elements and reduce waste generation.

16.7 Conclusion

In conclusion, the design and synthesis of ternary metal-based composites for coating applications in electrochemical reactions present a fascinating and promising field that merges material science, chemistry, and engineering. Through a multidisciplinary approach encompassing material selection, fabrication techniques, and desired coating properties, researchers aim to create coatings with enhanced functionalities tailored to meet the specific requirements of different electrochemical systems. The selection of metallic elements is a critical first step in designing ternary metal-based composites. Metals with diverse characteristics, including noble metals like platinum or palladium, transition metals such as nickel or iron, and rare earth elements, are chosen to achieve a balanced combination of properties such as catalytic activity, electrical conductivity, and corrosion resistance. For example, the combination of Au, TiO_2, and CdS in a ternary nanostructure demonstrated remarkable photocatalytic hydrogen production due to efficient electron transport facilitated by TiO_2.

Various synthesis methods, such as hydrothermal synthesis, are employed to fabricate ternary metal-based composite coatings. This method allows for controlled growth of nanoscale materials with specific compositions, sizes, and shapes, resulting in uniform and highly active coatings suitable for various electrochemical reactions. For example, the hydrothermal synthesis of graphene/FeS_2 composites demonstrated how graphene aids in nucleation and dispersion of FeS_2 particles. Characterization techniques such as SEM, TEM, XRD, and EDS are crucial for evaluating the structure, morphology, and composition of ternary metal-based composites. These techniques provide insights into the microstructure, distribution of components, and relationships between composition and morphology, as demonstrated in various studies.

The electrochemical performance and mechanisms of ternary metal-based coatings are vital for understanding their behavior and optimizing their functionality in electrochemical reactions. These coatings exhibit enhanced catalytic activity, charge transfer kinetics, and stability, contributing to improved efficiency and durability of electrochemical systems. By combining metals with different conductivities and optimizing the coating's microstructure, researchers can tailor the electrical conductivity and enhance the accessibility of active sites, thereby improving overall electrochemical performance. Ternary metal-based composites have shown significant potential in various electrochemical reactions, including energy storage, catalysis, corrosion protection, sensors, and electrochemical synthesis. These composites offer improved performance, stability, and versatility compared with traditional materials, making them promising candidates for a wide range of applications. For example, ZnO-based composites enhance the electrochemical performance of lithium-ion battery electrodes, as demonstrated by studies on NiO−ZnO/rGO and ZnO−NiO/graphene composites.

Future perspectives in the development of ternary metal-based coatings involve exploring new metal combinations, compositions, and synthesis techniques to tailor properties for specific applications. Integration into emerging technologies like flexible electronics and wearable devices presents exciting opportunities, while a deeper understanding of structure−property relationships and scalability challenges must be addressed. Long-term stability, environmental sustainability, and commercialization remain critical areas for future research and development.

In summary, ternary metal-based composites hold great promise for advancing electrochemical technologies, offering tailored properties to meet the evolving demands of various applications. Continued research and innovation in this field will drive the development of advanced coatings with improved performance, durability, and environmental sustainability, paving the way for a brighter future in electrochemical applications. In conclusion, this chapter delves into the comprehensive exploration of ternary metal-based composites, encompassing their design, synthesis, characterization, electrochemical performance, and various applications as coatings in electrochemical reactions. Ternary metal-based composites offer tremendous potential for enhancing the performance, durability, and efficiency of coatings in various electrochemical systems. The introduction highlighted the significance of electrochemical coatings and their role in optimizing the performance of electrochemical reactions. Ternary metal-based composites provide an opportunity to tailor the properties of coatings by carefully selecting and combining metallic elements. The subsequent sections discussed the design and synthesis of ternary metal-based composites, emphasizing the importance of material selection and fabrication techniques. The chapter highlighted electrochemical deposition, sol–gel synthesis, and physical vapor deposition as common methods for synthesizing these composites. Characterization techniques, such as SEM, TEM, XRD, FTIR, and Raman spectroscopy, were then presented as valuable tools for evaluating the structure, morphology, and composition of the coatings. The electrochemical performance and mechanisms of ternary metal-based coatings were thoroughly examined. The catalytic activity, charge transfer kinetics, and stability of these coatings were identified as crucial factors influencing their performance. The chapter emphasized the role of synergistic effects, high electrical conductivity, and optimized microstructure in enhancing the electrochemical behavior of ternary composites. Furthermore, the applications of ternary metal-based composites in electrochemical reactions were discussed. These composites were found to be advantageous in energy storage systems, fuel cells, corrosion protection, electrochemical sensors, and electrochemical synthesis. Their versatility and improved performance have opened up new opportunities for advancements in these fields. Lastly, the chapter presented future perspectives and challenges in the development of ternary metal-based coatings. The need for exploring novel metal combinations, integrating coatings into emerging technologies, understanding structure–property relationships, scaling up production, ensuring long-term stability, and addressing environmental sustainability were identified as key areas for future research.

In conclusion, the study of ternary metal-based composites for coating in electrochemical reactions offers exciting possibilities for enhancing electrochemical performance, optimizing energy storage systems, improving catalytic activity, providing corrosion protection, enabling sensitive sensors, and advancing electrochemical synthesis. By addressing the challenges and leveraging the opportunities, researchers can pave the way for the widespread implementation of ternary metal-based coatings, leading to significant advancements in electrochemical applications and contributing to a more sustainable and efficient future.

AI Disclosure

During the preparation of this work, the author(s) used ChatGpt to This book chapter incorporates artificial intelligence (AI) technologies for the sole purpose of grammar and line conversion. AI has been utilized in the following way: Grammar and Line Conversion: AI-powered tools have been employed exclusively for enhancing the clarity and readability of the text. These tools have been used to correct grammatical errors, improve sentence structure, and format text appropriately. After using this tool/service, the author(s) reviewed and edited the content as needed and take(s) full responsibility for the content of the publication.

References

Abdullah, K.T., Abdulrazzak, F.H., Hussein, F.H., 2020. Raman Spectroscopy & Fourier-Transform Infrared Spectroscopy for Ternary Composites Pt: TiO2/MWCNT.

Aljaafari, A., Parveen, N., Ahmad, F., Alam, M.W., Ansari, S.A., 2019. Self-assembled cube-like copper oxide derived from a metal-organic framework as a high-performance electrochemical supercapacitive electrode material. Nature Publishing Group, Saudi Arabia Sci. Rep. 9 (1). https://doi.org/10.1038/s41598-019-45557-6. Available from:http://www.nature.com/srep/index.html.

An, M., Li, L., Zhang, J., Zhao, L., Yang, C., 2019. Three-dimensional ordered hollow sphere composite Pt/TiO2-ZrO2: enhanced performance of photocatalytic degradation and photocatalytic hydrogen evolution. Springer Netherlands, China J. Nanoparticle Res. 21 (6). https://doi.org/10.1007/s11051-019-4564-8. Available from:https://link.springer.com/journal/11051.

Ansari, S.A., Khan, M.M., Ansari, M.O., Cho, M.H., 2015. Silver nanoparticles and defect-induced visible light photocatalytic and photoelectrochemical performance of Ag@m-TiO2 nanocomposite. Elsevier B.V., South Korea Sol. Energy Mater. Sol. Cell. 141, 162–170. https://doi.org/10.1016/j.solmat.2015.05.029. Available from:http://www.sciencedirect.com/science/journal/09270248/100.

Ansari, S.A., Yadav, H., Adeel, M., Yoo, K., Lee, J.J., 2019. Solvothermal growth of 3D flower-like CoS@FTO as high-performance counter electrode for dye-sensitized solar cell. Springer New York LLC, South Korea J. Mater. Sci. Mater. Electron. 30 (7), 6929–6935. https://doi.org/10.1007/s10854-019-01008-6. Available from:https://rd.springer.com/journal/10854.

Ansari, S.A., Khan, M.M., Ansari, M.O., Lee, J., Cho, M.H., 2014. Visible light-driven photocatalytic and photoelectrochemical studies of Ag-SnO2 nanocomposites synthesized using an electrochemically active biofilm. Royal Society of Chemistry, South Korea RSC Adv. 4 (49), 26013–26021. https://doi.org/10.1039/c4ra03448a. Available from:http://pubs.rsc.org/en/journals/journalissues.

Ali Ansari, S., Cho, M.H., 2017. Growth of three-dimensional flower-like SnS 2 on g-C 3 N 4 sheets as an efficient visible-light photocatalyst, photoelectrode, and electrochemical supercapacitance material. Sustain. Energy Fuels 1 (3), 510–519. https://doi.org/10.1039/C6SE00049E.

Ansari, S.A., Cho, M.H., 2016. Highly visible light responsive, narrow band gap TiO2 nanoparticles modified by elemental red phosphorus for photocatalysis and photoelectrochemical applications. Nature Publishing Group, South Korea Sci. Rep. 6. https://doi.org/10.1038/srep25405. Available from:http://www.nature.com/srep/index.html.

Ansari, S.A., Cho, M.H., 2017. Simple and large scale construction of MoS2-g-C3N4 heterostructures using mechanochemistry for high performance electrochemical supercapacitor and visible light photocatalytic applications. Nature Publishing Group, South Korea Sci. Rep. 7. https://doi.org/10.1038/srep43055. Available from:http://www.nature.com/srep/index.html.

Arsalan, M., Qiao, X., Awais, A., Wang, Y., Yang, S., Sheng, Q., Yue, T., 2021. Enhanced sensitive electrochemical sensor for simultaneous catechol and hydroquinone detection by using ultrasmall ternary Pt-based nanomaterial. John Wiley and Sons Inc, China Electroanalysis 33 (6), 1528–1538. https://doi.org/10.1002/elan.202100026. Available from: http://onlinelibrary.wiley.com/journal/10.1002/(ISSN)1521-4109.

Ashwini, R., Dileepkumar, V.G., Balaji, K.R., Viswanatha, R., Ravikumar, C.R., Srivastava, C., 2021. Mysore Sridhar Santosh, Ternary alkali metal chalcogenide engineered reduced graphene oxide (rGO) as a new class of composite (NaFeS2-rGO) and its electrochemical performance. Sensors International 2, 100125. https://doi.org/10.1016/j.sintl.2021.100125.

Badhe, R.A., Ansari, A., Garje, S.S., 2018. One-Pot synthesis of Pd-based ternary Pd@CdS@TiO2 Nanoclusters via a solvothermal route and their catalytic reduction efficiency toward toxic hexavalent chromium. American Chemical Society, India ACS Omega 3 (12), 18663–18672. https://doi.org/10.1021/acsomega.8b02924. Available from:http://pubs.acs.org/journal/acsodf.

Baruah, B., Kumar, A., 2018. PEDOT:PSS/MnO2/rGO ternary nanocomposite based anode catalyst for enhanced electrocatalytic activity of methanol oxidation for direct methanol fuel cell. Elsevier Ltd, India Synth. Met. 245, 74–86. https://doi.org/10.1016/j.synthmet.2018.08.009. Available from:http://www.journals.elsevier.com/synthetic-metals/.

Ben Saber, N., Mezni, A., Alrooqi, A., Altalhi, T., 2020. A review of ternary nanostructures based noble metal/semiconductor for environmental and renewable energy applications. Elsevier Editora Ltda, Saudi Arabia J. Mater. Res. Technol. 9 (6), 15233–15262. https://doi.org/10.1016/j.jmrt.2020.10.090. Available from:http://www.elsevier.com/journals/journal-of-materials-research-and-technology/2238-7854.

Cao, D., Fan, W., Kang, W., Wang, Y., Sun, K., Zhao, J., Xiao, Z., Sun, D., 2019. Metal–organic framework derived porous hollow ternary sulfide as robust anode material for sodium ion batteries. Elsevier Ltd, China Mater. Today Energy 12, 53–61. https://doi.org/10.1016/j.mtener.2018.12.013. Available from:https://www.journals.elsevier.com/materials-today-energy.

Caudillo-Flores, U., Kubacka, A., Berestok, T., Zhang, T., Llorca, J., Arbiol, J., Cabot, A., Fernández-García, M., 2020. Hydrogen photogeneration using ternary CuGaS2-TiO2-Pt nanocomposites. Elsevier Ltd, Spain Int. J. Hydrogen Energy 45 (3), 1510–1520. https://doi.org/10.1016/j.ijhydene.2019.11.019. Available from:http://www.journals.elsevier.com/international-journal-of-hydrogen-energy/.

Chauhan, M., Soni, K., Karthik, P.E., Reddy, K.P., Gopinath, C.S., Deka, S., 2019. Promising visible-light driven hydrogen production from water on a highly efficient CuCo2S4 nanosheet photocatalyst. Royal Society of Chemistry, India J. Mater. Chem. A 7 (12), 6985–6994. https://doi.org/10.1039/c9ta00391f. Available from:http://pubs.rsc.org/en/journals/journal/ta.

Cortie, M.B., McDonagh, A.M., 2011. Synthesis and optical properties of hybrid and alloy plasmonic nanoparticles. Chem. Rev. 111 (6), 3713–3735. https://doi.org/10.1021/cr1002529.

Deyab, N.M., Taha, M.M., Allam, N.K., 2022. A mesoporous ternary transition metal oxide nanoparticle composite for high-performance asymmetric supercapacitor devices with high specific energy. Royal Society of Chemistry, Egypt Nanoscale Adv. 4 (5), 1387–1393.

https://doi.org/10.1039/d1na00694k. Available from:http://pubs.rsc.org/en/journals/journalissues/na?_ga2.190536939.1555337663.1552312502-1364180372.1550481316#!issueidna001002&typecurrent&issnonline2516-0230.

Elemike, E.E., Onwudiwe, D.C., Wei, L., Chaogang, L., Zhiwei, Z., 2019. Noble metal–semiconductor nanocomposites for optical, energy and electronics applications. Sol. Energy Mater. Sol. Cell. 201.

Fang, J., Xu, L., Zhang, Z., Yuan, Y., Cao, S., Wang, Z., Yin, L., Liao, Y., Xue, C., 2013. Au@TiO2-CdS ternary nanostructures for efficient visible-light-driven hydrogen generation. ACS Appl. Mater. Interfaces 5 (16), 8088–8092. https://doi.org/10.1021/am4021654.

Fechler, N., Fellinger, T.P., Antonietti, M., 2012. Template-free one-pot synthesis of porous binary and ternary metal nitride@N-doped carbon composites from ionic liquids. Chem. Mater. 24 (4), 713–719. https://doi.org/10.1021/cm203667g.

Flomin, K., Diab, M., Mokari, T., 2017. Ternary hybrid nanostructures of Au-CdS-ZnO grown: via a solution-liquid-solid route using Au-ZnO catalysts. Royal Society of Chemistry, Israel Nanoscale 9 (42), 16138–16142. https://doi.org/10.1039/c7nr06382b. Available from: http://www.rsc.org/publishing/journals/NR/Index.asp.

Gopal, T.S., Alzahrani, K.E., Assaifan, A.K., Albrithen, H., Alodhayb, A., Muthuramamoorthy, M., Pandiaraj, S., Grace, A.N., 2022. Reduced graphene oxide supported MXene based metal oxide ternary composite electrodes for non-enzymatic glucose sensor applications. Nature Research, India Sci. Rep. 12 (1). https://doi.org/10.1038/s41598-022-24700-w. Available from:https://www.nature.com/srep/.

Gu, Z., Li, W., Chen, Y., Xia, X., Liu, H., 2020. Synthesis of the microspherical structure of ternary SiO x @SnO 2 @C by a hydrothermal method as the anode for high-performance lithium-ion batteries. Sustain. Energy Fuels 4 (5), 2333–2341. https://doi.org/10.1039/d0se00053a.

Gupta, S., Patel, M.K., Miotello, A., Patel, N., 2020. Metal boride-based catalysts for electrochemical water-splitting: areview. Wiley-VCH Verlag, United Kingdom Adv. Funct. Mater. 30 (1). https://doi.org/10.1002/adfm.201906481. Available from. http://onlinelibrary.wiley.com/journal/10.1002/(ISSN)1616-3028.

He, X., Hu, Y., Chen, R., Shen, Z., Wu, K., Cheng, Z., Peng, P., 2019. Foldable uniform GeOx/ZnO/C composite nanofibers as a high-capacity anode material for flexible lithium ion batteries. Chem. Eng. J. 360, 1020–1029. https://doi.org/10.1016/j.cej.2018.10.163.

Hsieh, C.T., Lin, C.Y., Chen, Y.F., Lin, J.S., 2013. Synthesis of ZnO@Graphene composites as anode materials for lithium ion batteries. Elsevier Ltd, Taiwan Electrochim. Acta 111, 359–365. https://doi.org/10.1016/j.electacta.2013.07.197. Available from:http://www.journals.elsevier.com/electrochimica-acta/.

Hu, Z., Cui, H., Li, J., Lei, G., Li, Z., 2020. Constructing three-dimensional Li-transport channels within the Fe3O4@SiO2@RGO composite to improve its electrochemical performance in Li-ion batteries. Elsevier Ltd, China Ceram. Int. 46 (11), 18868–18877. https://doi.org/10.1016/j.ceramint.2020.04.207. Available from:https://www.journals.elsevier.com/ceramics-international.

Huang, Y., Chen, X., Zhang, K., Feng, X., 2015. Preparation of graphene supported flower-like porous 3D ZnO-NiO ternary composites for high capacity anode materials for Li-ion batteries. Ceram. Int. 41 (10), 13532–13540. https://doi.org/10.1016/j.ceramint.2015.07.147.

Hussain, I., Sahoo, S., Mohapatra, D., Ahmad, M., Iqbal, S., Javed, M.S., Gu, S., Qin, N., Lamiel, C., Zhang, K., 2022. Recent progress in trimetallic/ternary-metal oxides nanostructures: misinterpretation/misconception of electrochemical data and devices. Elsevier

Ltd, Hong Kong Appl. Mater. Today 26. https://doi.org/10.1016/j.apmt.2021.101297. Available from:http://www.journals.elsevier.com/applied-materials-today/.

Idrees, M., Mukhtar, A., ur-Rehman, A.-, Abbas, S.M., Zhang, Q., Li, X., 2021. Transition metal nitride electrodes as future energy storage devices: a review. Elsevier Ltd, China Mater. Today Commun. 27. https://doi.org/10.1016/j.mtcomm.2021.102363. Available from. http://www.journals.elsevier.com/materials-today-communications/.

Imani, A.H., Ojani, R., Raoof, J.B., 2021. Facile synthesis of polyoxometalate-based composite with doped ternary NiCoFe cations as electrocatalyst for oxygen evolution reaction. Elsevier Ltd, Iran Int. J. Hydrogen Energy 46 (1), 449–457. https://doi.org/10.1016/j.ijhydene.2020.09.215. Available from:http://www.journals.elsevier.com/international-journal-of-hydrogen-energy/.

Iqbal, J., Bashir, S., Ansari, M.O., Jafer, R., Jilani, A., Mohammad, S., Ramesh, K., Ramesh, S., 2021. Ternary Nanocomposites for Supercapattery. Elsevier BV, pp. 141–173. https://doi.org/10.1016/b978-0-12-819897-1.00013-6.

Jo, Selvam, W.S.C., 2003. Dalton transactions Synthesis of GO supported Fe 2 O 3 –TiO 2 nanocomposites for enhanced visible-light photocatalytic applications.

Jose, J., Jose, S.P., Prasankumar, T., Shaji, S., Pillai, S., 2021. Emerging ternary nanocomposite of rGO draped palladium oxide/polypyrrole for high performance supercapacitors. Elsevier Ltd, India J. Alloys Compd. 855. https://doi.org/10.1016/j.jallcom.2020.157481. Available from:https://www.journals.elsevier.com/journal-of-alloys-and-compounds.

Kaneti, Y.V., Moriceau, J., Liu, M., Yuan, Y., Zakaria, Q., Jiang, X., Yu, A., 2015. Hydrothermal synthesis of ternary α-Fe2O3-ZnO-Au nanocomposites with high gas-sensing performance. Elsevier B.V., Australia Sens. Actuators, B 209, 889–897. https://doi.org/10.1016/j.snb.2014.12.065. Available from:https://www.journals.elsevier.com/sensors-and-actuators-b-chemical.

Katowah, D.F., Rahman, M.M., Hussein, M.A., Sobahi, T.R., Gabal, M.A., Alam, M.M., Asiri, A.M., 2019. Ternary nanocomposite based poly(pyrrole-co-O-toluidine), cobalt ferrite and decorated chitosan as a selective Co2+ cationic sensor. Elsevier Ltd, Saudi Arabia Compos. B Eng. 175. https://doi.org/10.1016/j.compositesb.2019.107175. Available from:https://www.journals.elsevier.com/composites-part-b-engineering.

Khan, A., Rashid, A., Younas, R., Ren, C., 2016. A chemical reduction approach to the synthesis of copper nanoparticles. Int. Nano Lett. 6 (1), 21–26. https://doi.org/10.1007/s40089-015-0163-6.

Kim, S.E., Woo, J.Y., Kang, S.Y., Min, B.K., Lee, J.K., Lee, S.W., 2016. A facile general route for ternary Fe2O3@TiO2@nanometal (Au, Ag) composite as a high-performance and recyclable photocatalyst. Korean Society of Industrial Engineering Chemistry, South Korea J. Ind. Eng. Chem. 43, 142–149. https://doi.org/10.1016/j.jiec.2016.07.060. Available from:http:www.sciencedirect.com/science/journal/1226086X.

Lee, Y., Kim, E., Park, Y., Kim, J., Ryu, W.H., Rho, J., Kim, K., 2018. Photodeposited metal-semiconductor nanocomposites and their applications. Chinese Ceramic Society, South Korea Journal of Materiomics 4 (2), 83–94. https://doi.org/10.1016/j.jmat.2018.01.004. Available from:https://www.journals.elsevier.com/journal-of-materiomics/.

Li, H., Han, X., Zhao, W., Azhar, A., Jeong, S., Jeong, D., Na, J., Wang, S., Yu, J., Yamauchi, Y., 2022. Electrochemical preparation of nano/micron structure transition metal-based catalysts for the oxygen evolution reaction. Royal Society of Chemistry, China Mater. Horiz. 9 (7), 1788–1824. https://doi.org/10.1039/d2mh00075j. Available from: http://pubs.rsc.org/en/journals/journal/mh.

Li, L., Zheng, Y., Zhang, S., Yang, J., Shao, Z., Guo, Z., 2018. Recent progress on sodium ion batteries: potential high-performance anodes. Royal Society of Chemistry, China Energy

Environ. Sci. 11 (9), 2310–2340. https://doi.org/10.1039/c8ee01023d. Available from: http://pubs.rsc.org/en/journals/journal/ee.

Li, L., Zhang, Y., Shi, F., Zhang, Y., Zhang, J., Gu, C., Wang, X., Tu, J., 2014. Spinel manganese-nickel-cobalt ternary oxide nanowire array for high-performance electrochemical capacitor applications. American Chemical Society, China ACS Appl. Mater. Interfaces 6 (20), 18040–18047. https://doi.org/10.1021/am5048653. Available from: http://pubs.acs.org/journal/aamick.

Li, S., Cai, J., Wu, X., Liu, B., Chen, Q., Li, Y., Zheng, F., 2018. TiO 2 @Pt@CeO 2 nanocomposite as a bifunctional catalyst for enhancing photo-reduction of Cr (VI) and photo-oxidation of benzyl alcohol. Elsevier B.V., China J. Hazard Mater. 346, 52–61. https://doi.org/10.1016/j.jhazmat.2017.12.001. Available from:http://www.elsevier.com/locate/jhazmat.

Lin, L., Miao, C., Weng, S., Ying, S., Chen, F., Ye, C., Zhuang, H., You, D., 2020. Ternary Pt–Au–FeOOH-decorated polyaniline nanocomposite for sensitive dopamine electrochemical detection. J. Electroanal. Chem. 877.

Liu, H., Feng, Y., Chen, D., Li, C., Cui, P., Yang, J., 2015. Noble metal-based composite nanomaterials fabricated via solution-based approaches. Royal Society of Chemistry, China J. Mater. Chem. A 3 (7), 3182–3223. https://doi.org/10.1039/c4ta05801a. Available from: http://pubs.rsc.org/en/journals/journalissues/ta.

Liu, M., Zhang, R., Chen, W., 2014. Graphene-supported nanoelectrocatalysts for fuel cells: synthesis, properties, and applications. American Chemical Society, China Chem. Rev. 114 (10), 5117–5160. https://doi.org/10.1021/cr400523y. Available from:http://pubs.acs.org/journal/chreay.

Liu, W., Niu, H., Yang, J., Cheng, K., Ye, K., Zhu, K., Wang, G., Cao, D., Yan, J., 2018. Ternary transition metal sulfides embedded in graphene Nanosheets as both the anode and Cathode for high-performance asymmetric supercapacitors. American Chemical Society, China Chem. Mater. 30 (3), 1055–1068. https://doi.org/10.1021/acs.chemmater.7b04976. Available from:http://pubs.acs.org/journal/cmatex.

Liu, Y., Feng, H., Luo, F., 2020. Quantitative analysis of the defects in CVD grown graphene by plasmon-enhanced Raman scattering. Elsevier Ltd, Spain Carbon 161, 153–161. https://doi.org/10.1016/j.carbon.2020.01.063. Available from:http://www.journals.elsevier.com/carbon/.

Lodhi, A.S., Iqbal, N., Noor, T., Zaman, N., Gao, J., 2022. ZIF -67 derived ternary NiMnCo -based nanoporous carbon material for methanol oxidation reaction. Int. J. Energy Res. 46 (12), 16736–16750. https://doi.org/10.1002/er.8335.

Luo, Y., Zhang, C., 2018. Pt-activated TiO2-MoS2 nanocomposites for H2 detection at low temperature. Elsevier Ltd, China J. Alloys Compd. 747, 550–557. https://doi.org/10.1016/j.jallcom.2018.03.068. Available from:

Ma, L., Pei, X.Y., Mo, D.C., Lyu, S.S., Fu, Y.X., 2018. Fabrication of NiO-ZnO/RGO composite as an anode material for lithium-ion batteries. Ceram. Int. 44 (18), 22664–22670. https://doi.org/10.1016/j.ceramint.2018.09.044.

Majumdar, D., 2021. Review on current progress of MnO2-based ternary nanocomposites for supercapacitor applications. Wiley-VCH Verlag, India Chemelectrochem 8 (2), 291–336. https://doi.org/10.1002/celc.202001371. Available from:http://onlinelibrary.wiley.com/journal/10.1002/(ISSN)2196-0216.

Mehmood, S., Zhao, X., Bhopal, M.F., Khan, F.U., Yang, Y., Wang, G., Pan, X., 2021. MoO2-Ni-graphene ternary nanocomposite for a high- performance room-temperature ethanol gas sensor. Appl. Surf. Sci. 554, 149595. https://doi.org/10.1016/j.apsusc.2021.149595.

Mummoorthi, G., Shajahan, S., Abu Haija, M., Mahalingam, U., Rajendran, R., 2022. Synthesis and characterization of ternary α-Fe2O3/NiO/rGO composite for high-performance supercapacitors. American Chemical Society, India ACS Omega 7 (31), 27390–27399. https://doi.org/10.1021/acsomega.2c02418. Available from:http://pubs.acs.org/journal/acsodf.

Nazal, M.K., Olakunle, O.S., Al-Ahmed, A., Merzougui, B., Abualkibash, A., Sultan, A., Yousaf, A.B., Zaidi, S.J., 2018. Precious metal free Ni/Cu/Mo trimetallic nanocomposite supported on multi-walled carbon nanotubes as highly efficient and durable anode-catalyst for alkaline direct methanol fuel cells. J. Electroanal. Chem. 823, 98–105. https://doi.org/10.1016/j.jelechem.2018.05.035.

Nouseen, S., Singh, P., Lavate, S., Chattopadhyay, J., Kuchkaev, A.M., Yakhvarov, D.G., Srivastava, R., 2022. Transition metal based ternary hierarchical metal sulphide microspheres as electrocatalyst for splitting of water into hydrogen and oxygen fuel. Elsevier B.V., India Catal. Today 397–399, 618–630. https://doi.org/10.1016/j.cattod.2021.05.019. Available from:http://www.sciencedirect.com/science/journal/09205861.

Ozkan, S., Petrov, V., Vasilev, A., Chernavskii, P., Efimov, M., Muratov, D., Pankina, G., Karpacheva, G., 2023. Formation features of polymer–metal–carbon ternary electromagnetic nanocomposites based on polyphenoxazine. Polymers 15.

Ozkan, S.Z., Karpacheva, G.P., Bondarenko, G.N., Topchiev, A.V., 2011. Phenoxazine polymers: synthesis and structure. Russ. Chem. Bull. 60 (8), 1651–1656. https://doi.org/10.1007/s11172-011-0247-z.

Pavitra, V., Soni, I., Praveen, B.M., Nagaraju, G., 2023. Brief review on carbon derivatives based ternary metal oxide composite electrode materials for lithium-ion batteries. International Association of Physical Chemists, India J. Electrochem. Sci. Eng. 13 (4), 605–640. https://doi.org/10.5599/jese.1470. Available from:https://pub.iapchem.org/ojs/index.php/JESE/article/download/1470/1793.

Quezada-Renteria, J.A., Ania, C.O., Chazaro-Ruiz, L.F., Rangel-Mendez, J.R., 2019. Influence of protons on reduction degree and defect formation in electrochemically reduced graphene oxide. Elsevier Ltd, Mexico Carbon 149, 722–732. https://doi.org/10.1016/j.carbon.2019.04.109. Available from:http://www.journals.elsevier.com/carbon/.

Rahman, M.U., Qazi, U.Y., Hussain, T., Nadeem, N., Zahid, M., Bhatti, H.N., Shahid, I., 2021. Solar driven photocatalytic degradation potential of novel graphitic carbon nitride based nano zero-valent iron doped bismuth ferrite ternary composite. Elsevier B.V., Pakistan Opt. Mater. 120. https://doi.org/10.1016/j.optmat.2021.111408. Available from:https://www.journals.elsevier.com/optical-materials.

Ramakrishnan, P., Lee, K.B., Sohn, J.I., 2021. Ternary metal-based inverse spinel oxide NiCrFeO4 nanoparticles as a highly efficient oxygen evolution catalyst. Elsevier B.V., South Korea Appl. Surf. Sci. 566. https://doi.org/10.1016/j.apsusc.2021.150653. Available from:http://www.journals.elsevier.com/applied-surface-science/.

Sarkar, S., Rawat, A., Das, T., Gaboardi, M., Chakraborty, S., Vinod, C.P., Peter, S.C., 2021. Structure-tailored non-noble metal-based ternary chalcogenide nanocrystals for Pt-like electrocatalytic hydrogen production. John Wiley and Sons Inc, India ChemSusChem 14 (15), 3074–3083. https://doi.org/10.1002/cssc.202100967. Available from:http://onlinelibrary.wiley.com/journal/10.1002/(ISSN)1864-564X.

Sawant, S.Y., Kim, J.Y., Han, T.H., Ansari, S.A., Cho, M.H., 2018. Electrochemically active biofilm-assisted biogenic synthesis of an Ag-decorated ZnO@C core-shell ternary plasmonic photocatalyst with enhanced visible-photocatalytic activity. Royal Society of Chemistry, South Korea New J. Chem. 42 (3), 1995–2005. https://doi.org/10.1039/c7nj03936k. Available from:http://pubs.rsc.org/en/journals/journal/nj.

Shi, W., Sahoo, Y., Zeng, H., Ding, Y., Swihart, M.T., Prasad, P.N., 2006. Anisotropic growth of PbSe nanocrystals on Au-Fe3O4 hybrid nanoparticles. Adv. Mater. 18 (14), 1889–1894. https://doi.org/10.1002/adma.200600685.

Shih, K.Y., Wei, J.J., Tsai, M.C., 2021. One-step microwave-assisted synthesis of PtNiCo/rGO electrocatalysts with high electrochemical performance for direct methanol fuel cells. MDPI AG, Taiwan Nanomaterials 11 (9). https://doi.org/10.3390/nano11092206. Available from. https://www.mdpi.com/2079-4991/11/9/2206/pdf.

Soni, K., Chauhan, M., Deka, S., 2019. Hydrothermally synthesized CuCo2S4 Nanosheets as an easily accessible and convenient heterogeneous catalyst for the sonogashira cross-coupling reactions. Frontiers Media S.A., India Frontiers in Materials 6. https://doi.org/10.3389/fmats.2019.00273. Available from:http://journal.frontiersin.org/journal/materials.

Srivastava, R., Chattopadhyay, J., Patel, R., Agrawal, S., Nouseen, S., Kumar, S., Karmakar, S., 2020. Highly efficient ternary hierarchical NiV2S4 nanosphere as hydrogen evolving electrocatalyst. Elsevier Ltd, India Int. J. Hydrogen Energy 45 (41), 21308–21318. https://doi.org/10.1016/j.ijhydene.2020.05.174. Available from:http://www.journals.elsevier.com/international-journal-of-hydrogen-energy/.

Sun, H., Yan, Z., Liu, F., Xu, W., Cheng, F., Chen, J., 2020. Self-Supported transition-metal-based electrocatalysts for hydrogen and oxygen evolution. Wiley-VCH Verlag, China Adv. Mater. 32 (3). https://doi.org/10.1002/adma.201806326. Available from:http://onlinelibrary.wiley.com/journal/10.1002/(ISSN)1521-4095.

Turunc, E., 2022. Synthesis, characterization and sensing application of HNTs@Pd-MoS2 ternary composite. Mater. Today Commun. 31, 103555. https://doi.org/10.1016/j.mtcomm.2022.103555.

Wang, J., Bao, W., Ma, L., Tan, G., Su, Y., Chen, S., Wu, F., Lu, J., Amine, K., 2015. Scalable preparation of ternary hierarchical silicon oxide-nickel-graphite composites for lithium-ion batteries. Wiley-VCH Verlag, China ChemSusChem 8 (23), 4073–4080. https://doi.org/10.1002/cssc.201500674. Available from:http://www.interscience.wiley.com/jpages/1864-5631.

Wang, T., Chen, H.C., Yu, F., Zhao, X.S., Wang, H., 2019. Boosting the cycling stability of transition metal compounds-based supercapacitors. Elsevier B.V., Australia Energy Storage Mater. 16, 545–573. https://doi.org/10.1016/j.ensm.2018.09.007. Available from: http://www.journals.elsevier.com/energy-storage-materials/.

Wu, M., Wang, Y., Wei, Z., Wang, L., Zhuo, M., Zhang, J., Han, X., Ma, J., 2018. Ternary doped porous carbon nanofibers with excellent ORR and OER performance for zinc-air batteries. Royal Society of Chemistry, China J. Mater. Chem. A 6 (23), 10918–10925. https://doi.org/10.1039/c8ta02416b. Available from:http://pubs.rsc.org/en/journals/journal/ta.

Xu, C., Kang, F., Li, B., Du, H., 2010. Recent progress on manganese dioxide based supercapacitors. J. Mater. Res. 25 (8), 1421–1432. https://doi.org/10.1557/jmr.2010.0211.

Yang, X., Fu, H., Wang, W., Xiong, S., Han, D., Deng, Z., An, X., 2020. Enhanced solar light photocatalytic performance based on a novel Au-WO3@TiO2 ternary core–shell nanostructures. Appl. Surf. Sci. 505, 144631. https://doi.org/10.1016/j.apsusc.2019.144631.

Yu, J., Li, Q., Li, Y., Xu, C.Y., Zhen, L., Dravid, V.P., Wu, J., 2016. Ternary metal phosphide with triple-layered structure as a low-cost and efficient electrocatalyst for bifunctional water splitting. Wiley-VCH Verlag, United States Adv. Funct. Mater. 26 (42), 7644–7651. https://doi.org/10.1002/adfm.201603727. Available from:http://onlinelibrary.wiley.com/journal/10.1002/(ISSN)1616-3028.

Zhang, D., Chang, H., Sun, Y., Jiang, C., Yao, Y., Zhang, Y., 2017. Fabrication of platinum-loaded cobalt oxide/molybdenum disulfide nanocomposite toward methane gas sensing at low temperature. Sens. Actuators, B 252, 624–632. https://doi.org/10.1016/j.snb.2017.06.063.

Zhang, D., Sun, Y., Jiang, C., Zhang, Y., 2017. Room temperature hydrogen gas sensor based on palladium decorated tin oxide/molybdenum disulfide ternary hybrid via hydrothermal route. Sens. Actuators, B 242, 15–24. https://doi.org/10.1016/j.snb.2016.11.005.

Zhang, D., Jiang, C., Zhou, X., 2018. Fabrication of Pd-decorated TiO2/MoS2 ternary nanocomposite for enhanced benzene gas sensing performance at room temperature. Elsevier B.V., China Talanta 182, 324–332. https://doi.org/10.1016/j.talanta.2018.01.064. Available from:

Zhang, J., Gu, P., Xu, J., Xue, H., Pang, H., 2016. High performance of electrochemical lithium storage batteries: ZnO-based nanomaterials for lithium-ion and lithium-sulfur batteries. Royal Society of Chemistry, China Nanoscale 8 (44), 18578–18595. https://doi.org/10.1039/c6nr07207k. Available from:http://www.rsc.org/publishing/journals/NR/Index.asp.

Zhang, M., Fu, D., 2019. An electrochemical sensor for dopamine detection based on ternary Pd-Au-P composites supported on PDDA/RGO. Electrochemical Science Group, China Int. J. Electrochem. Sci. 14 (10), 9909–9920. https://doi.org/10.20964/2019.10.51. Available from:http://www.electrochemsci.org/ESG.htm.

Zhang, R., Wang, D., Qin, L.C., Wen, G., Pan, H., Zhang, Y., Tian, N., Zhou, Y., Huang, X., 2017. MnCO3/Mn3O4/reduced graphene oxide ternary anode materials for lithium-ion batteries: facile green synthesis and enhanced electrochemical performance. Royal Society of Chemistry, China J. Mater. Chem. A 5 (32), 17001–17011. https://doi.org/10.1039/c7ta02874a. Available from:http://pubs.rsc.org/en/journals/journalissues/ta.

Zhou, X., Zi, Y., Xu, L., Li, T., Yang, J., Tang, J., 2021. Core-shell-structured prussian blue analogues ternary metal phosphides as efficient bifunctional electrocatalysts for OER and HER. American Chemical Society, China Inorg. Chem. 60 (15), 11661–11671. https://doi.org/10.1021/acs.inorgchem.1c01694. Available from:http://pubs.acs.org/journal/inocaj.

Zhuang, T.-T., Liu, Y., Sun, M., Jiang, S.-L., Zhang, M.-W., Wang, X.-C., Zhang, Q., Jiang, J., Yu, S.-H., 2015. A unique ternary semiconductor-(semiconductor/metal) nano-architecture for efficient photocatalytic hydrogen evolution. Angew. Chem. 127 (39), 11657–11662. https://doi.org/10.1002/ange.201505442.

Potential of conjugated polymers in nanostructured coatings for energy applications

17

Maria Federica De Riccardis
Italian National Agency for New Technologies, Energy and Sustainable Economic Development, Research Center of Brindisi, Brindisi, Italy

17.1 Introduction

Conjugated polymers have emerged as efficient materials for energy and electronics applications due to their unique combination of the electronic properties of metals and semiconductors with the mechanical plasticity of conventional polymers. Conjugated polymers belong to the broad class of conducting polymers and can be used in typical applications of conducting or semiconducting materials, such as shielding against electromagnetic interference, protective coatings against metal corrosion, light-emitting diodes, transistors, or photovoltaic solar cells. Their electrical activity is essentially due to the conjugation of π electrons along the polymer structure of carbon. In fact, the structure of conjugated polymers is characterized by alternating single and double bonds along the polymer backbone (conjugated bonds) with a delocalized π electron cloud over the backbone chain, or by aromatic rings such as those linked by carbon-carbon single bonds (Fig. 17.1). Linear conjugated polymers (e.g., polythiophene, polypyrrole, and polyaniline) have been intensively studied as active materials

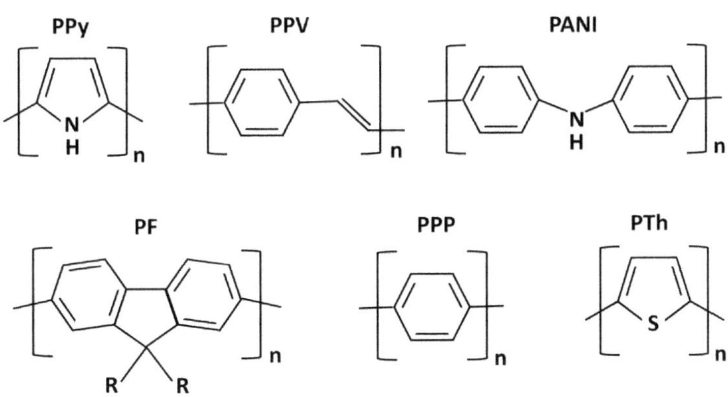

Figure 17.1 Structure of some conjugated polymer backbones. *PPy*, polypyrrole; *PPV*, poly(phenylenevinylene); *PANI*, polyaniline; *PF*, polyfluorene; *PPP*, poly(p-phenylene); *PTh*, polythiophene.

Multifunctional Nanostructured Coatings. https://doi.org/10.1016/B978-0-443-23683-9.00017-9

for electrochemical applications. One of the most interesting applications, and at the same time the most desirable in terms of limiting dependence on fossil fuels, is in energy conversion and storage systems. Despite their incredible properties, the use of conjugated polymers in this application is limited by their electrical conductivity, which is not high enough to allow high performance. The electrical conductivity of conjugated polymers is significantly increased by doping them with an electron donor or acceptor, as doping allows electrons to flow by creating new energy levels between the highest occupied molecular orbital (HOMO) and the lowest unoccupied molecular orbital (LUMO) (Namsheer and Rout, 2021). However, doping in conjugated polymers is more difficult to realize than in inorganic semiconductors because most dopants are molecules with different energy levels from the host polymers (Fig. 17.2). These molecular dopants are randomly distributed in the polymer matrix and bind to the hosts via weak but complex noncovalent interactions (Lu et al., 2021). Consequently, the n-doping level is limited by its incapacity to store electron charge stably and reversibly along the repeating units of the backbone, resulting in a reduced number of free electrons contributing to conductivity (Lu et al., 2021; Lu et al., 2020).

Several strategies have been employed to overcome this bottleneck. The introduction of secondary conductive phases, typically carbon-based and with a nanometric scale, such as CNTs, graphene, and nanowalls, can increase conductivity, but care

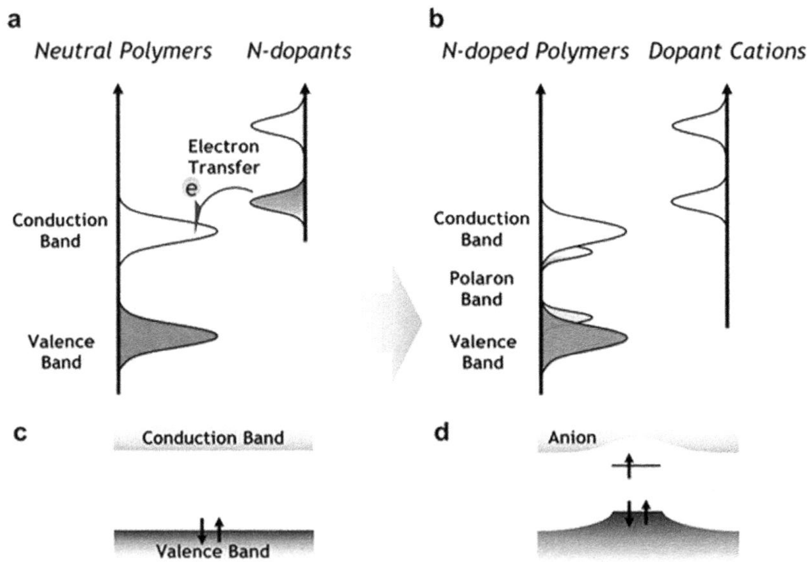

Figure 17.2 Scheme of the energy levels of neutral polymers and n-dopant (a), n-doped polymers and dopant cations (b), charged-induced energy levels in conjugated polymers from neutral state (c) to anion state (d).
From Lu, Y., Wang, J.-Y., Pei, J., 2021. Achieving efficient n-doping of conjugated polymers by molecular dopants. Acc. Chem. Res. 54(13), 2871–2883. https://doi.org/10.1021/acs.accounts.1c00223.

must be taken, especially during polymerization of the polymer matrix, to avoid uneven dispersion of the added nanophases and poor interface bonding. A more recent solution involves the development of composites with other robust materials, such as metal-organic frameworks (MOFs) (Qi et al., 2018). The excellent results obtained by combining the structural characteristics of MOFs or their derivatives with conjugated polymers allow not only to obtain good electrical properties, but also remarkable longevity when MOF-CP composites are used as supercapacitor (SC) electrodes (Yu et al., 2020), since the porous structure of MOFs-CPs limits the swelling and shrinking that occur during the operating cycles of the SCs (Siwach et al., 2024).

Enormous increases in electrical conductivity, up to 800% (Wang et al., 2023), can be achieved by modifying the molecular chains of conjugated polymers, for example by introducing amino and hydroxyl groups through active molecular polymerization (Zhang et al., 2022). This process produces an effect of hyperconjugation between the σ bonds of the polymer backbone and the π orbitals of the conjugated system. This interaction leads to a decrease in the energy gap between the HOMO and LUMO of the polymer.

To increase the electrical conductivity, π-conjugated polymers have been combined with nonconjugated redox polymers to obtain a "π-conjugated redox polymer" (Fig. 17.3). In this way, the weakness of each type of polymer, that is, the inability to store charges stably and reversibly of the π-conjugated polymers and the existence of limited redox active sites of the nonconjugated redox polymers, can be overcome and a high level of doping can be achieved (Jin et al., 2022; Liang et al., 2015).

For use in energy devices, the porosity of conjugated polymers is a very important issue. This is especially true for use as supercapacitor electrodes. For this reason, conjugated microporous polymers (CMPs) have attracted great interest from researchers who have sought to design a wide range of network-forming reactions and to synthesize building blocks with different properties and structures. CMPs are a subclass of porous materials that include covalent organic frameworks (COFs) and have an

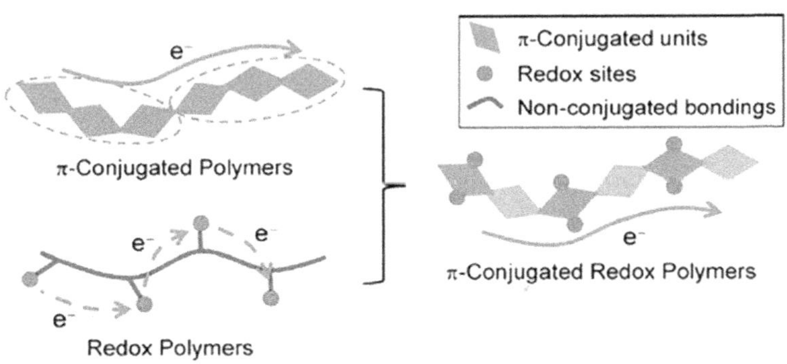

Figure 17.3 Scheme of the structural characteristics of π-conjugated polymers, redox polymers, and π-conjugated redox polymers.
From Liang, Y., Chen, Z., Jing, Y., Rong, Y., Facchetti, A., Yao, Y., 2015. Heavily n-dopable π-conjugated redox polymers with ultrafast energy storage capability. J. Am. Chem. Soc. 137(15), 4956–4959. https://doi.org/10.1021/jacs.5b02290.

extended π conjugation in combination with a permanently microporous backbone. The first example of CMP was the synthesis of microporous poly(aryleneethynylene) networks in 2007 (Jiang et al., 2007). Since then, a wide variety of molecular building blocks have been combined to control the functionality and structural diversity of CMPs, allowing them to be used in photo redox catalysis, energy storage, and biological and photocatalytic H_2 evolution (Lee and Cooper, 2020). Attempts have been made to transform bulk CMPs into nanostructured systems (CMP NPs) with the aim of obtaining nanostructures with high specific surface area, good thermal and chemical stability, and good electrical conductivity (MacFarlane et al., 2021; Kang et al., 2022). Some evidence of improved performance has been reported for CMPs combined with CNTs or graphene and used as cathodes in lithium-ion batteries (Alván et al., 2023; Huang et al., 2023; Li et al., 2022b; Lian et al., 2023) and SC electrodes (Alván et al., 2023; Jang et al., 2022), but the advantages of using 3D porous frameworks over 2D networks need to be investigated in more detail.

Extending the conjugation to the second dimension, like occurs in 2D COFs, could overcome the electrical conductivity limitations due to the short length of the conjugation and the point defects present on the linear conjugated polymers (Yu et al., 2020). Unfortunately, COFs have shown poor stability and limited durability in energy conversion devices, especially when operating in strong acid/alkaline electrolytes (Zhao et al., 2021). Therefore, conjugated polymers and 2D COFs have been strategically linked together to develop 2D conjugated polymer frameworks (CPFs), also known as 2D π-conjugated COFs, in which carbon is mainly bonded through imine bonds (C=N) (Fig. 17.4). These 2D carbon-rich conjugated frameworks are composed of atomically ordered planar networks of sp^2-carbon bonded to a nitrogen-doped backbone. This type of material is characterized by a nanoporous and layered structure that allows deep electrolyte infiltration and efficient mass transport through the pore channels, providing transport pathways for both ions and electrons along the stacking

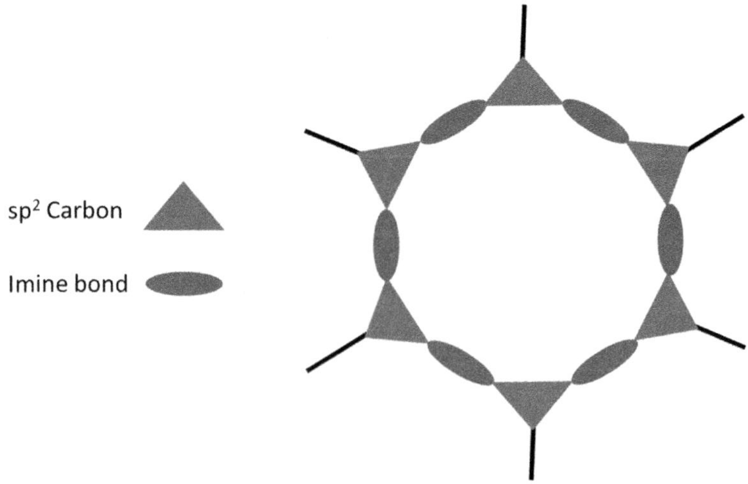

Figure 17.4 Structural configuration of conjugated polymer frameworks (CPFs).

directions and within the layers. In addition, the dimensionally stable framework structure ensures higher chemical and electrochemical stability than COFs (Xu et al., 2019b).

A new compound has been introduced in recent years. It can be considered a conducting polymer by any measure. It is polymeric graphitic carbon nitride (g-C_3N_4), which is composed of nitrogen and carbon and has a layered structure like graphite. It has a broadband gap (e.g., 2.5−3.0 eV) and is therefore characterized by photocatalytic activity in visible light (Hasija et al., 2019). Its potential in energy applications is enormous, especially as a photocatalyst for water splitting, and can be greatly enhanced by combining it with metals or oxide of transition metals (Alaghmandfard and Ghandi, 2022; Mazur et al., 2019).

From what has been described so far, it is evident that since the discovery of linear conducting polymers (Mac Diarmid and Shirakawa Nobel Prize in 2000), much research has been done to improve their performance. This chapter reports on the most recent applications of some new compounds related to conjugated polymers, used as active materials for supercapacitor electrodes, as electrocatalysts, and as materials for thermoelectric devices. For these applications, the conjugated polymer-based materials are used in the form of thin films or coatings applied to a substrate, which acts as the current collector of the device under consideration.

17.2 Supercapacitor electrodes

Supercapacitors (SCs) can be considered green energy storage devices because they are often based on cheap, sustainable, and abundant materials. They have a long lifetime and low maintenance costs, especially when compared to more common energy storage devices such as batteries. However, to be used on a large scale, SCs need to have not only high power density but also high energy density.

An SC consists of two electrodes with a separator between them, immersed in an electrolyte (Fig. 17.5). The active material responsible for the capacitance of the electrode is generally a thin film deposited on a current collector. The methods used to obtain the thin film of active material are the coating techniques usually used for large-scale production, characterized by ease of production and low cost: dip coating, spin coating, spray coating, and drop casting (Ramesh Peringath et al., 2023).

To be *super*, a capacitor must have a specific capacitance and a power density at least 100 times greater than those of conventional capacitors. These two characteristics

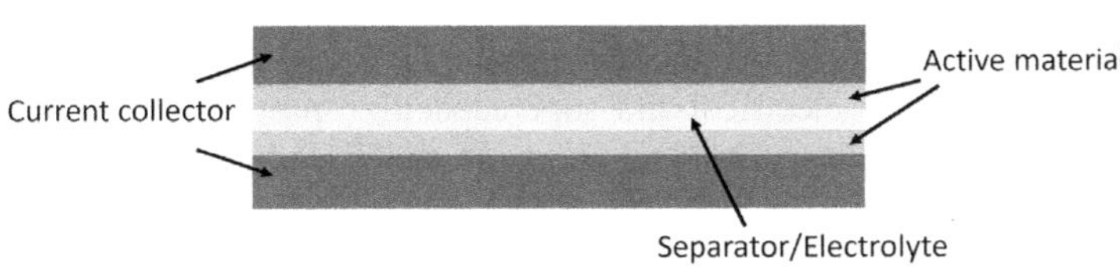

Figure 17.5 Scheme of a supercapacitor.

are essentially related to (i) the surface area of the electrode exposed to the electrolyte and (ii) the massive electrical resistance of the whole electrode system. It is therefore essential that the active material forming the SC electrode has a high specific surface area (SSA) and a high electrical conductivity, to enhance the specific capacitance and to minimize the potential drops that occur at the interface with the current collector.

The mechanisms underlying supercapacitive behavior are electrical double-layer capacitance (EDLC) or pseudocapacitance (PC). EDLC operates by reversible charge accumulation at the electrode-electrolyte interface and is proportional to the amount of charge stored. PC has a faradaic character and relies on fast reversible redox reactions at the surface of electroactive species. Maximizing the SSA allows for maximizing the exposure to the electrolyte, thus increasing the number of sites where charge can be stored, or redox reactions can take place. For both capacitance mechanisms, it is recommended that the active material of the electrode is highly porous. In terms of the electrical resistance of the electrode, it is important to ensure that the active material is as conductive as possible. For these reasons, CPs and especially CMPs are optimal candidates as active materials for SC electrodes. In the literature, many papers referred to numerous CMPs that showed excellent performance as SC electrodes, also when combined with nanocarbons (Dai et al., 2023; Gul et al., 2020; Zuo et al., 2024) or metal oxides (Kuznetsova et al., 2023). Table 17.1 shows some recent examples of electrodes based on CPs and the relative specific capacitance as reported.

Carbon nanotubes (CNTs) are often combined with CPs to increase the conductivity or to improve the structural properties of polymers. Zuo et al. (2024) adopted aminated multiwalled carbon nanotubes (NH2-MWCNT) as a template for the in-situ polymerization of polytriphenylamine (PTPA), which was synthesized via the Buchwald-Hartwig coupling reaction of tri(4-bromophenyl)amine and phenylenediamine. NH_2-MWNT was used because the $-NH_2$ group acts as a reactive site for coupling with aryl bromides in triphenylamine and in-situ polymerization of PTPA. The resulting active electrode material exhibited enhanced conductivity, high specific surface area, and elevated porosity. Compared with the original PTPA, the specific surface area of the core-shell PTPA@MWCNTs was significantly improved from 32 to 484 m^2/g. In addition, PTPA@MWNTs exhibited improved specific capacitance of 410 F/g in 0.5 M H_2SO_4 at a current of 10 A/g due to the hierarchical meso-micro pores, high redox activity, and electronic conductivity. In a symmetrical supercapacitor composed of PTPA@MWNTs electrodes, a capacitance of 216 F/g of total electrode material was measured and 71% of the initial capacitance was retained after 6000 cycles.

CNTs were used in combination with PANI to form nanostructured SC electrodes to prevent swelling, shrinkage, and fracture of CP due to the continuous doping and de-doping with intercalation and de-intercalation of electrolyte ions. The introduction of a small amount of a second phase, such as nanocarbons, suitable to avoid CP degradation without altering the electrical and supercapacitive properties and maintaining a low density, improves the mechanical behavior of CP and reduces the risk of short electrode life. In addition, the CP-CNT composite synergistically combines the two different charge storage mechanisms, the pseudo-capacitive behavior of CP and the EDL behavior of CNTs. CNTs were successfully combined with PANI in a layered

Table 17.1 examples of SC electrodes based on CPs.

Materials	Typology	Preparation/deposition Method	Electrolyte	Specific capacitance	References
PANI	Linear CP	Chemical synthesis/spin coating	H_3PO_4	13.22 mF/cm^2 @ 12.5 μA/cm^2	Ramesh Peringath et al. (2023)
PTh	Linear CP	Chemical synthesis/spin coating	H_3PO_4	3.3 mF/cm^2@ 12.5 μA/cm^2	Ramesh Peringath et al. (2023)
TPA-FC	CMP	Chemical synthesis	1M KOH	129 F/g @ 0.5 A/g	Samy et al. (2023)
PAQ	CMP	Buchwald–Hartwig coupling	0.5 M H_2SO_4	576 F/g @ 1 A/g	Liao et al. (2018)
PIBQ-1%rGO	CMP composite	Chemical synthesis	1 M H_2SO_4	63 F/g @ 1 A/g	Liao et al. (2018)
P(Th3CNTT–TPA)	CMP	Chemical synthesis/ Electropolymerization on ITO	0.1 M TBAPF6 in ACN	235 F/g @ 3 A/g	Topal et al. (2023)
Py-Ph-BDT	Redox-active CMP	Chemical synthesis	1.0 M KOH	712 F/g @ 0.5 A/g	Gaber et al. (2023)
PTPA@MWCNTs	CMP	Chemical synthesis	0.5 M H_2SO_4	410 F/g @ 10 A/g	Zuo et al. (2024)
Py-Ph-Pery	CMP	Chemical synthesis	1 M KOH	300 F/g @ 0.5 A/g	Singh et al. (2023)
pNTCDA-TAPM	CMP	Chemical synthesis	1 M Na_2SO_4	113.2 F/g @ 0.5 A/g	Dai et al. (2023)
pNTCDA-TAPT@ CNTs	CMP composite	Chemical synthesis	1 M Na_2SO_4	279.8 F/g @0.5 A/g	Dai et al. (2023)
PANI on CF	Linear CP composite	Chemical oxidative polymerization and pyrolysis	1 M H_2SO_4	224 F/g @ 0.5 A/g	Almeida and Ferreira (2020)
N/O-HPC	Linear CP	Electropolymerization	1 M H_2SO_4 0.5 M K_2SO_4 6 M KOH	419.8 F/g @ 0.2 A/g 179.8 F/g @ 0.2 A/g 343.2 F/g @ 0.2 A/g	Zhou et al. (2020)

Continued

Table 17.1 Continued

Materials	Typology	Preparation/deposition Method	Electrolyte	Specific capacitance	References
PANI-GO	Linear CP composite	Chemical oxidative synthesis	1 M H_2SO_4	264 F/g @ 1 A/g	Gul et al. (2020)
RGO/PEDOT/PANI	Linear CP composite	Electrodeposition	1 M H_2SO_4	535 F/g @1 A/g	Liu et al. (2020)
Hp-PANI	Hyperconiugated polymer	Chemical synthesis	1 M H_2SO_4	521 F/g @1 A/g	Zhang et al. (2022)

TPA-FC, tris(4-aminophenylamine)- ferrocene; *PAQ*, 3D polyaminoanthraquinone; *PANI*, polyaniline; *CF*, carbon fibers; *GO*, graphene oxide; *CNTs*, carbon nanotubes; *MWCNTs*, multiwalled carbon nanotubes; *PEDOT*, poly(3,4-ethylenedioxythiophene); *N/O-HPC*, Polyaniline-derived oxygen- and nitrogen-decorated hierarchical porous carbon; *PTh*, polythiophene; *PIBQ-1%rGO*, Quinone based imide conjugated microporous polymer-reductive graphene oxide composite; P(Th3CNTT−TPA):4-(2-(4-(bis(4-(thiophen-2-yl)phenyl)amino)phenyl)-5-(thiophen-2-yl)thieno [3,2-b]thiophen-3-yl) benzonitrile; *Py-Ph-BDT*, benzodithiophene-4,8-dione-pyrene-tethered; *PTPA*, polytriphenylamine; *Py-Ph-Pery*, perylene-3,4,9,10-tetracarboxylic acid dianhydride.

structure obtained by electrodeposition of PANI on a CNT substrate prepared by electrophoretic deposition (EPD) (De Riccardis and Re, 2019) (Fig. 17.6). A specific areal capacitance of 820 mF/cm^2 was measured at a current of 5 mA/cm^2(De Riccardis and Re, 2021).

The main candidates for EDLC electrode materials are porous carbons and nanocarbons. In addition to their porous structure, which increases the availability of active sites, porous carbons are characterized by high conductivity, high charge transport capability, and large surface area. When porous carbon materials are doped with heteroatoms such as oxygen, nitrogen, boron, or sulfur, their electrochemical performance as SC electrodes is further enhanced. CMPs are excellent precursors for porous carbons due to their inherent porosity, excellent stability, structurally controllable performance, and high specific surface area. A two-dimensional conjugated microporous polymer containing ketones was synthesized by a simple one-step coupling reaction of 4,4′-difluorodiphenylsulfone (DDS) and 4,4′-dihydroxybenzophenone (DHBP). After direct pyrolysis of CMP at 800°C, the resulting porous carbon exhibited a specific capacitance of 375 F/g at 0.5 A/g current (Xing et al., 2023).

An interesting property of CPs is plasticity, which can be useful in electronic applications where some flexibility of the SC is required, such as in wearable, smart, and portable electronic devices (Yan et al., 2023). Compared with traditional supercapacitors, in addition to the usual electrochemical performance, wearable SCs need to have good mechanical deformability and guarantee their performance even in a deformed state. A comprehensive review of the properties, preparation, and applications of portable SCs, mainly based on carbon and nanocarbon, is reported in (Yan et al., 2023).

Carbon fibers (Anand et al., 2020) or carbon cloths (Liu et al., 2020) used as substrates on which CPs are polymerized have good potential for flexible SC electrodes. A flexible binder-free hybrid mat was prepared by in-situ formation of PANI nanofibers on the surfaces of carbon nanofibers (CNFs) via chemical oxidation polymerization of aniline monomer (Anand et al., 2020). The nanostructured composite thus obtained ensures an enhanced charge/discharge rate and prevents structural deformation thanks to the interaction between the nanodimensional PANI and the porous CNFs through a π-bond. The PANI/CNF hybrid mat exhibits a large specific surface area, and a high gravimetric capacitance (493.75 F/g at 1 mA/cm^2), with electrochemical cyclic stability, as $>$ 90% retention of specific capacitance was measured after 5000 cycles.

CPs can also be used to prepare micro-electrodes for SC. Recently, skin-compatible and flexible conjugated polymer-based capacitor modules patterned by the parylene

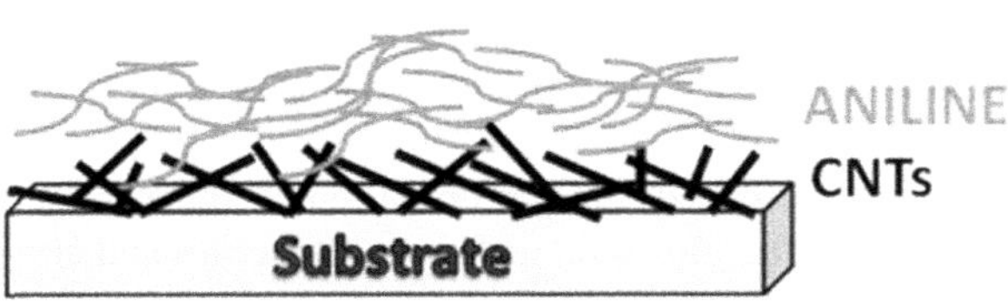

Figure 17.6 Scheme of PANI/CNTs nanocomposite prepared by electropolymerization and electrophoretic deposition. Adapted from De Riccardis, M. F., Re, M., 2019. Conducting nanocomposite coatings. In Nanomaterials-Based Coatings: Fundamentals and Applications, pp. 73−117, Elsevier. https://doi.org/10.1016/B978-0-12-815884-5.00004-1.

peel-off technique were fabricated (Girayhan Say et al., 2023). The resolution of the patterned devices was 20 μm, the electrode width was 180 μm, and the number of fingers was up to 20. The technique of designing polymer interdigitated electrodes on flexible parylene-C substrates has been shown to be applicable to two conjugated polymers, PEDOT:PSS and polythiophene functionalized with tetraethylene glycol (P(g42T-T)). The fabricated devices have shown to be mechanically stable and retain full charge storage capability after 1000 cycles with over 90% of initial capacity retained.

17.3 Electrocatalysts

In the future, the use of renewables will increase as their costs become competitive with fossil fuels. The technologies and catalytic processes on which renewables production is based is electrocatalysis, which makes extensive use of precious metals or their compounds. The technology of water splitting generates hydrogen and oxygen gas from electricity, through two half reactions that are hydrogen evolution reaction (HER) and oxygen evolution reaction (OER). Fuel cells are highly efficient devices capable of producing electrical energy with the electrochemical consumption of hydrogen, through hydrogen oxidation reaction (HOR) at the anode and oxygen reduction reaction (ORR) at the cathode. Metal-air batteries are an emerging technology requiring bifunctional ORR/OER processes. However, all electrochemical HER, OER, HOR, and ORR processes have sluggish dynamics that require high-performance electrocatalysts to reduce their overpotentials. Currently, Pt-, Ir- and Ru-based nanomaterials and their compounds show superior catalytic activities for hydrogen and oxygen electrochemical reactions. Unfortunately, the difficulty of their supply, especially at this historical moment, hinders the diffusion and commercialization of energy devices based on nonfossil fuels. Great efforts are therefore being made to study and research alternative materials that are abundant on Earth, easy to prepare, and have a catalytic activity comparable to that of noble metals. In theory, an excellent hydrogen and oxygen electrocatalyst should be rich in exposed active sites and have high electrical conductivity and mass transfer capability to facilitate electron and reactant transport. In recent years, many metal-based inorganic electrocatalysts, such as transition metal oxides, metal sulphides, metal nitrides, and metal carbides, have been investigated for efficient hydrogen and oxygen electrocatalysis (Deshmukh et al., 2023). Unfortunately, the limited presence of electrochemical active sites and poor electrical conductivity are responsible for their poor electrocatalytic activities compared with noble metal-based catalysts.

The quality of catalysts can be assessed mainly by some indicators, namely catalytic activity, selectivity, long-term stability, and cost. To improve these factors, many strategies have been adopted: (i) proper design of the electron density of the active sites, (ii) insertion of dopants or substituents to modify the peripheral environment of the active sites, (iii) adoption of special nanostructures with different geometries and crystalline degrees to make the active sites more exposed and thus accelerate the mass transport (Huang et al., 2021).

Carbon-based materials are promising electrocatalyst candidates having been shown to be suitable as bi- or even multifunctional electrocatalysts for zinc-air batteries (Xu et al., 2024), water splitting (Chi et al., 2024) and fuel cells (Li et al., 2024). In many of these cases, nanostructured carbon is associated with metal ions or heteroatoms and is often doped with nitrogen. In fact, the presence of heteroatoms in the carbon structure changes its electronic configuration and redistributes the charge in the carbon plane. In the case of nitrogen, which has an atomic radius similar to that of carbon, the doping does not cause the carbon framework to collapse.

CPs are optimal candidates as precursors for nitrogen-doped carbon because their conjugated structure makes them more resistant to carbonization than other polymers and allows the production of a graphitic structure. In addition, since nitrogen is intrinsically present in some CPs, such as PPy and PANI, the introduction of nitrogen atoms results in homogeneity. Finally, the conducting polymers can be doped or functionalized with metal-containing moieties. The presence of functional groups in the original CPs can favor the preparation of carbon-based nanocomposites with a synergistic catalytic effect (Li et al., 2022a).

In the preparation of carbon-based catalysts from CP precursors, special attention should be paid to the parameters of polymerization and subsequent pyrolysis to control the final composition and structure. In terms of composition, nonmetallic heteroatom-doped carbon, single-atom metal catalysts in carbon, and carbon-based hybrid materials can be prepared. For N-doped carbon, the most common bonding configurations are pyridinic N, pyrrolic N and graphitic N (Fig. 17.7) (Dai et al., 2015), which are particularly active for ORR (Lv et al., 2020; Singh et al., 2019; Cai et al., 2017) and OER (Cai et al., 2017; Jiang et al., 2019). The introduction of heteroatoms also

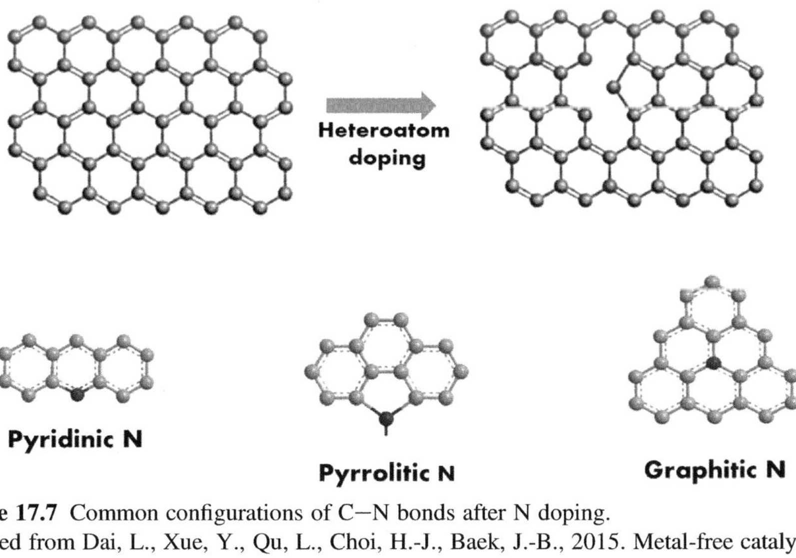

Figure 17.7 Common configurations of C—N bonds after N doping.
Adapted from Dai, L., Xue, Y., Qu, L., Choi, H.-J., Baek, J.-B., 2015. Metal-free catalysts for oxygen reduction reaction. Chem. Rev. 115(11), 4823—4892. https://doi.org/10.1021/cr5003563. Copyright 2015, American Chemical Society.

induces the presence of more chemically active sites and a high density of defects. Using models based on highly oriented pyrolytic graphite (HOPG), Guo et al. showed that the carbon atoms adjacent to pyridinic N are the active sites with Lewis's basicity where O_2 molecules are adsorbed as an initial step of ORR (Guo et al., 2016). In addition, the carbon adjacent to N-doping atoms can facilitate easier access to intermediates of H_2, thus generating hydrogen with a low overpotential in HER (Fang et al., 2017).

To enhance the electroactivity of N-doped carbon materials, a synergistic effect is achieved by incorporating transition metals in the form of single atoms (Bates et al., 2023; Zang et al., 2018) or nanoparticles (Mazur et al., 2019; Guo et al., 2019). Cobalt nanoparticles embedded in N-doped carbon have been demonstrated to have bifunctionality for both OER and ORR, even better than their oxidative counterparts (Kalusulingam et al., 2024).

As for the structure of the carbon-based catalysts, porous and hierarchical structures must be properly designed to have a large surface area, which allows the exposure of a large number of active sites and rapid electron transport. Many types of morphologies can be obtained during the synthesis of CPs: one-dimensional (1D) fibrous and tubular structure, two-dimensional (2D) sheet-like structure, three-dimensional (3D) microstructure, and even spherical core-shell structure (Fig. 17.8). It must be said, however, that the control of a structure subjected to pyrolysis is rather difficult. It is easier to achieve if the electrocatalyst preparation process does not include a high-temperature treatment, as this avoids the formation of aggregates. In addition, with nonpyrolysis polymer-based electrocatalysts, it is easier to define a structure-performance relationship, which is necessary to optimize the electrocatalysts.

Examples of electrocatalysts obtained from CMP without pyrolisis include nanopowders prepared using a CMP functionalized with a porphyrin moiety coordinated with Co(II) ions via the Suzuki C—C cross-coupling reaction (Bhunia et al., 2019). This provided an efficient electrocatalysts for OER. The tris-(4-aminophenyl)amine (TPA)-perylenediimide (PDI) CMP, which contains triphenylamine as the electron donor and PDI as the acceptor, showed excellent activity as a metal-free electrocatalyst for ORR due to its high conductivity and the existence of a charge-separated state (Bhattacharyya et al., 2019). A special nano-tentacular structure based on a novel CMP has been conferred on CNTs. The novel CMP (ST-CMP) was developed by rational design of the organic condensation reaction of squaric acid (denoted as S) and 1,3,5-tris(4-aminophenyl)benzene (denoted as T). The ST-CMP nanotentacles with strong polar sites provide a relatively large surface area, while the CNT network provides a fast electron transport path within the electrode. This articulated structure has shown efficacy in the catalytic conversion of lithium polysulphides (LiPs) and therefore potential applicability in Li—S batteries (Liu et al., 2021). CNTs were also decorated with an ethynyl-linked cobalt (II) porphyrin conjugated polymer, which was polymerized in situ on CNTs to create a 3D network of bimetallic central electrocatalysts for CO_2 reduction (Wang et al., 2021). Again, the absence of a pyrolysis process avoided the coalescence of the metal particle decoration on the CNTs and ensured the uniform disposition of the metal clusters, increased the electrochemically active surface area, and enhanced the interfacial electron transfer.

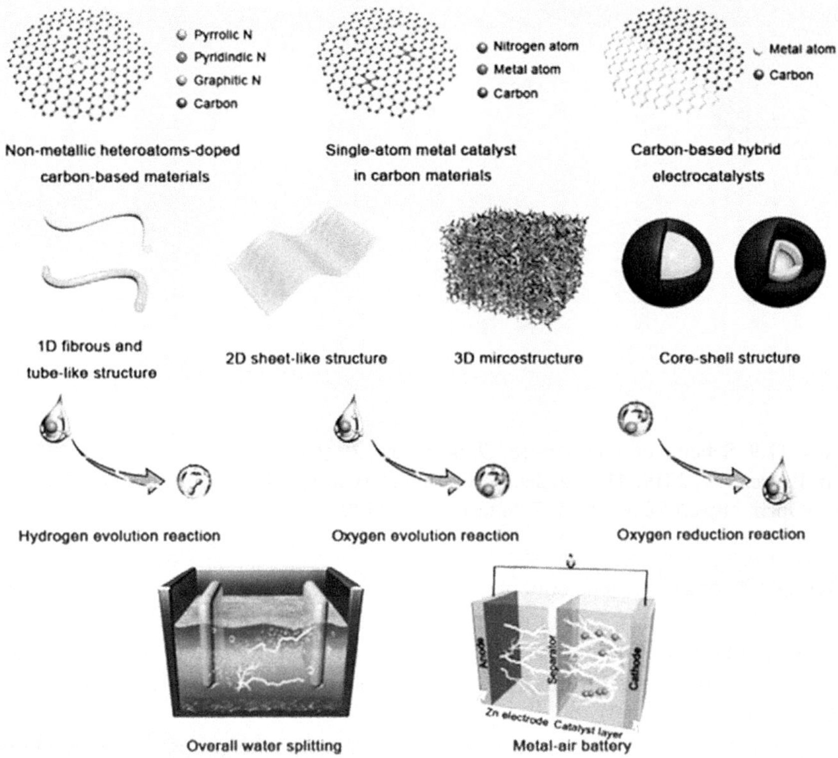

Figure 17.8 Possible composition and architecture of carbon-based materials with conducting polymers as precursors and their applications for hydrogen and oxygen electrocatalysis.
From Li, W., Wang, C., Lu, X., 2022. Conducting polymers-derived fascinating electrocatalysts for advanced hydrogen and oxygen electrocatalysis. Coord. Chem. Rev. 464, 214555. https://doi.org/10.1016/j.ccr.2022.214555.

17.4 Thermoelectric devices

The need to limit the use of energy from fossil sources has led researchers to explore ways of recovering energy from other sources, and energy harvesting was born with the aim of recovering energy lost in the environment or dispersed by natural sources, to convert it into electricity. The recovery of thermal energy is attractive, but the application of thermoelectric generators (TEGs) is limited to some low-power devices due to their low efficiency, although a value of 15% has recently been reported (Li et al., 2023).

Schematically, a TEG unit consists of a thermocouple with one leg of n-type material and one leg of p-type material connected by electrodes (Fig. 17.9). Heat is directly converted into electrical energy by the Seebeck effect since the temperature

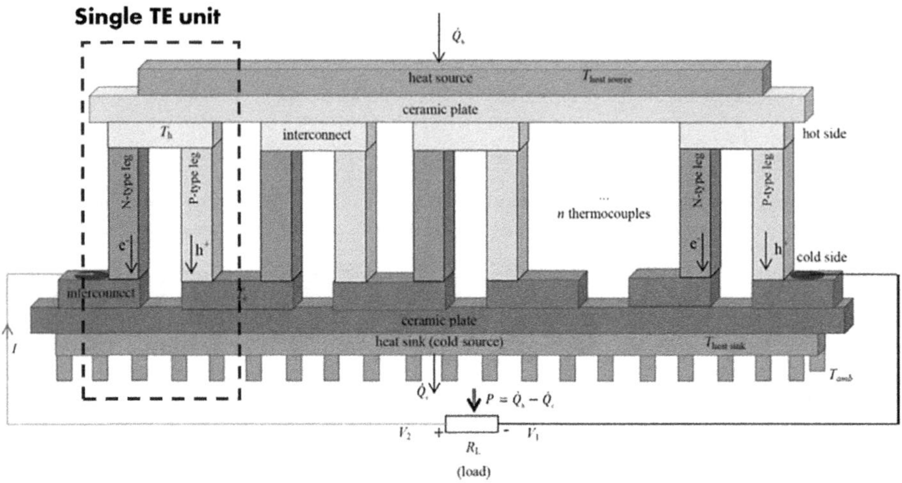

Figure 17.9 Scheme of a TE device. *Q*, heat flow; *P*, electric power.
From Enescu, D., 2019. Thermoelectric Energy Harvesting: Basic Principles and Applications, IntechOpen. https://doi.org/10.5772/intechopen.83495.

difference between a heat source and a heat sink connected by the legs creates a voltage difference and causes moving charges from the hot to the cold end of the thermoelectric material. The power generated by a single TEG unit is of the order of μW, but by connecting several hundred single TEG units, a power of several mW can be achieved (Enescu, 2019).

A parameter called Figure of Merit, ZT, quantifies the properties of a TEG. It is defined as

$$ZT = T\frac{\sigma S^2}{k}$$

T is the temperature, σ is the electrical conductivity of the carrier (electrons for n-type and holes for p-type materials), S is the Seebeck coefficient (positive for p-type and negative for n-type materials), and k is the thermal conductivity. To achieve high efficiency, a TE material should have a high Seebeck coefficient, high electrical conductivity, and low thermal conductivity. In general, S, σ, and k are interdependent, that is, an increase in σ will result in a decrease in S and an increase in k, with an overall detrimental effect on ZT.

A promising application of TEG is as portable flexible devices, since TE materials have long lifetimes, no mechanically moving parts, high portability, easy encapsulation, and no operating noise (Fan et al., 2021). Conjugated polymers seem to be the most suitable materials for this application, as they have not only mechanical properties appropriate for flexibility but also good electrical conductivity and low thermal conductivity. In addition, compared with inorganic TE materials, CPs have easy processability and wide tunability of their properties, allowing potential applications in miniaturized smart and healthcare systems (Fig. 17.10).

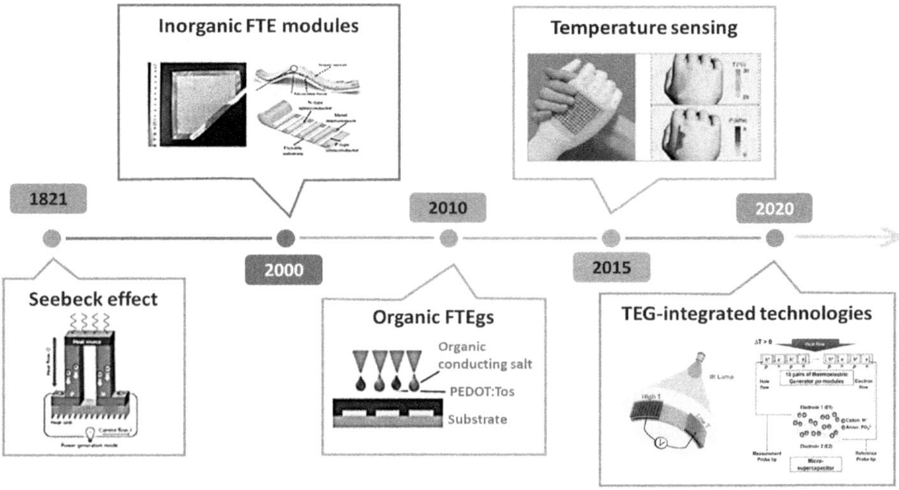

Figure 17.10 Roadmap of the representative application of TE devices.
From Fan, Z., Zhang, Y., Pan, L., Ouyang, J., Zhang, Q., 2021. Recent developments in flexible thermoelectrics: From materials to devices. Renew. Sustain. Energy Rev. 137, 110448. https://doi.org/10.1016/j.rser.2020.110448.

The most used linear conducting polymers, namely PEDOT, PANI, PPy, and PTh, are all p-type TE polymers, as holes are their main carriers. Among them, the PEDOT family is widely used in TEG because of its advantages: high transparency, good electrical conductivity, good water processability, and easy processability, which promote easy fabrication and large-scale production of printed/flexible TE devices.

The use of the complex PEDOT:PSS is widespread because poly(styrene sulphonate) (PSS) acts as a counterion for PEDOT and increases its solubility. Unfortunately, PSS reduces the intrinsic Seebeck coefficient and the electrical conductivity of PEDOT. Therefore, considerable efforts have been made by researchers to improve the power factor of PEDOT:PSS by removing excess PSS, improving the chain alignment, introducing some hybrid architectures, or controlling the oxidation level (Zhen et al., 2018). Recently, a high power factor (defined as σS^2) 141 $\mu W\ m^{-1}K^{-2}$ at 25°C, comparable to inorganic TE materials, has been obtained on a PEDOT:PSS film treated with triple post-treatments with formamide (CH_3NO), concentrated sulfuric acid (H_2SO_4), and sodium borohydride ($NaBH_4$) (Xu et al., 2019a). The CH_3NO and H_2SO_4 treatments contribute to the selective removal of excess PSS, while the H_2SO_4 treatment induces a conformational change in the PEDOT chains from benzenoid to quinoid. Finally, these treatments induce a structural modification of the PEDOT:PSS complex, producing highly ordered stacking lamellae that ensure a high carrier mobility (1786Scm^{-1}). The subsequent $NaBH_4$ treatment serves to optimize the TE properties by modulating the oxidation level.

The structure of PPy has also been modulated through functional blocks. Two different pendant groups, methyl (PBBTL-CH3) and phenyl (PBBTL-Ph), were combined with pyrrole[3,2-b:4,5-b ′]bis[1,4]benzothiazine (He et al., 2023). The π-π lamellar stacking distances and their orientations were influenced by these pendant groups (Fig. 17.11). The methyl group gives an exclusive edge-on orientation with smaller π-π lamellar stacking distances, whereas the phenyl group gives a mixed orientation with larger distances. As a result, $FeCl_3$-doped PBBTL- CH_3 has an electrical conductivity of about 36 S cm^{-1}, which is 38 times higher than that of $FeCl_3$-doped PBBTL-Ph. The power factor of PBBTL-CH_3 was also 13 times higher than that of PBBTL-Ph.

Unlike p-type polymers, there are few examples of conjugated polymers with efficient n-type TE properties. Polymers with LUMO energy levels below −4.0 eV are potential candidates for the development of polymers with efficient n-type TE properties. A π-conjugated polymer, PNDTI-BBT, consisting of naphtho[2,3-b:6,7-b’] dithiophenediimide (NDTI) and benzo[1,2-c:4,5-c’]-bis[1,2,5]thiadiazole (BBT) units was synthesized (Wang et al., 2017). PNDTI-BBT was then combined with two different alkyl side chains (DP and DT) to obtain two different polymers. Both polymers had a low-lying LUMO (−4.4 eV) and a large difference in transport and as well as in TE properties (in DP 3 times higher carrier mobility and 20 times higher conductivity and power factor than in DT), due to the different crystalline nature of the polymers in the thin film state. This demonstrates that by tuning the intermolecular

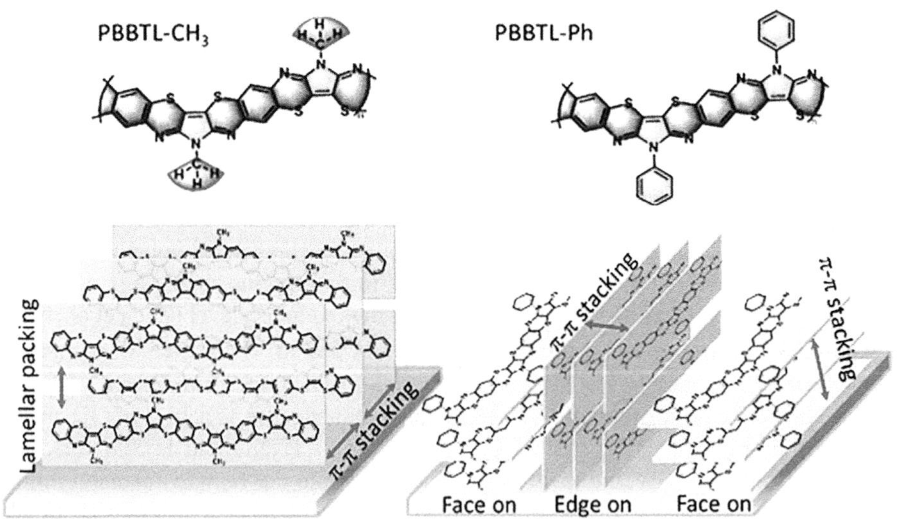

Figure 17.11 Scheme of the edge-on and the mixture of face-on and edge-on orientation in the PBBTL-Ch_3 and PBBTL-Ph complexes.
Adapted from He, Q., Wang, J., Dexter Tam, T. L., Zhang, X., Jiang, Z., Chien, S. W., Tham, N. N., Koh, X. Q., Soh, P. X., Meng, H., Huang, W., Xu, J., 2023. Thermoelectric performance enhancement of p-type pyrrolo[3,2-b:4,5-b′]bis[1,4]benzothiazine conjugated ladder polymer by pendant group engineering. ACS Mater. Lett. 5(10), 2829−2835. https://doi.org/10.1021/acsmaterialslett.3c00758. Copyright 2023, American Chemical Society.

interaction with alkyl side chains, both carrier mobility enhancement and doping effectiveness of conjugated polymers can be achieved in the thin film state.

The introduction of a secondary phase into a conjugated polymer matrix can increase the TE properties of the polymer, especially if the added filler is electrically conductive. Furthermore, by adding an n-type compound to a p-type conjugated polymer, a significant increase in the Seebeck coefficient can be obtained, as was observed in the case of MXene—an n-type TE material-incorporated into the p-type PEDOT: PSS (Guan et al., 2020). This may be due to the numerous boundaries created at the filler/polymer interfaces, which help to filter out the low-energy carriers. In addition, these interfacial boundaries also help to suppress phonon transport and therefore reduce the thermal conductivity via phonon scattering (Fan et al., 2021).

The use of CNTs as a secondary phase not only increases the electrical conductivity but also improves the mechanical properties of PANI (Yazdi et al., 2019). Moreover, the presence of the conducting CNTs has a double beneficial effect: (i) the electrical conductivity is improved due to better charge transport, and (ii) the carrier mobility is improved as the PANI chains are less compacted. This is due to the interactivity between CNTs and PANI through a strong π-π interaction. As a result, the power factor of PANI showed a significant increase when CNTs were added, changing from 0.002 to 0.3 $\mu W\ m^{-1}K^{-2}$(De Riccardis and Re, 2021).

17.5 Conclusions

Conjugated polymers and their derivatives have great potential for energy applications, particularly as materials for supercapacitor electrodes, electrocatalysts, and in thermoelectric devices. The most promising are conjugated microporous polymers, especially those combined with MOFs, because they have a large surface area coupled with interconnected porous networks and significant conductivity. They can also be engineered by appropriate design of chemical functionality. New future prospects for wide applications in the energy sector are opening up through the targeted nanostructuring and preparation of composite CPs with MXenes, metal nanoparticles, or carbon nanomaterials. For example, improvements in capacity, speed, and cycling performance of supercapacitors based on CMP composites can be achieved through functional molecular design and porosity improvement, as well as combination with nanomaterials in the form of nanoparticles or nanostructured carbons. In addition, the micro- and nanostructure can effectively optimize the performance of electrocatalysts through increased exposure and availability of active sites and increased activity. Finally, polymeric graphitic carbon nitrate (g-C_3N_4) seems to have great potential for energy storage and conversion, also due to the abundance of its constituent elements. It should not be underestimated that devices based on conjugated polymers can be integrated with other technologies and therefore used in other sectors such as sensors, photovoltaics, electromagnetic shielding, and photocatalysis. At this aim, the future trend of CP research should be toward a better understanding of the bonding mechanism between CP and MOF or other combined nanostructured components that are responsible for the variability in their structure. Detailed evaluation of this will lead to further progress in designing better-performing materials.

References

Alaghmandfard, A., Ghandi, K., 2022. A comprehensive review of graphitic carbon nitride (g-C3N4)−metal oxide-based nanocomposites: potential for photocatalysis and sensing. Nanomaterials 12 (2), 294. https://doi.org/10.3390/nano12020294.

Almeida, D.A.L., Ferreira, N.G., 2020. Fabrication of binary composites from polyaniline deposits on carbon fibers heat treated at three different temperatures: structural and electrochemical analyses for potential application in supercapacitors. Mater. Chem. Phys. 239, 122101. https://doi.org/10.1016/j.matchemphys.2019.122101.

Alván, D., Grieco, R., Patil, N., Mavrandonakis, A., Liras, M., Marcilla, R., 2023. Hybrid based on phenazine conjugated microporous polymer as a high-performance organic electrode in aqueous electrolytes. Batteries & Supercaps 6.

Anand, S., Ahmad, Md W., Saidi, A.K.A.Al, Yang, D.-J., Choudhury, A., 2020. Polyaniline nanofiber decorated carbon nanofiber hybrid mat for flexible electrochemical supercapacitor. Mater. Chem. Phys. 254, 123480. https://doi.org/10.1016/j.matchemphys.2020.123480.

Bates, J.S., Johnson, M.R., Khamespanah, F., Root, T.W., Stahl, S.S., 2023. Heterogeneous M-N-C catalysts for aerobic oxidation reactions: lessons from oxygen reduction electrocatalysts. Chem. Rev. 123 (9), 6233−6256. https://doi.org/10.1021/acs.chemrev.2c00424.

Bhattacharyya, S., Samanta, D., Roy, S., Radhakantha, V.P.H., Maji, T.K., 2019. In situ stabilization of Au and Co nanoparticles in a redox-active conjugated microporous polymer matrix: facile heterogeneous catalysis and electrocatalytic oxygen reduction reaction activity. ACS Appl. Mater. Interfaces 11 (5), 5455−5461. https://doi.org/10.1021/acsami.8b20610.

Bhunia, S., Bhunia, K., Chandra Patra, B., Das, S.K., Pradhan, D., Bhaumik, A., Pradhan, A., Bhattacharya, S., 2019. Efficacious electrochemical oxygen evolution from a novel Co(II) porphyrin/pyrene-based conjugated microporous polymer. ACS Appl. Mater. Interfaces 11 (1), 1520−1528. https://doi.org/10.1021/acsami.8b20142.

Cai, S., Meng, Z., Tang, H., Wang, Y., Tsiakaras, P., 2017. 3D Co-N-doped hollow carbon spheres as excellent bifunctional electrocatalysts for oxygen reduction reaction and oxygen evolution reaction. Appl. Catal. B Environ. 217, 477−484. https://doi.org/10.1016/j.apcatb.2017.06.008.

Chi, H., Deng, N., Wang, G., Feng, Y., Zeng, Q., Wang, Y., Peng, Z., Li, Q., Kang, W., Cheng, B., 2024. Controllable preparation, regulation mechanisms and development directions of multi-functional one-dimensional carbon nanomaterial as electrocatalysts for water splitting and zinc-air batteries. Int. J. Hydrogen Energy 51, 167−194. https://doi.org/10.1016/j.ijhydene.2023.09.294.

Dai, L., Xue, Y., Qu, L., Choi, H.-J., 2015. Jong-beom Baek, metal-free catalysts for oxygen reduction reaction. Chem. Rev. 115 (11), 4823−4892. https://doi.org/10.1021/cr5003563.

Dai, Y., Wang, Y., Xu, H., Li, X., Yan, X., Xu, X., 2023. Structure, morphology and energy storage properties of imide conjugated microporous polymers with different cores and the corresponding composites with CNT. Electrochim. Acta 441, 141820. https://doi.org/10.1016/j.electacta.2023.141820.

De Riccardis, M.F., Re, M., 2019. Conducting Nanocomposite Coatings Nanomaterials-Based Coatings: Fundamentals and Applications. Elsevier, Italy, pp. 73−117. https://doi.org/

10.1016/B978-0-12-815884-5.00004-1. Available from:http://www.sciencedirect.com/science/book/9780128158845.

De Riccardis, M.F., Re, M., 2021. Functional properties of nanocomposites based on polyaniline. Trends Phys. Chem. 21, 107–124.

Deshmukh, M.A., Park, S.-J., Thorat, H.N., Bodkhe, G.A., Ramanavicius, A., Ramanavicius, S., Shirsat, M.D., Ha, T.-J., 2023. Advanced energy materials: current trends and challenges in electro- and photo-catalysts for H2O splitting. J. Ind. Eng. Chem. 119, 90–111. https://doi.org/10.1016/j.jiec.2022.11.054.

Enescu, D., 2019. Thermoelectric Energy Harvesting: Basic Principles and Applications. IntechOpen, IntechOpen. https://doi.org/10.5772/intechopen.83495.

Fan, Z., Zhang, Y., Pan, L., Ouyang, J., Zhang, Q., 2021. Recent developments in flexible thermoelectrics: from materials to devices. Renew. Sustain. Energy Rev. 137, 110448. https://doi.org/10.1016/j.rser.2020.110448.

Fang, M., Dong, G., Wei, R., Ho, J.C., 2017. Hierarchical nanostructures: design for sustainable water splitting. Adv. Energy Mater. (23), 7. https://doi.org/10.1002/aenm.201700559.

Gaber, T.A., Ahmed, L.R., El-Mahdy, A.F.M., 2023. Efficient faradaic supercapacitor energy storage using redox-active pyrene- and benzodithiophene-4,8-dione-tethered conjugated microporous polymers. J. Mater. Chem. A 11 (36), 19408–19417. https://doi.org/10.1039/d3ta03198e.

Girayhan Say, M., Donahue, M.J., Kroon, R., Berggren, M., Engquist, I., 2023. Ultrathin polymer electrochemical microcapacitors for on-chip and flexible electronics. Org. Electron. 115, 106751. https://doi.org/10.1016/j.orgel.2023.106751.

Guan, X., Feng, W., Wang, X., Venkatesh, R., Ouyang, J., 2020. Significant enhancement in the Seebeck coefficient and power factor of p-type poly(3,4-ethylenedioxythiophene):poly(-styrenesulfonate) through the incorporation of n-type MXene. ACS Appl. Mater. Interfaces 12 (11), 13013–13020. https://doi.org/10.1021/acsami.9b21185.

Gul, H., Shah, A.-ul-H.A., Krewer, U., Bilal, S., 2020. Study on direct synthesis of energy efficient multifunctional polyaniline–graphene oxide nanocomposite and its application in aqueous symmetric supercapacitor devices. Nanomaterials 10 (1), 118. https://doi.org/10.3390/nano10010118.

Guo, D., Shibuya, R., Akiba, C., Saji, S., Kondo, T., Nakamura, J., 2016. Active sites of nitrogen-doped carbon materials for oxygen reduction reaction clarified using model catalysts. Science 351 (6271), 361–365. https://doi.org/10.1126/science.aad0832.

Guo, H., Feng, Q., Zhu, J., Xu, J., Li, Q., Liu, S., Xu, K., Zhang, C., Liu, T., 2019. Cobalt nanoparticle-embedded nitrogen-doped carbon/carbon nanotube frameworks derived from a metal–organic framework for tri-functional ORR, OER and HER electrocatalysis. J. Mater. Chem. A 7 (8), 3664–3672. https://doi.org/10.1039/c8ta11400e.

Hasija, V., Raizada, P., Sudhaik, A., Sharma, K., Kumar, A., Singh, P., Jonnalagadda, S.B., Thakur, V.K., 2019. Recent advances in noble metal free doped graphitic carbon nitride based nanohybrids for photocatalysis of organic contaminants in water: a review. Appl. Mater. Today 15, 494–524. https://doi.org/10.1016/j.apmt.2019.04.003.

He, Q., Wang, J., Tam, T.L.D., Zhang, X., Jiang, Z., Chien, S.W., Neng Tham, N., Qi Koh, X., Peng, X.S., Meng, H., Huang, W., Xu, J., 2023. Thermoelectric performance enhancement of p-type pyrrolo[3,2-b:4,5-b′]bis[1,4]benzothiazine conjugated ladder polymer by pendant group engineering. ACS Mater. Lett. 5 (10), 2829–2835. https://doi.org/10.1021/acsmaterialslett.3c00758.

Huang, B., Li, L., Tang, X., Zhai, W., Hong, Y., Hu, T., Yuan, K., Chen, Y., 2021. Pyrolysis-free polymer-based oxygen electrocatalysts. Energy Environ. Sci. 14 (5), 2789–2808. https://doi.org/10.1039/d1ee00306b.

Huang, T., Zhang, G., Chen, R., Lin, S., Zhou, H., Li, J., Chung, L.-H., Hu, X., Lin, Y., He, J., 2023. Donor−acceptor conjugated microporous polymer toward enhanced redox kinetics in lithium−sulfur batteries. ACS Appl. Mater. Interfaces 15 (17), 21075−21085. https://doi.org/10.1021/acsami.3c01558.

Jang, M., Cho, Y., Kim, Y., Hahn, M., Jung, D., Park, S.Y., Lee, W., Piao, Y., 2022. Redox-active conjugated microporous anthraquinonylamine-based polymer network grafted with activated graphene toward high-performance flexible asymmetric supercapacitor electrodes. Electrochim. Acta 434, 141315. https://doi.org/10.1016/j.electacta.2022.141315.

Jiang, J.X., Su, F., Trewin, A., Wood, C.D., Campbell, N.L., Niu, H., Dickinson, C., Ganin, A.Y., Rosseinsky, M.J., Khimyak, Y.Z., Cooper, A.I., 2007. Conjugated microporous poly(aryleneethynylene) networks. Angew. Chem. Int. Ed. 46 (45), 8574−8578. https://doi.org/10.1002/anie.200701595. Available from:http://onlinelibrary.wiley.com/journal/10.1002/(ISSN)1521-3773.

Jiang, H., Gu, J., Zheng, X., Liu, M., Qiu, X., Wang, L., Li, W., Chen, Z., Ji, X., Li, J., 2019. Defect-rich and ultrathin N doped carbon nanosheets as advanced trifunctional metal-free electrocatalysts for the ORR, OER and HER. Energy Environ. Sci. 12 (1), 322−333. https://doi.org/10.1039/c8ee03276a.

Jin, Z., Cheng, Q., Evans, A.M., Gray, J., Zhang, R., Tong Bao, S., Wei, F., Venkataraman, L., Yuan, Y., Nuckolls, C., 2022. π-Conjugated redox-active two-dimensional polymers as organic cathode materials. Chem. Sci. 13 (12), 3533−3538. https://doi.org/10.1039/d1sc07157b.

Kalusulingam, R., Ravi, K., Mathi, S., Mikhailova, T.S., Srinivasan, K., Biradar, A.V., Myasoedova, T.N., 2024. Bagasse derived N-doped graphitic carbon encapsulated cobalt nanoparticles as an efficient bifunctional catalyst for water splitting reactions. Colloids Surf. A Physicochem. Eng. Asp. 692, 133959. https://doi.org/10.1016/j.colsurfa.2024.133959.

Kang, S., Yoon, T.W., Kim, G.-Y., Kang, B., 2022. Review of conjugated polymer nanoparticles: from formulation to applications. ACS Appl. Nano Mater. 5 (12), 17436−17460. https://doi.org/10.1021/acsanm.2c04730.

Kuznetsova, L.S., Arlyapov, V.A., Plekhanova, Y.V., Tarasov, S.E., Kharkova, A.S., Saverina, E.A., Reshetilov, A.N., 2023. Conductive polymers and their nanocomposites: application features in biosensors and biofuel cells. Polymers 15 (18), 3783. https://doi.org/10.3390/polym15183783.

Lee, J.-S.M., Cooper, A.I., 2020. Advances in conjugated microporous polymers. Chem. Rev. 120 (4), 2171−2214. https://doi.org/10.1021/acs.chemrev.9b00399.

Li, W., Wang, C., Lu, X., 2022a. Conducting polymers-derived fascinating electrocatalysts for advanced hydrogen and oxygen electrocatalysis. Coord. Chem. Rev. 464, 214555. https://doi.org/10.1016/j.ccr.2022.214555.

Li, Z., Tan, J., Zhu, X., Xie, S., Fang, H., Ye, M., Shen, J., 2022b. High capacity and long-life aqueous zinc-ion battery enabled by improving active sites utilization and protons insertion in polymer cathode. Energy Storage Mater. 51, 294−305. https://doi.org/10.1016/j.ensm.2022.06.049.

Li, W., Poudel, B., Kishore, R.A., Amin, N., Liu, N., Zhang, Y., Priya, S., 2023. Toward high conversion efficiency of thermoelectric modules through synergistical optimization of layered materials. Adv. Mater. 35 (20). https://doi.org/10.1002/adma.202210407.

Li, X., Wang, D., Zha, S., Chu, Y., Pan, L., Wu, M., Liu, C., Wang, W., Mitsuzaki, N., Chen, Z., 2024. Active sites identification and engineering of M-N-C electrocatalysts toward oxygen reduction reaction. Int. J. Hydrogen Energy 51, 1110–1127. https://doi.org/10.1016/j.ijhydene.2023.07.161.

Lian, L., Li, K., Ren, L., Han, D., Lv, X., Wang, H.-G., 2023. Imine-linked triazine-based conjugated microporous polymers/carbon nanotube composites as organic anode materials for lithium-ion batteries. Colloids Surf. A Physicochem. Eng. Asp. 657, 130496. https://doi.org/10.1016/j.colsurfa.2022.130496.

Liao, Y., Wang, H., Zhu, M., Thomas, A., 2018. Efficient supercapacitor energy storage using conjugated microporous polymer networks synthesized from Buchwald–Hartwig coupling. Adv. Mater. 30 (12). https://doi.org/10.1002/adma.201705710.

Liang, Y., Chen, Z., Jing, Y., Rong, Y., Facchetti, A., Yao, Y., 2015. Heavily n-dopable π-conjugated redox polymers with Ultrafast energy storage capability. J. Am. Chem. Soc. 137 (15), 4956–4959. https://doi.org/10.1021/jacs.5b02290.

Liu, F., Xie, L., Wang, L., Chen, W., Wei, W., Chen, X., Luo, S., Dong, L., Dai, Q., Huang, Y., Wang, L., 2020. Hierarchical porous RGO/PEDOT/PANI hybrid for planar/linear supercapacitor with outstanding flexibility and stability. Nano-Micro Lett. 12 (1). https://doi.org/10.1007/s40820-019-0342-5.

Liu, J., Li, H., Wang, J., Zhang, Y., Luo, D., Zhao, Y., Li, Y., Yu, A., Wang, X., Chen, Z., 2021. Design zwitterionic amorphous conjugated micro-/mesoporous polymer assembled nanotentacle as highly efficient sulfur electrocatalyst for lithium-sulfur batteries. Adv. Energy Mater. 11 (40). https://doi.org/10.1002/aenm.202101926.

Lu, Y., Wang, J.Y., Pei, J., et al., 2021. Achieving efficient n-doping of conjugated polymers by molecular dopants. Acc. Chem. Res. 54, 2871–2883. https://doi.org/10.1021/acs.accounts.1c00223.

Lu, Y., Yu, Z.-D., Liu, Y., Ding, Y.-F., Yang, C.-Y., Yao, Z.-F., Wang, Z.-Y., You, H.-Y., Cheng, X.-F., Tang, B., Wang, J.-Y., Pei, J., 2020. The critical role of dopant cations in electrical conductivity and thermoelectric performance of n-doped polymers. J. Am. Chem. Soc. 142 (36), 15340–15348. https://doi.org/10.1021/jacs.0c05699.

Lv, Q., Wang, N., Si, W., Hou, Z., Li, X., Wang, X., Zhao, F., Yang, Z., Zhang, Y., Huang, C., 2020. Pyridinic nitrogen exclusively doped carbon materials as efficient oxygen reduction electrocatalysts for Zn-air batteries. Appl. Catal. B Environ. 261, 118234. https://doi.org/10.1016/j.apcatb.2019.118234.

MacFarlane, L.R., Shaikh, H., Diego Garcia-Hernandez, J., Vespa, M., Fukui, T., Manners, I., 2021. Functional nanoparticles through π-conjugated polymer self-assembly. Nat. Rev. Mater. 6 (1), 7–26. https://doi.org/10.1038/s41578-020-00233-4.

Mazur, D., Pariiska, O., Kurys, Y., 2019. Materials Today: Proceedings Ukraine Co-N-C electrocatalysts derived from nitrogen containing conjugated polymers for hydrogen evolution, p. 6. In: https://www.sciencedirect.com/journal/materials-today-proceedings.

Namsheer, K., Rout, C.S., 2021. Conducting polymers: a comprehensive review on recent advances in synthesis, properties and applications. RSC Adv. 11 (10), 5659–5697. https://doi.org/10.1039/d0ra07800j. Available from: http://pubs.rsc.org/en/journals/journal/ra.

Qi, K., Hou, R., Zaman, S., Qiu, Y., Xia, B.Y., Duan, H., 2018. Construction of metal–organic framework/conductive polymer hybrid for all-solid-state fabric supercapacitor. ACS Appl. Mater. Interfaces 10 (21), 18021–18028. https://doi.org/10.1021/acsami.8b05802.

Ramesh Peringath, A., Bayan, M.A.H., Beg, M., Jain, A., Pierini, F., Gadegaard, N., Hogg, R., Manjakkal, L., 2023. Chemical synthesis of polyaniline and polythiophene electrodes with excellent performance in supercapacitors. J. Energy Storage 73, 108811. https://doi.org/10.1016/j.est.2023.108811.

Singh, S.K., Takeyasu, K., Nakamura, J., 2019. Active sites and mechanism of oxygen reduction reaction electrocatalysis on nitrogen-doped carbon materials. Adv. Mater. 31 (13). https://doi.org/10.1002/adma.201804297.

Samy, M.M., Mohamed, M.G., Thomas, A., 2023. Conjugated microporous polymers based on ferrocene units as highly efficient electrodes for energy storage. Polymers 15, 1095. https://doi.org/10.3390/polym15051095.

Singh, P.N., Mohamed, M.G., Chaganti, S.V., Sharma, S.U., Ejaz, M., Lee, J.-T., Kuo, S.-W., 2023. Rational design of ultrastable conjugated microporous polymers based on pyrene and perylene units as high-performance organic electrode materials for supercapacitor applications. ACS Appl. Energy Mater. 6 (15), 8277–8287. https://doi.org/10.1021/acsaem.3c01391.

Siwach, P., Gaba, L., Dahiya, S., Punia, R., Maan, A.S., Singh, K., Ohlan, A., 2024. Recent progress in conjugated polymers composites with metal-organic frameworks as electrode materials for supercapacitors. Applied Surface Science Advances 19, 100555. https://doi.org/10.1016/j.apsadv.2023.100555.

Topal, S., Topal, S., Suna, G., Ustamehmetoğlu, B., Ozturk, T., Sezer, E., 2023. A microporous bifunctional electrochromic energy-storage polymer of thiophene, triphenylamine, and thienothiophene. Energy Technol. 11 (6). https://doi.org/10.1002/ente.202300015.

Wang, Y., Nakano, M., Michinobu, T., Kiyota, Y., Mori, T., Takimiya, K., 2017. Naphthodithiophenediimide–benzobisthiadiazole-based polymers: versatile n-type materials for field-effect transistors and thermoelectric devices. Macromolecules 50 (3), 857–864. https://doi.org/10.1021/acs.macromol.6b02313.

Wang, T., Xu, L., Chen, Z., Guo, L., Zhang, Y., Li, R., Peng, T., 2021. Central site regulation of cobalt porphyrin conjugated polymer to give highly active and selective CO2 reduction to CO in aqueous solution. Appl. Catal. B Environ. 291, 120128. https://doi.org/10.1016/j.apcatb.2021.120128.

Wang, Y., Chu, X., Zhang, H., Yan, C., Tian, G., Yang, W., Chen, X., Zhang, H., 2023. Hyperconjugated polyaniline delivering extraordinary electrical and electrochemical properties in supercapacitors. Appl. Surf. Sci. 628, 157350. https://doi.org/10.1016/j.apsusc.2023.157350.

Xing, Y., Dong, Y., Zhao, J., Zhang, Z., Wang, Y., 2023. Heteroatom-doped porous carbon derived from covalent organic framework for high performance supercapacitor applications. Carbon Letters 33 (6), 1723–1732. https://doi.org/10.1007/s42823-023-00553-9.

Xu, S., Wang, G., Biswal, B.P., Addicoat, M., Paasch, S., Sheng, W., Zhuang, X., Brunner, E., Heine, T., Berger, R., Feng, X., 2019b. A nitrogen-rich 2D sp 2-carbon-linked conjugated polymer framework as a high-performance cathode for lithium-ion batteries. Angew. Chem. Int. Ed. 58 (3), 849–853. https://doi.org/10.1002/anie.201812685.

Xu, S., Hong, M., Shi, X.-L., Wang, Y., Ge, L., Bai, Y., Wang, L., Dargusch, M., Zou, J., Chen, Z.-G., 2019a. High-performance PEDOT:PSS flexible thermoelectric materials and their devices by triple post-treatments. Chem. Mater. 31 (14), 5238–5244. https://doi.org/10.1021/acs.chemmater.9b01500.

Xu, C., Niu, Y., Ka-Man Au, V., Gong, S., Liu, X., Wang, J., Wu, D., Chen, Z., 2024. Recent progress of self-supported air electrodes for flexible Zn-air batteries. J. Energy Chem. 89, 110–136. https://doi.org/10.1016/j.jechem.2023.10.038.

Yan, Z., Luo, S., Qi, L., Wu, Z.-S., Liu, S.(F.), 2023. Recent advances in flexible wearable supercapacitors: properties, fabrication, and applications. Adv. Sci. https://doi.org/10.1002/advs.202302172.

Yazdi, M.K., Saeidi, H., Zarrintaj, P., Saeb, M.R., Mozafari, M., 2019. PANI-CNT Nanocomposites Fundamentals and Emerging Applications of Polyaniline. Elsevier, Iran, pp. 143–163. https://doi.org/10.1016/B978-0-12-817915-4.00009-9. Available from: http://www.sciencedirect.com/science/book/9780128179154.

Yu, M., Dong, R., Feng, X., 2020. Two-dimensional carbon-rich conjugated frameworks for electrochemical energy applications. J. Am. Chem. Soc. 142 (30), 12903–12915. https://doi.org/10.1021/jacs.0c05130.

Zang, W., Sumboja, A., Ma, Y., Zhang, H., Wu, Y., Wu, S., Wu, H., Liu, Z., Guan, C., Wang, J., Pennycook, S.J., 2018. Single Co atoms anchored in porous N-doped carbon for efficient Zinc–Air battery cathodes. ACS Catal. 8 (10), 8961–8969. https://doi.org/10.1021/acscatal.8b02556.

Zhang, H., Wang, Y., Chu, X., Xie, Y., Jiang, X., Huang, J., He, H., Yang, W., 2022. Dual-activity hyper-conjugated polymers delivering extraordinary electrical and electrochemical properties. SSRN Electron. J. https://doi.org/10.2139/ssrn.4185759.

Zhao, X., Pachfule, P., Thomas, A., 2021. Covalent organic frameworks (COFs) for electrochemical applications. Chem. Soc. Rev. 50 (12), 6871–6913. https://doi.org/10.1039/d0cs01569e.

Zhen, Y., Zeng, H., Zhu, Q., Xu, J., 2018. Recent advances of conducting poly (3,4-ethylenedioxythiophene): polystyrene sulfonate hybrids for thermoelectric applications. J. Mater. Chem. C 6, 8858–8873. https://doi.org/10.1039/C8TC01900B.

Zhou, F., Guan, S., Yan, Y., Pan, M., 2020. Polyaniline-derived nitrogen- and oxygen-decorated hierarchical porous carbons as an efficient electrode material for supercapacitors. J. Solid State Electrochem. 24 (4), 951–959. https://doi.org/10.1007/s10008-020-04545-x.

Zuo, H., Duan, J., Lyu, B., Lyu, W., Li, Y., Mei, X., Liao, Y., 2024. Carbon nanotube template-assisted synthesis of conjugated microporous polytriphenylamine with high porosity for efficient supercapacitive energy storage. Macromol. Rapid Commun. 45 (1). https://doi.org/10.1002/marc.202300238.

Index

'*Note:* Page numbers followed by "f" indicate figures and "t" indicate tables.'

A

Adsorption, oily wastewater treatment, 401
Advanced oxidation processes (AOPs), 380
Aeronautical engineering, 3–4
Aerosol assisted chemical vapor deposition (AACVD) process, 80f
Air flotation, oily wastewater treatment, 401
Alcoholic hepatitis, mesoporous silica-based nanostructures (MSNs), 258
Alkanethiol-based self-assembled monolayers (SAMs), 216–217
Amine functionalization, multifunctional nanostructure coatings (MNCs), 138
Antibacterial coatings, 446
Anticorrosion coatings, 37–38, 59, 326f
Antifouling, metal oxide-based nanocoatings, 49
Antifouling coatings, 59
Anti-icing coatings, 59, 347–351
Antimicrobial coatings, 59, 197
Antireflective coatings, 446
Antiscratch coatings, 59
Atherosclerosis diagnostics, mesoporous silica-based nanostructures (MSNs), 256–257
Atmospheric pressure plasma (APP), magnetic nanostructured materials, 210
Atomic force microscopy (AFM), conductive polymers (CPs), 375

B

Bio-catalysis applications
  enzymes, 221
  nanostructured coatings role, 222–223
  surface-coated magnetic nanostructured materials, 221–222
Biofouling, 344–345, 344f
Bio-inspired multifunctional coatings
  antifouling properties, 344–347
  anti-icing properties, 347–351
  bio-inspired slippery surfaces (SLIPS), 327–328
  butterfly wings, 332
  functional surfaces, 327–333
  gecko foot, 332–333
  lotus leaf, 329–330
  oil-water separation properties, 351–355
  pitcher plants, 333
  rice leaf, 331
  rose petals, 330–331
  self-cleaning characteristics
    biomimetic approaches, 334–335
    carbon steel slippery coating, 337–338
    cerium-based superhydrophobic coating, 336–337
    electrodeposition method, 337
    engineered POS@Al_2O_3@Epoxy superamphiphobic coating, 337f
    nanosecond laser treatment process, 338–339
    organic-inorganic hybrid strategy, 335–336
    spray-coating technique, 335–336
    superhydrophobicity, 334–335
  self-healing properties
    dual-action self-healing corrosion protection, 341
    epoxy-modified polyurethane composite coating (FD-EPU), 341–342
    extrinsic or intrinsic self-healing coatings, 339–340
    layered double hydroxides (LDHs), 341
    liquid-infused surface (LR), 343f
    revolutionary approach, 339
    self-healing coatings, 340

Bio-inspired multifunctional coatings (*Continued*)
slippery liquid-infused porous surface (SLIPS), 342–343
static contact angle analysis, 327
superhydrophobic surfaces, 327
traditional coatings, 333–334
water contact angle measurement, 327
water striders, 331–332
Biological functionalization, magnetic nanoparticles (MNPs), 216
Biological media utilization, oily wastewater treatment, 400
Biomass renewable energy, 467–468
Biomimetic coatings, magnetic nanostructured materials, 224
Biosensing modalities, 291
Biosensors, magnetic nanoparticle-based, 228
Butterfly wings, 332

C

Cancer treatment, mesoporous silica-based nanostructures (MSNs), 250–251
Carbide-based ceramics, 6
Cardiovascular diseases, mesoporous silica-based nanostructures (MSNs), 253–254
Cassie-Baxter model, 445–446
Cavity nanoengineering technology
extended, 308–314
high refractive index (HRI) Nd_2O_3–Ag nanohybrids, 303–304
hotspot-based plasmonic sensing, 303–304
nanocarbon florets (NCF), 306–307, 307f
plasmonic AgNPs, 304–306
plasmonic hotspots, 303–304
Purcel factor, 303–304
Centrifugation, oily wastewater treatment, 402
Ceramic-based nanocoatings, 5–6
Ceramic coatings, micro-/nanostructured biointerfaces, 182–183
Chemical etching method, 62–64, 66t
Chemical functionalization
magnetic nanoparticles (MNPs), 216
multifunctional nanostructure coatings (MNCs), 137–138
Chemical graft polymerization, multifunctional nanostructure coatings (MNCs), 139
Chemical vapor deposition (CVD), 16, 78–80, 81t, 97–98, 489
adsorption, 97
advantages and disadvantages, 97, 99t–100t
ceramics, 98
metal oxide-based nanocoatings, 40–41
multifunctional coatings, 197–198, 208
multifunctional nanostructure coatings (MNCs), 125–126
Coagulation, oily wastewater treatment, 402
Coalescence, oily wastewater treatment, 401
Coating application, 58
Cold spray (CS) technology
high-pressure cold spray (HPCS), 175–176
low-pressure cold spray (LPCS), 175
micro-/nano coatings, biomedical applications
ceramics, 182–183
composites, 183–185
kinetic energy perspective, 178
materials chemistry, 177–187
metals, 178–182, 181f
polymers, 185–187
tantalum coatings, 172–173
operational principles, 173–176, 176f
Conductive coatings, 59
Conductive polymers (CPs)
applications, 367
characteristics, 365
characterization, 371–375
charge carriers generation, 365–366
doping, 365
electrical conductivity, 373–374
electrochemical characterization, 371–372
electrochemical impedance spectroscopy, 372–373
electrochemical synthesis, 366–371
advantage, 367
deposition mechanisms, 368–371
disadvantages, 367–368
metal substrate, 368
environmental contaminants degradation, 377–384
organic molecules, 379–381
pesticides, 381–382

polluting and harmful gases, 378–379
toxic metals, 382–384
morphological characterization, 374–375
properties, 365
Conjugated microporous polymers (CMPs), 513–514
Conjugated polymer frameworks (CPFs), 514f
Conjugated polymers
applications, 511–512
conjugated microporous polymers (CMPs), 513–514
covalent organic frameworks (COFs), 514–515
doping, 511–512
electrical activity, 511–512
electrical conductivity, 513
electrocatalysts, 520–522
metal-organic frameworks (MOFs), 512–513
π-conjugated polymers, 513, 513f
polymeric graphitic carbon nitride, 515
structure, 511f
supercapacitors (SCs) electrodes, 515–520
thermoelectric devices, 523–527
Co-precipitation method, magnetic nanostructured materials, 205
Corrosion, 37–38, 39f
Corrosion inhibitors, metal oxide-based nanocoatings, 45–48
Corrosion process
anticorrosion coating, 326, 326f
categories, 325
economic burdens, 325
environmental factors, 325
measures, 325–326
Covalent organic frameworks (COFs), 514–515
Cyclic voltammetry, 370–371

D

Demulsification, oily wastewater treatment, 400
Dendrimers, 271
Deposition mechanisms, conducting polymers, 368–371
Diabetes diagnosis, mesoporous silica-based nanostructures (MSNs), 249–250
Dip coating approach, 107–108, 134
Dynamic glancing angle deposition (DGLAD), 127

E

Electrical conductivity/insulation, 13, 13t
Electrical double-layer capacitance (EDLC), 516
Electrical explosion of wires, magnetic nanostructured materials, 205
Electrocatalysts, 520–522
Electrochemical deposition, 80–84, 83t, 198
Electrochemical functionalization, magnetic nanostructured materials, 217
Electrochemical impedance spectroscopy (EIS)
conductive polymers (CPs), 372–373
ternary metal-based composites, 497
Electrochemical treatment, oily wastewater treatment, 400
Electrodeposition
magnetic nanostructured materials, 209
multifunctional nanostructure coatings (MNCs), 130
Electron beam evaporation (EBE), 102, 210
Electron beam lithography (EBL), magnetic nanostructured materials, 204
Electron cyclotron resonance (ECR) plasma, magnetic nanostructured materials, 210
Electrospinning technique, 73–77, 76t, 105, 132–134, 404
Electrospun nanofibrous membranes
electrospinning process, 404
factors impact, 404–405
nanofiber technology, 403
properties and applications, 405–406
Emulsion templating techniques, 268–269
Energy applications, 114–115
Energy-harvesting coatings, 197
Enhanced permeability and retention (EPR) effect, 246, 273
Extended cavity nanoengineering
AgAu nanohybrids synthesis, 312–314, 313f
Au-decorated SiO_2 NPs (AuSil) nanohybrids, 308–310, 309f
bio-inspired frugal science approach, 310
Hg^{2+} ions sensing, 310–312, 311f
LYCOAT biopolymer, 310
photo-plasmonic nanointerface, 312

F

Fabrication, multifunctional nanostructure coatings (MNCs), 124–136
Ferromagnetic materials, 200
Ferroptosis-related proteins, 278f
Flame-retardant technique, 110–111
Flocculation, oily wastewater treatment, 402
Fluoropolymers, self-cleaning coating, 450–451
Fourier transform infrared spectroscopy (FTIR), ternary metal-based composites, 494–496, 494f
Fullerene-like carbon nitride and carbon fluoride (FL-CNx and FL-CFx), nanostructured coatings, 126
Functional slippery lubricated coating (FSLC), 347, 348f

G

Galvanostatic technique, 369
Gas-phase deposition, magnetic nanostructured materials, 204
Gas-responsive membrane, oily wastewater treatment, 417
Gecko foot, 332–333
Geothermal energy, 470
Gradient coating nanostructure modification, 148–150
Grain refinement, 4–5
Graphene nanofibrous membrane (GPNM), 409–410
Gravity separation, oily wastewater treatment, 399

H

High-pressure cold spray (HPCS), 175–176
High-velocity oxy-fuel spraying (HVOF), 17
Hydrogen evolution reaction (HER), 520
Hydrophilic coatings, 446
Hydrothermal technique, 84–85, 217
Hydro turbines, 470–471
Hysteresis loss, 225–226

I

Ion beam irradiation, 151–152

J

Janus membrane, 411–412

K

Korea Advanced Institute of Science and Technology (KAIST), 245
Kretchmann-Raether (KR) configurations, 294

L

Laser ablation techniques, metal oxide-based nanocoatings, 41
Laser-assisted techniques, multifunctional coatings, 198
Laser cladding technique
  magnetic nanostructured materials, 209
  multifunctional nanostructure coatings (MNCs), 131–132
Laser pyrolysis, magnetic nanostructured materials, 205
Layer-by-layer (LbL) assembly, 127–129, 128f, 489, 490f
Layer-by-layer (LbL) deposition, 77–78, 79t
Layer-by-layer technique, 269–270
Lipid-based coatings, magnetic nanostructured materials, 224
Liposomes, 271
Lithography technique, 61–62, 63t
Liver-specific nanofiber scaffolds, 279f
Localized surface plasmon resonance (LSPR), 292
Lotus leaf, 329–330
Low-pressure cold spray (LPCS), 175

M

Magnetic hyperthermia, 225–226
Magnetic nanoparticles (MNPs)
  biocompatibility, 229
  biomedical applications, 223–229, 228f
  characterization, 207–208
  ferromagnetic materials, 200
  nanotoxicity and safety assessment, 230
  paramagnetic materials, 201–202
  properties, 207–208
  safety considerations and regulatory aspects, 229–230
  shapes and forms, 202f
  surface coating techniques, 204f, 208–221, 211t
    advantages and disadvantages, 217–221

biomedical and biocatalyst applications, 212t–213t
functionalization techniques, 215–217, 218t–220t
nanostructured coatings with 3D printing technology, 210–215
process parameters and coating materials, 214t–215t
synthesis methods, 203f
biological methods, 206
chemical methods, 205–206
physical methods, 203–205
Magnetic resonance imaging (MRI), 226–227
Magnetic-targeted drug delivery, 224–225
Magnetrons, 102
Marine engineering, 3–4
Mechanical milling method, magnetic nanostructured materials, 203
Membrane filtration, oily wastewater treatment, 402–403
Mesoporous silica-based nanostructures (MSNs)
for alcoholic hepatitis, 258
antimicrobial and antibacterial, 248–249
atherosclerosis diagnostics, 256–257
biodistribution, 246–247
biomedicine, 243f
cancer treatment, 250–251
for cardiovascular diseases, 253–254
categories, 242–245
as contrast agents, 254–255
development history, 242–245
diabetes diagnosis and treatment, 249–250
medical applications, 246–247
nanoformulations, 241–242
protein corona, 247
surface functionalization, 242
tuberculosis diagnostics, 257
for wound healing, 251–253
Metal-assisted chemical etching (MACE), 65f
Metal-based nanocoatings, 6
Metal-enhanced fluorescence (MEF), 293–294, 295f
Metal matrix nanocomposite-based coatings, 7
Metal-organic frameworks (MOFs), magnetic nanostructured materials, 224
Metal oxide-based nanocoatings
barrier properties, 42–43
biosynthesis method, 40, 41f
chemical vapor deposition (CVD) method, 40–41
as corrosion inhibitors, 45–48
corrosion protection, 48–49
design, 39–41
dispersibility, 48
laser ablation techniques, 41
nanoparticle synthesis, 40f
photodegradation resistance, 43–45
self-healing properties, 43
sol-gel method, 39–40
surface coatings, 41–45
thermal decomposition method, 41
Micelles, 270
Microbially influenced corrosion (MIC), 345
Micro-/nanostructured coatings
antipathogenic properties, 171–172
ceramics, 182–183
composites, 183–185
kinetic energy perspective, 178
materials chemistry, 177–187
for medicine, 171–172
metals, 178–182, 181f
osseointegration process, 171–172
polymers, 185–187
tantalum coatings, 172–173
Multifunctional coatings
antimicrobial coatings, 197
biomedical applications, 198
energy-harvesting coatings, 197
real-world biomedical applications, 199–200
self-healing coatings, 197
targeted drug delivery and controlled release, 198–199
techniques, 197–199
types, 196–197
Multifunctional fire protection nanocoating (MFNC), 111
Multifunctional nanocarriers, 273f
dendrimers, 271
infectious disease treatment, 272–274
inorganic nanoparticles, 272

Multifunctional nanocarriers (*Continued*)
- layer-by-layer technique, 269–270
- liposomes, 271
- micelles, 270
- nanocapsule structures, 269
- nanoprecipitation technique, 269–270
- polymersomes, 270–271

Multifunctional nanostructure coatings (MNCs)
- chemical functionalization, 137–138
- fabrication strategies, 124–136
- gradient coating modification, 148–150
- ion beam irradiation, 151–152
- layered materials functionalization, 142–143
- nanomaterials doping, 153
- nanoparticles functionalization, 141–142
- nanotexturing, 152–153
- plasma treatment modification, 150–151
- polymer modification, 138–141
- strain-engineered modification, 147–148
- structure modifications, 143–153
- surface tuning, 136–143
- template-assisted modification, 144–147

Multifunctional self-cleaning nanostructured coatings
- durability, 451–452
- materials used, 447–451
- self-cleaning coating types, 446
- solar photovoltaic (PV) devices, 441
  - antisoiling coatings and robotic cleaning, 443
  - capacity factors (CFs), 441–442, 442f
  - contaminant accumulations, 441
  - manual cleaning process, 443
  - nanomaterial coatings, 443–444
- wetting ability, 444–446

Multifunctional superhydrophobic coatings, 113–114, 113f

N

Nanocarbon florets (NCF), 306–307

Nanocoatings
- advantage, 3–4
- air filtering, pneumonia prevention, 276
- antimicrobial applications, 108–110
- antimicrobial properties, 12
- applicability, 3–4
- applications, 59, 108, 109f
- cardiovascular device surface modification, 274–276
- ceramic-based nanocoatings, 5–6
- characterization, 20, 21t
- customizable optical properties, 11–12
- diagnostics, 281
- drug and RNA delivery, 277–279
- drug screening, 277–279
- electrical conductivity/insulation, 13, 13t
- energy applications, 114–115
- enhanced adhesion, 10–11
- enhanced mechanical strength, 8
- environmental impact and sustainability, 20–21
- grain refinement, 4–5
- high thermal stability, 10
- improved corrosion resistance, 10
- metal-based nanocoatings, 6
- multifunctional smart nanocoatings, 110–111
- multifunctional superhydrophobic coatings, 113–114
- nanocomposite coatings, 7
- nano-scale thickness, 13–14, 14t
- polyurethane nanocomposite coating, 3–4
- preparation methods, 97–101
  - chemical vapor deposition (CVD), 97–98
  - dip coating, 107–108
  - electrospinning, 105
  - physical vapor deposition (PVD), 101
  - self-assembly, 105–107
  - sol-gel method, 103–105
  - spray coating, 103
  - sputtering, 101–102
- properties, 3–4
- skin repair, adhesion prevention, 276–277
- smart anticorrosion coatings, 112
- superior wear resistance, 8–9
- synthesis, 3–4, 14–20, 60–85, 60f
  - bottom-up synthesis, 70–85
  - chemical etching method, 62–64
  - chemical vapor deposition (CVD), 78–80
  - electrochemical deposition, 80–84
  - electrospinning technique, 73–77
  - hydrothermal technique, 84–85
  - layer-by-layer (LbL) deposition, 77–78
  - lithography technique, 61–62

plasma treatment, 64–67
research and development, 85–87
sol-gel method, 72–73
template technique, 67–70
top-down synthesis, 61–70
tailored surface functionality, 9, 9t
thickness, 58
tissue repair, 279–281
tooth and osteo implants, 276
transportation engineering, 3–4
types, 5–7
veterinary medicine, 281
wear-resistant coatings, 111–112
Nanocomposite coatings, 7
Nanofiber technology, 403
Nanofillers, 38–39, 96
Nanoparticle functionalization
magnetic nanoparticles (MNPs), 216
multifunctional nanostructure coatings (MNCs), 141–142
Nanopigments, 96
Nanoprecipitation technique, 269–270
Nanostructured composite materials, 87
Nanotechnology, 38–39, 57, 241
Nanotexturing, 152–153
Nitride-based ceramics, 5–6
Nonthermal plasma methods, magnetic nanostructured materials, 206

O

Oil-water separation properties, 351–355
Oily wastewater treatment
adsorption, 401
air flotation, 401
biological media utilization, 400
centrifugation, 402
coagulation and flocculation, 402
coalescence, 401
demulsification, 400
electrochemical treatment, 400
gravity separation, 399
membrane filtration, 402–403
membrane technology
Janus membrane, 411–412
oil-removing materials, 406–407
oil-removing membrane, 409–410
separation mechanism, 407–409
water-removing membrane, 410–411
wettability-regulated separation technology, 406–407
nanofibrous membrane design
carbon nanotubes (CNTs), 425–427
chemical and mechanical durability, 420–421
economic and environmental sustainability, 427–428
functionalization, 424–425
graphene oxide (GO), 425–427
Janus membrane (JM), 425–427
multifunctionality, 412–413
nanostructured membranes, 422–424
photocatalytic activity, 412–413
photocatalytic and fouling resistance, 419–420
real-world impact, 428–429
stimulus-responsive, 413–419
superhydrophilicity/superhydrophobicity, 412–413
thermo-responsive, 415–416
oily wastewater types, 397–398
traditional methods, 397–398
Oleophobic coatings, 446
Organic/inorganic micro-/nanostructured composites, 184–185
Organic molecules degradation, conductive polymers (CPs), 379–381
Oxide-based ceramics, 5–6
Oxygen evolution reaction (OER), 520

P

Peptide and protein nanostructures, 269–270
Peptide functionalization, magnetic nanoparticles (MNPs), 216
Pesticide degradation, conductive polymers (CPs), 381–382
Phospholipid-based self-assembled monolayers (SAMs), 216–217
Photocatalytic coatings, 446
Photocatalytic hydrogen production, 471
Photodegradation resistance, metal/metal oxide nanoparticles, 43–45
Photografting, multifunctional nanostructure coatings (MNCs), 139–141
Photo-responsive membranes, oily wastewater treatment, 414–415

pH-responsive membranes, oily wastewater treatment, 416–417
Physical vapor deposition (PVD), 16–17
  magnetic nanostructured materials, 208
  multifunctional nanostructure coatings (MNCs), 126–127
Pigments, 95
Pitcher plants, 333
Plasma-based techniques, magnetic nanostructured materials, 210
Plasma enhanced chemical vapor deposition (PECVD), magnetic nanostructured materials, 210
Plasma induced graft polymerization, multifunctional nanostructure coatings (MNCs), 139
Plasma spray coating, 461–462
Plasma treatment, 64–67, 69t
  magnetic nanoparticles (MNPs), 216
  nanomaterial modification, 150–151
Plasmonic biosensors, 292
Pneumonia prevention, 276
Poly(N-isopropylacrylamide) (PNIPAM), 415–416
Polymer-based nanostructures (PNMs)
  biomedical practices, 267
  multifunctional nanocarriers. *See* Multifunctional nanocarriers
  nanocoatings, 274–281
  polymeric nanocarriers, 268–269
  thiol and siloxane, 267–268
  universal coatings, 268
Polymer coating
  magnetic nanoparticles (MNPs), 216
  magnetic nanostructured materials, 217, 224
Polymeric materials, micro-/nanostructured biointerfaces, 185–187
Polymeric nanomaterials, self-cleaning coating, 448–450
Polymer matrix–based coatings, 7
Polymersomes, 270–271
Polyol method, magnetic nanostructured materials, 206
Polyurethane nanocomposite coating, 3–4
Potentiodynamic technique, 370
Potentiostatic technique, 369–370
Prism-based technique, 292–293
Protein corona, 246, 246f
Pseudocapacitance (PC), 516

R

Raman spectroscopy, ternary metal-based composites, 496–497, 496f
Renewable energy technology
  protective coatings and wind turbines, 463–467
  working environment, 467
Rice leaf, 331
Rose petals, 330–331

S

Scanning electron microscopy (SEM)
  conductive polymers (CPs), 374, 376f
  ternary metal-based composites, 491
Self-assembled monolayers (SAMs), magnetic nanoparticles (MNPs), 216–217
Self-assembly method, 18–20, 19t, 105–107
Self-cleaning coatings, 59, 196–197
Self-healing properties, metal oxide-based nanocoatings, 43, 49
Silica-based nanostructures (MSNs), biomedical applications, 245–247
Silica coatings, magnetic nanostructured materials, 224
Silicones, self-cleaning coating, 448
Silk fibroin polymers, 186–187
Slippery liquid-infused porous surface (SLIPS), 342–343, 351
Smart anticorrosion coatings, 112
Sol-gel method, 17, 18t, 39–40, 72–73, 74t
  alcohol-based methods, 103
  casting techniques, 104–105
  magnetic nanostructured materials, 205–206, 209
  metal alkoxides, 104
  multifunctional nanostructure coatings (MNCs), 129
  phases, 104
Solid-state spraying technique, 458–459
Soret colloids nano-engineering, 302f–303f
Spacer nanoengineering
  free space (FS) emission, 299–301
  gold nanocages (AuNCs), 299–301, 300f
  gold nanoparticles (AuNPs), 297, 301–303
  nanostars (NSs), 297
  plasmons, 297–299

reverse Kretschmann configuration, 299–301
soret colloids nano-engineering, 302f–303f
template-free nano-assemblies, 301
Spin coating technique, multifunctional nanostructure coatings (MNCs), 134–135
Spray coating technique, 17, 103
magnetic nanostructured materials, 209
multifunctional nanostructure coatings (MNCs), 130–131
Spray pyrolysis, magnetic nanostructured materials, 205
Sputtering technique, 101–102
Static contact angle analysis, 327
Stimulus-responsive nanofibrous membrane, oily wastewater treatment
gas-responsive membrane, 417
multiple-responsive membrane, 418–419
photo-responsive membranes, 414–415
pH-responsive membranes, 416–417
thermo-responsive membranes, 415–416
Strain-engineered nanostructure modification, 147–148
Supercapacitors (SCs) electrodes, 517t–518t
carbon nanofibers (CNFs), 519
carbon nanotubes (CNTs), 516–519
characteristics, 515–516
mechanisms, 516
micro-electrodes, 519–520
Superhydrophobic coatings, 446
Superhydrophobic surfaces, 327, 329f, 330t
Surface plasmon-coupled emission (SPCE) technology
nanostructured coatings, 296
cavity interface, 296
cavity nanoengineering technology, 303–307
extended cavity nanoengineering, 308–314
spacer nanoengineering, 297–303
optical configuration, 295–296
Surface plasmon resonance (SPR), 294–296

T

Template-assisted nanostructure synthesis
metal oxide nanostructure fabrication, 144–147, 146f
particle size and crystal structure, 144, 145t
porous surface, 144
Template technique, 67–70, 71t
Ternary metal-based composites
characterization techniques, 490–497
chemical deposition, 489
design and synthesis, 480–490
electrochemical performance and mechanisms, 497–498
electrochemical reactions, 498–499
facile colloidal process, 489
α-Fe_2O_3-ZnO-Au nanocomposites, 482–483, 484f
hematite, 486–487, 488f
hydrothermal synthesis method, 481–482
layer-by-layer (LbL) self-assembly process, 489
metal characteristics, 480
Pd-SnO_2-MoS_2 nanocomposites, 485–486, 486f
Pd-TiO_2-MoS_2 nanocomposites, 484–485, 485f
Pt-TiO_2-MoS_2 nanocomposites, 483–484, 485f
sol-gel process, 486
solvothermal methods, 487
sonochemical processes, 487–488
synthesis methods, 481–482
TiO_2@Pt@CeO_2 nanocomposite, 486, 487f
Thermal decomposition method, metal oxide-based nanocoatings, 41
Thermal spray coating techniques
adaptability, 461–462
advantages and disadvantages, 462t
coating material, 457–458, 462–463, 464t–466t
coating methods, 459–460
family tree diagram, 458f
geothermal energy, 470
high-velocity oxygen fuel (HVOF), 458–459
hydro turbines, 470–471
market-available coating process, 460–461, 461f
photocatalytic hydrogen production, 471
plasma spray coating, 461–462
renewable energy sources, 467–468
schematic representation, 460f
solar energy, 468–469

Thermal spray coating techniques (*Continued*)
solid-state spraying technique, 458–459
thermoelectric power generation, 472
wind energy, 469–470
Thermoelectric devices, 523–527
Thermoelectric power generation, 472
Thermo-responsive nanofibrous membranes, oily wastewater treatment, 415–416
Thiol functionalization, multifunctional nanostructure coatings (MNCs), 137
Tissue engineering (TE), magnetic nanoparticle-based, 229
Tissue repair, 279–281
Titanium dioxide nanoparticles, self-cleaning coating, 447
Titanium nitride (TiN), 5–6
Tooth and osteo implants, 276
Toxic metals degradation, conductive polymers (CPs), 382–384
Transmission electron microscopy (TEM)
conductive polymers (CPs), 374
ternary metal-based composites, 492, 492f
Transportation engineering, 3–4
Tuberculosis diagnostics, mesoporous silica-based nanostructures (MSNs), 257

V

Vacuum cold spray (VCS) technique, 103
Vapor deposition and patterning, magnetic nanostructured materials, 204
Veterinary medicine, 281

W

Water contact angle measurement, 327
Water-removing membrane, 410–411
Water striders, 331–332
Wear-resistant coatings, 111–112
Wettability-regulated separation technology, 406–407
Wind energy, 469–470
Wound healing, mesoporous silica-based nanostructures (MSNs), 251–253

X

X-ray diffraction (XRD), ternary metal-based composites, 492–494

Z

Zinc oxide nanoparticles, self-cleaning coating, 447–448

9780443236839